Actions for Bioenergy and Biofuels: A Sustainable Shift

Actions for Bioenergy and Biofuels: A Sustainable Shift

Editors

Idiano D'Adamo
Piergiuseppe Morone

MDPI • Basel • Beijing • Wuhan • Barcelona • Belgrade • Manchester • Tokyo • Cluj • Tianjin

Editors
Idiano D'Adamo
Control and Management Engineering
Sapienza University of Rome
Italy

Piergiuseppe Morone
Unitelma Sapienza—University of Rome
Italy

Editorial Office
MDPI
St. Alban-Anlage 66
4052 Basel, Switzerland

This is a reprint of articles from the Topic published online in the open access journal *Energies* (ISSN 1996-1073), *Sustainability* (ISSN 2071-1050) (available at: https://www.mdpi.com/topics/actions_for_bioenergy_and_biofuels).

For citation purposes, cite each article independently as indicated on the article page online and as indicated below:

LastName, A.A.; LastName, B.B.; LastName, C.C. Article Title. *Journal Name* **Year**, *Volume Number*, Page Range.

ISBN 978-3-0365-4065-8 (Hbk)
ISBN 978-3-0365-4066-5 (PDF)

Contents

About the Editors

Idiano D'Adamo is an Associate Professor at Sapienza University of Rome. He worked in the University of Sheffield, the National Research Council of Italy, Politecnico di Milano, University of L'Aquila and Unitelma Sapienza. In August 2015, he obtained the Elsevier Atlas Award. He received two Excellence Review Awards: Waste Management in 2017 and Resources Conservation and Recycling in 2018. During his academic career, Idiano D'Adamo published 96 papers in the Scopus database, reaching an h-index of 35. His current research interests are the bioeconomy, circular economy, renewable energy, sustainability and waste management. Idiano is among the 100,000 Top Scientists, with a global ranking in both 2019 and 2020. Idiano is Section Editor in Chief in Sustainability. He is Associate Editor in Global Journal of Flexible Systems Management and Frontiers in Sustainability. Idiano has participated in scientific research projects (Horizon 2020 "Star ProBio", Life "Force of the Future") and has collaborated with relevant national institutes (MATTM, CNBBSV, SVIMEZ).

Piergiuseppe Morone is Full Professor of Economic Policy at Unitelma Sapienza, with a strong interest in green innovation and sustainable circular bioeconomy, researching at the interface between innovation economics and sustainability transitions, an area of enquiry that has attracted growing attention over the last decade. His work regularly appears in prestigious innovation and environmental economics journals. He is the coordinator of the Bioeconomy in Transition Research Group (BiT-RG) and the director of the School of Sustainability Studies and Circular Economy (SUSTAIN). Moreover, he is/was involved in several European projects (including: H2020, BBI-JU, Life, Erasmus+, COST, Horizon Europe and CBE-JU), acting as a scientific coordinator, vice-chair and WP Leader. He was economic advisor to the Italian Minister of the Environment, Land and Sea Protection, till February 2021. He is now the Local Unit Scientific Coordinator in the LIFE EBP project (2020-2024), and a Scientific Committee member of the BIOVOICES and Biobridges projects (BBI JU CSAs). Piergiuseppe is Associate Editor of Cleaner and Circular Bioeconomy, member of the Editorial Boards of Current Opinion in Green and Sustainable Chemistry (ELSEVIER) and Open Agriculture (De Gruyter Open) and acts as a Guest Editor for various journal, including the Journal of Cleaner Production (ELSEVIER) and Sustainability (MDPI). Since April 2022, Piergiuseppe has been the vice-chair of the Circular Bio-based Europe Joint Undertaking (CBE JU) Scientific Committee.

Preface to "Actions for Bioenergy and Biofuels: A Sustainable Shift"

The pandemic period has led to a socioeconomic crisis that will have serious repercussions for production systems, as well as citizens' lives. This is furthered by the pollution caused by fossil fuels and concerns about damage to health and the environment. However, it is precisely in times of difficulty that it is necessary to identify a turning point and introduce a paradigmatic shift that can reconcile wellbeing with environmental and health protection. Increased citizen awareness, new employment outlets, the growing economic opportunities associated with the sustainable management of natural resources, as well as a growing respect for ecosystems, point to sustainability as a key driver of recovery. This Topic introduces the concept of sustainable hand, in which the point of connection between nature and man can occur if we adopt a systemic vision, which is participatory and oriented to social welfare. The use of natural resources, biofuels and bio-based products is one direction in which we can move. Are you ready for this green transition?

Idiano D'Adamo and Piergiuseppe Morone
Editors

MDPI

Editorial

Bioenergy: A Sustainable Shift

Idiano D'Adamo [1,*], Piergiuseppe Morone [2] and Donald Huisingh [3]

1   Department of Computer, Control and Management Engineering, Sapienza University of Rome, Via Ariosto 25, 00185 Rome, Italy
2   Department of Law and Economics, Unitelma Sapienza–University of Rome, Viale Regina Elena 295, 00161 Roma, Italy; piergiuseppe.morone@unitelmasapienza.it
3   Institute for a Secure and Sustainable Environment, University of Tennessee, 311 Conference Centre Building, Knoxville, TN 37996-4134, USA; dhuisingh@utk.edu
*   Correspondence: idiano.dadamo@uniroma1.it

Citation: D'Adamo, I.; Morone, P.; Huisingh, D. Bioenergy: A Sustainable Shift. *Energies* **2021**, *14*, 5661. https://doi.org/10.3390/en14185661

Received: 4 September 2021
Accepted: 7 September 2021
Published: 9 September 2021

Publisher's Note: MDPI stays neutral with regard to jurisdictional claims in published maps and institutional affiliations.

The European Commission emphasised that a bioeconomy is an economy that uses renewable biological resources from the land and sea (e.g., animals, crops, fish, forests and microorganisms) to produce energy, food and materials [1]. Consequently, the topic of bioenergy must be included within the sectors of the bioeconomy, which also includes traditional sectors (e.g., agriculture and forestry) and innovative sectors (e.g., manufacturing). Europe plays a key role in the evolving bioeconomy, but the performance of individual countries shows different values in terms of turnover, added value and personnel [2]. The aim is to support the development of these sectors globally in order to reach the 17 Sustainable Development Goals [3] with a special focus on circular bioeconomy [4].

Bioenergy, derived from biomass is applicable in all energy sectors (heat, electricity and transport). Human vulnerability to climate change—amplified by the COVID-19 pandemic—is a very serious threat to modern societies that requires urgent action. Given the complexity of the challenges, the efficient and sustainable usage of all resources is needed. The question we start with is: if people consume energy, are they responsible for environmental pollution? We believe that this is not the correct question to be posed, but certainly consumers are expected to engage in responsible behaviors. Similarly, should producers be expected to generate and use clean energy even if this leads to economic losses? Again, we believe this is not a valid question, yet producers are urged to make forward-looking choices when defining their production strategy—able to incorporate social and environmental (external) costs along with private costs.

Technological progress in transitioning to renewable energy-based systems is proving the greater competitiveness of these green resources that in various scenarios have become competitive with respect to gray resources, especially in the context of the fact that more than 10,000,000 people die each year due to air pollution, most of which is due to combustion of fossil fuels [5,6]. These and many climate-change related costs that are currently externalized must be integrated into our transition policies, practices and timetables.

Thus, we envisage two levels of analysis: first, all externalities should be properly identified; second, a sound and agreed upon methodology able to assign values to each of these externalities should be developed. However, this second phase is complex because not everything can always be translated into economic terms and because the value assigned may not be objective. While for the first problem, the solution may be to compare with comparable items, the second is simpler because it only requires alternative analyses to be carried out alongside the baseline ones.

The balance sheet associated with renewable energy tends to assign a positive value to externalities, but how high is this value? The social cost of carbon still has a low value, but in recent years it has grown significantly. Is the social component actually included in this externality? Unfortunately, we have to give a negative answer. It is precisely this absence that weighs on policy choices. How much do the exploitation of minors, the low rates of schooling, the serious poverty in which some families live, the lack of healthcare, the inefficiency of the Public Administration and some judicial criticalities really cost? For all these reasons, we can and should act now. This editorial focuses on the energy part which is only one component of this great revolution. Renewables should not compete with each other, as they are all required to replace fossil sources. The rapid advances in efficiency and falling prices of wind and solar power make them increasingly cost-effective. Sourcing and efficient usage of bio-energy based materials is also improving in many ways.

Analyzing bioenergy specifically, the International Energy Agency (IEA) concluded that a reduction in environmental impact occurs in the following scenarios: (i) "Biomass is grown sustainably or based on waste/residues"; (ii) "Converted to energy products efficiently"; and (iii) "Used to displace GHG-intensive fuels". Bioenergy includes several potential feedstocks (e.g., organic residues and waste; forestry and agriculture), production processes (e.g., fermentation; anaerobic digestion; gasification; pyrolysis; pelletisation; advanced biofuel processes; chipping and torrefaction), products (e.g., biodiesel; bioethanol; biogas; biomethane; renewable diesel; woodchips; pellets; pyrolysis oil; bio-synthetic gas; refuse derived-fuel and other advanced biofuels) and final energy use (e.g., biofuels for transport; combustion for electricity; combustion for heat and biomass-based materials and products) [7].

The literature has shown an exponential growth in the number of papers published on the topic of bio-resources for energy applications [8] and it has been documented that economic growth is influenced by the productivity of resources, including bioenergy [9]. Thus, it is evident that the growth of the bioenergy industry helps to reduce pollution and unemployment [10]. However, bio-resources have many other potential attributes that must be analyzed. Several authors have concluded that bioenergy can play an important role in the decarbonization of society. Thus, the uses and trade-offs of bioenergy must be explored from multiple perspectives.

It is essential to identify decision-making models that support both policy-makers and other stakeholders by comparing different alternatives [11]. The development of bioenergy must be integrated within emerging markets that support ecologically sound economic development of these territories based upon sustainable models [12]. In this direction, the comparison among countries is valuable in helping researchers to identify which business models have shown the main advantages [13].

In this context, it is essential to broaden and deepen the analyses of bio-energy materials that could be used to help to achieve sustainability objectives [14]; additionally, analyses must be performed that consider the bio-based system's sustainability by performing holistic analyses of the competing demands for and alternative uses of bio-based materials [15].

Analyses must be made on specific cases [16] as well as on local, regional, national and global bases for the short and long term future in the context of climate change, pandemics such as COVID-19 and one more giant factor.

What is that factor? It is that if the world's human population continues to increase at the current net increase of approximately 83 million people per year, the global population will increase from 7.9 billion to 9.7 billion by 2050 [17]. In that context, societies must factor in the increasing land that will be needed for food production and decide about the trade-offs between land for producing food and land for producing bio-based materials for non-dietary purposes.

Renewable energies are essential parts of the energy revolution in which the goal is to replace production from fossil fuels with those from renewable sources. In this framework, biomasses are the sources that can present the highest impact compared to other green sources, such as wind, photovoltaic and hydropower, but nevertheless energy consumption, both at the industrial and domestic level, are significant. Having a reduction due to energy efficiency interventions is certainly positive news; having a reduction due to a lower purchasing power indicates an economic decrease, and as such it cannot be welcomed. Biomasses are appropriate in the systems of supply and energy consumption at a local level.

This means that the production of energy from biomass obtained locally is significantly more sustainable than that obtained with biomass from other territories, in some cases even crossing national borders. This practice must be monitored and must include sustainability analyses that justify its application. In fact, the objective is to encourage the use of renewable sources to allow a circularity alongside an optimal trade-off between food production, and the production of bio-based energy resources. It must be remembered that the current transport sector contributes significant quantities of global emissions and the percentage of renewable energies in this sector has not yet achieved optimal levels [18].

Another delicate issue is certainly represented by the choice of bio-substrates. There is a perception that many of the available resources are not fully utilized, for example the proper use of forest trees, organic residues from agriculture, forestry or landscaping, or residues from the animal breeding sector. In addition, uncultivated (marginal) agricultural land could be used and the whole system could be optimized while ensuring an equitable and sustainable balance of the water–energy–food nexus.

The objective of future research should be focused on solving daily and longer-term problems; therefore, it is necessary to encourage collaboration among research centers and business structures to propose methodologies and solutions that will be able to develop answers to the food and bio-energy needs of society.

Unsustainable, intensive agricultural practices and burning forests and other vegetation are steadily increasing demands for bio-resources, and also causes serious environmental consequences. In this context, the European Commission highlighted the following points for reflection on the theme of bioenergy [19]:

(i)   Can the production of bioenergy penalize citizens because it has the effect of increasing the cost of raw materials, and therefore of the final price of some foods?
(ii)   Might the cultivation of some of these dedicated resources for energy purposes lead to labor exploitation?
(iii)   Can the demand for land on which to grow these dedicated resources be a threat to indigenous peoples?

These issues have a much broader perspective than the issue of bioenergy itself. They concern a model of sustainable and inclusive development that depends on an enhancement of the human component, particularly the collective, empowerment-oriented to identify and solve problems and not to amplify clashes that only benefit fossil fuel-based power lobbies. Education and training can play an essential role [20] and are manifested by university courses that have belatedly begun to be provided via public seminars and webinars.

Society can be empowered to demand answers from the political world to change the status quo, while at the same time, the great global challenges are not based solely on competitiveness, but on the reality that all of us must cooperate to help to ensure truly sustainable futures for our children's children's children!

It is undeniable that the economic component has always had a primary role in the choices of powerful institutional actors, and it is necessary to find a balance that does not create short-term profitability for the few monopolies to impede the transition to equitable, sustainable post-fossil-carbon societies globally, as soon as possible! We do not have time to waste!

The concept of Adam Smith's invisible hand could be readapted to that of a sustainable hand, which, within a market, seeks the social optimum, as this is the only truly sustainable approach, in the long term. For this to happen, new social norm paradigms models and customs must be developed and implemented to help to govern social interactions to truly sustainable futures.

In this sense, bio-based materials and products for all purposes, from food to other uses, are essential renewables that require changes in one's attitudes, visions, strategies and actions, based on a principle of sharing resources. However, this challenge is very complex—and becomes even more complex in those regions of the world characterized by armed conflicts and where there are dictatorial regimes. It is increasingly serious in regions where climate change is causing dramatic changes from flooding due to the sea level rising on the one side and to rapid expansion in desertification on the other side.

We have many challenges to address, but by co-working locally, regionally, nationally and globally, we can and will make urgently needed transitions. Bio-based production for food and other purposes is essential for humans and other life-forms on this planet.

Author Contributions: Conceptualization, I.D., P.M. and D.H.; writing—original draft preparation, I.D., P.M. and D.H.; writing—review and editing, I.D., P.M. and D.H. All authors have read and agreed to the published version of the manuscript.

Funding: This research received no external funding.

Conflicts of Interest: The authors declare no conflict of interest.

References

1. European Commission A New Bioeconomy Strategy for a Sustainable Europe. Available online: https://ec.europa.eu/info/research-and-innovation/research-area/environment/bioeconomy/bioeconomy-strategy_en (accessed on 7 September 2021).
2. D'Adamo, I.; Falcone, P.M.; Morone, P. A New Socio-economic Indicator to Measure the Performance of Bioeconomy Sectors in Europe. *Ecol. Econ.* **2020**, *176*, 106724. [CrossRef]
3. Dietz, T.; Börner, J.; Förster, J.J.; von Braun, J. Governance of the bioeconomy: A global comparative study of national bioeconomy strategies. *Sustainability* **2018**, *10*, 3190. [CrossRef]
4. Morone, P.; Imbert, E. Food waste and social acceptance of a circular bioeconomy: The role of stakeholders. *Curr. Opin. Green Sustain. Chem.* **2020**, *23*, 55–60. [CrossRef]
5. Vohra, K.; Vodonos, A.; Schwartz, J.; Marais, E.A.; Sulprizio, M.P.; Mickley, L.J. Global mortality from outdoor fine particle pollution generated by fossil fuel combustion: Results from GEOS-Chem. *Environ. Res.* **2021**, *195*, 110754. [CrossRef] [PubMed]
6. Pielke, R. Every Day 10,000 People Die Due to Air Pollution From Fossil Fuels. Available online: https://www.forbes.com/sites/rogerpielke/2020/03/10/every-day-10000-people-die-due-to-air-pollution-from-fossil-fuels/?sh=7f2aa322b6a7 (accessed on 9 July 2021).
7. IEA Bioenergy, a Sustainable Solution. Available online: https://www.ieabioenergy.com/bioenergy-a-sustainable-solution/ (accessed on 28 August 2021).
8. Ferrari, G.; Pezzuolo, A.; Nizami, A.-S.; Marinello, F. Bibliometric Analysis of Trends in Biomass for Bioenergy Research. *Energies* **2020**, *13*, 3714. [CrossRef]
9. Busu, M. Assessment of the Impact of Bioenergy on Sustainable Economic Development. *Energies* **2019**, *12*, 578. [CrossRef]
10. Alsaleh, M.; Abdulwakil, M.M.; Abdul-Rahim, A.S. Does Social Businesses Development Affect Bioenergy Industry Growth under the Pathway of Sustainable Development? *Sustainability* **2021**, *13*, 1989. [CrossRef]
11. Abdel-Basset, M.; Gamal, A.; Chakrabortty, R.K.; Ryan, M. Development of a hybrid multi-criteria decision-making approach for sustainability evaluation of bioenergy production technologies: A case study. *J. Clean. Prod.* **2021**, *290*, 125805. [CrossRef]
12. Carvalho, R.L.; Yadav, P.; García-López, N.; Lindgren, R.; Nyberg, G.; Diaz-Chavez, R.; Krishna Kumar Upadhyayula, V.; Boman, C.; Athanassiadis, D. Environmental Sustainability of Bioenergy Strategies in Western Kenya to Address Household Air Pollution. *Energies* **2020**, *13*, 719. [CrossRef]
13. Cicea, C.; Marinescu, C.; Pintilie, N. New Methodological Approach for Performance Assessment in the Bioenergy Field. *Energies* **2021**, *14*, 901. [CrossRef]
14. Pehlken, A.; Wulf, K.; Grecksch, K.; Klenke, T.; Tsydenova, N. More Sustainable Bioenergy by Making Use of Regional Alternative Biomass? *Sustainability* **2020**, *12*, 7849. [CrossRef]
15. D'Adamo, I.; Falcone, P.M.; Huisingh, D.; Morone, P. A circular economy model based on biomethane: What are the opportunities for the municipality of Rome and beyond? *Renew. Energy* **2021**, *163*, 1660–1672. [CrossRef]
16. Khawaja, C.; Janssen, R.; Mergner, R.; Rutz, D.; Colangeli, M.; Traverso, L.; Morese, M.M.; Hirschmugl, M.; Sobe, C.; Calera, A.; et al. Viability and Sustainability Assessment of Bioenergy Value Chains on Underutilised Lands in the EU and Ukraine. *Energies* **2021**, *14*, 1566. [CrossRef]
17. United Nations Population. Available online: https://www.un.org/en/global-issues/ (accessed on 2 September 2021).
18. D'Adamo, I.; Gastaldi, M.; Rosa, P. Assessing Environmental and Energetic Indexes in 27 European Countries. *Int. J. Energy Econ. Policy* **2021**, *11*, 417–423. [CrossRef]
19. European Commission Greening-Our-Energy-Supply. Available online: https://ec.europa.eu/info/food-farming-fisheries (accessed on 30 August 2021).
20. Ferrer-Balas, D.; Lozano, R.; Huisingh, D.; Buckland, H.; Ysern, P.; Zilahy, G. Going beyond the rhetoric: System-wide changes in universities for sustainable societies. *J. Clean. Prod.* **2010**, *18*, 607–610. [CrossRef]

sustainability

MDPI

Article

Corporate Power in the Bioeconomy Transition: The Policies and Politics of Conservative Ecological Modernization in Brazil

Mairon G. Bastos Lima [1,2]

1   Department of Space, Earth and Environment, Chalmers University of Technology, 412 96 Göteborg, Sweden; mairon.bastoslima@sei.org
2   Stockholm Environment Institute, 104 51 Stockholm, Sweden

Abstract: The bioeconomy transition is a double-edged sword that may either address fossil fuel dependence sustainably or aggravate human pressures on the environment, depending on how it is pursued. Using the emblematic case of Brazil, this article analyzes how corporate agribusiness dominance limits the bioeconomy agenda, shapes innovation pathways, and ultimately threatens the sustainability of this transition. Drawing from scholarship on power in agri-food governance and sustainability transitions, an analytical framework is then applied to the Brazilian case. The analysis of current policies, recent institutional changes and the case-specific literature reveals that, despite a strategic framing of the bioeconomy transition as a panacea for job creation, biodiversity conservation and local development (particularly for the Amazon region), in practice major soy, sugarcane and meatpacking conglomerates dominate Brazil's bioeconomy agenda. In what can be described as conservative ecological modernization, there is some reflexivity regarding environmental issues but also an effort to maintain (unequal) social and political structures. Significant agribusiness dominance does not bode well for smallholder farmers, food diversity or natural ecosystems, as major drivers of deforestation and land-use change (e.g., soy plantations, cattle ranching) gain renewed economic and political stimulus as well as greater societal legitimacy under the bioeconomy umbrella.

Keywords: agriculture; bioeconomy; agri-food systems; sustainability transitions; power relations; biofuels; Amazon; soy; sugarcane; value chains

Citation: Bastos Lima, M.G. Corporate Power in the Bioeconomy Transition: The Policies and Politics of Conservative Ecological Modernization in Brazil. *Sustainability* **2021**, *13*, 6952. https://doi.org/10.3390/su13126952

Academic Editor: Idiano D'Adamo

Received: 28 April 2021
Accepted: 17 June 2021
Published: 21 June 2021

Publisher's Note: MDPI stays neutral with regard to jurisdictional claims in published maps and institutional affiliations.

1. Introduction

A transition from a largely fossil-based to a renewable bio-based economy (a bioeconomy) has been increasingly seen as an imperative worldwide. Most global climate scenarios for staying under a 1.5 °C or 2 °C mean temperature increase foresee increased bioenergy use [1]. Climate change mitigation is, therefore, a key rationale behind bioeconomy promotion, but not the only one. Marine plastic pollution, too, reveals the urgent need to shift away from non-biodegradable goods and towards a circular economy based on renewable resources. Circularity, in this regard, includes adding value to by-products and reducing waste [2]. A question often left unasked, however, is who rules the circle. While most studies address the environmental and economic dimensions of circular bioeconomies, their social and political aspects have continuously been identified as a research gap [3].

Agri-food systems worldwide—understood as the entirety of supply chains, processing and retailing stages "from farm to fork"—have become increasingly consolidated over the past decades [4,5]. The bioeconomy, being significantly built on top of such systems, does not emerge over a level playing field; rather, the bioeconomy mostly adds to pre-existing agricultural and forestry sectors [6]. The novelty lies primarily in the new applications that such biomass starts having or in new value chains being created, while ideally also in the displacement of conventional fossil-based goods (e.g., fuels, plastics, lubricants). Bioeconomy proponents frequently hail the potentials for creating jobs, conserving biodiversity, and the overall social and environmental good that such a transition can spur [7–10]. Yet, it is fundamental not to take such ideal scenarios for granted, as power imbalances and vested interests can lead the bioeconomy onto tortuous ways.

This article explores some usually underexposed political dimensions of the bioeconomy transition. In particular, it assesses the role of power in steering bioeconomic development pathways to date, using Brazil as an in-depth case study. Political and economic power, notably from corporate agribusiness, largely shape the global agri-food system as we know it. Several studies have detailed how powerful actors broadly set agri-food agendas [11,12], mold policy incentive structures [13,14], establish dominant patterns of agricultural production (including the pervasive use of toxic chemicals such as glyphosate pesticides) [5], and broadly influence dietary patterns, food availability and consumer preferences [15]. These issues are critical not only from the perspective of democracy but also of sustainability, as corporate dominance limits governance agendas as well as shapes technological and innovation pathways [16].

The bioeconomy, nevertheless, despite being built mostly over and from such a consolidated agro-industrial system, is often approached gleefully, with its ample goodness assumed at face value [7,17]. Yet it is key to unpack how actors have been steering its development so far, off certain pathways and onto others, and the various ways power is used to effectively govern the bioeconomy. That means going beyond merely investigating the use of the concept of bioeconomy as a "master narrative" [18] to also assess the practices linked to it, i.e., policies coming now under its aegis and the concrete sectors that currently represent the bioeconomy's de facto building blocks. As we shall see, while creating some environmental benefits (notably from fossil fuel replacement), and despite continuous talk of its potential for poverty alleviation and native biodiversity valorization, the Brazilian bioeconomy remains primarily anchored on large sugarcane, soy and cattle agroindustrial conglomerates—major drivers of deforestation, other ecological impacts, and social exclusion [19–21]. How that dominance is achieved (and its workings) needs to be understood if such a pathway is to eventually change towards sustainability.

Based on a comprehensive review of policies, government reports, and specific literature on Brazil (in English and Portuguese), this analytical article is structured as follows. The next section develops a conceptual framework to study power manifestations in bioeconomy governance. It draws from work on the political economy of agri-food systems as well as scholarship on power in sustainability transitions. Section 3 analyzes Brazil's bioeconomy to date, reviewing its latest developments. Section 4 examines how different forms of corporate power have been used to steer the agenda, and Section 5 discusses the broader implications of that for the bioeconomy and sustainable development. Finally, Section 6 concludes the article with its key points and recommendations for further research.

2. Assessing Power in the Bioeconomy Transition: A Conceptual Framework

Although transition studies long remained focused on economic and technological dimensions, political interrogations have gained growing traction [22,23]. Analysts have observed that the governance of sustainability transitions does not escape the grips of politics; rather, their governance designs get embedded in pre-existing political conditions and often risk being captured by powerful actors [22,24]. Often, there are not just one but multiple possible transition pathways and competing visions, including of the bioeconomy and what it should accomplish [25,26]. Either out of self-interest, divergent values, different worldviews or any combination of those, different actors or advocates for particular pathways pursue distinct "policy beliefs." That refers to actors' contrasting visions of how the institutional setting should look like, and therefore how it should change or not change [27,28]. As such, there is a critical—and not to be overlooked—political element to governing bioeconomy transitions.

A key concept for such political analyses is power. Power is a multi-faceted concept that lacks a consensual and unambiguous definition. It has multiple dimensions. The most straightforward one is that of "power to", that is, the ability to get things done—or, more specifically in a governance setting, *"the capacity to mobilize resources and institutions to achieve a goal*" ([22], p. 516). However, it may not be suitable to regard power in an atomistic way, as if individual actors were isolated entities imbued with more or less power. Social and political contexts are made of relations, including power relations. That is where the dimension of "power over" comes to the fore, i.e., the ways through which one actor or social group can exert dominance over others. Some authors also speak of "power with", emerging from forms of cooperation that increase the means and possibilities of the ones involved [29].

In addressing power in the specific context of sustainability transitions, it is important to regard who gets empowered and who becomes disempowered through (the pursued) change [22]. In other words: who wins and who loses, whose capacities are increased or diminished. Understanding these processes is critical not only because we live in a world of marked inequalities that tend to be reproduced over time, but also because sustainability—by definition—is to entail improvements in social equity [30,31]. The bioeconomy is regularly presented as a socially benign transformation that can attain social as much as broader environmental good [7,8,32], yet it is pivotal to analyze the facts under that claim.

One framework identifies three forms of power in transitions: (a) innovative power, or the capacity to create new resources (e.g., technical innovations); (b) transformative power, or the ability to change institutional settings (changing policies, obtaining new incentives, etc.); and (c) reinforcive power, the capacity to reproduce—and, eventually, reinforce—existing institutions [22]. In a way, they all are framed as instances of "power to." However, the latter two also have clear dimensions of "power over", as control over agendas—to either reinforce or change existing institutions—denote a level of political dominance over other actors and competing advocacy. Those are different *ends* that may be sometimes pursued via the same type of means (e.g., lobbying for self-serving policy changes), and which change resulting configurations of who gets (further) empowered or disempowered in transitions [22].

An alternative three-part framework—building on a long tradition of power studies—focuses on the *ways* corporate power is exercised in agri-food governance [11]. First, *instrumental power* is when an actor mobilizes resources to directly accomplish a goal or impose its will over others, such as through political lobbying. Simply put, "*A has power over B to the extent that he can get B to do something that B wouldn't otherwise do.*" ([33], p. 203). However, that does not capture the exercise of power in the form of agenda-setting—what some authors refer to as *structural power* [11]. This "second face of power" builds on a longstanding observation that actors can also exert dominance by constraining others' range of choices and options, including limiting public or policy debates. As some have put it:

> "*[P]ower is also exercised when A devotes his energies to creating or reinforcing social and political values and institutional practices that limit the scope of the political process to public consideration of only those issues which are comparatively innocuous to A. To the extent that A succeeds in doing this, B is prevented, for all practical purposes, from bringing to the fore any issues that might in their resolution be seriously detrimental to A's set of preferences.*" ([34], p. 942)

Finally, a third face of power relates to how actors may shape the very views and wants of others, securing their consent and, thus, pre-empting political competition. This form of power encompasses various socially induced modifications of beliefs, attitudes or views—what some of the literature treats under the concept of *influence* [35]. As it has been argued:

> "*A may exercise power over B by getting him to do what he does not want to do, but he also exercises power over him by influencing, shaping or determining his very wants. Indeed, is it not the supreme exercise of power to get another or others to have the desires you want them to have—that is, to secure their compliance by controlling their thoughts and desires?*" ([36], pp. 23/27)

This more subtle form of power is extensively studied in the neo-Gramscian political science tradition, concerned with how actors may achieve hegemony through consent rather than coercion [37,38]. Through influential discourses, understood as ways of framing and apprehending the world, certain actors may come to shape what gets viewed as common sense and benefits from social legitimacy ([39], p. 9). That is regarded as *discursive power*, and some authors note how agribusiness corporations routinely use it to frame contentious issues (e.g., genetically modified organisms) favorably in the public discourse—which may, eventually, affect policymaking, regulations, and incentives [11,15]. Therefore, these various dimensions of power should not be seen as separate but intertwined. Preponderance and domination are usually achieved through a combination of the above, *"through an alignment of material, organizational and discursive formations which stabilize and reproduce relations of production and meaning"* ([40], p. 806).

Failure to address such a politics almost invariably is to the benefit of dominant actors, which tend to more or less silently reproduce and reinforce their privileged positions. Research, therefore, can *"denaturalize dominant constructions, in part by revealing their connection to existing power relations"*, whereas researchers can work *"to unmask these ideational structures of domination and to facilitate the imagining of alternative worlds"* ([41], p. 398).

Figure 1 illustrates this analytical framework based on the reviewed literature. The three concentric circles—or the range of such power relations—suggest the width of work of that power dimension. While the first is very specific to within a given economic sector and its materially related environs, the second pertains to a whole set of policy and politics surrounding the sector (in the bioeconomy's case, land-use policy, agricultural policy, energy policy, etc.), while the third theoretically reaches out to everyone from whom recognition and support are ultimately sought. This broadest reach frequently may be equated to public opinion and societal mores, which may in turn create social momentum for particular agendas, stir consumer behavior, and create political will towards one's desired goals. In other words, that is also about securing legitimacy, understood as *"a generalized perception or assumption that the actions of an entity are desirable, proper, or appropriate within some socially constructed system of norms, values, beliefs, and definitions"* ([42], p. 574). Working to not only conform to but also mold those broad social perceptions is therefore crucial, as actors can then enjoy a so-called social license to operate or to pursue one's (private) agendas under public approval [43].

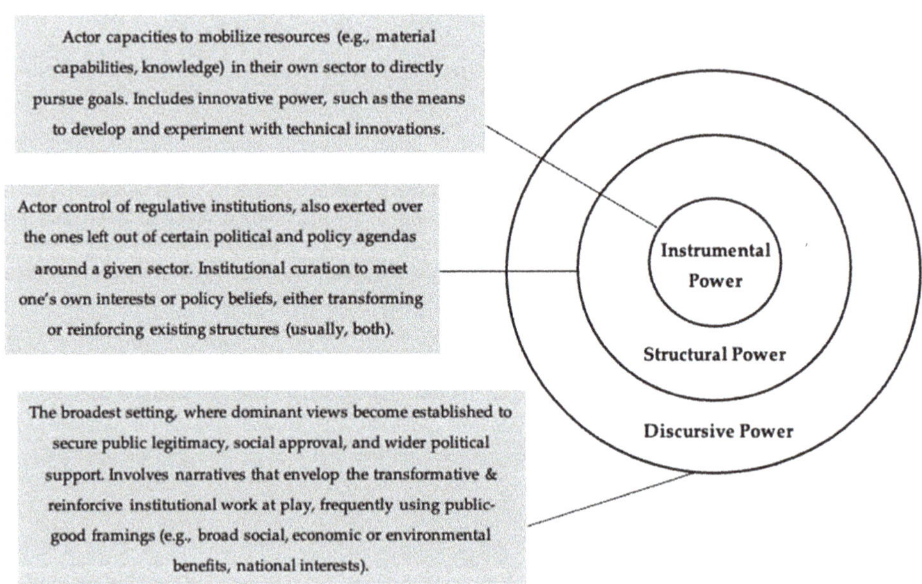

Figure 1. The range of power relations in governance.

3. Brazil: From Ethanol to the Bioeconomy

The bioeconomy has become a new framing concept rapidly gaining momentum in Brazil. Public and private actors linked to agribusiness, in particular, have warmly embraced it. Although the country lacks a unified bioeconomy strategy (as the EU and various European countries for example have), Brazilian scientists, government agencies and the business sector—notably well-established agroindustries such as the sugarcane one—have readily adopted the new umbrella term [9,17,44]. In a sense, the bioeconomy has been Brazil's prime way of engaging with ecological modernization, i.e., economic and technological modernization that seeks to address perceived environmental issues [28,45].

The bioeconomy is not an unambiguous concept—it is a field where multiple views compete, not the least over its definition [26]—and in Brazil it has been utilized quite liberally. On the one hand, some Brazilian scientists and government actors have utilized the term to refer mainly to novel bio-based value-chains, especially those designed to deliver goods or services that can replace fossil-based products [9,44]. A slight variation of that view focuses on biodiversity valorization, "biorefining" and value chain creation more generally (irrespective of whether new goods replace fossil-based products or not), particularly for the Amazon region [10,32]. On the other hand, however, some government and agribusiness actors refer to the bioeconomy in its broadest possible sense, to include not only well-established biofuel industries but *all* agricultural production [46,47]. Such an all-encompassing understanding is perhaps indicative of the appeal this new term has had among Brazilian agribusiness and shows its eagerness to frame itself under the bioeconomy umbrella.

In this article, the bioeconomy is regarded as all bio-based economic sectors beyond food and other conventional agricultural production (e.g., fibers, tobacco). This scope avoids overstretching the concept, while at the same time not limiting it to only the newest markets embraced after the term came in vogue. For bioenergy is a key—and still dominant—bioeconomy sector *avant la lettre*. As the following sections will show, it is primarily for bioenergy that crops traditionally regarded as food crops have been diverted, although their multiple uses have been expanding rapidly.

3.1. Bio-Based Add-Ons Have Sustained and Boosted Established Agribusiness

Brazil is the world's second-largest biofuel producer after the US and an active promoter of the bioeconomy worldwide. A long large-scale commercial experience with biofuels—with ethanol blending mandates being first introduced in the 1930s—and efficient agroindustries highly linked to the government have made Brazil the only country where more than 10% of the energy used in the transport sector comes from renewables [48]. Such an agro-efficiency, however, often obfuscates the power workings behind biofuel promotion in Brazil. Although the country boasts significant biodiversity, two Asian crops—sugarcane and soy—currently account for the bulk of the Brazilian bioeconomy. Most of the remainder, in turn, originates from animal fats that are by-products of large meatpacking industries. As such, there may be essential differences between the currently portrayed image and discourse of an Amazonia-friendly biodiverse bioeconomy and what a reality check can show.

Being the most established bio-based sector (aside from food and other traditional crop uses), biofuels dominate the bioeconomy in Brazil—as elsewhere. Sugarcane-based ethanol, used as a gasoline additive or replacement, represents 85% of the country's biofuel output [49]. Corn, which in Brazil is normally intercropped with soy, has nevertheless started also being used for ethanol production to "offload" fast-growing supplies and prevent a market glut that could depress corn prices [50]. Biodiesel, which replaces fossil diesel, in turn is chiefly produced from soybean oil (61%) or beef tallow (10.3%), followed by pork and chicken fat originating mostly from large meatpacking companies that in turn use soymeal for animal feed ([49], p. 14). One can see, therefore, how bioeconomy production to date—represented at commercial scale essentially by bioenergy—builds neatly on pre-existing agroindustrial conglomerates. In total, biofuel production claims as much as 37% of Brazil's vegetable oil supply and, on average, 65% of all its sugarcane, to then meet 23% of the country's transport energy needs ([51], p. 82).

Given that sugarcane is increasingly used also for electricity cogeneration, by processing the crushed remains (bagasse) after the sugar juice has been extracted, one can see that—despite its name—the crop has become far more an energy crop than a sugar crop in Brazil. Grown mostly in large-scale estates, some of them inherited and expanded from past centuries after the Portuguese introduced the crop to the country in 1530, the sector has deftly found new uses and downstream markets for it. The sugarcane agroindustry has navigated socio-technical changes—and, as we shall see, often actively promoting such changes—without significantly losing its power position but rather gaining further economic prominence in Brazil. In line with technological upgrading, however, it has mostly done away with manual laborers. While strenuous or even forced labor remained commonplace in the Brazilian sugarcane sector until at least the 2000s [52], now mechanized harvesting predominates.

Soy, in contrast, is a much newer "boom crop"—that is, a rapidly expanding cash crop embraced mostly for international markets ([53], p. 451)—that has made inroads across South America since the mid-20th century. If global soy production grew ten-fold between 1960 and 2016, today more than half of it originates from that continent [54]. Some speak of a process of increasing "soyzation", whereby soy has been replacing other land uses and gaining ever more economic relevance at the expense of other sectors in producing countries (notably Argentina, Brazil and Paraguay) [55]. In Brazil, the area dedicated to soy cultivation has been skyrocketing, having more than tripled—from 11 million hectares to over 36 million hectares—only between 1990 and 2020, largely at the expense of native ecosystems and smallholder farming [20,56]. Its biggest market is animal feed industries, primarily in China and Europe (which respectively import 68% and 13% of all Brazilian soy) [57]. Yet the bioeconomy, too, has offered a supplementary pull for the soy sector through ever-higher biodiesel blending mandates (see Section 4).

Table 1 shows how much of the supplies of some key agricultural commodities are destined to make biofuels in Brazil compared to their global averages. It indicates how some agroindustries have become deeply vested in bioeconomy promotion. Besides increasing market demand for products whose prices sometimes become undesirably low for the industry, the bioeconomy widens the range of possible downstream markets. That, in turn, grants producers the flexibility to choose what is most profitable and, sometimes, the possibility to switch back and forth according to price signals—as sugarcane mills routinely do between sugar and ethanol [6,58]. Some suggest such sectors have gone beyond single value chains to develop "value webs", where single crops lead to multiple interrelated strings (e.g., sugarcane-based electricity from bagasse being used to power ethanol production) [59].

Table 1. Agricultural commodity utilization for the Brazilian and global bioeconomies.

Agricultural Commodity Supplies	Brazil	Global Average (2016–2018)
Vegetable oil	37%	12.5%
Sugar	65%	21%
Coarse grains (corn and other cereals excluding wheat and rice)	4.5% *	13.4%

Data source: [49,60]. * This estimate refers exclusively to corn supplies, increasingly used as a secondary ethanol feedstock in Brazil [61].

Overall, biomass sources (including liquid biofuels and sugarcane electricity cogeneration) have met as much as 19% of Brazil's total energy consumption ([49], p. 27). Notably, bioenergy's growing relevance also represents an expanding share of private suppliers in a setting previously dominated by state-controlled energy companies. Energy supplying, therefore, is becoming increasingly privatized, even if to a degree those market shares are also being taken from transnational oil companies operating in Brazil. At any rate, given the significant consolidation of the sugarcane, soy or meatpacking sectors in the country, it can be said that its energy transition takes place essentially between giants—from the usual energy giants to agribusiness ones.

3.2. Opening Moves of the New Bioeconomy: In Whose Benefit?

As the bioeconomy broadens, there is growing momentum for expanding the realm of applications and products from the dominant agribusiness crops. Sugarcane has been increasingly used to produce renewable (though so far usually not biodegradable) polymers for so-called "green plastics." There are also prospects for using the crop to make solvents, lubricants, enzymes, cosmetics, and pharmaceutical products. The idea is not only to replace industrial inputs currently made from oil but generally to create novel and high value-added goods [59]. As it has come to be described, that is the idea of applying a "biorefinery" concept to agricultural crops, that is, extracting multiple components that can supply several markets [6].

Meanwhile, there is a growing appetite for widening Brazil's bioeconomy resource base to include, notably, Amazonian biodiversity as a basis for (bio)technological development and value chain creation [32]. A new landmark Biodiversity Law (13.123/2015) has paved the way for increased use of native flora and fauna by industry sectors (e.g., cosmetics, pharmaceuticals) while creating hurdles for public research institutions and less resourceful local actors. High entry costs in complex bureaucracy with technical and legal requirements have disproportionately impacted local actors such as indigenous communities or smallholder farmer associations vis-a-vis well-resourced corporate ones [62]. The number of private companies authorized to commercially utilize native Brazilian biodiversity under the new framework quintupled from 42 before the 2015 law to over 200 by 2018, as most stakeholders agree industry actors have been the new law's primary beneficiaries [63]. This growing dynamic, of course, raises fears that the bioeconomy may simply promote further corporate-led commodification of nature and accumulation by dispossession in the Amazon [64].

An additional piece of legislation has been a new framework in place since 2021 on payments for environmental services (Law 14.119/2021). Problematically, it requires only a self-made declaration into the online Rural Environmental Registry (*Cadastro Ambiental Rural*—CAR) as sufficient demonstration that one is supposedly entitled to the land and, therefore, to sell its environmental services (e.g., carbon credits). Although the CAR system has been designed for land-use change monitoring, in practice private landholders' entries have often dubbed as proofs of land tenure [65]. These entries have already been used for obtaining bank credit and, thus, the system has been conducive to pasture expansion at the expense of forests [66]. Overlapping claims have also been commonplace, with individual farmers and companies often utilizing CAR declarations to "grab" lands by registering as theirs plots that are under customary community use or even within protected areas and indigenous territories [20,65,67]. Such conflicts have been particularly salient in the Amazon, where traditional communities abound, tenure security is fragile, and land grabbing is rampant [68]. Environmental authorities are supposed to analyze and filter out undue claims, but deadlines for doing so have been continually extended over the years. In practice, as with regular amnesty to land grabbers (*grileiros*), Brazilian governments have continually shown leniency toward undue claimants and their utilization of CAR entries as suggestive of land rights in the meanwhile [65].

Despite a narrative of inclusive and sustainable development, Brazil's bioeconomy—old and new—seems therefore fit for dominant agribusiness actors. While indigenous people and other traditional communities have been recognized as the best in conserving forests in Latin America [69], it is unclear how they may gain or at least not experience further encroaching and dispossession. The following section analyzes how power workings of various kinds have shaped the Brazilian bioeconomy this way.

4. The Brazilian Bioeconomy and Its Powers That Be

4.1. Instrumental Forms of Power

Corporate agribusiness's instrumental power has manifested itself in at least three ways to shape Brazil's bioeconomy to date—both its policy agenda and concrete practices. First, agribusiness has abundantly used its vast material capabilities (e.g., financial resources, control over relevant technology) to expand its activities. That includes widening their market portfolio by pursuing technical innovations that suit their interests (i.e., novel goods produced from the crops they control). Power is also manifested—and increases—through sheer area expansion. Investments in soy's highly consolidated sector tripled its cropland area in Brazil between 1990 and 2020. That means that more of the country's land (and, at times, water) resources fall under an agribusiness that has no place for smallholders due to economies of scale requirements and where a handful of traders control the bulk of the market [20]. Corn, intercropped with soy, has accompanied that expansion and quickly become a sizeable ethanol feedstock [70]. Corporate agribusiness therefore utilizes its already-existing economic prowess, using multiple forms of "power to" (e.g., financial resources, control over land, technology) to get things done at scale and rapidly *be* the bioeconomy.

If, on the one hand, agribusiness actors disproportionately have the instrumental power to get things done on a practical level due to their economic prowess, on the other they have been stimulated by the bioeconomy, in a positive feedback loop [28,71]. Continual expansion means the economic power base of corporate agribusiness grows. That, in turn, allows for increasing say and leverage over agri-food and (other) biomass-based systems. For instance, while Brazil's sugarcane area has not expanded as much as soy's, it still doubled in the late 2000s during the country's latest ethanol boom [28]. The sector also continuously succeeds in conquering new markets (e.g., electricity) and developing new applications (e.g., cane-based "green plastics") [17,59]. In short, agribusiness uses its amassed power to create self-serving technological and innovation pathways, repeating in the bioeconomy what it does in the broader agri-food realm [16].

Second, besides using its material capabilities directly to expand its economic activities, agribusiness as an interest group has long had Brazil's most powerful parliamentary representation, too, and counts on notoriously successful lobbying [12,72]. Helping large agroindustries such as the sugarcane one has always been a key reason for the Brazilian government's creation of captive biofuel markets through blending mandates [28]. Such a form of political lobbying is a typical example of instrumental power use in agri-food governance [11], and its results yet another noxious effect of corporate dominance [16].

It is important to observe, however, that not all political lobbying relevant to shaping the bioeconomy relates specifically to it. Plenty of lobbying molds Brazil's broader agribusiness practices and, thus, shapes its bioeconomy production base indirectly. That includes, among others, land use policies that disproportionately benefit corporate agriculture [72,73] and lax rules on pesticide utilization suited for industrial monocultures [74]. Crucially, it also includes efforts to hinder collective land rights recognition, such as in indigenous territories that could block agribusiness expansion—as in the case of the sugarcane industry and the Guarani-Kaiowá people in Mato Grosso do Sul State [21].

The third prevalent form of instrumental power shaping Brazil's bioeconomy is corporate agribusiness's capacity to coercively displace competition that could eventually pursue alternative development pathways. Agribusiness's direct power over other actors is not limited to their sway over politicians or policymakers, it also affects the Judiciary. Even if often unconstitutional, law enforcement is regularly used to evict communities from disputed lands and to combat rural social movements that mobilize local resistance [75]. Between 2019 and 2020, a federal investigation arrested numerous judges that had colluded with large-scale farmers to favorably address land conflicts in the agricultural frontier region of Matopiba—an area of aggressive expansion of soy, Brazil's main biodiesel feedstock [76].

Besides manipulating the state apparatus to serve their interests, large agribusiness sometimes relies also on extra-legal—when not illegal or outright criminal—forms of coercion. That includes the forceful appropriation of land and water resources (land and water "grabbing") as well as intimidation and violence against local communities, rural social movements, or traditional populations in frontier regions [20]. Even if they are not always acknowledged as a form of power in governance, such direct actions in the end are critical for how agri-food systems—and the bioeconomy—become shaped.

While far from being an exhaustive list, Table 2 presents some key bioeconomy policies that have provided large agroindustry sectors with tax cuts, subsidized credit, public funding for R&D, biofuel blending mandates, as well as laws that facilitate business-controlled commodification of nature. As we shall see, those policies have sometimes responded to global trends (e.g., oil price hikes in the 1970s and 2000s, besides changing dynamics in international sugar markets in ethanol's case). Yet, *how* Brazil responded to such challenges—and who benefits from the chosen responses—are, in part, what indicates who holds power and how that power is manifested.

Table 2. A summary of Brazilian policies for bioeconomy promotion.

Year	Policy	Description
1931	E5 on imported gasoline	Mandatory blending of 5% of sugarcane-ethanol in all imported gasoline
1938	E5 all-across	E5 blending mandate extended to all gasoline, imported or not
1971	National Program of Sugarcane Improvement (*Planalsucar*)	Public R&D funding for sugarcane yield improvements
1975	Pro-Alcohol ethanol program (with E22)	Public funding for ethanol distilleries; mandatory 22% blending of ethanol in all gasoline
1979	Pro-Alcohol (Phase II)	Fiscal incentives for the automobile industry to produce cars running on 100% ethanol (E100)
2003	Flex-fuel cars	Fiscal incentives for the production and purchasing of cars able to run on any mixture of ethanol and gasoline
2004	National Program on Biodiesel Production and Use (*PNPB*)	Phase-in of mandatory biodiesel blending (B5 by 2013). Social Fuel Seal created as a certificate of smallholder inclusion, incentivized through preferential procurement and additional fiscal benefits.
2006	National Agroenergy Plan	Framework announcing public biofuel R&D and broad policy goals
2009	Sugarcane zoning policy introduced	Restriction of public credit eligibility to sugarcane cultivation outside ecologically sensitive biomes (e.g., the Amazon)
2009	Social Fuel Seal requirements hardened	Farming contracts between smallholder suppliers and biodiesel companies require approval by some rural worker union or collective organization
2014	New biodiesel blending mandates	Phase-in timeline for higher blends (B10 by 2018)
2015	Biodiversity Law (13.123/2015)	Legal framework for R&D and economic use of Brazilian biodiversity and its genetic resources
2017	National Biofuels Policy (*RenovaBio*)	Creation of a "decarbonization credits" market linked to carbon intensity reduction targets in Brazil
2018	New biodiesel blending mandates	Phase-in timeline for higher blends (B15 by 2023)
2019	Sugarcane zoning policy abolished	End of the area-based credit restrictions for sugarcane
2019	Social Fuel Seal requirements softened	Larger cooperatives become eligible as suppliers; end of the approval requirement by a rural worker union
2021	Payments for Environmental Services Law (14.119/2021)	Legal framework allowing payments for environmental services *even in untitled lands*, based on self-declaratory entries on the CAR registry

Sources: [28,77–79].

4.2. Structural Power

Structural power arguably emerges as an extension of what agribusiness can do through instrumental power. It then acquires an additional, analytically distinct facet. For instance, agrochemical companies that put together "technology packages" involving genetically modified seeds, fertilizers and pesticides keep great control over how crops such as soy and corn are produced [5]. Their initial instrumental power in creating those innovations then becomes also structural power in those sectors.

In Brazil's bioeconomy governance, agribusiness's structural power manifests itself in at least two complementary ways: (a) control over public institutions, building on path dependencies; and (b) agenda setting of overall agricultural development and specifically on bioeconomy pathways. Such an agenda-setting power grants corporate agribusiness the ability to exclude unwanted issues from sustainability debates, effectively preventing them from being addressed or even acknowledged in certain governance venues.

4.2.1. Agribusiness's Hold of Public Institutions

The policies displayed in Table 2 reveal how the Brazilian state has actively promoted biofuels (as the most developed bio-based sector) over three broad periods: the 1930s and the 1970s, before a resurgence in the 2000s' under a sustainability rationale. While part of the logic from the beginning was to reduce Brazil's oil import dependence and expenditures, those policies have also been mechanisms to support large agribusiness with public funding, credit, R&D investments, and new markets.

While rhetorically the emphasis is usually placed on the (globally benign) fossil-fuel replacement aspect of these policies, the adoption of sugarcane-ethanol, for one, has been equally—if not more—about supporting the sugarcane agroindustry in the face of low sugar prices and other market challenges [28,77,78]. The sugarcane sector had landed elites that for centuries enjoyed economies of scale and a privileged political position in the country. Such economies of scale and path dependency were associated with a measure of control over public institutions that has long made sugarcane Brazil's "favorite" feedstock, to the detriment of smallholders growing other crops [80]. Those public institutions then were transformed accordingly to create new incentives and structures that reinforced large agribusiness's hold over the country's ethanol sector and nascent bioeconomy. Once sustainability concerns came to the fore in the 2000s, a large sugarcane-ethanol sector was already prominent and could boast at least thirty years of commercial production experience. Similarly, once a biodiesel policy was introduced in 2004, the established soy sector managed to claim nearly all of the newly-created captive market despite the government's alleged smallholder inclusion goals.

Large agribusiness would become further empowered and, in time, support Brazil's growing shift to the right of the political spectrum [81,82]. In 2016, after President Rousseff's controversial impeachment, a new government promptly abolished the smallholder-oriented Ministry of Agrarian Development and, thereby, substantially cut funding and support for small-scale agriculture [83]. Corporate agribusiness would essentially crowd out those weaker players from Brazil's bioeconomy agenda. Since 2019, rule changes have allowed commercial soy-grower cooperatives to qualify as "family farmers" and thus benefit from the Social Fuel Seal policy originally conceived for smallholder inclusion and rural poverty alleviation [28]. In line with this and growing agribusiness representation in the federal government, recent government publications then started taking corn-and-soy endeavors from wealthy landholders as examples of biofuels originating from "family agriculture" ([70], p. 66). If anything, as a result of power feedback loops, consolidation in the hands of corporate agribusiness has only increased with Brazil's bioeconomy.

4.2.2. Setting the Agenda: What to Look at and How

Agenda-setting generally has two levels: (1) the definition of *what* comes onto the agenda and what is left out; and (2) questions of "attribute salience", i.e., *how* issues and actors are presented, which aspects are emphasized, and which ones are downplayed or obfuscated [84]. A broad literature on issue-framing points out how situations get recognized as problems only when successfully framed as such [85,86].

By pushing certain socio-economic and environmental issues out of Brazil's bioeconomy agenda, corporate agribusiness has been able to portray itself as sustainable and to offer its preferred bioeconomy pathways as a desirable option. Even though the bioeconomy agenda generally has sustainable development as a defining rationale and its very *raison d'être*, there are conspicuous absences. For instance, issues of agrobiodiversity loss, unsustainable water use by large-scale agriculture, land rights violations or widespread pesticide contamination—to name but a few—all are salient topics that nevertheless do not usually appear in Brazil's bioeconomy discussions [9,46,47]. For instance, the National Industries' Confederation flagship report on the bioeconomy extols sugarcane's multiplicity of products and boasts about the copious economic potentials from Brazil's biodiversity without ever acknowledging that monoculture expansion is a crucial driver of deforestation and overall biodiversity loss in the country [17].

In the same vein, as Brazil becomes the world's largest user of pesticides, it utilizes an abundance of known carcinogenic substances forbidden in Europe and other parts of the world. Yet, pesticides such as paraquat and other highly toxic substances have been applied on an increasingly large scale and gain speedy adoption in Brazil due to agribusiness's sway over the approval process [74]. Environmental and human contamination from increasing pesticide use in soy, corn and sugarcane crop fields are significant [20], but they are not acknowledged as a sustainability issue at the heart of a bioeconomy largely based on those monocultures. Arguably, the invisibility conferred to some of these environmental issues may suggest a level "solution aversion" in those dominant players, as their acknowledgement could bring the whole mainstream industrial agriculture into question [87].

Novel bioeconomy strands, too, seem poised to benefit business disproportionately. The 2015 biodiversity law empowering well-endowed industries instead of local communities is a case in point, and a look at the ways the CAR system has been used—and may come to be used, to draw payments for environmental services—is revealing of who is set to win the most and who may lose. As such laws privilege private land tenure over indigenous territory recognition or other collective land titling, the monetarization of "environmental services", and business-oriented rules for biodiversity use, they create a legal framework that secures disproportionate structural power to corporate actors. These actors indeed have eagerly benefited from further commodification of nature and control over natural resources, as seen. Meanwhile, indigenous and other Amazonian communities not only have been mostly left out of what so far has essentially been an effort to "mine the ecosystem" for more commodities, but they also stand to lose as these legal frameworks linked to bioeconomy promotion facilitate accumulation by dispossession in their environments [64,65].

Such a structural power of agribusiness in Brazil involves not only public but also private governance instances such as the various multi-stakeholder initiatives related to agriculture. Certification mechanisms such as the Round Table on Responsible Soy (RTRS) or Bonsucro (formerly Better Sugarcane Initiative) essentially represent agroindustry interests [88]. Even the more encompassing multi-stakeholder initiatives such as the Soy Working Group or the Cerrado Working Group, composed of environmental NGOs and commodity traders to govern agribusiness's "sustainable" expansion, tend to be busy mostly with the promotion of "best practices" and voluntary zero-deforestation commitments [89]. While that is often taken to represent "sustainable agriculture", this framing is deceptive as a plethora of environmental issues—and, thus, industrial monocultures' broad unsustainability—remain unchecked [90].

These various forms of structural power result in the exclusion of unwanted stakeholders, thorny issues, and competing narratives or alternative development pathways [20]. While endorsing and boosting conventional large-scale agriculture, this mainstream bioeconomy keeps alternative ways of rural development—as well as the local stakeholders that advocate for them—out of the agriculture sustainability debates and institutional spaces where such debates are held [89,91]. Through structural power, corporate agribusiness strategically makes local stakeholders, their concerns, views and visions invisible, even in the venues nominally dedicated to sustainability.

4.3. Discursive Power

As Gramsci put it, certain actors' political preeminence relies not only on material dominance but also on their portrayal as *"intellectual and moral leadership"*, from which emerges social legitimacy and, ultimately, consent ([92], pp. 182, 269). In Brazil, two key tactics underscore corporate agribusiness's discursive power over diverse social actors and the public at large: the portrayal of a corporate-controlled bioeconomy as a desirable, environmentally friendly endeavor; and the framing of "Brazilian" agribusiness as national champions whose successes and setbacks are tied to that of Brazil itself. As we shall see, the playing out of these two tactical discourses works to multiple effects and, eventually, also coalesce around a grand narrative of the bioeconomy as *the* way for Brazil to become a "great power".

Agribusiness in Brazil has a long history of associating its interests with that of the country in the public mind. Efforts, for instance, to collate agribusiness development with Brazilian pride, identity and self-image date as far back as the 1930s, with the promotion of banana plantations at the times of the famed "banana republics" [93]. It is not possible to understand Brazil's mainstream discourse around the bioeconomy without situating this in a longstanding effort by large agribusiness to earn public legitimacy, broad political support, and a social license to operate. On the backdrop of bioeconomy promotion is an overall framing of Brazil as a global protein breadbasket and, increasingly, also a supplier of other bio-based goods and (salable) environmental services [17]. Under a globally dominant neo-Malthusian food security narrative, the country is to fulfill a supposedly natural vocation as a major agricultural producer and meet the food demands of a growing world population [94,95]. In the minds of an increasingly evangelical popular base in Brazil's countryside, that operates as the country's "calling" to expand production with quasi-missionary zeal [96].

In the face of growing concerns about the "reprimarization" of Brazil's economy in recent years (that is, its growing economic dependence on agriculture and mining sectors in tandem with significant deindustrialization [97]), agribusiness has also been deft to portray itself as technologically advanced. In a broadly popular ongoing marketing campaign since 2019, the agricultural sector has portrayed itself as an "industry" to be regarded as the "wealth of the nation;" with prime-time commercial ads showing large-scale production, corporate agribusiness adopts a more informal shorthand—"agro"—and markets itself as being "tech", "pop", and "everything" (*o agro é tech, o agro é pop, o agro é tudo*) [98]. As such, a highly exclusive business group with considerable multinational capital poses as a national champion that should be the pride of Brazilians [99].

Environmental or human rights critiques hence become framed as outsiders' jealous attempts to undermine Brazil's development. Domestic NGOs who join that chorus are tarnished as a "fifth column" working for foreign interests. In Brazilian media coverage of international critiques against deforestation in Brazil—such as in the context of the EU-Mercosur trade agreement, which would see an increase in the exports of agricultural commodities from Brazil—it is routine for considerations to be made implying that, in truth, those are protectionist concerns and excuses due to fear of competition with the powerful Brazilian agriculture. While there may well be some truth to that, such a reasoning is tactically used to dismiss environmental critiques entirely [100]. Only "constructive" NGOs—which do not question the premises of Brazil's agribusiness expansion or bring thorny issues to the fore—are recognized as legitimate interlocutors and eventually welcomed in agri-food sustainability debates [89].

The bioeconomy strategically enters this setting as a way to boost both the "green" and the "high-tech" images corporate agribusiness wishes to confer to itself. As elsewhere in Latin America, the sector puts forth an "economic imaginary" grounded on crop-based technological developments and value-added from agriculture [101]. The bioeconomy, in Brazil's case, provides it with particularly "green" hues due to a growing emphasis on the economic potentials of Amazonian biodiversity. It is seen as offering nearly endless potential for value-chain creation with genetic improvements on "wild" foods and the commercialization of novel products through biotechnology (e.g., enzymes, pharmaceutical components, cosmetics) [17]. As elsewhere, technological progress is conflated with overall societal improvements, and a future is envisioned where scientific innovation supersedes all social problems and conflicts [102]. Noticeably, the bioeconomy here has little to do with replacing fossil-based products and more with sheer bio-based economic development. Some Brazilian scientists have, for instance, put forth an ambitious "Amazonia 4.0" agenda to promote economic and technological upgrading in the region, using its vast biodiversity to engender a socio-economic transformation [32]. The Bolsonaro administration's Ministry of Agriculture has endorsed the initiative and often pays lip service to such a bio-based economic development [103], but espousing deforestation activities all the while [104].

As Brazil has been on international headlines due to its soaring deforestation rates that tarnish agribusiness reputation and threaten to close export markets (notably in Europe), the bioeconomy also becomes part of an effort to "green" its image. Part of the sector's hope is that the bioeconomy may help Brazil join the Organisation for Economic Co-operation and Development (OECD) and dispel European hesitations regarding the EU-Mercosur free-trade agreement [105]. Finally, this effort blends with nationalistic grandeur to harness further public support for agribusiness and agro-technology also within Brazil. Drawing from an article in *The Economist* (possibly for increased legitimacy, as if it were merely reiterating what others abroad are saying), the National Industries Confederation has proposed that *"with bioeconomy development, there is a unique opportunity for Brazil to become one of the world's great powers."* ([17], p. 77) (see also [106]).

Once again, the tactic is to stir a sense of national pride where Brazil's fate is governed by the wishes and whims of corporate agribusiness. As some authors put it, *"such promises and the expectations they generate are performative: they act to build consensus around a techno-logical project; mobilize investment; enroll scientific, social, and economic actors; and construct a case for facilitative and supportive legislation"* ([102], p. 13). In Brazil's case, however, the bioeconomy's utopia of *"overcoming environmental, social, and economic challenges through biotechnological progress alone"* ([102], p. 12) includes also geopolitical achievements.

5. Discussion: Sustainable Development or Conservative Ecological Modernization?

Some authors have long noted that the bioeconomy, of which biofuels remain the leading sector to date, consists mainly of politically instituted markets. In other words, they are not markets that spontaneously arise out of consumer demand, but instead are created from "above" via directives, blending mandates, and other public policy determinations [107]. To a large extent, public policies have not only created such markets but also shaped them [28]. Some critics have, therefore, flagged the bioeconomy as "a political project" to reassert the interests of capital [102]. Other times, its expansion is rationalized as countries pursuing their national interests in an unregulated international space and sometimes with consequences beyond borders [108]. States have, indeed, been active—and some would say crucial—bioeconomy promoters worldwide [28,107]. Still, it is worth remembering that the state is neither an isolated entity nor a monolith; rather, the state usually expresses the will of competing interest groups and advocacy coalitions; it is an arena where some policy preferences become structured [109]. Therefore, understanding the inner workings of how bioeconomy politics takes place is of paramount importance.

Brazil's case shows a very supply-driven bioeconomy. Corporate agribusiness, amass-ing increasing land for a few "flex-crop" commodities (frequently at the cost of deforestation or smallholder displacement), mobilizes to develop more uses for them and thereby en-joy a wider variety of markets, greater demand, and better prices. The bioeconomy has therefore allowed such agribusiness actors to sustain and reinforce dominant positions, becoming all the more resilient in the face of international price volatility. The sugarcane sector, for one, has achieved that by transitioning seamlessly from a primary producer of sugar to becoming mostly dedicated to agroenergy—while retaining and expanding on the significant economic and political weight it historically inherited. The intercropped soy-corn duo, too, even if to a lesser extent, has also found in the bioeconomy new markets that help those producers modulate commodity prices.

Growing economic prowess, in turn, translates into political and social influence, shaping government agendas and the public discourse. Brazil's biodiversity may be broadly used as window-dressing for the bioeconomy, but in practice largeholder crops dominate the sector. The regard for biodiversity conservation or social inclusion remains entirely aspirational, if not deceptive. Table 3 summarizes the power uses through which such a corporate agribusiness dominance has been accomplished.

Table 3. Uses of power in Brazil's bioeconomy.

Power Typologies: Ends and Means	Instrumental	Structural	Discursive
Innovative	Technical innovations (e.g., biofuels, feedstock processing pathways, additional bio-products)	Institutional innovations for agenda-setting or privileged access to new markets or governance (e.g., private certification and governance instances, a credits market under the *RenovaBio* program.)	Innovative ideas and new framings (e.g., the "bioeconomy" label itself as a new framing for the well-established bioenergy industry)
Transformative	Political lobbying to secure new public policy incentives such as funding or tax cuts to agribusiness.	Changing agendas, excluding unwanted issues from the fore (e.g., abolition of sugarcane zoning, modification of rules that hitherto restricted large agribusiness, such as the Social Fuel Seal's flexibilization, virtually emptying it of its poverty-reduction rationale.)	The effort to transform the agri-food sustainability debate into a question only of efficiency and renewability, purposefully leaving out various social and environmental issues (e.g., land rights, water access, agrobiodiversity loss) from the public mind or the debate.
Reinforcive	Expansion of material capabilities (e.g., crop area, financial and technological resources) reinforcing Brazil's economic dependence on—and, thus, the political leverage of—corporate agribusiness in the country.	Creation of path dependencies around conventional, input-intensive and corporate-controlled monocultures (e.g., sugar economies giving rise to a dominant sugarcane-ethanol industry and, increasingly, entire sugarcane-based value webs as opposed to value webs from other crops.)	Legitimacy strengthening; the bioeconomy as a benign umbrella (a) portraying Brazilian agribusiness as a technologically advanced national champion, responding to growing concerns about "reprimarization" of the country's economy while (b) giving it "green" hues and shielding it from environmentalist critiques

In a neo-Gramscian sense, corporate agribusiness thus enjoys substantive hegemony in Brazil, achieved *"through the coercive and bureaucratic authority of the state, dominance in the economic realm, and the consensual legitimacy of civil society"* ([40], p. 806). The Brazilian bioeconomy clearly has no level playing field. Corporate agribusiness captures most if not all such new markets due to skewed material capabilities and power configurations. Often, indeed, entire bioeconomy segments (e.g., corn-based and sugarcane-based ethanol) serve primarily for the repurposing of supplies in captive markets. If others have noted that many of the emerging renewable energy regimes across the Americas have reproduced pre-existing inequalities [110], this analysis now exposes the details of their power workings and suggests that inequalities, as a consequence, are not just reproduced but have been widened.

A counter-hegemonic movement exists: various civil society organizations and rural social movements have coalesced around the banners of agroecology and food sovereignty to call for sustainable agri-food systems and, at times, an inclusive bioeconomy [28,91]. For instance, the concept of *alimergia*—a merger of the words "food" (*alimento*) and "energy" (*energia*) in Portuguese or Spanish—has been espoused by some of those critics of corporate agribusiness and advocates of locally-controlled farming [111]. Yet, if anything, these movements have become weaker since Brazilian politics shifted to the right and then to the far-right in the late 2010s [28]. It remains to be seen whether competing discourses and institutional agendas for the bioeconomy can eventually gain momentum.

Such an agribusiness-dominated bioeconomy represents, no doubt, a missed opportunity for sustainable development. There is little betterment to people's lives, a basic tenet of development [20,112], or social inclusiveness and poverty alleviation to speak of. Instead, the expansion of large-scale cash-cropping systems such as soy has been considered a driver of maldevelopment in Brazil, with numerous negative social impacts and resource dispossession for local populations [20]. As to sustainability, there is little of it in conventional input-intensive monocultures [113,114]. Even its "Brazilian" nationality is a misleading discursive tactic, as multinational corporations dominate the bulk of agribusiness in the country [99].

There is little development as such, entailing the overcoming of poverty and deprivation, but mostly technical and institutional innovations by and for those who already have power. This dynamic, in turn, reinforces their dominant positions in Brazil's overall agri-food governance. Innovative ideas such as Amazonia 4.0 are emptied of their hopes for "disruptive" or "transformative" change [32] and instead tamed into purely techno-economic upgrading and increased market opportunities for dominant actors. Such a pattern can be considered a case of *conservative modernization*, whereby political and social dominance structures largely remain in place or are even reinforced despite some economic and technological change [115,116]. As the bioeconomy is to a degree also a form of ecological modernization, which tries to reflect upon (some) environmental impacts and address them [117], what happens in Brazil can therefore be termed a form of conservative ecological modernization [28]. Not only are inequality matters tactically avoided, but environmental issues, too, are selectively addressed. Attention is limited to only relatively innocuous ones which do not require significant changes in the corporate agri-food system or threaten the dominant position of regime incumbents. Rather, these incumbents design the bioeconomy to precisely reinforce their positions of dominance. Without targeted incentives or redress measures that consider such pre-existing inequalities and skewed power relations in agri-food systems, the bioeconomy is therefore likely to expand on inequalities, entrench marginalization, and deepen exclusion.

6. Conclusions

This article has addressed the question of which stakeholders control the transition to circular bioeconomies—and how they do so. While the literature acknowledges that such social and political dimensions are important for understanding and steering these developments towards sustainability [2,6], this is among the first in-depth empirical studies on bioeconomy politics. The analysis shows that such questions are important not only in and of themselves but also because they bear consequences for the environmental and economic benefits of the transition.

Brazil's case shows that, while the bioeconomy has become an attractive umbrella term for environmentally-minded technical and institutional innovations, these have disproportionately benefited corporate agribusiness. Such already-dominant actors have resorted to instrumental, structural, and discursive power in multiple forms to shape bioeconomy policies and markets favorably. Under the guise of the public (environmental) good, they have transformed institutions and, ultimately, reinforced their dominant positions. That is problematic because, although closed-loop circular economies and bioeconomy value-addition are environmentally beneficial [3], socially inequitable production systems disempower vulnerable actors and concentrate economic benefits in a few hands, thus augmenting inequality. Moreover, such a corporate agribusiness dominance shapes technological and innovation pathways, limits potential developments to well-established sectors (e.g., sugarcane, soy), and expands on the environmental impacts these very systems create (e.g., freshwater depletion, deforestation). Such social equity considerations, therefore, are of paramount importance.

In Brazil, rather than sustainable development, the mainstream political project for its bioeconomy can be seen as a case of conservative ecological modernization. It promotes environmentally-minded technical and economic upgrading, but preserving social inequalities and reinforcing skewed power structures. Alas, being based primarily on industrial monocultures that coincidentally are also key drivers of water depletion, environmental pollution and land-use change, the bioeconomy further incentivizes these industries, their crops and production arrangements. It may, therefore, aggravate sustainability issues while precisely helping shield those agroindustries from environmental critiques. Portraying corporate agribusiness as national champions whose thriving would be equated to that of Brazil—in a bio-based, contemporary version of what General Motors once famously claimed to do for the United States—these actors also successfully earn further societal legitimacy, even while leading environmental destruction and social exclusion in the country.

Given that agri-food system consolidation has been a global trend, such an agribusiness dominance over emerging bioeconomies raises warning signs also to other countries, for in other similar contexts those powerful players may be similarly advantageously positioned to reap most benefits and set the bioeconomy agenda. It is important to reflect on who stands to benefit (and to lose) from the promotion of other circular economies, too. Further research is needed to understand such politics and how to make circular and bioeconomy promotion more equitable and sustainable.

The bioeconomy may well be an inevitable transition if fossil resources are to be phased away. Its potentials for inclusive and sustainable development remain in place, but these are not achievements to be taken for granted as a natural consequence of furthering bio-based sectors. It appears that redress efforts to correct for existing imbalances are needed if such new developments are not to fuel existing unsustainable systems under the guise of "green" progress. How to avoid such an elite capture through skillful bioeconomy governance, and how to effectively deliver on its social and environmental potentials, remain perhaps the most critical research and policy questions yet to be addressed.

Funding: This research received no external funding.

Institutional Review Board Statement: Not applicable.

Informed Consent Statement: Not applicable.

Data Availability Statement: Data sharing not applicable to this article.

Acknowledgments: The author would like to thank Carole-Anne Sénit, Karen Siegel and three anonymous reviewers for their helpful comments and suggestions on earlier versions of this article.

Conflicts of Interest: The author declares no conflict of interest.

References

1. IPCC. *Global Warming of 1.5 °C: An IPCC Special Report on the Impacts of Global Warming of 1.5 °C above Pre-Industrial Levels and Related Global Greenhouse Gas Emission Pathways, in the Context of Strengthening the Global Response to the Threat of Climate Change, Sustainable Development, and Efforts to Eradicate Poverty*; World Meteorological Organization: Geneva, Switzerland, 2018.
2. D'Adamo, I.; Falcone, P.M.; Morone, P. A New Socio-economic Indicator to Measure the Performance of Bioeconomy Sectors in Europe. *Ecol. Econ.* **2020**, *176*, 106724. [CrossRef]
3. D'Adamo, I.; Falcone, P.M.; Huisingh, D.; Morone, P. A circular economy model based on biomethane: What are the opportunities for the municipality of Rome and beyond? *Renew. Energy* **2021**, *163*, 1660–1672. [CrossRef]
4. McMichael, P. The land grab and corporate food regime restructuring. *J. Peasant. Stud.* **2012**, *39*, 681–701. [CrossRef]
5. Clapp, J. Explaining Growing Glyphosate Use: The Political Economy of Herbicide-Dependent Agriculture. *Glob. Environ. Chang.* **2021**, *67*, 102239. [CrossRef]
6. Bastos Lima, M.G. Toward Multipurpose Agriculture: Food, Fuels, Flex Crops, and Prospects for a Bioeconomy. *Glob. Environ. Politics* **2018**, *18*, 143–150. [CrossRef]
7. OECD (Organization for Economic Co-operation and Development). *The Bioeconomy to 2030: Designing a Policy Agenda*; OECD: Paris, France, 2009.
8. European Commission. *A Sustainable Bioeconomy for Europe: Strengthening the Connection Between Economy, Society and the Environment. Directorate General for Research and Innovation*; European Commission: Brussels, Belgium, 2018.

9.   MCTIC. *Plano De Ação Em Ciência, Tecnologia E Inovação Em Bioeconomia. Ministério Da Ciência, Tecnologia, Inovação E Comunicações*; Centro de Gestão e Estudos Estratégicos: Brasília, Brazil, 2018.
10.   Willerding, A.L.; Da Silva, L.R.; Da Silva, R.P.; Oliveira De Assis, G.M.; Monteiro De Paula, E.V.C. Estratégias para o desenvolvimento da bioeconomia no estado do Amazonas. *Estud. Avançados* **2020**, *34*, 145–165. [CrossRef]
11.   Clapp, J.; Fuchs, D. (Eds.) *Corporate Power in Global Agrifood Governance*; MIT Press: Cambridge, MA, USA, 2009.
12.   Søndergaard, N. Food regime transformations and structural rebounding: Brazilian state–agribusiness relations. *Territ. Politics Gov.* **2020**. [CrossRef]
13.   McMichael, P. A food regime genealogy. *J. Peasant. Stud.* **2009**, *36*, 139–169. [CrossRef]
14.   Bastos Lima, M.G.; Visseren-Hamakers, I.J.; Braña Varela, J.; Gupta, A. A reality check on the landscape approach to REDD+: Lessons from Latin America. *For. Policy Econ.* **2017**, *78*, 10–20. [CrossRef]
15.   Nestle, M. *Food Politics: How the Food Industry Influences Nutrition and Health*; University of California Press: Berkeley, CA, USA, 2013.
16.   Clapp, J. The problem with growing corporate concentration and power in the global food system. *Nat. Food* **2021**. [CrossRef]
17.   CNI. *Bioeconomia E a Indústria Brasileira*; Confederação Nacional da Indústria: Brasília, Brazil, 2020.
18.   Delvenne, P.; Hendrickx, K. The multifaceted struggle for power in the bioeconomy: Introduction to the special issue. *Technol. Soc.* **2013**, *35*, 75–78. [CrossRef]
19.   Piotrowski, M. *Nearing the Tipping Point: Drivers of Deforestation in the Amazon Region*; Inter-American Dialogue: Washington, DC, USA, 2019.
20.   Russo Lopes, G.; Bastos Lima, M.G.; Reis, T.N.P. Maldevelopment revisited: Inclusiveness and the impacts of soy expansion over Matopiba in the Brazilian Cerrado. *World Dev.* **2021**, *139*, 105316. [CrossRef]
21.   Ioris, A.A.R. Indigeneity and political economy: Class and ethnicity of the Guarani-Kaiowa. *Cap. Cl.* **2021**. [CrossRef]
22.   Avelino, F. Power in sustainability transitions: Analysing power and (dis)empowerment in transformative change towards sustainability. *Environ. Policy Gov.* **2017**, *27*, 505–520. [CrossRef]
23.   Kohler, J.; Geels, F.W.; Kern, F.; Markard, J.; Onsongo, E.; Wieczorek, A.; Alkemade, F.; Avelino, F.; Bergek, A.; Boons, F. An agenda for sustainability transitions research: State of the art and future directions. *Environ. Innov. Soc. Transit.* **2019**, *31*, 1–32. [CrossRef]
24.   Voss, J.; Bornemann, B. The politics of reflexive governance: Challenges for designing adaptive management and transition management. *Ecol. Soc.* **2011**, *16*, 9. [CrossRef]
25.   Scordato, L.; Bugge, M.M.; Fevolden, A.M. Directionality across diversity: Governing contending policy rationales in the transition towards the bioeconomy. *Sustainability* **2017**, *9*, 206. [CrossRef]
26.   Bugge, M.M.; Hansen, T.; Klitkou, A. What is the bioeconomy? A review of the literature. *Sustainability* **2016**, *8*, 691. [CrossRef]
27.   Sabatier, P.; Weible, C.M. The advocacy coalition framework: Innovations and clarifications. In *Theories of the Policy Process*, 2nd ed.; Sabatier, P., Ed.; Westview Press: Boulder, CO, USA, 2007; pp. 189–222.
28.   Bastos Lima, M.G. *The Politics of Bioeconomy and Sustainability: Lessons from Biofuel Governance, Policies and Production Strategies in the Emerging World*; Springer: Dordrecht, The Netherlands, 2021.
29.   Partzsch, L. 'Power with' and 'power to' in environmental politics and the transition to sustainability. *Environ. Politics* **2017**, *26*, 193–211. [CrossRef]
30.   Cook, S.; Smith, K.; Utting, P. *Green Economy or Green Society? Contestation and Policies for a Fair Transition. Social Dimensions of Green Economy and Sustainable Development*; Occasional Paper 10; United Nations Research Institute for Social Development: Geneva, Switzerland, 2012.
31.   UNCSD. *The Future We Want. A/RES/66/288*; United Nations Conference on Sustainable Development: Rio de Janeiro, Brazil, 2012.
32.   Nobre, I.; Nobre, C. The Amazonia third way initiative: The role of technology to unveil the potential of a novel tropical biodiversity-based economy. In *Land Use—Assessing the Past, Envisioning the Future*; Loures, L., Ed.; IntechOpen: London, UK, 2019. [CrossRef]
33.   Dahl, R. The concept of power. *Behav. Sci.* **1957**, *2*, 201–215. [CrossRef]
34.   Bachrach, P.; Baratz, M.M. Two faces of power. *Am. Polit. Sci. Rev.* **1962**, *56*, 947–952. [CrossRef]
35.   Willer, D.; Lovaglia, M.J.; Markovsky, B. Power and influence: A theoretical bridge. *Soc. Forces* **1997**, *76*, 571–603. [CrossRef]
36.   Lukes, S. *Power: A Radical View*; Palgrave Macmillan: London, UK, 2005.
37.   Cox, R. *Production Power and World Order: Social Forces in the Making of History*; Columbia University Press: New York, NY, USA, 1987.
38.   Morton, A.D. *Unravelling Gramsci: Hegemony and passive revolution in the global economy*; Pluto Press: London, UK, 2007.
39.   Dryzek, J. *The Politics of the Earth: Environmental Discourses*; Oxford University Press: Oxford, UK, 2005.
40.   Levy, D.L.; Egan, D. A neo-Gramscian approach to corporate political strategy: Conflict and accommodation in the climate change negotiations. *J. Manag. Stud.* **2003**, *40*, 803–829. [CrossRef]
41.   Finnemore, M.; Sikkink, K. Taking stock: The constructivist research program in international relations and comparative politics. *Annu. Rev. Polit. Sci.* **2001**, *4*, 391–416. [CrossRef]
42.   Suchman, M.C. Managing legitimacy: Strategic and institutional approaches. *Acad. Manag. Rev.* **1995**, *20*, 571–610. [CrossRef]
43.   Gehman, J.; Lefsrud, L.M.; Fast, S. Social license to operate: Legitimacy by another name? *Can. Public Adm.* **2017**, *60*, 293–317. [CrossRef]
44.   Nogueira, L.H.; Cantarella, H.; Souza, G.M.; Maciel Filho, R.; Cassinelli, L.F.D. Opinião—Bioenergia e Bioeconomia: É Preciso Manter o Rumo Certo. UNICA. Available online: https://unica.com.br/noticias/opiniao-bioenergia-e-bioeconomia-e-preciso-manter-o-rumo-certo/ (accessed on 27 April 2021).

45. Sadik-Zada, E.R. Natural resources, technological progress, and economic modernization. *Rev. Dev. Econ.* **2021**, *25*, 381–404. [CrossRef]
46. Oliveira e Silva, M.F.; Pereira, F.S.; Martins, J.V.B. A bioeconomia brasileira em números. *BNDES Set.* **2018**, *47*, 277–332.
47. Portugal, T. Programa Bioeconomia Brasil Sociobiodiversidade. Ministério da Agricultura, Pecuária e Abastecimento. Available online: https://www.gov.br/agricultura/pt-br/assuntos/camaras-setoriais-tematicas/documentos/camaras-setoriais/hortalicas/2019/58a-ro/bioeconomia-dep-saf-mapa.pdf (accessed on 27 April 2021).
48. REN21. *Renewables 2019 Global Status Report*; REN21 Secretariat: Paris, France, 2019.
49. EPE. *Balanço Energético Nacional: Ano Base 2019*; Ministério de Minas e Energia, Empresa de Pesquisa Energética: Rio de Janeiro, Brazil, 2020.
50. Peduzzi, P. Conab: Produção De Etanol A Partir Do Milho É Tendência Cada Vez Maior. Agência Brasil. Available online: https://agenciabrasil.ebc.com.br/economia/noticia/2019-05/producao-de-etanol-partir-do-milho-e-tendencia-maior-diz-conab (accessed on 30 March 2021).
51. EPE. *Balanço Energético Nacional: Ano Base 2018*; Ministério de Minas e Energia, Empresa de Pesquisa Energética: Rio de Janeiro, Brazil, 2019.
52. Novaes, J.R. Campeões de produtividade: Dores e febres nos canaviais paulistas. *Estud. Avançados* **2007**, *21*, 167–177. [CrossRef]
53. Vicol, M.; Pritchard, B.; Htay, Y.Y. Rethinking the role of agriculture as a driver of social and economic transformation in Southeast Asia's upland regions: The view from Chin State, Myanmar. *Land Use Policy* **2018**, *72*, 451–460. [CrossRef]
54. FAO. Faostat: Crops. Rome: Food and Agriculture Organization. Available online: http://www.fao.org/faostat/en/#data/QC/visualize (accessed on 30 March 2021).
55. Delvenne, P.; Vasen, F.; Vara, A.M. The "soy-ization" of Argentina: The dynamics of the "globalized" privatization regime in a peripheral context. *Technol. Soc.* **2013**, *35*, 153–162. [CrossRef]
56. Rausch, L.L.; Gibbs, H.K.; Schelly, I.; Brandão, A., Jr.; Morton, D.C.; Filho, A.C.; Strassburg, B.; Walker, N.; Noojipady, P.; Barreto, P. Soy expansion in Brazil's Cerrado. *Conserv. Lett.* **2019**, *12*, e12671. [CrossRef]
57. Trase. Trase Yearbook 2020: The State of Forest Risk Supply Chains. Available online: https://insights.trase.earth/yearbook/contexts/brazil-soy/ (accessed on 30 March 2021).
58. Borras, S.M.; Franco, J.C.; Isakson, S.R.; Levidow, L.; Vervest, P. The rise of flex crops and commodities: Implications for research. *J. Peasant. Stud.* **2016**, *43*, 93–115. [CrossRef]
59. Scheiterle, L.; Ulmer, A.; Birner, R.; Pyka, A. From commodity-based value chains to biomass-based value webs: The case of sugarcane in Brazil's bioeconomy. *J. Clean. Prod.* **2018**, *172*, 3851–3863. [CrossRef]
60. OECD/FAO. *Agricultural Outlook 2019–2028*; OECD: Paris, France, 2019.
61. Lopes, J. Conjuntura Atual Para a Produção De Etanol De Milho. Stone X Brasil—Mercados Agrícolas. Available online: https://www.mercadosagricolas.com.br/acucar-e-etanol/conjuntura-atual-para-a-producao-de-etanol-de-milho/ (accessed on 30 March 2021).
62. Kohlmann, G.; Ferreira, J.; Leitão, S.; Rossi, T. *Destravando a Agenda Da Bioeconomia: Soluções Para Impulsionar O Uso Sustentável Dos Recursos Genéticos E Conhecimento Tradicional No Brasil*; Instituto Escolhas: São Paulo, Brazil, 2021.
63. Maes, J. Biodiversidade: O Que Está Por Trás Da Batalha Sobre a Lei Ambiental Mais Complexa Do Brasil. Gazeta do Povo. 17 November 2021. Available online: https://www.gazetadopovo.com.br/politica/republica/biodiversidade-o-que-esta-por-tras-da-batalha-sobre-a-lei-ambiental-mais-complexa-do-brasil-aq3no2cl6totcp89y49p1sgqs/ (accessed on 13 April 2021).
64. Latorre, S.; Farrell, K.N.; Martínez-Alier, J. The commodification of nature and socio-environmental resistance in Ecuador: An inventory of accumulation by dispossession cases, 1980–2013. *Ecol. Econ.* **2015**, *116*, 58–69. [CrossRef]
65. Torres, M. Grilagem para principiantes: Guia de procedimentos básicos para o roubo de terras públicas. In *Perspectivas De Natureza: Geografia, Formas De Natureza E Política*; Marques, M.I.M., Ed.; Annablume: São Paulo, Brazil, 2018; pp. 285–314.
66. Jung, S.; Dyngeland, C.; Rausch, L.L.; Vang Rasmussen, L. Brazilian land registry impacts on land use conversion. *Agric. Appl. Econ. Assoc.* **2021**. [CrossRef]
67. Martins, H.; Nunes, S.; Souza, C., Jr. *CAR—Cadastro Ambiental Em Áreas Protegidas*; Imazon: Belém, Brazil, 2018.
68. Sparovek, G.; Reydon, B.P.; Pinto, L.F.G.; Faria, V.; De Freitas, F.L.M.; Azevedo-Ramos, C.; Gardner, T.; Hamamura, C.; Rajão, R.; Cerignoni, F. Who owns Brazilian lands? *Land Use Policy* **2019**, *87*, 104062. [CrossRef]
69. FAO; FILAC. *Forest Governance by Indigenous and Tribal Peoples: An Opportunity for Climate Action in Latin America and the Caribbean*; Food and Agriculture Organization of the United Nations (FAO): Santiago, Chile, 2021.
70. EPE. *Análise Da Conjuntura De Biocombustíveis: Ano 2019*; Ministério de Minas e Energia, Empresa de Pesquisa Energética: Rio de Janeiro, Brazil, 2019.
71. Cudlínova, E.; Sobrinho, V.G.; Lapka, M.; Salvati, L. New forms of land grabbing due to the bioeconomy: The case of Brazil. *Sustainability* **2020**, *12*, 3395. [CrossRef]
72. Kroger, M. Inter-sectoral determinants of forest policy: The power of deforesting actors in post-2012 Brazil. *For. Policy Econ.* **2017**, *77*, 24–32. [CrossRef]
73. De Toledo, P.M.; Dalla-Nora, E.; Vieira, I.C.G.; Aguiar, A.P.D.; Araujo, R. Development paradigms contributing to the transformation of the Brazilian Amazon: Do people matter? *Curr. Opin. Environ. Sustain.* **2017**, *26*, 77–83. [CrossRef]
74. Rocha, G.M.; Grisolia, C.K. Why pesticides with mutagenic, carcinogenic and reproductive risks are registered in Brazil. *Dev. World Bioeth.* **2019**, *19*, 148–154. [CrossRef]
75. Carter, M. The landless rural workers movement and democracy in Brazil. *Lat. Am. Res. Rev.* **2010**, *45*, 186–217. [CrossRef]

76. Procuradoria-Geral da República. Operação Faroeste: Corte Especial Do Stj Mantém Prisão Preventiva De Investigados Em Esquema De Venda De Sentenças. Available online: http://www.mpf.mp.br/pgr/noticias-pgr/operacao-faroeste-corte-especial-do-stj-mantem-prisao-preventiva-de-investigados-em-esquema-de-venda-de-sentencas (accessed on 27 April 2021).

77. Szmrecsányi, T.; Moreira, E.F.P. O desenvolvimento da agroindústria canavieira do Brasil desde a Segunda Guerra Mundial. *Estud. Avançados* **1991**, *11*, 57–79. [CrossRef]

78. Moreira, E.F.P. Evolução E Perspectivas Do Comércio Internacional De Açúcar E Álcool. Ph.D. Thesis, Pontifícia Universidade Católica de São Paulo, São Paulo, Brazil, 2007.

79. Hall, J.; Matos, S.; Severino, L.; Beltrão, N. Brazilian biofuels and social exclusion: Established and concentrated ethanol versus emerging and dispersed biodiesel. *J. Clean. Prod.* **2009**, *17*, S77–S85. [CrossRef]

80. Sakai, P.; Afionis, S.; Favretto, N.; Stringer, L.C.; Ward, C.; Sakai, M.; Weirich Neto, P.H.; Rocha, C.H.; Alberti Gomes, J.; De Souza, N.M. Understanding the Implications of Alternative Bioenergy Crops to Support Smallholder Farmers in Brazil. *Sustainability* **2020**, *12*, 2146. [CrossRef]

81. Otsuki, K. Ecological rationality and environmental governance on the agrarian frontier: The role of religion in the Brazilian Amazon. *J. Rural. Stud.* **2013**, *32*, 411–419. [CrossRef]

82. Søndergaard, N. Reforming in a democratic vacuum: The authoritarian neoliberalism of the Temer administration from 2016 to 2018. *Globalizations* **2021**, *18*, 568–583. [CrossRef]

83. Siegel, K.M.; Bastos Lima, M.G. When international sustainability frameworks encounter domestic politics: The sustainable development goals and agri-food governance in South America. *World Dev.* **2020**, *135*, 105053. [CrossRef]

84. McCombs, M. *Setting the Agenda*; Polity Press: Cambridge, MA, USA, 2014.

85. Kingdon, J.W. *Agendas, Alternatives, and Public Policies*; Pearson Longman: Harlow, UK, 1995.

86. Knaggard, A. The multiple streams framework and the problem broker. *Eur. J. Polit. Res.* **2015**, *54*, 450–465. [CrossRef]

87. Campbell, T.H.; Cay, A.C. Solution aversion: On the relation between ideology and motivated disbelief. *J. Pers. Soc. Psychol.* **2014**, *107*, 809–824. [CrossRef] [PubMed]

88. Hospes, O. Marking the success or end of global multi-stakeholder governance? The rise of national sustainability standards in Indonesia and Brazil for palm oil and soy. *Agric. Hum. Values* **2014**, *31*, 425–437. [CrossRef]

89. Bastos Lima, M.G.; Persson, U.M. Commodity-centric landscape governance as a double-edged sword: The case of soy and the Cerrado Working Group in Brazil. *Front. For. Glob. Chang.* **2020**, *3*, 27. [CrossRef]

90. Ofstehage, A.; Nehring, R. No-till agriculture and the deception of sustainability in Brazil. *Int. J. Agric. Sustain.* **2021**. [CrossRef]

91. Horlings, I.; Marsden, T. Rumo ao desenvolvimento espacial sustentável? Explorando as implicações da nova bioeconomia no setor agroalimentar e na inovação regional. *Sociologias* **2011**, *13*, 142–178. [CrossRef]

92. Gramsci, A. *Selections from the Prison Notebooks*; International Publishers: New York, NY, USA, 1971.

93. Rabelo, F. "Yes, nós temos bananas"? Uma análise de estereótipos brasileiros revisitados em eventos culturais e esportivos no Brasil. *Veredas Rev. Da Assoc. Int. De Lusit.* **2018**, *27*, 85–103. [CrossRef]

94. De Schutter, O. The political economy of food systems reform. *Eur. Rev. Agric. Econ.* **2017**, *44*, 705–731. [CrossRef]

95. Anderson, M.D.; Rivera-Ferre, M. Food system narratives to end hunger: Extractive versus regenerative. *Curr. Opin. Environ. Sustain.* **2020**, *49*, 18–25. [CrossRef]

96. Harris, B. Brazil's New Frontier is Transforming Its Fortunes—But at What Cost? *Financial Times*. Available online: https://www.ft.com/content/bc8a217f-804d-4b32-b2ea-e06e08e9eb7a (accessed on 30 March 2021).

97. Cooney, P. Reprimarization: Implications for the Environment and Development in Latin America: The Cases of Argentina and Brazil. *Rev. Radic. Political Econ.* **2016**, *48*, 553–561. [CrossRef]

98. Cardoso, A.S.R.; Sousa, R.A.D.; Reis, L.C. O agro é tech, é pop, é tudo: O (des) velar dessa realidade. *Geosul* **2019**, *34*, 836–857. [CrossRef]

99. Medina, G.; Santos, A.P. Curbing enthusiasm for Brazilian agribusiness: The use of actor-specific assessments to transform sustainable development on the ground. *Appl. Geogr.* **2017**, *85*, 101–112. [CrossRef]

100. Lahsen, M. Buffers against inconvenient knowledge: Brazilian newspaper representations of the climate-meat link. *Desenvolv. E Meio Ambiente* **2017**, *40*, 17–35.

101. Tittor, A. The key role of the agribusiness and biotechnology sectors in constructing the economic imaginary of the bioeconomy in Argentina. *J. Environ. Policy Plan.* **2021**. [CrossRef]

102. Goven, J.; Pavonne, V. The bioeconomy as political project: A Polanyian analysis. *Sci. Technol. Hum. Values* **2015**, *40*, 302–337. [CrossRef]

103. Brito, D. Projeto Amazônia 4.0 Sugere Utilização Da Tecnologia Para Exploração Sustentável Da Biodiversidade. Available online: https://www.gov.br/agricultura/pt-br/assuntos/noticias/amazonia-4-0-sugere-utilizacao-da-tecnologia-para-exploracao-sustentavel-da-biodiversidade (accessed on 30 March 2021).

104. Ferrante, L.; Fearnside, P.M. Brazil's new president and 'ruralists' threaten Amazonia's environment, traditional peoples and the global climate. *Environ. Conserv.* **2019**, *46*, 261–263. [CrossRef]

105. Rodrigues, M.J.; Bioeconomia Ajudará Brasil a Reduzir Dependência Externa E a Aumentar a Conservação. Agência De Notícias Cni—Confederação Nacional Da Indústria. Available online: https://noticias.portaldaindustria.com.br/noticias/sustentabilidade/bioeconomia-ajudara-brasil-a-reduzir-dependencia-externa-e-a-aumentar-a-conservacao/ (accessed on 30 March 2021).

106. The Economist. Deathwatch for the Amazon. Available online: https://www.economist.com/leaders/2019/08/01/deathwatch-for-the-amazon (accessed on 30 March 2021).
107. Pilgrim, S.; Harvey, M. Battles over biofuels in Europe: NGOs and the politics of markets. *Sociol. Res. Online* **2010**, *15*, 4. [CrossRef]
108. Bastos Lima, M.G.; Gupta, J. The extraterritorial dimensions of biofuel policies and the politics of scale: Live and let die? *Third World Q.* **2014**, *35*, 392–410. [CrossRef]
109. Jessop, B. *State Theory: Putting the Capitalist State in Its Place*; Polity Press: Cambridge, UK, 1990.
110. Backhouse, M.; Rodríguez, F.; Tittor, A. From a Fossil towards a Renewable Energy Regime in the Americas? Socio-Ecological Inequalities, Contradictions and Challenges for a Global Bioeconomy. Bioeconomy & Inequalities Working Paper No. 10. 2019. Available online: https://www.bioinequalities.uni-jena.de/sozbemedia/WorkingPaper10.pdf (accessed on 18 April 2021).
111. Patino, H.; Leal, M.; Ospina, B. Brazil: Associative production systems. Alimergia: Integrated food, environment and energy. In *Bioeconomy: New Framework for Sustainable Growth in Latin America*; Hodson de Jaramillo, E., Henry, G., Trigo, E., Eds.; Editorial Pontificia Universidad Javeriana: Bogotá, Colombia, 2019.
112. Sachs, W. *The Development Dictionary: A Guide to Knowledge as Power*; Palgrave Macmillan: New York, NY, USA, 2010.
113. IAASTD. *Agriculture at a Crossroads: Synthesis Report. International Assessment of Agricultural Knowledge, Science and Technology for Development*; Island Press: Washington, DC, USA, 2009.
114. IPBES. *The Global Assessment Report on Biodiversity and Ecosystem Services—Summary for Policymakers*; IPBES Secretariat: Bonn, Germany, 2019.
115. Moore, B., Jr. *Social Origins of Dictatorship and Democracy: Lord and Peasant in the Making of the Modern World*; Beacon Press: Boston, MA, USA, 1966.
116. Ioris, R.R.; Schneider, A. What is new in agribusiness in Brazil? The long path of Conservative Modernization in the Perpetual Country-of-the-Future. In *Frontiers of Development in the Amazon: Riches, Risks, and Resistances*; Ioris, A.A.R., Ed.; Lexington Books: Washington, DC, USA, 2020; pp. 107–134.
117. Mol, A.P.J. The environmental movement in an era of ecological modernisation. *Geoforum* **2000**, *31*, 45–56. [CrossRef]

Article

Analysis of Wood Chip Characteristics for Energy Production in Lithuania

Nerijus Pedišius, Marius Praspaliauskas, Justinas Pedišius and Eugenija Farida Dzenajavičienė *

Lithuanian Energy Institute, 44403 Kaunas, Lithuania; Nerijus.Pedisius@lei.lt (N.P.);
Marius.Praspaliauskas@lei.lt (M.P.); Justinas.Pedisius@lei.lt (J.P.)
* Correspondence: Farida.Dzenajaviciene@lei.lt

Abstract: Wood chips and logging residues currently comprise the largest share of biomass fuels used for heat generation in district heating plants and are provided by a variety of suppliers. Ash and moisture contents, as well as the calorific value, may vary considerably depending on the composition of the fuel, seasonality, location, and other factors. This paper provides the summarized results of the main characteristics of wood chip moisture and ash content and calorific value, experimentally tested for a significant range of samples. Chip samples were collected from two district heating companies and tested for a significant range of samples. Chip samples were collected from two district heating companies and tested for a 3-year period. The data on fuel chip prices were taken from the electronic wood chip trading platform. The tests were performed using standard express methods, where two sub-samples were taken and analyzed from every chip sample. It was determined that the moisture content of the wood chips varied from 35% to 45%, the calorific value from 18.4 to 19.6 MJ/kg, and the ash content from 0.5% to 4.5%. The calculated relative expanded uncertainty of the moisture content measurement was $\pm2.1\%$, of calorific value—$\pm1.5\%$, and of ash—$\pm1.0\%$. The repeatability of the results was estimated as the pooled standard deviation.

Keywords: solid biomass fuel; wood chips; moisture content; calorific value; ash content; pooled standard deviation

Citation: Pedišius, N.; Praspaliauskas, M.; Pedišius, J.; Dzenajavičienė, E.F. Analysis of Wood Chip Characteristics for Energy Production in Lithuania. *Energies* **2021**, *14*, 3931. https://doi.org/10.3390/en14133931

Academic Editor: Idiano D'Adamo

Received: 13 May 2021
Accepted: 29 June 2021
Published: 30 June 2021

Publisher's Note: MDPI stays neutral with regard to jurisdictional claims in published maps and institutional affiliations.

1. Introduction

Currently, the largest share of solid biomass used for heat generation in power plants consists of forest cutting and wood processing residues, which are usually prepared and supplied in the form of wood chips. Such fuel is a heterogeneous formation and may contain not only wood chips but also many different impurities, such as leaves, needles, soil clods, or other mineral substances.

Fuel moisture can also vary widely from 5 to 55% and affect other properties, such as calorific value. Accurate information on fuel properties is important in setting prices, assessing the amount of energy planned to be produced, and selecting the correct properties of fuel for efficient combustion. Fuel suppliers and energy producers often need information on how and within what limits the main parameters of the chips change: humidity, calorific value, and ash content depending on the composition of the chips, the seasonality of preparation, and the geographical location. This is necessary in order to make the proper decisions regarding the purchase, storage, and efficient combustion of fuel. Other important wood chip parameters, such as fraction size, the fraction of fines, and Cl and S quantities are often examined together to assess the suitability of the wood chips for the combustion installations.

Measurement and evaluation methods for wood chip fuel and analyses of main characteristics, such as moisture content, calorific value, and others, were surveyed in [1]. Accurate parameter setting requires complete methodological guidelines and standards. In addition, it should be noted that a very important factor influencing the accuracy of parameter setting is the number of samples. Such studies have been performed [1]. Here, it is shown how the error of moisture determination varies from the number of samples. It is also shown how the calorific value and chemical composition depend on the type of wood and its individual parts.

In general, initial studies on biomass for energy production have shown that it is a competitive fuel vs. fossil fuels and has a dry matter calorific value of around 17–21 MJ/kg for the wood [2]. Therefore, studies on the fuel, produced from solid biomass, and their characteristics have been carried out, and procedures for their determination have been improved, including research methods and the use of equipment [3,4], which were used for the development of respective standards for fuel classification and determination of its characteristics.

The main characteristics of fast vegetation agro-waste have also been investigated as potential fuels, but wood waste products (pellets, briquettes) were defined as qualitatively better in terms of moisture, ash content, and chemical composition [5].

As wood chips became the most important commercial product for energy generation, their quality became an object of increased focus. All stages of such fuel production from tree felling, storage, and preparation for efficient combustion are important.

It is considered that the felling and wood chip production season is important for fuel quality, as, during the season between February and May, the carbon content, calorific value, total moisture content, and density of wet wood chips decrease and ash content increases due to covering the woodpiles, thus, preventing air circulation [6], and fuel parameters deteriorate in dry or deadwood [7]. To prepare chips for combustion, the cutting residues must be piled up and chipped on the roadsides. The typical ash content of such chips is between 3% and 4%. To obtain the optimal moisture content (about 30%) and, thus, increase the calorific value of the fuel, these residues must be stored for about 5–7 months in the forest site before chipping, and the lowest moisture content is achieved in September. Thus, in terms of the best characteristics of wood chips: optimal cutting is at the clear-cut area after stacking; the storage period is important to prevent deterioration of fuel quality; summer months and early autumn are the most appropriate for forest cutting. The references envisage further research of the main characteristics to improve the energy efficiency of the combustion process and reduction emissions of harmful substances [8].

Logging residue storage in ventilated piles and the effect of the particle size on the storage were investigated, as particle size is important for conveyor transportation and proper burning of the chips. This can be achieved due to proper drying and maintaining appropriate microclimate conditions [9]. Fuel storage conditions also affect fuel moisture content and calorific values. It was found that, although the wood dries after 60 days, however, when stored for more than 18 months, the fuel quality characteristics deteriorate sharply, also due to exposure to molds and fungi [10]. Chips begin to decompose fast with a very high increase of moisture content and the share of a fine fraction [11]. Preference should be given to log drying [12]. Drying logging residue first and then forwarding the material to a windrow on a landing will ensure lower moisture content and better storage characteristics [13]. The temperature of the wood stack is a good indicator of the decomposition of logging waste chips, but the ash content and calorific value of dry matter vary slightly [14]. As a whole, many factors, such as storage method, biomass origin, size and shape of the fuel, and storage time, as well as temperature and humidity, simultaneously affect dry matter losses and must all be taken into account. Logs dried for two weeks produce fewer fines while chipping compared to one-week dried logs [15].

When using wood chips or other wood fuels for cogeneration or special applications in industry, the reduction of fuel moisture by drying using various technologies has been investigated: the lower calorific value of wood fuels is 9.72 MJ/kg at 40% moisture content, however, calorific value can be increased to 14.76 MJ/kg after drying the fuel to 15% moisture content [16]. Minimization of ash content in wood chips has a significant impact on heat and fuel production economy and ash handling costs. Studies in Central European countries show that reducing ash content in wood chips is important as the storage of slag after incineration is an important environmental issue. Chip screening can improve chip quality by reducing ash content and eliminating unwanted size classes [17].

The composition of wood chips can vary considerably as this fuel is inhomogeneous and may contain significant amounts of needles, bark, and minerals [18]. The size of wood chips also depends on the density of felling waste, which can vary up to 58%. Chip size affects the movement of chips on a screw conveyor in the boiler house. The additional mechanical energy from 1 kJ/kg to 5 kJ/kg to feed chips is needed [19]. Chip additives, such as pinecones, can also improve fuel quality by increasing calorific value up to 21.16 MJ/kg for spruce and 19.41 MJ/kg for pine, which exceeds the calorific value of chips produced from these wood types [20].

In practice, the main characteristics have the greatest impact on the quality of wood chips. In a Baltpool biofuel market study commissioned for an accredited laboratory, wood chip samples to define moisture and ash content showed that the moisture content values varied between 13.5% and 19.9% and ash content values—between 1.0% and 14.4%, and those values were higher at higher outdoor temperatures and humidity [21]. The dependence of calorific value on the moisture of the same wood species was investigated, where the calorific value of dry wood at 0% humidity reached 18.802–20.224 MJ/kg, and at 50% humidity—calorific value reached only 9.74–6.36 MJ/kg [22]. Storage costs accounted for about 4.8% and transportation costs for about 23.2% [23].

Following from the review, the parameters of the chips depend on many factors, such as the composition of the raw material and the conditions and time of its preparation. Although these factors have begun to be explored, efforts are still insufficient, while their importance is growing due to the increasing need of renewable energy resources. At the same time, the importance of measurement accuracy in estimating the total cost of production and use of renewable fuels, as well as the volumes of harmful emissions to the environment [24], is increasing.

Increasing the production of wood chips from certified raw materials and using them for energy generation, as well as improving energy production technologies using biofuels of various qualities can lead to expansion of the use of bio-resources for green energy, which helps us to make progress towards a sustainable and circular economy and decarbonization of the municipal thermal energy sector [25].

Socio-economic indicators for the bioeconomy (SEIB) have been proposed to assess the sustainability of bio-economic sectors in Europe, including the bio-energy sector. They could be used to measure the impact of policy strategies on the specific performances evaluating the contribution of single bio-based sectors (especially in bio-energy). The monitoring and assessment of indicators, related with management practices, is required for European countries and would serve as example on a global scale. Lithuania, according to this evaluation of the European average, is in the middle of MS [26].

The present investigation is focused on the determination of the main characteristics (moisture content, calorific value, and ash content) of wood chips supplied to boiler houses for heat production and their changes due to seasonality and geographical location in Lithuania. Data for a period of 3 years, as well as the accuracy of their determination using express methods, are analyzed.

2. Materials and Methods

2.1. Data Collection

To determine the characteristics of wood chips (moisture content, calorific value, and ash content) supplied to district heating companies (DH), chip samples were selected from two district heating companies to define geographical differences. Companies were located in the northwest (Mažeikiai DH company—Company M) and southwest (Kaišiadorys DH company—Company K) of Lithuania, the distance between being about 250 km. Wood chip samples (approximately 600–700 g) were selected and submitted by the staff of the wood chip supplier and the DH company following the procedure provided in their contract. Two sub-samples from each sample were prepared for testing procedures.

Figure 1 shows two sub-samples sets (a) and samples drying in the oven (b).

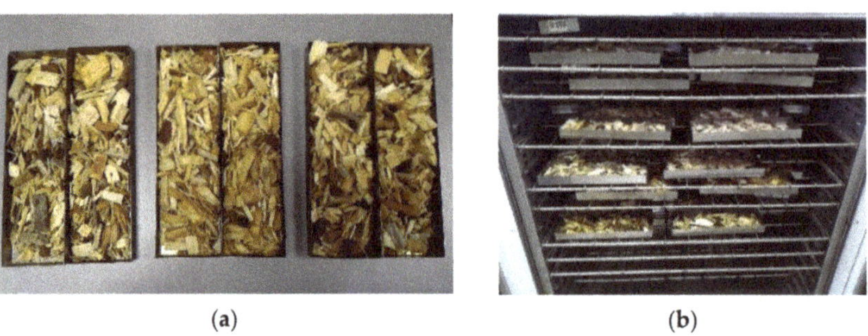

(a) **(b)**

Figure 1. Examples of samples: (**a**) two sub-sample sets and (**b**) samples drying in the oven.

The elemental composition of the chips was not determined for each sample. However, a comparison of the CHNSO road measurement averages with the averages of similar elements of firewood (alder, ash wood, oak, birch) oak and black alder briquettes [5], presented in Table 1, shows that there are no large differences between the main elements, and there was no task to identify detailed differences in composition.

Table 1. Comparison of the composition of main elements for wood chips, firewood, and briquettes, %.

Type of Biomass	Carbon (C)	Hydrogen (H)	Nitrogen (N)	Sulphur (S)	Ash	Oxygen (O)
Chips	53.3	5.8	0.22	0.045	3.2	37.5
Firewood, briquettes	49.3	6.0	0.29	0.02	0.9	43.5

The data on chip prices in two regions under investigation for the period of test performed during the years 2018–2020 were collected in statistics of Biomass Exchange—the electronic wood chip trading platform.

2.2. Testing Equipment and Methods

As in practice, the tests were performed using standard express methods, where only two sub-samples were taken and analyzed from one chip sample, and it was not possible to repeat them. Separate chapters of this paper discuss the additional uncertainty inputs introduced by this method and the application of the total standard deviation to assess the scatter.

Moisture content was determined in two sub-samples (Figure 1). Before the test wood chips sub-sample was weighed, it was sieved through the 31.5 mm sieve to homogenize the sub-sample. The total test sub-sample in a layer did not exceed 1 g of matter per cm^2. Two sub-samples were dried at 105 °C until the constant mass was achieved.

Dried wood chips were ground to obtain a nominal particle top size of 1 mm. Ash content was determined in two test sub-samples, minimum of 1 g of test sub-sample. Weighed samples were placed into a furnace and heated for approximately 4 h in two steps (250 °C and 550 °C). Cooled dishes with samples were weighed.

Calorific value was determined also in two test sub-samples. The sample was tested in a pellet form. Prepared sawdust was pressed with the hydraulic press at a force of about 10 t, having a diameter of about 13 mm and a mass of (1.0 ± 0.2) g. The repeatability limit for two wood chip samples could not exceed more than 140 J/g.

The uncertainties presented in Table 2 were achieved for homogenous samples, and tests could be repeated many times to define the standard deviation of achieved results.

Table 2. Test apparatus and methods.

No	Measured Characteristic and Unit	Method Used	Apparatus, Type	Expanded Measurement Uncertainty *
1	Ash content, %	Solid biofuels. Determination of ash content. EN ISO 18112:2016	Nabertherm LVT/9/11/P330 Mettler Toledo XS205DU/M	($\pm$0.30%)
2	Lower calorific value of dry fuel, kJ/kg	Solid biofuels. Determination of calorific value. EN ISO 18125:2017	Calorimeter IKA C 5000 Mettler Toledo XS205DU/M	($\pm$0.70%)
3	Moisture content, %	Solid biofuels. Determination of moisture content. Oven dry method. Part 1: Total moisture. Reference method EN ISO 18134-1:2016	BINDER FD 115 Nr. 13-18110 0021 Mettler Toledo XP 2003 SDR	($\pm$0.40%)

* The best measurement option. Expanded measurement uncertainty is defined from [27,28] by multiplying the standard uncertainty by the coverage factor, which is determined by estimating the repeatability of the measurement results and the effective number of degrees of freedom.

2.3. Data Analysis and Processing

Due to a large number of samples and the need to determine the moisture and ash content and calorific value in a limited time, simplified procedures, so-called express methods, have been adopted in practice, where a minimum number of two samples can be analyzed. There is no regulation on the threshold of the assessment of the main characteristics of the wood chip, which is important in resolving disputes between purchasers and suppliers of the wood chip, where several laboratories may be involved to determine the moisture content of the spare chip sample, the results of which may differ.

The numbers of measurements for this investigation are as follows:

- 1317 measurements for Company M and 1453 measurements for Company K performed on moisture content during the period 2018–2020;
- 1320 measurements for Company M and 383 measurements for Company K performed on calorific value during the period 2018–2020; and
- 1498 measurements for Company M and 467 measurements for Company K were performed on ash content during the period 2018–2020.

The figures represent the total numbers of samples provided to the laboratory during the testing period, which is needed for the more precise definition of pooled standard deviation [27].

Since the moisture content of biofuels in routine tests is only determined from two sub-samples of a fuel sample, the repeatability of the results is estimated as the pooled standard deviation [28] according to Equation (1):

$$S_p = \sqrt{\frac{1}{N-K}\sum_{i=1}^{N}(n_i - 1)\times s_i^2},\tag{1}$$

where S_p—pooled standard deviation;

s_i—standard deviation per sample assessed;

n_i—number of measurements/sub-samples per sample;

N—total number of measurements;

K—number of samples.

The standard deviation per fuel sample is determined [28] according to Equation (2):

$$s_i = \sqrt{2(x_i - \overline{x})^2},\tag{2}$$

where x_i—measured values of two sub-samples;

$\overline{x}$—an average of the two measured values.

Equation (1) was used for repeatability measurement of all parameters under the test.

3. Results and Discussion

3.1. Moisture Content of Wood Chips

Moisture affects the energy value of the fuel, combustion process efficiency, and temperature achievable during the combustion process. Its lower calorific value is because some of the heat released during combustion is used to evaporate the moisture. During the cooling process of flue gas, the water vapor in the fuel can condense and cause corrosion of the economizer and heat exchange surfaces. Moisture lowers the temperature in the furnace, and as a result, combustion conditions deteriorate.

Experimental measurements of the total moisture of the chip samples provided from the heat supply companies M and K revealed that the change of this parameter correlates very well with annual seasonal changes in regional climate indicators. As is seen in Figure 2, in the summer from May to September, the minimum moisture content of the wood chips reached (30 ± 5)%. Maximum wood chip moisture up to (45 ± 5)% was observed in the period from December to April. Although only 3 years were analyzed, it can be said that the range of variation in the moisture content of the supplied wood chips varied slightly, as well as the annual changes in environmental conditions. Short-term changes in air humidity or precipitation did not change the basic pattern of wood chip moisture variation. These factors, together with the shortcomings of the express method used for moisture measurement, determined the deviations of the moisture data from the basic pattern curve and data dispersion.

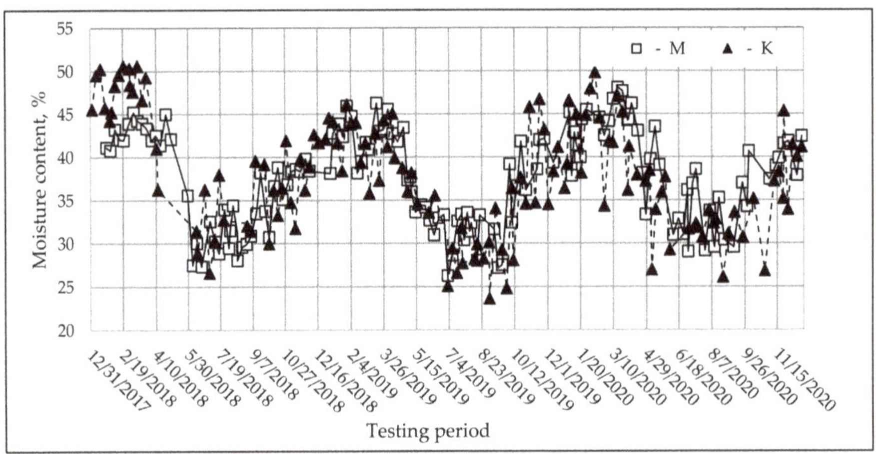

Figure 2. Daily average values of moisture content of wood chips supplied to companies M and K.

By increasing the averaging time from one day (Figure 2) to one month (Figure 3), the impacts of short-term and random factors disappeared, and the changes of the seasonal environmental conditions became more transparent and reflected the moisture content of the main share of wood chips supplied to DH.

As expected, the geographical location of the sites did not affect the moisture content of the chips, as the distance between the sites was only (250–300) km, and there was no strict boundary between the regions of origin of the wood chips.

To evaluate the uncertainty on a determination of the total moisture of the wood chip sample by taking the test results of two sub-samples, an additional analysis of the results on the evaluation of the total moisture of 396 sub-samples out of 198 samples was performed. It showed (Figure 4) that the absolute values of the standard deviation of the moisture content values of two sub-samples from one sample were distributed according to the normal distribution, and at 95% confidence level, the standard deviation of these values varied within wide limits and reached up to 4.5%. This means that the total uncertainty in the determination of the total moisture content of the wood chips is strongly influenced by the sampling procedure, the inhomogeneity of the samples, and the non-uniformity of the moisture distribution in the individual fractions. These factors make an important contribution to the total uncertainty of the moisture assessment in biofuels for two sub-samples, as other contributions related to the measurement and test conditions made a smaller and constant contribution of $\pm$ 0.4% (Table 2).

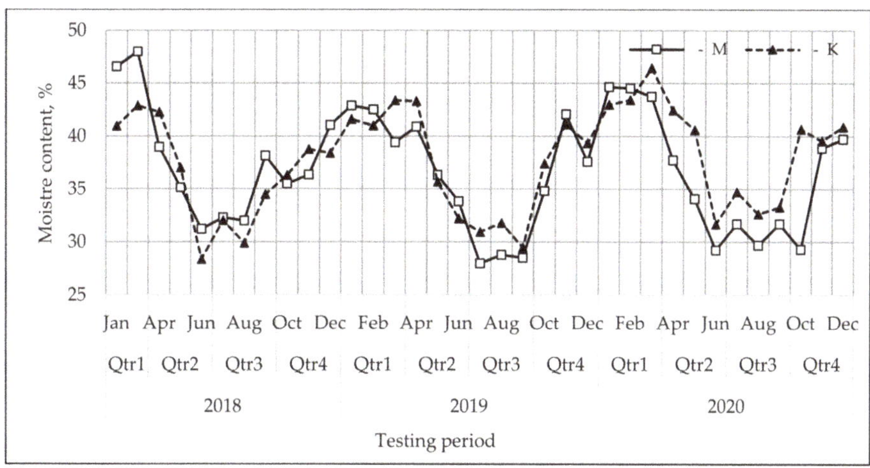

Figure 3. Monthly average values of moisture content of wood chips supplied to companies M and K.

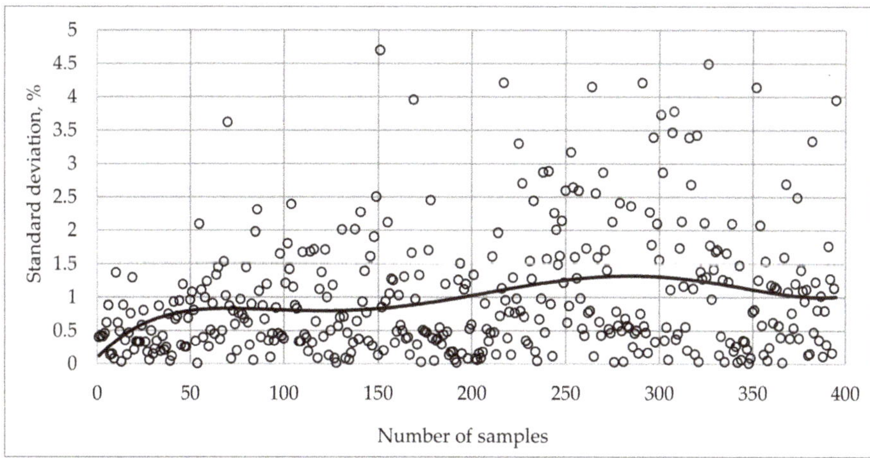

Figure 4. Standard deviation from mean values for two sub-samples of the same sample and mean trendline curve.

The data presented in Figure 4 show that the mean of the standard deviations is close to 1% and to the values of the pooled standard deviations calculated using Equation 1 and presented in Table 3.

Table 3. The values of pooled standard deviation, calculated from the measured value of 50 samples.

Sample	0–50	51–100	101–150	151–200	201–250	251–300	301–350	351–400
Sp, %	0.58	1.07	1.14	1.24	1.5	1.67	1.71	1.37

Thus, when estimating the scatter of the total wood chip moisture measurement results by the total standard deviation, the expanded moisture measurement uncertainty reached ±2.1%. In this case, the contributions to the total uncertainty associated with the accuracies of the apparatus given in Table 2 and the scatter of the results are summed arithmetically, i.e., the sum of their squares is not calculated. Deviations from individual measurements to achieve 95% confidence, which is important in resolving potential legal disputes, must be considered separately.

3.2. Calorific Value of Wood Chips

A comparison of the daily average variation of the lower (net) calorific values of dry (moisture-free) chips supplied to companies M and K in 2018–2020 is presented in Figure 5.

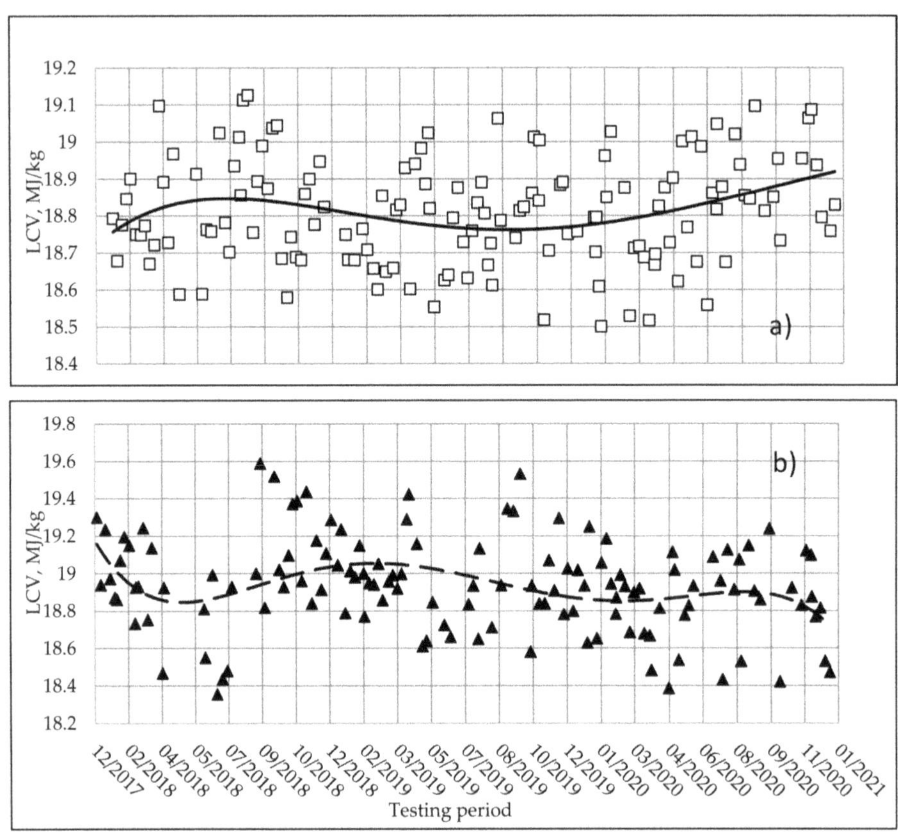

Figure 5. Daily average lower calorific values (LCV) and mean trendlines curves: (**a**) company M; (**b**) company K.

It can be seen that the calorific value of the chips used in company M was quite stable over the whole 3-year period and reached (18.8 ± 0.3) MJ/kg. At company K, a little higher mean calorific value and dispersion were observed, that of (18.9 ± 0.4) MJ/kg. It also confirmed that the geographical location of the sites did not affect the calorific values of wood chips.

When determining the uncertainty of the calorific value measurement, it was observed that the difference between the calorific values of the two sub-samples in one sample did not exceed 0.14 MJ/kg. This value can be considered as an indicator of the scatter of measurement results on calorific value and can be considered a contribution to the total uncertainty of the measurement result, together with the contribution of 0.7% provided by the calorimeter and scales (Table 2). As the uncertainty in the determination of the moisture content of the general analysis was small compared to the two main inputs indicated, the calculated relative expanded uncertainty of the calorific value measurement was ± 1.5%.

3.3. Ash Content of Wood Chips

The ash content of pure wood is low and reaches up to 0.5%. However, logging waste often contains large amounts of various impurities, such as leaves, needles, soil, or other minerals, that increase the amount of ash, which has undesirable effects: it reduces the calorific value of the fuel and accelerates the wear of fuel combustion equipment. Higher amounts of ash increase the costs of ash storage and disposal.

The comparison of the daily average values of biofuel ash content in the companies M and K in 2018–2020 is shown in Figure 6.

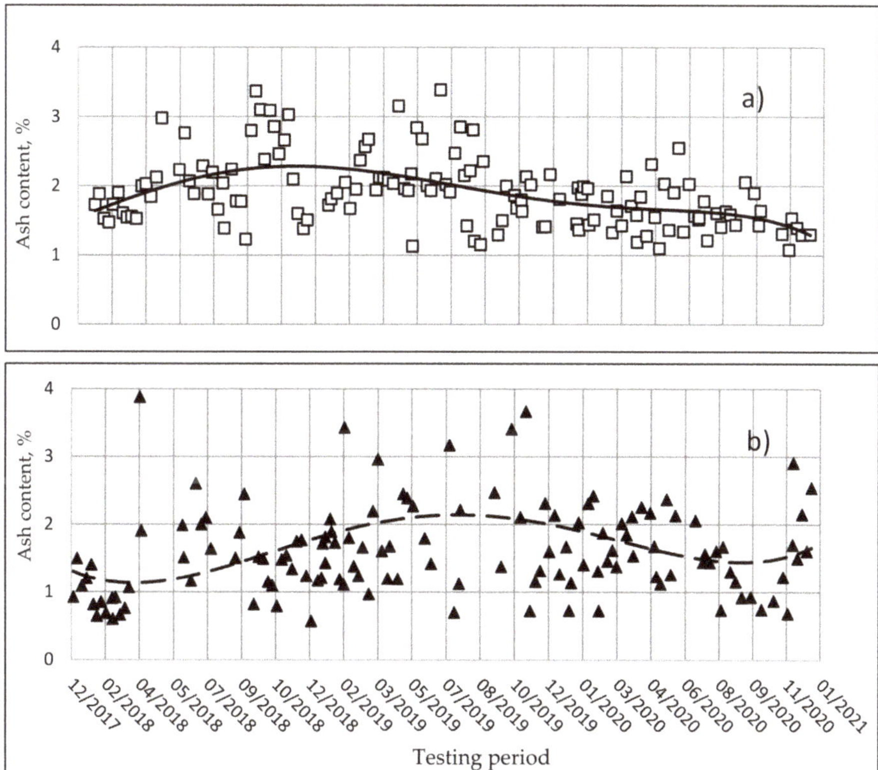

Figure 6. Daily average of ash content values and trendlines curves: (**a**) company M; (**b**) company K.

The distribution of ash content of wood chips according to the established classes showed that up to 90% of the ashes in the analyzed samples fell in the range of biofuel ash classes from A0.7 to A3.0. The calculated relative expanded uncertainty of the ash value measurement was ±1.0%.

3.4. Biomass Fuel Prices

Prices of wood chips purchased at biofuel exchange auctions in regional counties, where companies M and K were located during the period 2018–2020, are presented in Figure 7.

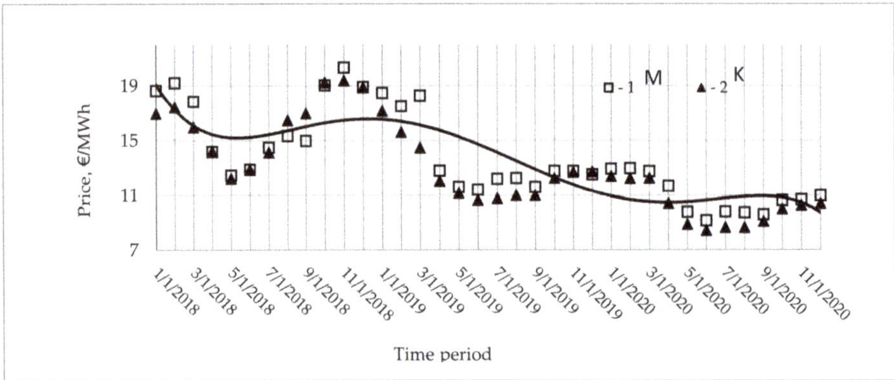

Figure 7. Changes of chip fuel prices at biofuel exchange during period 2018–2020 in counties, where companies M and K are located.

Auction prices showed both seasonal and slight geographical differences. The geographical differences were reflected via the prices of chips sold through the biofuel exchange and did not exceed 10%. Total tendency showed that this was related to delivery distance, not quality. Seasonal differences did not so much reflect fuel quality as seasonal fluctuations in demand and supply as prices fell during the summer when fuel consumption fell to baseline needs. During the heating season, when demand increased, biofuel prices also rose. On the other hand, prices continued to rise depending on climatic conditions—during warmer winters, the prices were lower than in cold winters.

These data also show that chip fuel prices have fallen significantly in the recent couple of years due to low-cost imported fuels. However, in terms of the prospects, the rise in the price of biomass is expected due to several reasons: the increasingly strong competition with the growing furniture and pulp industries for high-quality chips with lower moisture content, as those industries have been expanding rapidly in Lithuania over the past few years; competition due to growing demand for biomass fuel on national as well as international biomass markets.

3.5. Discussion

The presented results show that the wood chips supplied to boilers with an average capacity of up to 10 MW operating in DH systems are of relatively high and stable quality. The moisture content of such fuels changes regularly depending on seasonal climate changes and affects the calorific value of the fuel accordingly. Such impact was also observed in the study [16].

In our case, the high calorific value and low ash content of the chips were due to the fact that amounts of impurities, in the form of small particles with diameter lower than 3.15 mm, did not exceed 10% [21], and the production—storage—use time of the raw material was relatively short, usually not exceeding one year. This prevents the raw material from biodegrading. Such chips, the moisture of which can change in the prescribed pattern over the year, are the most suitable for medium-capacity boilers with furnaces adapted to wet fuels.

When wood chips with low impurities are used as fuel, their calorific value and ash content depend little on the type of wood used to produce chips in central and northern Europe [8,16,20]. However, as the practice of chip preparation in Latvia [6] showed, when storing green cutting waste in piles from half a month to 4 years and producing chips in February–May 2012, their relative moisture, calorific value, and ash content varied within the respective range (20.6–73.1)%, (15.7–19.7) MJ/kg, and (1.5–23.3)%. It is a consequence of the long storage of the raw material in piles under field conditions that determines the wider limits of variation in ash content and moisture content of the raw material.

Decreasing wood chip quality is observed in the results of our study, as high-quality local raw material resources are declining not only due to the growing needs for heating but also due to the increased use of wood for other purposes. This suggests that further intense rates of wood chip use could inevitably lead to serious violations of the sustainable use of renewable fuels based mainly on wood. Therefore, increased attention needs to be paid to the improvement of technologies for the preparation and incineration of wood waste and mixtures with other types of biofuels, as well as separate agro-wastes.

Nearly 80% of heat was produced from wood chips in 2019 in Lithuania. Transformation of bio-wastes into green municipal energy helps to make progress towards introducing circular and sistainable economies [25]. Although Lithuania is still considered an "in-between" EU country according to the socio-economic indicator for the bioeconomy [26], improvement of waste management, together with renewable energy management at the regional or municipal levels, paves the way to the increase of implementation of sustainable processes based on resource circularity.

4. Conclusions

A large amount of data for a period of 3 years from 2018 to 2020 on wood chips, supplied to the two district heating companies in Lithuania, and their parameters, such as moisture, ash content, and the calorific value, were analyzed to assess the dispersion and dependence of these parameters on seasonality and geographical factors.

It was determined that the change of total wood chips moisture was fully consistent with annual seasonal changes in regional climate indicators. In the cold season, the moisture content of wood chip felling residues was about $(45 \pm 5)\%$ and reached its highest values from December to April, and in the warm season, it decreased to $(30 \pm 5)\%$, reaching the minimum values in the period from May to September.

It was also determined that the mean lower calorific value of the wood chips was quite stable over the whole 3-year (2018–2020) period and reached (18.85 ± 0.35) MJ/kg.

The ash content values ranged from 0.5 to 4.0%. These parameters indicated that the wood chips used were still relatively clean, with a small content of small particles that did not exceed 10% of the total weight. Geographic locations did not show tangible differences between these parameters.

Despite the use of express methods, the relative expanded uncertainties of moisture content, lower calorific value, and ash content determination were $\pm 2.1\%$, $\pm 1.5\%$, and $\pm 1.0\%$, respectively.

Such parameters indicated that the wood chips used were still relatively clean, with a small content of small particles that did not exceed 10% of the total weight. This was ensured by the requirements of Baltpool to provide fuel of a quality that would guarantee efficient combustion and would not adversely affect the equipment. At this stage, it could be argued that the principles of sustainability were sought to be maintained. However, it cannot be guaranteed that this will be the case in the future if chips are used for heating at such a rate. Unless the incineration of household waste occurs in modern power plants, the insulation of buildings and various energy saving measures can all make a significant contribution to reducing the incineration of wood chips and maintaining sustainability.

Current rather stable and low chip prices lead to lower heating prices for consumers, while their seasonal and geographic variations depend more on demand/supply and the distance of delivery. However, the rise in the price of biomass is expected due to the growing strong competition with the furniture and pulp industries, growing biomass fuel national and international demand, as well as changing EU policy on the role of biomass and supply-chain emissions, which include increasing atmospheric CO_2 and the rated pace of global warming.

Author Contributions: Conceptualization, N.P., methodology, N.P. and M.P., investigation, M.P. and E.F.D., resources, M.P., data curation, J.P., writing—original draft preparation—N.P., M.P. J.P. and E.F.D., writing, review and editing, N.P., M.P., J.P. and E.F.D., visualization, J.P., supervision, N.P. All authors have read and agreed to the published version of the manuscript.

Funding: This research received no external funding.

Data Availability Statement: Data available on request due to privacy restrictions.

Conflicts of Interest: The authors declare no conflict of interest.

References

1. Nurmi, J. Measurement and evaluation of wood fuel. *Biomass Bioenergy* **1992**, *2*, 157–171. [CrossRef]
2. McKendry, P. Energy production from biomass (part 1): Overview of biomass. *Bioresour. Technol.* **2002**, *83*, 37–46. [CrossRef]
3. Alakangas, E. Analysis of Particle Size of Wood Chips and Hog Fuel—ISO/TC 238. Research Report. VTT-R-02834-12. VTT: Jyväskylä, Finland, 2012. Available online: https://www.vttresearch.com/sites/default/files/julkaisut/muut/2012/VTT-R-02834-12.pdf (accessed on 14 January 2021).
4. Alakangas, E.; Hurskainen, M.; Laatikainen-Luntama, J.; Korhonen, J. *Properties of Indigenous Fuels in Finland*; VTT Technology 272; VTT: Jyväskylä, Finland, 2016; 251p.
5. Praspaliauskas, M.; Pedišius, N.; Čepauskienė, D.; Valantinavičius, M. Study of chemical composition of agricultural residues from various agro-mass types. *Biomass Convers. Biorefin.* **2020**, *10*, 937–948. [CrossRef]
6. Gruduls, K.; Bārdule, A.; Zālītis, T.; Lazdiņš, A. Characteristics of wood chips from logging residues and quality influencing factors. *Res. Rural Dev.* **2013**, *2*, 49–54.
7. Barrette, J.; Pothier, D.; Duchesne, I. Lumber and wood chips properties of dead and sound black spruce trees grown in the boreal forest of Canada. *Forestry* **2015**, *88*, 108–120. [CrossRef]
8. Moskalik, T.; Gendek, A. Production of Chips from Logging Residues and Their Quality for Energy: A Review of European Literature. *Forests* **2019**, *10*, 262. [CrossRef]
9. Jirjis, R. Storage and drying of wood fuel. *Biomass Bioenergy* **1995**, *9*, 181–190. [CrossRef]
10. Gejdoš, M.; Lieskovský, M.; Slančík, M.; Němec, M.; Danihelová, Z. Storage and Fuel Quality of Coniferous Wood Chips. *BioResources* **2015**, *10*, 5544–5553. [CrossRef]
11. Pari, L.; Scarfone, A.; Santangelo, E.; Gallucci, F.; Spinelli, R.; Jirjis, R.; Del Giudice, A.; Barontini, M. Long term storage of poplar chips in Mediterranean environment. *Biomass Bioenergy* **2017**, *107*, 1–7. [CrossRef]
12. Manzone, M. Energy and moisture losses during poplar and black locust logwood storage. *Fuel Process. Technol.* **2015**, *138*, 194–201. [CrossRef]
13. Nurmi, J. The storage of logging residue for fuel. *Biomass Bioenergy* **1999**, *17*, 41–47. [CrossRef]
14. Hofmann, N.; Mendel, T.; Schulmeyer, F.; Kuptz, D.; Borchert, H.; Hartmann, H. Drying effects and dry matter losses during seasonal storage of spruce wood chips under practical conditions. *Biomass Bioenergy* **2018**, *111*, 196–205. [CrossRef]
15. Van der Merwe, J.P.; Ackerman, P.; Pulkki, R.; Längin, D. The Impact of Log Moisture Content on Chip Size Distribution When Processing Eucalyptus Pulpwood. *Croat. J. For. Eng.* **2016**, *37*, 297–307.
16. Lieskovsky, M.; Jankovsky, M.; Trenčiansky, M.; Merganič, M.; Dvořák, J. Ash Content vs. the Economics of Using Wood Chips for Energy: Model Based on Data from Central Europe. *BioResources* **2017**, *12*, 1579–1592. [CrossRef]
17. Huber, C.; Kroisleitner, H.; Stampfer, K. Performance of a Mobile Star Screen to Improve Woodchip Quality of Forest Residues. *Forests* **2017**, *8*, 171. [CrossRef]
18. Nurek, T.; Gendek, A.; Roman, K. Forest Residues as a Renewable Source of Energy: Elemental Composition and Physical Properties. *BioResources* **2019**, *14*, 6–20. [CrossRef]
19. Rackl, M.; Günthner, W.A. Experimental investigation on the influence of different grades of wood chips on screw feeding performance. *Biomass Bioenergy* **2016**, *88*, 106–115. [CrossRef]
20. Aniszewska, M.; Gendek, A. Comparison of heat of combustion and calorific value of the cones and wood of selected forest trees species. *For. Res. Pap.* **2014**, *75*, 231–236. [CrossRef]
21. Baltpool (Biomass Exchange). 2017. Available online: https://www.baltpool.eu/dl/~wp-content~uploads~2017~10~Tyrimai_EN_2017-09-25.pdf/Comparative-research-of-biomass-chips-samples-2017-09-25 (accessed on 14 January 2021).
22. Spirchez, C.; Lunguleasa, A.; Pruna, M.; Gaceu, L.; Forfota, R.; Dumitrache, I. Research on the Calorific Value of the Hardwood Species. *Bull. Eng.* **2019**, *12*, 111–114.
23. Kofman, P.D. Wood Ash. Processing/Products No. 43. 2016. © COFORD. 2016. Available online: http://www.coford.ie/media/coford/content/publications/projectreports/cofordconnects/cofordconnectsnotes/00675CCNPP43Revised091216.pdf (accessed on 14 January 2021).
24. Allen, J.; Browne, M. Logistics management and costs of biomass fuel supply Costs of biomass fuel supply. *Int. J. Phys. Distrib. Logist. Manag.* **1998**, *28*, 463–477. [CrossRef]
25. D'Adamo, I.; Falcone, P.M.; Huisingh, D.; Morone, P. A circular economy model based on biomethane: What are the opportunities for the municipality of Rome and beyond? *Renew. Energy* **2021**, *163*, 1660–1672. [CrossRef]

26. D'Adamo, I.; Falcone, P.M.; Morone, P. A New Socio-economic Indicator to Measure the Performance of Bioeconomy Sectors in Europe. *Ecol. Econ.* **2020**, *176*, 106724. [CrossRef]
27. Statistical How To. Pooled Standard Deviation. Available online: https://www.statisticshowto.com/pooled-standard-deviation (accessed on 14 January 2021).
28. Beckert, S.F.; Domeneghetti, G.; Bond, D. Use of Pooled Standard Deviation of Paired Samples in Calculating the Measurement Uncertainty by the Monte Carlo Method. In Proceedings of the 16th International Congress of Metrology, Paris, France, 7–10 October 2013; p. 03003.

Review

Crop Residue Management in India: Stubble Burning vs. Other Utilizations including Bioenergy

Gaurav Kumar Porichha [1,*], Yulin Hu [2], Kasanneni Tirumala Venkateswara Rao [1] and Chunbao Charles Xu [1,*]

[1] Department of Chemical and Biochemical Engineering, Western University, London, ON N6A 5B9, Canada; venkat.kasanneni@gmail.com

[2] Faculty of Sustainable Design Engineering, University of Prince Edward Island, Charlottetown, PE C1A 4P3, Canada; yhu268@uwo.ca

* Correspondence: gauravporichha21@gmail.com (G.K.P.); cxu6@uwo.ca (C.C.X.)

Abstract: In recent studies, various reports reveal that stubble burning of crop residues in India generates nearly 150 million tons of carbon dioxide (CO_2), more than 9 million tons of carbon monoxide (CO), a quarter-million tons of sulphur oxides (SO_X), 1 million tons of particulate matter and more than half a million tons of black carbon. These contribute directly to environmental pollution, as well as the haze in the Indian capital, New Delhi, and the diminishing glaciers of the Himalayas. Although stubble burning crop residue is a crime under Section 188 of the Indian Penal Code (IPC) and the Air and Pollution Control Act (APCA) of 1981, a lack of implementation of these government acts has been witnessed across the country. Instead of burning, crop residues can be utilized in various alternative ways, including use as cattle feed, compost with manure, rural roofing, bioenergy, beverage production, packaging materials, wood, paper, and bioethanol, etc. This review article aims to present the current status of stubble-burning practices for disposal of crop residues in India and discuss several alternative methods for valorization of crop residues. Overall, this review article offers a solid understanding of the negative impacts of mismanagement of the crop residues via stubble burning in India and the other more promising management approaches including use for bioenergy, which, if widely employed, could not only reduce the environmental impacts of crop residue management, but generate additional value for the agricultural sector globally.

Keywords: agricultural residue; stubble burning; alternative management practices; valorization

Citation: Porichha, G.K.; Hu, Y.; Rao, K.T.V.; Xu, C.C. Crop Residue Management in India: Stubble Burning vs. Other Utilizations including Bioenergy. *Energies* **2021**, *14*, 4281. https://doi.org/10.3390/en14144281

Academic Editor: Idiano D'Adamo

Received: 31 May 2021
Accepted: 8 July 2021
Published: 15 July 2021

Publisher's Note: MDPI stays neutral with regard to jurisdictional claims in published maps and institutional affiliations.

1. Introduction

Stubble burning is a practice where fire is purposely put to the stubble which remains after grains, such as paddy, wheat, rice, corn, etc., have been harvested. This represents an important source of atmospheric aerosol and gas emissions, hence having a potential effect on the global air quality and environmental chemistry. Open-field biomass burning is a longstanding method for land clearing and improvements in land use to dispose of living and dead vegetation, used globally. It has been estimated that humans account for nearly 90% of biomass combustion, although only a small portion of natural fires are responsible for the overall amount of vegetation burnt [1]. Over the past few decades, biomass burning has increased worldwide. In India alone, the total amount of crop residue and the burnt was estimated to be 516 million tones and 116 million tonnes (Mt), respectively, in the year of 2017–2018, approximately generating 176.1 Mt CO_2, 10 Mt of CO, 0.31 Mt CH_4, 0.008 Mt N_2O, 0.151 Mt NH_3, 0.814 Mt NMVOC, 0.453 Mt $PM_{2.5}$ (particulate matter) and 0.936 Mt PM_{10} [2]. Stubble burning has many environmental impacts and consequences, compared with alternatives such as ploughing stubble back in the field or harvesting them for industrial purposes. However, there are inadequate data on the impacts of crop stubble burning. Extrapolation has been commonly used in estimating the pollution factors in the database of farm residues, which may result in high uncertainty in the emission figures. It is well known that, due to agricultural field burning during the harvest season, air quality is greatly affected. Aerosol and gaseous pollutant source profiles from an agricultural fire are needed to assess their contribution to ambient air quality. As described above, agricultural field burning has created many environmental problems, utilizations of crop residues for such as cattle feed, compost with manure, rural roofing, biomass fuel, beverage production, packaging materials, wood, paper, and bioethanol, etc., should be explored and promoted. In the following sections, the detrimental environmental impacts of open burning of the agricultural residues are discussed in detail and current approaches for managing these crop residues are also presented.

This paper aims to present an overview of the practice of stubble burning of crop residues in India, its effects on the environment and health, and discuss some alternatives to stubble burning for valorization of crop residues.

2. Practices in India

India is a farming nation with many farming practices in step with agro-climatic zones. Rice, paddy and wheat cropping patterns are among the extensive farming practices in the states of Haryana, Punjab, Rajasthan, and western Uttar Pradesh. These regions are also infamous for burning the straw and stubble after the harvesting season. The state Punjab crosses India Pakistani border and is also called a 'breadbasket' because it produces two-thirds of India's food grains. Even though the government increasingly restricted the practice after 1990s, each year in late September and October, farmers from Punjab and Haryana in particular burn an estimation of 35 million tons of crop residue from their paddy fields after harvesting [3]. This practice serves as a low-cost method of getting rid of the straw and reduces the turnaround time between harvesting and sowing for the second (winter) crop. Figure 1 shows the crop-wise distributions of crop production, residue generated, and residue burnt in India for the year of 2018 [4].

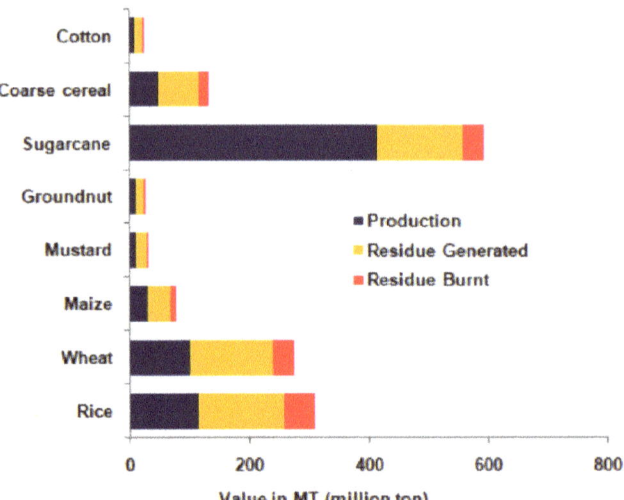

Figure 1. Crop-wise distributions of crop production, residue generated, and residue burnt in India for the year 2018 [4].

Burning the residue leads to the loss of nutrients and resources. Apart from deteriorating the ambient air quality, flaming stubbles causes soil nutrient loss of organic carbon (3850 million kg), nitrogen (59 million kg), phosphorus (20 million kg), and potassium (34 million Kg), and discharges large volumes of various air pollutants such as CO_X, CH_4, NO_X, SO_X, and particulate matters (PM_{10} and $PM_{2.5}$) [5]. The burning of straw and stubble is still a major disposal method in India, although the government of India has taken a few steps to prohibit the practice in recent years. The National Green Tribunal (NGT), located in the capital of India, imposed a ban on the flaming of straw and stubble in the neighbouring four states (Haryana, Rajasthan, Punjab, and Uttar Pradesh) to New Delhi, which contribute the significant air pollution during the early winter [6]. The government encourages farmers to utilize the straw and stubbles for alternative practices like mulching or in situ incorporation rather than burning. This agricultural waste can be used for animal fodder, generation of electricity, growing mushroom, and paper industry.

It has been reported that crop stubble burning contributes to one-quarter of the air pollution that blankets the entire capital city in the winters almost every year. In November 2018, the Delhi pollution index climbed to 12 times higher than the upper limit for healthy air. According to the United States embassy in New Delhi, the air quality in central Delhi in the morning was termed as "unhealthy," being more than three times the permitted level. The federal government was strongly criticized after the US-based United Airlines suspended flights to New Delhi due to its pollution. Apart from the stubble burning, the entire country celebrates the Diwali festival with fireworks, which also contributes to a lot of air pollution, especially in New Delhi. The Supreme Court of India passed a judgment in early 2019 to regulate the emission of gases from different sources, including the combustion of crop stubble and garbage as well as emissions from motor vehicles. However, most of the farmers in North India, as well as in other parts of the country, are not well educated and they do not become aware of air pollution and its threats to human life and the environment. Thus, the farmers use fire on agricultural waste, not deliberately, as they are the first victims of smoke inhalers. The North Indian farmers, especially from Punjab and Haryana, are in a helpless situation because there are no viable alternatives available to them to clear the fields but burning. All that threatens human health to such an extent is wrong and needs to be prevented. For implementation of the prohibition on the combustion of crop stubbles, a centralized, designated and accountable authority should be established to execute a comprehensive strategy to solve this problem with concrete, time-limited goals.

3. Effects of Stubble Burning

3.1. Environmental Effects

The burning of crop residues generates various environmental issues. The most adverse effects of crop residue burning embody the emission of greenhouse gases (GHGs) that contribute to global climate changes. In addition to that, enhanced levels of PM and other air pollution that cause health hazards, loss of diversity of agricultural land, and the deterioration of soil fertility [7]. The burning of the crop stubble in an open field influences soil fertility, eroding the sum of soil nutrients.

3.1.1. Air Pollution

Crop residue burning produces various air pollutants like GHG emissions, CO, NH_3, NO_X, SO_X, non-methane organic compound (NMHC), volatile organic compounds (VOCs), semi-volatile organic compounds (SVOCs) and PM [3], leading to the loss of organic carbon, nitrogen, and alternative nutrients, which otherwise might have preserved in soil. In Punjab, approximately 22 Mt of CO_2, 0.92 Mt of CO, and 0.03 Mt of SO_2 is generated from around 15 Mt of rice residues on an annual basis [8]. Similarly, according to a study by Jain et al., GHG emissions account for 91.6% of total air emissions caused by the burning of 98.4 Mt of crop residue and the remaining 8.4% are CO, NO, NMHCs and SVOCs [3]. Stubble burning also leads to emission of aerosols [9]. As also reported by Gadde et al. [10], the open burning of rice straw in India, Thailand and the Philippines results in severe air emissions of SO_2, NOx, CO_2, CO, and CH_4.

The major emissions of polluting gases and PM, as well as aerosols and trace gases as a result of crop residue burning, are listed in Tables 1 and 2. The PM released from the burning of crop residues is 17 times higher than that of the emissions from various other sources like motor vehicles, waste incineration, and industrial waste [11]. Intrinsically, the crop residue burning among the northwest vicinity of the country contributes to a considerable amount of about 200 organic carbon compounds in terms of the national emissions budget [12]. Street et al. [13], have anticipated that about 730 Mt of biomass was burned annually in Asian countries, and among them, India is in 18th position. Crop residue burning will increase the PM within the atmosphere and contribute to temperature change considerably. The fact that the fine black and also brown carbon (primary and secondary) would change sun light absorption and hence contribute to the global climatic change [10,14–16].

Table 1. Major pollutants released into the atmosphere during crop residue burning [8].

Category	Pollutant	Source
Particulate matters	$PM_{2.5}$ and PM_{10}	Condensation after combustion of gases and incomplete combustion of organic matters
	PM_{100}	Incomplete combustion of in-organic materials, particles on burnt soil
Gases	CO	Incomplete combustion of organic matters
	CH_4	Incomplete combustion of organic matters
	O_3	A secondary pollutant formed due to the reaction of nitrogen oxide and hydrocarbon
	NO, NO_2 N_2O	Oxidation of fuel-N or N_2 in the air at high temperatures
	Polycyclic aromatic hydrocarbons (PAHs)	Incomplete combustion of organic matters

Table 2. Emission levels of air pollutants during harvesting season in Haryana and Punjab (Source: Delhi Pollution Control Committee [DPCC], 2016).

Pollutants	Area in Delhi	Current Level ($\mu g/m^3$)	Permissible Limit ($\mu g/m^3$)
$PM_{2.5}$	Punjabi Bagh	650	60–80
PM_{10}	Punjabi Bagh	1000	60–80
CO	IGI Airport	6.3	2–4
SO_2	IGI Airport	29.8	60–80
NO_X	Anand Vihar	167	60–80

Usually, PM within the air is classified as $PM_{2.5}$ and PM_{10} in terms of its particle size ($PM_{2.5}$ is fine particles with a diameter <2.5 µm and PM_{10} is coarser, with a diameter <10 µm). Lightweight PM materials will remain suspended within the air for an extended time and might travel a prolonged distance with the wind [17,18]. PM pollution worsens under some climatic conditions, when the lightweight particles stay in air for an extended time causing severe air pollution. The annual contribution of $PM_{2.5}$ from the burning of paddy residue within the Patiala district of the geographic region was estimated to be around 60 to 390 mg/m^3 [19]. With the onset of cooler weather in November, the smoke, mixed with fog, dust, and industrial pollution, forms a thick haze. In the season if there is a lack of wind, the thick haze would continue for many days, as was the case throughout November 2017. Many major cities, including New Delhi, Lucknow, and Kanpur, faced elevated levels of pollution [20].

According to the United Nations, the permissible levels of $PM_{2.5}$ within the air is 10 µg/m^3, while India's National Air Quality commonplace allows the permissible level of $PM_{2.5}$ to be about at 40 µg/m^3. However, the capital territory of the urban center recorded a mean of 97 µg/m^3, which is double that of any other Indian place and 10 times more than that of the United Nations guidelines [19].

As well known, the emission of toxic gases from burning of the crop residue could lead to coughing asthma, emphysema, bronchitis, irritation of the eye, an opacity of the corneas, and skin disorders. Inhaling of PM can lead to intensifying persistent cardiac and pulmonary ailments and is related to the premature deaths of people who are already suffering from these illnesses [21]. About half of the world's population now lives in urban areas, which facing sever air pollution issues that adversely affects human health through the cardiovascular and respiratory systems [22]. Air pollution results in metabolism diseases like eye irritation, bronchitis, asthma, etc. Increasing individuals' sickness mitigation expenses and, additionally, poignant ones' operating capability. Annually, 3.3 million people are dying prematurely due to air pollution around the world. If air emissions continue to rise, this number will double by 2050 [21]. The Organization for Economic Cooperation and Development (OECD) estimates that in Delhi NCR alone, air pollution contributes to approximately 20,000 premature deaths and this number is expected to increase to 30,000 by 2025 and to 50,000 by 2050 (OECD, 2016). Table 2 shows that current pollutant emission levels in most areas of Delhi are way off the permissible limits.

The burning of crop waste also puts in danger of the survival of animals that produce milk. Air pollution can lead to animal death, as high CO_2 and CO levels in the blood can alter normal haemoglobin leading to death. More than 60,000 people who live in rice-growing areas are vulnerable to air pollution as a result of rice stubble burning, according to Singh et al. [8].

Detrimental compounds such as polyhalogenated organic compounds, namely polychlorinated dibenzodioxins, peroxyacetyl nitrate, polyaromatic hydrocarbons, polychlorinated biphenyls, and polychlorinated dibenzofurans, a family of organic compounds commonly called 'furans', are all classes of chemical emitted from open burning of farm residue [21]. These atmospheric pollutants are potentially toxic, and some are teratogenic, mutagenic, or carcinogenic in nature. The burning of crop straw and stubble has severe negative impacts on health. Pregnant women and infants are most prone to suffer adverse effects due to stubble burning pollutants. Respiratory inhalation of suspended $PM_{2.5}$ prompts asthma and can even worsen symptoms of bronchial attacks [6].

Open burning in the field also affects the lifespan of animals, birds and insects. Burning from time to time also causes poor visibility and increases the impact of road accidents. House surveys show that paddy husk burning causes several other problems as well as increasing health costs in polluted lands. Whether or not households are conscious of the harmful effects of waste burning, the answer would be 'yes' for about 90% of households; however, almost no household would take precautionary action to tackle pollution-related diseases [5].

Eye irritation and congestion within the chest are the two major issues faced by the bulk of people suffering from the stubble burning pollution. Metabolism allergic reaction, asthma attack and cartilaginous tube issues are the smoke connected chronic and non-chronic diseases that affected house members. Within the harvesting season, the affected families have to see doctors or use home medication for temporary relief from irritation/itching in eyes, respiration problems and similar alternative smoke connected issues. Sometimes they need to be hospitalized for 3 to 4 days, with extra expenditure incurred. On average, households spend a lot and suffer a lot from non-chronic metabolism diseases like coughing, a problem in respiration, irregular heartbeat, itchiness in eyes minimized respiratory organ performance, etc., throughout the year in particular during the months of crop husk burning [23]. The state government, from time to time, advises farmers not to set their field blazing. This is publicized in the native newspapers to create individual awareness of the adverse effects of crop husk burning. The administration even makes such announcements by loudspeakers within the villages. However, farmers who store the husk rather than burning it are not given any incentive from the administration, and farmers are not privy to the provision of alternative techniques to burning. The majority of farmers would be interested in adopting other practices if the state government offers enough resources.

In addition to the emission of air pollutants and the associated health effects to living organism, there is a continuous deterioration of soil fertility because of burning, which will be discussed in the following Section 3.1.2.

3.1.2. Soil Fertility

According to the Department of Agriculture, Government of Punjab, the soils of Punjab typically contain low nitrogen content, low to medium phosphorus, and moderate to high potassium. Besides, the organic carbon in the soil has decreased to very low, and insufficient levels and organic manure and crop residue have not been properly applied. Production of 7 t/ha rice and 4 t/ha wheat extract more than 300 kg of nitrogen, 30 kg of phosphorus, and 300 kg of potassium from the soil per hectare. The burning of crop residues contributes to the depletion of soil organic carbon, according to the Department of Soil, Punjab Agricultural University. Moreover, CO_2 and soil nitrogen balance changes quickly, and nitrogen is converted into nitrate, leading to depletion of 0.824 million tons of nitrogen-phosphorus-potassium (NPK) from the soil annually.

In addition, repeated burns can diminish by more than 50% the bacterial population. Long-term burning also reduces the amount of 0–15 cm soil loss along with loss of total nitrogen, biomass, and potentially mineralized nitrogen and organics. The burning of agriculture residues raises the soil temperature and causes depletion of the microorganism and flora population. The residue burning will increase the dirt temperatures to just about 35.8–42.2 °C at 10 mm depth [11], and semi-permanent effects will even reach up to 15 cm of the highest soil. Furthermore, the frequent burning reduces the nitrogen and carbon content in the soil and kills the microflora and fauna, which are useful to the soil, and additionally removes a significant portion of organic matter. With crop residue burning, the carbon-nitrogen equilibrium of the soil can be totally lost [5,24]. According to NPMCR [25], open incineration of 1 tonne of stubble would result in the loss of all organic carbon, 5.5 kg of nitrogen (N), 2.3 kg of phosphorous (P), 25 kg of potassium (K) and 1.2 kg of sulphur (S) in the soil. If the crop residue is kept within the soil itself, it will enrich the soil with C, N, P and K additionally.

Mandal et al.'s study [26] revealed that the burning of rice and wheat residues contributes to a loss of about 80% of nitrogen, 25% of phosphorus, 21% of potassium and 4-60% of soil sulphur, although it does destroy unwanted bugs and diseases borne by the soil.

Corp residue burning also contributes to a depletion in the crop essential nutrients. Around 25% of nitrogen and phosphorus are kept in crop residues, making 50% of sulphur and 75% of cereals potassium intake viable nutrient sources [13]. According to Singh et al. [8], there was a loss of 2400 kg of carbon, 35 kg of nitrogen, 3.2 kg of phosphorus, 21 kg of potassium, and 2.7 kg of sulphur from burning of rice residues in 1 ha in Punjab between 2001 and 2002. As presented in Table 3, burning of rice residue led to almost complete loss of carbon and nitrogen, and about 20–60% loss of P, K and S.

Table 3. Nutrient losses as a result of burning of rice residue in Punjab [8].

Nutrient	Concentration in Straw (g/kg)	Percentage Lost in Burning	Loss (kg/ha)
C	400	100	2400
N	6.5	90	35
P	2.1	25	3.2
K	17.5	20	21
S	0.75	60	2.7

4. Alternative Methods to Open Burning

In order to implement sound selections of alternative crop residue management methods, it is necessary to scientifically perceive the short and temporary effects of various crop residue management practices and to develop new residue management technologies that are cost-efficient and environmentally acceptable.

Crop residue management choices should be measured on the premise of productivity, gain, and environmental impact. These criteria would overlap with those employed in the approach of ecological intensification for intensive crop production systems aiming to fulfill the increasing demand for food, feed, fiber, and fuel, while meeting acceptable standards of environmental quality [27].

4.1. In Situ Incorporation

Although there are different alternatives to stubble burning, currently only two options are available to farmers, either to integrate the remains of crop stubble in situ or to burn it within the field. Farmers do not favour in situ incorporation because the stubble takes a long time to break down into the soil. According to the Department of Agriculture of Punjab, less than 1% of farmers implement in situ incorporation of crop stubble.

As per Singh et al. [28], however, the crop yield was significantly lower if the rice residue was added immediately before seeds are sown due to inorganic nitrogen immobilization and its adverse effect because of nitrogen deficiency. In a few studies, rice stubble was incorporated in the soil during the first 1–3 years, 30 days before wheat planting, and the wheat yield was found to decrease. Yet rice stubble incorporation in later years had no effect on the production of wheat crops.

According to another study by Sidhu and Beri [29], the incorporation of rice residues in the soil could be the best alternative practice (Table 4). A six-year study period showed that the production of subsequent wheat and rice crops was not adversely affected if the rice residue was introduced into the soil between 10 to 40 days before the sowing of the wheat crop. The resulting rice residue had no residual effect on the paddy straw combined with wheat, producing similar yields of rice and wheat with different residual management practices, including burning, elimination and integration [28,30]. Singh et al. [31] reported that paddy straw was introduced to the soil three weeks before wheat sowing and the wheat yield increased significantly in clay loam soil but in sandy loam soil. It was also reported that there was no adverse impact of in situ incorporation of crop stubble on the subsequent grain outputs according to research by Sharma et al. [32,33].

Table 4. Comparison of impacts of different residue management practices on soil properties in Ludhiana, Punjab [29].

Soil Property	Crop Residue Management		
	Incorporated	Removed	Burned
Total P (mg/kg)	612	420	390
Total K (mg/kg)	18.1	15.4	17.1
Olsen P (mg/kg)	20.5	17.2	14.4
Available K (mg/kg)	52	45	58
Available S (mg/kg)	61	55	34

Similar results reported that the incorporation of rice residue 3 weeks prior to sowing the wheat crop increased the amount of wheat only in clay loam soil and not in sandy loam soil [28]. This study also showed that organic carbon increased by 14–29% when the crop residues were incorporated in the soil. Incorporation of rice residue into the soil within 30 days before sowing wheat crops led to lower yields of grape wheat relative to those when the rice residue was burned [34]. Moreover, rice stubble incorporation into the soil has a beneficial effect on physical, chemical, and biological soil properties such as pH, organic carbon, the ability of water retention and bulk soil density. According to Mandal et al. [26], the impacts on the physicochemical properties of the soil over 7 years of various crop residue management practices (incorporated, removed and burned) are comparatively shown in Table 5. Both Tables 4 and 5 clearly show that methods of handling the rice residue for soil nutrient conservation are in the following order: in situ incorporation > removal of the rice residue from the land > stubble burning.

Table 5. Impacts of different residue management practices on soil properties [26].

Physiochemical Properties of the Soil	Crop Residue Management		
	Incorporated	Removed	Burned
pH	7.7	7.6	7.6
EC (dSM^{-1})	0.18	0.13	0.13
Organic C (%)	0.75	0.59	0.69
Available N (kg/ha)	154	139	143
Available P (kg/ha)	45	38	32
Available K (kg/ha)	85	56	77
Total N (kg/ha)	2501	2002	1725
Total P (kg/ha)	1346	924	858
Total K (kg/ha)	40480	34540	38280

4.2. Mulching

Mulching is the protective covering of the soil using sawdust, compost, or paper to reduce evaporation, prevent erosion, control the weeds, enrich the soil, or cleanse fruit. Mulching usually requires biomass transfer from the field before the soil is puddled, and then the biomass returns once the soil is prepared [35]. After harvesting the rice in India, some farmers do not drain their land on early rice crop and keep the field flooded during a short transition to the late rice crop transplant. Placing straw from early rice as mulch in lengths along the transplanting path for late rice ensures that the soil is moist enough to allow late rice transplants, controls the growth of weed and prevents rice rotting, while other Indian rice farmers have attempted to avoid flooded rice systems that allow crop residue to be retained on the surface [36]. Farmers either throw rice seedlings or seed germinated rice directly in many northern states such as Uttar Pradesh. Residues of crops are maintained on the soil surface in these systems. During plant establishment, the soil is saturated or flooded and weeds are managed by herbicides. Some farmers use relay crops to sow rice in wheat fields before mixing harvests. During the rice crop, the standing wheat stubble gradually decays. In order to save water and enhance N performance, a rice system for rice production (GCRPS) covers the soil under non-flooded conditions with soil covered by rice straw paw during development, but the yield of grain was often lower than in flooded rice [37].

A reduced or no-tillage system (Figure 2) makes it fairly easy to preserve the residue on the soil like a mulch by merely holding it onto the field during harvesting, when the residue does not need to be removed and added until tillage.

Figure 2. No-tillage farming (https://smallfarms.cornell.edu/2016/01/no-till-permanent-beds/ (accessed on 8 July 2021)).

Farmers in rice-based cultivation systems in India typically practice no sowing of winter crops, including wheat, barley, and rapeseed. Surface seeding of wheat is performed in ~60% of the rice-wheat system in Southern India by making a little hole in the earth (dibbling) followed by mulching with rice residue (4–6 t/ha). As the time span between maize harvest and the wheat crop sowing is relatively short (20 to 45 days) in India, it is difficult to enforce activity to manage weeds and to minimize evaporation in the fallow while not flooding the next crop [38]. Maize residue is spread to the mulch immediately following the harvest. For farmers who grow rice rather than wheat after rice, this technique is attractive. Farmers in the Indo-Gangetic Plains in north-western India have been gradually using reduced and no-tillage for wheat since the end of the 1990s and it results in substantial cost savings by reducing the fuel and labour use. Early seeding, particularly after the late rice harvest, also makes it easier in the eastern part of the Indo-Gangetic Plain to achieve potential yield advantages. Since the end of the 1990s, the region of reduced and non-smoking wheat has grown exponentially in the Indo-Ganges plain to an estimated 20% to 30% or 2–3 million hectares in 2006. However, the non-sowing of weeds after combine-harvesting of rice encounters some difficulties, e.g., accumulation of residues in the furrow openers, problems with the friction of the driving wheel of the drill, difficulties with loose straw fertilizer metering systems and uniform sowing depth due to frequent leaving of the drill to clear blockages [39]. Many methods for direct sowing into rice residue are being tested to address the problem of clogging and 'hair pinning' when the straw bends but is not cut and buried, which leaves the seed on the surface. These include the double and three-disk systems [40], the paw thrower and stubble chopper, although, to date, none has been successfully implemented.

Tillage mulching is a method to retain enough residue remains on the surface of the soil to significantly reduce erosion. It is used in low-soil lands. The residue can also be used as mucus to the following non-flooded crop (Figure 3). This method is not used by many farmers because it requires that the residue be partially removed from the field and returned after planting [41]. For farmers with limited land holdings and ample labour, however, this approach can be easier. In conventionally tilled fields in the rice-wheat system in South Asia, rice residue mulching is carried out only occasionally [42].

Figure 3. Tillage method of mulching (www.livinghistoryfarm.org (accessed on 8 July 2021)).

4.3. Composting

Composting is a biological process where organic waste is converted into compost that can be used as a fertilizer by microorganisms under controlled aerobic conditions [43]. A composting technique is usually used for the management of off-field residues where the compost generated is not returned to the field, while it can also be implemented in fields (in situ composting) [44].

In situ, rice straw is stacked up at threshing locations [45] as an instance of composting, where the straw decomposes gradually, mostly aerobically, and then at the beginning of the next season the compost can be dispersed into the soil as a fertilizer. The drawbacks of this operation include the establishment of suitable habitat for rodent pests and the undesirable presence of immobilized residual N. In China, another form of in situ composting is practiced in which wheat or barley residue is buried in ditches parallel to the rice transplant [46]. Crop residue can be composted alone or with other organic materials such as animal manure. The resulting compost can then be collected and added to the soil as a fertilizer. It is an appealing alternative to stubble burning owing to its ability to turn waste in a farm into a valuable fertilizer product by simply packing crop residues into piles or pits for a long time [47]. The composting method used in Indian rural areas often employs a passive aeration system with aeration holes and the treatment time is commonly within two to three months. Composting technology requires the input of the labour force but does not entail capital investment nor advanced machinery and infrastructure, which can be particularly attractive to small farms with sufficient labour resources.

4.4. Happy Seeder Machines

Although crop residue retained in the field plays a positive role in recovering soil quality and reducing environmental pollution caused by stubble burning, seeding of wheat in the field with rice residue retained was a challenge until the development of a happy seeder machine recently. The 'Happy Seeder', incorporating mulching and drilling of stubble in a single unit, is a promising new method [48], where the stubble is cut and gathered before seeding, and the cut stubble is then deposited as a mulch behind the seed sower.

The technology evolution has led to invention of a device known as the Combo Happy Seeder equipment (Figure 4) [49]. Several Happy Seeder machines have been built including the nine-row Turbo Happy Seeder (v.2). The 9-row Turbo Happy Seeder (vers.2) with a weight of 506 kg can be driven by a 33.6 kW tractor, and it has a working capacity of 0.3 ha h^{-1} for seeding wheat in the field with rice residue retained [50]. The adoption of happy seeder machine is facing some constraints including its low machine operating window (25 days a year), limited machine efficiency, inability for wet straw operation, and the lack of spreaders on a combined harvester. According to the National Academy of Agricultural Sciences (2017), rice residue can be managed by concurrent use of a super straw management system (SMS) fitted combines and turbo happy seeder. More importantly, profit analysis showed that Happy Seeder systems are more profitable than other alternatives for crop residue management [51]. Sowing wheat with the combination of happy seeders has an operating cost 50–60% lower than that with a traditional seeder [8].

Figure 4. Happy seeder equipment [49].

4.5. Bioenergy

Another promising residue management method is the production of bioenergy from crop residues. For example, crop residues can be used for liquid biofuel production, i.e., conversion of biomass to biopower or electricity or liquid fuel [52,53]. The most common biofuel produced from crop residues is cellulose-based ethanol, and the production involves enzymatically breaking down the polysaccharide within the straw into its element sugars, which will then be fermented into ethyl alcohol. Biopower is generated by direct combustion of straw alone (direct firing), or together with another fuel (co-firing) [54]. Another alternative method to direct combustion is through gasification, in which straw is gasified by air or steam to generate a fuel-gas mixture of N_2, H_2, CO, and CO_2 followed by combustion of fuel-gas mixture to generate electricity.

Bioenergy can also be generated from crop residues via anaerobic digestion to produce biogas mainly CH_4, which is collected and combusted to generate electricity [55]. For example, Gross et al. [56] applied cattle and buffalo manure and crop residues as the feedstock to produce biogas by anaerobic digestion, which was accompanied by the production of organic fertilizer. In another study, Abdelsalam et al. [57] produced biogas from the mixture of agricultural waste (e.g., potato peels, lettuce leaves, and peas peels) and livestock manure by co-digestion, and it was observed that the highest CH_4 and biogas production yield of 6610.2 and 1,23, 55 mL, respectively, were achieved using the co-feedstock of manure and lettuce leaves.

A state-wise surplus crop residue biomass potential in India is illustrated in Figure 5 [58]. As shown in the Figure, Uttar Pradesh produces the largest amount of surplus residues (40 MT) followed by Maharashtra (31 MT) and Punjab (28 MT), which can be used for bioenergy production. As suggested by the previous literature, the state ranking ranges from the lowest in western Bengal (679 MJ to the highest in Punjab (16,840 MJ) [58].

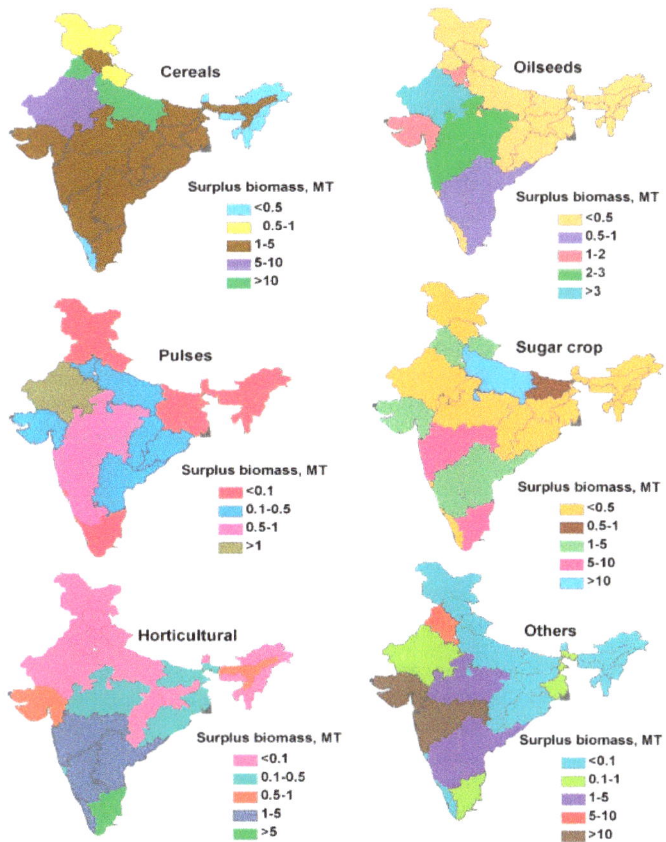

Figure 5. State-wise surplus crop residue biomass potential in India [58].

4.5.1. Case Study of Generation of Electricity from Agricultural Residues

The Jalkheri, Fatehgarh Sahib District thermal plant is the first plant in India using bioenergy sourced from agricultural and forestry residues [5]. The plant uses rice husk, wood chips, and stalks from different crops, such as paddy, wheat, etc. In June 1992, Bharat Heavy Electricals Limited (BHEL) commissioned this plant to use rice straw at the cost of Rs. 47.2 crores from Punjab State Electricity Board (PSEB). Originally, some of the teething problems were investigated by small-scale experiments, and a process of 10 MW was realized by modifications of an existing boiler to accept biological feedstocks like rice husk, wood chips, etc. The plant was leased and run on a sustainable basis in its full capacity of 10 MW. The Punjab Biomass Power, Bermaco Energy, Archean Granites and Gammon Infrastructure Projects Limited in Punjab have also built another 10–15 MW agricultural waste power plant. The plant uses local farm wastes, including rice paw and sugar cane bagasse. The annual total biomass consumption is approximately 120,000 tonnes of biomass obtained from that region. Punjab annually produces approximately 20–25 million tonnes of rice straw, conventionally disposed of by burning. As the technology is developed, this straw waste can now be used to generate electricity. The plant will provide 15,000 farmers with additional income from supplying agricultural waste. The development of this plant will be an important milestone for protecting the environment and creating new jobs and revenues by turning farm waste into bioenergy or green electricity.

In contrast to other biomass to produce bioenergy, using crop residues for large-scale production of bioenergy, there are several challenges to address with respect to both efficiency and economy. For instance, crop residues commonly contain a high content of alkaline ash which would pose operating issues (corrosion and deposition) in boilers for electricity generation. Cost-wise, since the crop residues are bulky, so the feedstock transportation and processing (crushing/pelletizing) costs would be high for centralized large-scale power plants. While it might not make it profitable for large bioenergy plants fuelled by crop residue, it can be adopted as small-scale energy suppliers for households and smaller communities. For instance, digestive biogas can be easily used as a bioenergy source for households.

4.5.2. Shortcomings and Ways to Overcome Them

Although there is a rapidly increasing trend to use wheat, rice husk for energy [59], biomass transport is among the main costs for energy use due to its bulky nature. To address this shortcoming, decentralized energy systems could offer a chance to use biomass to meet local demands for heat and electricity. For instance, rice husk can be used within the rice mill to some extent for energy, thereby reducing the total energy consumption of the entire plant. Rice millers may preferably sell the husk to a power plant operator. Energy suppliers may manage their own rice millers to supply the husk for power generation. A new trend can be that the rice millers themselves produce electricity and then sell electricity to a grid.

As mentioned previously, crop residue's transport costs are a significant constraint in its use as a source of bioenergy. Transport distances farer than 25–50 km are usually not cost-effective, depending on the local infrastructure. For large scale applications, straw can be packed in the field into bales or briquettes, making transportation more viable to the power generation site. The logistics of a supply chain for straw, although complicated, should be established for large-scale applications of crop residues for bioenergy generation.

5. Government Support and Policies

India is a country rich in legislation concerning pollution. Scientists, engineers, environmentalists, and government officials are also aware of the harmful consequences of the practice of stubble burning of agricultural residues on human health, soil, soil fertility and the environment. There are 11 major pollution control laws in India and many different regulations for implementation of these laws [60]. However, in order to avoid the burning of the straw, Section 144 of the Code is called upon by the Government to prohibit paddy burning but is difficult to implement, likely due to insufficient efforts having been made to increase the awareness of farmers about the serious impacts of stubble burning practices [19]. Nevertheless, the government must play more active roles in implementing all measures or practices planned or suggested by the various government or non-government groups, environmental scientists, and activists at the ground level in order to put an end to this damaging activity of stubble burning.

5.1. Steps Were Taken by the Government

Instead of working on solutions, the government has not even come out with the final version of its much-touted National Clean Air Plan (NCLAP) yet. The present Prime Minister of India, Narendra Modi and his cabinet ministers signed off on the plan in early 2018. The expenditure is far less than the $600 million per year that National Institution for Transforming India (NITI Aayog), a government policy advisory group, had initially proposed to the government. As per the proposed plan, money will be given to growers in three states bordering Delhi–Punjab, Haryana, and Uttar Pradesh to subsidize 80% of the cost of machinery for extracting crop residues from the fields, so as to avoid burning. Farmers are welcoming the plan since most of them cannot afford the machinery on their own. While this is an important step, it will depend on how quickly the scheme is rolled out at a scale that can make a difference. Other proposals put forward to deal with the stubble include purchasing crop residue by the state electricity company NTPC as a fuel in its coal-fired power stations.

5.2. Potential Future Strategies

(1)   Providing farmers with incentives not to burn crop residue outdoors.
(2)   Facilitation of maximum land cover using agricultural conservation practices.
(3)   Promoting the sustainable, environmentally friendly, and cost-effective use of surplus crop residues for generating bioenergy in power plants.
(4)   Crop residues should be classed as recycled fertilizers, and their use as fertilizers or amendments should receive government support.
(5)   Increasing subsidy rates for farmers who retain and utilize their crop residues.
(6)   There should not be free power as the same policy has resulted in the installation of high-powered tube wells that draw water from deep within the earth.
(7)   Promoting in situ management of crop residues by fast decay by chemical or biological means and mulching by mechanical means.
(8)   Promoting the use of machines such as double disks, zero tillage and happy seeders.
(9)   Valorization of crop residues for useful products, for example, compost, organic manure and biochar as a renewable fuel for power generation or as a soil amendment to improve soil health and fertility.
(10) Increasing the awareness of farmers on the serious impacts of the open field burning practice.

6. Discussion

Considering the detrimental influence of the lack of proper management practice, the Indian Government should advise farmers on alternative solutions to open field burning of crop residues, in order to reduce the toxic clouds over Delhi at the very least. For instance, to stop burning stubble altogether, many farmers from Punjab are pleased with the Happy Seeder, a system that can seed crops while minimizing stubble burning.

In fact, none of the alternative methods to open burning are perfect, but it is still noteworthy to implement them in a proper way considering geographic location, transportation, economic feasibility, etc.

When used as feedstock for livestock, bioenergy, and industrial raw materials, crop residues will be a bioresource of a higher economic value. However, in some regions complications in crop residue management remain diverse. The government should promote and provide alternatives in order to stop stubble burning, but simply legally prohibiting incineration of crop residues may not be successful because farmers do not have proper information about its consequences on soil, human, and animal health. Although farmers are aware of the adverse effects of paddy stroke on the farm, these are limited by the lack of economically viable and acceptable machinery and options for paddy residue disposal. Nevertheless, alternatives should be promoted, such as gasification of crop residues for fueling boilers, transformation into briquettes, etc. Other alternative management methods such as using Happy Seeder, zero-till machine, dual disc coulters, straw choppers, and agricultural conservation activities are also promising, as they could reduce the stubble burning in the rotation of rice-wheat. Encouraging organic recycling practices will reduce air pollution and convert wastes into resources, while machinery availability is currently the major obstacle for this management method and the other obstacle is lack of crop residue-fueled power plants.

Very recently, D'Adamo et al. [61,62] compared the status and performance of the EU countries in the bio-economy and circular economy and assessed them using a new socio-economic indicator for the bio-economy (SEIB), and performed a techno-economic analysis on integration of an effective management of renewable energy and municipal organic wastes. They demonstrated that the transformation of municipal organic waste into clean bio-energy, particularly through the generation of bio-methane, would close the loop and help societies make progress toward becoming circular economies, which can contribute to decarbonizing the transport sector. We believe that the results of their analyses may be applicable not only for utilization of municipal organic wastes but also agricultural wastes that are currently under-utilized but being disposed of by stubble burning in India as well as in many other Asian countries.

Although many alternative treatment methods have been developed and proven to be helpful for preventing the mismanagement of crop residues to some extent, the following practices are still recommended: (i) social awareness needs to be strengthened regarding the detrimental effects of open burning of agricultural residue; (ii) conservation agriculture techniques to maximize land cover need to be implemented; (iii) the conversion of crop residue into biofuels, thereby enhancing air and soil quality and halting greenhouse gas emissions, needs to be promoted; (iv) methods of in situ decomposition of crop residue by using chemical, biological, and mechanical approaches should be developed.

7. Conclusions

Undoubtedly, the lack of proper management of abundant crop residue has had an adverse influence on the environment and human health not only in India but also in the world. Agricultural field burning has created many environmental problems, particularly causing a threat to the soil fertility and the emission of toxic gases such as CO_2, CO, SO_2, $PM_{2.5}$ and PM10. Consequently, a variety of alternative approaches should be considered as substitutes for open field burning, e.g., in situ incorporation, mulching, composting, Happy Seeder machines, and bioenergy use. In conclusion, this article provides an overall understanding of the adverse impacts of open burning of crop residues in terms of ecology and environment and more promising alternative management practices for the crop residues, which, if widely employed, could not only reduce the environmental impacts of crop residue management, but generate additional value for the agricultural sector globally.

Author Contributions: Conceptualization, G.K.P. and C.C.X.; formal analysis, G.K.P.; investigation, G.K.P.; resources, G.K.P. and C.C.X.; writing—original draft preparation G.K.P.; writing—review and editing, C.C.X., K.T.V.R., Y.H.; supervision, C.C.X.; project administration, C.C.X.; funding acquisition, C.C.X. All authors have read and agreed to the published version of the manuscript.

Funding: Natural Sciences and Engineering Research Council of Canada (NSERC), DG Grant.

Institutional Review Board Statement: Not applicable.

Informed Consent Statement: Not applicable.

Acknowledgments: This work was supported through the Biomass Canada Cluster (BMC) that is funded through Agriculture and Agri-Food Canada's AgriScience Program and industry partners. The authors also acknowledge the financial support from Natural Sciences and Engineering Research Council of Canada (NSERC) through the Discovery Grant awarded to C. Xu.

Conflicts of Interest: The authors declare no conflict of interest.

References

1. Paul, J.; Meinrat, O. Biomass Burning in the Tropics: Impact on Atmospheric Chemistry and Biogeochemical Cycles. *Science* **1990**, *250*, 1669–1678.
2. Venkatramanan, V.; Shah, S.; Rai, A.K.; Prasad, R. Nexus between crop residue burning, bioeconomy and sustainable development goals over north-western India. *Front. Energy Res.* **2021**, 26. [CrossRef]
3. Badarinath, K.V.S.; Chand, T.R.K.; Prasad, V.K. Agriculture crop residue burning in the Indo-Gangetic Plains—A study using IRS-P6 AWiFS satellite data. *JSTOR* **2006**, *91*, 1085–1089.
4. Sahu, S.K.; Mangaraj, P.; Beig, G.; Samal, A.; Pradhan, C.; Dash, S.; Tyagi, B. Quantifying the high resolution seasonal emission of air pollutants from crop residue burning in India. *Environ. Pollut.* **2021**, *286*, 117165. [CrossRef]
5. Parmod, K.; Surender, K.; Laxmi, J. *Socioeconomic and Environmental Implications of Agricultural Residue Burning*; Springer: New Delhi, India, 2015.
6. Siddiqui, N.A.; Tauseef, S.M.; Bansal, K. *Advances in Health and Environment Safety*; Springer Nature Singapore Pte Ltd.: Singapore, 2016.
7. Jin, Y.; Bin, C. Global warming impact assessment of a crop residue gasification project- A dynamic LCA perspective. *Appl. Energy* **2014**, *122*, 269–279.
8. Singh, R.P.; Dhaliwal, H.S.; Humphreys, E.; Sidhu, H.S.; Manpreet, S.; Yadvinder, S.; Blackwell, J. Economic Evaluation of the Happy Seeder for Rice-Wheat Systems in Punjab, India. Available online: https://ageconsearch.umn.edu/record/5975/files/cp08si03.pdf (accessed on 10 May 2021).
9. Venkataraman, C.; Habib, G.; Kadamba, D.; Shrivastava, M.; Leon, J.F.; Crouzille, B.; Boucher, O.; Streets, D.G. Emissions from open biomass burning in India: Integrating the inventory approach with high-resolution Moderate Resolution Imaging Spectroradiometer (MODIS) active-fire and land cover data. *Glob. Biogeochem. Cycles* **2006**, 2. [CrossRef]
10. Gadde, B.; Sébastien, B.; Christoph, M.; Savitri, G. Air pollutant emissions from rice straw open field burning in India, Thailand and the Philippines. *Environ. Pollut.* **2000**, *157*, 1554–1558. [CrossRef] [PubMed]
11. Jitendra, S.V.; Ishan, K.; Kundan, P.; Deepanwita, G.N.; Polash, M. India's Burning Issues of Crop Burning Takes a New Turn. Down Earth 2017. Available online: https://www.downtoearth.org.in/coverage/agriculture/river-of-fire-57924 (accessed on 10 May 2021).
12. Prabhat, K.G.; Gupta, S.S.; Nahar, S.; Dixit, C.K.; Singh, D.P.; Sharma, C.; Tiwari, M.K.; Raj, K.G.; Garg, S.C. Residue burning in rice–wheat cropping system: Causes and implications. *Curr. Sci.* **2004**, *87*, 1713–1717.
13. Streets, D.G.; Bond, T.C.; Carmichael, G.R.; Fernandes, S.D.; Fu, Q.; He, D.; Klimont, Z.; Nelson, S.M.; Tsai, N.Y.; Wang, M.Q.; et al. An Inventory of Gaseous and Primary Aerosol Emissions in Asia in the Year 2000. *J. Geophys. Res.* **2003**, *108*, 8809–8823. [CrossRef]
14. Washenfelder, R.A.; Attwood, A.R. Biomass burning dominates brown carbon absorption in the rural southeastern United States. *J. Geophy. Res.* **2015**, *42*, 653–664. [CrossRef]
15. Hatch, L.E.; Luo, W.; Pankow, J.F.; Yokelson, R.J.; Stockwell, C.E.; Barsanti, K.C. Identification and Quantification of Gaseous Organic Compounds Emitted from Biomass Burning using Two-Dimensional Gas Chromatography-time-of-flight Mass Spectrometry. *Atomspheric Chem. Phys. Atomos. Chem. Phys.* **2015**, *15*, 1865–1899. [CrossRef]
16. Jiang, H.H.; Alexander, L.F.; Avi, L.; Jin, Y.C.; Zhang, H.F.; Roya, B.; Lin, Y.H. Brown Carbon Formation from Nighttime Chemistry of Unsaturated Heterocyclic Volatile Organic Compounds. *Environ. Sci. Technol. Lett.* **2019**, *6*, 184–190. [CrossRef]
17. Singh, C.P.; Sushma, P. Characterization of residue burning from agricultural system in India using space-based observations. *J. India. Soc. Remote Sens.* **2011**, *39*, 423–429. [CrossRef]
18. Jain, N.; Bhatia, A.; Pathak, H. Emission of Air Pollutants from Crop Residue Burning in India. *Aerosol. Air Qual. Res.* **2014**, *14*, 422–430. [CrossRef]
19. Lohan, S.K.; Jat, H.S.; Arvind, K.Y.; Sidhu, H.S.; Jat, M.L.; Madhu, C.; Jyotsna, K.P.; Sharma, P.C. Burning issues of paddy residue management in northwest states of India. *Renew. Sust. Energ. Rev.* **2018**, *81*, 693–706. [CrossRef]
20. Allen, A.; Voiland, A. NASA Earth Observatory, Haze Blankets Northern India. Available online: https://earthobservatory.nasa.gov/images/91240/haze-blankets-northern-india (accessed on 1 May 2021).
21. Kaliramana, R.S.; Sangwan, A.; Parveen, S. Burning of Crop Residues Harmful Effects in Soil Ecosystem and Potential Solutions, Effective Management of Paddy Residue/Wheat Stubble. *Adv. Agric. Sci.* **2019**, *19*, 99–124.
22. Mario, J.M.; Luisa, T.M. Megacities and atmospheric pollution. *J. Air. Waste Manag.* **2004**, *54*, 644–680.
23. Parmod, K.; Surender, K. *Economic Impact of Air Pollution from Agricultural Residue Burning on Human Health*; Springer International Publishing: New York, NY, USA, 2016.

24. Singh, Y.; Gupta, R.K.; Jagmohan, S.; Gurpreet, S.; Gobinder, S.; Ladha, J.K. Placement effects on paddy residue decomposition and nutrient dynamics on two soil types during wheat cropping in paddy-wheat system in north western India. *Nutr. Cycling. Agroecosyst.* **2010**, *88*, 471–480. [CrossRef]

25. NPMCR. Available online: http://agricoop.nic.in/sites/default/files/NPMCR_1.pdf (accessed on 6 March 2019).

26. Mandal, K.G.; Arun, K.M.; Kuntal, M.H.; Kali, K.B.; Prabir, K.G.; Manoranjan, M. Rice residue management options and effects on soil properties and crop productivity. *Food Agric. Environ.* **2004**, *2*, 224–231.

27. Gangwar, K.S.; Singh, K.K.; Sharma, S.K.; Tomar, O.K. Alternative tillage and crop residue management in wheat after rice in sandy loam soils of Indo-Gangetic plains. *Soil. Till. Res.* **2006**, *88*, 242–252. [CrossRef]

28. Singh, Y. Crop Residue Management in Rice-Wheat Cropping System. In *Abstracts of Poster Sessions. 2nd International Crop Science Congress*; National Academy of Agricultural Sciences: New Delhi, India, 1996; Volume 43, Available online: https://www.researchgate.net/profile/Bijay-Singh-3/publication/315515730_Management_of_Crop_Residues_in_Rice-Wheat_Cropping_System_in_the_Indo-Gangetic_Plains/links/58d34be3aca2723c0a777e7b/Management-of-Crop-Residues-in-Rice-Wheat-Cropping-System-in-the-Indo-Gangetic-Plains.pdf (accessed on 10 May 2021).

29. Sidhu, B.S.; Beri, V. Experience with managing rice residues in intensive rice-wheat cropping system in Punjab. *Conserv. Agric. Status Prospect.* **2005**, 55–63.

30. Walia, S.S. Effect of management of crop residues on soil properties in rice-wheat cropping system. *Environ. Ecol.* **1995**, *13*, 503–507.

31. Singh, Y.; Singh, B. Efficient management of primary nutrients in the rice-wheat system. *J. Crop. Prod.* **2001**, *4*, 23–85. [CrossRef]

32. Sharma, H.L.; Rajendra, P. Effect of applied organic manure, crop residues and nitrogen in rice-wheat cropping system in northwestern Himalayas. *HJAR* **1985**, *11*, 63–68.

33. Sharma, H.L.; Singh, C.M.; Modgal, S.C. Use of organics in rice-wheat crop sequence. Indian. *J. Agric. Sci.* **1987**, *57*, 163–168.

34. Verma, T.S.; Bhagat, R.M. Impact of rice straw management practices in yields, nitrogen uptake and soil properties in a wheat-rice rotation in northern India. *Fertil. Res.* **1992**, *33*, 97–106. [CrossRef]

35. Olaf, E. Crop residue mulching in tropical and semi-tropical countries: An evaluation of residue availability and other technological implications. *Soil. Till. Res.* **2002**, *67*, 115–133.

36. Peeyush, S.; Vikas, A.; Sharma, R.K. Impact of tillage and mulch management on economics, energy requirement and crop performance in maize–wheat rotation in rainfed subhumid inceptisols, India. *Eur. J. Agron.* **2011**, *34*, 46–51.

37. Acharya, C.L.; Sharma, P.D. Tillage and mulch effects on soil physical environment, root growth, nutrient uptake and yield of maize and wheat on an Alfisol in northwest India. *Soil. Till. Res.* **1994**, *32*, 291–302. [CrossRef]

38. Jeffrey, P.M.; Purnendu, N.S.; Wesley, W.W.; Daniel, S.M.; Jonathan, F.W.; William, R.H.; Philip, H.; Robert, R.; Blaine, R.H. No-tillage and high-residue practices reduce soil water evaporation. *Calif. Agric.* **2012**, *66*, 55–61.

39. Hari, R.; Yadvinder, S.; Saini, K.S.; Kler, D.S. Agronomic and economic evaluation of permanent raised beds, no tillage and straw mulching for an irrigated maize-wheat system in northwest India. *Exp. Agric.* **2012**, *48*, 21–38.

40. Rickman, J.F. Manual for Laser Land Levelling. In *Rice-Wheat Consortium for the Indo-Gangetic Plains*; CIMMYT: México-Veracruz, Mexico, 2002.

41. Unger, P.U.; Stewarta, B.A.; Parrb, J.F.; Singh, R.P. Crop residue management and tillage methods for conserving soil and water in semi-arid regions. *Soil. Till. Res.* **1991**, *20*, 219–240. [CrossRef]

42. Zaman, A.; Choudhuri, S.K. Water use and yield of wheat under unmulched and mulched conditions in laterite soil of the Indian Sub-continent. *J. Agron. Crop. Sci.* **1995**, *175*, 349–353. [CrossRef]

43. Sunita, G.; Lata, N.; Patel, V.B. Quality evaluation of co-composted wheat straw, poultry droppings and oil seed cakes. *Biodegradation* **2009**, *20*, 307–317.

44. Kumar, S. Composting of municipal solid waste. *Crit. Rev. Biotechnol.* **2011**, *31*, 112–136. [CrossRef] [PubMed]

45. Ponnamperuma, F.N. Straw as a source of nutrients for wetland rice. *Org. Matter Rice* **1984**, 117–136.

46. Peng, S.B.; Roland, J.B.; Huang, J.I.; Jian, C.Y.; Ying, B.Z.; Zhong, X.H.; Wang, G.H.; Zhang, F. Strategies for overcoming low agronomic nitrogen use efficiency in irrigated rice systems in China. *Field. Crops. Res.* **2006**, *96*, 37–47. [CrossRef]

47. Misra, R.V.; Roy, R.N.; Hiraoka, H. *On-Farm Composting Methods*; Land and Water Discussion Paper 2; UN-FAO: Rome, Italy, 2003.

48. Blackwell, J.; Sidhu, H.S.; Dhillon, S.S.; Prashar, A. The Happy Seeder concept- a solution to the problem of sowing into heavy stubble residues. *Aust. J. Exp. Agric.* **2003**, *47*, 5–6.

49. Sims, B.; Kienzle, J. Mechanization of conservation agriculture for smallholders: Issues and options for sustainable intensification. *Environments* **2015**, *2*, 139–166. [CrossRef]

50. Sidhu, S.; Manpreet, S.; Blackwell, J.; Humphreys, E.; Bector, V.; Yadvinder, S.; Malkeet, S.; Sarbjit, S. Development of the Happy Seeder for direct drilling into combine harvested rice. *Aust. J. Exp. Agric.* **2008**, *127*, 159–170.

51. Kaur, A.; Grover, D.K. Happy Seeder Technology in Punjab: A Discriminant Analysis. *IJE&D* **2017**, *13*, 123–128.

52. Kishore, V.V.N.; Bhandari, P.M.; Gupta, P. Biomass energy technologies for rural infrastructure and village power- opportunities and challenges in the context of global climate change concerns. *Energy Policy* **2004**, *32*, 801–810. [CrossRef]

53. Priya, C.; Bisaria, V.S. Simultaneous Bioconversion of Cellulose and Hemicellulose to Ethanol. *Crit. Rev. Biotechnol.* **1998**, *18*, 295–331.

54. Jenkins, B.M.; Baxter, L.L.; Miles, T.R. Combustion properties of biomass. *Fuel. Process. Technol.* **1998**, *54*, 17–46. [CrossRef]

55. Venkateswara, P.R.; Baral, S.S.; Dey, R.; Mutnuri, S. Biogas generation potential by anaerobic digestion for sustainable energy development in India. *Renew. Sustain. Energy. Rev.* **2010**, *14*, 2086–2094.
56. Gross, T.; Breitenmoser, L.; Kumar, S.; Ehrensperger, A.; Wintgens, T.; Hugi, C. Anaerobic digestion of biowaste in India municipalities: Effects on energy, fertilizers, water and the local environment. *Resour. Conserv. Recycl.* **2021**, *170*, 105569. [CrossRef]
57. Abdelsalam, E.M.; Samer, M.; Amer, M.A.; Amer, B.M.A. Biogas production using dry fermentation technology through co-digestion of manure and agricultural wastes. *Environ. Dev. Sustain.* **2021**, *23*, 8746–8757. [CrossRef]
58. Moonmoon, H.; Das, D.; Baruah, D.C. Bioenergy potential from crop residue biomass in India. *Renew. Sustain. Energy. Rev.* **2014**, *32*, 504–512.
59. Rajeev, K.S.; Vikram, J.S.; Raveendran, S.; Parameshwaran, B.; Kanakambaran, U.J.; Kuttavan, V.S.; Kuni, P.R.; Ashok, P. Lignocelulosic ethanol in India: Prospects, challenges and feedstock availability. *Bioresour. Technol.* **2010**, *101*, 4826–4833.
60. Gathala, M.K.; Ladha, J.K.; Vivak, K.; Yashpal, S.S.; Virender, K.; Paradeep, K.S.; Sheetal, S.; Himanshu, P. Tillage and Crop Establishment Affects Sustainability of South Asian Rice–Wheat System. *ASA* **2011**, *103*, 961–971. [CrossRef]
61. D'Adamo, I.; Falcone, P.M.; Morone, P. A New Socio-economic Indicator to Measure the Performance of Bioeconomy Sectors in Europe. *Ecol. Econ.* **2020**, *176*, 106724. [CrossRef]
62. D'Adamo, I.; Falcone, P.M.; Huisingh, D.; Morone, P. A circular economy model based on biomethane: What are the opportunities for the municipality of Rome and beyond? *Renew. Energy* **2021**, *163*, 1660–1672. [CrossRef]

 energies

MDPI

Article

1,2—Propanediol Production from Glycerol Derived from Biodiesel's Production: Technical and Economic Study

Juan B. Restrepo [1], Carlos D. Paternina-Arboleda [2] and Antonio J. Bula [3],*

[1] Programa de Ingeniería Química, Facultad de Ingeniería, Universidad del Atlántico, Carrera 30 # 8-49, Puerto Colombia C.P. 081001, Colombia; juanrestrepo@mail.uniatlantico.edu.co
[2] Departamento de Ingeniería Industrial, Universidad del Norte, Km 5 Vía Puerto Colombia, Barranquilla C.P. 080001, Colombia; cpaterni@uninorte.edu.co
[3] Departamento de Ingeniería Mecánica, Universidad del Norte, Km 5 Vía Puerto Colombia, Barranquilla C.P. 080001, Colombia
* Correspondence: abula@uninorte.edu.co

Abstract: For every nine tons of produced biodiesel, there is another ton of glycerol as a byproduct. Therefore, glycerol prices dropped significantly worldwide in recent years; the more significant biodiesel production is, the more glycerol exists as a byproduct. glycerol prices also impact the biodiesel manufacturing business, as it could be sold according to its refinement grade. The primary objective of this work was to evaluate the economic potential of the production of 1,2-propanediol derived from the biodiesel produced in Colombia. A plant to produce 1,2-propanediol via catalytic hydrogenation of glycerol in a trickle-bed reactor was designed. The plant comprised a reaction scheme where non-converted excess hydrogen was recycled, and the heat generated in the reactor was recovered. The reactor effluent was sent to a separation train where 98% *m/m* purity 1,2-propanediol was attained. Capital and operational costs were estimated from the process simulation. The net present value (NPV) and the modified internal return rate (MIRR) of the plant were used to assess the viability of the process. Their sensitivity to key input variables was evaluated to find the viability limits of the project. The economic potential of the 1,2-propanediol was calculated in USD 1.2/kg; for the base case, the NPV and the MIRR were USD 54.805 million and 22.56%, respectively, showing that, for moderate variations in products and raw material prices, the process is economically viable.

Keywords: 1,2 propanediol; biodiesel; glycerol as a byproduct; Colombia case-study; trickle-bed reactor

Citation: Restrepo, J.B.; Paternina-Arboleda, C.D.; Bula, A.J. 1,2—Propanediol Production from Glycerol Derived from Biodiesel's Production: Technical and Economic Study. *Energies* **2021**, *14*, 5081. https://doi.org/10.3390/en14165081

Academic Editor: Idiano D'Adamo

Received: 12 May 2021
Accepted: 19 July 2021
Published: 18 August 2021

Publisher's Note: MDPI stays neutral with regard to jurisdictional claims in published maps and institutional affiliations.

1. Introduction

The current models of production and consumption are affecting the world, producing inequality and environmental damage. Studies show that the increment of biobased products is more effective in the reduction of greenhouse gases (GHG) [1,2]; moreover, using an alternative to fossil resource that fulfills the same technical requirements as its fossil counterpart is defined as adding value [1]. Biodiesel is an attractive added value bioresource. Since it is carbon-neutral and compatible with the petroleum-based fuel supply infrastructure, it can reduce the transport sector's greenhouse gas (GHG) emission, estimated as 15% of the total GHG emissions [3].

Developing countries such as Colombia implemented laws such as 693 from 2001 and 939 from 2004, which ruled blends between biofuels as well as gasoline and biofuels mixtures for diesel engines; these laws promoted new business opening in bioethanol and biodiesel production enterprises. As of today, Colombia's biodiesel production capacity is estimated at 591 tons per year. Between 2009 and 2013, Colombia's biodiesel production grew 197.1%. An increasing biodiesel production trend across the globe was also observed [4], but its negative impact on markets such as glycerol was evident.

Biodiesel production is commonly done via catalyst-based transesterification. In this type of reaction, for every 9 tons of produced biodiesel, there is another ton of glycerol as a byproduct [5]; therefore, glycerol prices dropped significantly worldwide [6]. The larger the biodiesel production is, the more glycerol there is as a byproduct. Lower glycerol prices also impact biodiesel production, as it could be sold depending on its refinement [7,8]. The opportunity to make biodiesel production plants more profitable by generating added value products using glycerol as a raw material source would impact the biofuels market, making it nearly as profitable as Petro fuels [9]. Increasing the availability of carbon-neutral commodity chemicals and contributing to the implementation of circular economy biorefineries [1].

Glycerol can be used to obtain several commodity chemicals, especially but not exclusively by chemo-catalytic methods [10–12]. Glycerol can be catalytically oxidized using Pd, Pt, and Au as catalysts to obtain dihydroxyacetone (DHA) [13,14], hydroxypyruic acid (HYPA) [14], and glyceraldehyde (GLYALD) [15,16], which can serve as final products or intermediaries to other commodity chemicals such as formic acid and lactic acid [11]. It can also be dehydrated to obtain acrolein, a chemical intermediate to produce acrylic acid esters and polymers [17]. In the presence of metallic catalysts and hydrogen, the selective hydrogenolysis of glycerol can also be carried on producing 1,2-propanediol [18] or 1,3-propanediol [19,20].

The 1,2-propanediol is one of the most promising platform chemicals, as it is a raw material for polyester unsaturated resins, a solvent, an antifreeze agent, and an additive in food and cosmetics [18,21,22]. As shown in [8], the calculated sales price/total production cost ratio of 1,2-propanediol is 1.57, making it a high added value product compared with acrolein (1.34) and PHB (1.42). Therefore, the production of 1,2-propanediol from glycerol derived from biodiesels production would add value to the biodiesel production chain and become an alternative source of carbon-neutral commodity chemicals for several industries. It would to some degree, reduce the carbon footprint of well-established processes that are already buying petroleum-derived 1,2 propanediol in Colombia.

This work is the technical and the economic analysis of a 1,2-propanediol production plant that uses glycerol, a byproduct of Colombian biodiesel's production. Technological aspects such as reaction technology selection, separation train design, and non-converted reactants recycle are addressed, while their impact on the process capital and the operational cost is calculated.

2. Materials and Methods

2.1. Raw Materials and Products

2.1.1. Glycerol

The (1,2,3-propanetriol) can be found as a component of different fatty acids, which are present naturally, forming triglycerides. At least 10% of total fatty acids are glycerol. Glycerol is found on many industrial products, and it is practically on almost every modern day life product. Figure 1 shows the consumption proportion from different industries regarding glycerol as raw material. Therefore, the glycerol market is vast, and it can be purchased at different concentrations or grades. The glycerin obtained from the biodiesel production line is called raw glycerin. It can be refined to reach United States Pharmacopeia (USP) standards or the Food Chemical Codex (FCC) standards for pharmaceutical or food industry products.

Raw glycerin composition is rich in methanol, water (12% max), soaps, and 80% to 88% glycerol. On technical grade commercially traded items, glycerin is a clear liquid substance with a 98% minimum glycerol content, and no water, methanol, or salts are expected to be present. USP glycerin is a highly refined substance that complies with the USP standards monitored and controlled by the United States Food and Drugs Administration (FDA) (SRS Engineering, 2013).

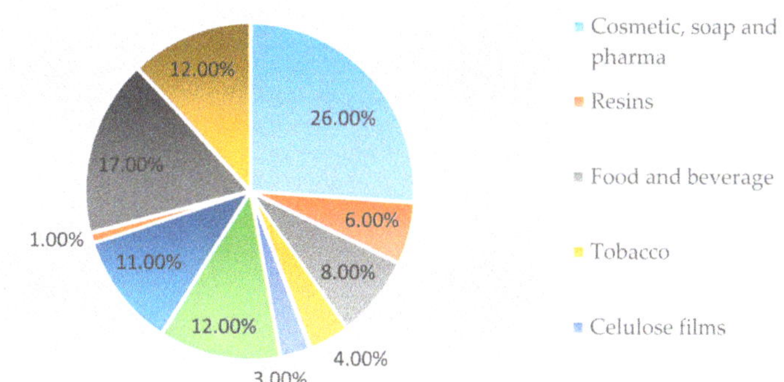

Figure 1. Glycerol consumption distribution (%) [21].

2.1.2. Hydrogen

Hydrogen worldwide production is as high as 54 million tons. In total, 95% of the total output comes from production facilities that use it as raw material in their production lines. Only 5% is sold but only to nearby facilities [22]. Most of the hydrogen also comes from fossils fuels such as natural gas. However, production processes are already tested to obtain hydrogen from biomass [23].

2.1.3. 1,2-Propanediol

Additionally, known as propylene glycol, it has a market value estimated at 1.58 million tons per year. It is the primary raw material on the polyester unsaturated resin manufacturing business used in construction and transport companies. At least 75% of total polyester unsaturated resins are reinforced with fiberglass and substances to reinforce plastics, making them highly resistant and light-weight [24]. Propylene glycol is also used as a solvent in anti-freezing industry, freezing media in the food industry, and even polyglycol industry responsible for the hydraulic braking systems industry. Recent studies indicated the rising concerns on ethylene glycol and its toxicity; therefore, a rise in the propylene glycol demand as a substitute for ethylene glycol could be expected.

2.2. Glycerol Catalytic Conversion to 1,2-Propanediol

Glycerol is the primary raw material for 1,2-propanediol, ethylene glycol, and other sub-products such as lactic acid, acetol, and prop-2-enal as well as degradation products such as methanol and carbon dioxide, methane, and propanol [25]. The dehydration–hydrogenation reaction seen in Figure 2 was achieved in the presence of Cu-based catalysts [26–28]. In [29], two adsorption sites were assumed in the reaction mechanism, and a Langmuir–Hinshelwood kinetics was proposed. Therefore, for the reactor design, the kinetic model expressed by Equations (1) and (2) extracted from [28] was used.

$$r_1 = \frac{k_1 b_G C_G}{1 + b_G c_G + b_A c_A + b_P c_P} \tag{1}$$

$$r_2 = \frac{k_1 b_A c_A b_H P_H}{\left(1 + b_G c_G + b_A c_A + b_P c_P\right)\left(1 + \sqrt{b_H P_H}\right)^2} \tag{2}$$

where r_1 is the acetol formation rate, r_2 is 1,2-propanediol formation rate, c is molar concentration, b the adsorption constant, P_H is hydrogen pressure, and the sub-indexes A, G, P, and H stand for acetol, glycerol, 1,2-propanediol, and hydrogen, respectively. The reaction and the adsorption constants were modeled according to Equations (3) and (4), and the parameters of Table 1 obtained in [28] were used.

$$k_i = k_i^0 \exp\left[-\frac{E_i}{R_g T}\right], \, i = 1, 2, \ldots,$$ (3)

$$b_j = b_i^0 \exp\left[-\frac{Q_j}{R_g T}\right], \, j = A, G, P, H$$ (4)

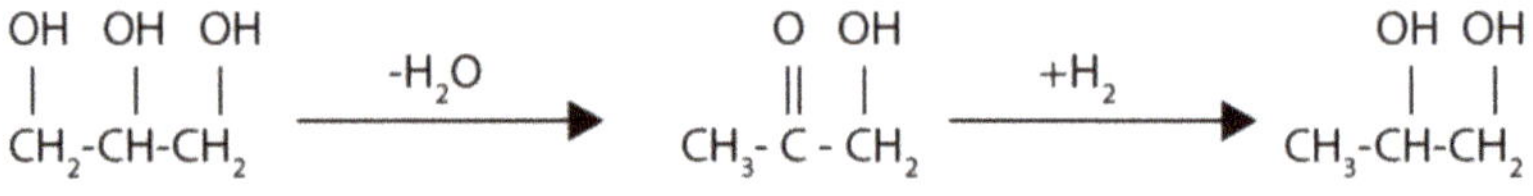

Figure 2. Two-step mechanism proposed by Dasari et al. [25].

Table 1. Kinetics parameters from Zhou et al. [28].

Parameter	Pre-Exponential	Activation Energy
k_1	1.54×10^4	86.56
k_2	7.16×10^3	57.8
b_A	2.22×10^{-3}	36.42
b_G	8.73×10^{-3}	25.94
b_P	5.8×10^{-3}	25.77
b_H	1.86×10^{-5}	36.24

Table 1 constants k_i and b_i are based on $\left(\mathrm{mol \cdot g^{-1} \cdot s^{-1}}\right)$ and $\left(\mathrm{m^3 \, mol^{-1}}\right)$, respectively, b_h is on $\mathrm{MPa^{-1}}$, and the activation energy is on kJ/mol. These constants were experimentally obtained for a Cu-ZnO-Al_2O_3 catalyst on a 1:1:0.5 molar proportion and a 0.17 mm particle size by Zhou et al. [28]. The selectivities used in the present for these reaction conditions were 93.4% for 1,2-propanediol, 1.2% for acetol, 2.7% for ethylene glycol, and 2.7% for other compounds such as methanol, ethanol, and propanol [28].

2.3. Reactor Technology Selection and Design

There is a significant industrial background for the trickle bed reactor (TBR) technology for hydrogenolysis, hydrodesulfurization, hydrocracking, and hydro-refinery processes [30,31]. Therefore, this reactor type was selected as the most suitable technology to obtain 1,2-propanediol. The reactor operates using a liquid and gas phase percolation through a fixed catalytic bed on a co-current operation (ideal operation is assumed). Thus, the gaseous phase is the continuous one, while the liquid phase creates a thin film that descends and covers the surface of the catalyst particles. This kind of reactor enables continuous operation, high pressure operating conditions, and a more considerable length to diameter (L/D) ratio.

Conversely, low mass transfer coefficients limit the reaction process as the catalyst gets partially covered with the liquid [31]. The reactor was simulated with the RStoic model in Aspen Plus using the selectivity mentioned above, operating adiabatically at 5 MPa and an outlet temperature of 243.588 °C. The calculations of the mass transfer limitations and the reactor sizing were done using Matlab® with the methodologies developed in [31,32]. For the solution of the model presented in Equations (5) and (6), catalysts deactivation mechanism was assumed (coke formation) as 14 h according to [33].

$$\frac{dF_G}{dx} = \eta_{CE}\rho_c(1 - \varepsilon_B)r_1$$ (5)

$$\frac{dF_P}{dx} = [\eta_{CE}r_2 + (1 - \eta_{CE})\backslash r_2^*]\rho_c(1 - \varepsilon_B)$$ (6)

where F_P, and F_G are the 1,2-propanediol, glycerol mole flows inside the reactor, η_{CE} is the wetting of the catalyst surface calculated as in [31], and ε_B is the bed voidage calculated for the HIFUEL W220 [34] from Alfa Aesar, which is the commercially available catalyst more similar to what is described in [28].

2.4. Design and Simulation Methodology and Considerations

The economic analysis for the 1,2-propanediol produced from biodiesel's glycerol and the conceptual process design were performed using Aspen Plus and Aspen Economic Analyzer software. The thermodynamic model used for liquid phase simulation was non-random two liquids [35], and for the gas phase, the Redlich–Kwong model [36].

The process comprises two stages—the reaction and the heat recovery stages—where the glycerol is converted, the excess hydrogen is recovered, and the purification stage occurs, where the 1,2 propanediol is recovered with 99% purity.

The reaction and heat recovery process diagram is shown in Figure 3. The outlet temperatures on heat exchangers HX1 and HX2 were optimized to reach the maximum energy recovery. Feeding proportion rates between glycerol and hydrogen were set to 5:1 as suggested in the literature to achieve the 1,2-propanediol formation [25] successfully. As the hydrogen proportion is five times larger than glycerol, considering that the stoichiometric ratio is 1:1, the excess of hydrogen must be recycled. Therefore, the reaction process design considered heat exchangers and flash separators to recover a fraction of the heat generated during the reaction step and use it as heating media on the reactor's inlet stream. A stoichiometric reactor was selected for the simulation, assuming 100% glycerol conversion and the selectivity mentioned earlier.

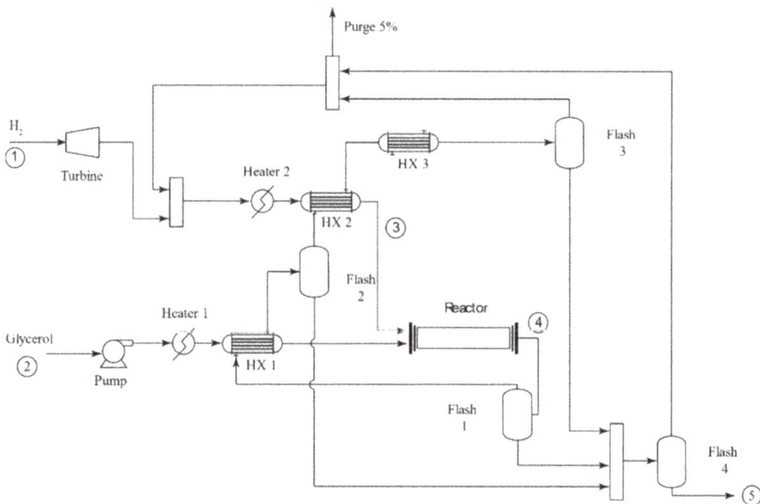

Figure 3. Reaction process flow diagram.

As the reactor effluent was rich in excess hydrogen, successive cooling and condensate recovery in flashes 1, 2, 3, and 4 allowed the recovery of the excess hydrogen that was recirculated and combined with the fresh feed. The reactor scheme was designed using a hierarchical approach [37,38] combined with optimization of key temperatures [39,40] to achieve an effective recovery of the heat generated in the reactor.

Figure 4 illustrates the distillation process. After the reaction process step was completed, hydrogen traced, and the other two most volatile compounds were removed from reaction products at the first distillation tower. In the next distillation tower, the lightest compounds such as acetol and propanol were separated from ethylene glycol. However, as

1,2-propanediol has a mid-boiling point, thus it could be found at both the top of the distillation tower (stream 7) and the bottom of the distillation tower (stream 6). In order to refine 1,2-propanediol up to 99% purity, it was necessary to take stream 7 to a distillation tower where volatile compounds such as acetol were separated from 1,2-propanediol, which came out of the bottom of the distillation tower. Stream 6 was also taken to a 65 stages distillation tower to isolate ethylene glycol from 1,2-propanediol; in this step, 1,2-propanediol came out of the top of the distillation tower. Both streams were then mixed and then cooled to 25 C in stream 8 to reach 99% purity.

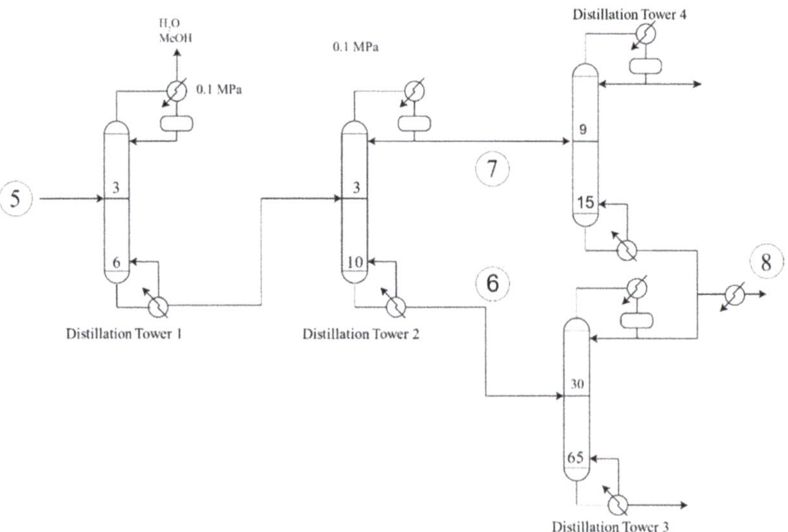

Figure 4. Separation process.

Distillation was first designed using the Winn-Underwood-Gilliland equations, which are included on the DSTWU unit from Aspen Plus. Once the initial design parameters were estimated, the RadFrac module from Aspen Plus was used, and the parameters were adjusted to fulfill the design specifications more precisely. The distillation train was designed based on non-sharp distillations to separate the ethylene glycol and the 1,2-propanediol which have close boiling points [41,42].

Production Rate Based on the Colombia Case Study

A private association in Colombia in which some of its primary responsibilities include the national alcohol sustainable production supporting, enabling, and designing the regulations required for biofuels production and usage. This federation is called Fedebiocombustibles; according to them, biodiesel production during 2017 was 460.121 tons [43]. As stated earlier, for every 9 kg of produced biodiesel, another kilogram of glycerol is produced. Thus, an average production of 4.260 tons of glycerol every month was estimated. In this paper, it is assumed as a calculation base that 3.924 kg/h of glycerol will enter our reaction process.

Another critical assumption is the glycerol quality, which was selected as raw glycerin. After the preliminary isolation and purification process, salts and other compounds are separated from our inlet stream, leaving 84% glycerol, 1% methanol, and 14% water as total stream composition. Hydrogen comes as a pure gas from an available storage tank at 1.000 psi of pressure, and the hydrogen pressure must be reduced as low as 734.8 psi via turbo expansion [28].

2.5. Economic Analysis Methodology and Considerations

Economic analysis was done based on material and energy balances from process simulation for a plant that processes 3.924 kg/h of glycerol to produce 2514.84 kg/h of 1,2-propanediol. Aspen economic analyzer was selected as the calculation engine. Initial parameters fed to the software were aligned with Colombia's economic benchmarks using US dollars as reference coin and a 10-year operation period, with 52 weeks/year and 24 h/day. An annual interest rate of 17%, straight-line depreciation, and a 34% tax rate were also considered to model project economics outcomes. Labor costs of USD 2.14/h for the operator's labor time and USD 4.29/h for supervisors were used. Power cost was estimated at USD 0.03044/kWh, and other utilities such as water (USD 1.252/m^3) and low-pressure steam (USD 8.18/ton) were also included [8,44]. A glycerin cost of USD 0.2/kg and a hydrogen cost at 1000 psi of USD 2/kg were selected as the base case for calculations. USD 1.54/kg for the 1,2-propanediol price was assumed considering historical data presented in [45] and information from local importers in private communications.

The economic assessment of the process was performed based upon the concept of money time value under the discounted cash flow (DCF) methodology [1,46]. Based on this methodology, the net present value (NPV) and the modified internal return rate (MIRR) were selected as the plant's profitability indices. The following models were used to calculate the profitability indices:

$$NPV = \sum_{t=0}^{n} \frac{C_t}{(1+r)^t} \tag{7}$$

$$MIRR = \left(\frac{FV}{PV}\right)^{\frac{1}{t}} - 1 \tag{8}$$

where C_t is the net cash flow, r is the discount rate, FV is the future value of the cash flows at the reinvestment rate, PV is the present value of the negative cash flows at the finance rate, t is the period, and n is the number of periods of the analysis. The economic analysis was performed based on the variation of the net present value (NPV) and the modified internal return rate (MIRR) with the baseline prices of glycerin, hydrogen, and 1,2-propanediol. The catalyst price was estimated fixed at USD 95/kg, yet the refresh rate of the catalyst and its regeneration remain uncertain, and most of the consulted studies are still in early stages [18,33,47,48] for the specific reaction; therefore, the refresh rate of the catalysts was introduced as an input variable to calculate the viability of the project.

3. Results

As a result of applying the methodology, the plant simulations were followed by an optimization procedure to adjust the system to design specifications. The data obtained from the plant simulation economic analysis revealed the economic potential of the process and its sensitivity to relevant input variables. The lack of information impeded the catalysts' deactivation simulation; therefore, the catalyst refresh rate was added as an economic input variable, and the plant's NPV was analyzed.

3.1. Reactor Design

With the same methodology used in [31,32], the total catalyst required to reach a 100% conversion was calculated. Equations (1)–(4) with the parameters reported in Table 1 were used as the trickle bed reactors' kinetic model described by differential Equations (5) and (6). Figure 5 shows the results from the simulation used to design the reactor, where 3700 kg was calculated as the necessary mass of catalysts to carry on the reaction.

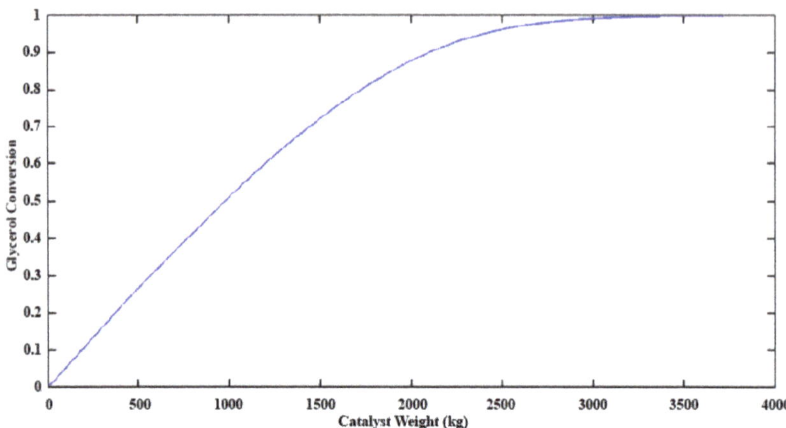

Figure 5. Catalyst's mass required to reach 100% glycerol conversion.

3.2. Simulation and Design Results

Table 2 shows process stream compositions and temperatures for the simulated process. It was observed that, even though the distillation system includes four distillation towers, stream 8, which is the final product stream, had a 99% mass fraction for 1,2-propanediol but still had a 1% mass fraction of ethylene glycol.

Table 2. Simulation results for Figures 3 and 4 flowsheets.

Parameter	Streams *							
	1	2	3	4	5	6	7	8
Mass Flow (kg/h)	87.62	3924.00	411.91	4335.91	3997.72	2433.55	206.59	2514.84
Temperature (°C)	25.00	25.00	200.00	243.59	176.35	187.42	130.28	25.00
Pressure (bar)	68.95	1.00	49.78	50.00	50.00	1.00	1.00	1.00
Vapor Frac.	1.00	0.00	1.00	0.89	0.00	0.00	0.00	0.00
Mass Fraction								
1,2-Propanediol	0	0	0.00009	0.58685	0.63638	0.97500	0.80418	0.98885
Ethylene glycol	0	0	0	0.01383	0.01500	0.02373	0.01060	0.00897
Acetol	0	0	0.00007	0.00734	0.00791	0.00127	0.08766	0.00218
1-Popanol	0	0	0.00291	0.01516	0.01447	1.17766×10^{-9}	0.00098	1.87134×10^{-9}
Water	0	0.15	0.01548	0.29239	0.30863	2.00252×10^{-7}	0.09631	5.00583×10^{-7}
H_2	1	0	0.97673	0.06631	6.64086×10^{-5}	0	0	0
Methanol	0	0.01	0.00472	0.01811	0.01754	2.5741×10^{-12}	0.00028	7.1212×10^{-12}
glycerol	0	0.84	0	0	0	0	0	0
Cost $/hr	96.58102	502.9981						

* The stream numbers correspond to Figures 3 and 4.

3.3. Estimated Capital Cost

The total project cost of capital was USD 15,456,582.64, out of which USD 6,757,832 were associated with direct cost and can be seen in Table 3, USD 1,796,000 were related to indirect construction and installation cost, and USD 7,094,051 were administrative and non-field costs. The total cost included process unit procurement and installation costs, instrumentation, pipelines, steel procurement, civil and electrical works, and isolation where required.

Table 3. Detailed direct costs of the plant in USD.

	Labor Cost	Mat Cost	Total Cost
Equipment	135,772.95	2,573,980.00	2,709,752.95
Piping	477,683.95	333,634.47	811,318.42
Civil	104,195.42	140,486.82	244,682.24
Steel	16,730.19	94,990.92	111,721.11
Instruments	286,258.38	1,162,905.89	1,449,164.27
Electrical	129,173.18	973,076.62	1,102,249.80
Insulation	138,412.94	125,151.70	263,564.64
Paint	47,330.64	18,047.51	65,378.15
Total Direct Field Costs	1,335,557.65	5,422,273.92	6,757,831.58

Table 4 shows the cost of all the process equipment in the plant. As expected, the most expensive fixed costs were the reactor (including catalyst) and the distillation train. Nevertheless, the catalyst was not a fixed cost, as it must be refreshed periodically. The catalyst's stability was a concern, as it is a key factor for the plant's operation.

Table 4. Process equipment, as labeled in Figures 3 and 4, cost in USD.

	Total Direct Cost	Equipment Cost		Total Direct Cost	Equipment Cost
Heater 1	48,300.00	10,000.00	Distillation Tower 1	456,100.00	94,400.00
Heater 2	53,300.00	11,900.00	Distillation Tower 2	430,000.00	81,400.00
HX 1	60,200.00	12,400.00	Distillation Tower 3	2,110,200.00	1,500,100.00
HX 2	55,400.00	10,500.00	Distillation Tower 4	397,500.00	92,600.00
HX 3	76,300.00	12,500.00	Cooler	64,400.00	13,600.00
Flash 1	123,300.00	32,300.00	Turbine	183,100.00	92,600.00
Flash 2	109,600.00	32,300.00	Pump	90,700.00	54,900.00
Flash 3	121,900.00	32,300.00	Reactor	140,900.00	32,300.00
Flash 4	121,700.00	32,300.00	Catalyst	362,700.00	349,280.00

3.4. Estimated Profit and Operational Cost

Operational cost reached USD 10,002,707.23/year; raw materials, operational labor, utilities, and general and administrative costs were also included in the income statement calculation.

As it can be observed in Table 5, raw materials were as high as 85.73% of the total operating cost.

Table 5. Manufacturing unit operational cost.

Cost Items	Annual Cost (USD)	Cost Contribution
Raw materials	8,575,486.53	85.3%
Labor	72,735.30	0.73%
Maintenance	202,631.90	2.03%
Utilities	255,044.79	2.55%
Operating charges	18,183.83	0.18%
Plant Overhead	137,683.60	1.38%
Administrative cost	740,941.28	7.41%

Due to the heat recovery proposed early in the design phase, fluid services cost was lower than administrative cost. If the total production was sold, the net income could reach USD 38,490,503.50/year, enough to cover the cost of capital by two. 1,2 Propanediol economic potential was calculated at USD 1.2/kg.

Table 6 shows the cost of heating and cooling in the exchangers of the plant. A factor that allowed us to diminish the heating utility expenses recycled the heat generated in the reactor, which was recovered using HX1 and HX2 with a combined heat duty of

400.1052 kJ/s. The cooling water mass flow was 406,604.837 kg/h, and the steam flow was 5665.7344 kg/h for the heat duties reported in Table 6.

Table 6. Heating and Cooling duty of the plant.

Total heating duty	kJ/s	2744.6214
Total cooling duty	kJ/s	2336.7594
Net duty (Total heating duty—Total cooling duty)	kJ/s	407.8620
Total heating cost flow	US $/h	24.0990
Total cooling cost flow	US $/h	1.7834
Net cost (Total heating cost + Total cooling cost)	US $/h	25.8824

3.5. Sensitivity Analysis

The economic viability of the project is dependent on the cash flow of the plant. The net present value (NPV) of the project is sensitive to economic changes such as the variability of the raw materials cost and the product's sales price. For the base raw material costs and product price, the NPV of the project for a 10-year period is USD 54,805 million, and the modified internal rate of return (MIRR) was 22.56%.

Figures 6 and 7 show the variations of the NPV and the MIRR as raw materials prices varied from the baseline case established in Section 2.5.

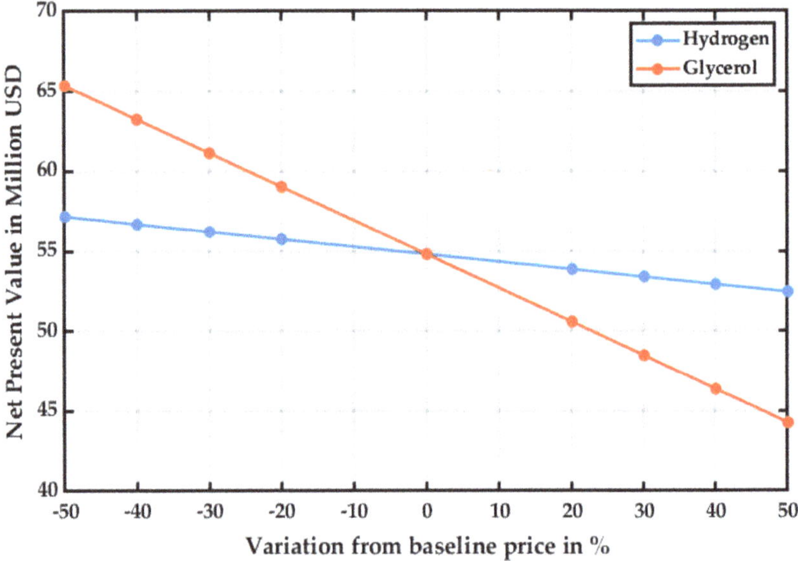

Figure 6. Sensitivity analysis for the NPV of the project as the raw material cost varies.

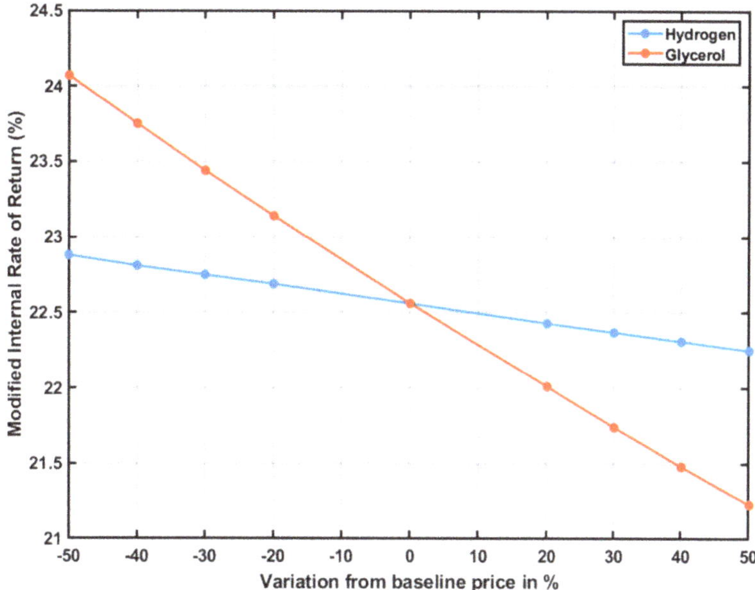

Figure 7. Sensitivity analysis for the MIRR of the project as the raw material cost varies.

Figure 8 shows the sensitivity of the project to the sales price of the product. The actual price of 1,2-propanediol allowed the project to be profitable, yet the project was most sensitive to the high prices of the product. The MIRR presented a nonlinear variation with the product price, which was not observed with the raw materials.

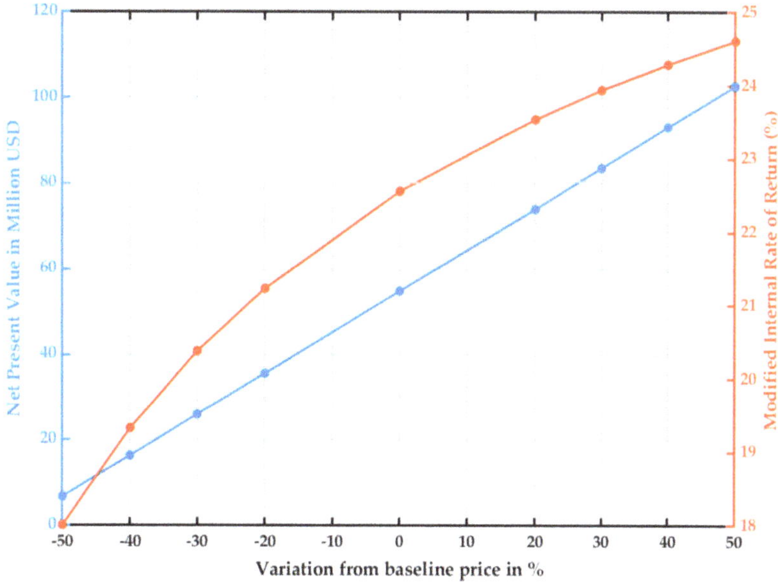

Figure 8. Sensitivity analysis for NPV and MIRR of the project as the sales price of 1,2-propanediol varies.

Figure 9 shows how the NPV varies as the catalysts must be replaced as it deactivates. The deactivation amount is expressed in spent catalyst per year/reactor's catalyst capacity.

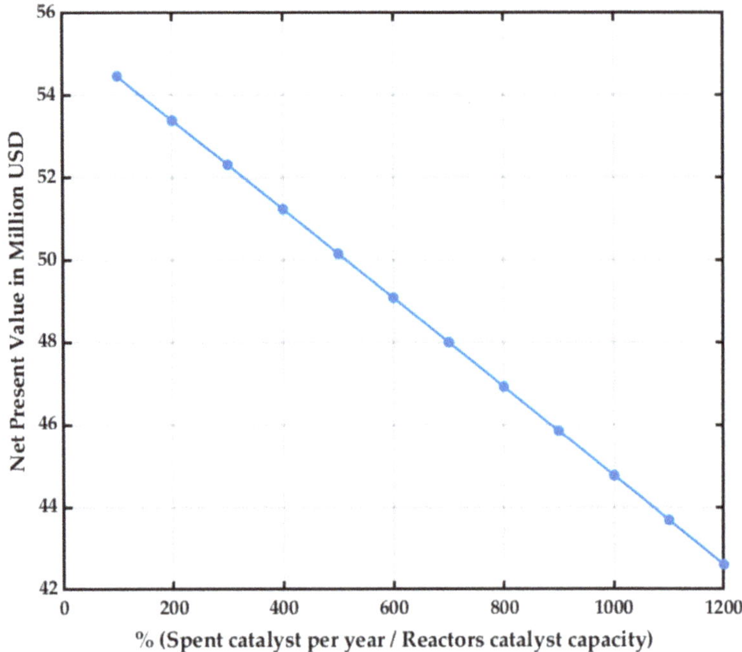

Figure 9. Sensitivity analysis for the catalysts refresh rate.

4. Discussion

This work explored a promising way to take advantage of the residual glycerol from biodiesel production based on the Colombian industrial context. The conceptual design and the economic analysis of the production of 1,2-propanediol were performed without neglecting the byproduct formation. As a result, a small ethylene glycol fraction was formed. Its higher boiling point altered the distillation column train design strategy for a non-sharp distillation, which resulted in efficient recovery of 1,2-propanediol. Consequently, operational and capital costs would rise; this portrays a more realistic economic analysis of the selected catalytic and the reaction technology in the separation train. With all these considerations, the profit margin remains attractive for investment.

The reactor technology selection led to a three-phase reactor in which a complex mass transfer process took place. Therefore, a much more complicated model was used for the sizing of the reactor, leading to the calculation of a 3700 kg catalyst load, which would allow a high conversion of glycerol, avoiding its presence in the separation train which would raise the utility costs, as glycerol normal boiling point is 290 °C.

The high temperature of the reactor allowed the recovery of 400.1052 kJ/s used for heating the inlet streams; this accounts for a 12% reduction in the heating costs compared to the total heating duty of the plant without the thermal energy recycle. The excess hydrogen required for the reaction to take place was recycled with high purity.

The capital cost estimation led to a USD 5,456,582.64 where the process equipment cost account was 17%, and electrical, instrumentation, and piping costs were 21.8%. This means that the initial investment was significant, and with a process start-up period of 20 weeks, the process reaches the equilibrium point after the second year.

Raw materials accounted for 85.73% of the operational costs, and all the plant's income was calculated based on the 1,2-propanediol as the only product. Therefore, these were the key input values for economic analysis. The net present value (NPV) measured the economic potential and the risk of commercialization of the current process; the positive NPV meant that the process was profitable. With a base NPV of USD 54.805 million for a 10-year period, the sensitivity analysis showed that the plant profitability depended on the sale price of 1,2-propanediol followed by the glycerol cost, which highly depended on its availability. The modified internal return rate MIRR assumed that any positive cash flow was reinvested at the external rate of return, making this a more accurate way to measure the viability of the project. As with NPV, the tendencies were the same; the plant was highly dependent on the variability of the 1,2-propanediol price followed by the glycerol price.

A non-accounted factor during the design due to the lack of mathematical models was the catalyst's refresh and regeneration rates. In consequence, the impact of the catalyst's refresh rate was analyzed. In conclusion, the plant is not viable (NPV < 0) if the catalyst has to be totally replaced every seven days, much less than expected of a commercial catalyst. This factor highly affects the plant's economy, and further research must be done to establish its deactivation mechanism and exact cost.

5. Conclusions

The conceptual design and the economic analysis of a process are potent tools for possible solutions to existing problems—in this case, the surplus glycerol produced by the biodiesels industry in a country such as Colombia. Nevertheless, many assumptions induce uncertainty to the result in this analysis. In this case, logistics costs analysis was not considered, as there is not enough information to predict the distribution chain of glycerol in the country, especially for a facility as big as the designed one, with the country's biodiesel production scattered across the territory. The availability of data on the reaction also limits the reach of the predictions; most of the available kinetics were obtained at a laboratory scale under particular conditions. Yet, an effort was made to use the intrinsic kinetics combined with the mass transfer equations to dimension the reactor as accurately as possible. Although the deactivation kinetics of the chosen $Cu/Zn/Al_2O_3$ is not available, it was possible to consider its economic effect on the plant. Accurate calculations are possible, especially in the separation where most of the utilities are spent, and, therefore, energetic predictions are considered reliable. The estimates of operational costs, raw material prices, and product prices are also accurate, as they are done based on the literature and well-known analysis methodologies and software.

With the above-mentioned limitations, the potential of the 1,2 Propanediol production as an alternative for re-valuing raw glycerin is high, since the production costs and the revenues exceed capital costs in the long term, even under moderate variations on feedstock costs. In the short term, the project would start being profitable after the second year. Since glycerol production is increasing worldwide, all alternatives should be considered to produce added value processes that make the biodiesel refinery more profitable and promote carbon-neutral alternatives to petroleum-derived products.

Author Contributions: Conceptualization, J.B.R. and A.J.B.; methodology, J.B.R.; investigation, J.B.R. and C.D.P.-A.; writing—original draft preparation, J.B.R.; writing—review and editing, J.B.R., C.D.P.-A. and A.J.B.; supervision, A.J.B.; project administration, C.D.P.-A.; funding acquisition, C.D.P.-A. All authors have read and agreed to the published version of the manuscript.

Funding: This research was funded through the project "Proyectos I+D Para Promover el Desarrollo y la Transferencia Tecnológica De Cadenas Productivas Agroindustriales y la Implementación de Tecnologías de Ultima Generación para el Procesamiento de Biocombustibles en el Departamento del Atlántico" Grant: "Convenio UNINORTE - Gob Atlántico: 0103_2013_000014", through the general royalties System of Colombia (Sistema General de Regalías). The APC was funded by Universidad del Norte Barranquilla-Colombia.

Acknowledgments: Especial thanks to the Universidad del Atlántico in Barranquilla for the ongoing support and funding of Juan B. Restrepo's research in biofuels.

Conflicts of Interest: The authors declare no conflict of interest.

References

1. D'Adamo, I.; Falcone, P.M.; Huisingh, D.; Morone, P. A circular economy model based on biomethane: What are the opportunities for the municipality of Rome and beyond? *Renew. Energy* **2021**, *163*, 1660–1672. [CrossRef]
2. D'Adamo, I.; Falcone, P.M.; Morone, P. A new socio-economic indicator to measure the performance of bioeconomy sectors in Europe. *Ecol. Econ.* **2020**, *176*, 106724. [CrossRef]
3. Alhindawi, R.; Abu Nahleh, Y.; Kumar, A.; Shiwakoti, N. Projection of Greenhouse Gas Emissions for the Road Transport Sector Based on Multivariate Regression and the Double Exponential Smoothing Model. *Sustainability* **2020**, *12*, 9152. [CrossRef]
4. Raturi, A.K. Renewables 2019 Global Status Report. Available online: https://www.ren21.net/gsr-2019/ (accessed on 17 April 2021).
5. Karinen, R.S.; Krause, A.O.I. New biocomponents from glycerol. *Appl. Catal. A Gen.* **2006**, *306*, 128–133. [CrossRef]
6. Quispe, C.A.G.; Coronado, C.J.R.; Carvalho, J.A., Jr. Glycerol: Production, consumption, prices, characterization and new trends in combustion. *Renew. Sustain. Energy Rev.* **2013**, *27*, 475–493. [CrossRef]
7. Chiu, C.-W.; Dasari, M.A.; Sutterlin, W.R.; Suppes, G.J. Removal of residual catalyst from simulated biodiesel's crude glycerol for glycerol hydrogenolysis to propylene glycol. *Ind. Eng. Chem. Res.* **2006**, *45*, 791–795. [CrossRef]
8. Posada, J.A.; Rincón, L.E.; Cardona, C.A. Design and analysis of biorefineries based on raw glycerol: Addressing the glycerol problem. *Bioresour. Technol.* **2012**, *111*, 282–293. [CrossRef]
9. Lari, G.M.; Pastore, G.; Haus, M.; Ding, Y.; Papadokonstantakis, S.; Mondelli, C.; Pérez-Ramírez, J. Environmental and economical perspectives of a glycerol biorefinery. *Energy Environ. Sci.* **2018**, *11*, 1012–1029. [CrossRef]
10. Zhou, C.-H.C.; Beltramini, J.N.; Fan, Y.-X.; Lu, G.Q.M. Chemoselective catalytic conversion of glycerol as a biorenewable source to valuable commodity chemicals. *Chem. Soc. Rev.* **2008**, *37*, 527–549. [CrossRef]
11. Len, C.; Delbecq, F.; Corpas, C.C.; Ramos, E.R. Continuous flow conversion of glycerol into chemicals: An overview. *Synthesis* **2018**, *50*, 723–741. [CrossRef]
12. Muraza, O. Peculiarities of glycerol conversion to chemicals over zeolite-based catalysts. *Front. Chem.* **2019**, *7*, 233. [CrossRef] [PubMed]
13. Walgode, P.M.; Faria, R.P.V.; Rodrigues, A.E. A review of aerobic glycerol oxidation processes using heterogeneous catalysts: A sustainable pathway for the production of dihydroxyacetone. *Catal. Rev. Sci. Eng.* **2020**, 1–90. [CrossRef]
14. Yang, L.; Li, X.; Chen, P.; Hou, Z. Selective oxidation of glycerol in a base-free aqueous solution: A short review. *Chin. J. Catal.* **2019**, *40*, 1020–1034. [CrossRef]
15. Dimitratos, N.; Porta, F.; Prati, L. Au, Pd (mono and bimetallic) catalysts supported on graphite using the immobilisation method: Synthesis and catalytic testing for liquid phase oxidation of glycerol. *Appl. Catal. A Gen.* **2005**, *291*, 210–214. [CrossRef]
16. Dimitratos, N.; Porta, F.; Prati, L.; Villa, A. Synergetic effect of platinum or palladium on gold catalyst in the selective oxidation of D-sorbitol. *Catal. Lett.* **2005**, *99*, 181–185. [CrossRef]
17. Katryniok, B.; Paul, S.; Dumeignil, F. Recent developments in the field of catalytic dehydration of glycerol to acrolein. *Acs Catal.* **2013**, *3*, 1819–1834. [CrossRef]
18. Zhao, H.; Zheng, L.; Li, X.; Chen, P.; Hou, Z. Hydrogenolysis of glycerol to 1,2-propanediol over Cu-based catalysts: A short review. *Catal. Today* **2019**, *355*, 84–95. [CrossRef]
19. Nakagawa, Y.; Shinmi, Y.; Koso, S.; Tomishige, K. Direct hydrogenolysis of glycerol into 1,3-propanediol over rhenium-modified iridium catalyst. *J. Catal.* **2010**, *272*, 191–194. [CrossRef]
20. Kurosaka, T.; Maruyama, H.; Naribayashi, I.; Sasaki, Y. Production of 1,3-propanediol by hydrogenolysis of glycerol catalyzed by Pt/WO$_3$/ZrO$_2$. *Catal. Commun.* **2008**, *9*, 1360–1363. [CrossRef]
21. Bondioli, P. Overview from oil seeds to industrial products: Present and future oleochemistry. *J. Synth. Lubr.* **2005**, *21*, 331–343. [CrossRef]
22. Ball, M.; Wietschel, M. The future of hydrogen–opportunities and challenges. *Int. J. Hydrogen Energy* **2009**, *34*, 615–627. [CrossRef]
23. Holladay, J.D.; Hu, J.; King, D.L.; Wang, Y. An overview of hydrogen production technologies. *Catal. Today* **2009**, *139*, 244–260. [CrossRef]
24. Bridgwater, A.V.; Chinthapalli, R.; Smith, P.W. *Identification and Market Analysis of Most Promising Added-Value Products to Be Co-Produced with the Fuels*; Aston University: Birmingham, UK, 2010.
25. Dasari, M.A.; Kiatsimkul, P.-P.; Sutterlin, W.R.; Suppes, G.J. Low-pressure hydrogenolysis of glycerol to propylene glycol. *Appl. Catal. A Gen.* **2005**, *281*, 225–231. [CrossRef]
26. Balaraju, M.; Rekha, V.; Prasad, P.S.S.; Prasad, R.B.N.; Lingaiah, N. Selective hydrogenolysis of glycerol to 1,2 propanediol over Cu–ZnO catalysts. *Catal. Lett.* **2008**, *126*, 119–124. [CrossRef]
27. Wang, S.; Liu, H. Selective hydrogenolysis of glycerol to propylene glycol on Cu–ZnO catalysts. *Catal. Lett.* **2007**, *117*, 62–67. [CrossRef]
28. Zhou, Z.; Li, X.; Zeng, T.; Hong, W.; Cheng, Z.; Yuan, W. Kinetics of Hydrogenolysis of Glycerol to Propylene Glycol over Cu-ZnO-Al$_2$O$_3$ Catalysts. *Chin. J. Chem. Eng.* **2010**, *18*, 384–390. [CrossRef]
29. Restrepo, J.B.; Bustillo, J.A.; Bula, A.J.; Paternina, C.D. Selection, Sizing, and Modeling of a Trickle Bed Reactor to Produce 1,2 Propanediol from Biodiesel Glycerol Residue. *Processes* **2021**, *9*, 479. [CrossRef]
30. Mary, G.; Chaouki, J.; Luck, F. Trickle-Bed Laboratory Reactors for Kinetic Studies. *Int. J. Chem. React. Eng.* **2009**, *7*. [CrossRef]

31. Ranade, V.V.; Chaudhari, R.V.; Gunjal, P.R. *Trickle Bed Reactors*; Elsevier: Amsterdam, The Netherlands, 2011; ISBN 978-0-444-52738-7.
32. Xi, Y.; Holladay, J.E.; Frye, J.G.; Oberg, A.A.; Jackson, J.E.; Miller, D.J. A kinetic and mass transfer model for glycerol hydrogenolysis in a trickle-bed reactor. *Org. Process Res. Dev.* **2010**, *14*, 1304–1312. [CrossRef]
33. Panyad, S.; Jongpatiwut, S.; Sreethawong, T.; Rirksomboon, T.; Osuwan, S. Catalytic dehydroxylation of glycerol to propylene glycol over Cu–ZnO/Al$_2$O$_3$ catalysts: Effects of catalyst preparation and deactivation. *Catal. Today* **2011**, *174*, 59–64. [CrossRef]
34. HiFUEL®Base Metal Water Gas Shift Catalysts. Available online: https://www.alfa.com/media/HiFUEL_Base_Metal_Water_Gas_Shift_Catalysts.pdf (accessed on 17 April 2021).
35. Renon, H.; Prausnitz, J.M. Local compositions in thermodynamic excess functions for liquid mixtures. *AIChE J.* **1968**, *14*, 135–144. [CrossRef]
36. Redlich, O.; Kwong, J.N.S. On the thermodynamics of solutions. V. An equation of state. Fugacities of gaseous solutions. *Chem. Rev.* **1949**, *44*, 233–244. [CrossRef] [PubMed]
37. Turton, R.; Bailie, R.C.; Whiting, W.B.; Shaeiwitz, J.A. *Analysis, Synthesis and Design of Chemical Processes*; Pearson Education: London, UK, 2008; ISBN 0132459183.
38. Douglas, J.M. *Conceptual Design of Chemical Processes*; McGraw-Hill: New York, NY, USA, 1988.
39. El-Halwagi, M.M. *Sustainable Design through Process Integration: Fundamentals and Applications to Industrial Pollution Prevention, Resource Conservation, and Profitability Enhancement*; Butterworth-Heinemann: Oxford, UK, 2017; ISBN 0128098244.
40. Cardona, C.A.; Sanchez, O.J.; Gutierrez, L.F. *Process Synthesis for Fuel Ethanol Production*; CRC Press: Boca Raton, FL, USA, 2009; ISBN 1439815984.
41. Petlyuk, F.B. *Distillation Theory and Its Application to Optimal Design of Separation Units*; Cambridge University Press: Cambridge, UK, 2004; ISBN 113945563X.
42. Westerberg, A.W. The synthesis of distillation-based separation systems. *Comput. Chem. Eng.* **1985**, *9*, 421–429. [CrossRef]
43. Fedebiocombustibles. Available online: http://www.fedebiocombustibles.com/v3/estadistica-mostrar_info-titulo-Biodiesel.htm (accessed on 17 April 2021).
44. Posada, J.A.; Cardona, C.A.; Cetina, D.M. Bioglicerol como materia prima para la obtención de productos de valor agregado. In *CARDONA, CA Avances Investigativos en la Producción de Biocombustibles*; Artes Graficas Tizán: Manizales, Colombia, 2009; pp. 103–127.
45. Pandia, R. Markets and Potential for Non-polymer Derivatives of Propylene. In Proceedings of the Olefins Asia 2014, Shanghai, China, 6–7 March 2014.
46. Cucchiella, F.; D'Adamo, I.; Gastaldi, M.; Miliacca, M. A profitability analysis of small-scale plants for biomethane injection into the gas grid. *J. Clean. Prod.* **2018**, *184*, 179–187. [CrossRef]
47. Hirunsit, P.; Luadthong, C.; Faungnawakij, K. Effect of alumina hydroxylation on glycerol hydrogenolysis to 1,2-propanediol over Cu/Al$_2$O$_3$: Combined experiment and DFT investigation. *RSC Adv.* **2015**, *5*, 11188–11197. [CrossRef]
48. Du, Y.; Wang, C.; Jiang, H.; Chen, C.; Chen, R. Insights into deactivation mechanism of Cu–ZnO catalyst in hydrogenolysis of glycerol to 1,2-propanediol. *J. Ind. Eng. Chem.* **2016**, *35*, 262–267. [CrossRef]

Article

Synthesis, Characterization, and Synergistic Effects of Modified Biochar in Combination with α-Fe$_2$O$_3$ NPs on Biogas Production from Red Algae *Pterocladia capillacea*

Mohamed A. Hassaan [1,*], **Ahmed El Nemr** [1], **Marwa R. Elkatory** [2], **Safaa Ragab** [1], **Mohamed A. El-Nemr** [3] **and Antonio Pantaleo** [4]

[1] Marine Pollution Lab, National Institute of Oceanography and Fisheries (NIOF), Alexandria 21556, Egypt; ahmedmoustafaelnemr@yahoo.com (A.E.N.); safaa_ragab65@yahoo.com (S.R.)

[2] Advanced Technology and New Materials Research Institute, City for Scientific Research and Technological Applications, Alexandria 21934, Egypt; marwa_elkatory@yahoo.com

[3] Department of Chemical Engineering, Faculty of Engineering, Minia University, Minia 61519, Egypt; mohamedelnemr1992@yahoo.com

[4] Department of Agriculture and Environmental Sciences, Bari University, 70121 Bari, Italy; antonio.pantaleo@uniba.it

* Correspondence: mhss95@mail.com

Abstract: This study is the first work that evaluated the effectiveness of unmodified (SD) and modified biochar with ammonium hydroxide (SD-NH$_2$) derived from sawdust waste biomass as an additive for biogas production from red algae *Pterocladia capillacea* either individually or in combination with hematite α-Fe$_2$O$_3$ NPs. Brunauer, Emmett, and Teller, Fourier transform infrared, thermal gravimetric analysis, X-ray diffraction, transmission electron microscopy, Raman, and a particle size analyzer were used to characterize the generated biochars and the synthesized α-Fe$_2$O$_3$. Fourier transform infrared (FTIR) measurements confirmed the formation of amino groups on the modified biochar surface. The kinetic research demonstrated that both the modified Gompertz and logistic function models fit the experimental data satisfactorily except for 150 SD-NH$_2$ alone or in combination with α-Fe$_2$O$_3$ at a concentration of 10 mg/L. The data suggested that adding unmodified biochar at doses of 50 and 100 mg significantly increased biogas yield compared to untreated algae. The maximum biogas generation (219 mL/g VS) was obtained when 100 mg of unmodified biochar was mixed with 10 mg of α-Fe$_2$O$_3$ in the inoculum.

Keywords: *Pterocladia capillacea*; biochar; sawdust; modified sawdust; α-Fe$_2$O$_3$ NPs; biogas

Citation: Hassaan, M.A.; El Nemr, A.; Elkatory, M.R.; Ragab, S.; El-Nemr, M.A.; Pantaleo, A. Synthesis, Characterization, and Synergistic Effects of Modified Biochar in Combination with α-Fe$_2$O$_3$ NPs on Biogas Production from Red Algae *Pterocladia capillacea*. *Sustainability* **2021**, *13*, 9275. https://doi.org/10.3390/su13169275

Academic Editor: Idiano D'Adamo

Received: 13 July 2021
Accepted: 16 August 2021
Published: 18 August 2021

Publisher's Note: MDPI stays neutral with regard to jurisdictional claims in published maps and institutional affiliations.

1. Introduction

In the recent decades, renewable energy has gained significance largely because of the sources from which it comes [1,2]. Waste-to-energy technologies help process or dispose of less waste while also generating electricity, encouraging the move away from fossil fuels [1,2]. Biomass resources represent an opportunity for sustainable development in bio-based industries, which encompass sectors as diverse as agriculture, food, bio-based chemicals, bio-energy, bio-based textiles, and forestry [3]. In general, initial studies on biomass for energy production have shown that it is a competitive fuel vs. fossil fuels and has a dry matter calorific value of around 17–21 MJ/kg [4]. Biogas may be created from a wide range of sources, as long as they contain organic material. Among these sources are seaweeds, municipal sewage, manure, agricultural waste, and waste dumps [5,6]. The gas composition may vary depending on the source, but methane will always be the significant component [6]. Numerous studies have been conducted to optimize the anaerobic digestion (AD) performance and energy efficiency of biogas-producing technologies to meet global demand for a stable and clean energy source. For example, Europe was striving to achieve one-fifth of renewable energy by 2020 by improving the energy efficiency of existing systems [1]. Given the nature of organic waste, various strategies for increasing the digestibility of these waste materials have been identified, including co-digestion, pre-treatments, and the use of carbonaceous additives to stimulate microbial activity and reduce the inhibitory concentration of certain by-products [2–6]. El Nemr et al. [7] mentioned that *P. capillacea* is a common marine biomaterial in the Mediterranean, where a considerable amount is habituated on rocks on the coast and in shallow water each year, and it possesses impressive metal adsorption properties. Hassaan et al. [8] also stated that *P. capillacea* is a marine red alga in which the chlorophyll pigment phycoerythrin obscures the chlorophyll. They are invariably multicellular and range from tiny to moderate size, with a hollow frond with a cartilaginous texture.

Among the approaches mentioned above, carbonaceous additives are the most practicable for commercial application, particularly in landfills, due to their simplicity and lack of infrastructure modification [5,9,10]. Carbonaceous additions have been demonstrated to be helpful due to their good effect on biogas generation, widespread availability, and inexpensive application [10]. For example, activated carbon (AC) can be synthesized at a low cost by steam activating char, a by-product of woody biomass gasification [11]. Nowadays, AC has been successfully used as an additive in AD to boost the process efficiency in wastewater treatment plants [12,13]. Among the several technologies used to improve biogas, AC has been identified as one of the most economically viable [13]. Biochar is a carbon-rich substance that is formed when biomass is thermally decomposed in the absence of oxygen [14]. It is created in a variety of ways, including pyrolysis (300–700 °C; N2; atmospheric pressure) and hydrothermal carbonization (170–250 °C; water above saturated pressure) [14]. The amount of energy required to produce biochar varies according to the type of biomass used. Around 160 MJ would be required for an efficient pyrolysis process using the wood biomass employed in this investigation [15,16].

Recent studies have demonstrated that including biochar into the AD of food waste improved the biogas output [17–19]. Sunyoto et al. [18] noticed a 41.6% increase in CH_4 synthesis when they added pine sawdust biochar (generated at 650 °C) to AD of aqueous carbohydrate food waste prepared from white bread. The addition of 8.3 g/L biochar to food waste resulted in a higher methane yield (from 55 to 78%), whereas 33.3 g/L biochar resulted in the lowest output. The biochar, it is claimed, increases the surface area available for colonization by the AD's microbial flora and functions as an adsorbent for chemicals such as limonene and ammonia [20] that would otherwise hinder the AD's performance. Wang et al. [19] heated vermicompost-based biochar to 500 °C and reported that the biochar worked as a buffer and boosted CH_4 generation due to the inclusion of 15–20% (w/w) biochar. Meyer-Kohlstock et al. [17] reported that addition of holm oak residue biochar (produced at 650 °C) to municipal biowaste increased the CH_4 production per kilogram of organic dry matter (ODM) of 5% (257–272 NL/kgODM) with a biochar content of 5% (w/w) and 3% (252–267 NL/kgODM) with a biochar content of 10% (w/w).

Iron nanoparticles can be utilized as an electron donor to convert carbon dioxide (CO_2) to CH4 [21], alter the type of hydrolysis fermentation, and increase the acetic acid content [22–25]. Similarly, Farghali et al. [26] observed that adding 20 and 100 mg/L Fe_2O_3 NPs increased biogas and CH_4 generation by 9 and 105%, respectively, compared to using only cattle dung. Additionally, the addition of 20 and 100 mg/L Fe_2O_3 NPs resulted in a 53.02 and 57.93% reduction in H_2S, respectively. Hassanein et al. [27] discovered that supplementing poultry litter with 100 mg/L Fe NPs (30.0 to 80.9 nm) and 15 mg/L Fe_3O_4 NPs (94.3 to 400 nm) increased CH_4 production by 29.1 and 27.5%, respectively, as compared to poultry litter alone. Additionally, Yu et al. [28] discovered that adding 10 g/L of Fe NPs (5–100 nm) enhances CH_4 output from sludge by 46.1%. While Su et al. [29] discovered that 0.10 wt % of Fe NPs (20 nm) boosts CH_4 and biogas production from sludge by 9.1 and 30.4%, respectively; this concentration of Fe NPs also significantly reduces H_2S production by 98%.

Rasapoora et al. [4] studied the effects of biochar and activated carbon on biogas generation and the results showed that by using 20 g/L biochar, a significant increase occurred in the rate of AD for all types of biochar, as confirmed by the thermogravimetric results. The physical properties of the additives, including electrical conductivity and surface area, were found to influence only the rate of AD process and not the biogas production yield [4]. Biochar showed more promising biogas generation results than activated carbon due to its ability to adsorb ammonia nitrogen [4]. Wambugu et al. [5] studied the role of biochar in anaerobic digestion-based biorefinery for food waste, and the results showed that the biogas volume produced by the treatments with the brewery residue hydrochar and treated waste wood pyrochar was lower than the amount of biogas produced by the control with only food waste. These study results indicate that the type of biochar and trace elements concentration in biochar plays a key role in determining the effectiveness of the biochar in enhancing biogas production from food waste.

Biochar modification can enhance the properties associated with the porosity and functional groups and has been identified as an effective way to improve adsorption capacity [30]. The functional groups and surface charge of biochar mainly influence the immobilization of heavy metals, which also depend upon environmental conditions and the type of metals studied for remediation [31]. Recent research has been conducted to determine the role of trace elements and unmodified biochar in the AD of food waste in anaerobic systems, respectively. However, this research did not examine the impact of biochar in conjunction with iron oxide nanoparticles in AD. Additionally, the role of biochar in the continuing AD process has not been reported. Thus, this work aimed to: (i) synthesize, characterize, and modify biochar; (ii) determine the potential of various forms of biochar to increase biogas production, and (iii) determine the influence of biochar in combination with α-Fe$_2$O$_3$ on biogas production. To the authors' knowledge, this is the first study to examine the influence of modified biochar by ammonification on the production of biogas from algae, both singly and in combination with Fe$_2$O$_3$ NPs.

2. Materials and Methods

2.1. Collection of Red Algae P. capillacea

Red alga *P. capillacea* was obtained from the Mediterranean coast near Alexandria, gently cleaned with water to eliminate contaminants, and washed multiple times with distilled water and dried in an oven. *P. capillacea* was dried, processed, crushed to a particle size of approximately 0.5 mm, and stored until usage. According to the literature [32,33], the dry matter has been calculated. By ashing the ground dried samples overnight in a muffle furnace at 550 °C, the ash content was measured. Carbon and nitrogen content was determined from energy dispersive X-ray analysis.

2.2. Preparation of Unmodified and Modified Sawdust Raw Materials

The precursors used for the preparation of biochar was sawdust collected from an Egyptian local wood carpentry workshop. It was rinsed numerous times using tap water. Clean sawdust was dried in an oven at 105 °C and then ground and crushed. The crushed sawdust was cooked in a refluxed system for 2 h using a Soxhlet containing 250 g in a 1000 mL solution of 50% H$_2$SO$_4$ (99.999%), then filtered and washed with distilled water until the wash solution became neutral, followed by washing with ethanol. The final result of biochar was dried at 70 °C and then weighed. Modified SD (SD-$_{NH2}$) was prepared by boiling the sawdust (25 g) for 2 h in a refluxed system utilizing a Soxhlet in a 100 mL solution of 25% NH$_4$OH (25%), followed by filtration and washing with distilled water and ethanol. The final charcoal product was dried in an oven set to 70 °C [34,35]. H$_2$SO$_4$ and NH$_4$OH were obtained from Aldrich Chemicals, Milwaukee, WI, USA.

2.3. Characterization and Measurement

The following techniques were used to characterize the samples of α-Fe$_2$O$_3$ NPs and biochar: Fourier transform infrared (FTIR) spectroscopy (platinum ATR) model V-100 VERTEX70, Germany, in the wavenumber range (400–4000 cm^{-1}) with resolution values of (4 cm^{-1}) and 16 scan, X-ray diffractograms (XRD) using a Bruker Meas Srv (D2-

208219)/D2-2082019 diffractometer operating at 30 kV, 10 mA with a Cu tube ($\lambda = 1.54$), with a 2θ range of (5–80) for biochar and from (15–80) for α-Fe$_2$O$_3$. Individually, the prepared green nanostructure α-Fe$_2$O$_3$ was characterized using Raman (the sample was exposed to this beam for 1 s at a power of 10 mW and an aperture of 25 1000 mm; three distinct points were measured, and displacement occurred between 100 and 1400 cm^{-1}). Transmission electron microscope (TEM) (JEOL, Model JSM 6360LA, Tokyo, Japan), and PSA (the Malvern Mastersizer 3000 is a compact optical system that uses laser diffraction to measure particle size distribution) were used. Thermal analyses of biochar were conducted using SDT650-Simultaneous Thermal Analyzer equipment in the temperature range of room temperature to 900 °C, with a ramping temperature of 5 °C per minute.

2.4. Inoculum and Substrates Preparation

Cow excrement was collected from a slaughterhouse in Alexandria, Egypt, sealed in a plastic bag, and stored in a plastic box container until the next day. The cow excrement was diluted 1:1 (w/v) with water.

2.5. Biogas Tests

Laboratory tests were conducted on reactors in similar digesters of cylindrical syringes [36–38]. The syringes were reversed directly onto the reactor lid [39]. A plastic syringe was used to sample the fuel equipped with a three-way valve and re-injected into the waste. In all tests, 100 mL glass syringes were applied. As feedstock, 1.5 g of milled *P. capillacea* (dried weight) was used. In each syringe, 20 g (wet weight) of each manure or activated sludge was applied to the untreated and treated *P. capillacea*. For 10 min, the working volume was flushed with N$_2$. For each anaerobic degradation set-up, three replicates were performed. Until no apparent methane was produced, the inoculum was pre-incubated for three days. At 37 °C with continuous shaking at 150 rpm, the digesters were incubated. Table 1 offers an overview of the substrates used in batch experiments to estimate the *P. capillacea* biogas yield.

Table 1. Overview of substrates and pretreatment processes used to estimate of the biogas yield of *P. capillacea* in batch experiments.

Experiment	Pretreatment	Incubation Temp. (°C)	I/S Ratio
Batch 1	Manure + algae untreated	37 ± 1	20:1.5
Batch 2	Manure + Algae (Fe 10 mg/L)	37 ± 1	20:1.5
Batch 3	Manure + Algae (SD 50 mg/L)	37 ± 1	20:1.5
Batch 4	Manure + Algae (SD 100 mg/L)	37 ± 1	20:1.5
Batch 5	Manure + Algae (SD 150 mg/L)	37 ± 1	20:1.5
Batch 6	Manure + Algae (SD 50 mg/L + Fe 10 mg/L)	37 ± 1	20:1.5
Batch 7	Manure + Algae (SD 100 mg/L + Fe 10 mg/L)	37 ± 1	20:1.5
Batch 8	Manure + Algae (SD 150 mg/L + Fe 10 mg/L)	37 ± 1	20:1.5
Batch 9	Manure + Algae (SD +NH$_2$ 50 mg/L)	37 ± 1	20:1.5
Batch 10	Manure + Algae (SD + NH$_2$ 100 mg/L)	37 ± 1	20:1.5
Batch 11	Manure + Algae (SD + NH$_2$ 150 mg/L)	37 ± 1	20:1.5
Batch 12	Manure + Algae (SD + NH$_2$ 50 mg/L + Fe 10 mg/L)	37 ± 1	20:1.5
Batch 13	Manure + Algae (SD + NH$_2$ 100 mg/L + Fe 10 mg/L)	37 ± 1	20:1.5
Batch 14	Manure + Algae (SD + NH$_2$ 150 mg/L + Fe 10 mg/L)	37 ± 1	20:1.5

2.6. Green Synthesis of α-Fe2O3 Nanoparticles

Twenty grams of dried *P. capillacea* was added to 1.5 L of water and cooked on a hot plate for 20 min at 80 °C. Filtration and storage of the extract solution at 4 °C were performed. Five grams of Fe(NO$_3$)3.9H$_2$O was dissolved in 30 mL *P. capillacea* extract and then diluted to 50 mL with distilled water in a beaker. After adding a 6M NaOH solution dropwise to the stirring mixture at room temperature, the product was washed with water and dried at 80 °C for 2 h before being calcified at 450 °C for 2 h [36,40,41].

2.7. Kinetics Study and Statistical Analysis

Numerous researchers have used the nonlinear regression models, and the modified Gompertz and logistic function models. Equations (1) and (2) were applied to determine the cumulative biogas production [42–44]. In order to compare the accuracy of the studied models, (R^2) was calculated using Excel 2010 methods and Origin 2020b.

$$M = Pb \times \exp\left\{-\exp\left[\frac{Rm.e}{Pb}(\lambda - t) + 1\right]\right\} \tag{1}$$

$$M = Pb/((1 + \exp\{4.Rm.(\lambda - t))/pb + 2) \tag{2}$$

3. Results

3.1. Chemical Compositions of P. capillacea

As shown in Table 2, the VS content of the investigated *P. capillacea* is 83.99%. Table 2 shows a C/N ratio of about 5.89%. The majority of the literature recommends a working C/N ratio of between 20 and 30, with a maximum of 25, for anaerobic bacterial growth in the AD system [45], which is still significantly higher than the measured value for *P. capillacea*.

Table 2. The relative values of different substrates.

Proximate Tests	*P. capillacea*	Manure
DM%	83.97	80.67
Ash%	16.01	15.33
VS%	83.99	84.66
C%	40.02	48.95
N%	6.79	4.16
C/N	5.89	11.76

3.2. Characterization of Green α-Fe$_2$O$_3$ NPs

3.2.1. Fourier Transform Infrared Spectrum (FTIR)

The FTIR spectrum of hematite nanoparticles depicted in Figure 1 demonstrates a series of absorption bands ranging from 400 to 900 cm^{-1}. The Fe–O vibrational bands of α-Fe$_2$O$_3$ are roughly 627, 580, and 485 cm^{-1} in this region [46–50]. The vibrational band at 977 cm^{-1} is due to longitudinal absorptions, but the bands at 538 and 439 cm^{-1} are due to the transverse absorption of a hematite structure. These bands are seen in Figure 1's FTIR spectrum [51,52]. The FTIR spectrum of the α-Fe$_2$O$_3$ sample exhibits no additional vibrational bands due to the hydration and organic phase used as a capping agent being completely removed after 600 °C calcination.

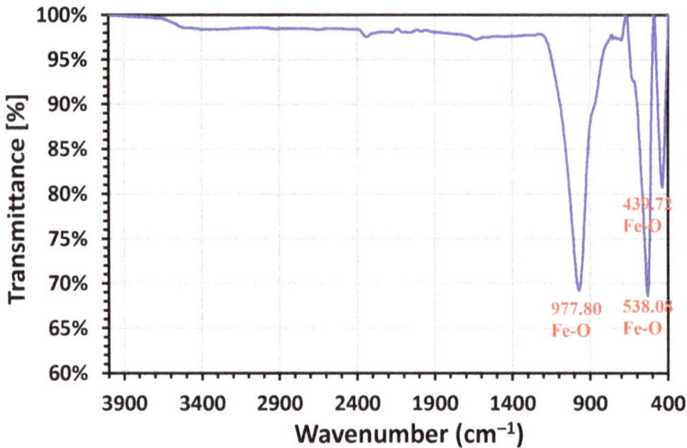

Figure 1. FTIR spectrum of α-Fe$_2$O$_3$ NPs.

3.2.2. Raman Spectroscopy

The Raman spectrum of hematite α-Fe$_2$O$_3$ is depicted in Figure 2. It is devoid of peaks associated with maghemite or magnetite. The A1g modes are associated with the 214 and 567 cm^{-1} peaks [53]. The remaining four peaks at 278, 390, 430, and 616 cm^{-1} are assigned to the Eg modes [53]. This indicates that heating the initial Fe to 600 °C for four hours will completely convert it to hematite. The data obtained from the FTIR was confirmed by the results obtained from the Raman analysis which proved the formation of the hematite as reported in literature [46–53].

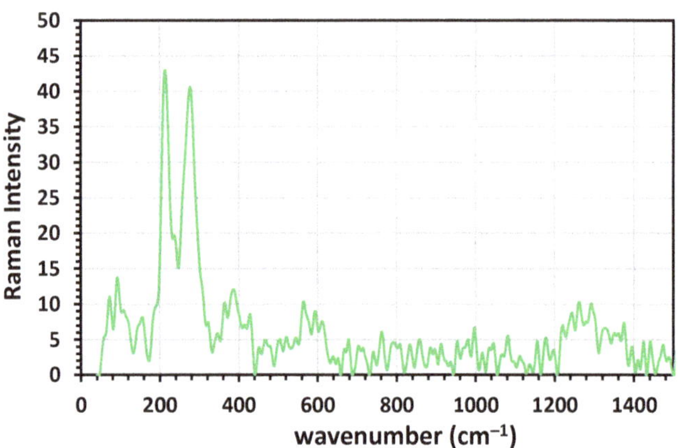

Figure 2. Raman spectrum of α-Fe$_2$O$_3$ NPs.

3.2.3. X-ray Diffraction (XRD)

XRD was used to characterize the crystalline structure of the produced α-Fe$_2$O$_3$ nanoparticles. As illustrated in Figure 3, diffraction patterns correspond to the crystallographic planes (012), (104), (110), (113), (024), (116), (018), (214), and (300) of rhombohedral phase α-Fe$_2$O$_3$, were assigned for 2theta of 24.106, 33.149, 35,577, 40.888, 49.446, 54.119, 56.517, 62.252, and 64.013, respectively, based on the standard COD card (No. 9,000,139) α-Fe$_2$O$_3$ hematite, which confirms the synthesis of α-Fe$_2$O$_3$ nanoparticles. This demonstrates that α-Fe$_2$O$_3$ hematite nanoparticles may be synthesized using such a simple and environmentally friendly method.

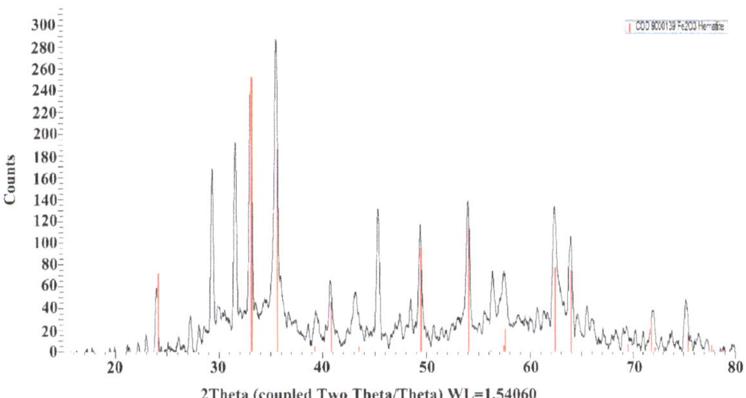

Figure 3. X-ray diffractograms of α-Fe_2O_3 NPs.

3.2.4. Transmission Electron Microscopy (TEM)

The TEM micrograph (Figure 4) revealed that the particles were oval and pyramid-shaped, suggesting that they are self-assembled into a large spindle with pores via electrostatic and/or van der Waals forces and aggregated as a result of algal extract solvating and capping the nanoparticles [54]. The particle diameters range from 5.6 to 16.8 nm.

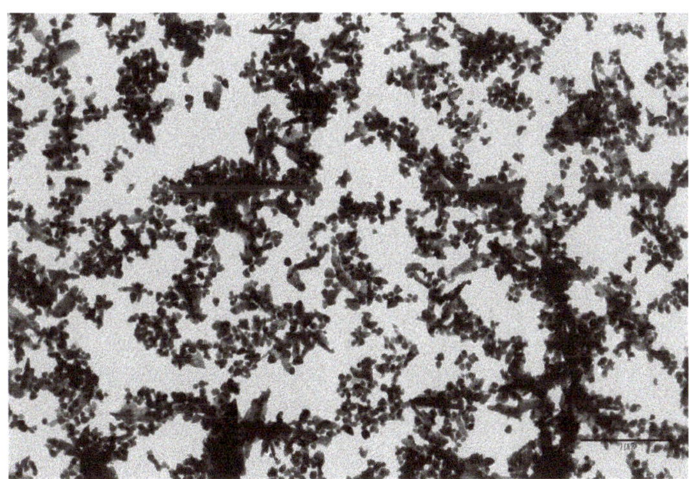

Figure 4. TEM of α-Fe_2O_3 NPs.

3.2.5. Particle Size Analyzer (PSA) and BET Analysis of the Surface Area

The size distribution is shown in Figure 5 and is defined by PSA for α-Fe_2O_3 NPs, as a result of detecting α-Fe_2O_3 hematite in two ranges of size. The range from 6 to 8 nm using a 10° test angle has small size particles and another range of particle sizes around 122 to 691 nm using a 90° test angle; the results indicate the widest size distribution of the α-Fe_2O_3 nanoparticles. In addition, the dominant sizes of the hematite were about 421 nm. The BET analysis (Table 3) of green α-Fe_2O_3 shows that the surface area and average pore size of the synthesized magnetite (α-Fe_2O_3) nanoparticles were 29.29 m^2/g and 11.92 nm, respectively.

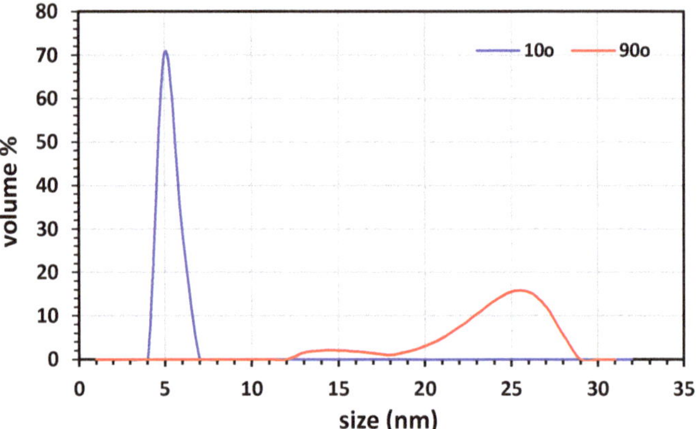

Figure 5. PSA of the α-Fe$_2$O$_3$ NPs.

Table 3. BET surface area and porosity of green α-Fe$_2$O$_3$ NPs.

Sample	BET Surface Area (m^2/g)	Mean Pore Diameter (nm)	Total Pore Volume (cm^3/g)
Green α-Fe$_2$O$_3$ NPs	29.29	11.92	0.087

3.3. Characterization of Biochar

3.3.1. Fourier Transform Infrared Spectra (FTIR)

The FTIR spectrum (Figure 6) was used to qualitatively analyze the chemical structures of biochar modified with H$_2$SO$_4$ and NH$_4$OH. Both samples' FTIR spectra exhibit some similarities. The band between 3388 and 3203 cm^{-1} corresponds to the –OH and –NH in SD and SD-NH$_2$ biochars, respectively [16,55]. C–H stretching can be attributed to the adsorption peak at 2921 cm^{-1}. The significant adsorption peak at 1701 cm^{-1} can be attributed to the carboxyl group's C=O stretching, which was absent in SD and completely absent in SD-NH$_2$ biochars [16,55]. In both SD and SD-NH$_2$ biochars, the band 1581 cm^{-1} corresponds to the C=O stretching vibration. In SD and SD-NH$_2$ biochars, the peak at 1209–1176 cm^{-1} represents a rise in C–O–C, while at (1029–1033, 784–792, 615–626) cm^{-1} represents a Si–O–Si stretching.

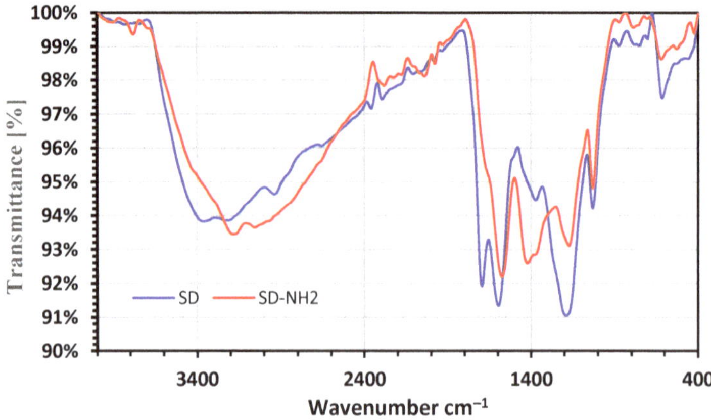

Figure 6. FTIR analysis of SD and SD-NH$_2$.

3.3.2. X-ray Diffraction (XRD)

Figure 7 shows the XRD of the SD and SD-NH$_2$ biochar. The broad peak in the region of 2θ = 10–30 is indexed as C (002) diffraction peak indicating an amorphous carbon structure with randomly oriented aromatic sheets. There are sharp peaks around 2θ = 27 and 43.65. In the case of SD-NH$_2$, sharp peaks around 2θ = 25.8, 43.6, and 63.9 correspond to the miscellaneous inorganic components mainly constituted of quartz and albite, within the structure of SD, which indicated that the original feedstocks were rich in Si, which can be manifested by the Si-O-Si stretching band from FTIR spectra [55].

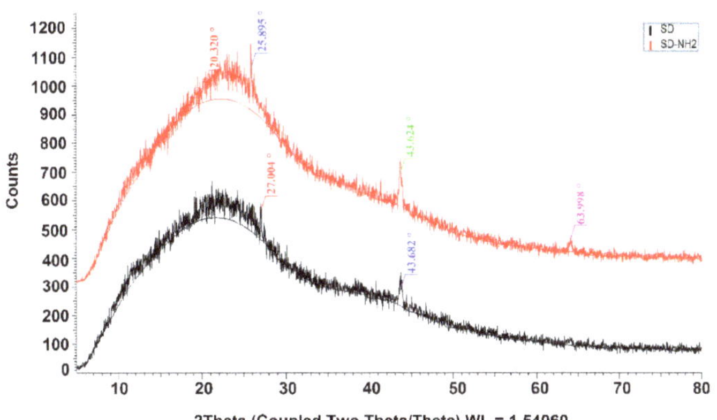

Figure 7. XRD analysis of SD and SD-NH$_2$.

3.3.3. Thermal Analysis (TGA)

The decomposition of sawdust biochar occurs in three steps, whereas the SD-NH$_2$ biochars decompose in two steps, as illustrated in Figure 8. The first step occurred between 50 and 150 °C as a result of the loss of surface-bound water and moisture present in the sample, resulting in a weight loss of 6.7, 13.21, and 6.7%, respectively, for SD and SD-NH$_2$ [7,34,35]. The second phase results in a 56.30% weight loss at 150–350 °C and a 3.9% weight loss at 150–275 °C for SD and SD-NH$_2$. The third phase results in a 22.55% weight loss at 350–1000 °C and a 30.03% weight loss at 275–1000 °C, for SD and SD-NH$_2$, respectively [7,34,35]. The weight retained by SD-NH$_2$ biochars and the percentages of 2.96% obtained, as well as the DTA curves, indicate that the SD-NH$_2$ biochar amine-modified sample exhibits more stability than the SD biochar [7,34,35]. This enhances the susceptibility to consume SD during anaerobic digestion, which explains why the cumulative amount of biogas produced was greater when SD biochar was used rather than SD-NH$_2$ biochars.

3.3.4. BET Analysis of the Surface Area

The parameters of SD and SD-NH$_2$ biochars, including their BET-specific surface area, total pore volume, and mean pore diameter, are reported in Table 4. Interestingly, the modification enhanced the surface area of SD (2.913) and SD-NH$_2$ (3.19) biochars.

Table 4. BET surface area and porosity of biochar.

Sample	BET Surface Area (m^2/g)	Mean Pore Diameter (nm)	Total Pore Volume (cm^3/g)
SD	2.913	16.874	0.01220
SD-NH$_2$	3.190	8.370	0.00668

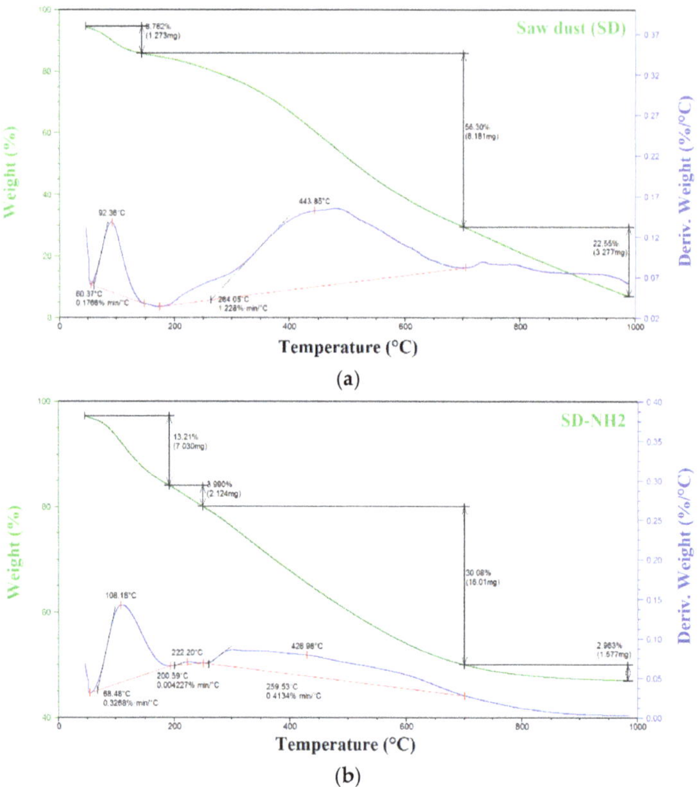

Figure 8. TGA analysis of (**a**) SD and (**b**) SD-NH$_2$.

3.4. Impact of Pretreatment on Anaerobic Digestion by Batch

For 40 days, the experimental results of biogas output yields were collected and are shown in Figure 9. When the treated *P. capillacea* was treated with an SD dose (100, 50 mg/L), the average biogas production yield was marginally increased compared to the biogas production yield without biochar treatment, as shown in Figure 9.

A major positive effect on the production of biogas ($p < 0.05$) was achieved when *P. capillacea* was treated with 100 mg SD combined with 10 mg α-Fe$_2$O$_3$. The dosage of 100 mg SD combined with 10 mg α-Fe$_2$O$_3$ produces higher biogas yield with 219 mL/g VS for *P. capillacea* combined with manure. It is also worth mentioning that when *P. capillacea* was treated with unmodified SD with different dosages of 50, 100, and 150 mg/L, the biogas was increased more than the control sample, which produces a higher biogas yield with 171, 205, and 169.5 mL/g VS, respectively. It is also clear that modified SD-NH$_2$ with all higher dosages, except 100 and 150 mg/L, has inhibitory effects on the biogas production.

Biogas output tests have been completed when, as seen in Figure 10, the regular production of biogas is <1% of the total production of most of the tests conducted. It is clear that the biogas output of *P. capillacea* treated with 50 mg/L of SD-NH$_2$ is around 169.5 mL/g VS, which is equal to the biogas yield formed by 100 mg/L dosage of unmodified biochar and is higher than the biogas produced from the untreated algae (control), which yields 138 mL/g VS. This may indicate that the little dosages of SD-NH$_2$ may have no detrimental effect on methane yield of biogas production and the addition effect of biochar in biogas is dosage dependent.

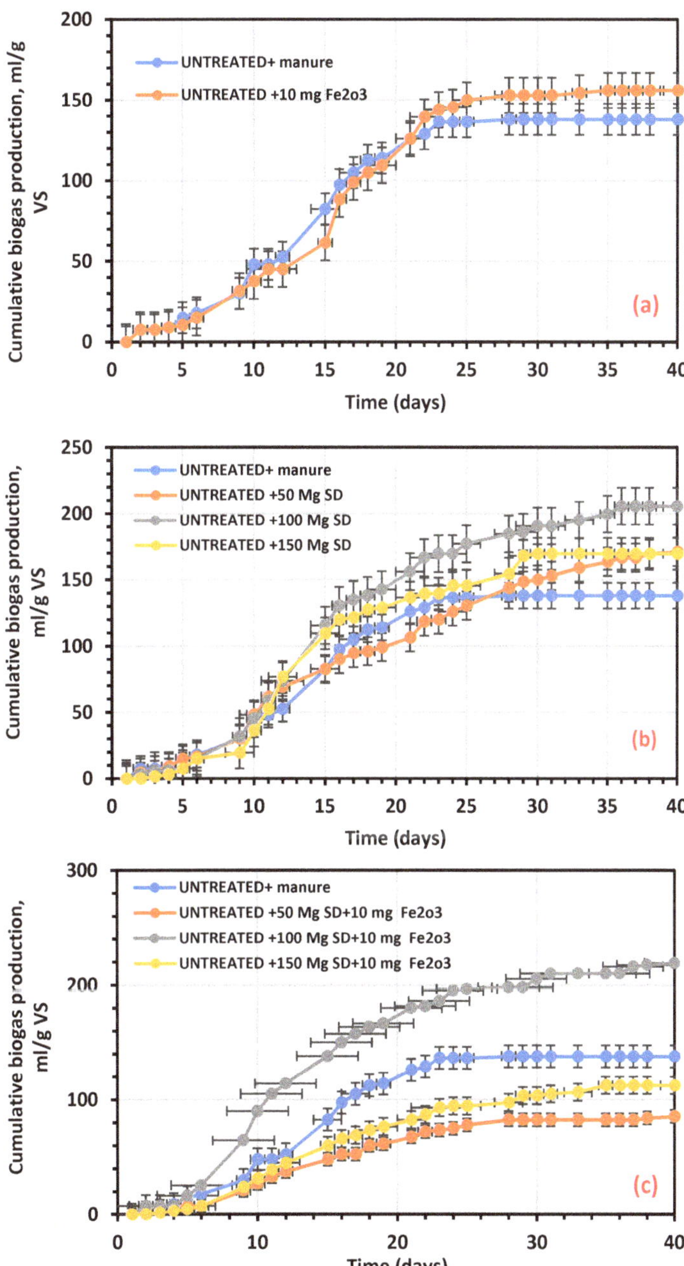

Figure 9. *Cont.*

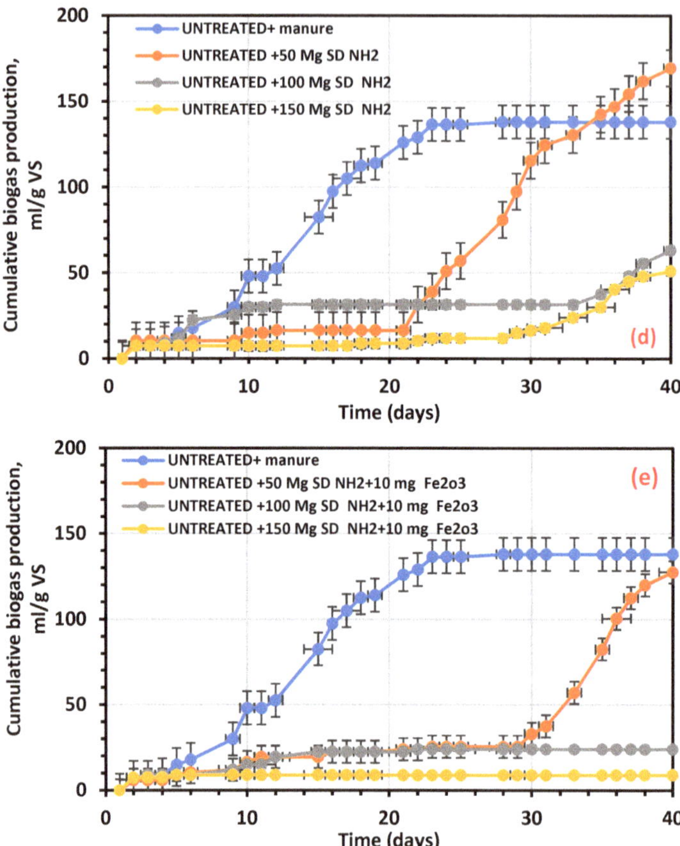

Figure 9. Average production of cumulative net biogas (mL/g VS) using (**a**) raw and untreated + raw pretreated with α-Fe$_2$O$_3$, (**b**) raw and raw pretreated with SD, (**c**) raw and raw pretreated with SD-NH$_2$, (**d**) raw and combination of α-Fe$_2$O$_3$ NPs and SD, and (**e**) raw and combination of SD-NH$_2$ with α-Fe$_2$O$_3$ NPs.

Due to the possibility of ammonia accumulation during anaerobic digestion of food waste, its use in industrial biogas facilities is restricted. Sheng et al. [56] investigated the effect of ammonia and nitrate on biogas production from food waste via anaerobic digestion. They discovered that lower ammonia concentrations (1.544 g L^{-1}) had no detrimental effect on methane yield, whereas higher TAN concentrations (>3.78 g L^{-1}) resulted in severe inhibition of methanogenesis. These findings corroborate our findings that greater doses of SD-NH$_2$ restrict the formation of biogas from *P. capillacea*. This inhibitory effect became worse and gave more inhibitory effect when the *P. capillacea* was treated with SD-NH$_2$ with a combination with α-Fe$_2$O$_3$ with a dosage of 10 mg/L and the biogas yield was 127, 24, and 9 mL/g VS. The reason for this inhibitory effect may be explained by the releasing of more nitrogen which will be converted into ammonia in the digester according to the following equation:

$$Fe_2O_3 \text{ (s)} + 2\,NH_3 \text{ (aq)} \rightarrow 2\,Fe \text{ (s)} + N_2 \text{ (g)} + 3\,H_2O \tag{3}$$

The release of N$_2$ can alter the C/N ratio, and changes in the C/N ratio can change the pH of a slurry [57]. Increased carbon content results in increased carbon dioxide creation and a lower pH value. In contrast, increased nitrogen content results in increased ammonia gas production, which may raise the pH to the detriment of microorganisms [57].

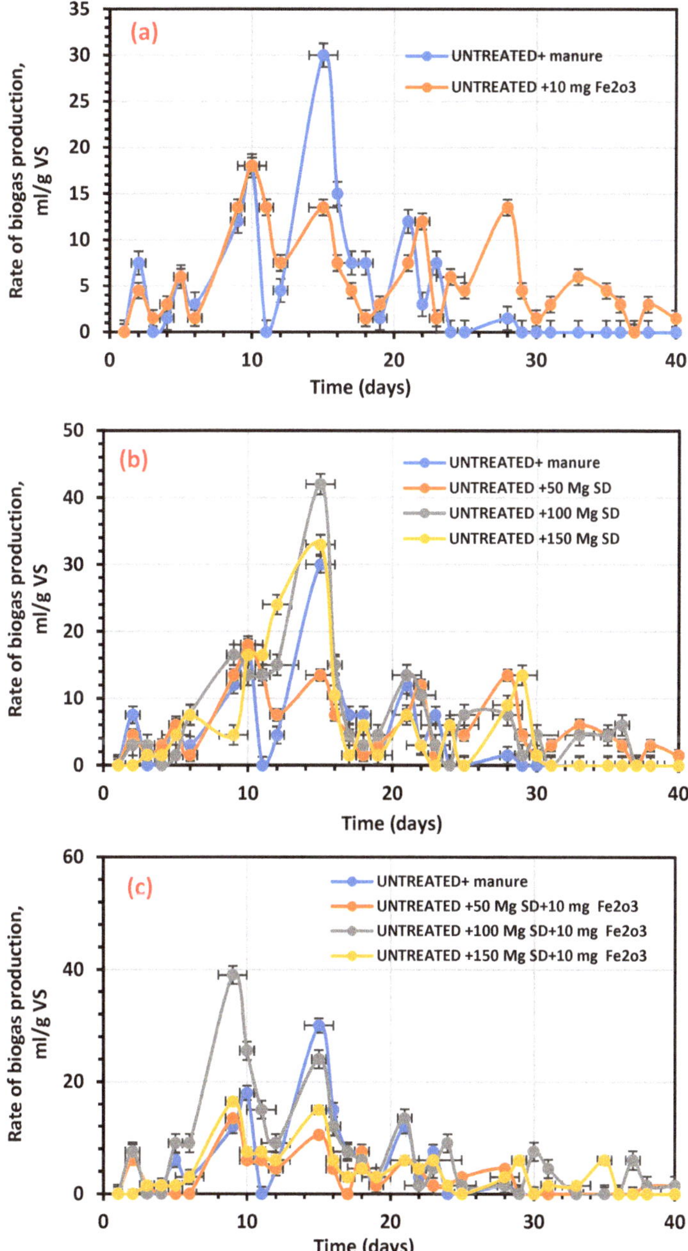

Figure 10. *Cont.*

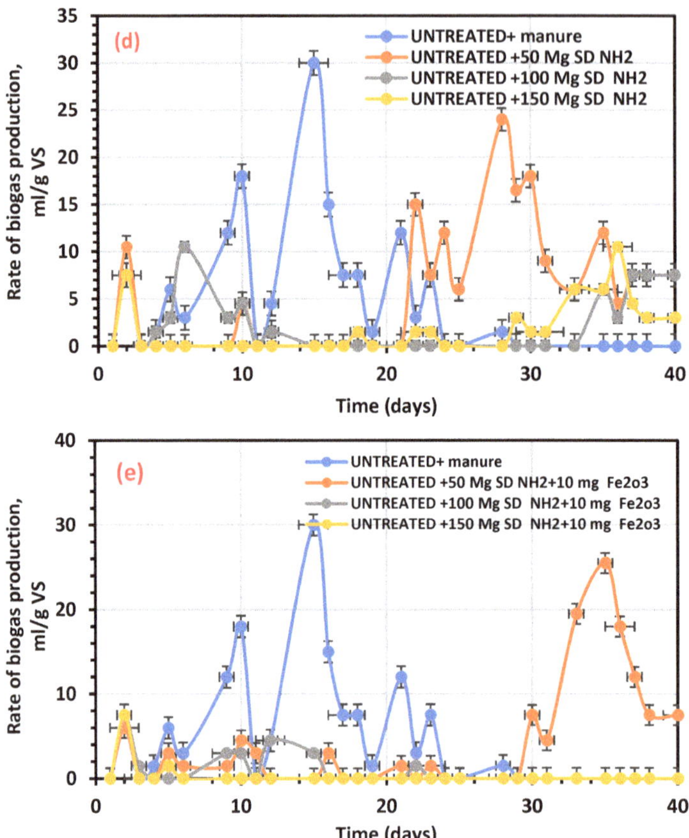

Figure 10. Average rate of the daily production of the biogas using (**a**) raw and untreated + raw pretreated with α-Fe$_2$O$_3$, (**b**) raw and raw pretreated with SD, (**c**) raw and raw pretreated with SD-NH$_2$, (**d**) raw and combination of α-Fe$_2$O$_3$ NPs and SD, and (**e**) raw and combination of SD-NH$_2$ with α-Fe$_2$O$_3$ NPs.

The addition of 10 mg/L α-Fe$_2$O$_3$ resulted in a 12% increase in biogas production over control (138 mL/g VS), which is consistent with the results obtained by Abdelwahab et al. [58], who stated that the addition of 15, 30, and 60 mg/L Fe NPs increases specific biogas production startup by 124.7, 85.1, and 40.3%, respectively. The addition of SD in combination with 10 mg/L α-Fe$_2$O$_3$ increased biogas generation by 40% compared to the control and by 29% compared to the individual SD treatment. These findings are consistent with those of Abdelsalam et al. [59], who discovered that adding 5, 10, and 15 mg/L Fe NPs (20 nm) increases biogas production startup by 250.7, 270.3, and 264.5%, respectively, as compared to using only cattle dung. This discrepancy in results, however, could be explained by the size and concentration of Fe NPs. The size, concentration, and type of NPs have a significant effect in biogas production [38,59]. In our study, the TEM analysis showed α-Fe$_2$O$_3$ NPs in the range of 5–16 nm, which is near the range studied by Abdelsalam [59] (20 nm), which may confirm that the small size of α-Fe$_2$O$_3$ NPs may have a good impact on methanogenesis bacteria. Cumulative specific biogas production results demonstrated a substantial increase ($p < 0.05$) in cumulative biogas production when α-Fe$_2$O$_3$ NPs and unmodified SD were added to control manure alone, as seen in Figure 9. In this study, the highest biogas yield of 219 mL/g VS was still lower than the results obtained by Hassaan et al. [12], who studied ozonation pretreatment's effect on the biogas production from Ulva lactuca with biogas yield around 499 mL/g VS with higher ozonation time (30 min), but it is still higher than the biogas yield attained by Hassaan et al. [12] when using ozonation time (10 min).

This work can be a good addition to support the circularity of resources due to it having a significant effect on the circular bioeconomy by using red algae *P. capillacea* as a source of biomass and also to synthesize green α-F_2O_3. Another way of using biomass to enhance the bioeconomy is by using the SD waste to synthesize biochar and that will lead to reducing pressure on the environment, and to the creating of new green industries and jobs and boosting economic growth.

3.5. Kinetic Study

The data of the gas production kinetic study have been summarized in Tables 5 and 6. It is reported that the Gompertz and logistic feature models matched well the experimental findings, except for 150 SD-NH_2 individually or in combination with α-Fe_2O_3 with a dosage of 10 mg/L. For the logistic feature model and the modified Gompertz model, Rm of 14.53 and 12.16 mL/g VS, respectively, of biogas production when algae was treated with unmodified SD were observed [42,60]. The modified Gompertz and logistic models' functional received λ values of 0.1 and 0.29 days, respectively. Our work's value is extremely low compared to previously published values [38,61], for both modified Gompertz and logistic function models. The model's reliability was tested by plotting the calculated values for biogas production against the observed values (Figures 11 and 12). Tables 5 and 6 additionally include statistical indicators (R^2) to help visualize the kinetics study. According to Nguyen et al. [42], the higher R^2 values (0.999 and 0.994) for modified Gompertz and logistic feature models, respectively, indicated a more appropriate kinetic model. Both models in our analysis have a superior R^2 of 0.997, which is near the same values attained by [12,38].

Table 5. Data of kinetic analysis using the modified model of Gompertz.

	R^2	Predicted P (ml/g VS)	Differences (%)	Rmax mL/g VS.day	λ (Day)
SD					
untreated	0.991	142.31	2.69	10.94	0.19
50 SD	0.993	180.94	0.879	13.32	0.10
100 SD	0.996	203.95	1.86	12.16	0.16
150 SD	0.989	168.92	0.822	11.60	0.19
Modified SD-NH_2					
untreated	0.991	142.31	2.69	10.94	0.19
50 SD-NH_2	0.975	220.78	1.62	27.34	0.11
100 SD-NH_2	0.691	52.12	29.16	8.40	0.058
150 SD-NH_2 *	0.91	51.41	0.804	891	0.0036
α-Fe_2O_3 (10 mg/L)					
untreated	0.991	142.31	2.69	10.94	0.19
Fe 10 mg/L	0.989	162.73	3.06	12.76	0.16
Combined SD-α-Fe_2O_3 (10 mg/L)					
untreated	0.991	142.31	2.69	10.94	0.19
50 SD + 10 mg/L	0.995	86.91	0.025	11.11	0.14
100 SD + 10 mg/L	0.995	213.07	3.32	9.81	0.17
150 SD + 10 mg/L	0.997	113.92	0.763	12.06	0.14

Table 5. Cont.

	R^2	Predicted P (ml/g VS)	Differences (%)	Rmax mL/g VS.day	λ (Day)
		SD			
Combined Modified SD-NH$_2$—α-Fe$_2$O$_3$(10 mg/L)					
untreated	0.991	142.31	2.69	10.94	0.19
50 SD-NH$_2$ + 10 mg/L *	0.92	133.31	4.55	753	0.0044
100 SD-NH$_2$ + 10 mg/L	0.955	24.50	2.05	4.92	0.17
150 SD-NH$_2$ + 10 mg/L *	−10	3.3	63	8.85	0

* Fit Status = Failed.

Table 6. Data of kinetic analysis using the logistic model.

	R^2	Predicted P (ml/g VS)	Differences (%)	Rmax mL/g VS.day	λ (Day)
		SD			
untreated	0.997	139.50	1.03	13.22	0.28
50 SD	0.984	169.02	3.07	16.49	0.17
100 SD	0.991	196.99	4.30	14.53	0.25
150 SD	0.983	164.34	3.10	13.68	0.29
Modified SD-NH$_2$					
untreated	0.997	139.50	1.03	13.22	0.28
50 SD-NH$_2$	0.983	183.28	0.78	28.75	0.21
100 SD-NH$_2$	0.677	300.46	19.09	86.68	0.03
150 SD-NH$_2$ *	0.916	51.97	1.55	136.01	0.08
α-Fe$_2$O$_3$ (10 mg/L)					
untreated	0.997	139.50	1.03	13.22	0.28
Fe 10 mg/L	0.995	157.82	0.983	15.26	0.25
CombinedSD-α-Fe$_2$O$_3$(10 mg/L)					
untreated	0.997	139.50	1.03	13.22	0.28
50 SD + 10 mg/L	0.995	84.00	2.05	13.88	0.22
100 SD + 10 mg/L	0.986	207.57	5.30	12.22	0.25
150 SD + 10 mg/L	0.988	109.36	3.16	14.77	0.22
Combined Modified SD-NH$_2$—α-Fe$_2$O$_3$(10 mg/L)					
untreated	0.997	139.50	1.03	13.22	0.28
50 SD-NH$_2$ + 10 mg/L *	0.925	134.30	5.33	121.29	0.10
100 SD-NH$_2$ + 10 mg/L	0.959	24.25	1.00	7.48	0.24
150 SD-NH$_2$ + 10 mg/L *	−0.07	4.5	50	-	0

* Fit Status = Failed.

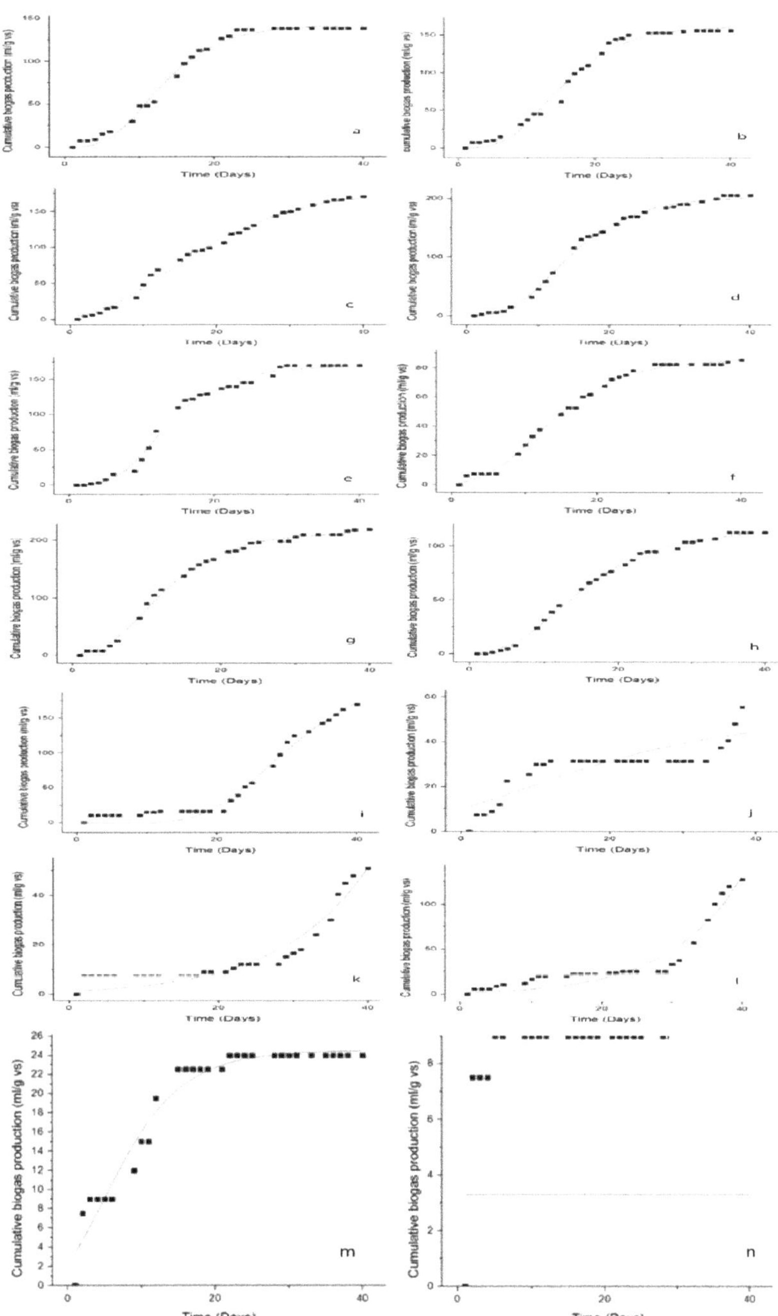

Figure 11. Cumulative biogas yield from Gompertz model, (**a**) untreated, untreated + α-Fe$_2$O$_3$ (10 mg/L) (**b**), 50, 100, 150 mg/L SD + untreated (**c–e**), 50, 100, 150 mg/L SD + α-Fe$_2$O$_3$ (10 mg/L) + untreated (**f–h**), 50, 100, 150 mg/L SD-NH$_2$ + untreated (**i–k**), 50, 100, 150 mg/L SD-NH$_2$ + α-Fe$_2$O$_3$ (10 mg/L) + untreated (**l–n**).

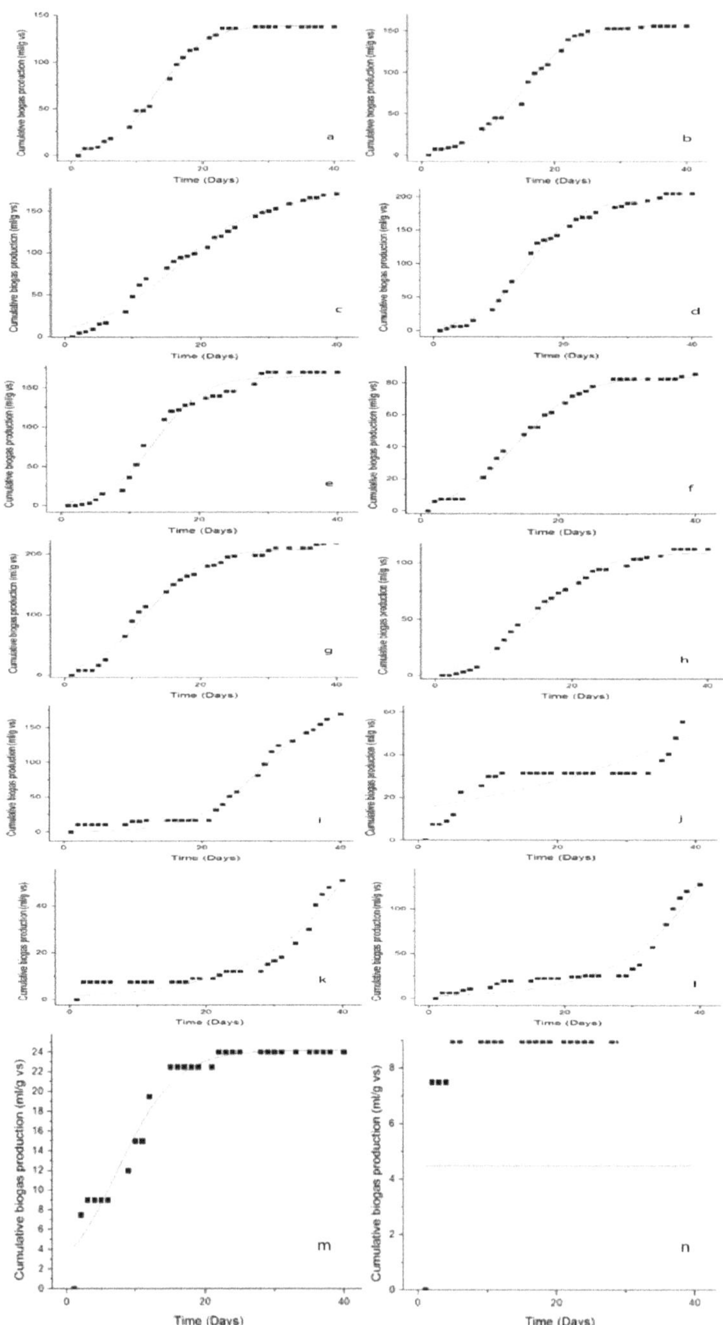

Figure 12. Cumulative biogas yield from logistic model, (**a**) untreated, untreated + α-Fe$_2$O$_3$ (10 mg/L) (**b**), 50, 100, 150 mg/L SD + untreated (**c–e**), 50, 100, 150 mg/L SD + α-Fe$_2$O$_3$ (10 mg/L) + untreated (**f–h**), 50, 100, 150 mg/L SD-NH$_2$ + untreated (**i–k**), 50, 100, 150 mg/L SD-NH$_2$ + α-Fe$_2$O$_3$ (10 mg/L) + untreated (**l–n**).

4. Conclusions

This study is the first work that studies the impact of nanoparticles with biochar on biogas production from seaweeds. The biomass of the red algae *P. capillacea* was pretreated with two different types of biochar either individually or combined with α-Fe$_2$O$_3$ for enhancing biogas production in this work. As a result, the unmodified biochar SD with all different dosages increased the biogas ability of the studied red algae *P. capillacea* compared to untreated *P. capillacea*. On the other hand, the modified biochar SD-NH$_2$ has an inhibitory effect on biogas production for higher dosages (100 and 150 mg/L). TEM, Raman, FTIR, PSA, and XRD confirmed the synthesis of α-Fe$_2$O$_3$ NPs. When *P. capillacea* pretreated with α-Fe$_2$O$_3$ alone and combined with 100 mg/L SD, the biogas increased by 12 and 40%, respectively. The updated Gompertz model and the logistic function model ($R^2 = 0.997$) were appropriate models to match the calculated biogas production and could be used more reasonably to characterize the kinetics of the AD phase. Moreover, from our results and literature, the type of biochar, trace elements concentration in biochar, and the modification method play a key role in determining the effectiveness of the biochar in enhancing biogas production. The compatibility of the *P. capillacea* bioprocess, the emission of biogas, techno-economic analysis, and compositional analysis of the used seaweeds should also be studied in a future study. On the basis of our data, the bio-energy is one of the major renewable energy types that demands substantial financial investment. In order to meet the challenge, the Egyptian government should make a significant contribution to the construction of more biogas and biomethane production plants in the coming years.

Author Contributions: M.A.H. conceived the research, carried out the experimental and theoretical work, and drafted the report. A.E.N. conceived the research, supervised and assisted with the study, as well as writing the manuscript. The unmodified and modified biochars were prepared by Eng. M.A.E.-N., M.R.E. assisted with the manuscript's writing and performed some analysis. S.R. assisted with writing and work discussion. A.P. was in charge of supervising research and assisting with theoretical work and publication. All authors have read and agreed to the published version of the manuscript.

Funding: This research received no external funding.

Conflicts of Interest: The authors declare no conflict of interest.

Abbreviations

AD	Anaerobic digestion
NPs	Nanoparticles
Fe$_2$O$_3$ NPs	Hematite nanoparticles
SD	Sawdust
SD+NH$_2$	Modified sawdust with NH$_4$OH
FTIR	Fourier transform infrared
XRD	X-ray diffractograms
TEM	Transmission electron microscopy
EDX	Energy dispersive X-ray spectroscopy
BET	Brunauer–Emmett–Teller
TGA	Thermo gravimetric analysis
TS	Total solids
Rm	The maximum biogas production rate
VS	Volatile solids
λ	The lag phase time (days)

References

1. Baena-Moreno, F.M.; Malico, I.; Marques, I.P. Promoting Sustainability: Wastewater Treatment Plants as a Source of Biomethane in Regions Far from a High-Pressure Grid. A Real Portuguese Case Study. *Sustainability* **2021**, *13*, 8933. [CrossRef]
2. D'Adamo, I.; Falcone, P.M.; Huisingh, D.; Morone, P. A circular economy model based on biomethane: What are the opportunities for the municipality of Rome and beyond? *Renew. Energy* **2021**, *163*, 1660–1672. [CrossRef]

3. D'Adamo, I.; Falcone, P.M.; Morone, P. A new socio-economic indicator to measure the performance of bioeconomy sectors in Europe. *Ecol. Econ.* **2020**, *176*, 106724. [CrossRef]

4. Pedišius, N.; Praspaliauskas, M.; Pedišius, J.; Dzenajavičienė, E.F. Analysis of Wood Chip Characteristics for Energy Production in Lithuania. *Energies* **2021**, *14*, 3931. [CrossRef]

5. Romero-Güiza, M.; Vila, J.; Mata-Alvarez, J.; Chimenos, J.; Astals, S. The role of additives on anaerobic digestion: A review. *Renew. Sustain. Energy Rev.* **2016**, *58*, 1486–1499. [CrossRef]

6. Wambugu, C.W.; Rene, E.R.; van de Vossenberg, J.; Dupont, C.; van Hullebusch, E.D. Role of biochar in anaerobic digestion based biorefinery for food waste. *Front. Energy Res.* **2019**, *7*, 14. [CrossRef]

7. El Nemr, A.; El Sikaily, A.; Khaled, A.; Abdelwahab, O. Removal of toxic chromium from aqueous solution, wastewater and saline water by marine red alga *Pterocladia capillacea* and its activated carbon. *Arab. J. Chem.* **2015**, *8*, 105–117. [CrossRef]

8. Hassaan, M.A.; Hosny, S.; ElKatory, M.R.; Ali, R.M.; Rangreez, T.A.; El Nemr, A. Dual action of both green and chemically synthesized zinc oxide nanoparticles: Antibacterial activity and removal of Congo red dye. *Desalin. Water Treat.* **2021**, *218*, 423–435. [CrossRef]

9. Ali, R.M.; Hassaan, M.A.; Elkatory, M.R. Towards Potential Removal of Malachite Green from Wastewater: Adsorption Process Optimization and Prediction. In *Materials Science Forum*; Trans. Tech. Publications Ltd.: Stafa-Zurich, Switzerland, 2020; Volume 1008, pp. 213–221.

10. Zhang, L.; Zhang, J.; Loh, K.-C. Activated carbon enhanced anaerobic digestion of food waste-Laboratory-scale and Pilot-scale operation. *Waste Manag.* **2018**, *75*, 270–279. [CrossRef]

11. Maneerung, T.; Liew, J.; Dai, Y.; Kawi, S.; Chong, C.; Wang, C.-H. Activated carbon derived from carbon residue from biomass gasification and its application for dye adsorption: Kinetics, isotherms and thermodynamic studies. *Bioresour. Technol.* **2016**, *200*, 350–359. [CrossRef]

12. Hassaan, M.A.; El Nemr, A.; Elkatory, M.R.; Eleryan, A.; Ragab, S.; El Sikaily, A.; Pantaleo, A. Enhancement of Biogas Production from Macroalgae Ulva latuca via Ozonation Pretreatment. *Energies* **2021**, *14*, 1703. [CrossRef]

13. Skouteris, G.; Saroj, D.; Melidis, P.; Hai, F.I.; Ouki, S. The effect of activated carbon addition on membrane bioreactor processes for wastewater treatment and reclamation—A critical review. *Bioresour. Technol.* **2015**, *185*, 399–410. [CrossRef]

14. Aguilera, P.; Ortiz, F.G. Techno-economic assessment of biogas plant upgrading by adsorption of hydrogen sulfide on treated sewage–sludge. *Energy Convers. Manag.* **2016**, *126*, 411–420. [CrossRef]

15. Cha, J.S.; Park, S.H.; Jung, S.-C.; Ryu, C.; Jeon, J.-K.; Shin, M.-C. Production and utilization of biochar: A review. *J. Ind. Eng. Chem.* **2016**, *40*, 1–15. [CrossRef]

16. Man, Y.; Wang, B.; Wang, J.; Slaný, M.; Yan, H.; Li, P.; El-Naggar, A.; Shaheen, S.M.; Rinklebe, J.; Feng, X. Use of biochar to reduce mercury accumulation in Oryza sativa L: A trial for sustainable management of historically polluted farmlands. *Environ. Int.* **2021**, *153*, 106527. [CrossRef]

17. Meyer-Kohlstock, D.; Haupt, T.; Heldt, E.; Heldt, N.; Kraft, E. Biochar as additive in biogas-production from bio-waste. *Energies* **2016**, *9*, 247. [CrossRef]

18. Sunyoto, N.M.S.; Zhu, M.; Zhang, Z.; Zhang, D. Effect of biochar addition on hydrogen and methane production in two-phase anaerobic digestion of aqueous carbohydrates food waste. *Bioresour. Technol.* **2016**, *219*, 29–36. [CrossRef]

19. Wang, D.; Ai, J.; Shen, F.; Yang, G.; Zhang, Y.; Deng, S.; Zhang, J.; Zeng, Y.; Song, C. Improving anaerobic digestion of easy-acidification substrates by promoting buffering capacity using biochar derived from vermicompost. *Bioresour. Technol.* **2017**, *227*, 286–296. [CrossRef]

20. Lü, F.; Luo, C.; Shao, L.; He, P. Biochar alleviates combined stress of ammonium and acids by firstly enriching Methanosaeta and then Methanosarcina. *Water Res.* **2016**, *90*, 34–43. [CrossRef]

21. Karri, S.; Sierra-Alvarez, R.; Field, J.A. Zero valent iron as an electron-donor for methanogenesis and sulfate reduction in anaerobic sludge. *Biotechnol. Bioeng.* **2005**, *92*, 810–819. [CrossRef]

22. Zhang, Y.B.; Jing, Y.W.; Quan, X.; Liu, Y.W.; Onu, P. A built-in zero valent iron anaerobic reactor to enhance treatment of azo dyewastewater. *Water Sci. Technol.* **2011**, *63*, 741–746. [CrossRef]

23. Liu, Y.W.; Zhang, Y.B.; Quan, X.; Chen, S.; Zhao, H.M. Applying an electric field in a built-in zero valent iron-anaerobic reactor for enhancement of sludge granulation. *Water Res.* **2011**, *45*, 1258–1266. [CrossRef] [PubMed]

24. Zhang, Y.B.; Jing, Y.W.; Zhang, J.X.; Sun, L.F.; Quan, X. Performance of a ZVIUASB reactor for azo dye wastewater treatment. *J. Chem. Technol. Biotechnol.* **2011**, *86*, 199–204. [CrossRef]

25. Zhang, J.; Zhang, Y.; Quan, X.; Liu, Y.; An, X.; Chen, S.; Zhao, H. Bioaugmentation and functional partitioning in a zero valent ironanaerobic reactor for sulfate-containing wastewater treatment. *Chem. Eng. J.* **2011**, *174*, 159–165. [CrossRef]

26. Farghali, M.; Andriamanohiarisoamanana, F.J.; Ahmed, M.M.; Kotb, S.; Yamashiro, T.; Iwasaki, M.; Umetsu, K. Impacts of iron oxide and titanium dioxide nanoparticles on biogas production: Hydrogen sulfide mitigation, process stability, and prospective challenges. *J. Environ. Manag.* **2019**, *240*, 160–167. [CrossRef]

27. Hassanein, A.; Lansing, S.; Tikekar, R. Impact of metal nanoparticles on biogas production from poultry litter. *Bioresour. Technol.* **2019**, *275*, 200–206. [CrossRef]

28. Yu, B.; Huang, X.; Zhang, D.; Lou, Z.; Yuan, H.; Zhu, N. Response of sludge fermentation liquid and microbial community to nano zero-valent iron exposure in a mesophilic anaerobic digestion system. *RSC Adv.* **2016**, *6*, 24236–24244. [CrossRef]

29. Su, L.; Shi, X.; Guo, G.; Zhao, A.; Zhao, Y. Stabilization of sewage sludge in the presence of nanoscale zero-valent iron (nZVI): Abatement of odor and improvement of biogas production. *J. Mater. Cycles Waste Manag.* **2013**, *15*, 461–468. [CrossRef]

30. Deng, Y.; Li, X.; Ni, F.; Liu, Q.; Yang, Y.; Wang, M.; Ao, T.; Chen, W. Synthesis of Magnesium Modified Biochar for Removing Copper, Lead and Cadmium in Single and Binary Systems from Aqueous Solutions: Adsorption Mechanism. *Water* **2021**, *13*, 599. [CrossRef]

31. Mandal, S.; Pu, S.; Shangguan, L.; Liu, S.; Ma, H.; Adhikari, S.; Hou, D. Synergistic construction of green tea biochar supported nZVI for immobilization of lead in soil: A mechanistic investigation. *Environ. Int.* **2020**, *135*, 105374. [CrossRef]

32. Saitİzgi, M.; Saka, C.; Baytar, O.; Saraçoğlu, G.; Şahin, Ö. Preparation and characterization of activated carbon from microwave and conventional heated almond shells using phosphoric acid activation. *Anal. Lett.* **2018**, *52*, 772–789. [CrossRef]

33. Guo, J.; Song, Y.; Ji, X.; Ji, L.; Cai, L.; Wang, Y.; Zhang, H.; Song, W. Preparation and characterization of nanoporous activated carbon derived from prawn shell and its application for removal of heavy metal ions. *Materials* **2019**, *12*, 241. [CrossRef] [PubMed]

34. El-Nemr, M.A.; Abdelmonem, N.M.; Ismail, I.M.A.; Ragab, S.; El Nemr, A. The efficient removal of the hazardous Azo Dye Acid Orange 7 from water using modified biochar from Pea peels. *Desalin. Water Treat.* **2020**, *203*, 327–355. [CrossRef]

35. El-Nemr, M.A.; Abdelmonem, N.M.; Ismail, I.M.A.; Ragab, S.; El Nemr, A. Removal of Acid Yellow 11 Dye using novel modified biochar derived from Watermelon Peels. *Desalin. Water Treat.* **2020**, *203*, 403–431. [CrossRef]

36. Hassaan, M.A.; Pantaleo, A.; Tedone, L.; Elkatory, M.R.; Ali, R.M.; El Nemr, A.; Mastro, G.D. Enhancement of biogas produc-tion via green ZnO nanoparticles: Experimental results of selected herbaceous crops. *Chem. Eng. Commun.* **2019**, *208*, 1–14.

37. Amirante, R.; Demastro, G.; Distaso, E.; Hassaan, M.A.; Mormando, A.; Pantaleo, A.M.; Tamburrano, P.; Tedone, L.; Clodo-veo, M.L. Effects of ultrasound and green synthesis ZnO nanoparticles on biogas production from Olive Pomace. *Energy Procedia* **2018**, *148*, 940–947.

38. Hassaan, M.A.; Pantaleo, A.; Santoro, F.; Elkatory, M.R.; De Mastro, G.; El Sikaily, A.; Ragab, S.; El Nemr, A. Techno-Economic Analysis of ZnO Nanoparticles Pretreatments for Biogas Production from Barley Straw. *Energies* **2020**, *13*, 5001. [CrossRef]

39. Remigi, E.U.; Buckley, C.A. *Co-Digestion of High Strength/Toxic Organic Effluents in Anaerobic Digesters at Wastewater Treatment Works*; Water Research Commission: Pretoria, South Africa, 2016.

40. Salah, H.; Elkatory, M.R.; Fattah, M.A. Novel zinc-polymer complex with antioxidant activity for industrial lubricating oil. *Fuel* **2021**, *305*, 121536. [CrossRef]

41. Ahmmad, B.; Leonard, K.; Islam, M.S.; Kurawaki, J.; Muruganandham, M.; Ohkubo, T.; Kuroda, Y. Green synthesis of mesoporous hematite (α-Fe$_2$O$_3$) nanoparticles and their photocatalytic activity. *Adv. Powder Technol.* **2013**, *24*, 160–167. [CrossRef]

42. Nguyen, D.D.; Jeon, B.-H.; Jeung, J.H.; Rene, E.R.; Banu, J.R.; Ravindran, B.; Vu, C.M.; Ngo, H.H.; Guo, W.; Chang, S.W. Thermophilic anaerobic digestion of model organic wastes: Evaluation of biomethane production and multiple kinetic models analysis. *Bioresour. Technol.* **2019**, *280*, 269–276. [CrossRef]

43. Kafle, G.K.; Chen, L. Comparison on Batch Anaerobic Digestion of Five Different Livestock Manures and Prediction of Bio-chemical Methane Potential (BMP) Using Different Statistical Models. *Waste Manag.* **2016**, *48*, 492–502. [CrossRef]

44. Donoso-Bravo, A.; Pérez-Elvira, S.; Fdz-Polanco, F. Application of simplified models for anaerobic biodegradability tests. Evaluation of pre-treatment processes. *Chem. Eng. J.* **2010**, *160*, 607–614. [CrossRef]

45. Pang, Y.Z.; Liu, Y.P.; Li, X.J.; Wang, K.S.; Yuan, H.R. Improving Biodegradability and Biogas Production of Corn Stover through Sodium Hydroxide Solid State Pretreatment. *Energy Fuels* **2008**, *22*, 2761–2766. [CrossRef]

46. Lassoued, A.; Dkhil, B.; Gadri, A.; Ammar, S. Control of the shape and size of iron oxide (a-Fe2O3) nanoparticles synthe-sizedthrough the chemical precipitation method. *Results Phys.* **2017**, *7*, 3007–3015. [CrossRef]

47. de Jesús Ruíz-Baltazar, Á.; Reyes-López, S.Y.; de Lourdes Mondragón-Sánchez, M.; Robles-Cortés, A.I.; Pérez, R. Eco friendly synthesis of Fe3O4nanoparticles: Evaluation of their catalytic activity in methylene blue degradation by kinetic adsorption models. *Results Phys.* **2019**, *12*, 989–995. [CrossRef]

48. Morel, M.; Martínez, F.; Mosquera, E. Synthesis and Characterization of magnetite nanoparticles from mineral magnetite. *J. Magn. Magn. Mater.* **2013**, *343*, 76–81. [CrossRef]

49. Tadic, M.; Panjan, M.; Damnjanovic, V.; Milosevic, I. Magnetic properties of hematite (a-Fe2O3) nanoparticles prepared by hydrothermal synthesis method. *Appl. Surf. Sci.* **2014**, *320*, 183–187. [CrossRef]

50. Soliman, E.A.; Elkatory, M.R.; Hashem, A.I.; Ibrahim, H.S. Synthesis and performance of maleic anhydride copolymers with alkyl linoleate or tetra-esters as pour point depressants for waxy crude oil. *Fuel* **2018**, *211*, 535–547. [CrossRef]

51. Ali, R.M.; Elkatory, M.R.; Hamad, H.A. Highly active and stable magnetically recyclable CuFe2O4 as a heterogenous catalyst for efficient conversion of waste frying oil to biodiesel. *Fuel* **2020**, *268*, 117297. [CrossRef]

52. Ramesh, R.; Ashok, K.; Bhalero, G.M.; Ponnusamy, S.; Muthamizhchelvan, C. Synthesis and properties of a-Fe$_2$O$_3$ nanorods. *Cryst. Res. Technol.* **2010**, *45*, 965–968. [CrossRef]

53. Chamritski, I.; Burns, G. Infrared-and Raman-active phonons of magnetite, maghemite, and hematite: A computer simulation and spectroscopic study. *J. Phys. Chem. B* **2005**, *109*, 4965–4968. [CrossRef]

54. Jiang, X.C.; Yu, A.B.; Yang, W.R.; Ding, Y.; Xu, C.X.; Lam, S. Synthesis and growth of hematite nanodiscs through a facile hydrothermal approach. *J. Nanopart. Res.* **2010**, *12*, 877–893. [CrossRef]

55. Slaný, M.; Jankovič, L'.; Madejová, J. Structural characterization of organo-montmorillonites prepared from a series of primary alkylamines salts: Mid-IR and near-IR study. *Appl. Clay Sci.* **2019**, *176*, 11–20. [CrossRef]

56. Sheng, K.; Chen, X.; Pan, J.; Kloss, R.; Wei, Y.; Ying, Y. Effect of ammonia and nitrate on biogas production from food waste via anaerobic digestion. *Biosyst. Eng.* **2013**, *116*, 205–212. [CrossRef]
57. Dioha, I.J.; Ikeme, C.H.; Nafi'u, T.; Soba, N.I.; Yusuf, M.B.S. Effect of carbon to nitrogen ratio on biogas production. *Int. Res. J. Nat. Sci.* **2013**, *1*, 1–10.
58. Abdelwahab, T.A.M.; Mohanty, M.K.; Sahoo, P.K.; Behera, D. Impact of iron nanoparticles on biogas production and effluent chemical composition from anaerobic digestion of cattle manure. *Biomass Convers. Biorefin.* **2020**, 1–13. [CrossRef]
59. Abdelsalam, E.; Samer, M.; Attia, Y.A.; Abdel-Hadi, M.A.; Hassan, H.E.; Badr, Y. Influence of zero valent iron nanoparticles and magnetic iron oxide nanoparticles on biogas and methane production from anaerobic digestion of manure. *Energy* **2017**, *120*, 842–853. [CrossRef]
60. Deepanraj, B.; Sivasubramanian, V.; Jayaraj, S. Experimental and Kinetic Study on Anaerobic Digestion of Food Waste: The Effect of Total Solids and PH. *J. Renew. Sustain. Energy* **2015**, *7*, 063104. [CrossRef]
61. Mao, C.; Wang, X.; Xi, J.; Feng, Y.; Ren, G. Linkage of Kinetic Parameters with Process Parameters and Operational Conditions during Anaerobic Digestion. *Energy* **2017**, *135*, 352–360. [CrossRef]

 *sustainability*

Article

Effects of Variable Weather Conditions on Baled Proportion of Varied Amounts of Harvestable Cereal Straw, Based on Simulations

Alfredo de Toro [1],*, Carina Gunnarsson [2], Nils Jonsson [2] and Martin Sundberg [2]

[1] Independent Researcher, 172 40 Stockholm, Sweden
[2] Department of Agriculture and Food, Research Institutes of Sweden (RISE), 750 07 Uppsala, Sweden; carina.gunnarsson@ri.se (C.G.); nils.jonsson@ri.se (N.J.); martin.sundberg@ri.se (M.S.)
* Correspondence: alfredo.de.toro@telia.com; Tel.: +46-73-073-3249

Abstract: All harvestable cereal straw cannot be collected every year in regions where wet periods are probable during the baling season, so some Swedish studies have used 'recovery coefficients' to estimate potential harvestable amounts. Current Swedish recovery coefficients were first formulated by researchers in the early 1990s, after discussions with crop advisors, but there are no recent Swedish publications on available baling times and recovery proportions. Therefore, this study evaluated baling operations over a series of years for representative virtual farms and machine systems in four Swedish regions, to determine the available time for baling, baled straw ratio and annual variation in both. The hourly grain moisture content of pre-harvested cereals and swathed straw was estimated using moisture models and real weather data for 22/23 years, and the results were used as input to a model for simulating harvesting and baling operations. Expected available baling time during August and September was estimated to be 39–49%, depending on region, with large annual variation (standard deviation 22%). The average baling coefficient was estimated to be 80–86%, with 1400 t·year^{-1} harvestable straw and 15 t·h^{-1} baling capacity, and the annual variation was also considerable (s.d. 20%).

Keywords: bioenergy; biofuels; sustainability; renewable; cereal straw; recovered; weather; simulation; Sweden; baling time; baling proportion; baling coefficient

Citation: Toro, A.d.; Gunnarsson, C.; Jonsson, N.; Sundberg, M. Effects of Variable Weather Conditions on Baled Proportion of Varied Amounts of Harvestable Cereal Straw, Based on Simulations. *Sustainability* **2021**, *13*, 9449. https://doi.org/10.3390/su13169449

Academic Editors: Idiano D'Adamo and Piergiuseppe Morone

Received: 21 June 2021
Accepted: 18 August 2021
Published: 23 August 2021

Publisher's Note: MDPI stays neutral with regard to jurisdictional claims in published maps and institutional affiliations.

1. Introduction

Cereal straw is a sustainable and renewable resource that has been used for various purposes for centuries [1] Considerable amounts of this residue are available in many regions (e.g., [1–4]), where it can be used in the energy sector as part of the green energy supply and in efforts to mitigate climate change (e.g., [5,6]). Extensive re-use of biomass, including straw and waste resources, would also generate other beneficial effects, such as (a) progress towards a biobased economy centred on biological and renewable resources (e.g., [7–10]), (b) creation of job opportunities in rural areas, strengthening the rural economy and promoting sustainable development (e.g., [5,6,9,11,12]), (c) raw material for biobased industries (e.g., biobased textiles) and (d) raw material for second-generation biofuels (e.g., biomethane, bioethanol) or cogeneration of heat and electricity (e.g., [5,13–16]). Use of biomass and wastes in this way to contribute to a circular economy would reduce overexploitation of non-renewable resources and decrease greenhouse gas emissions (e.g., [7,9,16–18]).

The total amounts of crop residues produced in the European Union-27 (EU-27) are significant. Based on data for the period 1997–2008, Scarlat et al. [3] estimated total production to be 258 M dry tonnes/year. However, not all harvestable straw can be removed, as local conditions may require all or part of the straw to be incorporated into the soil, to maintain or improve soil organic matter content and cultivation properties.

According to Scarlat et al. [3], there is a sustainable collectable amount of 111 M dry tonnes of crop residues/year on average in the EU-27, when considering environmental and harvesting constraints, with the annual amount ranging from 86 to 133 M tonnes over a 10-year period. A proportion of the collectable total, around 28 M tonnes, is used in animal production or for purposes other than energy, leaving an estimated 87 M tonnes per year available for bioenergy purposes [3]. In a review of multiple studies, Kretschmer et al. [1] reported similar estimates, i.e., 50–127 M dry tonnes/year of agricultural residues (mainly cereal straw) available for bioenergy purposes or new uses in the EU-27. However, this potential is unevenly distributed and mainly concentrated in France, Germany, Poland, Romania, Italy, Hungary and Spain [1].

The significant amounts of agricultural residues available in the EU-27 represent an important resource for supplying renewable and sustainable bioenergy. Based on data for the period 1997–2008, Scarlat et al. [3] estimated that these residues can make a potential contribution of 3.2% on average to final energy consumption. Hence, agricultural residues can contribute significantly to achieving the EU-27 target of "at least 32% energy from renewable sources in the Union's gross final consumption of energy in 2030" [6,19].

In Sweden, the potential amount of straw available for fuel purposes is estimated at approximately 1 million tonnes per year, which corresponds to 3–4 TWh or 300,000–400,000 m^3 diesel [20]. This study focused mainly on bioenergy and biofuel applications using this renewable and sustainable agricultural residue.

However, straw as a fuel has the following major limitations:

- It is a bulky material, even when compacted
- It is a biological, hygroscopic and degradable material that needs to be stored at moisture content <18% (wet basis, w.b.) to avoid spoilage due to mould growth
- The collection period is short (a few weeks), particularly in the Nordic countries
- Annual supply varies, mainly due to yearly yield differences and recovery difficulties due to wet weather conditions during baling
- Weather variations mean that the time available for cereal harvesting and straw recovery operations varies between years
- The annual amounts of straw expected to be collected are uncertain and therefore planning its use is difficult at both farm and regional levels [21,22].

Approximately one-third of days during the Swedish baling period (August–September) are rainy ($\geq$0.5 mm) in the main cereal production regions and there is great monthly variation. Average air relative humidity is approximately 81% and the variation is also considerable (data from SMHI, period 1990–2018, own compilation [23]). This means that the Swedish weather conditions for straw baling are troublesome, particularly in some years.

As all harvestable straw cannot be collected every year, due to climate factors, delayed crop maturity, lack of time or resources due to other farm operations etc., Henriksson & Stridsberg proposed in 1992 [24] that the harvestable straw amounts in Sweden be multiplied by a specific factor, a 'recovery coefficient' (0.4 to 0.8 depending on the cereal crop and region), to estimate potential quantities that can be harvested.

These recovery coefficients were developed after discussions with crop advisors and were based on cultivation conditions and machine systems existing in the 1980s. Later Swedish studies have used them at the farm or national level (e.g., [21,24–28]). An advantage of recovery coefficients is that they are easy to use when harvest index (ratio between grain yield and aboveground biomass) or grain/straw ratio and grain yield are known, which is the case in Sweden. The drawbacks of recovery coefficients are their poor correlation with cereal yield [20,22], in addition, yields are subject to considerable annual variation in Sweden [29].

The straw recovery process is complex and dependent on multiple factors. Some factors are more or less well known at the farm level, e.g., crop grown, expected straw yield, crop area, number of fields, field size, available machines and their capacity, human resources, etc. An unknown but important consideration for harvesting and straw recovery operations is climate, a less predictable factor, particularly in regions where wet periods can be expected. Grain and straw harvesting operations require low moisture contents of the grain and straw, but the requirements are higher for straw as drying of bales is difficult to carry out at a reasonable cost.

Weather variation also leads to considerable differences in the time available to perform harvesting and baling operations, and weather conditions are difficult to determine and may vary between and within days. However, some knowledge of the expected time available for work is important when planning field operations or when new systems need to be dimensioned (e.g., [30,31]). To avoid the use of an uncertain average available time and, in parallel, to capture possible interactions between crops, machine system, weather and other factors, researchers have developed simulation models where field operations are replicated for a series of years, either on a daily or hourly basis, using historical weather data (e.g., [21,30–39]).

Straw recovery consists of several steps (baling, bale collection/loading, transport, unloading and storage), which differ depending on the system used. The most uncertain link in this chain is baling, as it requires a low straw moisture content, which in turn is weather-dependent and thus an unpredictable factor.

No assessment of harvestable cereal straw recovery operations under Swedish conditions has been published in recent years. Therefore this study evaluated baling operations over a series of years for representative farm conditions and machine systems in four regions of Sweden with a straw surplus (Västmanland, Östergötland, Västra Götaland and Skåne). The objectives were to determine the available baling time, baled straw recovery ratio and annual variation in these, using simulation and real weather data, thereby contributing to greater utilisation of this resource, supporting bioenergy systems towards sustainability goals.

2. Materials and Methods

2.1. Outline

The following steps were taken to achieve the objectives of the study (Figure 1):

- An existing model [30] was applied for predicting hourly grain moisture content of standing ripe winter wheat during 22/23 harvesting seasons
- A second existing model [40,41] was used for estimating the hourly moisture content of swathed cereal straw during the same 22/23 seasons
- Predicted moisture content data for grain and straw were used as input to a simulation model for cereal harvesting operations, which included a module for straw baling. The operations were simulated for 22/23 harvesting seasons on an hourly basis
- Several baling parameters were evaluated on representative virtual farms for the four Swedish regions in terms of their effects on amounts of baled straw, baled straw ratio and annual variation in both.

2.2. Weather Data

The moisture content of ripe standing cereal grain and swathed straw is largely determined by the weather, which in turn affects when harvesting and baling operations can be carried out. The simulation models used in this study to estimate moisture contents require data on the following weather variables on an hourly basis: temperature, precipitation, relative humidity, global radiation, wind speed and cloudiness. These data were obtained from weather stations of the Swedish Meteorological and Hydrological Institute (SMHI [23]) and the Swedish Ordnance Survey [42] for the period from 16 July to 15 October, 1995/1996–2017. For the region of Västmanland, the weather data were downloaded from stations near Västerås (59°36′ N, 16°32′ E), for Östergötland from stations near Linköping (58°24′ N, 15°37′ E), for Västra Götaland from stations near Skara (58°23′ N 13°26.3′ E) and for Skåne from stations near Hörby (55°51′ N, 13°39′ E). Global radiation data were completed using the STRÅNG model [43].

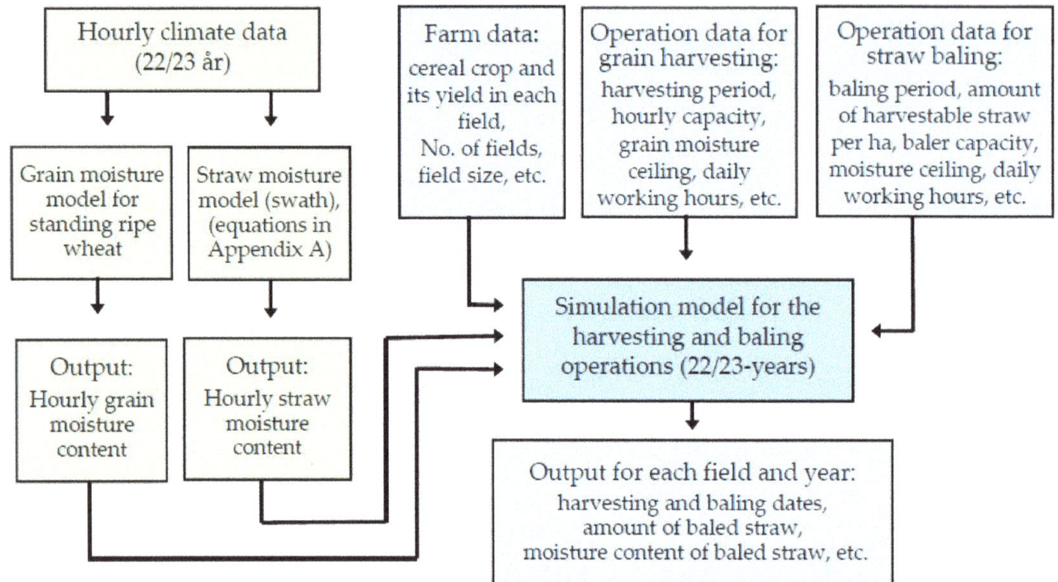

Figure 1. Flowchart of the procedure used to estimate the yearly amount of baled straw on a farm.

2.3. Estimation of Hourly Moisture Content of Swathed Cereal Straw

A simulation model was used to predict the hourly moisture content of swathed straw. The model was developed under Swedish conditions and is described in detail by Nilsson & Karlsson [40] and Nilsson & Bernesson [41]. The calculations were made in a spreadsheet computer application, employing the equations presented in Appendix A. The results were verified in a field experiment with swathed straw from winter wheat in the Uppsala region (59°51′ N, 17°38′ E) in August 2019 and are depicted in Figure 2.

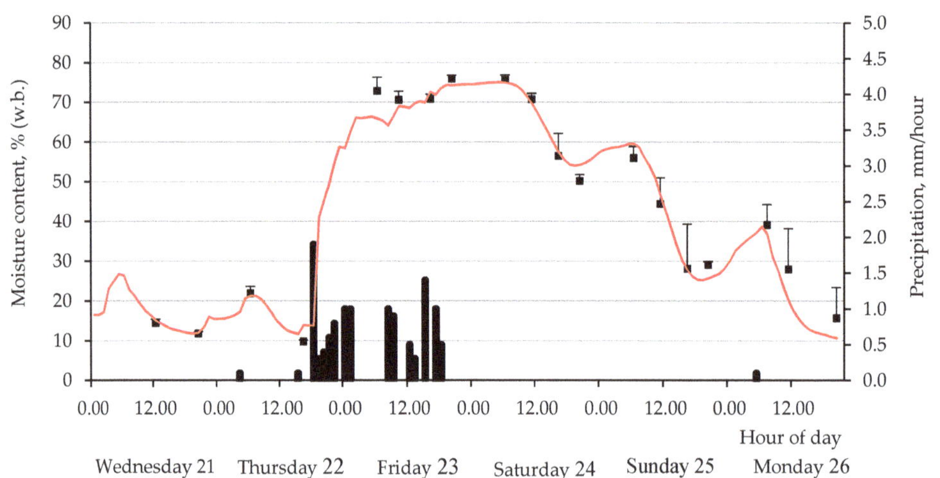

Figure 2. Comparison of moisture content (%, wet basis) of swathed wheat straw obtained with the simulation model proposed by Nilsson & Karlsson [40] and Nilsson & Bernesson [41] using as parameters: b = 0.45 and c = 0.35 (continuous line) and values measured (squares) in a field experiment in Uppsala in August 2019. The error bar on the measured moisture contents represents one standard deviation ($n = 3$) for each determination. The secondary axis shows precipitation, drawn as bars.

The model estimates the equilibrium moisture content of swathed straw based on the adsorption (wetting) process, estimating the 18% (w.b.) moisture content equilibrium at 85–90% air relative humidity. At this humidity level, there are risks of mould growth and heat generation in stored bales [44–46]. The moisture content at equilibrium with 80% air relative humidity (16% moisture content according to the model), which corresponded to 18% (w.b.) moisture content for the desorption (drying) process [47,48], was used as the moisture ceiling for the baling operation. For further details, see Appendix A.

In a comparison of equilibrium moisture content in yellow and grey straw (i.e., straw that had been exposed to precipitation) from several cereal crops harvested with a straw walker or axial-flow combine harvester, Nilsson & Bernesson [41] found small differences.

All water content values in this study are expressed on a wet basis (w.b.) and all amounts of straw are expressed in kilograms (kg) or metric tonnes (t), unless otherwise stated.

2.4. Simulation Model for Harvesting and Straw Baling Operations

The event-driven model used for simulating harvesting operations [30,31,38] simulates the operation hourly for many years with specific farm conditions (e.g., number of fields, crops, crop area, amount of straw per ha, harvesting capacity, current precipitation, working hours etc.) (Figure 1).

Annual maturation dates of the cereal crops were calculated for each field with a procedure based on daily temperature and photoperiod according to Angus et al. [49]. If the estimated maturity date for a crop fell outside the intervals shown in Table 1, it was assumed that the field maturity date would be the nearest value in the range shown in the table. This was done to avoid extreme values in the simulation model.

Table 1. Expected maturation period for the main cereal crops in the four Swedish regions studied.

Region/County	Winter Wheat	Spring Wheat	Spring Barley	Oats
Västmanland	30 July–20 August	21 August–12 September	13 August–3 September	15 August–5 September
Östergötland	28 July–18 August	19 August–9 September	10 August–1 September	13 August–3 September
Västra Götaland	28 July–18 August	19 August–9 September	10 August–1 September	13 August–3 September
Skåne	21 July–11 August	13 August–3 September	7 August–28 August	9 August–30 August

The harvesting model, whose functional unit was one hectare, was extended with a module that simulated baling operations, using as input the hourly straw moisture content estimated with a separate model [40,41].

Harvesting and baling operations were simulated for representative virtual farms in the four Swedish regions, with varied harvestable straw quantities for 22/23 years in terms of amounts baled and baled proportions.

2.5. Virtual Farms

The virtual farms were located in the regions of Västmanland, Östergötland, Västra Götaland and Skåne. Table 2 shows the distribution of cereal crops by area on the virtual farms, which corresponds to the county level, and Table 3 shows the standard yields of the main cereal crops for the county.

Table 2. Cereal crop area distribution (%) on virtual farms in the four Swedish regions studied *.

Region/County	Winter Wheat	Spring Wheat	Spring Barley	Oats	Total
Västmanland	31	12	32	25	100
Östergötland	66	5	19	10	100
Västra Götaland	36	6	24	34	100
Skåne	51	4	40	5	100

* Values based on the crop relative area over four years (2015–2018) at the county level, data from the Swedish Board of Agriculture (Jordbruksverket [50]).

Table 3. Grain yield (kg·ha^{-1}) for crops grown on virtual farms in the four Swedish regions studied *.

Region/County	Winter Wheat	Spring Wheat	Spring Barley	Oats
Västmanland	5573	4316	4617	4149
Östergötland	6829	4228	5088	4035
Västra Götaland	6189	3747	4926	4433
Skåne	7722	5200	5998	4900

* Values based on standard yields for 2017 and 2018 for the respective county, data from the Swedish Board of Agriculture (Jordbruksverket [51]).

Straw/grain yield ratio (Table 4) was used to estimate amounts of harvestable straw per hectare for the main cereal crops in the different regions (Table 5).

Table 4. Straw/grain yield ratio for the main cereal crops.

	Winter Wheat	Spring Wheat	Spring Barley	Oats
Straw: grain ratio [20] *	0.6	0.66	0.37	0.52

* Cutting height 20 cm.

Table 5. Estimated harvestable straw quantities (kg·ha^{-1}) for the main cereal crops in the four Swedish regions studied *.

Region/County	Winter Wheat	Spring Wheat	Spring Barley	Oats
Västmanland	3344	2848	1708	2157
Östergötland	4097	2790	1883	2098
Västra Götaland	3713	2473	1823	2305
Skåne	4633	3432	2219	2548

* Straw quantities based on grain yields (kg·ha^{-1}) in Table 3, multiplied by the straw/grain ratio for each crop in Table 4.

The effective baling time required to bale 1400 tonnes was approximately 10 days with 15 t·h^{-1} baling capacity and 9 h·day^{-1} working time. In a baling period of approximately 45 days, it was most likely that there were a sufficient number of available days for completing the operation in most years.

2.6. Premises and Input Data for Simulating Harvesting and Baling Operations

The following main assumptions, parameters and input data were used in the models for simulating harvesting and baling operations on the virtual farms:

- Harvestable amount of straw: 600–2800 t·year^{-1}
- Baling capacity (square bales): 15 t·h^{-1}
- Daily harvesting capacity: 6% of cereal crop area and 22% grain moisture ceiling for operating (parameters for this operation under Swedish conditions [30])
- 20 days as a maximum after crop maturation to perform harvesting operations on individual fields to avoid unreasonable delays in the simulations due to precipitation or other reasons (ripening periods are given in Table 1)
- Daily schedule for harvesting and baling operations: 11.00–20.00 h, including weekends
- Grain drying capacity was assumed not to be a limiting factor
- 2 days as a minimum waiting time between harvesting and baling operations, for swathed straw moisture content to reach equilibrium with air relative humidity and for weeds to dry
- 16 days as a maximum between harvesting and baling operations on individual fields
- A baling period until September 15, 18, 18 and 22 for Västmanland, Östergötland, Västra Götaland and Skåne, respectively

- Baling operations carried out by contractors. To replicate a real farm that does not own a large baler and hires a baler only when there are sufficient amounts of dry straw, the following requirements were set to start the operation on a certain day:
  - At least 60 tonnes sufficiently dry swathed straw ($\leq$18% w.b.), i.e., approx. 4 h effective baling time
  - Straw moisture content $\leq$18% for at least 4 h on the first day of a baling batch; on the following days the operation continued as long as dry straw was available
- Number of fields in the virtual farms: 32
- Individual field data on expected ripening period according to Table 1
- Maximum precipitation of 0.1 mm in the current simulation hour, resulting in approximately 1.6% higher straw moisture content at around 18% (w.b.) moisture content.

3. Results

3.1. Predicted Moisture Content of Swathed Cereal Straw

The hourly moisture content of swathed straw was estimated using a simulation model and historical weather data for 22/23 years (for further details, see Section 2.3). Figure 3 presents the mean available time for baling when straw moisture content did not exceed 18% (w.b.) during daytime for the four regions. The regional estimation can be considered the expected available baling time.

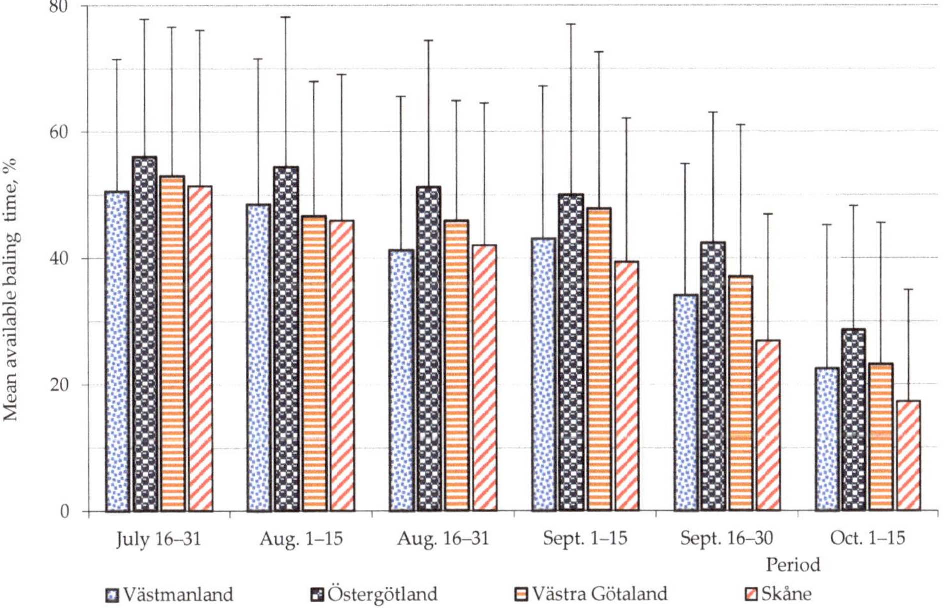

Figure 3. Mean available baling time (i.e., expected proportion of time when the moisture content of swathed straw did not exceed 18% (w.b.)) during baling hours (11.00–21.00 h), by 15-day periods in the Västmanland, Östergötland, Västra Götaland and Skåne regions from July 16 to October 15. Error bars indicate one standard deviation (n = 22/23 years) in percentage annual time available. Values based on simulation with hourly weather data from Västerås and Skara 1996–2017 and from Linköping and Malmö 1995–2017.

The available time decreased from approximately 50–55% in the second half of July to approximately 20% during the first two weeks of October for all regions except Östergötland, which showed 5–10% higher available baling time. The annual variation for 15-day periods was large, as indicated by the standard deviation (error bars in Figure 3) (22% on average considering all regions and 15-day periods).

Figure 4 shows the estimated available baling time per year in August and September for the Skåne region. In some years (e.g., 1998) this time was very short throughout the whole season, while in other years (e.g., 1995, 2001, 2006 and 2007) it was very short during one month. The yearly variation was even larger for 15-day periods. The other regions studied displayed a similar pattern as in Figure 4. Variation in available baling time makes straw recovery an uncertain process, demanding high baling capacity in Sweden, particularly during years with unfavourable weather conditions.

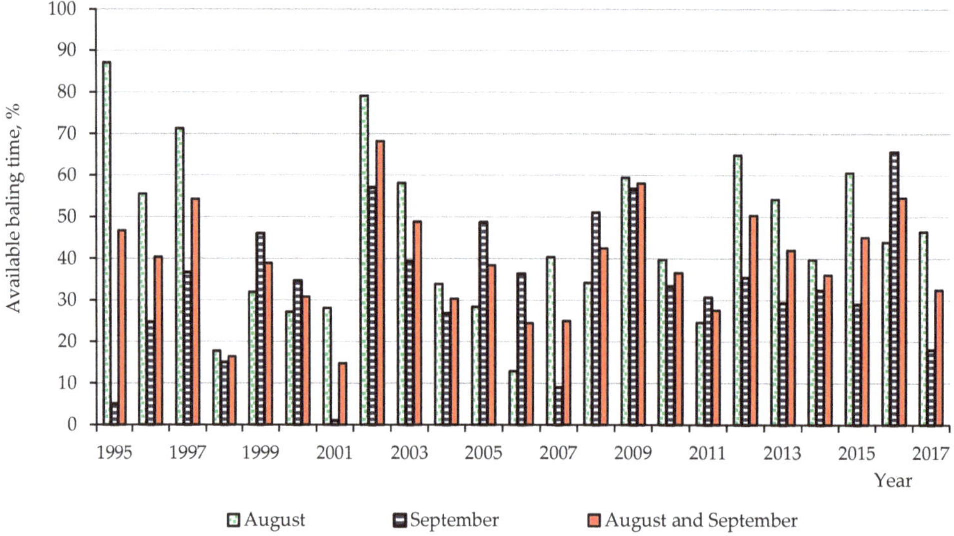

Figure 4. Estimated available baling time per year, i.e., the proportion of time when straw moisture content did not exceed 18% (w.b.) in August and September and mean values for both months during baling hours (11.00–21.00 h) for the Skåne region in the period 1995–2017. Values based on simulation and hourly weather data, most from the Hörby weather station.

3.2. Estimated Straw Baling Coefficient

Figure 5 depicts the estimated average percentage of baled straw with varied amounts of harvestable straw and a baling capacity of 15 t·h^{-1} for representative virtual farms in each region. The highest estimated percentages of baled straw were at 600 and 1000 t·year^{-1}. With quantities of harvestable straw above 1000 t·year^{-1} the baled share decreased, especially with amounts higher than 1800 t. However, the highest baled proportions estimated meant low annual baler utilisation (600 tonnes requires approx. 40 h of operation). A balance between high amounts of harvestable straw, a baled proportion of 80% or higher and high annual baler utilisation occurred at 1400 t·year^{-1}, on a virtual farm with approximately 570, 410, 520 and 400 hectares of cereal crops in Västmanland, Östergötland, Västra Götaland and Skåne, respectively. The estimated baled percentage with 1400 t·year^{-1} of harvestable straw in these regions was 84, 86, 82 and 80%, respectively (Figure 5).

With an amount of 1400 t·year^{-1} harvestable straw, the standard deviation, in this case, a measure of variation in annual baled percentage, was 21, 22, 18 and 22%, respectively. The Östergötland region showed more favourable conditions for straw baling, as reflected in the higher proportion of baled straw (Figure 5).

The average percentage of the annually baled straw with 1400 t·year^{-1} of harvestable straw was considered as the baling coefficient for the respective region, as these percentages were estimated for representative regional farm conditions and cereal crop distributions. They corresponded to approximately 90 h or 10–11 days of effective baling work when 100% of the 1400 t was baled. Note that these coefficients apply only under the conditions and restrictions specified in Section 2.6.

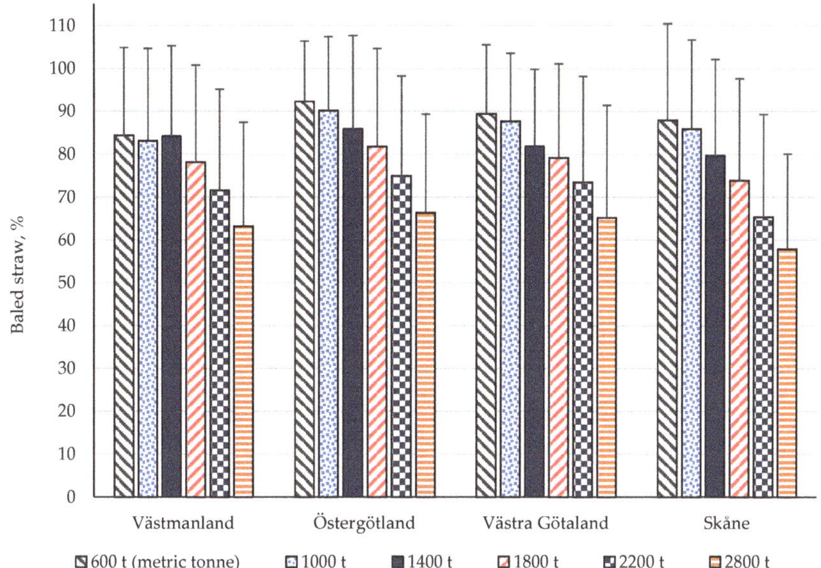

Figure 5. Mean estimated annual percentage of baled straw with varied amounts of harvestable straw for virtual farms located in the regions of Västmanland, Östergötland, Västra Götaland and Skåne, with a baling period until September 15, 18, 18 and 22, respectively, and baling capacity of 15 t·h^{-1} (one baler). The error bars indicate one standard deviation of the annual baled straw percentage (n = 22/23). Values based on simulations and hourly weather data (for further details see Figure 1 and Sections 2.4 and 2.6).

The calculated proportion of baled straw varied for the individual crops (Figure 6). It was highest for winter wheat (>80%) which also showed less annual variation, and lowest for spring wheat (67–76% depending on the region, with large annual variation).

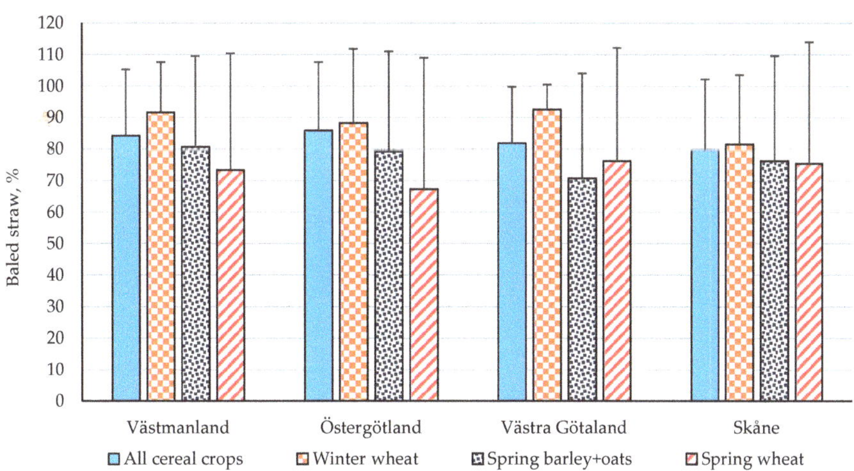

Figure 6. Mean estimated annual percentage of baled straw (baling coefficient) for all cereal crops (winter wheat, spring barley, oats and spring wheat) and for individual crops (spring barley and oats were considered as one crop) on virtual farms located in Västmanland, Östergötland, Västra Götaland and Skåne, with 1400 t·year^{-1} harvestable straw, a baling period up to September 15, 18, 18 and 22, respectively, and baling capacity of 15 t·h^{-1} (one baler). The error bars represent one standard deviation of the annual baled percentages. Values based on simulations for 22/23 seasons (for further details see Figure 1 and Sections 2.4 and 2.6).

Figure 7 depicts the quartile distributions of annual baled percentage for the four regions. In 50% of years, 90% or more of the harvestable straw was baled, while in 75% of years more than 60% was baled. Lower amounts of baled straw mainly occurred during a few years, with very low percentages during one or two years, as shown in Figure 8 for the Skåne region

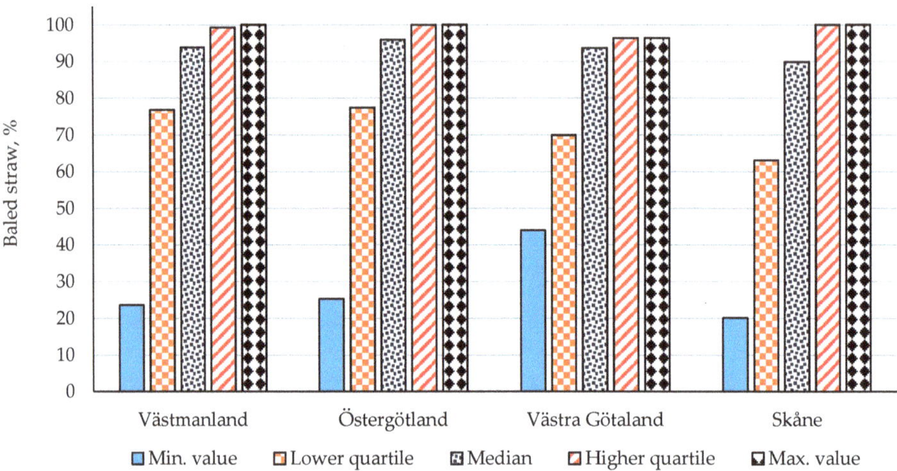

Figure 7. Quartile distributions of estimated annual baled straw percentage for virtual farms located in Västmanland, Östergötland, Västra Götaland and Skåne with 1400 t·year^{-1} harvestable straw, a baling period up to September 15, 18, 18 and 22, respectively, and baling capacity of 15 t·h^{-1} (one baler). Values based on simulations for 22/23 seasons. For further details, see Figure 1 and Sections 2.4 and 2.6.

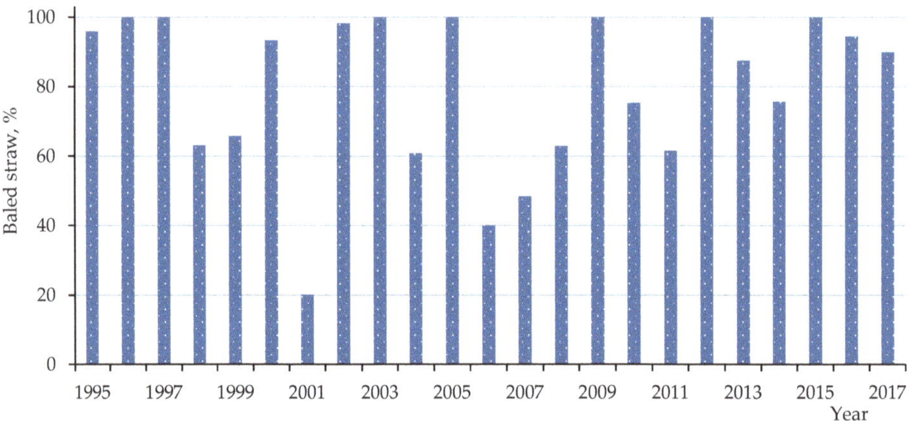

Figure 8. Estimated percentage of baled straw per year for a virtual farm located in the Skåne region with 1400 t·year^{-1} harvestable straw and a cereal crop distribution representative for the region (Table 2), the baling capacity of 15 t·h^{-1} (one baler) and a baling period up to September 22.

3.3. Moisture Content of Baled Straw

The median moisture content of the baled straw was within the range of 12–13% (w.b.) for all regions. The higher quartiles were in the range of 13–14%, indicating that the majority of the baled fields were baled at straw moisture content lower than 15%. At this moisture content, straw is in equilibrium with air relative humidity of about 70% for the drying (desorption) process [47,48,52]. At 70% humidity, mould development is greatly reduced even at optimal temperatures for microbe growth (about 20 °C) [45]. For further information, see Appendix A.

3.4. Sensitivity Analysis

A sensitivity analysis involving reducing baler accessibility and baling capacity was made, to assess how these factors affected amounts of baled straw. Baling operations were simulated for a representative virtual arable farm in the Skåne region with 1400 t·year^{-1} harvestable straw.

3.4.1. Limited Baler Accessibility

Figure 9A displays the average percentage of straw baled annually with varied accessibility of a baler to start a batch of operating days. Baler accessibility was reduced for each field by a random parameter in the simulation model. The baled amount decreased with lower accessibility, particularly when it was 60% or lower. However, the annual variation was considerable, as indicated by the standard deviation of the baled straw in Figure 9A. The reductions were not evenly distributed annually and mainly occurred in years with little available baling time.

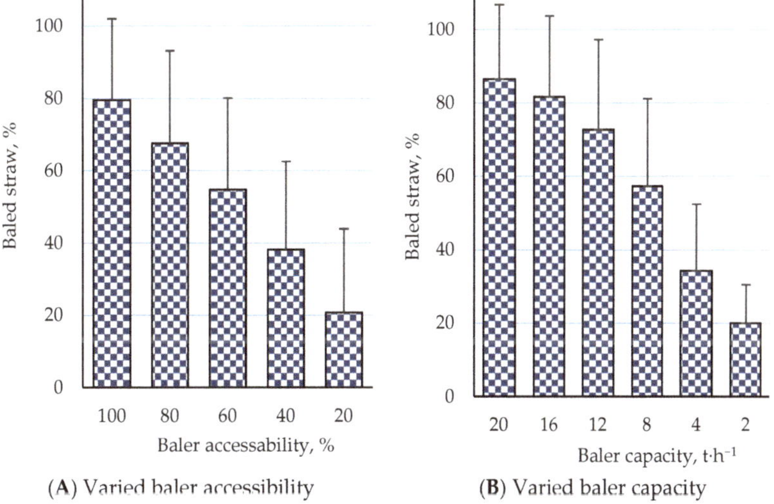

(**A**) Varied baler accessibility   (**B**) Varied baler capacity

Figure 9. Mean percentage of straw baled annually with (**A**) varied baler accessibility to start a batch of baling days and baling capacity of 15 t·h^{-1} and (**B**) varied baler capacity and 100% accessibility to a baler for a representative arable farm in the Skåne region with 1400 t·year^{-1} harvestable straw and a baling period until September 22. The error bars represent one standard deviation of annual baled percentage (n = 23 years), for further details see Figure 1 and Sections 2.4 and 2.6.

3.4.2. Reduced Baler Capacity

Figure 9B presents the average percentage of straw baled annually with varied baling capacity. The percentage only decreased considerably when capacity was significantly reduced, specifically to 4 or 8 t·h^{-1}. Note that baler accessibility was set to 100% in the model when a value was required. As with baler accessibility, the decreases were not evenly distributed between years. For several baling seasons, the amount baled was close to 100% with the baling capacity of 8 t·h^{-1}. This can be explained by the high available baling time in some years in the Skåne region (see Figure 4). The effective number of working days with this capacity was around 20 days (1400 tonnes of straw/(8 t·h^{-1} × 9 h·day^{-1})). Therefore during years with favourable weather conditions, most of the straw could be baled.

4. Discussion

4.1. Weather Conditions

The average proportion of rainy days ($\geq$0.5 mm precipitation) in August and September 1990–2018 was approximately one-third in the four Swedish regions analysed, although the annual variation was considerable. Fortunately, very problematic years with little available baling time only occurred in a few years, as clearly illustrated in Figure 4 for the Skåne region. The Västra Götaland region showed poorer weather conditions but had a more even crop distribution (Table 2), allowing harvesting and baling operations to be spread over more days.

Most of the weather data used were downloaded from one station for each region and assumed to be valid for the whole region. However, all four regions are rather large and daily weather may vary from site to site. This adds uncertainty to the estimated available baling times for particular locations, years or months. Nevertheless, the estimated averages from 22/23 years for the regions did not differ greatly (Figure 3), and hence a similar pattern can be expected within a region in the long term.

4.2. Moisture Content Prediction Models for Standing Mature Wheat and Swathed Cereal Straw

Drying and wetting processes are complicated and depend on multiple factors, so different simplifications and assumptions must be made if a model is to be useful and the amount of input data manageable [41].

To increase the reliability of estimates of swathed straw moisture content, a model developed and validated in Sweden for crops and weather conditions similar to those in the four regions studied was utilised [40,41]. The harvesting model used was also developed under Swedish conditions [30,38] so its results should be reliable for the regions.

In this study, it was assumed that the moisture content in swathed straw was evenly distributed, which is not always the case under real field situations. Swath parameters may vary a great deal, e.g., swath thickness, amount of weeds, straw lying on stubble or soil, site exposure to rain, dew, wind, solar radiation etc. All these factors lead to an uneven straw moisture content distribution in swathed straw in regions with spells of unfavourable weather conditions during the baling season, making their prediction the weakest link in this study.

The estimated moisture content median and higher quartile for baled straw were in the range 12–13% and 13–14% (w.b.), respectively, for all four regions, i.e., much lower than the 18% (w.b.) moisture ceiling for operations. Abawi [36] analysed harvesting operations and found that the moisture content of ripe standing wheat fluctuates around a certain average, depending on prevailing weather conditions. Weather variation means a mixture of "good" and "bad" days, making moisture contents tend to an average. However, weather variation also includes periods of persistent inclement weather in some years.

4.3. Farm Premises and Assumptions for Simulating Baling Operations

The harvestable straw quantities per hectare used in this study were based on grain yields (Table 3), but the actual annual yield variation is considerable. A yield calculation for the analysed crops and regions over the 10-year period 2010–2019 showed a yield range of 1470–3800 kg·ha^{-1}, with a coefficient of variation of 10–23% (data from Swedish Board of Agriculture, own compilation (Jordbruksverket [50]. Similar yield variations have been found in another study based on 50-year statistics for cereal yields in Sweden, with the annual differences even larger at the farm level [29].

An equally wide range of estimated recoverable residues:grain yield ratios (0.8–1.6 for wheat, 0.8–1.3 for barley) has been reported by Glithero et al. [22], who concluded that there is no clear relationship between harvested grain:straw yield for wheat in their study area (England). Based on five-year data for Swedish cereal crops, Nilsson & Bernesson [20] reported varied harvestable straw:grain ratios (20 cm stubble) of 0.41–0.96 for winter wheat cultivars and 0.29–0.46 for barley.

In this study, quantities from 2.2 to 4.6 t·ha^{-1} harvestable straw were assumed, depending on crop and region. These values are similar to estimates of harvestable straw for wheat and barley (median 2.5 and 2.3 t·ha^{-1}, respectively) for the 2010 harvest in England [22] from straw yield experiments in Denmark [53], from a recovery study of wheat residues in Sweden [54] and for harvestable cereal straw in Finland with a cutting height of 20 cm [55].

4.4. Available Baling Time

Mean available baling time during working hours (11.00–21.00 h) was estimated to be approximately 44% in August and September for the four regions, with higher values for August than September and higher values for Östergötland than other regions (Figure 3). Unfortunately, moisture content values for swathed straw are not reported in the Swedish literature, so comparisons were not possible.

With a straw recovery period of about 40–50 days, the expected number of suitable days for baling varied from 18–22 days per year, which suggests that around 3000 t·year^{-1} of straw could theoretically be baled on average over a series of years with a baling capacity of 15 t·h^{-1}. In practice, about half this amount is usually recovered at that baling capacity [21,56,57], which is in line with the results from this study (for further details, see Section 3.2).

4.5. Baling Coefficient

When estimating average baled percentages with varying quantities of straw for representative farm conditions in each region (see Figure 5), it was assumed that at least 80% of the harvestable straw was baled on average for a series of years to be considered a baling coefficient. However, there is no objective criterion for this proportion and a lower percentage could be accepted, e.g., around 60%, which occurred at 2800 t·year^{-1} harvestable straw, leading to a lower baling capacity requirement in relation to straw amount. A similar baled proportion of 65% with 2200 t·year^{-1} as shown in Figure 5 was reported by Nilsson & Bernesson [21] for a straw amount of approximately 2000 t·year^{-1}.

The average proportion of baled straw for the whole farm was estimated at 84, 86, 82 and 80% for the Västmanland, Östergötland, Västra Götaland and Skåne region, respectively, with an amount of 1400 t·year^{-1} harvestable straw, 15 t·h^{-1} baling capacity and a baling period until mid-September. These coefficients are at least 5% higher than those originally reported by Henriksson & Stridsberg [24] following discussions with crop advisors. This discrepancy can be explained in part by the fact that the advisors based their estimates on experiences from the 1980s, with cultivation conditions, farm sizes, machine systems, climate and so on for that decade.

In this paper, we explicitly state the method, premises and parameters with which the coefficients were estimated, so that the values can be adjusted if farm conditions or premises deviate, e.g., amount of straw, baling period duration, moisture content ceiling for operating, baling capacity, working hour per day, the minimum expected baled straw quantities in seasons with poor weather conditions, etc.

4.6. Baled Proportions with a High Amount of Harvestable Straw

Figure 10 shows the quartile distribution of baled percentage for the virtual farms in the four regions with an amount of 2800 t·year^{-1} harvestable straw and baling capacity of 15 t·h^{-1}. The higher quartile denotes that in at least 25% of years, it was possible to bale more than 80% except in the Skåne region. Comparing Figure 7 with Figure 10 shows that the proportion of baled straw decreased for half the years by approximately 30% with an amount of 2800 t·year^{-1} (median values). The reductions were not evenly distributed over the years, mainly occurring in 50% of years (Figures 7, 8 and 10). Nilsson & Bernesson [21] arrived at a similar conclusion.

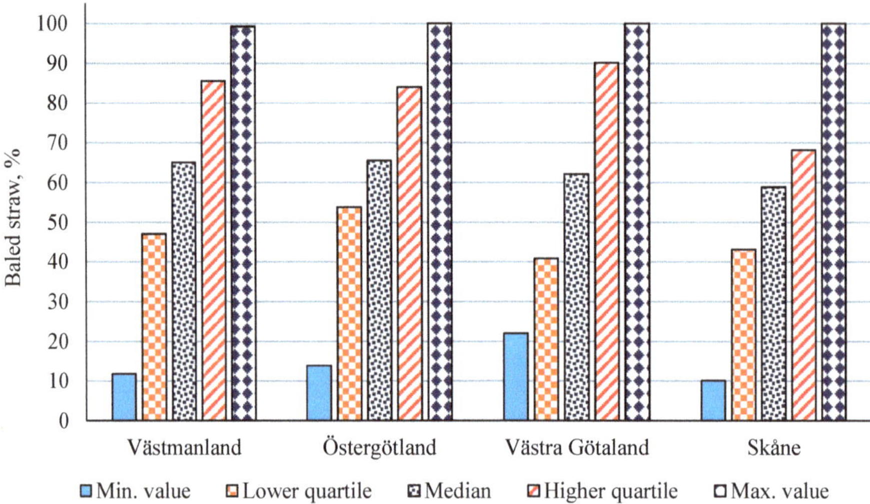

Figure 10. Quartile distributions of annual baled straw percentage for virtual farms located in Västmanland, Östergötland, Västra Götaland and Skåne with 2800 t·year^{-1} harvestable straw, a baling period up to September 15, 18, 18 and 22, respectively, and baling capacity of 15 t·h^{-1}. Values based on simulations for 22/23 seasons, for further details see Section 2.4 and Figure 1.

On the other hand, if a farmer will accept the collection of a lower proportion (e.g., 60% on average) of the 2800 t·year^{-1} harvestable straw during 50% of years, then 15 t·h^{-1} baling capacity is adequate, assuming the conditions and constraints described in Section 2.6.

4.7. Validity of Results

In general, models allow general patterns and trends of complex scenarios to be explained and effects of main parameters and variables of a process to be quantified, so new knowledge can be acquired. In addition, it is possible to capture interactions between the main factors influencing a system, which are difficult to visualise with analysis. However, model results are usually difficult to apply to specific cases and do not necessarily give a completely correct picture of reality, mainly due to general assumptions, simplifications and limited input data.

In this study, harvesting and baling operations were simulated for virtual farms in four Swedish regions, to assess the influence of weather conditions on amounts and proportions of baled straw. It was assumed that each farm represented the standard conditions for the region. However, it is well known that no two farms have similar conditions in terms of crop distribution, operation priorities, access to a baler, machine system, soils, swath properties, annual variations in straw yield, etc. The complexity of the drying and wetting process of swathed straw adds further uncertainty in the estimation of straw moisture content.

Considering the above limitations, the main conclusions of this study are likely to be mainly valid for regions with comparable climate and agricultural conditions to the four Swedish regions (central and southern parts of the country, not strictly geographically).

4.8. Equation for Rough Estimation of Performance of a Straw Baling System

A system in which a high proportion of harvestable straw is baled requires the number of available days for operating to be equal to or greater than the number of effective baling days that the work requires during a collecting season, i.e.:

$$A_{days} \geq E_{days} \tag{1}$$

where:

A_{days}: number of available days for straw baling.

E_{days}: number of effective days that the baling operation requires for a given amount of straw; and where:

$$A_{days} = Pe_{length} * A_{time} * Af_{baler} * Rf_{low} * Mf_{matching} \tag{2}$$

$$\text{e.g., } A_{days} = 45 * 0.45 * 0.9 * 0.8 * 0.6 = 9 \text{ days}$$

where:

Pe_{length}: period length for straw baling operations (e.g., 45), days

A_{time}: available expected time proportion for straw baling (e.g., 0.45 for the regions studied, see Figure 3)

Af_{baler}: access factor to a baler, (0.1 if the chance of getting a baler when needed is 10%, 1.0 if a baler is always available when needed)

Rf_{low}: reducing factor for baling a higher straw proportion than average (0.5 to 1.0: 0.5 for baling a higher straw proportion than average a 1 for mean available baling time according to Figure 3)

$Mf_{matching}$: matching factor to compensate restrictions of the baling system, e.g., not enough straw ready to be baled on an available baling day (0.3 if the baling system operates with difficulties and 0.8 if everything goes smoothly, without bottlenecks); and

$$E_{days} = \frac{H_a * St_{amount}}{Ba_{cap} * Wh_{day}} \tag{3}$$

$$\text{e.g., } E_{days} = \frac{450 * 2.5}{15 * 9} = 8.3 \text{ days}$$

where:

H_a: area to be baled, ha

St_{amount}: amount of straw per hectare, $t \cdot ha^{-1}$

Ba_{cap}: baling capacity, $t \cdot h^{-1}$

Wh_{day}: working hours per day, $h \cdot day^{-1}$

The results of Equation (2) were compared with estimates from other Swedish studies. Considering only weather factors, Lundin [58] (Cited by Nilsson [59]) estimated 40, 36, 41 and 55 available days for straw recovery in Västmanland, Östergötland, Västra Götaland and Skåne, respectively. These values are twice as high as the estimates in the present study because the mean available baling time was estimated at approximately 45% (Figure 3), which is consistent with information from Nilsson & Bernesson [21], C. Gunnarsson [56] and F. Johansson [57].

Brundin [60] (Cited by Nilsson [59]) developed a similar formula to Equation (2) to estimate the number of available days for straw collection but obtained higher estimates than those in this study. However, his equation does not take into account limiting factors caused by the machine system. Even so, any collecting system requires machines, whose parameters and limitations affect the amounts of straw collected.

5. Conclusions

By analysing the straw recovery process in regions where wet periods are probable during the baling season, this study provided important data support for increasing the efficiency of operations, enabling higher proportions of straw to be recovered and helping bioenergy systems move towards greater sustainability.

This study simulated 22/23 years of baling operations for representative virtual farms in four Swedish regions on an hourly basis, using historical weather data. The main conclusions, which should be valid for regions with similar climate and agricultural conditions to the regions assessed, were as follows:

- The estimated available baling time during working hours (11.00–21.00 h) was 39–49%, depending on the region. The time decreased from around 50% at the beginning of August to 30–40% at the end of September. The annual variation was large, with a standard deviation of around 22% in all regions.
- A reasonable balance between a large amount of harvestable straw and a high proportion of baled straw (over 80% on average) was reached at 1400 t·year^{-1} harvestable straw with the baling capacity of 15 t·h^{-1} (one baler) and a baling period to mid-September.
- An 80% proportion of baled straw is not standard. A lower ratio, e.g., around 60% on average, would be reached at 2800 t·year^{-1} harvestable straw with the same baling capacity.
- A rough estimate showed that approximately 25% of days in the baling season were effective baling days on average. Matching factors between sufficiently dry swathed straw amount and the baling system set restrictions on utilising most of the available baling time.
- In one to two baling seasons out of 10, the proportion of baled straw was reduced to about 60% or less, even for an "optimised" system with the baling capacity of 15 t·h^{-1}, 1400 t·year^{-1} harvestable straw and a baling period of 45 days per year.
- A baling system where a high proportion of harvestable straw is baled in most years requires the number of available working days for the operation to be greater than the average number of effective baling days that a given amount of straw demands with a certain baling capacity.
- The most uncertain item of this study was the prediction of changes in straw moisture content for the swathed straw, due to the complexity of straw wetting and drying processes.

Author Contributions: Conceptualization and methodology, C.G., N.J. and M.S.; software and data curation, A.d.T.; writing—original draft preparation A.d.T. and M.S.; writing—review and editing C.G., N.J. and M.S.; supervision, C.G.; project administration, C.G.; funding acquisition, C.G. All authors have read and agreed to the published version of the manuscript.

Funding: This research was funded by the Swedish Energy Agency, project grant number 45910-1 "Residues from cereal production for the Swedish biobased industry".

Institutional Review Board Statement: Not applicable.

Informed Consent Statement: Not applicable.

Data Availability Statement: Not applicable.

Acknowledgments: Authors would like to thank Fredrik Johansson, Executive Manager and Agricultural Contractor, for the valuable information on straw handling under practical conditions and, the Researcher, Martin Knicky (RISE) for his contribution in the straw moisture determinations in the field experiment.

Conflicts of Interest: The authors declare no conflict of interest.

Appendix A. Model for Estimating Moisture Content of Swathed Cereal Straw

The baling requirement of low straw moisture content means that weather has a great influence on when this operation can be carried out. Several researchers have used models to investigate how weather affects moisture contents of straw or hay in the field (e.g., [40,41,61–64]).

A model developed in Sweden [40,41] was used in this study. It divides straw moisture into bound and free water according to an idea of Atzema) [64]. Bound water follows relative air humidity with some delay, and increases when air relative humidity is higher than the equilibrium moisture content of straw, or vice versa. Free water is related to dew and precipitation so that the potential evapotranspiration (E) equation can be used to estimate the effects of drying and wetting processes. If E > 0 free water evaporates, and when E < 0 water vapour condenses to dew. Table A1 presents the equations applied in this study for estimating the moisture content of swathed cereal straw. A complete model description can be found in Nilsson & Karlsson [40] and Nilsson & Bernesson [41].

Table A1. Moisture model equations for swathed cereal straw *.

Process	Condition	Equations (Time Step = One Hour)	Value of Parameter
Quantity of bound water at equi-librium moisture content	$q_{eq} \leq 0.30\, q_{dm}$	$q_{eq}(t) = q_{dm} \left(\dfrac{-A}{\ln RH(t)} \right)^{1/B}$	A = 582.3 B = 2.69
Quantity of bound water	$q_b(t) \geq 0$ and $q_b(t) \leq 0.25\, q_{dm}$	$q_b(t) = \dfrac{q_{eq}(t) + q_{eq}(t-1) + q_{eq}(t-2)}{3}$	
Change in dew water due to dew **	$E < 0$	$\Delta q_d(t) = E(t)$	
Change in dew water due to evaporation **	$E \leq 0$	$\Delta q_d(t) = -E(t)$	
Quantity of water due to dew	$q_d(t) \geq 0$	$q_d(t) = q_d(t-1) + \Delta q_d(t)$	
Change in precipitation water due to absorption	$P > 0$	$\Delta q_{pa}(t) = I_c t(t)(1 - e^{-\frac{bP}{I_c(t)}})$ where $I_c(t) = q_{dm}(M_{cMax} - M_c(t-1))$	b = 0.45 M_{cMax} = 3.5
Change in precipitation water due to evaporation **	$E > 0$	$\Delta q_{pe}(t) = E(t)\left(1 - e^{-cq_p(t-1)}\right)$	c = 0.35
Quantity of water due to precipitation	$q_p(t) \geq 0$	$q_p(t) = q_p(t-1) + \Delta q_{pa}(t) - \Delta q_{pe}(t)$ and: if $(q_b(t) + q_d(t) + q_p(t)) > 3.5\, q_{dm}$ then $q_p(t) = 3.5\, q_{dm} - q_b(t) - q_d(t)$	
Actual moisture content, decimal (dry base)	$M_c(t) \leq 3.5$	$M_c(t) = \dfrac{q_b(t) + q_d(t) + q_p(t)}{q_{dm}}$	

*: For further information on the model, see Nilsson & Karlsson [40] and Nilsson & Bernesson [21,41]. **: If weather data are available, the potential evapotranspiration computation may be facilitated by programmes obtainable on the internet (e.g., [65]).

where:

$q_{eq}(t)$:   quantity of water at equilibrium moisture content at time t, kg·m^{-2}

q_{dm}:   quantity of dry matter (straw), kg·m^{-2}

RH(t):   air relative humidity at time t, decimal

$q_b(t)$:   quantity of bound water at time t, kg·m^{-2}

$q_{eq}(t)$, $q_{eq}(t-1)$, $q_{eq}(t-2)$:   quantity of bound water at time t, t − 1, t − 2, kg·m^{-2}

$\Delta q_d(t)$:   change in absorbed or dried dew water per time step (one hour), kg·m^{-2} h^{-1}

E(t):   potential evapotranspiration of water at time t, mm h^{-1} (or kg·m^{-2} h^{-1});

$q_d(t)$:   quantity of dew water at time t, kg·m^{-2}

$\Delta q_{pa}(t)$:   change in precipitation water due to absorption per time step (one hour), kg·m^{-2} h^{-1}

$I_c(t)$:   interception storage capacity at time t, kg·m^{-2}

P: precipitation, mm/h (or kg·m^{-2} h^{-1})

M_{cMax}:   maximal moisture content of straw: 3.5 (dry base) (empirical determination)

$\Delta q_{pe}(t)$:   change in precipitation water due to evaporation per time step (one hour), kg·m^{-2} h^{-1}

$q_p(t)$:   quantity of precipitation water at time t, kg·m^{-2}

$M_c(t)$:   actual moisture content of straw at time t, decimal (dry base).

Parameter "b" in Table A1 is related to the hourly increase of water from precipitation, i.e., how much water straw absorbs from it. The higher the value, the higher the estimated absorbed water. In this study, the parameter "b" was set to 0.45 in the model, which is the value used by Nilsson & Bernesson [21].

Parameter "c" (Change of precipitation water due to evaporation), is related to hourly moisture decrease from precipitationI, i.e., the higher the value, the faster the estimated drying process due to evaporation. Nilsson & Bernesson [21], in their study on *Dynamic simulation of handling systems*, set this parameter to 0.85, but in their report on straw *Moisture characteristics*, values from 0.35–1.05 are given by Nilsson & Bernesson, [41]. All parameter values in this range showed a high correlation between measured straw moisture content and simulated values. In this study, the parameter "c" was set to a lower value (0.35), which is more in line with the field experiment conducted in Uppsala in August 2019 (Figure 2), to ensure that the model did not overestimate the drying process after precipitation.

The parameter variations between different studies indicate that the drying or wetting processes for swathed straw are complex, depending on many factors (including weather, swath properties, weed amount, swath compaction, nitrogen fertilisation, fungal treatment, etc.) that are difficult to capture in a simulation model.

As a hygroscopic material, straw dries out or moistens if the straw moisture content is not in equilibrium with surrounding air relative humidity (without considering rain or dew). In Sweden, the recommendation is to bale cereal straw at a maximum of 18% (w.b.) moisture content, to reduce the risk of mould development and heat generation. At this humidity, the straw moisture content is in equilibrium with air relative humidity of between 80 and 90%, depending on temperature and whether the moisture equilibrium was reached by drying (desorption) or moistening (adsorption). In experiments comparing the processes, the equilibrium moisture content of wheat straw has been found to be 1.5–2% lower for the moistening process within a certain range of air relative humidity, due to the so-called hysteresis effect [47,48]. Thus, an 18% moisture content is reached at lower air relative humidity when the straw is drying than wetting.

A Swedish study examining the drying process of wheat and barley straw reported that an 18% (w.b.) equilibrium moisture content was reached at 80% relative humidity at 10 °C and 83% relative humidity at 20 °C [52].

The model used in this study predicts equilibrium moisture content of straw based on the humidification process, resulting in 18% (w.b.) moisture content relating to 85–90% air relative humidity [41]. If a baling operation is carried out when the straw moisture content is in equilibrium with such a high relative humidity level (85–90%), there is a risk of mould development in bales stored at temperatures higher than 5 °C [45], which occur commonly during the baling season in Sweden (August–September). At lower relative humidity, for example, 80%, mould development is strongly inhibited at temperatures of 10 °C or lower [44–46]. These temperatures are not uncommon from the end of August in the regions analysed in this study. Accordingly, the moisture content that is in equilibrium with 80% air relative humidity, i.e., 18% (w.b.), was used as the moisture ceiling for straw baling operations in the drying process but 16% was used in the model that based moisture estimation on the moistening process. Thus available baling time was unlikely to be overestimated, which would probably lead to a higher proportion of baled straw for the baling systems analysed.

References

1. Kretschmer, B.; Allen, B.; Hart, K. *Mobilising Cereal Straw in the EU to Feed Advanced Biofuel Production*; Institute for European Environmental Policy: Bruxelles, Belgium, 2012. Available online: http://minisites.ieep.eu/assets/938/IEEP_Agricultural_residues_for_advanced_biofuels_May_2012.pdf (accessed on 20 August 2021).
2. Edwards, R.A.H.; Šúri, M.; Huld, M.A.; Dallemand, J.F. GIS-Based Assessment of Cereal Straw Energy Resource in the European Union. In Proceedings of the 14th European Biomass Conference & Exhibition, Biomass for Energy, Industry and Climate Protection, Paris, France, 17–21 October 2005. Available online: https://www.academia.edu/22384487/GIS_based_assessment_of_cereal_straw_energy_resource_in_the_European_Union?from=cover_pag (accessed on 20 August 2021).
3. Scarlat, N.; Martinov, M.; Dallemand, J.F. Assessment of the availability of agricultural crop residues in the European Union: Potential and limitations for bioenergy use. *Waste Manag.* **2010**, *30*, 1889–1897. [CrossRef] [PubMed]
4. Eurostat. Agricultural Production—Crops. Statistics Explained. 2020. Available online: https://ec.europa.eu/eurostat/statistics-explained/index.php?title=Agricultural_production_-_crops#Cereals (accessed on 13 August 2021).
5. European Commission. Greening Our Energy Supply—The Role of Bioenergy from Forestry and Agriculture. 2010. Directorate General for Agriculture and Rural Development. Available online: https://ec.europa.eu/info/sites/default/files/food-farming-fisheries/sustainability_and_natural_resources/documents/leaflet-greening-our-energy-supply_en.pdf (accessed on 13 August 2021).
6. Directive (EU) 2018/2001 of the European Parliament and of the Council of 11 December 2018 on the Promotion of the Use of Energy from Renewable Source. Available online: https://eur-lex.europa.eu/legal-content/EN/TXT/HTML/?uri=CELEX:32018L2001&from=EN#d1e1665-82-1 (accessed on 13 August 2021).
7. Scarlat, N.; Dallemand, J.F.; Monforti-Ferrario, F.; Nita, V. The role of biomass and bioenergy in a future bioeconomy: Policies and facts. *Environ. Dev.* **2015**, *15*, 3–34. [CrossRef]
8. Morone, P. Sustainability Transition towards a Biobased Economy: Defining, Measuring and Assessing. *Sustainability* **2018**, *10*, 2631. [CrossRef]
9. D'Adamo, I.; Falcone, P.M.; Morone, P. A New Socio-economic Indicator to Measure the Performance of Bioeconomy Sectors in Europe. *Ecol. Econ.* **2020**, *176*, 106724. [CrossRef]
10. Bastos Lima, M.G. Corporate Power in the Bioeconomy Transition: The Policies and Politics of Conservative Ecological Modernization in Brazil. *Sustainability* **2021**, *13*, 6952. [CrossRef]
11. Eziyi, I.; Krothapalli, A. Sustainable Rural Development: Solar/Biomass Hybrid Renewable Energy System. *Energy Procedia* **2014**, *57*, 1492–1501. [CrossRef]
12. Spinelli, L.; Pari, L.; Magagnotti, N. New biomass products, small-scale plants and vertical integration as opportunities for rural development. *Biomass Bioenergy* **2018**, *115*, 244–252. [CrossRef]
13. Ekman, A.; Wallbergb, O.; Joelsson, E.; Börjesson, P. Possibilities for sustainable biorefineries based on agricultural residues—A case study of potential straw-based ethanol production in Sweden. *Appl. Energy* **2013**, *102*, 299–308. [CrossRef]
14. Yamakawa, C.K.; Qin, F.; Mussatto, S.I. Advances and opportunities in biomass conversion technologies and biorefineries for the development of a bio-based economy. *Biomass Bioenergy* **2018**, *119*, 54–60. [CrossRef]
15. AGROinLOG. Success Case of the Integration of a Logistics Centre into an Agro-Industry of the Grain-Milling and Feed Sector. Integrated Biomass Logistics Centres for the Agro-Industry, 2020. Available online: http://agroinlog-h2020.eu/wp-content/uploads/2020/06/AGROinLOG_D5.7_Success-case-IBLC-Sweden_FV_30042020_RISE.pdf (accessed on 13 August 2021).
16. D'Adamo, I.; Falcone, P.M.; Huisingh, D.; Morone, P. A circular economy model based on biomethane: What are the opportunities for the municipality of Rome and beyond? *Renew. Energy* **2021**, *163*, 1660–1672. [CrossRef]
17. Sheldon, R.A. Biocatalysis and biomass conversion: Enabling a circular economy. *Philos. Trans. R. Soc. A* **2020**, *378*, 20190274. [CrossRef]
18. Sherwood, J. The significance of biomass in a circular economy. *Bioresour. Technol.* **2020**, *300*, 122755. [CrossRef]
19. European Commission. The Revised Renewable Energy Directive. Factsheet: Renewable Energy Directive. Available online: https://ec.europa.eu/energy/sites/default/files/documents/directive_renewable_factsheet.pdf (accessed on 13 August 2021).
20. Nilsson, D.; Bernesson, S. *Halm som Bränsle. Del 1: Tillgångar och Skördetidpunkter (Straw as Fuel. Part 1: Available Resources and Harvest Times)*; Department of Energy and Technology, Swedish University of Agricultural Sciences: Uppsala, Sweden, 2009. Available online: https://pub.epsilon.slu.se/4854/1/nilsson_d_et_al_100630.pdf (accessed on 20 August 2021).

21. Nilsson, D.; Bernesson, S. *Halm som Bränsle. Del 3: Dynamisk Simulering av Hanteringssystem (Straw as Fuel. Part 3: Dynamic Simulation of Handling Systems)*; Department of Energy and Technology, Swedish University of Agricultural Sciences: Uppsala, Sweden, 2010. Available online: https://pub.epsilon.slu.se/4852/1/nilsson_d_et_al_100630_2.pdf (accessed on 20 August 2021).

22. Glithero, N.J.; Wilson, P.; Ramsden, S.J. Straw use and availability for second generation biofuels in England. *Biomass Bioenergy* **2013**, *55*, 311–321. [CrossRef]

23. SMHI. Ladda ner Meteorologiska Observationer. (Swedish Meteorological and Hydrological Institute. Download Meteorological Observations). Available online: https://www.smhi.se/data/meteorologi/ladda-ner-meteorologiska-observationer/#param=airtemperatureInstant,stations=all (accessed on 10 June 2021).

24. Henriksson, A.; Stridsberg, S. *Möjligheter att Använda Halmeldning till Energiförsörjningen i Södra Sverige (The Potential of Using Straw as a Fuel for the Support of Energy in the Agricultural Area of Southern Sweden)*; Department of Agricultural Engineering, Swedish University of Agricultural Sciences: Uppsala, Sweden, 1992. Available online: https://pub.epsilon.slu.se/3883/ (accessed on 20 August 2021).

25. Lindqvist, J. Potential för Biobränsleproduktion i Uddevalla Kommun (Biofuel, Bioenergy, Potential, Municipality of Uddevalla). Master's Thesis, Department of Technology and Society, Lund University, Lund, Sweden, 2007. Available online: https://lup.lub.lu.se/luur/download?func=downloadFile&recordOId=4468201&fileOId=4469151 (accessed on 20 August 2021).

26. Broberg, A. Potential för Biogasproduktion i Västra Götaland. Hushållningssällskapet (Potential of biogas production in Västra Götaland. The Rural Economy and Agricultural Societies), 2009. Available online: https://docplayer.se/134805689-Potential-for-biogasproduktion-i-vastra-gotaland.html (accessed on 10 June 2021).

27. Gustafsson, E. Lokal Handlingsplan för Bioenergi—Tingsryds Kommun. Energikontor Sydost (Local Handling Plan for Bioenergy—Municipality of Tingsryds. Energy Division), 2012. Available online: http://static.wm3.se/sites/2/media/25367_Lokal_handlingsplan_f%C3%B6r_bioenergi_Tingsryd_2012-05-09.pdf?1418646636 (accessed on 10 June 2021).

28. Sternberg, H. Underlag för Potentialberäkningar av Förnybar Energi (Basic Data for Potential Renewable Energy Estimations). County Administrative Board of Dalarna, Report 2013:03, 2013. Available online: https://www.lansstyrelsen.se/download/18.4df86bcd164893b7cd92d381/1534488489669/2013-03Underlagf%C3%B6rpotentialber%C3%A4kningaravf%C3%B6rnybarenergi.pdf (accessed on 10 June 2021).

29. De Toro, A.; Eckersten, H.; Nkurunziza, L.; von Rosen, D. *Effects of Extreme Weather on Yield of Major Arable Crops in Sweden*; Department of Energy and Technology, Swedish University of Agricultural Sciences: Uppsala, Sweden, 2015. Available online: https://pub.epsilon.slu.se/12606/8/deToro_a_etal_151109.pdf (accessed on 20 August 2021).

30. De Toro, A.; Gunnarsson, C.; Lundin, G.; Jonsson, N. Cereal harvesting—Strategies and costs under variable weather conditions. *Biosyst. Eng.* **2012**, *111*, 429–439. [CrossRef]

31. Gunnarsson, C.; de Toro, A.; Jonsson, N.; Lundin, G. Spannmålsskörd—Strategier och Kostnader vid Varierande Väderlek. In *JTI-Rapport. Lantbruk & industri, No. 403*; Jordbrukstekniska Institutet: Uppsala, Sweden, 2012. Available online: https://www.diva-portal.org/smash/get/diva2:959461/FULLTEXT01.pdf (accessed on 20 August 2021).

32. Donaldson, G.F. Allowing for Weather Risk in Assessing Harvest Machinery Capacity. *Am. J. Agric. Econ.* **1968**, *50*, 24–40. [CrossRef]

33. Van Kampen, J.H. Optimizing Harvesting Operations on a Largescale Grain Farm. Ph.D. Thesis, Wageningen University, Wageningen, The Netherlands, 30 May 1969. Available online: https://edepot.wur.nl/192243 (accessed on 20 August 2021).

34. Philips, P.R.; O'Callaghan, J.R. Cereal harvesting—A mathematical model. *J. Agric. Eng. Res.* **1974**, *19*, 415–433. [CrossRef]

35. Van Elderen, E. Models and techniques for scheduling farm operations: A comparison. *Agric. Syst.* **1980**, *5*, 1–17. [CrossRef]

36. Abawi, G.Y. A Simulation Model of Wheat Harvesting and Drying in Northern Australia. *J. Agric. Eng. Res.* **1993**, *54*, 141–158. [CrossRef]

37. Sokhansanj, S.; Mani, S.; Bi, X. Dynamic simulation of Mcleod Harvesting System for Wheat, Barley and Canola Crops. In Proceedings of the ASAE/CSAE Annual International Meeting, Ottawa, ON, Canada, 1–4 August 2004; Paper No. 048010. Available online: https://elibrary.asabe.org/abstract.asp?aid=17710 (accessed on 20 August 2021).

38. De Toro, A. *Assessment of Field Machinery Performance in Variable Weather Conditions Using Discrete Event Simulation*; Swedish University of Agricultural Sciences: Uppsala, Sweden, 2004; Volume 462, Available online: https://pub.epsilon.slu.se/553/ (accessed on 20 August 2021).

39. Nawi, N.M.; Chen, G.; Zare, D. The effect of different climatic conditions on wheat harvesting strategy and return. *Biosyst. Eng.* **2010**, *106*, 493–502. [CrossRef]

40. Nilsson, D.; Karlsson, S. A model for the field drying and wetting processes of cutflax straw. *Biosyst. Eng.* **2005**, *92*, 25–35. [CrossRef]

41. Nilsson, D.; Bernesson, S. *Halm som Bränsle. Del 2: Fuktegenskaper (Straw as Fuel. Part 2: Moisture Characteristics)*; Department of Energy and Technology, Swedish University of Agricultural Sciences: Uppsala, Sweden, 2009. Available online: https://pub.epsilon.slu.se/4854/ (accessed on 20 August 2021).

42. LantMet. Lantmet Väderstationer. (Ordnance Survey: Weather Stations). Available online: https://www.slu.se/fakulteter/nj/om-fakulteten/centrumbildningar-och-storre-forskningsplattformar/faltforsk/vader/lantmet/ (accessed on 10 June 2021).

43. SMHI. Extracting STRÅNG Data. (Swedish Meteorological and Hydrological Institute). Available online: http://strang.smhi.se/extraction/index.php (accessed on 10 June 2021).

44. Lehmann, D. Verlustvorgänge und schimmelbildung bei der trocknung und lagerung von halmfutter (Processes of decomposition and mold growth during drying and storage of straw forage). *Landtech. Forshung* **1971**, *19*, 180–187.
45. Sedlbauer, K. *Prediction of Mould Fungus Formation on the Surface of and Inside Building Components*; Fraunhofer Institute for Building Physics: Stuttgart, Germany, 2001. Available online: https://www.ibp.fraunhofer.de/content/dam/ibp/ibp-neu/de/dokumente/dissertationen/ks_dissertation_etcm45-30729.pdf (accessed on 20 August 2021).
46. Ramberg, D.T.; Jeppson, P.; Wang, A. Simulation of Relative Humidity and Temperature and Modeling of Mold Growth in Exterior Walls Insulated with Straw Bales in a Southern Swedish climate. In Proceedings of the BS2015: 14th Conference of International Building Performance Simulation Association, Hyderabad, India, 7–9 December 2015. Available online: http://www.ibpsa.org/proceedings/BS2015/p2859.pdf (accessed on 10 June 2021).
47. Hedlin, C.P. Sorption isotherms of five types of grain straw at 70 degrees F. *Can. Agric. Eng.* **1967**, *9*, 37–39. Available online: https://nrc-publications.canada.ca/eng/view/ft/?id=efb99a49-f9d7-4050-b4ca-8f155722704e (accessed on 10 June 2021).
48. Almeida, G.; Rémond, R.; Perré, P. Hygroscopic behaviour of lignocellulosic materials: Dataset at oscillating relative humidity variations. *J. Build. Eng.* **2018**, *19*, 320–333. [CrossRef]
49. Angus, J.F.; Mackenzie, D.H.; Morton, R.; Schafer, C.A. Phasic development in field crops. II. Thermal and photoperiodic responses of spring wheat. *Field Crop. Res.* **1981**, *4*, 269–283. [CrossRef]
50. Jordbruksverket. Hektar-och Totalskörd Efter län och Gröda. År 1965–2020. (Swedish Board of Agriculture. Hectare Yields and Total Production: County and Crop. Year 1965–2020). Available online: https://statistik.sjv.se/PXWeb/pxweb/sv/Jordbruksverkets%20statistikdatabas/Jordbruksverkets%20statistikdatabas__Skordar/JO0601J02.px/ (accessed on 19 August 2021).
51. Jordbruksverket. Normskörd i kg/ha efter län, Gröda och typ av Normskörd. År 2003–2021. (Swedish Board of Agriculture. Standard Yields in kg/ha per County, Crop, and Type of Standard Yield, Year 2003–2021). Available online: https://statistik.sjv.se/PXWeb/pxweb/sv/Jordbruksverkets%20statistikdatabas/Jordbruksverkets%20statistikdatabas__Skordar__Normskord/JO0602A01.px/?rxid=5adf4929-f548-4f27-9bc9-78e127837625 (accessed on 19 August 2021).
52. Swain, G. Desorption Equilibrium Moisture Content (EMC) of Straw. JTI—Swedish Inst. of Agric. and Envir. Eng., 1985, Report No. 63. Available online: http://www.diva-portal.org/smash/get/diva2:959801/FULLTEXT01.pdf (accessed on 10 June 2021).
53. Larsen, S.U.; Bruun, S.; Lindedam, J. Straw yield and saccharification potential for ethanol in cereal species and wheat cultivars. *Biomass Bioenergy* **2012**, *45*, 239–250. [CrossRef]
54. Suardi, A.; Stefanoni, W.; Bergonzoli, S.; Latterini, F.; Jonsson, N.; Pari, L. Comparison between Two Strategies for the Collection of Wheat Residue after Mechanical Harvesting: Performance and Cost Analysis. *Sustainability* **2020**, *12*, 4936. [CrossRef]
55. Hakala, K.; Heikkinen, J.; Sinkkob, T.; Pahkala, K. Field trial results of straw yield with different harvesting methods, and modelled effects on soil organic carbon. A case study from Southern Finland. *Biomass Bioenergy* **2016**, *95*, 8–18. [CrossRef]
56. Gunnarsson, C.; (Research Institutes of Sweden (RISE), Gothenburg, Sweden; Carina.gunnarsson@ri.se). Personal communication, 2018.
57. Johansson, F.; (Industrigatan 1, 523 78 Trädet, Sweden). Executive Manager/Agricultural Marketing Agent, Lundby Maskin Station. Personal communication, 2020.
58. Lundin, G. *Underlag för Bestämning av Leveranssäkerhet och Erforderlig Maskininsatsvid Halmbärgning (Basic Data for Determination of Delivery Security and Required Machine Input for Straw Recovery)*; Unpublished material; 1984.
59. Nilsson, D. *Bärgning, Transport, Lagring och Förädling av Halm till Bränsle; Metoder, Energibehov, Kostnader (Harvesting, Transport, Storage and Upgrading of Straw as a Fuel—Methods, Energy Needs, Costs)*; Department of Energy and Technology, Swedish University of Agricultural Sciences: Uppsala, Sweden, 1991. Available online: https://pub.epsilon.slu.se/3893/1/nilsson_d_090904.pdf (accessed on 19 August 2021).
60. Brundin, S. *Fastbränsle från Jordbruket. Kostnader för Halm- och Gräsbränslesystem (Solid Fuels from Agriculture. Straw and Grass Feedstock Costs for Energy)*; Swedish University of Agricultural Sciences: Uppsala, Sweden, 1988.
61. Brück, I.G.M.; Van Elderen, E. Field drying of hay and wheat. *J. Agric. Eng. Res.* **1969**, *14*, 105–116. [CrossRef]
62. Thompson, N. Modelling the field drying of hay. *J. Agric. Sci.* **1981**, *97*, 241–260. [CrossRef]
63. Smith, E.A.; Duncan, E.J.; McGechan, M.B.; Haughey, D.P. A model for the field drying of grass in windrows. *J. Agric. Eng. Res.* **1988**, *41*, 251–274. [CrossRef]
64. Atzema, A.J. A model for the drying of grass with real time weather data. *J. Agric. Eng. Res.* **1992**, *53*, 231–247. [CrossRef]
65. Ref-ET: Software. Reference Evapotranspiration Calculator; Version 4.1. University of Idaho, 2016. Available online: https://www.uidaho.edu/cals/kimberly-research-and-extension-center/research/water-resources/ref-et-software (accessed on 10 June 2021).

 *sustainability*

Article

Pyrolysis of Solid Digestate from Sewage Sludge and Lignocellulosic Biomass: Kinetic and Thermodynamic Analysis, Characterization of Biochar

Aleksandra Petrovič [1,*], Sabina Vohl [1], Tjaša Cenčič Predikaka [2], Robert Bedoić [3], Marjana Simonič [1], Irena Ban [1] and Lidija Čuček [1]

[1] Faculty of Chemistry and Chemical Engineering, University of Maribor, Smetanova ul. 17,
    2000 Maribor, Slovenia; sabina.vohl@um.si (S.V.); marjana.simonic@um.si (M.S.); irena.ban@um.si (I.B.);
    lidija.cucek@um.si (L.Č.)
[2] IKEMA d.o.o., Institute for Chemistry, Ecology, Measurements and Analytics, Lovrenc na Dravskem polju 4,
    2324 Ptuj, Slovenia; tjasa@ikema.si
[3] Faculty of Mechanical Engineering and Naval Architecture, University of Zagreb, Ul. Ivana Lučića 5,
    10000 Zagreb, Croatia; robert.bedoic@fsb.hr
*   Correspondence: aleksandra.petrovic@um.si

Citation: Petrovič, A.; Vohl, S.; Cenčič Predikaka, T.; Bedoić, R.; Simonič, M.; Ban, I.; Čuček, L. Pyrolysis of Solid Digestate from Sewage Sludge and Lignocellulosic Biomass: Kinetic and Thermodynamic Analysis, Characterization of Biochar. *Sustainability* **2021**, *13*, 9642. https://doi.org/10.3390/su13179642

Academic Editor: Idiano D'Adamo

Received: 21 July 2021
Accepted: 24 August 2021
Published: 27 August 2021

Abstract: This study investigates the pyrolysis behavior and reaction kinetics of two different types of solid digestates from: (i) sewage sludge and (ii) a mixture of sewage sludge and lignocellulosic biomass—*Typha latifolia* plant. Thermogravimetric data in the temperature range 25–800 °C were analyzed using Flynn–Wall–Ozawa and Kissinger–Akahira–Sunose kinetic methods, and the thermodynamic parameters (ΔH, ΔG, and ΔS) were also determined. Biochars were characterized using different chemical methods (FTIR, SEM–EDS, XRD, heavy metal, and nutrient analysis) and tested as soil enhancers using a germination test. Finally, their potential for biosorption of NH_4^+, PO_4^{3-}, Cu^{2+}, and Cd^{2+} ions was studied. Kinetic and thermodynamic parameters revealed a complex degradation mechanism of digestates, as they showed higher activation energies than undigested materials. Values for sewage sludge digestate were between 57 and 351 kJ/mol, and for digestate composed of sewage sludge and *T. latifolia* between 62 and 401 kJ/mol. Characterizations of biochars revealed high nutrient content and promising potential for further use. The advantage of biochar obtained from a digestate mixture of sewage sludge and lignocellulosic biomass is the lower content of heavy metals. Biosorption tests showed low biosorption capacity of digestate-derived biochars and their modifications for NH_4^+ and PO_4^{3-} ions, but high biosorption capacity for Cu^{2+} and Cd^{2+} ions. Modification with KOH was more efficient than modification with HCl. The digestate-derived biochars exhibited excellent performance in germination tests, especially at concentrations between 6 and 10 wt.%.

Keywords: digestate; pyrolysis; kinetics; thermogravimetric analysis; biochar characterization; germination test; biosorption

1. Introduction

The continuous growth of the human population is correlated with an increase in primary energy consumption, where the main sources of energy are (still) of fossil origin, and are responsible for the majority of greenhouse gas (GHG) emissions into the atmosphere [1]. Renewable energy sources, such as solar [2], wind, geothermal, hydropower energy, and energy recovered from biomass and different wastes [3], are promising alternatives to fossil fuels, offering solutions to the above challenges. Renewable energy is environmentally friendly [4] and more sustainable than non-renewable energy [5].

Many researchers address issues related to energy resources and energy recovery in their studies. Some deal with waste-to-energy recovery and sustainable waste management [6], others attempt to find the right programming approach, with an optimal mix of

117

power generation for socioeconomic sustainability [7]. Others aim for progress towards circular economy models that optimize the use of renewable energy (e.g., biomethane from waste) [8]. Promoting the production of renewable resources and converting them into valuable products and bioenergy to satisfy sustainable development is also the goal of the European Bioeconomy Strategy, which was accepted by the European Commission [9]. The bioeconomy aims to replace non-renewable resources with bio-based alternatives, emphasizing the introduction of bio-based energy and material to reduce environmental risks [10].

One of the biggest environmental challenges, in addition to increasing energy consumption, is the problem of large quantities of sewage sludge generated during the operation of wastewater treatment plants as a result of increasing demand for clean water [11]. The most common disposal processes for sewage sludge are landfills, agricultural applications, and incineration [1]. Alternative, more environmentally friendly processes should be developed, due to stricter regulations and the environmental impacts associated with sewage sludge. Since sewage sludge has a relatively high calorific value and organic matter content, its waste-to-energy valorization technologies, such as anaerobic digestion [12], hydrothermal carbonization [13] and pyrolysis, are gaining attention [14].

Lignocellulosic biomass is recognized as one of the most sustainable alternative energy sources that contributes considerably to the reduction of GHG emissions [15]. Different types of lignocellulosic biomass can be used for energy recovery. The plant *Typha latifolia* (cattail) is one of the lignocellulosic feedstocks with high potential for energy recovery due to its special characteristics, such as high carbon content, high C:N ratio, and high yield due to rapid growth [16]. The *T. latifolia* plant grows on marginal lands and wetlands around the world, making it a low-cost biomass resource. Despite the significance of *T. latifolia*, there are not many studies related to energy recovery from it. Ciria et al. made an assessment of its potential as a biomass fuel by thermochemical characterization of a wetland produced biomass [17], while Grosshans studied the compression of cattail into compacted fuel products, wherein the combustion experiments showed that its calorific value is comparable to that of commercial wood pellets [18]. Hu et al. studied the potential of *T. latifolia* for methane production by anaerobic mono-digestion [19], and its performance in biogas production when co-digested with sewage sludge, including nutrient recovery from the obtained digestate, has also been investigated in an earlier study by the authors of the current work [20]. The efficiency of hydrothermal carbonization to produce hydrochar from cattail [21] and cattail digestate [22], and liquefaction processes to produce bio-oil [23] were also examined. Ahmad et al. studied the pyrolytic behavior of cattail and its thermal degradation process [24].

Various processes have been applied to convert biomass into energy, such as thermochemical processes including incineration, pyrolysis, torrefaction, hydrothermal carbonization and liquefaction [4], and biological processes, such as anaerobic digestion [19]. Of these technologies, pyrolysis and anaerobic digestion are among the most promising methods for conversion of sewage sludge [11], as well as lignocellulosic biomass into valuable products [15]. The coupling of anaerobic digestion and pyrolysis in an integrated process provides an opportunity to obtain higher bioenergy recovery compared to single processes [25], especially when using lignocellulosic biomass [15].

Anaerobic digestion is the process in which biomass, with the help of anaerobic microorganisms, is converted into biogas, mostly methane, which can be used for heat and/or electricity generation [11] or upgraded to biomethane [26]. The enormous potential biomethane production represents a sustainable way towards the decarbonization of the transport sector [27]. Anaerobic digestion and methane production can be enhanced by pretreatment of feedstocks or the addition of natural enzymes and microorganisms, such are those in cattle rumen fluid [19]. A by-product of anaerobic digestion is digestate, which can be applied as fertilizer as it contains valuable nutrients, such as nitrogen, phosphorus and potassium for plant growth, although the possible presence of pathogens and heavy metals could limit such application [28]. The separation of digestate into solid and liquid parts is also possible, where the solid part can be used in the pyrolysis process [25]. Pyrolysis is an endothermic process that occurs in an inert atmosphere, during which biomass is converted into three fractions: char, oil, and a gaseous fraction that is represented mainly by CO_2, H_2 and CO [15]. During the degradation process, the organic matter undergoes a series of complex reactions, generating volatile products and condensed molecules, which finally leads to char formation [29]. The pyrolysis process and the characteristic of the products depend on various factors, especially the pyrolysis temperature, and the type and composition of biomass used [30]. Cellulose and hemicellulose in the lignocellulosic materials mostly contribute to bio-oil production, while lignin mainly contributes to biochar formation [31]. Lignocellulosic feedstocks usually require pre-treatment to enhance the pyrolysis efficiency, where chemical, thermal, or biological methods can be applied [15]. Pyrolysis is a particularly promising technology for sewage sludge management due to the reduction of sewage sludge volume, stabilization of heavy metals in the solid residue [32], and elimination of pathogens [33]. Compared to combustion, pyrolysis appears to be less polluting as most hazardous trace elements are retained in the biochar [34].

Thermogravimetric analysis (TGA) is widely employed to investigate the behavior of biomass during pyrolysis and the related degradation mechanisms [35]. The pyrolysis behavior of various biomasses and waste materials has been explored in detail using TGA, such as that of sewage sludge, animal manure [36], rice husks [33], miscanthus [37], and others. To determine the kinetic and thermodynamic parameters of the pyrolysis reaction, iso-conversional methods, such as the Flynn–Wall–Ozawa (FWO), Kissinger–Akahira–Sunose (KAS), and Friedman methods could be applied, besides model-fitting methods, such as the Coats–Redfern method [38]. The advantage of iso-conversional methods is that they do not require prior knowledge of the reaction mechanism [4]. Various studies have attempted to describe the kinetic and thermodynamic behavior of sewage sludge pyrolysis [39–41], while only a few were dedicated to the pyrolysis kinetics of sewage sludge digestates [34,42,43]. The FWO and Vyazovkin kinetic models have been used to determine the activation energy of pyrolysis of sewage sludge digestate or co-digestate of sewage sludge and grease waste, although the thermodynamic parameters have not been determined [42]. In another study, a FWO model was applied to describe the combustion of sewage sludge digestate [43]. The nth-order reaction model was used to calculate the activation energy and pre-exponential factor for sewage sludge digestate pyrolysis and combustion [34]. In contrast to sewage sludge digestates, kinetic studies dedicated to lignocellulosic digestates [25,44] and swine manure digestate [45] are more widely available.

Although pyrolysis is primarily intended for energy valorization, it has the added benefit of char production as a valuable carbon product. In recent years, a number of studies have been published on the characterization of biochars [46], the impact of feedstock type [47–49], and the operation conditions [50], including the pyrolysis temperature [51,52] on the properties and quality of the resulting biochars. Special attention was paid to the study of the impact of pyrolysis conditions on the toxicity and environmental safety of potentially toxic elements (heavy metals and others) in the biochars [53]. Sewage sludge and its solid digestate are promising feedstocks for the production of low-cost biochar that can be applied for various purposes, such as adsorbent or soil enhancer [46]. To improve the quality of biochar, sewage sludge can be co-pyrolyzed with other organic biomass, for example manure [36], rice husks [54], or any other biomass.

Biochar can be used for a variety of purposes, such as carbon sequestration [55], soil improvement as a fertilizer, pollution remediation, and with proper modification it can be used as a catalyst or supercapacitor [56]. Regarding the biosorption potential of biochars, studies reveal that sewage sludge derived biochars are effective in adsorption of phosphorus [57], ammonium and heavy metals [52], polycyclic aromatic hydrocarbons (PAHs), emerging organic pollutants (EOPs) [56], and other micropollutants from wastewater [35]. Since the sorption ability of sewage sludge-derived biochars may be relatively low compared to that of other biochars, modifications, such as chemical treatment can be applied to improve their sorption capacity. Modification with KOH improved the biosorption of heavy metals by sewage sludge digestate-derived biochar [58], HCl, and $FeCl_3$ were effective in modification of wheat straw biochar tested for ammonium biosorption [59], while the impregnation of sewage sludge biochar with Mg, Ca, Al, Cu, and Fe demonstrated the better sorption ability of phosphorus [60]. N-doped biochars proved to be successful in removing emerging organic pollutants [56].

Biochar also contains significant amounts of micro- and macro-nutrients, which makes it valuable as a soil amender. Several researchers have studied the potential of sewage sludge biochars as soil amenders, and obtained quite diverse results, from a negative influence on plant growth due to heavy metal toxicity [61], to a positive impact due to nutrient enrichment of the soil [62,63]. Therefore, each biochar should be carefully evaluated for its own specific characteristics before being used for a particular purpose.

Research Motivation and Paper Organization

Several factors motivated us to conduct the research reported on in this paper; the literature review revealed that there are knowledge gaps in many of the areas mentioned above. For example, there are no studies on the pyrolysis of *T. latifolia* digestate or its co-digestate with other biomass, such as sewage sludge, and no studies on the co-pyrolysis of this plant with other biomass. The data for kinetic and thermodynamic parameters are also lacking. In addition, there is limited information about the potential of sewage sludge digestate derived biochars for soil improvement and their impact on seed germination, and none about biochar derived from *T. latifolia*. Furthermore, biosorption studies with sewage sludge biochars are usually performed with only one ion species, while studies with different types of ions and different biochar modifications are less common. In order to fill the knowledge gaps mentioned above, this study investigated the thermogravimetric behavior of two solid fractions of digestates obtained from anaerobic digestion. The first digestate was obtained from mono-digestion of sewage sludge, while the second digestate was obtained from co-digestion of sewage sludge and *Typha latifolia* (1:1 ratio). For comparison, the analysis of undigested feedstocks was carried out. Kinetic analysis was performed by applying two iso-conversional methods, KAS and FWO. The obtained biochars were characterized by several analytical methods, wherein elemental, heavy metal, and nutrient analysis, FTIR, SEM–EDS, and XRD analyses were performed. Moreover, experiments were conducted on the further applicability of the digestate derived biochars. The fertility potential of cress seeds exposed to different biochar concentrations was studied, and the adsorption potential for biosorption of NH_4^+, PO_4^{3-}, Cu^{2+}, and Cd^{2+} ions by unmodified and chemically modified biochars was evaluated.

Several novelties are introduced by this work. To the best of the authors' knowledge, the pyrolysis kinetics of digestate composed of sewage sludge and the lignocellulosic plant *T. latifolia* were investigated for the first time. Thermodynamic parameters, such as ΔH, ΔG, and ΔS were determined as well, which cannot be found in the literature for this kind of digestate. A significant novelty is represented by the data obtained in the germination and biosorption tests, especially those from biochar modification, which bring valuable information on the possible use of the obtained biochars in agriculture, water treatment, and for other purposes.

The paper is organized as follows: Section 1 presents the study background and motivation for the research. The materials and methods used in the experiments and the kinetic models used in the kinetic analysis are presented in Section 2. In Section 3, the results of the characterization of feedstocks and products (including the results of biosorption and germination tests), as well as the results of thermogravimetric, kinetic, and thermodynamic analyses are discussed. Section 4 summarizes the main conclusions of the work and presents some directions for future research.

2. Materials and Methods

In this section, first, the methods for sample preparation and characterization methods are presented; further, the procedure for TGA is presented, and the models used in kinetic and thermodynamic analyses are introduced. Finally, the procedures for the biosorption and cress seed germination tests using digestate-derived biochars are described.

2.1. Preparation and Characterization of Feedstocks and Products

TGA experiments were conducted on two different solid fractions of digestates, digestate from mono-digestion of sewage sludge and from co-digestion of sewage sludge and the *T. latifolia* plant. In addition, raw sewage sludge and *T. latifolia* were analyzed for comparison.

2.1.1. Feedstocks Preparation

The solid fractions of digestates were obtained from anaerobic digestion experiments performed in 1 L batch reactors under mesophilic conditions (42 °C) with a retention time of 50 days. Digestate D1 was obtained from mono-digestion of sewage sludge, while digestate D2 was from co-digestion of sewage sludge and *Typha latifolia* plant (cattail). The composition of the samples on a dry matter basis (d.m.) used in the anaerobic digestion from which digestates D1 and D2 were obtained is shown in Table S1 in the supplementary material.

The ratio between substrate and inoculum was 1:1 (15:15 g on a dry matter basis). To both samples, 50 mL of cattle rumen fluid was added to promote fermentation and degradation of the lignocellulosic components. The mixtures were diluted with a buffer solution [64] to achieve a dry matter content of 6 wt.% in each reactor. The dewatered sewage sludge sample was collected from a local municipal wastewater treatment plant with tertiary biological treatment of wastewater with the capacity of 68,000 PE. *Typha latifolia* was gathered near a small river in the eastern part of Slovenia and cut into small pieces (0.5 cm × 0.5 cm). The inoculum was obtained from a biogas plant producing biogas from poultry manure. Cattle rumen fluid was acquired from a nearby slaughterhouse. The results of biogas production by anaerobic digestion are presented in our previous work [20]. After the anaerobic digestion process was stopped, the obtained digestates were separated into two parts, liquid and solid fractions, by centrifugation (Eppendorf 5804 R centrifuge, 7500 rpm, 8 min). The solid fractions of digestates D1 and D2 were dried at 105 °C in a laboratory dryer until constant weight, then ground and stored in a desiccator until further use in the thermogravimetric study. Undigested sewage sludge (SS) and *T. latifolia* were likewise dried and ground before being used in TGA.

2.1.2. Characterization of Feedstocks and Biochars

The basic characteristics of the feedstocks (digestates D1 and D2, raw sewage sludge, and *T. latifolia* plant) and their biochars were determined, such as proximate, ultimate, and heavy metal analyses. Moisture and dry matter content were determined according to the corresponding standard [65]. Ash content was determined as mass percentage of residues after combustion of the samples at 800 °C in a furnace for 4 h. Volatile matter (VM) was determined by measuring the weight loss after combustion of the samples in a furnace at 900 °C for 1 h. The fixed carbon (FC) was calculated as:

$$\text{FC (wt.\%)} = 100 - \text{VM} - \text{Ash} \tag{1}$$

Higher heating value (HHV) was determined experimentally by combustion of the samples in a bomb calorimeter calibrated by combustion of certified benzoic acid [66]. Besides experimental values, the theoretical HHV values were also calculated. Several correlation models were established to estimate the HHV of biomass using the proximate values of biomass. The following equation was used in this study [67]:

$$HHV\left(\frac{\text{MJ}}{\text{kg}}\right) = 0.3491 \cdot \text{C} + 1.1783 \cdot \text{H} + 0.1005 \cdot \text{S} - 0.1034 \cdot \text{O} - 0.0151 \cdot \text{N} - 0.0211 \cdot \text{Ash} \tag{2}$$

where C, H, N, O, S, and Ash are the dry basis weight percentages of carbon, hydrogen, nitrogen, oxygen, sulfur, and ash in the solid samples.

The Elemental Analyzer PerkinElmer Series II 2400 was used to determine the carbon, hydrogen, nitrogen, and sulfur contents. The oxygen content was calculated as:

$$O = 100 - C - H - N - S - Ash \ (all \ in \ wt.\%) \tag{3}$$

Before and after TGA, the content of heavy metals and elements K^+, Ca^{2+}, and Mg^{2+} was measured in the samples by inductively coupled plasma–optical emission spectrometry, ICP–OES [68]. The pH value of the biochars was determined as the pH value of the solution containing biochar at the mixing ratio biochar/deionized water = 1:20 (w/v).

The feedstocks and biochars were characterized using Fourier-transform infrared spectroscopy (FTIR) to study the functional groups present on the sample surface. For FTIR analysis, each dry sample was mixed with KBr (at a ratio of 1:30) and pressed into tablet form. The FTIR spectra were then recorded using a Shimadzu IRAffinity FTIR spectrophotometer (Japan). Scanning Electron Microscopy (SEM) analysis and X-ray powder diffraction (XRD) analysis were additionally used to characterize the biochars. XRD analysis was performed at room temperature with parameters of 2Theta from $10°$ to $70°$ and a scan rate of $0.033° \ s^{-1}$ (XRD, D-5005 diffractometer of Bruker Siemens manufacturer, Karlsruhe, Germany). SEM combined with Energy Dispersive X-ray Spectroscopy (EDS) analysis was used to identify the surface morphology and composition of the biochars. The specimens were observed using a Sirion 400 FEI scanning electron microscope equipped with an energy dispersive microanalysis system (EDS Oxford INCA 350). The surface area of biochars and the pore size were determined using N_2 adsorption by means of a Micromeritics Tristar II 3020 porosimeter. The Brunauer–Emmett–Teller (BET) method was used for surface area determination and the Barrett, Joyner, and Halenda (BJH) model for pore size distribution.

2.2. Thermogravimetric Analysis (TGA) and Pyrolysis of Feedstocks

TGA was performed on the following feedstocks: digestates D1 and D2, undigested sewage sludge (SS) and *T. latifolia* plant (TLP). Biochar derived from undigested sewage sludge was designated as "B-SS", biochar from *T. latifolia* plant as "B-TLP", biochar from digestate D1 as "B-D1", and biochar from digestate D2 as "B-D2" (see Table 1).

Table 1. Feedstocks and experimental conditions used in the thermogravimetric analysis.

Feedstock	Biochar Mark	Pyrolysis Conditions	Tested Heating Rates
Sewage sludge digestate (D1) *	B-D1	25–800 °C, inert atmosphere (nitrogen flow of 100 mL/min)	15, 30, and 100 °C/min
Digestate of sewage sludge mixture with *T. latifolia* plant (D2) *	B-D2		
Sewage sludge (SS)	B-SS		
T. latifolia plant (TLP)	B-TLP		

* Solid fraction of digestate.

The TGA studies were carried out using the TGA/SDTA851e thermogravimetric analyzer (Mettler Toledo) in the temperature range from 25 to 800 °C under an inert atmosphere, ensured by a constant nitrogen flow of 100 mL/min. Samples weighing about 25 ± 1 mg, were exposed to the slow pyrolysis process at the following heating rates β: 15, 30, and 100 °C/min. These heating rates were chosen to cover as wide a range of the "slow pyrolysis" area as possible and promote the formation of solid biochar as the main product, rather than the liquid product normally formed at higher heating rates. From the results of TGA the TG curves (mass weights vs. temperatures) and derivative (DTG) curves were constructed using the MS Excel software tool.

For the germination and biosorption tests and characterization studies (XRD, FTIR, and SEM–EDS analyses), the biochar samples were obtained by pyrolysis of feedstocks at 800 °C in a tube furnace under an inert atmosphere at a heating rate of 15 °C/min. The biochars, after achieving the desired temperature, were kept in a furnace for another 30 min under the same conditions. After cooling to room temperature, the biochars were stored in a desiccator until further use.

2.3. Kinetic and Thermodynamic Analysis

The kinetic study was performed using the Kissinger–Akahira–Sunose (KAS) and Flynn–Wall–Ozawa (FWO) models. The thermodynamic analysis was further carried out based on the obtained kinetic parameters from the KAS and FWO models.

2.3.1. Kinetic Models

For the kinetic analysis of the thermogravimetric data obtained in this study, the KAS and FWO models were used, since they are less susceptible to errors than differential iso-conversional methods, such as the Friedman method [69]. Since both models have been explained in detail in the literature, only the final expression of the temperature integral is presented here. The FWO kinetic model, which uses the Doyle equation to approximate the temperature integral, is described by Equation (4) [70]:

$$\ln[\beta] = \ln\left[\frac{A \cdot E_\alpha}{R \cdot g(\alpha)}\right] - 5.331 - 1.052\frac{E_\alpha}{R \cdot T} \tag{4}$$

The KAS kinetic model [37] is given by Equation (5):

$$\ln\left[\frac{\beta}{T^2}\right] = \ln\left[\frac{R \cdot A}{E_\alpha \cdot g(\alpha)}\right] - \frac{E_\alpha}{R \cdot T} \tag{5}$$

To determine the kinetic parameters for the selected conversion point (α), the left sides of Equations (4) and (5) were plotted on the y-axis against the $(-1/RT)$ on the x-axis. The activation energy E_α was then calculated from the value of the slope of the linear plots using the KAS and FWO methods.

Since iso-conversional methods are often limited to estimate the pre-exponential factor A and predict the reaction model, Kissinger developed a model-free non-isothermal equation to determine the pre-exponential factor [41], described by Equation (6):

$$A = \left[\beta \cdot E_\alpha \cdot \exp\left(\frac{E_\alpha}{RT_p}\right)\right] / \left(RT_p^2\right) \tag{6}$$

2.3.2. Thermodynamic Parameters

The thermodynamic parameters of biomass decomposition such as the change in enthalpy ΔH (kJ/mol), Gibbs free energy ΔG (kJ/mol), and entropy ΔS (kJ/mol·K), can be calculated based on the previously obtained kinetic parameters using Equations (7)–(9) [54]:

$$\Delta H = E_\alpha - RT \tag{7}$$

$$\Delta G = E_\alpha + RT_p \ln\left(\frac{K_B T_p}{hA}\right) \tag{8}$$

$$\Delta S = \frac{\Delta H - \Delta G}{T_p} \tag{9}$$

where K_B represents Boltzmann constant (1.381×10^{-23} J/K), T_p represents peak temperature of the DTG curve (K) at a given heating rate, and h represents the Planck constant (6.626×10^{-34} Js) [71].

2.4. Cress Seed Germination Test

Since seed germination is a critical step in a plant's life cycle [72], the cress seed germination test was conducted to investigate the potential of digestate-derived biochars (B-D1 and B-D2) for use as soil enhancers and to evaluate their toxicity to plants. To examine the response of plants to the obtained biochars according to the corresponding standard [73], 10 cress seeds (*Lepidium sativum* L.) were placed in each petri dish containing peat and then exposed to different concentrations of biochars for 72 h under controlled conditions (25 °C, absence of light). The following concentrations of biochars B-D1 and B-D2 were tested: 2, 6, 10, and 15 wt.%. Experiments were performed in triplicate for each concentration. A control sample containing water-soluble fertilizer with essential macronutrients ($N:P_2O_5:K_2O = 15:10:20$, concentration of 1.5 g/L) was also prepared to compare the results. Based on the results of the growth test, the root length (RL) index was calculated using the equation described by Chemetova et al. [74] and Munoo-Liisa vitality (MLV) index, using the equation given by Maunuksela et al. [75].

2.5. Adsorption Tests

The biosorption potential of the digestate-derived biochars (B-D1 and B-D2) was evaluated by an adsorption test. The adsorption of NH_4^+, PO_4^{3-}, Cd^{2+}, and Cu^{2+} ions was studied at pH 7 and at constant temperature (22 °C). The pH 7 was chosen because the pH of aquatic solutions or wastewater is usually close to the neutral value. Experiments were conducted in 100 mL conical flasks containing 0.05 g of biochar and 50 mL of water solution with the initial ion concentration of 50 mg/L. The flasks were placed on an orbital shaker and shaken at 200 rpm for 24 h. The Cd^{2+} and Cu^{2+} contents in the solution were determined by ICP–OES, while the NH_4^+ and PO_4^{3-} contents were determined spectrophotometrically using the standards DIN 38 406-E5-1 [76] and SIST EN ISO 6878:2004 [77]. Before analyses, samples were filtered through 0.45 µm filters.

To enhance biosorption capacity, the biochars were chemically modified with 2 mol/L KOH or HCl solution. For this purpose, 2 g of biochar was exposed to 50 mL of modification solution, which was shaken for 2 h. After modification, the biochar was rinsed several times with distilled water and dried at 105 °C before being used in the adsorption tests. Three different types of each biochar (each in two parallel runs) were tested for adsorption: unmodified, HCl modified, and KOH modified biochar. The removal efficiency of particulate ion species from the aqueous solution and the amount of ion adsorbed on the biochar (biosorption capacity) were calculated from the reduction of ion concentration in the solution using standard equations described in Tang et al. [78].

3. Results and Discussion

In this section, the proximate and ultimate analyses of the feedstock materials are presented and the results of the TGA are introduced. Further, the results of the kinetic analysis by applying KAS and FWO kinetic models are presented, and the thermodynamic parameters are noted. Finally, the results of the characterization of the pyrolysis products are described.

3.1. Characterization of Feedstock Materials

The results of proximate and ultimate analyses, heavy metals, and other parameters for the feedstocks used in this study (sewage sludge, *T. latifolia* plant, and digestates D1 and D2) are shown in Table 2.

In general, the digestates have higher ash content and lower volatile matter, carbon, and hydrogen content than undigested SS and TLP. Solid digestates D1 and D2 contained 58% and 60% volatiles, while undigested SS contained 71% and TLP 79%. The lower percentage of volatiles is a consequence of pre-treatment with the anaerobic digestion. Ash content was highest in the digestates (D1–36%, D2–31%), lower in SS (18%) and lowest in TLP (7%). On the other hand, TLP contained the highest content of fixed carbon (13%), followed by SS (11%) and both digestates (D1–7%, D2–9%).

Table 2. Proximate, ultimate, and heavy metal analysis of the feedstocks.

Parameter	Sewage Sludge (SS)	*T. latifolia* Plant (TLP)	Digestate D1 [a]	Digestate D2 [a]
Dry matter (wt.%)	17.42	19.77	12.16	14.38
Moisture content (wt.%)	82.58	80.23	87.84	85.62
Volatile matter (wt.%) [b]	70.99	79.42	57.87	60.18
Ash (wt.%) [b]	18.48	7.31	35.66	31.08
FC (wt.%) [b]	10.52	13.27	6.47	8.74
HHV (MJ/kg)	19.91	17.02	12.75	12.64
$HHV_{theoretical}$ (MJ/kg)	19.93	20.50	13.01	13.30
C (wt.%)	42.30	45.79	31.90	34.83
H (wt.%)	6.75	7.10	4.15	3.73
N (wt.%)	8.10	3.63	4.35	3.58
S (wt.%)	1.16	0.49	1.35	1.09
O (wt.%)	23.21	35.68	22.59	25.69
H/C [c]	1.91	1.86	1.56	1.29
O/C [c]	0.41	0.58	0.53	0.55
N/C [c]	0.16	0.07	0.12	0.09
P (wt.%)	2.61	0.57	1.01	0.86
Ca (wt.%)	2.04	1.45	5.35	4.87
Mg (wt.%)	0.82	0.32	0.34	0.26
K (wt.%)	0.82	3.44	1.15	1.18
Si (wt.%)	0.14	0.04	0.09	0.05
Fe (wt.%)	0.88	0.18	0.31	0.21
Cd (mg/kg d.m.)	1.03	1.05	<1	<1
Cr_{total} (mg/kg d.m.)	45.15	1.42	38.58	30.69
Cu (mg/kg d.m.)	173.61	6.24	165.59	153.10
Ni (mg/kg d.m.)	25.35	1.68	17.10	13.07
Pb (mg/kg d.m.)	26.66	2.10	15.85	12.04
Zn (mg/kg d.m.)	740.79	26.73	596.36	489.40

[a] Solid fraction, [b] on a dry basis, [c] molar ratio.

Elemental analysis revealed higher content of C in raw SS and TLP (42 and 46%) than in digestates (D1-32%, D2-35%). Undigested SS contained around 8% N, while TLP and digestates contained about half of this. The content of sulfur in the samples was low. The content of heavy metals was highest in the case of undigested SS, while digestates D1 and D2 contained slightly lower content of heavy metals. Among the heavy metals detected, the highest concentrations belonged to Zn and Cu, with Ni, Pb, Cr, and Cd also detected. Otherwise, the raw SS satisfies the limit values of heavy metals set in the Slovenian decree on the use of sewage sludge in agriculture [79] and, thus, could be used for agricultural purposes. TLP exhibited low content of heavy metals, and a high content of K^+ ions. The SS and digestates were rich in nitrogen as well as other nutrients, such as P, Mg, and Ca. Therefore, they could potentially be used as alternative sources for nutrient recovery or as soil enhancers. Further comparison of the digestates D1 and D2 revealed that digestate D2 contained lower content of H, N, and S elements, a lower amount of ash and heavy metals, and higher content of carbon, fixed carbon, and volatile matter.

The calorific value, i.e., higher heating value (HHV) reflects the amount of energy that can be released from a form of biomass when it is subjected to combustion, therefore, the determination of HHV is important as it provides valuable information regarding the bioenergy potential of the biomass [80]. The experimental HHV of both digestates was ~13 MJ/kg, which is lower than the HHV of sewage sludge (20 MJ/kg) and TLP (17 MJ/kg). Both sewage sludge and TLP show similar calorific values as reported in the literature. Values between 11 and 22 were reported for SS [81], and the value of 18 MJ/kg was found for TLP [24]. The values are comparable with the values of other energy crops such as miscanthus (19 MJ/kg) and wheat straw (16 MJ/kg) [81]. Regarding the HHV of SS digestates, different values were found, from relatively high, 17 MJ/kg (Aragon-Briceno 2017), 18 MJ/kg [34] and 16 MJ/kg [43], to relatively low, 13 MJ/kg [42]. The theoretical HHV were also calculated using Equation (2) developed by Channiwala and Parikh [67]. The agreement between the experimental and theoretical values for SS and the digestate samples was good, while for TLP, the difference between the values was more significant.

However, based on their properties, the tested feedstocks have promising potential to be applied in further thermal degradation processes for energy recovery. Since they have quite diverse compositions, the characteristics and quality of the final products can vary greatly.

3.2. Thermogravimetric Analysis

3.2.1. Analysis of TG and DTG Curves

Figure 1 shows TG and DTG curves for the analyzed feedstocks: digestates D1 and D2, raw sewage sludge and lignocellulosic *T. latifolia*. The curves for three different heating rates are shown: 15, 30, and 100 °C/min. The curve of TG represents the mass loss with respect to temperature and the curve of DTG represents the rate of mass loss with respect to temperature at a chosen heating rate.

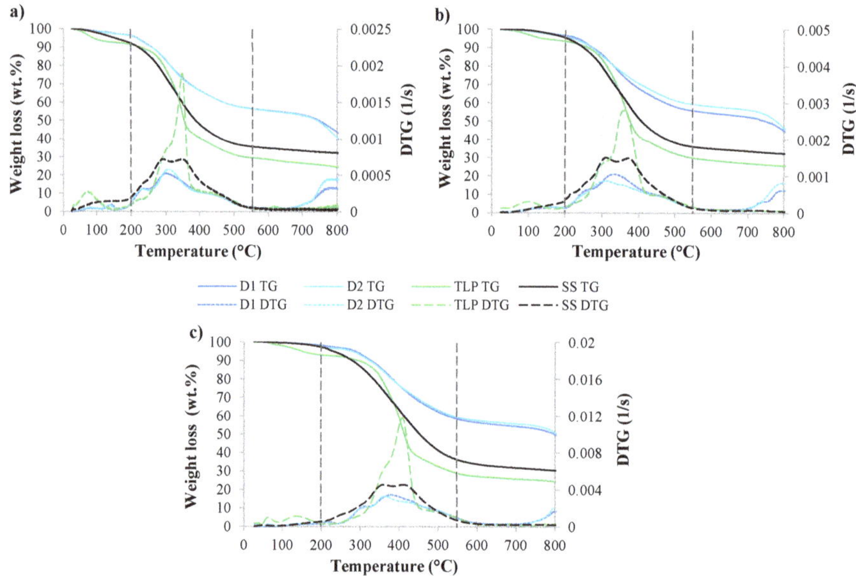

Figure 1. TG and DTG curves of sewage sludge (SS), *T. latifolia* plant (TLP) and digestate samples D1 and D2 at heating rates: (**a**) 15 °C/min, (**b**) 30 °C/min, and (**c**) 100 °C/min.

Generally, the TG and DTG curves of the selected samples show similar characteristics at all three heating rates. The degradation of SS and TLP starts at lower temperatures (~150 °C) than the degradation of digestates (~200 °C), and it also ends earlier, at around 650 °C. The digestate samples show lower weight loss than the raw SS or TLP, implying that the AD pre-treatment has a significant effect on the thermal degradation of the biomass. The overall weight loss was highest for the TLP (75.1 wt.% on average), lower for SS (68.4 wt.%), and lowest for digestates (D1–54.1 wt.% and D2–54.6 wt.%). As TLP loses more weight than SS or digestates, more volatile matter is decomposed, so higher oil and gas yields than biochar yields are expected for this feedstock. The digestates gave the highest residue (45 wt.%) and showed quite similar TG and DTG profiles despite different feedstock compositions.

The thermogravimetric data revealed that the decomposition of the tested feedstocks occurred in three main stages (see Figure 1 and Table 3). The first stage (stage I) is attributed to mass loss due to dehydration of the low boiling fractions, mainly evaporation of intracellular water from the samples [82]. This occurred at a temperature interval between 25 and 200 °C. In this stage, the digestates lose approximately 3 wt.% of weight, while undigested SS and TLP around 4.5 wt.% and 6.5 wt.%. The main decomposition step, active pyrolysis (stage II), takes place in the temperature range of 200–550 °C, with most organic matter volatilized in this step. The greatest weight loss for all samples was observed in this stage (on average 40 wt.% for digestates, 59.7 wt.% for SS and 54.6 wt.% for TLP). The weight loss of SS and TLP was faster than the weight loss of digestates. The weight loss at this stage can be ascribed to the degradation of carbohydrates, hemicellulose and cellulose [80]. In the case of SS feedstock, thermal degradation of amino acids and proteins also occurred, which originates mainly from the bacteria present in the SS [14]. According to Hung et al. [28], the remaining solid residue at the end of the second stage could contain large amounts of inorganic minerals, such as calcite ($CaCO_3$) and calcium phosphates ($Ca_3(PO_4)_2$, $Ca_5(PO_4)_3(OH)$, and others). This could also apply to this study, as the presence of these components in biochars was later confirmed by XRD analysis. The last stage, passive pyrolysis (stage III), occurred at temperatures between 550 and 800 °C, where the degradation of high-temperature thermally stable components, such as lignin components happened. In contrast to SS and TLP, the digestates in this stage showed significant weight loss even at the highest temperatures (700–800 °C), which was associated with the deep decomposition of digestates, such as refractory organic matter, inorganic matter and char residues [12]. Decomposition of calcium carbonate and other minerals has been reported to occur in this temperature range as well [36].

The differences in the degradation mechanisms of the studied feedstocks are more evident from the DTG profiles. Their shape indicates that biomass decomposition incorporates more than one step. The DTG curves of the SS revealed two main overlapping peaks, the first (~300 °C) being associated with lipid degradation, while the second (~380 °C) is related to carbohydrate decomposition. The DTG curves of the TLP exhibited typical patterns of thermal degradation of lignocellulosic materials, as also observed in the case of camel grass [82], sawdust [14], or rice straw [29]. Lignocellulosic biomass usually consists of cellulose, hemicellulose, lignin, extractives, and a small portion of inorganic mineral matter [83]. The highest peak (~380 °C) corresponds to cellulose decomposition, which occurs between 325 and 400 °C with levoglucosan as the main pyrolysis product [83]. The shoulder before that peak (at ~300 °C) is related to hemicellulose pyrolysis, which takes place between 250 and 350 °C and is represented by xylan [84]. The long tail at higher temperatures is attributed to the decomposition of lignin, which is the most difficult to degrade because it consists of aromatic rings, e.g., benzene rings, connected with ether bonds, which are more stable and degrade in a wider temperature range, between 160 and 900 °C [5]. In the DTG profiles of the digestates, the peaks of cellulose and hemicellulose are less emphasized, indicating the lower content of these compounds in the digestates, which is due to their degradation during the AD process.

Table 3. Weight loss during different decomposition stages and characteristics of DTG curves (peak temperature T_p and maximum value of the derivative curve DTG_{max}) for the tested samples.

Sample	Heating Rate (°C/min)	Weight Loss (wt.%)			Total Weight Loss (wt.%)	Final Residue (wt.%)	T_p (°C)	DTGmax (1/s)
		Dehydration (Stage I)	Active Pyrolysis (Stage II)	Passive Pyrolysis (Stage III)				
Digestate D1	15	2.70	40.44	13.33	56.47	43.53	299	5.31×10^{-4}
	30	3.12	41.47	10.60	55.18	44.82	332	1.06×10^{-3}
	100	2.09	41.79	6.83	50.71	49.29	380	3.44×10^{-3}
Digestate D2	15	2.63	40.49	16.72	59.85	40.15	306	5.84×10^{-4}
	30	3.51	37.19	13.78	54.48	45.52	312	8.79×10^{-4}
	100	3.18	38.47	7.89	49.54	50.46	362	3.32×10^{-3}
Sewage sludge (SS)	15	7.05	56.79	3.96	67.80	32.20	292	7.17×10^{-4}
	30	4.23	59.34	3.94	67.51	32.49	311	1.51×10^{-3}
	100	3.38	63.01	3.45	69.85	30.15	359	5.28×10^{-3}
T. latifolia plant (TLP)	15	7.22	54.96	13.37	75.55	24.45	347	1.89×10^{-3}
	30	5.60	55.85	12.91	74.36	25.64	359	2.80×10^{-3}
	100	7.17	55.45	12.81	75.44	24.56	408	1.18×10^{-2}

Detailed characteristics of the DTG curves, including the pyrolysis peak temperatures (T_p) and the maximum values of the DTG curves (DTG_{max}) for the analyzed samples are presented in Table 3. The pyrolysis peak temperatures and DTG_{max} values were highest for TLP, while the other three SS-based feedstocks showed lower, but comparable values. For all feedstocks, a shift in T_p for about 60 °C was observed when the heating rate was increased from 15 to 100 °C, reflecting that the heating rate affects the T_p, and the pyrolysis process. The maximal value of DTG at a heating rate of 100 °C/min for the chosen sample was higher than that at 15 °C/min, suggesting that the heating rate enhances the thermal decomposition rate of the sample. This applies to all samples. Comparison of SS and digested SS (sample D1) showed that AD caused an increase in T_p and DTG_{max} values. Similar observations regarding the effect of AD on these two parameters were found in one of the previous studies [12].

The findings associated with the degradation of the feedstocks used in this study are in agreement with the findings on the thermal degradation of sewage sludge [14,40], SS digestate [43], *T. latifolia* [24], as well as grass and its digestate [85].

3.2.2. The Influence of the Heating Rate

The heating rate plays an important role in the pyrolysis process, since the rate of change of heat affects the characteristics of pyrolysis products, especially biochar characteristics, such as porosity, surface area, volatile compound content, and biochar yield [31]. Therefore, the optimum heating rate for each material should be determined to obtain the products with desired properties.

Increasing the heating rate from 15 °C/min to 100 °C/min resulted in an absolute decrease in the weight loss of the digestates, by about 10 wt.% for digestate D2 and 6 wt.% for digestate D1. On the other hand, the heating rate has little effect on the weight loss of TLP and SS, as the differences in weight loss were almost negligible (~1%). Thus, the biochar yield increases at higher heating rates for the digested samples, but remains almost the same for the raw samples, which could be attributed to the AD pre-treatment affecting the composition of the materials. During AD pre-treatment, components, such as cellulose and hemicellulose, were degraded; therefore, the digestates lost less weight during pyrolysis than the raw samples, which is reflected in a higher biochar yield for these samples. However, the heating rates used in this study represent a slow pyrolysis process that yields less gases and produce more biochar [83].

3.3. Kinetic Analysis

The knowledge of reaction dynamics and kinetic parameters is essential for the design of a pyrolysis process [82]. In this study, two iso-conversional methods were applied in the kinetic analysis, the Kissinger–Akahira–Sunose (KAS) and Flynn–Wall–Ozawa (FWO) models. To determine the kinetic parameters, activation energy (E_α), and pre-exponential factor (A), the linear fit plots were first constructed for all tested samples (digestates D1 and D2, sewage sludge and *T. latifolia*) using the KAS and FWO kinetic models, as shown on Figure S1 in the supplementary material. For TLP and SS, data for conversion values (α) between 0.1 and 0.9 were considered in the calculations, while for digestates D1 and D2, the data in the conversion range of 0.1–0.8 were applied. The data below or above these conversion degrees were excluded due to high fluctuations and low correlation coefficients. The correlation coefficients R^2 were slightly higher when the FWO kinetic model was used, but in general the values for both models were quite close. The correlation coefficients for linear plots of the *T. latifolia* plant were >0.92, for sewage sludge >0.98, for digestate D1 > 0.88 and for digestate D2 > 0.79. Both models showed good agreement with the data representing SS or the TLP sample, while in the case of digestates the correlations were high up to a conversion level 0.5, afterwards they apparently decreased.

3.3.1. Activation Energy (E_α)

Activation energy is a barrier that must be overcome before a chemical reaction is occurred. It determines the reactivity of a material, sensitivity of a reaction rate, and is proportional to material stability [29]. The values of the activation energies E_α calculated from the slopes of the linear plots at each degree of conversion are presented in Figure 2. The error bars represent confidence intervals with a confidence level of 95%.

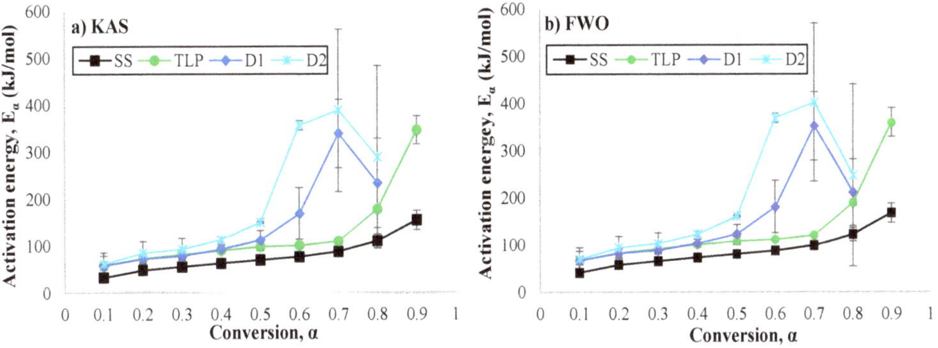

Figure 2. Activation energy E_α as a function of conversion degree calculated according to: (**a**) the KAS and (**b**) FWO models.

The activation energy E_α determined with the KAS and FWO kinetic models varied strongly with the conversion level, with significant differences found between digestate and raw samples.

In both kinetic models, E_α for SS and TLP samples increased gradually with increasing conversion level. Above the conversion level of 0.7 (stage III), a more significant increase was noticed, corresponding to the decomposition of lignin and proteins in the biomass. High E_α values at higher conversion degrees were also reported for pyrolysis of SS and its co-pyrolysis with rice husks [33]. The maximum value of 167 kJ/mol for SS and 359 kJ/mol for TLP was calculated at a conversion level of 0.9 by the FWO model. The KAS model gave lower values (SS–154 kJ/mol, TLP–346 kJ/mol). The differences in activation energies between the models comes from the different approximations used to solve the temperature integral. For the digestates D1 and D2, the E_α values increased slowly up to the conversion level of 0.4, after which a huge increase was observed, and the highest E_α values for both digestates were calculated at a conversion level of 0.7. From this point on, the values declined. The increase in E_α indicates endothermic reactions while the decrease is associated with exothermic reactions [86]. The decrease in E_α at a higher degree of conversion may be ascribed to the porous structure of the intermediate formed, which increases diffusion, the release of volatiles, and further decomposition with metal, thus catalyzing the degradation process [69]. Besides, the formation of biochar is also reflected in a decrease of activation energy [82]. The E_α for digestate D1 varied in the range of 66–351 kJ/mol for the KAS model and from 57 to 339 kJ/mol for the FWO model. The E_α for digestate D2, with a more complex composition, were generally higher (FWO model: 70–401 kJ/mol, KAS model: 62–388 kJ/mol). The higher values of E_α in the case of digested compared to undigested biomass were most likely the consequence of stabilization of biomass during the AD process. Anaerobic digestion promotes several biochemical reactions in the biomass in which the organic material is converted into methane and carbon dioxide [3], therefore digestates after AD contained less organic material and higher content of inorganic material, e.g., minerals, which impacts thermal degradation and causes an increase in E_α at higher conversion levels. It seems that minerals and inorganic matter originated from SS act as a barrier and hinder the diffusion of heat and the release of degraded volatiles, causing E_α to increase. Similar results were observed in one of the previous studies where minerals in manure feedstock also caused the increase of E_α [29]. Different E_α at different conversions illustrate the multi-step complex reaction mechanism of thermal decomposition of the analysed samples, depicted by the progressive change of E_α with conversion [87]. In particular, digestates are composed of various constituents with different reactivities resulting from the differences in chemical nature and inherent structure of the constituent components and, therefore, each constituent contributes to the overall E_α [69].

The higher E_α values in the case of TLP compared to SS may arise from higher content of cellulose and lignin in this sample. The same is true for digestate D2, which likewise contained TLP, which is reflected in the higher E_α due to the more complex structure of the sample due to the presence of lignocellulosic components.

Higher E_α values were reported for cellulose than for hemicellulose in previous studies [5], while lignin was characterized by both lower and higher values than cellulose, depending strongly on the feedstock type. Thus, the strong increase in E_α values at conversion levels above 0.7 for TLP could be related to lignin degradation. The E_α values of both digestates and TLP were very close to each other up to conversion point 0.5, from that point on, the differences become larger. According to the results, pyrolysis of SS is the reaction that proceeds most easily, followed by pyrolysis of TLP, while the highest barrier has to be overcome in the pyrolysis of digestates, particularly digestate D2.

The literature review regarding E_α revealed that the E_α values for the feedstocks analysed in this study are comparable to those reported for similar feedstocks. A detailed comparison of activation energies and other kinetic and thermodynamic parameters, which will be discussed in detail in the following sections, is presented in Table 4. For SS, the E_α values in a wider range were reported, between 46 and 232 kJ/mol [69], while for *T. latifolia* a narrow area was stated, 135–204 kJ/mol [24]. For SS digestate, the values ranged between 49 and 198 kJ/mol in one of the studies [34], and from 90–227 kJ/mol in another [42]. The upper limit of E_α values for SS digestate obtained in this study is higher than in the case of other digestates, but it must be considered that the conversion range for reported E_α could be different. No data for E_α of digestates composed of SS and lignocellulosic biomass can be found in the literature. Nevertheless, some correlations could be made with swine manure digestate [45], corn stover digestate [88], roadside grass digestate [85], and other lignocellulosic rich digestates [25,44]. As shown in Table 4, the E_α varied greatly with the type of feedstock. A comparison for some other lignocellulosic materials (para grass, camel grass, castor residue, canola residue, etc.) is also carried out.

Table 4. Comparison of kinetic and thermodynamic parameters calculated for the analyzed feedstocks (digestates D1 and D2, sewage sludge, and *T. Latifolia*) with data from the literature.

Feedstock	E_α (kJ/mol)	A (1/s)	ΔH (kJ/mol)	ΔG (kJ/mol)	ΔS (J/mol·K)	Kinetic Model	Ref.
Digestate D1 (sewage sludge)	66–351	6.73×10^3–3.80×10^{30}	61–347	160–168	(−185)–(327)	FWO	This study
Digestate D2 (sewage sludge + *T. latifolia* plant)	70–401	1.37×10^4–5.43×10^{34}	65–397	161–170	(−180)–(406)	FWO	This study
SS digestate	90–227	/	/	/	/	FWO	[42]
SS digestate	49–198	5.49×10^1–7.92×10^{14}	/	/	/	nth-order reaction model	[34]
Sewage sludge co-digested with grease waste (ratio 95:5)	132–226	/	/	/	/	FWO	[42]
Swine manure digestate	179–223	2.55×10^{16}–1.45×10^{20}	179–219	143–147	(54)–(127)	FWO	[45]
Lignocellulosic biomass digestate	75–175	1.83×10^{-2}–9.74×10^9	/	/	/	FWO	[44]
Lignocellulosic biomass digestate	130–230	1.05×10^2–7.83×10^{15}	/	/	/	Starink model-free method	[25]
Roadside grass digestate	30–175	6.74×10^{-3}–1.59×10^{15}	/	/	/	KAS	[85]
Corn stover digestate	99–331	10^{12}–10^{22}	/	/	/	DAEM [a]	[88]
Sewage sludge	41–167	2.12×10^1–4.85×10^{13}	36–163	161–167	(−233)–(4)	FWO	This study
Sewage sludge	63–323	3.22×10^4–5.78×10^{26}	70–318	85–90	(−90)–(650)	FWO	[41]
Sewage sludge	46–232	1.02×10^9–3.97×10^{19}	41–227	53–295	(−151)–(63)	FWO	[69]
Sewage sludge	48–82	1.34×10^1–5.92×10^5	11–134	/	/	Coats and Redfern	[89]
Sewage sludge	75–292	/	/	/	/	FWO	Wang 2020 [33]
Sewage sludge	200–400	10^{15}–10^{25}	/	/	/	DAEM [a]	[1]
T. latifolia plant	67–359	2.18×10^3–4.83×10^{28}	62–354	174–183	(−195)–(290)	FWO	This study
T. latifolia plant	135–204	7.6×10^9–7.9×10^{15}	130–199	171–173	(−70)–(45)	FWO	[24]
Para grass	152–242	3.06×10^{11}–2.26×10^{19}	113–237	169–173	(−98)–(111)	FWO	[80]
Camel grass	85–193	1.77×10^5–4.70×10^{14}	79–188	174–178	(−159)–(23)	FWO	[82]
Chicken manure	149–288	1.00×10^6–1.00×10^{14}	165–170	158–175	(−8)–(12)	FWO	[29]
Castor residue	102–216	3.06×10^8–6.26×10^{18}	97–211	151–154	(−97)–(101)	FWO	[86]
Canola residue	129–391	6.5×10^9–3.4×10^{27}	136–385	158–212	(−51)–(284)	DAEM [a]	[71]

[a] Distributed activation energy model.

3.3.2. The Pre-Exponential Factor (A)

The values of pre-exponential factors (A) for the analyzed feedstocks, calculated by the KAS and FWO models, are presented in Table 5. The pre-exponential factor describes the solid phase reaction dynamics and reaction chemistry, which is an essential factor for the optimization of biomass pyrolysis and is directly related to the material structure [69]. In general, the pre-exponential factors showed the same variational trend as E_α. For example if E_α increases with the conversion level, then A increases as well. The values of A calculated with the FWO model ranged between 12×10^1–4.85×10^{13} 1/s for SS and between 2.18×10^3–4.83×10^{28} 1/s for TLP. For these two samples, the values were in almost the whole conversion range, except for the highest conversions below 10^9 1/s, which could mainly indicate a surface reaction. On the other hand, if the reactions are not surface dependent, low A values may also indicate a closed complex [86]. The pre-exponential factors for SS and TLP are comparable to the pre-exponential factors of similar feedstocks reported in the literature, while a much wider range for the A values was calculated for digestates D1 and D2, as for the digestates from other studies (see Table 4). The explanations could be found in the presence of SS in the digestate samples and the more complex composition of digestates, since digestates contain both organic and inorganic material. The A values calculated with the FWO model ranged for digestate D1 between 6.73×10^3–3.80×10^{30} 1/s and for digestate D2 between 1.37×10^4–5.43×10^{34} 1/s. The KAS model gave similar results. The A values for digestates D1 and D2 were at lower conversion levels $<10^9$ 1/s, while at conversion levels above 0.4 they were $>10^9$ 1/s. This behavior indicates a multi-phase reaction due to the complex nature of these feedstocks, where degradation is slower and the reactions require more energy and a higher rate of molecular collisions [45]. Therefore, higher values of A indicate a simple complex [86].

3.3.3. Kinetic Compensation Effect

To characterize the dependence of E_α and lnA on the conversion degree, the kinetic compensation effect is frequently used [70]. The relation between the pre-exponential factors (lnA) and activation energy (E_α) for the tested feedstocks is presented in the supplementary material, in Figure S2. For all samples, the linear relationship between Arrhenius parameters was observed in the case of both kinetic models (KAS and FWO), which can be expressed as follows: $lnA = aE_\alpha + b$. This reflects that there exists a compensation effect between E_α and lnA during pyrolysis, where the constants a and b refer to the compensation coefficients [90]. The correlation coefficients R^2 for the linear fit plots for digestate D1 were >0.97 and for digestate D2 > 0.93. For SS and *T. latifolia*, the R^2 were >0.99. High correlation coefficients indicate that the KAS and FWO kinetic models are suitable for describing the pyrolysis data of the tested feedstocks in the chosen conversion range.

3.4. Thermodynamic Analysis

The values of thermodynamic parameters (enthalpy ΔH, Gibbs free energy ΔG, and entropy ΔS) for sewage sludge, *T. latifolia* and solid digestates D1 and D2, calculated at DTG peak temperatures (heating rate of 15 °C/min) using the KAS and FWO methods, are shown in Table 5.

3.4.1. Enthalpy (ΔH)

The enthalpy ΔH for all feedstocks changed significantly with the conversion level and followed a similar trend as the activation energy E_α (Table 5). The ΔH for SS ranged between 36 and 163 kJ/mol for the FWO and 27–150 kJ/mol for the KAS methods, while for TLP it ranged from 62–354 kJ/mol for the FWO and 53–341 kJ/mol for the KAS methods. The KAS model gave lower values in all cases. Positive values of ΔH indicated an endothermic process, implying that an external source of energy needs to be provided to convert the biomass to its transition state [4]. The ΔH for TLP in the literature ranged from 130–199 kJ/mol [24], while for SS it ranged from 11 kJ/mol [89] to 318 kJ/mol [41] depending on the conversion level. As shown in Table 5, the lowest ΔH values at the specific conversion point were calculated in the case of SS, followed by TLP and digestate D1. The feedstock with the highest ΔH was digestate D2 (65–397 kJ/mol, calculated by the FWO method). For comparison, the ΔH of the swine manure digestate in one of the earlier studies ranged between 179 and 219 kJ/mol [45]. Otherwise, ΔH represents the total energy required for pyrolysis of biomass and its conversion into final products such as biogas, bio-oil and biochar [41]. Therefore, digestate D2 requires the highest amount of energy to be provided for the formation of the final products compared to other samples. The ΔH differed from E_α at each conversion point by 4.70 kJ/mol for SS, 5.15 kJ/mol for TLP, 4.76 kJ/mol for digestate D1, and 4.82 kJ/mol for digestate D2. The difference between E_α and ΔH indicates the possibility of the pyrolysis reaction occurring (Rasam et al., 2020). Small differences indicate that only a small amount of additional energy (~5 kJ/mol) is required to form the final product.

Table 5. Thermodynamic parameters (A, ΔH, ΔG, ΔS) of pyrolysis of sewage sludge, *T. latifolia* and solid digestates D1 and D2 calculated at the heating rate of 15 °C/min.

	FWO Method					KAS Method				
α	A (1/s)	R^2	ΔH (kJ/mol)	ΔG (kJ/mol)	ΔS (J/mol·K)	A (1/s)	R^2	ΔH (kJ/mol)	ΔG (kJ/mol)	ΔS (J/mol·K)
Digestate D1										
0.1	6.73×10^3	0.99	61.46	167.53	−185.36	9.14×10^2	0.98	52.63	168.21	−201.97
0.2	2.19×10^5	0.99	77.03	166.52	−156.39	2.62×10^4	0.99	67.51	167.11	−174.06
0.3	9.87×10^5	1.00	83.80	166.14	−143.89	1.06×10^5	0.99	73.77	166.72	−162.42
0.4	2.67×10^7	0.99	98.76	165.40	−116.46	2.67×10^6	0.99	88.29	165.91	−135.63
0.5	2.09×10^9	0.98	118.67	164.56	−80.21	1.90×10^8	0.98	107.70	165.01	−100.14
0.6	5.10×10^{14}	0.93	175.86	162.75	22.91	4.17×10^{13}	0.92	164.27	163.07	2.10
0.7	3.80×10^{30}	0.93	346.57	159.59	326.76	2.74×10^{29}	0.93	334.23	159.76	304.89
0.8	3.38×10^{17}	0.92	206.03	162.02	76.91	4.14×10^{19}	0.88	228.43	161.54	116.90
Digestate D2										
0.1	1.37×10^4	0.92	65.47	169.53	−179.58	1.95×10^3	0.90	56.74	170.17	−195.76
0.2	2.47×10^6	0.95	89.12	168.14	−136.36	3.09×10^5	0.94	79.61	168.65	−153.66
0.3	1.66×10^7	0.96	97.86	167.71	−120.55	1.87×10^6	0.95	87.83	168.20	−138.71
0.4	1.37×10^9	1.00	118.25	166.83	−83.85	1.43×10^8	1.00	107.78	167.26	−102.64
0.5	4.92×10^{12}	1.00	156.39	165.53	−15.77	4.64×10^{11}	1.00	145.35	165.88	−35.42
0.6	5.21×10^{31}	1.00	363.45	161.55	348.42	4.38×10^{30}	1.00	351.68	161.71	327.85
0.7	5.43×10^{34}	0.97	396.52	161.14	406.20	3.63×10^{33}	0.97	383.64	161.30	383.71
0.8	4.49×10^{20}	0.79	242.63	163.47	136.61	2.79×10^{24}	0.82	283.96	162.73	209.22
Sewage sludge (SS)										
0.1	2.12×10^1	0.99	35.80	167.48	−233.13	2.83×10^0	0.98	27.43	168.57	−249.89
0.2	9.99×10^2	1.00	52.28	165.88	−201.12	1.15×10^2	1.00	42.95	166.72	−219.12
0.3	6.06×10^3	1.00	60.14	165.27	−186.13	6.31×10^2	0.99	50.29	166.05	−204.94
0.4	3.43×10^4	1.00	67.76	164.75	−171.71	3.34×10^3	0.99	57.53	165.46	−191.09
0.5	1.90×10^5	1.00	75.32	164.28	−157.51	1.71×10^4	0.99	64.68	164.95	−177.52
0.6	8.76×10^5	0.99	82.13	163.90	−144.77	7.30×10^4	0.99	71.09	164.54	−165.44
0.7	1.17×10^7	0.99	93.72	163.31	−123.22	9.07×10^5	0.99	82.28	163.89	−144.49
0.8	2.24×10^9	0.99	117.38	162.30	−79.53	1.57×10^8	0.99	105.39	162.79	−101.61
0.9	4.85×10^{13}	0.99	162.77	160.82	3.46	2.86×10^{12}	0.99	149.86	161.19	−20.07
T. latifolia plant (TLP)										
0.1	2.18×10^3	0.94	61.54	182.65	−195.40	4.00×10^2	0.92	53.47	183.32	−209.49
0.2	5.89×10^4	1.00	77.43	181.55	−167.99	8.05×10^3	1.00	67.81	182.19	−184.54
0.3	3.72×10^5	1.00	86.39	181.02	−152.68	4.71×10^4	1.00	76.34	181.62	−169.86
0.4	2.36×10^6	1.00	95.44	180.54	−137.30	2.83×10^5	1.00	85.06	181.10	−154.94
0.5	1.24×10^7	1.00	103.57	180.13	−123.53	1.42×10^6	0.99	92.94	180.66	−141.53
0.6	2.31×10^7	1.00	106.65	179.99	−118.33	2.56×10^6	1.00	95.84	180.51	−136.62
0.7	1.48×10^8	1.00	115.82	179.58	−102.88	1.60×10^7	1.00	104.83	180.08	−121.40
0.8	1.37×10^{14}	1.00	184.27	177.27	11.29	1.39×10^{13}	1.00	172.82	177.59	−7.70
0.9	4.83×10^{28}	0.99	353.61	173.98	289.81	3.93×10^{27}	0.99	340.87	174.17	268.96

3.4.2. Gibbs Free Energy (ΔG)

The Gibbs free energy ΔG, also called free enthalpy, reflects the total energy increase of the system for the formation of the activated complex and thus shows bioenergy potential of the biomass [86]. The ΔG calculated by the FWO method for SS, digestate D1 and digestate D2 were in the range of 161–167 kJ/mol, 160–168 kJ/mol, and 161–170 kJ/mol, respectively. According to the results presented in Table 5, these three feedstocks have very similar bioenergy potential. The *T. latifolia* plant showed the highest ΔG values among all feedstocks (174–183 kJ/mol) and had the highest bioenergy potential. In contrast to the enthalpy ΔH, the Gibbs free energy ΔG was quite stable and showed only little variation with the conversion degree. The values calculated with the KAS model were very similar. The Gibbs free energies of the tested feedstocks are comparable to those for SS, TLP and other lignocellulosic materials in the literature (see Table 4).

3.4.3. Entropy (ΔS)

The entropy ΔS of a system represents the degree of disorder in a reaction system, and in the context of pyrolysis it reflects the degree of arrangement of carbon layers in biochar samples [38]. The ΔS for digestate D1 ranged from -185 to 327 J/(mol·K) and for digestate D2 from -180 to 406 J/(mol·K). These values are in agreement with the ΔS values of swine manure digestate, sewage sludge, canola residue and para grass (Table 4). Values were mostly negative at lower conversion levels and positive at higher levels (Table 5). The occurrence of both negative and positive values reflects that the thermal conversion of the digestates D1 and D2 is more complex than the conversion of SS and TLP, both of which had negative ΔS throughout the conversion range (with one exception at the highest conversion point of 0.9 for TLP). At the conversion points with negative ΔS, the ΔG values were higher than ΔH, suggesting that a significant fraction of heat energy provided to the system is excess or free energy [38]. The occurrence of negative ΔS and positive ΔG values implies that thermal decomposition of biomass is a non-spontaneous process [91]. Negative ΔS values illustrate a more organized structure of the activated complex (product) compared to the feedstock and that the degree of disorder of the activated complex is lower compared to the feedstock, therefore the reactivity is low with long reaction times [29]. On the other hand, a positive ΔS indicates that the material is far from its thermodynamic equilibrium and the reactivity is high with short reaction times [54].

3.5. Characterization of Biochars

The properties of biochar have great influence on its further use and depend on various parameters, such as type of feedstock, temperature [55], heating rate [53], and residence time [31]. The obtained biochars were characterized by chemical analysis, elemental analysis, XRD, FTIR, and SEM–EDS analyses, the results of which are presented below. The results of the cress seed germination test and biosorption experiments performed on digestate derived biochars are also presented.

3.5.1. Chemical Characteristics of Biochars

The pyrolysis temperature significantly affects the distribution and properties of the final products [30]. Therefore, the chemical composition of the produced biochars was determined, focusing on elemental, heavy metal and nutrient analysis. The parameters varied depending on the type of feedstock. Comparison of the parameters between the biochars (Table 6) and feedstocks (Table 2) showed that the content of elements H, N, O and S in the biochars decreased due to the degradation of organic material. The content of C decreased in all biochars except TLP biochar, where it increased. The opposite trend in carbon content suggests that the pyrolysis mechanisms of TLP and the other three feedstocks that contained sewage sludge differed. Yin et al. [92] found a similar trend in the pyrolysis of SS and walnut shell. The decrease of C content in SS biochars was also noticed by other researchers [32]. The decrease in N content in TLP biochar was lower than that in biochars derived from SS and SS digestate. The explanation could be found in the chemistry of nitrogen in the feedstocks, as nitrogen is more volatile than the other nutrients and the concentrations may change differently depending on the biomass type and the chemistry of its binding [83]. Total nitrogen decreased mainly due to the loss of volatile nitrogen species (NH_4 and/or NO_3), which tend to convert to stable pyridine compounds at high pyrolysis temperatures [78].

Table 6. Chemical characteristics of the obtained biochars.

Parameter	Biochar			
	B-SS	B-TLP	B-D1	B-D2
Biochar yield (wt.%)	32.49	25.64	44.82	45.52
HHV (MJ/kg)	11.84	12.92	11.61	12.15
Ash (wt.%) [a]	56.22	41.06	70.20	64.78
C (wt.%)	37.10	53.42	24.33	29.03
H (wt.%)	0.58	1.47	0.18	0.31
N (wt.%)	5.28	3.33	1.39	1.48
S (wt.%)	0.05	0.21	0.92	0.59
O (wt.%)	0.77	0.51	2.98	3.81
H/C [b]	0.19	0.33	0.09	0.13
O/C [b]	0.02	0.01	0.09	0.10
N/C [b]	0.12	0.05	0.05	0.04
P (wt.%)	7.94	1.67	2.11	1.72
Ca (wt.%)	6.60	11.57	8.60	12.37
Mg (wt.%)	2.34	1.78	0.74	0.58
K (wt.%)	2.51	1.78	2.04	3.35
Si (wt.%)	0.13	0.04	0.09	0.05
Fe (wt.%)	1.49	0.28	0.54	0.52
Cd (mg/kg d.m.)	<1	<1	<1	<1
Cr_{total} (mg/kg d.m.)	60.45	6.40	62.31	47.58
Cu (mg/kg d.m.)	300.77	11.00	369.27	283.25
Ni (mg/kg d.m.)	33.78	3.00	29.77	26.64
Pb (mg/kg d.m.)	51.88	1.40	25.09	20.73
Zn (mg/kg d.m.)	2352.42	56.90	1179.38	939.61
pH	9.36	10.89	11.05	11.22

[a] On a dry basis, [b] Molar ratio.

Along with the decrease of C, H, N, and O, the molar ratios of H/C, O/C, and N/C also decreased in biochars. The H/C ratio together with volatile organic matter (VOM) content could be used as a parameter for the carbonization degree of biochar [32], because lower H/C ratio and VOM content indicate greater carbonization. In this study, digestate derived biochars showed the lowest H/C ratios. The lower ratios of H/C and O/C also indicated higher aromaticity and a less hydrophilic biochar surface [78]. Biochars with higher aromaticity are more resistant to decomposition and could be retained in the soil longer [93]. The O/C ratio of digestate derived biochars was higher than that of undigested SS and TLP, implying that digestate biochars contained more oxygen-containing functional groups. Digestate derived biochars showed very similar ratios, although the composition of their feedstocks differed. The changes in H/C and O/C also indicate the occurrence of dehydrogenative polymerization and dehydration polycondensation during pyrolysis, with significant loss of oxygen and aliphatic hydrogen [94]. The H/C and O/C ratios of the biochars from this study are consistent with the ratios of the sludge-based biochars obtained in other studies [51,78]. The decrease in N/C ratio in biochar mainly resulted from the reduction of N-related functional groups [94]. The biochar yield was highest for pyrolysis of digestates (44.8% for B-D1 and 45.5% for B-D2), lower for SS (32.5%), and lowest for TLP pyrolysis (25.6%). Biochar yields and higher heating values listed in Table 6 are comparable with the data for SS [95], and other bio-waste [50] given in the literature. The ash content in the biochars increased compared to the feedstocks. Digestate derived biochars contained more ash and had higher pH than biochars obtained from raw biomass. Ash content data are important because the ash plays an important role in biochar properties, such as surface area, pore volume, aromaticity, carbon stability, and sorption capacity [51]. The higher heating value of biochars was lower compared to that of the feedstocks.

The biochars showed alkaline characteristics as their pH value ranged from 9.4 to 11.2. The alkaline characteristics come from the release of alkali salts from the pyrolytic structure and organic nitrogen present as amine functionalities, which transforms into pyridine-like compounds [93]. The presence of metal oxides and minerals also leads to higher pH of the biochar, and high pH of the biochar ensures the safety of heavy metal leaching [52]. Biochars obtained from SS in other studies also showed alkaline properties, especially those obtained at higher pyrolysis temperatures, as pH increases with increasing temperature [96]. The content of heavy metals and macronutrients (P, Ca, Mg, K) in the biochars increased due to mass loss because of thermal degradation. Similar observations were also noted by Liu et al. [96]. However, potentially toxic elements in biochar, such as heavy metals, are usually transformed from bioavailable fraction into a more stable form during thermal conversion [53].

The content of heavy metals in biochar D2 was lower than that in biochar D1. This is most likely due to the co-digestion of SS with TLP, which caused the reduction of total heavy metal content in biochar D2. Co-digestion also improved the content of C and reduced the ash content in the biochar. The higher organic matter content and lower heavy metal content in biochar D2 indicated the higher quality of this biochar. Since biochar obtained from SS or other lignocellulosic biomass has been extensively studied, while there is a lack of knowledge on biochar derived from various digestates, the digestate derived biochars in this study were subjected to further characterization studies, biosorption experiments, and fertility tests.

3.5.2. FTIR Analysis

The FTIR spectra of the feedstocks (sewage sludge, *T. latifolia*, and digestates D1 and D2) and the corresponding biochars (B-SS, B-TLP, B-D1, B-D2) obtained after pyrolysis of the feedstocks at 800 °C, are presented in Figure 3a,b, respectively. Before pyrolysis, several peaks were common for the tested feedstocks (Figure 3a). A broad peak in the range 3500–3100 cm^{-1} corresponds to the vibrations of hydroxyl groups (–OH) of water molecules and carbohydrates [97]. The vibrations of N-H groups also appear in this area due to presence of amines and amides. Peaks between 3000 and 2800 cm^{-1} indicate the aliphatic (–CH_x) vibrations. The peak at 1647 cm^{-1} represents aromatic C=C vibrations and peaks at around 1400 cm^{-1} are attributed to aliphatic groups –CH_2 and –CH_3 [87]. The peak at 1073 cm^{-1} represents C–O and P-O bonds [30]. Lignin in the raw *T. latifolia* plant is represented by C=C aromatic vibrations (1653 cm^{-1}), while hemicellulose and cellulose are represented by C=O (1765 cm^{-1}), C-H (1375 cm^{-1}), C-O-C (1240 and 1160 cm^{-1}), C–O (1056 cm^{-1}) and C-H vibrations (896 cm^{-1}) [14]. Peaks at 777 cm^{-1} and 669 cm^{-1} are associated with aromatic hydrogen. SS and digestates shown the common peak at 1560 cm^{-1} associated with amide (-CO-NH-) originated from sewage sludge proteins [78]. The bands between 1550 and 1400 cm^{-1} are related to nitrogen compounds (N-H and N-O), while the peaks in the 400–600 cm^{-1} range are from metal-oxygen bonding [98]. The sharp peak at 871 cm^{-1} could correspond to calcium carbonate. The main differences between digested and undigested feedstocks are related to the AD pre-treatment that destroys the complex lignocellulose structure, which is reflected in the reduction of the intensity of some peaks. For example, the peak representing the C-O-C group (1240 cm^{-1}) of hemicellulose and the linkages between hemicelluloses and lignin [87] is lower in the digestates. The results are in agreement with the findings of previous studies, where AD likewise caused a decrease in carbohydrates, protein (amide) compounds, fats, and lipids on the one hand, and an increase in aromatic compounds and polysaccharide groups (C-O) in the digestates on the other [12].

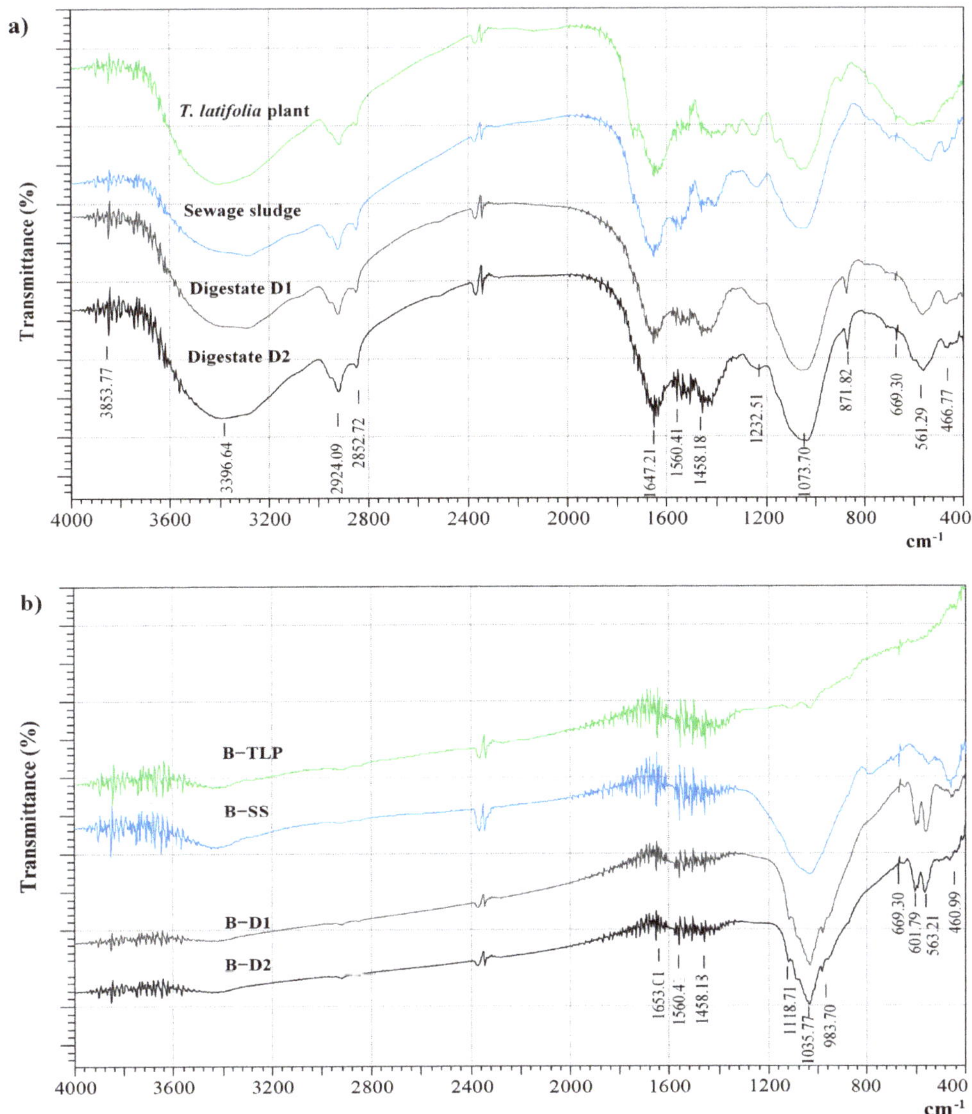

Figure 3. FTIR spectra of sewage sludge, *T. latifolia* and solid fraction of digestates D1 and D2 before pyrolysis (**a**) and biochars obtained from these feedstocks after pyrolysis (**b**).

The FTIR spectra of the biochars (Figure 3b) reflect significant changes in chemical bonds and functional groups after pyrolysis of the feedstocks. The basic functional groups representing organic components, such as hydroxyl (-OH), amine (-NH) and aliphatic groups (-CH$_x$), have almost disappeared, while the intensity of aromatic C=C ring stretching vibrations increases slightly. The disappearance of aliphatic groups in the biochars proved that the alkane groups were involved in the carbonization process [97], revealing that organic fatty hydrocarbons were converted into aromatic structures or decomposed into methane, carbon dioxide, and other gases during pyrolysis [51]. The disappearance of the majority of peaks in the case of TLP biochar illustrated deep decomposition of this sample due to the high pyrolysis temperature. The FTIR spectra of the SS biochar and digestate derived biochars were very similar. They showed a sharp peak at 1036 cm^{-1}, which besides Si-O-Si vibrations also represents the vibrations of the PO_4^{3-} group [47]. The absorption peak at 984 cm^{-1} refers to Al–O bonds [98]. Significant peaks were also detected between 400 and 600 cm^{-1} reflecting vibrations of different oxides and silicates, such as Fe-O, Mg-O, Si-O-Si, and Si-O-Al vibrations [99]. Small peaks between 600 and 800 cm^{-1} could be assigned to aromatic and hetero-aromatic compounds [93]. The differences in the composition of digestates did not essentially affect the functional groups of the biochars, as the FTIR spectra of biochars B-D1 and B-D2 are very similar. According to the FTIR analysis, the SS-based biochars contain functional groups that could cooperate in the adsorption process, and therefore could potentially be used as adsorbents for various ions from wastewater.

3.5.3. SEM–EDS Analysis

SEM images of digestate derived biochars (B-D1 and B-D2) are shown in the supplementary material in Figure S3a,b, respectively. The biochars consisted of irregular grains of various compositions and had a rough surface with porous structure containing small holes and pits on the surface. Differences in the composition of digestates had no special effect on the morphology of the biochars. The structure was consistent with the data regarding the specific surface area. The EDS spectra of biochars B-D1 (Figure S3c) and B-D2 (Figure S3d) revealed high contents of C, O, Si, P, K, Ca, and Mg in the samples, and Cl, Na, Al, and Fe were also detected. However, the contents of these elements varied among the samples. For example, biochar B-D2 contained a higher amount of C and lower amounts of heavy metals, while in biochar B-D1 Cu was also found. The results of the EDS analysis mainly agree with the results of the elemental and chemical analysis.

3.5.4. XRD Analysis

X-ray diffraction (XRD) analysis was performed to identify the crystalline phases in the digestate derived biochars. The XRD diffractograms of biochars B-D1 and B-D2 are presented in the supplementary material (Figure S4a,b). The XRD analysis revealed that the biochars have similar mineral compositions despite differences in the composition of feedstocks. Nevertheless, some differences in the contents of mineral phases were found. The main crystalline phases in biochar samples B-D1 and B-D2 were attributed to calcium phosphates, with hydroxyapatite (with variable Cl content) and whitlockite (with possible presence of Na) as the main representatives. Silicates were present in the form of mineral quartz (SiO_2) and Al silicates (with variable Mg content). Ca-Mg-carbonates (Mg calcite) were also identified in the biochars, although Mg ions could be substituted by Fe ions, which were likewise present in biochars. Traces of other phases, such as iron oxides (hematite—Fe_2O_3, magnetite—Fe_3O_4), aluminum oxide, and pure carbon phases were detected as well. The obtained biochars have similar mineral characteristics as biochars obtained from sewage sludge [39] or sewage sludge digestate [58] in other studies.

3.5.5. The Potential of Digestate-Derived Biochars for Use as a Soil Enhancer

Biochars contain a range of macro- and micro-nutrients, making them valuable as soil amenders to enhance plant growth and to sustain and increase crop yield [83]. The potential of the digestate derived biochars (B-D1 and B-D2) for use as a soil enhancer was evaluated by performing a cress seed germination test. The results of the root length *(RL)* index and Munoo-Liisa vitality *(MLV)* index obtained after cress seeds (*Lepidium sativum* L.) were exposed to different concentrations of biochars B-D1 and B-D2 (2%, 6%, 10% and 15%) for 72 h, are shown in Figure 4.

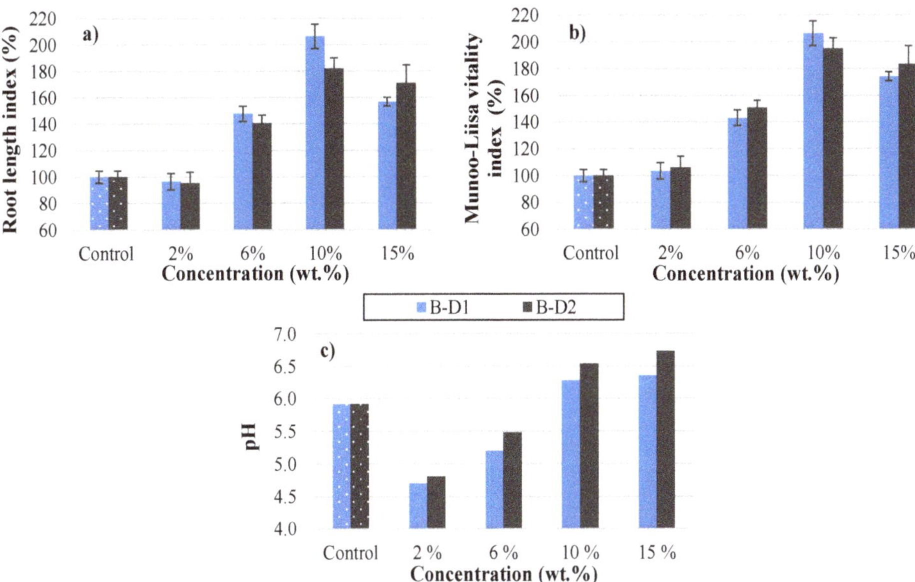

Figure 4. The results of the cress seed germination test performed with biochars B-D1 and B-D2 and dependence on the concentration of biochars: (**a**) the root length index, (**b**) Munoo-Liisa vitality index, and (**c**) pH of the soil.

The *RL* and *MLV* indexes of the control sample are given for comparison. The best results of *RL* and *MLV* indexes for both biochars were achieved when using 10 wt.% concentration of biochar. *RL* is expressed as the percentage difference of the root length of the tested material compared to the root length of the control sample. The highest *RL* index for biochar B-D1 was 206% and for biochar B-D2 182%. Both significantly exceeded the *RL* index of the control sample (100%). The highest *MLV* index, that compares the germination rate and the average lengths of roots in the test and control samples, was 206% for biochar B-D1, while for biochar D2 it was 195% (at 10 wt.% concentration). The concentration of 2 wt.% gave the worst results in both cases, even lower than the control sample, and therefore it is too low. At 15 wt.% concentration, the *RL* and *MLV* indexes of both biochars decreased, especially those of biochar B-D1. This could be connected with the phytotoxic effect of the biochars on the cress seeds. Nevertheless, the values of the *RL* and *MLV* indexes were still higher than in the case of the control sample. The phytotoxicity could occur due to the higher content of heavy metals in the biochars, especially Zn and Cu. High heavy metal concentrations negatively affect plant growth and biomass yield, and the toxic effect of heavy metals and their bioaccumulation in the plants is one of the major problems in the application of SS biochars as a soil amenders [63]. For example, Song et al. reported the problem of accumulation of Zn and Cu in garlic root and bulb [62]. The content of bioavailable heavy metals in the biochars can be efficiently reduced by selecting higher pyrolysis temperatures [100]. In addition, biochars derived at higher pyrolysis temperatures were reported to promote wheat growth more than biochars derived at lower temperatures [72]. Since the biochars in this study were obtained at relatively high temperature, this could be one of the reasons for their good performance and low toxicity.

Biochar B-D1 generally gave better results than biochar B-D2, which is a consequence of the different compositions of these two samples. Biochar B-D1 contained more P and Mg, while biochar B-D2 contained more N, K, and Ca. The advantage of using biochar B-D2 instead of biochar B-D1, despite lower RL and MLV indexes, is the lower content of heavy metals in this biochar. According to the results, the biochars obtained from digestates D1 and D2 have good potential to be used in agriculture as alternative sources of nutrients for plant growth. The optimum concentration for both biochars is around 10 wt.%.

The results of the germination test are comparable to the results of similar tests performed for SS biochars in previous studies, while the results cannot be compared with those for SS-digestate derived biochars, as these studies are quite rare. The concentrations of SS biochars of up to 5 wt.% were found to be efficient in a wheat seed germination test [72]. The same concentration was used in the cultivation of cucumber seeds [61]. In another study, 10 wt.% of SS biochar was optimal for improving cucumber growth, with cucumbers absorbing small proportions of potentially toxic elements from the biochar [101]. The positive effects of biochar addition to soil were likewise observed by Rehman et al. [63].

The soil pH values at the tested biochar concentrations ranged from 4.7 to 6.4 for biochar B-D1 and from 4.8 to 6.7 for biochar B-D2. The soil containing biochar B-D2 had a higher pH due to the higher pH of this biochar (the pH of B-D1 was 11.05 and that of B-D2 was 11.22, at biochar/water ratio of 1:20). Since the biochars have alkaline characteristics, no additional chemicals were added to the soil to ensure optimum pH for plant growth, i.e., between 5.5 and 6.5, according to the standard [73]. Besides improved soil fertility due to pH amendment, there are also some other benefits of using biochar in agriculture; it can increase the amount of bacterial biomass in the soil [102], improve the quality of nutrient-deficient soils, retain nutrients (especially N in permeable soils), improve carbon sequestration, supplement nitrogen fixation, and reduce bioaccumulation of heavy metals and polycyclic aromatic hydrocarbons (PAHs), which improves crop productivity [103]. However, since each biochar has unique characteristics, its fertility potential and phytotoxic effects should be carefully evaluated. Although digestate-derived biochars shows promising potential for use as a soil enhancer, further studies on the leaching of heavy metals and their accumulation in plants should be conducted to evaluate the possibility of their actual use for this purpose.

3.5.6. Biosorption Potential of Digestate-Derived Biochars

Biochars have specific properties such as large specific surface area, porous structure, and enriched functional groups which make them suitable as adsorbents for the removal of various pollutants from wastewater [103]. They can also be physically or chemically modified to produce so-called activated carbons, which have higher surface area and lower ash content [35]. Chemical activation also reduces mineral matter, activates carbonaceous materials, and increases the number of surface functional groups, which provides better cation and anion exchange properties [31]. In this study, digestate-derived biochars (B-D1 and B-D2) and their modifications were tested as biosorbents for the adsorption of NH_4^+, PO_4^{3-}, Cd^{2+} and Cu^{2+} ions from a water solution at an initial concentration of 50 mg/L (pH 7). To enhance the biosorption capacity, the biochars were chemically modified by KOH or. HCl. The results of the biosorption experiments performed with HCl modified, KOH modified or unmodified biochars B-D1 and B-D2 are shown in Figure 5.

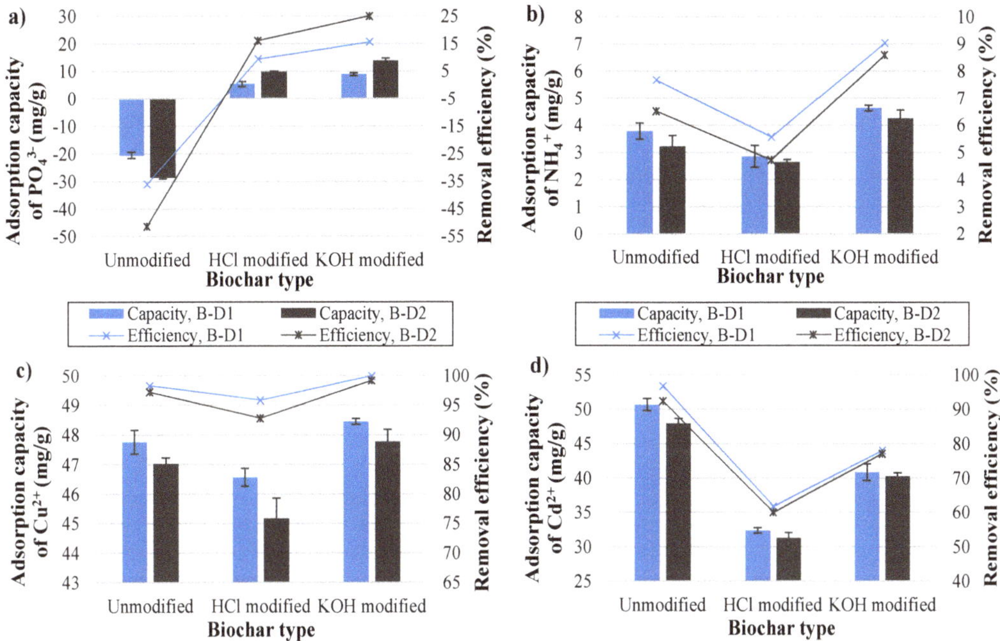

Figure 5. The results of the adsorption tests with the biochars B-D1 and B-D2 for: (**a**) PO_4^{3-} adsorption, (**b**) NH_4^+ adsorption, (**c**) Cu^{2+} adsorption, and (**d**) Cd^{2+} adsorption.

Both biochars showed relatively low adsorption capacities for NH_4^+ and PO_4^{3-} ions, although PO_4^{3-} adsorption was slightly better. In general, biochar B-D1 showed higher affinity for NH_4^+ ions, while biochar B-D2 for PO_4^{3-} ions. The best results for PO_4^{3-} adsorption were achieved with KOH modified biochars, then with HCl modified biochars, while unmodified biochar even gave negative biosorption capacities (Figure 5a). This could be explained by the leaching effect of PO_4^{3-} ions from biochars [104]. In the case of the modified biochars, leaching was not possible because PO_4^{3-} ions were already leached during modification with HCl or KOH. However, the highest biosorption capacities for PO_4^{3-} ions were obtained in the case of KOH modified biochar, 8.91 mg/g for biochar B-D1 and 13.89 mg/g for biochar B-D2. The corresponding removal efficiencies were 15.6% (B-D1) and 24.9% (B-D2). The binding of ions on the biochar surface can take place in several ways, via physical adsorption, chemical adsorption, electrostatic interaction, precipitation, complexation of ions and ion exchange process [103]. Lewis acid-base interactions, electrostatic interactions, and ligand exchange are mentioned as among the most common controlling mechanisms for PO_4^{3-} removal with SS biochars [105], which could also be applicable to this study.

Regarding the adsorption capacities of biochars for PO_4^{3-} ions, the values in the literature vary considerably, as the removal efficiency is closely related to the biochar type, modification and pH of the solution. Yin et al. [92] reported capacities of around 50 mg/g for SS biochar at a PO_4^{3-} conc. of 50 mg/L, while Xu et al. [106] achieved an adsorption capacity of 15.2 mg/g at an initial conc. of 80 mg/L. The PO_4^{3-}-P adsorption capacity of the dolomite-modified SS biochar was 19.9 mg/g (conc. of 50 mg/L) [57], while Ca-rich SS biochar showed a capacity of 27.4 mg/g at a PO_4^{3-} conc. of 40 mg/L [60]. A much lower P uptake was reported in another study, less than 1 mg/g was adsorbed at an initial concentration of 50 mg/L, but it must be considered that the biochar was not modified [49]. On the other hand, the adsorption capacity on pyrolusite-activated SS biochar was 10.8 mg/g, (conc. of 50 mg/L) [97].

The adsorption of NH_4^+ ions was likewise most efficient when KOH-modified biochars were used (Figure 5b). Capacities of 4.62 mg/g (9.1% removal efficiency) for biochar B-D1 and 4.25 mg/g (8.6%) for biochar B-D2 were achieved. Modification with HCl negatively affected the biosorption capacity of NH_4^+ ions, as the removal efficiencies were lower than in the case of unmodified biochars. Interestingly, in contrast to this study, the modification of wheat straw biochar using a combination of HCl and $FeCl_3$ increased the efficacy of the biochar in treating ammonium-contaminated wastewater [59]. Otherwise, the biosorption capacities for NH_4^+ ions achieved here were slightly higher than in other studies due to the KOH modification of the biochar. The NH_4^+ adsorption capacity of unmodified biochars is generally low, <20 mg/g [107]. In addition, sewage sludge biochars generally have a lower biosorption capacity for NH_4^+ ions than biochars derived from other organic materials. The biosorption capacity of 1.4 mg/g was achieved when SS biochar was used for the adsorption of NH_4^+ ions at a conc. of 80 mg/L [78]. Yin et al. [92] reported even lower capacities, 0.6 mg/g (at a conc. of 50 mg/L), while co-pyrolysis of SS with walnut shells improved the biosorption capacity up to 3 mg/g. The results of NH_4^+ biosorption capacity from this study are much closer to the biosorption capacities obtained by biochars derived from different wetland plant species, which have capacities between 0.8 and 5.5 mg/g at the same conc. of NH_4^+ ions (50 mg/L) [104].

However, there are several factors that affect the biosorption potential of biochar and its affinity for certain ionic species. It has been reported that the specific surface area of the adsorbent is one of the factors that significantly affects NH_4^+ adsorption, while it does not affect PO_4^{3-} adsorption [92], because the surface area of biochar is mostly negatively charged. The specific surface area, average pore size, and pore volume of biochar B-D1 were equal to 32.9290 m^2/g, 7.4885 nm, and 0.0563 cm^3/g, respectively, while biochar B-D2 had higher surface area of 60.0527 $\pm$ 0.5038 m^2/g, lower pore size (6.6904 nm), and higher pore volume (0.0788 cm^3/g). Although the surface area of biochar B-D2 was higher than that of biochar B-D1, its adsorption capacity for NH_4^+ ions was lower. However, the specific surface area of the tested biochars is comparable to the specific surface areas reported in the literature. For example, a specific surface area of 101.9 m^2/g was measured for biochar derived from manure digestate (at 800 °C) [28], while a specific surface area between 15 and 89 m^2/g was reported for sewage sludge biochars derived at high temperatures (600–900 °C) [48,51,100]. Higher pyrolysis temperatures (>700 °C) generated more pores and higher surface area due to high aromaticity caused by thermal decomposition of lignocelluloses and volatilization of inorganic minerals [28]. The surface area values of lignocellulose derived biochars are generally in a larger range of 2–500 m^2/g than those of SS, because the compact nature of sewage sludge restrict the formation of developed porosity structures [100].

Besides the surface area, the oxygen-containing surface functional groups have a great influence on the adsorption capacity of NH_4^+ ions, especially alkyl and carboxyl groups form chemical or electrostatic interactions with NH_4^+ ions [92]. Therefore, biochars with higher O/C ratios could have a higher NH_4^+ adsorption capacity [104]. The O/C ratio of unmodified biochars B-D1 and B-D2 was almost the same (0.09 and 0.10), therefore, it could not significantly impact the adsorption capacity. On the other hand, the coexistence of P, Mg and different metal elements on the biochar surface also contributes to NH_4^+ removal. Since biochar B-D1 contained higher amounts of metals, as well as P and Mg, than biochar B-D2, this could explain its better performance in NH_4^+ biosorption. The presence of surface functional groups and metals on the biochar surface is likewise crucial for the adsorption efficiency of PO_4^{3-} ions, as ligand exchange could occur between metal oxides and PO_4^{3-} ions [97]. Furthermore, elements such as Ca, Si, Al, Fe, Ca, and Mg could serve as active sites and react with PO_4^{3-} through complexation or formation of precipitates, with Mg and Ca in particular significantly promoting PO_4^{3-} adsorption due to strong divalent cation bridging [107]. A higher Ca/P ratio of biochar B-D2 compared to biochar B-D1 reflects the higher adsorption capacity of biochar B-D2 for PO_4^{3-} ions. Both biochars had a similar Mg/P ratio, so its influence was less significant.

The binding between functional groups and selected ions is also highly affected by the pH of the solution. The optimal pH value for NH_4^+ adsorption was reported in the range of 7–9, while PO_4^{3-} could be adsorbed in a wider pH range, between 4 and 9 [92]. Lower pH values cause the protonation of functional groups on the biochar surface and the removal efficiency of NH_4^+ could therefore be lower [107]. This could explain the lower capacities achieved when HCl modified biochar was used, as it lowered the pH of the solution (despite initial adjustment) compared to KOH modified biochar, which increased the pH of the solution. Another reason for the better performance of KOH modified biochar compared to HCl modified biochar is most likely due to the higher increase in the specific surface area of the biochar, as alkali treatment could significantly increase the specific surface area [108], which is one of the major factors affecting NH_4^+ adsorption.

The results of adsorption of heavy metals (Cu^{2+} and Cd^{2+}) performed at the same experimental conditions as adsorption of NH_4^+ and PO_4^{3-} ions are presented in Figure 5c,d. KOH modified biochars were found to be the most efficient for the biosorption of Cu^{2+} ions, followed by unmodified biochars. The highest biosorption capacity for Cu^{2+} ions was achieved with biochar B-D1, 48.45 mg/g (removal efficiency >99%). Biochar B-D2 showed very similar biosorption capacities. Modification of biochars with HCl has a negative effect on Cu^{2+} biosorption, as well as on Cd^{2+} biosorption. The explanation for the better performance of the KOH modified and unmodified biochars over the HCl modified biochars could be connected with the alkalinity properties of these biochars. However, it is interesting to note that the KOH modification decreases the pH of the biochars. For biochar B-D1, the decrease from pH 11.05 (unmodified biochar) to pH 9.87, and for biochar B-D2 from 11.22 to 9.79 was observed. In the case of HCl treatment, the pH value decreased to 4.60 for biochar B-D1 and 4.22 for biochar B-D2. Modification of biochars by KOH also brings several advantages, it increases the number of hydroxyl groups on the surface, dissolves ash, condenses organic matter in the biochar [58] and produces a larger surface area with higher H/C, N/C, and lower O/C ratios [31]. Modification with KOH was also found to be successful in other studies. Wongrod et al. [58] reported enhanced Pb^{2+} sorption when SS digestate biochar was treated with KOH solution. On the other hand, acid modification removes impurities, such as heavy metals, and introduces the acidic functional groups on the surface of the biochars, but in some cases, it may also decrease the surface area [102], which could be one of the explanations for the lower performance of HCl modified biochars in adsorbing heavy metals in this study.

In contrast to Cu^{2+} ions, the highest biosorption capacities for Cd^{2+} ions were achieved with unmodified biochars, followed by KOH modified biochars. The maximum biosorption capacity of 50.67 mg/g was calculated for biochar B-D1 (97% removal efficiency) and 47.92 mg/g (92%) for biochar B-D2. The differences in biosorption capacities of unmodified and modified biochars were higher in the case of Cd^{2+} ions than Cu^{2+} ions. Biochar B-D1 generally has a better affinity for heavy metals than biochar B-D2. However, the tested biochars exhibit higher biosorption capacity for biosorption of heavy metals than for NH_4^+ or PO_4^{3-} ions. This could be related to the presence of mineral phases such as aluminosilicate, quartz, calcite and metal oxides on the biochar surface, which promote the sorption of metals [108]. The main mechanisms responsible for the heavy metal adsorption on biochar include complexation with oxygen-containing functional groups (-OH, -COOH), coordination of heavy metals with π electrons in unsaturated bonds (-CH, C=O and C=N), precipitation with different minerals such as PO_4^{3-} and ion exchange with positively charged ions such as K^+, Ca^{2+}, Na^+ and Mg^{2+} [48]. In particular, π-electrons in biochars with the aromatic structure have been reported to have a strong potential to bind heavy metals [93].

The adsorption capacities for Cu^{2+} and Cd^{2+} ions obtained in this study were slightly higher than those reported in other works, but the comparison is difficult because the biosorption properties of biochars depend highly on the pyrolysis temperature and type of modification, while the initial ion concentrations also varied. For Cu^{2+} adsorption by SS biochar, one of the studies reported a capacity of 5.3 mg/g (initial conc. of Cu^{2+}

100 mg/L) [109], while another reported 11 mg/g [110]. Biosorption capacities of up to 89 mg/g for Cu^{2+} and 93 mg/g for Cd^{2+} were achieved with hydroxyapatite-modified sewage sludge biochar, and the capacity lower than 15 mg/g with unmodified biochar at initial conc. of 100 mg/L [111]. Similar removal capacities for SS biochar of around 20 mg/g were obtained at a Cd^{2+} concentration of 50 mg/L in studies performed by Chen et al. [52] and Gao et al. [48]. When SS biochar obtained by an electromagnetic induction heating method was used, a biosorption capacity of 32.3 mg/g for Cd^{2+} ions (100 mg/L) was achieved [32]. Compared with the above results, the biochars from this study showed good adsorption performance for Cu^{2+} and Cd^{2+} ions, so their application as biosorbents for the removal of heavy metals from wastewater could be possible. Further experiments on multi-metal biosorption or biosorption of multiple pollutants need to be performed in addition to the experiments with real wastewater.

4. Conclusions

In this work, kinetic and thermodynamic analyses of two types of solid digestate subjected to pyrolysis process were presented: (i) sewage sludge digestate and (ii) digestate obtained from co-digestion of sewage sludge and lignocellulosic biomass—specifically the plant *T. latifolia*. Pyrolysis of raw SS and TLP was performed as well for the comparison. Based on the experimental results, the following conclusions were made:

- Thermogravimetric analysis revealed that the digestate samples had lower weight loss than raw SS or TLP due to pre-treatment with AD and gave higher biochar yield. E_α values were higher for digestate than for raw samples. The maximum values were obtained for digestate composed of a mixture of SS and TLP. The KAS and FWO models showed excellent matching for raw materials, while for digestates, lower correlations were observed, most likely because of heterogeneous constitution, which influenced the pyrolysis process. Variation of the thermodynamic parameters (ΔH, ΔG and ΔS) indicated that the degradation of digestates is more complex than degradation of SS or TLP. TLP with the highest ΔG values exhibited the highest bioenergy potential.
- Chemical characterization of the biochars revealed high nutrient content and, thus, good prospects for their further utilization. Biochars performed very well with regard to biosorption of heavy metals (Cu and Cd), while biosorption of NH_4^+ and PO_4^{3-} ions was less efficient. Modification of biochars with KOH significantly improved their biosorption ability for all ionic species, whereas HCl modification was found to be efficient only in the case of PO_4^{3-} adsorption. Germination tests with cress seeds showed that digestate-derived biochars can be used as soil amenders at a concentration of up to 10 wt.%. SS digestate-derived biochar showed better performance than the biochar derived from a digestate mixture of SS and TLP, the advantage of the latter being its lower heavy metal content.

Depending on biochar properties and the results obtained, the digestate-derived biochars can be used in various fields, such as soil conditioning and agriculture, pollution remediation, and in modified form for other purposes. This work contributes to sustainability by promoting the circularity of bioresources by using a by-product of anaerobic digestion (digestate of sewage sludge and *T. latifolia* biomass) to synthesize biochar, a valuable product that can be used as a biofuel. The use of biochars for various other purposes also follows the bioeconomy approach and represents a major step towards sustainability.

Limitations and Directions for Future Studies

Despite the extensive work done in this study, there are some limitations and knowledge gaps that open a new path for future research.

For example, only basic thermogravimetric experiments and kinetic analysis were performed in this study, the results of which cannot provide all the information needed for a complete understanding of the pyrolysis process of the selected feedstocks and, thus, experiments at a larger, pilot scale should be performed. The operating conditions, the changes in the biochar characteristics and its yield, the formation of other phases, such as the gas and liquid phases, and other parameters, should also be studied in more detail. The effects of various pollutants, including heavy metals, and the addition of various catalysts on the pyrolysis process and biochar quality could be studied. Due to the global shortage of phosphorus fertilizers, the possibility of phosphorus recovery should also be investigated.

This study does not address the economic aspects of biochar production from *T. latifolia* and its digestate; thus, the assessment of operating costs, economic viability, and other risks of using this plant in the pyrolysis process could be investigated. The environmental impact is also an important issue in the pyrolysis of *T. latifolia* in combination with SS. A study on the biosorption of pollutants from real wastewater by the obtained biochars would also be interesting, as the behavior of biochars in real wastewater and the biosorption efficiency are likely to be significantly different from the results obtained using the model water in this study. Before the actual use of the obtained biochars for agricultural purposes, germination experiments with other plant species, such as potato or cabbage, would be necessary to evaluate the possible accumulation of heavy metals from biochar in the plants and their fruits, and other changes in plant growth. Another interesting area for further study is the modification of biochar to improve its adsorption and catalytic ability, and to produce biochar-based catalysts or supercapacitors.

Supplementary Materials: The following are available online at https://www.mdpi.com/article/10.3390/su13179642/s1, Table S1: Composition of the digestate mixtures D1 and D2, Figure S1: Linear fit plots for FWO (a–d) and KAS method (e–h) to determine activation energy values for digestate D1 (a,e), digestate D2 (b,f), sewage sludge (c,g), and *T. latifolia* (d,h), Figure S2: Linear fit plots for the compensation effects between the pre-exponential factors $ln(A/f(\alpha))$ and the activation energy E_α for: (a) digestate D1, (b) digestate D2, (c) sewage sludge, and (d) *T. Latifolia*, Figure S3: SEM images of biochars B-D1 (a) and B-D2 (b), and EDS spectra of biochars B-D1, (c) and B-D2 (d), Figure S4: The XRD diffractograms of biochars B-D1 (a) and B-D2 (b).

Author Contributions: Conceptualization, A.P.; data curation, S.V. and R.B.; formal analysis, S.V. and R.B.; investigation, A.P., S.V., and T.C.P.; methodology, A.P.; resources, M.S. and I.B.; software, R.B.; supervision, L.Č.; validation, T.C.P.; visualization, A.P.; writing—original draft, A.P.; writing—review and editing, M.S., I.B., and L.Č. All authors have read and agreed to the published version of the manuscript.

Funding: The authors would like to acknowledge the Slovenian Ministry of Education, Science, and Sport (project no. C3330-19-952041, OP20.04349), Slovenian Research Agency (research core funding no. P2-0412 and P2-0032) and the Slovenia-Croatia bilateral project Interdisciplinary Research on Variable Renewable Energy Source and Biomass in Clean and Circular Economy (BIOVARES) for financial support.

Institutional Review Board Statement: Not applicable.

Informed Consent Statement: Not applicable.

Data Availability Statement: The data presented in this study are available in supplementary material.

Acknowledgments: The authors acknowledge the IKEMA d.o.o. Institute for their support in chemical analytics and sewage sludge management, and Muzafera Paljevac and Peter Krajnc from the Laboratory of Organic and Polymer Chemistry and Technology from the Faculty of Chemistry and Chemical Engineering, University of Maribor, Slovenia, for providing help in chemical analyses.

Conflicts of Interest: The authors declare no conflict of interest.

References

1. Soria-Verdugo, A.; Goos, E.; Morato-Godino, A.; García-Hernando, N.; Riedel, U. Pyrolysis of biofuels of the future: Sewage sludge and microalgae—Thermogravimetric analysis and modelling of the pyrolysis under different temperature conditions. *Energy Convers. Manag.* **2017**, *138*, 261–272. [CrossRef]
2. Zainol Abidin, M.A.; Mahyuddin, M.N.; Mohd Zainuri, M.A.A. Solar Photovoltaic Architecture and Agronomic Management in Agrivoltaic System: A Review. *Sustainability* **2021**, *13*, 7846. [CrossRef]
3. Mehariya, S.; Patel, A.K.; Obulisamy, P.K.; Punniyakotti, E.; Wong, J.W.C. Co-digestion of food waste and sewage sludge for methane production: Current status and perspective. *Bioresour. Technol.* **2018**, *265*, 519–531. [CrossRef]

4. Rasam, S.; Moshfegh Haghighi, A.; Azizi, K.; Soria-Verdugo, A.; Keshavarz Moraveji, M. Thermal behavior, thermodynamics and kinetics of co-pyrolysis of binary and ternary mixtures of biomass through thermogravimetric analysis. *Fuel* **2020**, *280*, 118665. [CrossRef]

5. Sobek, S.; Werle, S. Kinetic modelling of waste wood devolatilization during pyrolysis based on thermogravimetric data and solar pyrolysis reactor performance. *Fuel* **2020**, *261*, 116459. [CrossRef]

6. Cucchiella, F.; D'Adamo, I.; Gastaldi, M. Sustainable waste management: Waste to energy plant as an alternative to landfill. *Energy Convers. Manag.* **2017**, *131*, 18–31. [CrossRef]

7. Khan, M.F.; Pervez, A.; Modibbo, U.M.; Chauhan, J.; Ali, I. Flexible Fuzzy Goal Programming Approach in Optimal Mix of Power Generation for Socio-Economic Sustainability: A Case Study. *Sustainability* **2021**, *13*, 8256. [CrossRef]

8. D'Adamo, I.; Falcone, P.M.; Huisingh, D.; Morone, P. A circular economy model based on biomethane: What are the opportunities for the municipality of Rome and beyond? *Renew. Energy* **2021**, *163*, 1660–1672. [CrossRef]

9. European Commission. *A New Bioeconomy Strategy for a Sustainable Europe*; European Comission: Brussels, Belgium, 2018.

10. D'Adamo, I.; Falcone, P.M.; Morone, P. A New Socio-economic Indicator to Measure the Performance of Bioeconomy Sectors in Europe. *Ecol. Econ.* **2020**, *176*, 106724. [CrossRef]

11. Cao, Y.; Pawłowski, A. Sewage sludge-to-energy approaches based on anaerobic digestion and pyrolysis: Brief overview and energy efficiency assessment. *Renew. Sustain. Energy Rev.* **2012**, *16*, 1657–1665. [CrossRef]

12. Li, X.; Chen, L.; Dai, X.; Mei, Q.; Ding, G. Thermogravimetry–Fourier transform infrared spectrometry–mass spectrometry technique to evaluate the effect of anaerobic digestion on gaseous products of sewage sludge sequential pyrolysis. *J. Anal. Appl. Pyrolysis* **2017**, *126*, 288–297. [CrossRef]

13. Ahmed, M.; Andreottola, G.; Elagroudy, S.; Negm, M.S.; Fiori, L. Coupling hydrothermal carbonization and anaerobic digestion for sewage digestate management: Influence of hydrothermal treatment time on dewaterability and bio-methane production. *J. Environ. Manag.* **2021**, *281*, 111910. [CrossRef] [PubMed]

14. Alvarez, J.; Amutio, M.; Lopez, G.; Bilbao, J.; Olazar, M. Fast co-pyrolysis of sewage sludge and lignocellulosic biomass in a conical spouted bed reactor. *Fuel* **2015**, *159*, 810–818. [CrossRef]

15. Feng, Q.; Lin, Y. Integrated processes of anaerobic digestion and pyrolysis for higher bioenergy recovery from lignocellulosic biomass: A brief review. *Renew. Sustain. Energy Rev.* **2017**, *77*, 1272–1287. [CrossRef]

16. Hu, Z.-H.; Yu, H.-Q.; Yue, Z.-B.; Harada, H.; Li, Y.-Y. Kinetic analysis of anaerobic digestion of cattail by rumen microbes in a modified UASB reactor. *Biochem. Eng. J.* **2007**, *37*, 219–225. [CrossRef]

17. Ciria, P.; Solano, M.; Soriano, P. Role of Macrophyte *Typha latifolia* in a Constructed Wetland for Wastewater Treatment and Assessment of Its Potential as a Biomass Fuel. *Biosyst. Eng. BIOSYST ENG* **2005**, *92*, 535–544. [CrossRef]

18. Grosshans, R. Cattail (*Typha* spp.) Biomass Harvesting for Nutrient Capture and Sustainable Bioenergy for Integrated Watershed Management. Ph.D. Thesis, University of Manitoba, Winnipeg, MB, Canada, 2014.

19. Hu, Z.-H.; Yue, Z.-B.; Yu, H.-Q.; Liu, S.-Y.; Harada, H.; Li, Y.-Y. Mechanisms of microwave irradiation pretreatment for enhancing anaerobic digestion of cattail by rumen microorganisms. *Appl. Energy* **2012**, *93*, 229–236. [CrossRef]

20. Petrovič, A.; Simonič, M.; Čuček, L. Nutrient recovery from the digestate obtained by rumen fluid enhanced anaerobic co-digestion of sewage sludge and cattail: Precipitation by $MgCl_2$ and ion exchange using zeolite. *J. Environ. Manag.* **2021**, *290*, 112593. [CrossRef]

21. Jaruwat, D.; Udomsap, P.; Chollacoop, N.; Eiad-ua, A. Preparation of carbon supported catalyst from cattail leaves for biodiesel fuel upgrading application. *Mater. Today Proc.* **2019**, *17*, 1319–1325. [CrossRef]

22. Zhang, B.; Joseph, G.; Wang, L.; Li, X.; Shahbazi, A. Thermophilic anaerobic digestion of cattail and hydrothermal carbonization of the digestate for co-production of biomethane and hydrochar. *J. Environ. Sci. Health Part A* **2020**, *55*, 230–238. [CrossRef]

23. Arun, J.; Gopinath, K.P.; Sivaramakrishnan, R.; Madhav, N.V.; Abhishek, K.; Ramanan, V.G.K.; Pugazhendhi, A. Bioenergy perspectives of cattails biomass cultivated from municipal wastewater via hydrothermal liquefaction and hydro-deoxygenation. *Fuel* **2021**, *284*, 118963. [CrossRef]

24. Ahmad, M.S.; Mehmood, M.A.; Taqvi, S.T.H.; Elkamel, A.; Liu, C.-G.; Xu, J.; Rahimuddin, S.A.; Gull, M. Pyrolysis, kinetics analysis, thermodynamics parameters and reaction mechanism of *Typha latifolia* to evaluate its bioenergy potential. *Bioresour. Technol.* **2017**, *245*, 491–501. [CrossRef]

25. Bartocci, P.; Tschentscher, R.; Stensrød, R.; Barbanera, M.; Fantozzi, F. Kinetic Analysis of Digestate Slow Pyrolysis with the Application of the Master-Plots Method and Independent Parallel Reactions Scheme. *Molecules* **2019**, *24*, 1657. [CrossRef] [PubMed]

26. Bedoić, R.; Jurić, F.; Ćosić, B.; Pukšec, T.; Čuček, L.; Duić, N. Beyond energy crops and subsidised electricity—A study on sustainable biogas production and utilisation in advanced energy markets. *Energy* **2020**, *201*, 117651. [CrossRef]

27. D'Adamo, I.; Falcone, P.M.; Gastaldi, M.; Morone, P. RES-T trajectories and an integrated SWOT-AHP analysis for biomethane. Policy implications to support a green revolution in European transport. *Energy Policy* **2020**, *138*, 111220. [CrossRef]

28. Hung, C.-Y.; Tsai, W.-T.; Chen, J.-W.; Lin, Y.-Q.; Chang, Y.-M. Characterization of biochar prepared from biogas digestate. *Waste Manag.* **2017**, *66*, 53–60. [CrossRef]

29. Xu, Y.; Chen, B. Investigation of thermodynamic parameters in the pyrolysis conversion of biomass and manure to biochars using thermogravimetric analysis. *Bioresour. Technol.* **2013**, *146*, 485–493. [CrossRef] [PubMed]

30. Jin, J.; Li, Y.; Zhang, J.; Wu, S.; Cao, Y.; Liang, P.; Zhang, J.; Wong, M.H.; Wang, M.; Shan, S.; et al. Influence of pyrolysis temperature on properties and environmental safety of heavy metals in biochars derived from municipal sewage sludge. *J. Hazard. Mater.* **2016**, *320*, 417–426. [CrossRef] [PubMed]

31. Rangabhashiyam, S.; Balasubramanian, P. The potential of lignocellulosic biomass precursors for biochar production: Performance, mechanism and wastewater application—A review. *Ind. Crop. Prod.* **2019**, *128*, 405–423.

32. Xue, Y.; Wang, C.; Hu, Z.; Zhou, Y.; Xiao, Y.; Wang, T. Pyrolysis of sewage sludge by electromagnetic induction: Biochar properties and application in adsorption removal of Pb(II), Cd(II) from aqueous solution. *Waste Manag.* **2019**, *89*, 48–56. [CrossRef]

33. Wang, C.; Bi, H.; Lin, Q.; Jiang, X.; Jiang, C. Co-pyrolysis of sewage sludge and rice husk by TG–FTIR–MS: Pyrolysis behavior, kinetics, and condensable/non-condensable gases characteristics. *Renew. Energy* **2020**, *160*, 1048–1066. [CrossRef]

34. Folgueras, M.B.; Alonso, M.; Díaz, R.M. Influence of sewage sludge treatment on pyrolysis and combustion of dry sludge. *Energy* **2013**, *55*, 426–435. [CrossRef]

35. Gao, N.; Kamran, K.; Quan, C.; Williams, P.T. Thermochemical conversion of sewage sludge: A critical review. *Prog. Energy Combust. Sci.* **2020**, *79*, 100843. [CrossRef]

36. Ruiz-Gómez, N.; Quispe, V.; Ábrego, J.; Atienza-Martínez, M.; Murillo, M.B.; Gea, G. Co-pyrolysis of sewage sludge and manure. *Waste Manag.* **2017**, *59*, 211–221. [CrossRef] [PubMed]

37. Cortés, A.M.; Bridgwater, A.V. Kinetic study of the pyrolysis of miscanthus and its acid hydrolysis residue by thermogravimetric analysis. *Fuel Process. Technol.* **2015**, *138*, 184–193. [CrossRef]

38. Mallick, D.; Poddar, M.K.; Mahanta, P.; Moholkar, V.S. Discernment of synergism in pyrolysis of biomass blends using thermogravimetric analysis. *Bioresour. Technol.* **2018**, *261*, 294–305. [CrossRef]

39. Tang, S.; Zheng, C.; Yan, F.; Shao, N.; Tang, Y.; Zhang, Z. Product characteristics and kinetics of sewage sludge pyrolysis driven by alkaline earth metals. *Energy* **2018**, *153*, 921–932. [CrossRef]

40. Magdziarz, A.; Werle, S. Analysis of the combustion and pyrolysis of dried sewage sludge by TGA and MS. *Waste Manag.* **2014**, *34*, 174–179. [CrossRef] [PubMed]

41. Zaker, A.; Chen, Z.; Zaheer-Uddin, M.; Guo, J. Co-pyrolysis of sewage sludge and low-density polyethylene—A thermogravimetric study of thermo-kinetics and thermodynamic parameters. *J. Environ. Chem. Eng.* **2020**, *9*, 104554. [CrossRef]

42. González Arias, J.; Gil, M.; Fernández, R.; Martinez, J.; Fernández, C.; Papaharalabos, G.; Gómez, X. Integrating anaerobic digestion and pyrolysis for treating digestates derived from sewage sludge and fat wastes. *Environ. Sci. Pollut. Res.* **2020**, *27*, 32603–32614. [CrossRef]

43. Sanchez, M.E.; Otero, M.; Gómez, X.; Morán, A. Thermogravimetric kinetic analysis of the combustion of biowastes. *Renew. Energy* **2009**, *34*, 1622–1627. [CrossRef]

44. Akor, C.I.; Osman, A.I.; Farrell, C.; McCallum, C.S.; John Doran, W.; Morgan, K.; Harrison, J.; Walsh, P.J.; Sheldrake, G.N. Thermokinetic study of residual solid digestate from anaerobic digestion. *Chem. Eng. J.* **2021**, *406*, 127039. [CrossRef]

45. Vuppaladadiyam, A.K.; Liu, H.; Zhao, M.; Soomro, A.F.; Memon, M.Z.; Dupont, V. Thermogravimetric and kinetic analysis to discern synergy during the co-pyrolysis of microalgae and swine manure digestate. *Biotechnol. Biofuels* **2019**, *12*, 170. [CrossRef] [PubMed]

46. Chen, Y.; Wang, R.; Duan, X.; Wang, S.; Ren, N.; Ho, S.-H. Production, properties, and catalytic applications of sludge derived biochar for environmental remediation. *Water Res.* **2020**, *187*, 116390. [CrossRef] [PubMed]

47. Xu, X.; Cao, X.; Zhao, L.; Sun, T. Comparison of sewage sludge- and pig manure-derived biochars for hydrogen sulfide removal. *Chemosphere* **2014**, *111*, 296–303. [CrossRef] [PubMed]

48. Gao, L.-Y.; Deng, J. H.; Huang, G.-F.; Li, K.; Cai, K.-Z.; Liu, Y.; Huang, F. Relative distribution of Cd^{2+} adsorption mechanisms on biochars derived from rice straw and sewage sludge. *Bioresour. Technol.* **2019**, *272*, 114–122. [CrossRef]

49. Melia, P.M.; Busquets, R.; Hooda, P.S.; Cundy, A.B.; Sohi, S.P. Driving forces and barriers in the removal of phosphorus from water using crop residue, wood and sewage sludge derived biochars. *Sci. Total Environ.* **2019**, *675*, 623–631. [CrossRef]

50. Yousaf, B.; Liu, G.; Ubaid Ali, M.; Abbas, Q.; Liu, Y.; Ullah, H.; Imtiyaz Cheema, A. Decisive role of vacuum-assisted carbonization in valorization of lignin-enriched (Juglans regia-shell) biowaste. *Bioresour. Technol.* **2021**, *323*, 124541. [CrossRef]

51. Fan, J.; Li, Y.; Yu, H.; Li, Y.; Yuan, Q.; Xiao, H.; Li, F.; Pan, B. Using sewage sludge with high ash content for biochar production and Cu(II) sorption. *Sci. Total Environ.* **2020**, *713*, 136663. [CrossRef]

52. Chen, T.; Zhang, Y.; Wang, H.; Lu, W.; Zhou, Z.; Zhang, Y.; Ren, L. Influence of pyrolysis temperature on characteristics and heavy metal adsorptive performance of biochar derived from municipal sewage sludge. *Bioresour. Technol.* **2014**, *164*, 47–54. [CrossRef]

53. Yousaf, B.; Liu, G.; Abbas, Q.; Ali, M.U.; Wang, R.; Ahmed, R.; Wang, C.; Al-Wabel, M.I.; Usman, A.R.A. Operational control on environmental safety of potentially toxic elements during thermal conversion of metal-accumulator invasive ragweed to biochar. *J. Clean. Prod.* **2018**, *195*, 458–469. [CrossRef]

54. Naqvi, S.R.; Hameed, Z.; Tariq, R.; Taqvi, S.A.; Ali, I.; Niazi, M.B.K.; Noor, T.; Hussain, A.; Iqbal, N.; Shahbaz, M. Synergistic effect on co-pyrolysis of rice husk and sewage sludge by thermal behavior, kinetics, thermodynamic parameters and artificial neural network. *Waste Manag.* **2019**, *85*, 131–140. [CrossRef] [PubMed]

55. Ullah, H.; Abbas, Q.; Ali, M.U.; Amina; Cheema, A.I.; Yousaf, B.; Rinklebe, J. Synergistic effects of low-/medium-vacuum carbonization on physico-chemical properties and stability characteristics of biochars. *Chem. Eng. J.* **2019**, *373*, 44–57. [CrossRef]

56. Başer, B.; Yousaf, B.; Yetis, U.; Abbas, Q.; Kwon, E.E.; Wang, S.; Bolan, N.S.; Rinklebe, J. Formation of nitrogen functionalities in biochar materials and their role in the mitigation of hazardous emerging organic pollutants from wastewater. *J. Hazard. Mater.* **2021**, *416*, 126131. [CrossRef]

57. Li, J.; Li, B.; Huang, H.; Lv, X.; Zhao, N.; Guo, G.; Zhang, D. Removal of phosphate from aqueous solution by dolomite-modified biochar derived from urban dewatered sewage sludge. *Sci. Total Environ.* **2019**, *687*, 460–469. [CrossRef] [PubMed]

58. Wongrod, S.; Simon, S.; Guibaud, G.; Lens, P.N.L.; Pechaud, Y.; Huguenot, D.; van Hullebusch, E.D. Lead sorption by biochar produced from digestates: Consequences of chemical modification and washing. *J. Environ. Manag.* **2018**, *219*, 277–284. [CrossRef] [PubMed]

59. Wang, S.; Ai, S.; Nzediegwu, C.; Kwak, J.-H.; Islam, M.S.; Li, Y.; Chang, S.X. Carboxyl and hydroxyl groups enhance ammonium adsorption capacity of iron (III) chloride and hydrochloric acid modified biochars. *Bioresour. Technol.* **2020**, *309*, 123390. [CrossRef] [PubMed]

60. Saadat, S.; Raei, E.; Talebbeydokhti, N. Enhanced removal of phosphate from aqueous solutions using a modified sludge derived biochar: Comparative study of various modifying cations and RSM based optimization of pyrolysis parameters. *J. Environ. Manag.* **2018**, *225*, 75–83. [CrossRef] [PubMed]

61. Nuagah, M.B.; Boakye, P.; Oduro-Kwarteng, S.; Sokama-Neuyam, Y.A. Valorization of faecal and sewage sludge via pyrolysis for application as crop organic fertilizer. *J. Anal. Appl. Pyrolysis* **2020**, *151*, 104903. [CrossRef]

62. Song, X.D.; Xue, X.Y.; Chen, D.Z.; He, P.J.; Dai, X.H. Application of biochar from sewage sludge to plant cultivation: Influence of pyrolysis temperature and biochar-to-soil ratio on yield and heavy metal accumulation. *Chemosphere* **2014**, *109*, 213–220. [CrossRef]

63. Rehman, R.A.; Rizwan, M.; Qayyum, M.F.; Ali, S.; Zia-ur-Rehman, M.; Zafar-ul-Hye, M.; Hafeez, F.; Iqbal, M.F. Efficiency of various sewage sludges and their biochars in improving selected soil properties and growth of wheat (*Triticum aestivum*). *J. Environ. Manag.* **2018**, *223*, 607–613. [CrossRef] [PubMed]

64. Angelidaki, I.; Alves, M.; Bolzonella, D.; Borzacconi, L.; Campos, J.; Guwy, A.; Kalyuzhnyi, S.; Jenicek, P.; Van Lier, J. Defining the biomethane potential (BMP) of solid organic wastes and energy crops: A proposed protocol for batch assays. *Water Sci. Technol.* **2009**, *59*, 927–934. [CrossRef] [PubMed]

65. SIST EN 14346:2007, *Characterization of Waste—Calculation of Dry Matter by Determination of Dry Residue or Water Content*; Slovenian Institute for Standardization (SIST): Ljubljana, Slovenia, 2007.

66. SIST-TS CEN/TS 16023:2014, *Characterization of Waste—Determination of Gross Calorific Value and Calculation of Net Calorific Value*; Slovenian Institute for Standardization (SIST): Ljubljana, Slovenia, 2014.

67. Channiwala, S.A.; Parikh, P.P. A unified correlation for estimating HHV of solid, liquid and gaseous fuels. *Fuel* **2002**, *81*, 1051–1063. [CrossRef]

68. EN 16170:2016, *Sludge, Treated Biowaste and Soil—Determination of Elements Using Inductively Coupled Plasma Optical Emission Spectrometry (ICP-OES)*; European Committee for Standardization: Brussels, Belgium, 2016.

69. Naqvi, S.R.; Tariq, R.; Hameed, Z.; Ali, I.; Taqvi, S.A.; Naqvi, M.; Niazi, M.B.K.; Noor, T.; Farooq, W. Pyrolysis of high-ash sewage sludge: Thermo-kinetic study using TGA and artificial neural networks. *Fuel* **2018**, *233*, 529–538. [CrossRef]

70. Yao, Z.; Yu, S.; Su, W.; Wu, W.; Tang, J.; Qi, W. Kinetic studies on the pyrolysis of plastic waste using a combination of model-fitting and model-free methods. *Waste Manag. Res.* **2020**, *38*, 77–85. [CrossRef]

71. Tahir, M.H.; Çakman, G.; Goldfarb, J.L.; Topcu, Y.; Naqvi, S.R.; Ceylan, S. Demonstrating the suitability of canola residue biomass to biofuel conversion via pyrolysis through reaction kinetics, thermodynamics and evolved gas analyses. *Bioresour. Technol.* **2019**, *279*, 67–73. [CrossRef]

72. Kong, L.; Liu, J.; Han, Q.; Zhou, Q.; He, J. Integrating metabolomics and physiological analysis to investigate the toxicological mechanisms of sewage sludge-derived biochars to wheat. *Ecotoxicol. Environ. Saf.* **2019**, *185*, 109664. [CrossRef]

73. SIST EN 16086-2:2012, *Soil Improvers and Growing Media—Determination of Plant Response—Part 2: Petri Dish Test Using Cress*; Slovenian Institute for Standardization (SIST): Ljubljana, Slovenia, 2012.

74. Chemetova, C.; Fabião, A.; Gominho, J.; Ribeiro, H. Range analysis of Eucalyptus globulus bark low-temperature hydrothermal treatment to produce a new component for growing media industry. *Waste Manag.* **2018**, *79*, 1–7. [CrossRef]

75. Maunuksela, L.; Herranen, M.; Torniainen, M. Quality assessment of biogas plant end products by plant bioassays. *Int. J. Environ. Sci. Dev.* **2012**, *3*, 305. [CrossRef]

76. DIN 38 406-E5-1, *Photometric Determination of Ammonium Nitrogen by Sodium Dichloroisocyanurate and Sodium Salicylate*; DIN—German Institute for Standardization: Berlin, Germany, 1983.

77. SIST EN ISO 6878:2004, *Water Quality—Determination of Phosphorus—Ammonium Molybdate Spectrometric Method*; ISO: Geneva, Switzerland, 2004.

78. Tang, Y.; Alam, M.S.; Konhauser, K.O.; Alessi, D.S.; Xu, S.; Tian, W.; Liu, Y. Influence of pyrolysis temperature on production of digested sludge biochar and its application for ammonium removal from municipal wastewater. *J. Clean. Prod.* **2019**, *209*, 927–936. [CrossRef]

79. Government of the Republic of Slovenia. Decree on the use of sludge from municipal sewage treatment plants in agriculture. In *Official Gazette of the Republic of Slovenia, No. 62/08*; The Official Gazette of the Republic of Slovenia Ltd., Public Company: Ljubljana, Slovenia, 2008.

80. Ahmad, M.S.; Mehmood, M.A.; Al Ayed, O.S.; Ye, G.; Luo, H.; Ibrahim, M.; Rashid, U.; Arbi Nehdi, I.; Qadir, G. Kinetic analyses and pyrolytic behavior of Para grass (*Urochloa mutica*) for its bioenergy potential. *Bioresour. Technol.* **2017**, *224*, 708–713. [CrossRef]

81. Syed-Hassan, S.S.A.; Wang, Y.; Hu, S.; Su, S.; Xiang, J. Thermochemical processing of sewage sludge to energy and fuel: Fundamentals, challenges and considerations. *Renew. Sustain. Energy Rev.* **2017**, *80*, 888–913. [CrossRef]

82. Mehmood, M.A.; Ye, G.; Luo, H.; Liu, C.; Malik, S.; Afzal, I.; Xu, J.; Ahmad, M.S. Pyrolysis and kinetic analyses of Camel grass (*Cymbopogon schoenanthus*) for bioenergy. *Bioresour. Technol.* **2017**, *228*, 18–24. [CrossRef] [PubMed]

83. Kan, T.; Strezov, V.; Evans, T.J. Lignocellulosic biomass pyrolysis: A review of product properties and effects of pyrolysis parameters. *Renew. Sustain. Energy Rev.* **2016**, *57*, 1126–1140. [CrossRef]

84. Shen, D.; Jin, W.; Hu, J.; Xiao, R.; Luo, K. An overview on fast pyrolysis of the main constituents in lignocellulosic biomass to valued-added chemicals: Structures, pathways and interactions. *Renew. Sustain. Energy Rev.* **2015**, *51*, 761–774. [CrossRef]

85. Bedoić, R.; Ocelić Bulatović, V.; Čuček, L.; Ćosić, B.; Špehar, A.; Pukšec, T.; Duić, N. A kinetic study of roadside grass pyrolysis and digestate from anaerobic mono-digestion. *Bioresour. Technol.* **2019**, *292*, 121935. [CrossRef]

86. Kaur, R.; Gera, P.; Jha, M.K.; Bhaskar, T. Pyrolysis kinetics and thermodynamic parameters of castor (*Ricinus communis*) residue using thermogravimetric analysis. *Bioresour. Technol.* **2018**, *250*, 422–428. [CrossRef] [PubMed]

87. Rasool, T.; Kumar, S. Kinetic and Thermodynamic Evaluation of Pyrolysis of Plant Biomass using TGA. *Mater. Today Proc.* **2020**, *21*, 2087–2095. [CrossRef]

88. Zhang, D.; Wang, F.; Yi, W.; Li, Z.; Shen, X.; Niu, W. Comparison study on pyrolysis characteristics and kinetics of corn stover and its digestate by TG-FTIR. *BioResources* **2017**, *12*, 8240–8254. [CrossRef]

89. Gao, N.; Li, J.; Qi, B.; Li, A.; Duan, Y.; Wang, Z. Thermal analysis and products distribution of dried sewage sludge pyrolysis. *J. Anal. Appl. Pyrolysis* **2014**, *105*, 43–48. [CrossRef]

90. Hu, Y.; Wang, Z.; Cheng, X.; Ma, C. Non-isothermal TGA study on the combustion reaction kinetics and mechanism of low-rank coal char. *RSC Adv.* **2018**, *8*, 22909–22916. [CrossRef]

91. Naqvi, S.R.; Ali, I.; Nasir, S.; Ali Ammar Taqvi, S.; Atabani, A.E.; Chen, W.-H. Assessment of agro-industrial residues for bioenergy potential by investigating thermo-kinetic behavior in a slow pyrolysis process. *Fuel* **2020**, *278*, 118259. [CrossRef]

92. Yin, Q.; Liu, M.; Ren, H. Biochar produced from the co-pyrolysis of sewage sludge and walnut shell for ammonium and phosphate adsorption from water. *J. Environ. Manag.* **2019**, *249*, 109410. [CrossRef]

93. Huang, H.-j.; Yang, T.; Lai, F.-y.; Wu, G.-q. Co-pyrolysis of sewage sludge and sawdust/rice straw for the production of biochar. *J. Anal. Appl. Pyrolysis* **2017**, *125*, 61–68. [CrossRef]

94. Xie, S.; Yu, G.; Li, C.; Li, J.; Wang, G.; Dai, S.; Wang, Y. Treatment of high-ash industrial sludge for producing improved char with low heavy metal toxicity. *J. Anal. Appl. Pyrolysis* **2020**, *150*, 104866. [CrossRef]

95. Arauzo, P.J.; Atienza-Martínez, M.; Ábrego, J.; Olszewski, M.P.; Cao, Z.; Kruse, A. Combustion Characteristics of Hydrochar and Pyrochar Derived from Digested Sewage Sludge. *Energies* **2020**, *13*, 4164. [CrossRef]

96. Liu, Q.; Fang, Z.; Liu, Y.; Liu, Y.; Xu, Y.; Ruan, X.; Zhang, X.; Cao, W. Phosphorus speciation and bioavailability of sewage sludge derived biochar amended with CaO. *Waste Manag.* **2019**, *87*, 71–77. [CrossRef] [PubMed]

97. Yao, S.; Wang, M.; Liu, J.; Tang, S.; Chen, H.; Guo, T.; Yang, G.; Chen, Y. Removal of phosphate from aqueous solution by sewage sludge-based activated carbon loaded with pyrolusite. *J. Water Reuse Desalination* **2017**, *8*, 192–201. [CrossRef]

98. Ren, N.; Tang, Y.; Li, M. Mineral additive enhanced carbon retention and stabilization in sewage sludge-derived biochar. *Process Saf. Environ. Prot.* **2018**, *115*, 70–78. [CrossRef]

99. Wang, Q.; Li, J.; Poon, C.S. Using incinerated sewage sludge ash as a high-performance adsorbent for lead removal from aqueous solutions: Performances and mechanisms. *Chemosphere* **2019**, *226*, 587–596. [CrossRef]

100. Xing, J.; Li, L.; Li, G.; Xu, G. Feasibility of sludge-based biochar for soil remediation: Characteristics and safety performance of heavy metals influenced by pyrolysis temperatures. *Ecotoxicol. Environ. Saf.* **2019**, *180*, 457–465. [CrossRef] [PubMed]

101. Xie, S.; Yu, G.; Ma, J.; Wang, G.; Wang, Q.; You, F.; Li, J.; Wang, Y.; Li, C. Chemical speciation and distribution of potentially toxic elements in soilless cultivation of cucumber with sewage sludge biochar addition. *Environ. Res.* **2020**, *191*, 110188. [CrossRef] [PubMed]

102. Wang, J.; Wang, S. Preparation, modification and environmental application of biochar: A review. *J. Clean. Prod.* **2019**, *227*, 1002–1022. [CrossRef]

103. Singh, S.; Kumar, V.; Dhanjal, D.S.; Datta, S.; Bhatia, D.; Dhiman, J.; Samuel, J.; Prasad, R.; Singh, J. A sustainable paradigm of sewage sludge biochar: Valorization, opportunities, challenges and future prospects. *J. Clean. Prod.* **2020**, *269*, 122259. [CrossRef]

104. Cui, X.; Hao, H.; He, Z.; Stoffella, P.J.; Yang, X. Pyrolysis of wetland biomass waste: Potential for carbon sequestration and water remediation. *J. Environ. Manag.* **2016**, *173*, 95–104. [CrossRef]

105. Almanassra, I.W.; McKay, G.; Kochkodan, V.; Ali Atieh, M.; Al-Ansari, T. A state of the art review on phosphate removal from water by biochars. *Chem. Eng. J.* **2021**, *409*, 128211. [CrossRef]

106. Xu, G.; Zhang, Z.; Deng, L. Adsorption Behaviors and Removal Efficiencies of Inorganic, Polymeric and Organic Phosphates from Aqueous Solution on Biochar Derived from Sewage Sludge of Chemically Enhanced Primary Treatment Process. *Water* **2018**, *10*, 869. [CrossRef]

107. Zhang, M.; Song, G.; Gelardi, D.L.; Huang, L.; Khan, E.; Mašek, O.; Parikh, S.J.; Ok, Y.S. Evaluating biochar and its modifications for the removal of ammonium, nitrate, and phosphate in water. *Water Res.* **2020**, *186*, 116303. [CrossRef]

108. Wongrod, S.; Simon, S.; van Hullebusch, E.D.; Lens, P.N.L.; Guibaud, G. Changes of sewage sludge digestate-derived biochar properties after chemical treatments and influence on As(III and V) and Cd(II) sorption. *Int. Biodeterior. Biodegrad.* **2018**, *135*, 96–102. [CrossRef]
109. Zhou, D.; Liu, D.; Gao, F.; Li, M.; Luo, X. Effects of Biochar-Derived Sewage Sludge on Heavy Metal Adsorption and Immobilization in Soils. *Int. J. Environ. Res Public Health* **2017**, *14*, 681. [CrossRef]
110. Matheri, A.N.; Eloko, N.S.; Ntuli, F.; Ngila, J.C. Influence of pyrolyzed sludge use as an adsorbent in removal of selected trace metals from wastewater treatment. *Case Stud. Chem. Environ. Eng.* **2020**, *2*, 100018. [CrossRef]
111. Chen, Y.; Li, M.; Li, Y.; Liu, Y.; Chen, Y.; Li, H.; Li, L.; Xu, F.; Jiang, H.; Chen, L. Hydroxyapatite modified sludge-based biochar for the adsorption of Cu^{2+} and Cd^{2+}: Adsorption behavior and mechanisms. *Bioresour. Technol.* **2021**, *321*, 124413. [CrossRef] [PubMed]

MDPI

Article

Value Proposition of Different Methods for Utilisation of Sugarcane Wastes

Ihsan Hamawand [1,*], Wilton da Silva [2], Saman Seneweera [3] and Jochen Bundschuh [4]

[1] Wide Bay Water, Fraser Coast Regional Council, Urangan 4655, Australia
[2] Department of Physics, Federal University of Campina Grande, Campina Grande 58429-900, Brazil; wiltonps@uol.com.br
[3] National Institute of Fundamental Studies, Kandy 20000, Sri Lanka; seneweera@gmail.com
[4] Centre for Crop Health, School of Civil Engineering and Surveying, University of Southern Queensland, Toowoomba 4350, Australia; Jochen.Bundschuh@usq.edu.au
* Correspondence: ihsan.hamawand@gmail.com or Ihsan.Hamawand@frasecost.gov.qld.au; Tel.: +61-04668-97659

Abstract: There are four main waste products produced during the harvesting and milling process of sugarcane: cane trash, molasses, bagasse and mill mud–boiler ash mixture. This study investigates the value proposition of different techniques currently not being adopted by the industry in the utilisation of these wastes. The study addresses the technical challenges and the environmental impact associated with these wastes and comes up with some recommendations based on the recent findings in the literature. All the biomass wastes such as bagasse, trash (tops) and trash (leaves) have shown great potential in generating higher revenue by converting them to renewable energy than burning them (wet or dry). However, the energy content in the products from all the utilisation methods is less than the energy content of the raw product. This study has found that the most profitable and challenging choice is producing ethanol or ethanol/biogas from these wastes. The authors recommend conducting more research in this field in order to help the sugar industry to compete in the international market.

Keywords: sugarcane wastes; bioenergy; utilisation; mill mud; value proposition; ethanol

Citation: Hamawand, I.; da Silva, W.; Seneweera, S.; Bundschuh, J. Value Proposition of Different Methods for Utilisation of Sugarcane Wastes. *Energies* **2021**, *14*, 5483. https://doi.org/10.3390/en14175483

Academic Editor: Idiano D'Adamo

Received: 22 July 2021
Accepted: 29 August 2021
Published: 2 September 2021

Publisher's Note: MDPI stays neutral with regard to jurisdictional claims in published maps and institutional affiliations.

1. Introduction

One of the greatest challenges of the 21st century is to meet the global energy demand while maintaining environmental sustainability. The possible decline in fuel supply, together with the detrimental release of greenhouse gases, has led to an urgent need to identify sources of renewable fuel/energy. The Australian government was expecting to achieve a renewable energy target of 10% of total energy production by 2020 [1,2]. However, Australia has yet to reach 2% renewable energy production. Likewise, India had similar renewable targets to achieve by 2020 [3]. To achieve this target, biofuel was identified to play a major role [2]. However, the use of food crops for biofuel production has been widely criticised because the demand for food is still increasing, and the need will be increased by a further 50% by the middle of the century [4]. Thus, there is an urgent need to develop more sustainable alternative biofuel feedstocks that do not impact global food production [5]. These challenges represent the key drivers for this article.

Sugarcane is identified as a main target in Australia for biofuel production. Other annual crops such as maize, sorghum and wheat have also been identified as potential candidates for biofuel production in Australia because these crops are widely grown. Sugarcane bagasse, a byproduct after sugar extraction, remains in large quantities and is used to generate electricity for the sugar factory operation by burning wet bagasse. Bagasse is the fibrous residue, which is left over after crushing and extraction of the plant juice. In Australia, the amount of bagasse produced each year just from sugarcane is around

151

10 million tons [6]. The net theoretical calorific value produced from wet bagasse is around 7588 kJ/kg. Currently, bagasse is burned wet in the mills' boilers [7]. This practice causes a significant loss of energy. About 52% of energy is lost in evaporating the moisture in bagasse. The conversion efficiency of heat to electricity in this process is low; the energy efficiency of a conventional thermal power station is typically 33% to 48% [6]. This means around 52% to 67% of the energy content in the fuel is lost, and this also releases a large quantity of CO_2 into the atmosphere.

Australian bagasse fibre mainly consists of 43% and 25% of cellulose and hemicellulose, respectively. The remaining constituents are ashes and waxes. Cellulose and hemicellulose content are around 70%; this makes it a potential candidate for bioethanol production. Bagasse represents a low-cost raw material when compared to wood. Its estimated value at the sugar mills is typically around AUD 40 per dry ton [8]. If energy can be produced from bagasse by converting it to ethanol, additional income can be generated. The solid waste generated from the fermentation is almost 50% of the bagasse, and this can be used as fuel for a boiler to generate steam and electricity.

In Australia, crop residues from maize, sorghum, wheat and other small grain crops are largely used as mulch after harvesting. For example, the harvest index of modern grain crops falls within the range of 0.4 to 0.6, which is the ratio of harvestable yield to biomass ratio [9,10]. On this basis, the remaining agricultural waste, except waste from cereal crops, accounts for nearly 250 million metric tons of lignocellulose biomass globally. This can be used for bioethanol production, and this can further facilitate the production of a large amount of energy needed globally. Another advantage of using crop biomass for bioethanol is the low lignin content compared to perennial trees and higher cellulose to hemicellulose ratio [11].

The structure of lignocellulosic biomass is determined by the cell wall structure. Understanding biomass composition between different species and genotypes will provide important insights into potential bioethanol production efficiency. For example, the relative carbohydrate and lignin ratio plays an important role in the deconstruction of plant cell walls [12]. A detailed understanding of carbohydrate composition, protein concentration and also macro and micronutrient concentrations are essential for promoting microbial growth, fermentation and fermentation efficiency.

Converting lignocellulose to fermentable monosaccharides is a technical challenge. This process is hindered by many physio-chemical, structural and compositional factors [13]. Various pre-treatment techniques, such as steam explosion, alkaline, diluted and concentrated acid and ammonia are introduced to overcome these difficulties [14]. It was suggested that one or more of these pre-treatments might be required prior to enzymatic hydrolysis to make the cellulose more accessible to the enzymes [1].

Australia ranks third after Brazil and Thailand in supplying raw sugar. Queensland produces approximately 95 per cent of the sugar in Australia, and the remaining five per cent is produced by New South Wales. There are approximately 4000 cane farming businesses supplying 24 mills. Seven milling companies own the milling industry. Around 75 per cent of this industry is currently foreign-owned [15,16].

In 2012/13, QSL's Discretionary Pool returned AUD 438.55 per ton of sugar. A total of just over 30.5 million tons of sugarcane was crushed during the 2013 season with an average cane yield of 83 tons per hectare, resulting in more than 4 million tons of sugar being produced [15,16]. Table 1 shows the total production of sugar and byproducts from the sugar mills in Australia.

Table 1. Main product and byproducts of sugar cane mill.

Products	Unit	Production Annually	Unit *	Production
		Main Product		
Sugar	metric tonne	4,360,000	tonne/hectare	11.75
Cane	metric tonne	30,500,000	tonne/hectare	82.21
		Byproducts		
Bagasse	tonne/tonne sugar	2.294	tonne/hectare	26.954
Trash	tonne/tonne sugar	1.63	tonne/hectare	19.1
Molasses	tonne/tonne sugar	0.229	tonne/hectare	2.695
Mill mud **	tonne/tonne sugar	0.140–0.419	tonne/hectare	1.644–4.932

* 371,000 ha harvested, ** Mill mud quantity varies depend on the seasons.

Around 660 kg of solid residues are produced for each milled ton of cane (wet basis) processed. The energy content of the residues from this industry could mean about 85 million tons of oil equivalent [17]. Figure 1 shows the trend of sugarcane production in the world from 1997 to 2010. It shows a sharp increase in the production of sugar cane after 2006. In 2010, the world's sugarcane agro-industry processed more than 1685×10^6 tons of sugarcane. Currently, Brazil is the world's leader among the sugarcane-producing countries. In 2019, the total production of sugarcane raised to 2016 million tons; Australia produces 1.7% of the world's production [18].

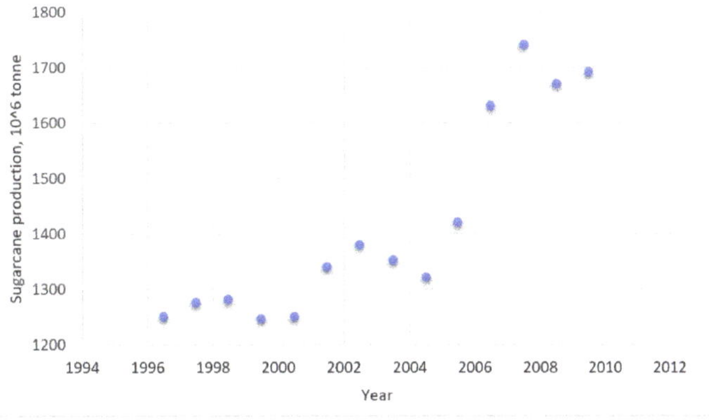

Figure 1. World sugarcane production from 1997 to 2010 [17].

Around 80 per cent of the sugar produced in Australia is exported. Australian major competitors are mainly from Asian Countries, especially Thailand. Thailand exports to Indonesia, Japan, South Korea and China. Countries such as South Korea, Japan and Indonesia import approximately 75 per cent of their sugar from Australia [15,16]. This study aimed to explore the feasibility of a variety of methods in the utilisation of the wastes from the sugar industry in order to generate other revenue streams to help the industry in Australia compete in the international market. Based on the recent findings in the literature, recommendations were made to either support the approach in some of this literature and/or suggest different ways of utilisation based on the authors' interpretation. From this study, it was shown that the milling industries in most countries, including Australia, are not in favor of exploring new methods due to a lack of knowledge related to the commercialisation aspects. In this study, a value proposition analysis was carried out to show the economic and environmental benefits of some of the new utilisation techniques suggested. This study showed, despite the technical issues with some of these methods, the potential of adopting one or more of these technologies is feasible.

2. Methodology

The approach in this article included setting the question of the research, conducting a targeted search tailored to the question, article screening, critical appraisal, data extraction, examination and carrying out calculations. Different database sources were identified to be used in the review, including scientific trusted websites and reports, and only the literature published in English was considered. All of the literature retrieved were screened for relevance, reliability and relevant subjects (country, scale, crop: sugarcane wastes).

The research question was focused on the quality and quantity of biomass waste produced in sugarcane farming and processing. The data collected were analysed, processed and presented in tables and graphs; this included sugarcane productivity, sugarcane waste byproducts, benefits and/or impacts on the environment. The methods and conversion factors of sugarcane to biofuel and other products were presented and considered in the economic calculation.

The main concept of this article is to show the economic advantage of converting waste biomass solid to products. Despite the fact that most of the data are from Australia, the analysis and calculation can be applied anywhere in the world.

3. Waste Utilisation/Recovery

Several byproducts are produced from the milling of cane stalks: wastewater, trash, molasses, bagasse, mill mud and boiler ash. Bio-dunder is another byproduct produced in mills that process molasses to produce ethanol. The utilisation of these byproducts in Australia has remained the same for decades. Bagasse (wet product) is mainly used to fuel the mill boilers and generate electricity; mill mud or filter mud and ash are mixed and mainly used as soil ameliorants or, to a lesser extent, as plant nutrients not far away from the mill. Molasses is mostly used as food for animals and, recently, ethanol production. Sugarcane trash is mostly leftover in the field, and in some places, it is burned in the field before harvesting [19].

3.1. Bagasse

Bagasse is the expended cane fibre that remains after extracting sugar juice. Bagasse is considered a renewable fuel and can also be used as stock feed. Sugar mills burn bagasse to generate electricity and steam for the operation. The mills actually generate more electricity than their need. In Australia, around 400 GWh was fed to the grid in 2012 [16].

Bagasse generally contains 44–53% moisture, 1–2% soluble solids, 1–5% insoluble solids and the remaining is lignocellulosic fibre. A typical constitutive analysis of Australian bagasse fibre on a dry basis shows around 43% cellulose content, as shown in Table 2. The reported composition of bagasse varies because its composition depends upon the growth conditions of the plant, the plant tissue and the age at harvesting [8].

Table 2. A typical constitutive analysis of Australian and world bagasse fibre on a dry basis [8].

Bagasse Constituent	Australia, Weight per Cent	World, Weight per Cent
Cellulose	43	34–47
Hemicellulose	31	24–29
- Xylose	27	–
- Arabinose	4	–
Lignin	23	18–28
Extractives	1	–
Ash	2	–

Bagasse after crushing and extraction processes will be in the form of fibres. The lengths of these fibres vary between 1 and 25 mm. Cellulose is surrounded by lignin and hemicellulose. The hemicellulose provides an interpenetrating matrix for the cellulose microfibrils, while lignin is incorporated into the spaces around the fibrillary elements [8]. Based on moisture content of 52% and the quantity produced per hectare, the potential dry biomass available from bagasse in Australia is around 4.8 million tons. The quantity of each component in bagasse is presented in Figure 2.

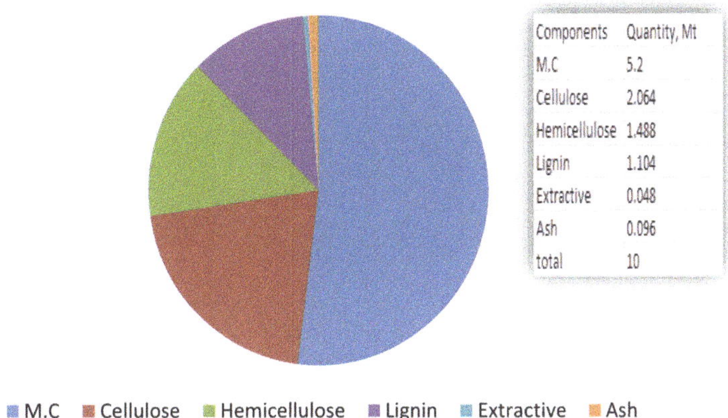

Figure 2. Quantity of each component in bagasse (Mt) in Australia.

The practice of using bagasse as a source of renewable energy reduces Australia's greenhouse gas emissions by over 1.5 million tons annually [16]. Due to the cellulosic nature of the bagasse, there are other methods of utilisation of this byproduct rather than burning it, such as manufacturing of paper pulp and fibreboard, animal feed, manufacturing energy pellets, fermenting to produce ethanol, biogas, butanol or hydrogen and building products.

3.2. Molasses

Molasses is a black syrup that remains after passing the boiled sugar syrup through a centrifugation process. Around 50 per cent of the molasses produced in Australia is exported. The remainder is used in many applications such as stock feed to make industrial alcohol (ethanol), rum and carbon dioxide.

The amount of molasses produced as a byproduct of sugarcane processing is around one million tons. The moisture content and total sugar content of this byproduct are around 23% and 63%, respectively. Molasses composition varies between crushing season months, milling regions and years. The full composition of molasses is presented in Table 3.

Table 3. Mean molasses composition from surveys of NSW and Queensland sugar mills [20].

Molasses Component	NSW (1997–2001)	Queensland (1997–2001)
Dry matter (g/kg)	769 ± 5.9	765 ± 1.0
Total sugars (g/kg DM)	651 ± 9.6	637 ± 1.4
- Reducing sugars (g/kg DM)	214 ± 10.6	183 ± 1.8
- Sucrose (g/kg DM)	436 ± 6.9	454 ± 1.3
Ash (g/kg DM)	164 ± 8.1	176 ± 1.1
Other organic matter (g/kg DM)	184 ± 10.3	187 ± 1.3
Calculated metabolisable energy (MJ/kg DM)	10.6 ± 0.10	10.5 ± 0.02

Around 670,000 tons of Australian molasses (56% of the whole production) was exported in 1996 at prices less than those achieved selling into the domestic market [21]. Molasses used to feed beef cattle have the lowest price per ton. The prices of molasses is around USD 100/t, compared to whole cottonseed (USD 160/t), cottonseed meal (USD 380/t), urea (USD 520/t) and alkali-treated bagasse (USD 154/t) [20]. Molasses at USD 90/ton equates to (processed) barley at USD 122/ton in the ration as a source of metabolisable energy (ME) and provides a cost advantage when barley exceeds USD 122/tonne, or whole cottonseed (WCS) exceeds USD 143 [21]. In 2019 the price of exported molasses increased to USD 120/ton [22].

Typically, Australian molasses is 76.5% DM, has an ME value of 11.0 MJ/kg, crude protein of 5.0% and is high in some minerals. Sugars contribute approximately 65% of the solids, of which sucrose accounts for 70% [21]. Table 4 presents a complete analysis of the molasses produced in Bundaberg.

Table 4. Bundaberg molasses analysis [23].

Specifications	Values
TDN, %	62–65
Dry Matter, %	75.0
Total Sugars, %	50.0
Sucrose, %	35.0
Protein, %	3–5
Calcium, %	1.15
Phosphors, %	0.07
Magnesium, %	0.61
Potassium, %	5.19
Sodium, %	0.1
Chlorine, %	2.98
Sulphur, %	0.73
Copper, mg/kg	11.0
Zinc, mg/kg	11.6
Manganese, mg/kg	82.4
Iron, mg/kg	246.0
Energy, MJ.ME/kg	10.29

3.3. Cane Trash

The plant tops and dry leaves that are left on the field after harvesting are known in the sugarcane industry as sugarcane trash. In the past, farmers used to burn this biomass in the field in order to maintain the cultivation practices of the ratoons. In 1976, a very wet season in North Queensland prompted the re-introduction of green cane harvesting after a gap of more than 30 years. Green cane harvesting is associated with trash blanketing, spreading leaves and other plant residues in a thick layer of mulch over the ground, aiming mainly at the conservation of the soil and water [24].

It was found that the practice of maintaining the post-harvest sugarcane residues on the field has an agronomic benefit plus improved flexibility in harvesting. With the green cane approach, harvesting is still possible without worry about unfavorable weather for burning, such as wet weather and unfavorable wind conditions. Zero tillage often occurs with green cane harvesting, which enhances the movement of farm machinery in wet weather. Moreover, this practice contributes to the protection of soil against erosion, the reduction in variation in the soil temperature (protection from direct radiation), an increase in the biological activity, higher rate of water infiltration (less evapotranspiration), improvement in weed control and increase in the soil carbon stock and nutrients cycling. However, there are also some negative effects associated with this practice. The large amounts of trash left in the field may contribute to the reduction in ratoon sprouting, increased risk of fire, greater incidence of sugarcane pest and disease, and difficulties in mechanised cultivation. Harvesting green costs more because cutting rates are lower (60% to 70%), and losses during harvesting are much higher than for burnt cane. Finally, burning the cane before or the trash after harvesting causes an environmental issue due to the large smoke produced [25,26].

In Brazil, currently, around 85% of sugarcane area is mechanically green harvested, which produces cane trash between 10 to 30 t dry biomass/ha/year [25]. Over 85% of Queensland's sugarcane is now harvested green. In the Burdekin and in northern New South Wales, the adoption of green harvesting is restricted due to the inability of current machines to harvest large cane plants and two-year crops at commercial rates. In some other districts, this practice is inhibited due to older machinery [26].

There are possibilities of using sugarcane trash as fuel for boilers to cogenerate electricity and in the production of second-generation ethanol. In order to gain the benefit of green harvesting and reduce its disadvantages, the question is how much trash (tops and leaves) can be removed from a sugarcane field that can be used for other purposes [25].

A study by Franco et al. (2013) [25] showed that the tops and dry leaves contain different nutrients and moisture; around 80% of N, P and K are derived from tops. It is important to consider this fact when collecting trash. This study showed that more nutrients would be recycled, and fewer mineral fertilizers might be used for sugarcane production if tops are left in the field. It also predicted that for second-generation ethanol production, the pre-treated dry leaves are superior to the tops. Franco et al. suggested that it is more viable to leave the tops in the field and recover parts of the dry leaves for bio-fuel production.

Analysis of the fresh trash samples showed that 67% of trash is composed of tops (12.8 t/ha), and 33% is dry leaves. The studies with post-harvest residues of sugarcane trash in Brazil and in Australia showed the amounts of trash produced per hectare did not vary [25]. Tables 5–7 show the nutrient content, lignocellulosic materials content, and potential amount of trash produced in Australia (both tops and leaves).

Table 5. Trash produced per hectare and its nutrient content.

	Fresh Matter, %	Fresh Matter, t/ha	Moisture, %	N, g/kg	K, g/kg	P, g/kg	Ca, g/kg	Mg, g/kg	S, g/kg
Tops	67	12.8	62	7.5	12.4	0.86	6.8	1.7	1.5
Dry leaves	33	6.3	9.2	3.4	1.8	0.17	5.3	2.5	1.5
Total/average	100	19.1	44.6	4.75	5.29	0.39	5.79	2.24	1.5

Table 6. Composition of trash in regard to lignocellulosic materials and ash content.

		Ash	Lignin	Cellulose	Hemicelluloses
Tops	%	4.7	21.7	39.7	32.0
	t	84,813	391,586	716,403	577,454
Dry leaves	%	4.7	22.7	40.8	28.7
	t	99,746	481,754	865,885	609,091

Table 7. Amount of tops and leaves produced in Australia.

Byproduct	Produced, t/ha	Land Harvested, ha	Produced, t	Moisture Content, %
Trash, tops	12.8	371,000	4,748,800	62
Trash, leaves	6.3	371,000	2,337,300	9.2
Total/average	19.1	371,000	7,086,100	44.6

4. Products from Sugarcane Wastes

4.1. Ethanol

4.1.1. From Molasses

Bioethanol and biodiesel are the only non-fossil liquid transport fuels currently of significance on a global scale. In 2007, the world production of biofuels exceeded 111 million L/d, which is sufficient to cover 1.5% of total road transport fuel use. Biofuel production is forecast by EIA to grow by about 8.6% annually to approximately 938 million L/d in 2030, increasing to 5.5% of total liquid fuel consumption [6]. Global production of ethanol is expected to increase by 2030 to 132 billion L [27].

Ethanol can be produced from a variety of sugarcane feedstocks, including juice, molasses and crystal sugar. Molasses (final) contains around 50% sugar. Ethanol can be used in several products, including perfume, toiletries, cleaning products and shoe polish, as well as being used as a fuel. In Australia, industrial-grade ethanol is manufactured from final molasses at Sarina by CSR Distilleries Operations Pty Ltd. and at Rocky Point by the ethanol plant attached to Rocky Point Mill. Bundaberg Distillery uses ethanol from fermented molasses to make rum. Bio-dunder is a waste product of the process and is used as a fertilizer on cane farms. Table 8 shows ethanol plants in Australia, the feedstock and the capacity of each plant.

Table 8. Ethanol plants in Australia [28].

Ethanol Plant	Feed Stock	Location	Owner	Installed Capacity, ML
Dalby Bio-Refinery	Red Sorghum	South, QLD	Dalby Bio-Refinery Pty Ltd.	80
Manildra Ethanol Plant	Waste Starch	Coastal, NSW	Manildra Group	300
Sarina Distillery	Molasses	Central, QLD	Sucrogen	60
		Total Capacity (ML)		440

In the fermentation of molasses or sugarcane juice, sucrose is hydrolysed to hexoses (glucose and fructose) which are fermented to ethanol, as shown in Equations (1) and (2) [6].

$$C_{12}H_{22}O_{11} + H_2O \rightarrow 2C_6H_{12}O_6 \tag{1}$$

$$C_6H_{12}O_6 \rightarrow 2C_2H_5OH + 2CO_2 \tag{2}$$

The most used feedstock to produce ethanol in the cane sugar industry is final molasses. The production of final molasses in Australia is around one million tons per annum.

The updated Gay–Lussac equation for the fermentation of sugars to ethanol is as follows [29]:

Reaction: $C_6H_{12}O_6 \rightarrow 2C_2H_5OH + 2CO_2$
Molar mass balance (kmol $\times$ kg/kmol) 1$\times$180.16 2$\times$46.07 2$\times$44.01
Mass balance (%) 100.00 51.14 48.86

The maximum theoretical yield of ethanol, based on molar mass balance, is 51.14, and CO_2 is 48.86 mass units produced per 100 mass units of dextrose. A series of experiments carried out by Pasteur in 1989 demonstrated that the maximum practical yield could not be more than 48.40 mass units of ethanol per 100 mass units of dextrose. This is because some of the dextroses are consumed in side reactions necessary for ethanol synthesis. There are many products from the side reactions such as glycerol, succinic acid, acetic acid and others. The yields between 88% and 94% are considered good in practice. The distillation efficiency is another factor that should be considered; distillation efficiencies are usually in the order of 98.5% or higher [29].

Table 9 shows the potential of production of ethanol from molasses in Australia, considering the variables involved in the process, such as conversion and distillation efficiencies.

The residue from the fermentation of final molasses can be concentrated and is called CMS (condensed molasses soluble). Using final molasses for ethanol production will affect animal feed, so CMS can be used to provide a substitute for final molasses in some animal feeds [8].

4.1.2. From Bagasse/Trash

Bagasse

Bagasse is lignocellulosic biomass that can be used for biofuel production such as ethanol. Sugarcane bagasse is an economically viable and very promising raw material for bioethanol and biomethane production. Converting cellulose to fermentable monosaccharides is a technical challenge. This process is hindered by many physico-chemical, structural and compositional factors. Pre-treatments are introduced to overcome these difficulties, such as with steam explosion, alkaline, diluted and concentric acid and ammonia.

One or some of these pre-treatments may be required prior to enzymatic hydrolysis in order to make the cellulose more accessible by the enzymes [1].

Table 9. Potential of ethanol production form final molasses in Australia.

Production Steps	Estimated
Molasses, t	1,000,000
Sugar content, %	50
Sugar (hexoses), t	500,000
Fermentable to Ethanol, %	48.40
Ethanol potential, t	242,000
Distillation efficiency, %	98.5
Ethanol lost in the distillation, t	3630 (1.5%)
Ethanol lost as side products, t	13,700 (51.14 − 48.4 = 2.74%)
Ethanol produced, t; ML; L/t molasses	238,370; 302; 302
For fermentation yield of 88% *, t; L/t molasses	209,766; 266
CO$_2$ produced, t	244,300
Total gain, selling molasses, (AUD 100/t)	AUD 100,000,000/year
Energy content, kJ/kg; Total energy GJ	29,677; 6,225,226
Total gain, based on AUD 3.93/GJ Based on 1 tonne of black coal, which can produce 28 GJ energy at a price of AUD 110, this means AUD 3.93 per GJ.	AUD 24,465,137/year
Based on AUD 1/L price of ethanol	AUD 266,000,000/year

Ethanol density: 789 kg/m3, * O'Hara, 2010.

The hydrolysis reactions for cellulose can be described by Equation (3), which results in the production of glucose monomers. However, the hydrolysis of hemicellulose, as shown in Equation (4), produces pentose monomers (five-carbon sugars), xylose and arabinose [8].

$$(C_6H_{10}O_5)_n + nH_2O \rightarrow nC_6H_{12}O_6 \tag{3}$$

$$(C_5H_8O_4)_n + nH_2O \rightarrow nC_5H_{10}O_5 \tag{4}$$

As shown in Equations (3) and (4) above, in the hydrolysis reaction of cellulose, the molecular weight increases by 11.1% and for hemicelluloses by 13.6%. Hydrolysis of cellulose is very important for ethanol production. It is glucose, not cellulose, that can be consumed by the fermenting bacteria. Cellulose is very stable under many chemical conditions because it has a crystalline structure due to the dense packing of cellulose chains. Cellulose is not soluble in water, many organic solvents, weak acids and weak bases.

Bagasse contains around 43% cellulose (dry basis) and has a complex structure, and for these reasons, pre-treatment is required. The harsh nature of the pre-treatment processes may lead to the formation of several degradation products, which may reduce hexose and pentose yields. Moreover, these products can be inhibitory to the organisms involved in the fermentation of the sugars to ethanol. These degradation products include furfural, 5-hydroxymethylfurfural, levulinic acid, formic acid, and acetic acid [8].

The crystalline nature of cellulose restricts the achievable theoretical yield of glucose from cellulose hydrolysis. While hemicellulose can only be hydrolysed to pentoses using mild acid, glucose can be fermented at very high efficiencies using conventional fermentation organisms. Fermentation of pentoses by yeasts and other organisms happens at a slow rate. Currently, the focus of research is on improving enzyme and fermentation organism effectiveness and minimising the formation of degradation products.

In Australia, there is in excess of 10 million tons of bagasse potentially available for the manufacture of ethanol. Bagasse represents a low-cost raw material when compared to wood, and its estimated value to the sugar mills is typically around AUD 40 per dry ton. Currently, bagasse is burned to produce steam and electricity [7].

A study by O'Hara (2010) stated that an ethanol yield of around 340 L/t dry fibre could be achieved. This consists of about 260 L/t dry fibre from the cellulose component and 80 L/t dry fibre from the hemicellulose component of the fibre [8]. The potential of ethanol produced from the dry fibre in bagasse is shown in Table 10.

Table 10. Potential of ethanol production from bagasse in Australia.

Bagasse Content	In Wet Solid, %	Amount in 10 Mt, Mt	Ethanol Potential, L/t	Amount of Ethanol, L in 3.552 Million Tonne Dry Fibre
Cellulose	20.64	2.064	340	1,207,680,000
Hemicellulose	14.88	1.488		
Lignin	11.04	1.104		Leftover solid
Assume 20% leftover solid from cellulose and 50% from hemicellulose.				
Leftover dry solids (0.2064 × 0.2 + 0.1488 × 0.5 + 0.1104 = 0.226 kg/kg wet bagasse)				

In order to convert the crystalline structure of cellulose to an amorphous form, high temperature and pressure are required (>300 °C and 25 MPa). Generally, there are two methods to hydrolyse cellulose: chemically and enzymatically. The chemical method involves using concentrated strong acids and high temperature and pressure. This method results in toxic byproducts, which will affect the fermentation step and the fermenting bacteria. The enzymatic method is a milder treatment and seems to be much favorable to hydrolyse cellulose.

Table 11 shows the amount of energy produced from bagasse compared to bagasse-based ethanol. The gain can be around 171 million dollars in favour of ethanol production. It is worth mentioning that the number of wet solids leftover from this process is around 0.226 kg/kg wet bagasse, which can be used as boiler fuel. If ethanol is sold for AUD 1/L, then the total revenue from ethanol can be above AUD 171 million.

Table 11. Energy production from wet bagasse compared to bagasse-based ethanol.

Sugarcane Waste	Energy, kJ/kg	Conversion	Energy Content in 1 kg, kJ	Comments
Wet Bagasse	7588	—	7588	Expensive to transport and store
Ethanol	29,677	0.268 kg ethanol/kg wet bagasse	7953 + 3990 = 11,943 kJ	Easy to transport and store. The 0.226 kg solid/kg wet bagasse leftover can produce 3990 kJ when burned wet (17,659 kJ/kg × 0.226 kg)
Net Energy gain			4355 kJ/kg	
Net energy gain for 10 million tonnes bagasse			43,550,000 GJ	
One tonne of black coal can produce 28 GJ energy at a price of AUD 110, this means AUD 3.93 per GJ. The saving is around AUD 171,151,500/year				

Trash

Franco et al. (2013) [25] reported that sugarcane trash, both tops and dry leaves, could be pre-treated hydrothermally. Hydrothermal treatment was carried out at 190 °C for 10 min reaction time and 1:10 (m:v) solid–liquid ratio. The pre-treated samples then underwent enzymatic hydrolysis. The percentage and amount of cellulose, hemicellulose and lignin in the trash (dry basis), both the tops and leaves, were estimated and presented in Table 12.

Table 12. Composition and amount of trash in regard to lignocellulosic materials after hydrothermal treatment.

	Tops, %	Amount, t	Dry Leaves, %	Amount, t
Cellulose	27.1	489,031	33.6	713,082
Hemicelluloses	8.8	158,799	7.7	163,414
Lignin	15.8	285,117	18.4	390,497

The hydrothermal pre-treatment aims to remove a large fraction of hemicellulose and the partial fraction of lignin. Glucose yields obtained after 72 h interval of enzymatic hydrolysis of trash subjected to hydrothermal pre-treatment are presented in Tables 13 and 14.

These processes considerably increase cellulose conversion to sugars. Based on the conversion of glucose to ethanol of 51.1% as reported by Lavrack (2003) [29], the potential amount of ethanol that can be produced from the trash from both the tops and the leaves is estimated as shown in Tables 13 and 14.

Table 13. Potential of ethanol production from trash/tops.

Bagasse Content	Amount, t	Conversion to Glucose after Hydrolysis, %	Amount of Glucose, t	Ethanol Potential, Conversion%	Amount of Ethanol, t	Amount of Ethanol kg/t Wet Tops
Cellulose	489,031	63.6 ± 1.7	311,023	51.1	158,933	33.5
Hemicellulose	158,799	—	—	—	—	—
Lignin	285,117	—	—	—	—	—
Solid leftover, t (37% Cellulose + Hemicellulose + Lignin)	624,859	Solid leftover is 0.1315 kg/kg wet tops				

Table 14. Potential of ethanol production from trash/leaves.

Bagasse Content	Amount, t	Conversion to Glucose after Hydrolysis, %	Amount of Glucose, t	Ethanol potential, Conversion%	Amount of Ethanol, t	Amount of Ethanol kg/t Wet Leaves
Cellulose	713,082	61.0 ± 1.0	434,980	51.1	222,274	95.1
Hemicellulose	163,414	—	—	—	—	—
Lignin	390,497	—	—	—	—	—
Solid leftover, t (39% Cellulose + Hemicellulose + Lignin)	832,014	Solid leftover is 0.356 kg/kg wet leaves				

The net energy gain from the produced ethanol and the solid remains after the fermentation process are estimated in Table 15. In Table 15, the potential energy produced from wet trash burned in the boiler is compared to the energy produced from converting the trash to ethanol. The saving can be around AUD 134,626,192/year; note that this does not include the capital and operation costs for producing ethanol from the trash. As shown in Table 15, the leaves can produce more ethanol compared to the tops, and this supports the suggestion made by Franco et al. (2013) [25] that the tops can be left at the field and the leaves used for alcohol production. If ethanol is sold for AUD 1 then the total revenue from ethanol (162.8 L/t wet tops and leaves) can be around AUD 380 million (tops AUD 158,933,000 and leaves AUD 222,274,000).

Table 15. Energy production from wet trash compared to ethanol-trash-based.

Sugarcane Waste	Energy, kJ/kg	Conversion, kg/kg Wet	Energy Content in 1 kg Wet, kJ	Comments
Wet trash	7588	—	7588	Expensive to transport and store
Ethanol from tops	29,677	0.0335	994 + dry solid (0.1315 × 17,659) = 3316	Easy to transport and store
Ethanol from leaves	29,677	0.095	2819 + dry solid (0.356 × 17,659) = 9105	Easy to transport and store
Net Energy gain			4834 kJ/kg wet	
Net energy gain for 7,086,100 tonne wet tops plus leaves			34,256,028 GJ	
One tonne of black coal can produce 28 GJ energy at a price of AUD 110, and this means AUD 3.93 per GJ. The saving is around AUD 134,626,192/year				

4.2. Other Alcohols

4.2.1. ABE—Bagasse/Trash

Butanol, an industrial solvent, can also be produced from renewable resources such as molasses, corn, wheat straw (WS), corn stover/fibre and other agricultural byproducts. Butanol is a superior fuel to ethanol. The fermentation process of butanol produces acetone–butanol–ethanol (ABE) with a typical ratio of 3:6:1 [30].

A batch fermentation experiment run with glucose (59.4 g/L) as a substrate for 72 h produced 21.37 g/L total ABE (0.31 g/L·h). The conversion of glucose to ABE was around 36% (21.37/59.4). When pre-treated wheat straw was used as a substrate, the optimum production of ABE was achieved, the ratio of acetone, butanol and ethanol in the fermentation broth was found to be 3:6:1, around 2.72 g/L acetone, 6.05 g/L butanol and 0.59 g/L ethanol in 50 h [30,31]. While in a study by Qureshi et al. 2008 [31], the productivity of 0.36 g/L·h and 0.77 g/L·h was observed in normal and highly active cultures, respectively. C. beijerinckii P260 was used as a hydrolysate to produce butanol from waste solid in integrated fermentations.

In a study by Jonglertjunya et al. (2013) [32], butanol production from the fermentation of sugarcane bagasse was considered using Clostridium sp. The bagasse was first pre-treated mechanically by a ball mill to reduce its particle size and then by acid hydrolysis at different temperatures. The results showed that 24 h butanol fermentation of sugarcane bagasse hydrolysate by Clostridium beijerinckii (TISTR 1461) provided the highest butanol concentration of 0.27 g/L. The bagasse (dry)-to-solvent ratio used was 1:10. However, ethanol and acetone were observed to be very low (<0.05 g/L).

In another study, Congcong Lu (2011) [33], showed that butanol could be produced from various agricultural bio-wastes using selected mutant strains of C. beijerinckii. Corn fiber, cassava bagasse, wood pulp and sugarcane bagasse were investigated as potential feedstocks for butanol production from ABE fermentation. In batch fermentation, 12.7 g/L and 15.4 g/L ABE were obtained using corn fibre hydrolysate and cassava bagasse hydrolysate, respectively. For wood pulp hydrolysate and sugarcane bagasse hydrolysate, 11.35 g/L and 9.44 g/L ABE, respectively, were produced [33]. By assuming the same conversion rate of the cellulose in the tops, leaves and bagasse, Tables 16 and 17 show the potential amount of ABE produced and the revenue that can be generated from these wastes.

Table 16. Potential amount of ABE produced from trash and bagasse.

		Cellulose Amount, t	Convert to Glucose, Average, %	Glucose Convert to ABE, %	Amount of ABE, t	Amount Of ABE, Kg/Kg Wet Waste
Trash	tops	489,031	61	36	107,391	0.0226
	leaves	713,082	61	36	156,592	0.0669
Bagasse		2,064,000	61	36	453,254	0.0453

Table 17. Energy production from wet trash and bagasse compared to ABE-trash/bagasse based.

Sugarcane Waste		Energy, ABE kJ/kg	Amount of ABE, kg/kg Wet Waste	Energy Content, kJ/kg Wet Solid	Total Amount, t	Total Energy, GJ
ABE ABE 3:6:1 Average Energy = (3 × A + 6 × B + 1 × E)	Tops	31,377	0.0226	709 + dry solid (0.094 × 17,659) = 2369 *	4,748,800	11,249,650
	Leaves	31,377	0.0669	2099 + dry solid (0.24 × 17,659) = 6337 *	2,337,300	14,811,844
	Bagasse	31,377	0.0453	1421 + dry solid (0.26 × 17,659) = 6012 *	10,000,000	60,123,400

* solidleftover is Hemicellulose+ Lignin, kg dry solid/kg wet solid × kJ/kg dry solid = kJ/kg wet solid.

4.2.2. Furfural—Bagasse/Trash

Furfural is produced from several agricultural wastes such as corncobs, cottonseed hull, oat hull, bran, sawdust, bagasse and rice hull. The most readily available and potentially low-cost raw material for furfural production is bagasse. These biomasses contain the polysaccharide hemicellulose (pentosan); when hemicellulose is heated with sulfuric acid, it undergoes hydrolysis to yield the monosaccharide xylose. Under the same conditions of heat and acid, xylose undergoes dehydration to provide furfural [34].

Polysaccharide hemicellulose is available in many plant materials. It is a polymer of sugars that contains five carbon atoms each. When heated with sulfuric acid, it undergoes hydrolysis to yield sugars, principally xylose. Under the same conditions of heat and acid, xylose and other five-carbon sugars undergo dehydration, losing three water molecules to become furfural: $[C_5H_{10}O_5 \rightarrow C_5H_4O_2 + 3H_2O]$.

In a study by Uppal et al. (2008) [34], the experiments were carried out to show furfural production from sugarcane bagasse. Bagasse samples were treated with different acids at variable concentrations, solid–liquid ratio of 1:15 at 110 °C and 1.05 kg/cm^2 steam pressure for different reaction times of 30, 60 and 90 min. The study showed that 2% H_2SO_4 and 90 min reaction provided the highest yield. The highest amount of furfural was 14%.

Furfural can be converted to furfural alcohol with flammable hydrogen gas and requires pressure equipment. Based on a study by Wondu Business and Technology Services (WBTS; 2006) [35], around 9.2% furfural and 4.2% acetic acid could be produced from hydrolysis and dehydration of foliage wood (dry basis).

By considering the conversion rate presented by WBTS (2006) [35], Table 18 shows the amount of furfural and acetic acid that can be produced from the hemicellulosic portion of trash and bagasse.

Table 19 shows the revenue that can be made from investing in the production of furfural and acetic acid from these cellulosic byproducts. Note that the capital and operating costs associated with this process are not included.

Table 18. Potential amount of furfural produced from trash and bagasse.

		Hemicellulose Amount, t	Convert to Furfural, Yield, %	Amount of Furfural, t	g/kg Wet	Convert to Acetic Acid, Yield, %	Amount of Acetic Acid, t	g/kg Wet
Trash	tops	158,799	9.2	14,609	3.1	4.2	6669	1.4
	leaves	163,414	9.2	15,034	6.4	4.2	6863	2.9
Bagasse		1,488,000	9.2	136,896	13.7	4.2	62,496	6.3

Table 19. Potential revenue of furfural produced from trash and bagasse.

Liquid Products		Amount of Furfural, t	Price, AUD/t *	Revenue Generated, AUD	Amount of Acetic Acid, t	Price, AUD/t	Revenue Generated, AUD
Trash	tops	14,609	1200	17,530,800	6669	550	3,667,950
	leaves	15,034	1200	18,040,800	6863	550	3,774,650
Bagasse		136,896	1200	164,275,200	62,496	550	34,372,800
Solid left after fermentation		Energy, kJ/kg dry solid	Amount of dry solid, t (cellulose + lignin)		Total, GJ		Revenue
Trash	tops	17,659	774,149		13,670,697		53,725,839
	leaves	17,659	1,103,579		19,488,101		76,588,239
Bagasse		17,659	3,168,000		55,943,712		219,8587,88
One tonne of black coal can produce 28 GJ energy at a price of AUD 110, this means AUD 3.93 per GJ.							
Total revenue from the liquids produced and the solids remained							
Trash	tops			74,924,589			
	leaves			98,403,689			
Bagasse				418,506,788			

* assumption made AUD = USD.

4.3. Hydrogen—Bagasse/Trash

An investigation by Tanisho et al. (1996) [36], hydrogen production from anaerobic fermentation of sugar showed a conversion efficiency of 2.4 mol H_2/mol sugar. The direct fermentation of sugarcane juice for the production of H_2 does not seem feasible; it will be a major diversion from sugar or ethanol production. The available energy is around 1/1.35 of that in ethanol, and it comes at a price around 1.3-fold that of producing ethanol [37].

Lignocellulosic materials such as bagasse can be used to produce hydrogen after pre-treatment. There are two pre-treatment methods for sugar extraction from bagasse biomass, and in both methods, Enterobacter aerogenes MTCC 2822 is used: (a) hot water treatment (10% w/v, lasting 2 h,110 °C); the resulting extract contained ~1% sucrose (w/w), while in (b), bagasse is treated by acid hydrolysis (HCl 1N, 80 °C, 2 h) wherein the sugar yield increased to 2.5% (w/w).

Table 20 shows the potential of hydrogen production from generated sugar and the byproducts of sugarcane processing. In the case of bagasse as a substrate, the water extraction route seems to be a better alternative; although sugar recovery is limited to 1% (w/w), the advantage is complete recovery, and the spent bagasse can be reused as fuel or composting. Based on the study by Tanisho et al. (1996) [36], theoretical hydrogen production from sugar can be presented by the following equation.

Reaction: $C_6H_{12}O_6 \rightarrow 2.4\,H_2 + $ others
Molar mass balance (kmol × kg/kmol) 1 × 180.16 2.4 × 2
Mass balance (kg) 180.16 4.8

Table 20 shows the production of 26.6 kg H_2 for each ton of sugar; this means a conversion of 2.6% only. Table 21 shows that converting sugar to hydrogen is not a feasible option; the revenue generated from sugar-based hydrogen is only 3.8% of the revenue generated from selling the sugar.

Table 20. Potential of production of hydrogen from sugar, molasses, bagasse and trash [36,38].

Substrate		H2 Production	H2 Production	Sugar and Byproduct Production, t	Total H_2 Production, mm^3
Sugar		2.4 mol H_2/mol sugar *	313.4 m^3/t sugar	4,360,000	1366.4
Molasses (53% sugar)		2.4 mol H_2/mol sugar *	166.1 m^3/t molasses	998,440	165.8
Bagasse	Acid hydrolysis route, sugar 2.5% (*w/w*)	0.072 m^3H_2/kg of dry bagasse	72 m^3/t dry bagasse	5,000,000	360
	Water extraction route, sugar 1% (*w/w*)	0.055 m^3H_2/kg of dry bagasse	55 m^3/t dry bagasse	5,000,000	275
Trash Water extraction route, sugar 1% (*w/w*)	Tops	0.055 m^3H_2/kg of dry trash	55 m^3/t dry tops	1,804,544	99.25
	Leaves	0.055 m^3H_2/kg of dry trash	55 m^3/t dry leaves	2,122,268	116.7

Gas density of hydrogen: 0.085 kg/m^3 (at 1.013 Bar and 15 °C), * hydrogen production of 1.85 mol H_2/mol hexose [39].

Table 21. Potential of hydrogen production from bagasse.

	mm^3, kg	Energy Content MJ/kg	Total Energy GJ	Gain, AUD
Sugar	1366.4; 116,144,000	142	16,492,448	64,815,320
Molasses (53% sugar)	165.8; 14,093,000	142	2,001,206	7,864,739
H2 from bagasse, acid hydrolysis	360; 30,600,000	142	4,345,200	17,076,636
H2 from bagasse, water extraction	275; 23,375,000	142	3,319,250	13,044,652
H2 from Trash—tops, water extraction	99.25; 8,440,000	142	1,198,480	4,710,026
H2 from Trash—leaves, water extraction	116.7; 9,920,000	142	1,408,640	5,535,955
Solid left after fermentation	Energy, kJ/kg dry solid	Amount of dry solid, t (hemicellulose + lignin)	Total, GJ	Revenue
Bagasse, Water extraction	17,659	2,592,000	45,772,128	179,884,463
Trash, water extraction — Tops	17,659	969,040	17,112,277	67,251,250
Trash, water extraction — Leaves	17,659	1,090,845	19,263,231	75,704,501

One tonne of black coal can produce 28 GJ energy at a price of AUD 110, and this means AUD 3.93 per GJ.

Sugarcane Trash is assumed to have a similar conversion to bagasse in order to estimate the hydrogen production potential for this byproduct. The potential amounts of energy and the revenue from this energy that can be produced from sugar, molasses, bagasse and trash are summarised in Table 21. The table shows the value of the hydrogen produced based on the cost of one GJ of energy from coal.

Based on the price of molasses of USD 100/t, the revenue generated from molasses-based hydrogen is only 7.8% of the molasses value in the market. Both options of producing hydrogen from sugar and molasses are not feasible based on the current conversion ratio and the available technologies in this regard.

Based on the water extraction method for producing hydrogen from the bagasse and trash, Table 21 shows the revenue that can be generated. In these cases, only cellulose was assumed to be consumed in the process.

4.4. Ethanol and Biogas—Bagasse/Trash

Anaerobic digestion (AD) of sugarcane waste is a promising strategy because the digestate could still be used to partially replace mineral fertilizers in addition to the production of biogas. Biogas could be upgraded to biomethane and sold as a new energy produced by the sugarcane plants. However, implementing a large-scale biogas plant using sugarcane waste as a substrate has many challenges. The C:N ratio of bagasse and trash are around 116:1 and 83:1, respectively. This shows a significant lack of nitrogen in these wastes. Moreover, not all the carbon content in these lignocellulosic substrates is available for degradation, such as lignin. In addition, sulfur and phosphorus contents are low. A co-digestion strategy is required to balance the macronutrients of the sugarcane waste and add nitrogen to balance the C:N ratio [40]. The results by Janke et al. (2015) [40] showed that hydrogen peroxide pre-treatment (1 h, 25 °C, 7.35% v/v) provides better results than lime pre-treatment, the total production of methane from the separated liquor and from the enzymatic hydrolysis (3.5 FPU/g, 25 CBU/g) of the solid residue achieved around 49.1 g methane/kg dry bagasse (74.8 L/kg). In addition, this process produced, from the other products (liquor from the enzymatic hydrolysis), around 201.5 g ethanol/kg dry bagasse and 112.7 g recovered lignin. Biochemical potential (BMP) tests were carried out under mesophilic conditions with the addition of a solution of macroelements (source of N, P, Mg, Ca, K), a solution of oligoelements, a solution of bicarbonate (buffer solution) and inoculum from an anaerobic digester.

In Australia, the price of electricity is around 30 cents/kWh for households, and it is around 15–26 cents for large commercial businesses. A biogas project in Australia can generate electricity for large industrial users for 7–10 cents/kWh. Such a project will generate more income from renewable energy certificates (RECs) and savings by recycling the generated heat. For example, a biogas plant that produces 26.5 million m^3 of biogas (52% methane) annually will be able to generate 55 million kWh of electricity. This means that each one cubic meter of biogas can generate around 2.1 kWh of electricity worth 21 cents of electricity (10 cents/kWh). With further development in the field of pre-treatment of the substrates fed to the anaerobic digester, biogas yield and quality can be enhanced, and this may lead to higher economic benefit for the biogas plant owners [37].

Based on the price of electricity, the revenue from sugarcane waste can be around AUD 8.1 billion for the bagasse and trash (Table 22).

Table 22. Biogas production from sugarcane wastes (bagasse and trash).

Waste Type	Fuel Type	Methane Production, m^3 Methane/Ton of Waste on Dry Weight Basis	Total Amount, t	Electricity Generated, kWh/m^3	Price, AUD/kWh; AUD/L	Value, AUD
Bagasse	Biogas	74.8	5,000,000	2.1	0.1	78,540,000
Trash		74.8	3,926,812	2.1	0.1	61,682,362
			Total			140,222,362
Bagasse	Ethanol	0.255	5,000,000	–	1.0	1,275,000,000
Trash		0.255	3,926,812	–	1.0	1,001,337,060
			Total			2,276,337,060
			Overall Total			2,416,559,422

4.5. Dry Pellets—Bagasse/Trash

Bagasse is the fibrous residue of the cane stalk left after crushing and extraction of the juice. It consists of fibres, water and relatively small quantities of soluble solids—mostly sugar. The average composition of mill-run bagasse is as follows: fibre (including ash) 48%, moisture 50% and soluble solids 2% [41].

The calorific value (CV) of bagasse is provided by the formula [41]:

$$\text{Net CV} = 18\,309 - 31.1\,S - 207.3\,W - 196.1\,A \text{ (expressed in kJ/kg).}$$

where; S = soluble solids % bagasse, W = moisture % bagasse, and, A = ash % bagasse,
    In case of dry material;
    If W = 0, S = 2 and A = 3, then the net CV of bone-dry bagasse = 17,659 kJ/kg.
    In case of wet material;
    If W = 50, S = 2 and A = 1.4 then the net CV of mill run bagasse = 7588 kJ/kg.
    The ratio of the energy content of dry to wet bagasse is 2.33. This means that by drying the bagasse, there is a potential to raise the energy produced by 2.3 times for each kilogram burned.

The energy required for evaporating the water from the wet solid can be calculated as follows:

The amount of energy required to evaporate 1 kg of water at a medium drying temperature of 150 °C is;

$$Q = m \, [\text{Cwater} \times (\text{To} - \text{Ti}) + \lambda]$$

$$Q = 1 \text{ kg } [4.1868 \text{ KJ}/(\text{kg water.K}) \times (150 - 25) \,^{\circ}\text{C} + 2260 \text{ KJ}/\text{kg}] = 2783 \text{ kJ}$$

Most of this energy, around 80%, can be recovered by a heat exchanger (2226.7 kJ). Table 23 shows the energy required and the potential energy that can be recovered from burning dry bagasse.

Table 23. Energy required for drying bagasse.

	Energy for Drying, kJ/kg Water	Total Energy for Drying, GJ	Recovered, kJ/kg (80%)	Total Recovered, GJ	Net Lost, GJ
Dry Bagasse	2783	13,915,000	2226.7	11,133,400	2,781,600

The net energy difference between wet bagasse and dry bagasse can be around 14,925,400 GJ which means a saving of AUD 37,859,262, as shown in Table 24.

Table 24. Net energy difference between wet and dry bagasse.

	Amount, t	Energy Content, kJ/kg	Total, GJ	Lost Due to Drying GJ	Total, GJ
Wet Bagasse	10,000,000	7588	75,880,000	—	75,880,000
Dry Bagasse	5,000,000	17,659	88,295,000	2,781,600	85,513,400
	Net energy gain from drying bagasse				9,633,400
One tonne of black coal can produce 28 GJ energy at a price of AUD 110, and this means AUD 3.93 per GJ. The saving is around AUD 37,859,262					

By using a similar approach for sugarcane trash, the revenue that can be made from burning wet trash can be around AUD 35 million. This can be increased by AUD 5 million when the trash is dried before burning it in the boilers (Table 25). However, these figures do not include the capital and operation costs of collecting, storing and hand selling the trash from the field.

Table 25. Net energy difference between wet and dry trash.

		Amount, t	Energy Content, kJ/kg	Total, GJ	Lost Due to Drying GJ	Total, GJ
Wet trash	Tops	4,748,800	4472	21,239,150	—	21,239,150
	Leaves	2,337,300	15,417	36,036,421	—	36,036,421
Dry trash	Tops	1,804,544	17,659	31,866,442	1,639,950	30,226,492
	Leaves	2,122,268	17,659	37,477,137	119,772	37,357,365
	Net energy gain from drying Trash				Tops	8,987,342
					Leaves	1,320,944
One tonne of black coal can produce 28 GJ energy at a price of AUD 110, and this means AUD 3.93 per GJ. By collecting and burning the trash (tops), the revenue generated can be as much as AUD 35,320,254. By drying and burning the trash (leaves), the revenue can be increased by AUD 5,191,309.						

Superheated steam drying may be suitable for drying these wastes as they are non-sensitive materials and save energy. There are many advantages for superheated steam drying that have been reported in the literature, one of these is the ability to recycle the steam through the process [42]. The only energy leaving the system in this case is the extra vapor that has been evaporated from the materials under drying, which saves energy.

Dust collection in superheated steam drying is much simpler by passing the excess vapor through condensers [43,44]. Pronyk et al. (2004) [45] illustrated that the use of superheated steam as a drying medium for non-temperature sensitive products could lead to energy savings as high as 50–80% over the use of hot air or flue gases. These savings can be achieved due to higher heat transfer coefficients and the increased drying rates in the constant and falling periods if the steam temperature is above the inversion temperature. Some valuable volatile organic compounds generated from the drying material can be recovered and separated by a condenser [46,47]. Some applications involving steam drying are drying of fuels and biofuels with high moisture contents prior to combustion in a boiler and cattle feed exemplified by sugar beet pulp, lumber, paper pulp, paper and sludge [48].

4.6. Pyrolysis and Gasification—Bagasse/Trash

Pyrolysis and gasification are thermochemical processes where the gases and liquids produced can be used as energy. These processes produce a combination of gases such as methane (CH_4), hydrogen (H), carbon monoxide (CO), carbon dioxide (CO_2) and light hydrocarbons. Moreover, biochar and tar (bio-oil) are produced. These products have high commercial value and can be produced from the pyrolysis of bagasse. The pyrolysis of sugarcane bagasse occurs in the absence of oxygen and can be accomplished in two ways, with or without carrier gas (nitrogen). The literature emphasises that it is very difficult to perform this process without the use of carrier gas (N_2); this occurs because of the formation of soot. Gasification takes place in the presence of oxygen in the form of air, pure oxygen or steam, and it is not therefore necessary to use a carrier gas. The gasification process can be defined as a process of partial combustion because it uses an amount of air less than that required stoichiometrically [49].

A company called Agri-Therm built the first bubbling fluidised bed mobile pyrolysis unit on the market. The cost of the mobile unit is estimated to be AUD 1.5 million, processing 3600 dry tones (20% MC) per year. The biochar produced from bagasse is expected to be of lower technical quality and can be valued at the lower end of the price range (AUD 250 per ton). Energy costs are minimal as the non-condensable gases, and entrained bio-oil from the pyrolysis are used to provide an energy self-sustaining operation [50].

Table 26 shows the results of a study by Figuero et al. (2013) [49]; the percentages of gas, tar and char produced at different reaction temperatures in a pyrolysis reaction. The table shows clearly that the increase in temperature favours the production of gas. Table 27 shows the percentages of gas, tar and char produced at different reaction times in a gasification reaction.

Table 26. Gas composition at different temperatures—pyrolysis [49].

		500 °C	600 °C	700 °C	800 °C	900 °C
	Char, wt%	45.4	41.9	32.9	27.6	26.4
	Tar, wt%	22.1	22.8	24.1	20.9	20.6
	Gas, wt%	32.5	35.3	43.0	51.5	53.1
	H_2	–	3	42	49	42
	CO	8	10	–	–	19
Gas, mole	CH_4	22	29	35	29	18
	CO_2	70	56	23	22	21
	Butane	–	1	–	–	–
	Ethane	–	1	–	–	–

Table 27. Gas composition at 900 °C and different reaction times—gasification [49].

		20 min	40 min	60 min	80 min	120 min
	Char, wt%			7		
	Tar, wt%			27		
	Gas, wt%			66		
Gas, mole	H_2	52	54	57	59	60
	CO	19	19	18	17	16
	CH_4	5	6	5	4	4
	CO_2	24	21	20	20	20
	Butane	–	–	–	–	–
	Ethane	–	–	–	–	–

Tables 28 and 29 show the conversion from mole per cent to weight per cent for the gases produced during the two processes.

Table 28. Converting mole per cent to weight per cent—pyrolysis at 900 °C.

		900 °C, kmole	Mwt, kg/kmole	Wt, kg	Wt %
Gas, mole	H_2	42	2	84	4.6
	CO	19	28	532	29.1
	CH_4	18	16	288	15.8
	CO_2	21	44	924	50.5
	Butane	–	–	–	–
	Ethane	–	–	–	–
		100		1828	100

Table 29. Converting mole per cent to weight per cent—gasification at 900 °C.

		60 min, kmole	Mwt, kg/kmole	Wt, kg	Wt %
Gas, mole	H2	57	2	114	7.2
	CO	18	28	504	31.9
	CH4	5	16	80	5.1
	CO2	20	44	880	55.8
	Butane	–	–	–	–
	Ethane	–	–	–	–
		100		1578	100

Table 30 shows the potential amount of char, tar and gases produced during the pyrolysis and gasification processes for each tonne of wet bagasse.

Table 31 shows the potential energy generated from the gasses produced during the pyrolysis and gasification processes. It is worth mentioning here that the capital and operating costs for these processes are not included in the calculations of the revenues. Based on the revenue generated for each kilogram of dry solid bagasse, the revenue that can be generated from pyrolysis and gasification of cane trash is estimated in Table 32.

Table 30. Amount of char, tar and gases produced during the pyrolysis and gasification processes.

		Pyrolysis			Gasification		
		Wt% at 900 °C	Amount, kg/t dry Bagasse	Amount, kg/t wet Bagasse	Wt% at 900 °C and 60 min	Amount, kg/t Dry Bagasse	Amount, kg/t Wet Bagasse
Char		26.4	264	132	7	70	35
Tar		20.6	206	103	27	270	135
Gas		53.1	531	265.5	66	660	330
	H_2	4.6	24.4	12.2	7.2	47.5	23.75
	CO	29.1	154.5	77.25	31.9	210.5	105.25
Gas	CH_4	15.8	83.9	41.95	5.1	33.7	16.85
	CO_2	50.5	268.2	134.1	55.8	368.3	184.15
	Butane	–	–	–	–	–	–
	Ethane	–	–	–	–	–	–

Table 31. The potential energy and revenue from pyrolysis and gasification.

Utilisation Method	Amount, kg/t Wet Bagasse	Total Amount, t	Energy Content, kJ/kg	Total Energy, GJ	Value, AUD
		Pyrolysis at 900 °C for 1 h			
Char (AUD 250/t)	132		–	–	250,000,000
Tar	103		22,100	22,763,000	89,458,000
H_2	12.2	10,000,000	142,000	17,324,000	68,083,320
CO	77.25		10,100	7,802,250	30,662,842
CH_4	41.95		55,500	23,282,250	91,499,242
CO_2	134.1		–	–	–
Total *					529,700,000
		Gasification at 900 °C and 60 min			
Char (AUD 250/t)	35		–	–	87,500,000
Tar	135		—	—	—
H_2	23.75	10,000,000	142,000	33,725,000	132,539,250
CO	105.25		10,100	10,630,250	41,776,882
CH_4	16.85		55,500	9,351,750	36,752,377
CO_2	184.15		–	–	–
Total *					298,568,000

* The energy required for these processes is not included.

Table 32. The potential revenue from pyrolysis and gasification of cane trash based on AUD/kg generated from bagasse/cane trash.

Bagasse			Dry Solid, kg	Revenue, AUD	Rate, AUD/kg
Bagasse		Pyrolysis	5,000,000	529,700,000	106
		Gasification	5,000,000	298,568,000	60
	Trash		Rate, AUD/kg	Dry solid, kg	Revenue, AUD
Trash—Tops		Pyrolysis	106	1,804,544	191,281,664
		Gasification	60	1,804,544	108,272,640
Trash—Leaves		Pyrolysis	106	2,122,268	224,960,408
		Gasification	60	2,122,268	127,336,080

4.7. Pulp and Paper—Bagasse/Trash

In Australia, in 2003, an estimated 1.45 million tons of pulp were transformed into paper grades, of which 357,000 tones were imported [7]. Table 33 shows the quantity of bagasse produced in Australia and the potentially available quantity of de-pithed bagasse. The fibre content of the cane is 13%; 35% of the dry bagasse fibre is required to be removed as pith. Assuming 45% pulp yield, then chemical pulp produced from bagasse can be as much as 438,750 tons of pulp per year. The separated solid and pith would be returned to the sugar mill to generate energy [7]. The total gain, assuming no loss in the solid during the process, can be around AUD 777,137,455/year, as shown in Table 34. As the trash (top) is similar to the bagasse, the same pulp production rate of 0.088 kg pulp/kg dry solid is assumed to calculate the potential revenue from trash (tops). As shown in Table 34, the revenue that can be generated from the trash (tops) can be as much as 290 million dollars. The trash (leaves) were not considered for pulp due to its unsuitable structure for this process.

Table 33. Potential pulp production from Bagasse in Australia.

	Total Quantity, t	Fibre Content, %	Total Fibre, t	Fibre Is Required to Be Removed as Pith, %	Pulp Yield, %	Potential Pulp, t
Bagasse	10,000,000	13	1,300,000	35	45	438,750
	Total quantity, t	Solid remained after removing the fibre, %	Total solid remained, t	Solid remained as pith	Solid remained based on converted yield, t	Total remained after removing the potential pulp, t
Solid remained, dry, wet	10,000,000	37	3,700,000	455,000	250,250	4,405,250; 9,405,250

Table 34. Potential gain from pulp production from Bagasse and trash in Australia.

	Product	Amount	Price	Total Revenue, AUD
Bagasse	Pulp *	438,750 t	1132 AUD/t **	496,665,000
	Wet Solid, fuel for the boiler (50% MC)	9,405,250 t ***, 71,367,037 GJ	3.93 AUD/GJ	280,472,455
	Total ***			777,137,455
	Product	Production rate, kg/kg dry solid	Total, t	Total produced, t	Total revenue, AUD
Trash—tops	Pulp *	0.088	1,804,544	158,799	179,761,455
	dry Solid, fuel for the boiler	0.88	1,804,544	1,587,998	110,206,904
	Total ***				289,968,359

* Does not include the processing of bagasse to produce pulp. ** retrieved from http://www.indexmundi.com/commodities/?commodity=wood-pulp¤cy=aud (access on 15 April 2021). *** Assume no solid losses during the process.

There are several technical challenges related to the manufacturing process of pulp from bagasse [7]; the following are some of these technical issues:

- Storage: bagasse is a seasonal byproduct from the milling process; it is produced in huge quantities in a short period of time. Large storage facilities and long storage time is required in order for the pulping process to proceed. Special methods of storage are required as bagasse is prone to biological activity. This may lead to color and fibre degradation, as well as loss of fibre properties;
- De-Pithing: bagasse normally contains 30–35% fine, thin walled and low cellulose content cells, which should be removed from the pulping process. These cells are called "Pith Cells", the presence of such fibres can result in higher consumption of chemicals, poor draining pulp and reduced scattering power in mechanical pulps. An enhanced de-pithing process is required to reduce economic losses;
- Silica content: bagasse contains high quantities of silica compared to woody fibre sources such as eucalyptus. Silica is a major issue in the pulping process. The removal, chemical recovery or other reliable methods are required to make the operation practical.

4.8. Product from Sugar—Ethanol

Theoretically, around 48.4% of the sugar can be converted to ethanol. Table 35 shows that converting sugar to ethanol can generate extra revenue of around AUD 974,175,411/year.

Table 35. Ethanol production from sugar.

Products	Production, t/y	Fermentable Ethanol, %	Ethanol, t/Year	Energy Content, kJ/kg	Total Energy, GJ
Sugar	4,360,000	48.4	2,110,240	29,677	62,625,592
	Production, t/y	Price AUD/t		Total AUD	
Sugar	4,360,000	390 *		1,700,400,000	
Based on AUD 3.93/GJ the total gain from converting sugar to ethanol will be around AUD 246,118,576/year or based on AUD 1/l ethanol, the revenue is around AUD 2,674,575,411/year					

* http://www.rabobank.com.au/Research/Documents/Agribusiness_monthly/2013/Agri_Monthly_Jun-2013.pdf (access on 15 April 2021).

5. Sustainability of Sugarcane Production

There are 100 million people on the globe who make a livelihood from the sugarcane industry. The industry needs to overcome significant obstacles to accomplish voluntary sustainability standard (VSS) compliant sugarcane. This includes control pollution and greenhouse gas emission, enhancing producer profitability and providing a healthy, safe environment and respect for the right of the laborers [51].

The European Union imported more than 43 million liters of ethanol produced from Brazilian sugarcane. Brazil produces ethanol from sugar, currently this leading to exploitation and destruction of Amazon and Pantanal native vegetation [52]. This is happening in other countries; for example, Australia is currently working to reverse the impact of nutrient runoff on the Great Barrier Reef, which has been an issue for a long time [26]. Adding to these issues, artificial sugar can threaten this industry and direct its bath to less profitable products of biofuel from sugar.

This article addresses two of the important factors to accomplish sustainability in this industry: enhancing profitability and reducing greenhouse emissions. One way to execute this is by utilising the byproducts (rather than sugar) from this industry, such as bagasse, molasses and cane trash, to produce biofuel and other products that have a higher value. The technology is still progressing to make biofuel from lignocellulosic material economically feasible [53]. Moreover, the use of sugarcane byproducts does not conflict with land use for agricultural products. A small portion of the byproduct biomass can be left on the ground for soil enhancement purposes, and a larger portion can be utilised in the production of biofuel.

6. Recent Development in Sugarcane Industry

Worldwide sugarcane production in 2020/2021 reached 1,889,268,880 tons produced per year. The largest sugarcane producer in the world is Brazil, with 768,678,382 tons per year. The second highest producer is India, with 348,448,000 tons yearly production. Both countries, Brazil and India, produce 59% of the world's total. Australia ranks eighth in the world with 34,403,004, and the production increased by around 4 million tons in the last decade [54]. Worldwide sugar production in the 2019–2020 crop year was approximately 166.18 million metric tons. The global sugar production in 2020/2021 was approximately 179 million metric tons [55]. Sugar production worldwide, by leading country, is presented in Figure 3; the figures are in a million metric tons. Based on the sugar production figures, bagasse, trash, molasses and mill mud can be calculated as they represent percentages of the sugar production.

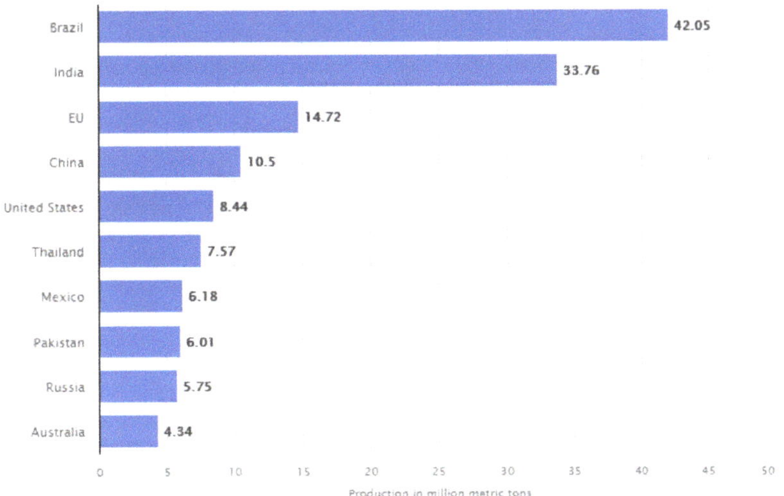

Figure 3. Worldwide sugar production by the ten leading countries [55].

In 2019 the price (in USD) of exported molasses increased to USD 120/ton [22]. Recent prices in 2021 for different energy sources are as following: ethanol USD 0.586/L, coal USD 0.171/kg, natural gas 0.121/m^3, and bagasse USD 29–35/dry ton.

Based on Table 1 and the percentages of each of the byproducts produced per tone of sugar, Table 36 shows the expected global byproducts productions in 2020/2021. Following the same method/concept above, an estimation of biofuel production from these byproducts can be derived. By 2030, global ethanol production is expected to increase to 132 billion L. Currently, maize contributes to about 60% of the ethanol produced, sugarcane 25%, wheat 3%, molasses 2% and the remainder from other grains, cassava or sugar beets [27].

Table 36. Estimated global production of sugarcane byproducts is 2021.

Byproduct	Produced, Tone
Bagasse	9575.87
Trash	6804.3
Molasse	955.92
Mill mud	584.40–592.75

In a study by Strain and Akshaya [56], they showed 74 ± 3 g/L of fermentable sugar from pre-treated bagasse, steam-exploded acid bagasse can be achieved, in addition to bioethanol production of 34 ± 2 g/L during a fermentation period of 36 h [56]. The theoretical amount of bioethanol that can be produced from sugar (glucose) is 0.51 g of ethanol per 1 g of glucose. In a recent study, bioethanol production of 82.1 g/L was reported. Sulphite pre-treated momentary pine slurry (25%, w/w) was used as a substrate. Pre-hydrolysis was performed at 50 °C for 24 h and 200 rpm, and this was then followed by fermentation at 28°C or 35 °C using 5 g/L dry inoculums of S. cerevisiae. It is also reported the production of bioethanol was increased from lignocellulosic materials after pre-treatment in feedstocks with 3.5% H_2SO_4 at 121 °C for 30 min, followed by enzymatic hydrolysis [57].

Currently, wood chips and logging residues represent the largest share of biomass fuels; depending on its composition (ash and moisture contents, and calorific value), its energy content may vary considerably. In general, the moisture content of the wood chips varies from 35% to 45%, which has a calorific value of 18.4 to 19.6 MJ/kg and ash content from 0.5% to 4.5% [58]. Lignocellulosic biomass (trash, bagasse and other agro-residues) generally consists of the combination of lignin (26–31%), hemicellulose (25–32%) and cellulose (41–46%). Methane, hydrogen, ethanol, butanol or other forms of fuels can be generated from lignocellulosic biomass after going through intensive thermal and/or chemical pre-treatments. A renewable supply of bioenergy can then be created by the biological systems [59].

The authors of [60] designed and investigated a system for biomass-based hydrogen production integrated with the organic Rankine cycle. The aim was to predict the performance of the system, under various operating conditions, regarding hydrogen production yield and electricity generation. Different types of biomass such as wood chips, manure and sorghum were compared under the same operating condition. Wood chips demonstrated a maximum hydrogen yield of 11.59 mol/kg. After optimising the system using a genetic algorithm based on the response surface model, a hydrogen yield of 39.31 mol/kg was shown to have achievable [60].

Biogas production in anaerobic digestion can be increased by carrying out the best pre-treatment techniques, trace metal additive and control process conditions such as temperature, pH, C/N and volatile solid percentage. It was reported that thermochemical pre-treatment of bagasse enhanced its anaerobic biodegradation. Moreover, the pre-treatment of biomass with ultrasonic improved biogas production to 71%, the range used was 31–93 W h/L. This has contributed to increasing the solubility of cellulose and hemicelluloses in the feedstock [61].

The recent reporting on pyrolysis yield and product distribution is presented in Table 37. It can be seen that slow pyrolysis targets biochar, fast pyrolysis targets bio-oil, while intermediate pyrolysis targets both [62].

Table 37. Different pyrolysis processes and products' yield [62].

Property	Slow	Intermediate	Fast	Flash
Heating rate (°C/s)	1.1–1	1–10	10–200	>1000
Feed size (mm)	5–50	1–5	<1	<0.5
Reaction temperature (°C)	400–500	400–650	850–1250	>1000
Vapor residence time (s)	300–550	0.5–20	0.5–10	<1
Feed water content (%)	Up to 40	Up to 40	<<10	<<10
Biooil yield (%)	20–50	35–50	60–75	60–75
Biochar yield (%)	25–35	25–40	10–25	10–25
Gas yield (%)	20–50	20–30	10–30	10–30

To develop a global-scale biofuel production from organic trash, a biomass trash collection and effective regulatory strategies need to be established. A study in 2021 [59] discussed the advances in the manufacturing of biofuels from agro-residues utilising the novel technologies currently available. It is obvious that the efficiency and yield of biofuel from sugarcane byproducts (or other agro-residues) have not progressed effectively in the last few decades because it is limited by the chemical/biological reaction and the content of the agricultural byproducts. The only progress is in the processes; nowadays, processes to produce biofuel from biomass become more energy-efficient, have less labor required and more efficient in utilising chemicals/enzymes.

Biofuel yield from biomass has limitations; a development of the current technologies is required to ease the pre-treatment process to overcome the limitations and make the concept of biorefinery commercially viable [63]. The transition to bioeconomy can be pursued by either addressing fossil fuel dependence sustainably or showing a human-activity impact on the environment [64].

The European Commission supports the conversion of renewable biomass into value-added products and bioenergy. In October 2018, a new bioeconomy strategy aimed at promoting a sustainable Europe was commenced. However, not much was conducted to analyse the impacts of bioeconomy sectors development [65]. Subsidies bioenergy (biomethane) for usage as vehicle fuel instead of natural gas for the potential reduction in emissions is essential for the development of the bioenergy industry. The subsidies must be accompanied by other development such as the construction and operation of new fueling facilities and the increase in vehicles derived by biomethane [66].

In this study, the aim was to introduce the investors in this field, manufacture second-generation biofuel from agricultural biomass/byproducts, to the limitation of the production. Despite the advance in chemical, thermal and biological technology, these only will contribute to producing the maximum amount of biofuel limited by the chemical/biological reaction and the content of the raw materials. The yield for each of the biofuel discussed in this article is based on the chemical formula of the raw materials.

The investor should account for the capital and operational cost, and carry out the Net Present Value (NPV) calculation to justify any investment. Investing in this field is volatile due to changes in regulation, price of the raw materials and final products.

7. Conclusions and Recommendations

Table A1 (Appendix A) and Figures 4–6 summarise the findings of this study by comparing the potential energy produced from each byproduct. In many cases, the capital and operating costs for the technology used in utilising the wastes were not included in the estimates. The value of energy is based on the energy content and price of coal. Moreover, the market prices of some of the products were not included. The sugarcane milling industry produces one main product, sugar and four byproducts: molasses, bagasse, cane trash and mill mud/ash mixture. Sugar is an edible product, and it is not recommended to be used for any other purposes even though an extra one billion dollars can be made from converting sugar to ethanol.

Molasses is a byproduct that can be used for human consumption and animal feeds and to produce ethanol due to its high sugar content (around 50–70%). Based on the current market value of molasses, it is recommended that molasses be converted to ethanol to generate revenue of around AUD 165 million. The current practice of burning bagasse in the milling boiler has the potential to generate about AUD 38 million when first dried, around AUD 1.2 billion per year when converted to ethanol and AUD 1.35 billion when converted to ethanol/biogas. It is worth mentioning that fermenting bagasse comes with many technical challenges, such as the pre-treatment processes. Fermenting bagasse to ABE is a good option compared to burning it; however, the revenue generated is much lower than converting it to ethanol. When fermenting bagasse to furfural, acetic acid and hydrogen are not feasible options because the revenues generated from these processes are lower than burning bagasse. Pyrolysis, gasification and production of pulp are all feasible options but still way below ethanol or ethanol/biogas production. The same also applies to sugarcane trash.

As shown in Figure 5, the product that has the highest energy content is the dry bagasse (drying energy is subtracted from the net energy) followed by wet bagasse, products from pyrolysis and gasification. Figure 5 does not consider the leftover solid from these processes. For example, if the energy from the leftover solid from the ethanol production process is added to the energy content of the produced ethanol, the total energy will be approximately 80% of the energy content in the dry bagasse. However, if the energy consumed during the production of ethanol from bagasse is considered, the net energy of this process will significantly be lower than that of dry bagasse. It is obvious that the value of the final product depends on the energy consumed during the processes of conversion. Despite the fact that the amount of ethanol/biogas produced is less than the other products, as shown in Figure 6, these two products have the potential to produce the highest revenue Figure 4.

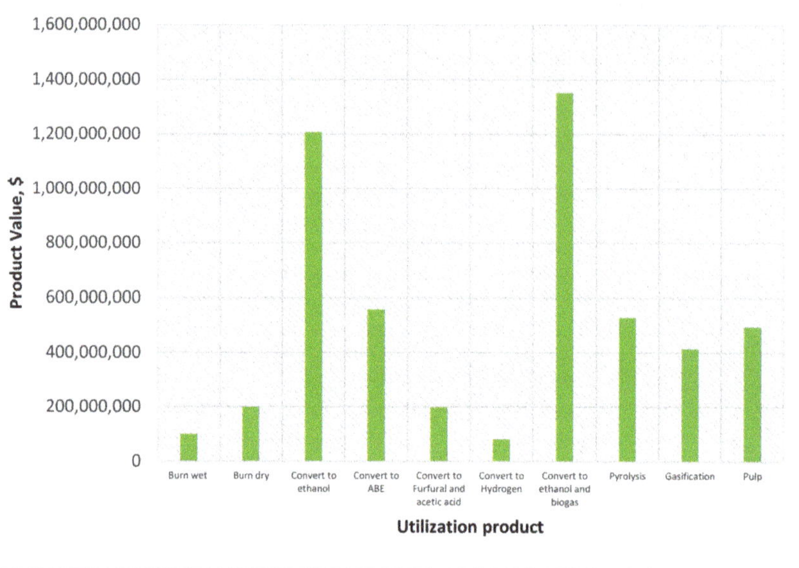

Figure 4. Potential revenue from different utilisation processes of bagasse, the leftover solid was not considered in the calculation.

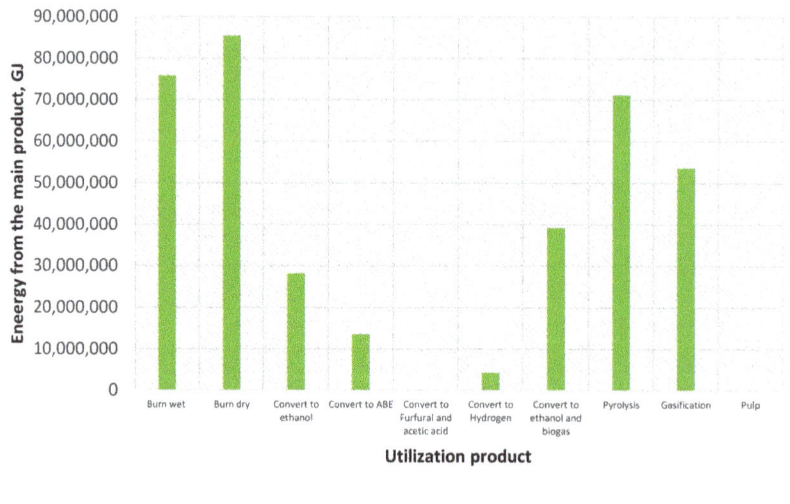

Figure 5. Energy content of the products from different utilisation processes of bagasse, the leftover solid was not considered in the calculation.

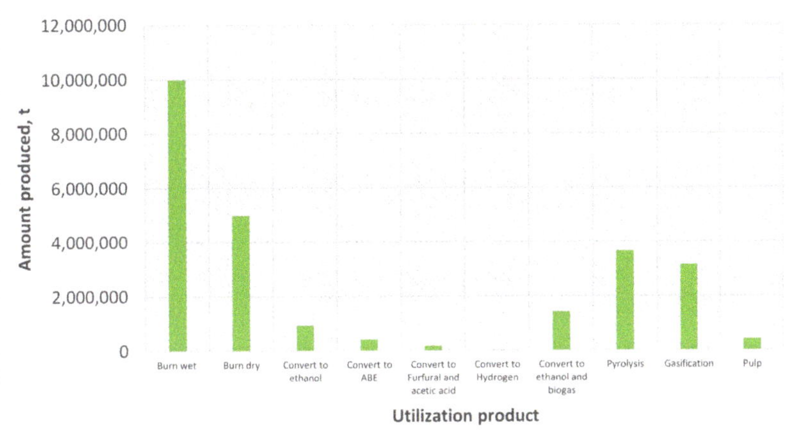

Figure 6. Amount of the products from different utilisation processes of bagasse, the leftover solid was not considered in the calculation.

Author Contributions: Conceptualisation, methodology, validation, formal analysis, investigation, resources, data curation and writing—original draft preparation: I.H. Review, editing and visualisation: W.d.S., S.S. and J.B., All authors have read and agreed to the published version of the manuscript.

Funding: This research received no external funding.

Institutional Review Board Statement: Not applicable.

Informed Consent Statement: Not applicable.

Data Availability Statement: This study did not report any data.

Conflicts of Interest: The authors declare no conflict of interest.

Appendix A

Table A1. Value proposition for different scenarios of utilising the wastes from sugarcane milling industry.

Utilisation Method	Amount, kg/t Material	Total Amount, t/Year	Energy Content, kJ/kg	Total Energy, GJ	Value, AUD
Sugar and Molasses					
Sugar (AUD 390/t)	—	4,360,000	—	—	1,700,400,000
Sugar to Ethanol (based on energy content)	484	2,110,240	29,677	62,625,592	246,118,576
Sugar to Ethanol (AUD 1/L)	484	2,110,240	—	—	2,674,575,411
Molasses (AUD 100/t)	—	1,000,000	—	—	100,000,000
Molasses—Ethanol (based on energy content)	209	209,766	29,677	6,225,226	24,465,137
Molasses—Ethanol (AUD 1/L)	209	209,766	—	—	265,863,000
Bagasse					
Burn Wet	—	10,000,000	7588	75,880,000	298,208,400
Burn Dry (energy for drying included)	500	5,000,000	17,659	85,513,400	336,067,662

Table A1. *Cont.*

Utilisation Method		Amount, kg/t Material	Total Amount, t/Year	Energy Content, kJ/kg	Total Energy, GJ	Value, AUD
Sell dry (AUD 40/t)		500	5,000,000	—	—	200,000,000
Ethanol	Ethanol (Energy)	268.3	952,859	29,677	28,277,996	111,132,526
	Ethanol (AUD 1/L)	268.3	952,859	—	—	1,207,680,000
	dry solid	226	2,260,000	17,659	39,909,340	156,843,706
ABE	ABE (energy)	45.3	435,000	31,377	13,648,995	53,640,550
	ABE (AUD 1/L)	45.3	435,000	—	—	559,259,259
	dry solid	260	2,600,000	17,659	45,913,400	180,439,662
Furfural + acetic acid	Furfural (AUD 1200/t)	13.7	136,896	—	—	164,275,200
	Acetic acid (AUD 550/t)	6.3	62,496	—	—	34,372,800
	Dry solid	0.317	3,168,000	17,659	55,943,712	219,8587,88
Hydrogen, Water extraction	H2 (energy)	3.06	30,600	142,000	4,345,200	17,076,636
	H2 (AUD 2.7/kg)	3.06	30,600	—	—	82,620,000
	Dry solid	260	2,600,000	17,659	45,913,400	180,439,662
Biogas + Ethanol	Biogas (energy)	86.02	430,100	22,000	9,462,200	37,186,446
	Biogas (Electricity)	86.02	430,100	—	—	78,540,000
	Ethanol (energy)	201.2	1,005,975	29,677	29,854,320	117,327,477
	Ethanol (AUD 1/L)	201.2	1,005,975	—	—	1,275,000,000
Pyrolysis at 900 °C for 1 h		—	3,664,000	—	71,171,500	529,703,000
Gasification at 900 °C		—	3,158,500	—	53,707,000	415,819,000
Pulp	Pulp (AUD 1132/t)	0.044	438,750	–	–	496,665,000
	Remain solid	0.94	9,405,250	7588	71,367,037	280,472,455
Total						777,137,455
Burn Wet	Tops	—	4,748,800	4472	21,239,150	83,469,859
	Leaves	—	2,337,300	15,417	36,036,421	141,623,134
Burn Dry + drying energy included	Tops	0.38	1,804,544	17,659	30,226,492	118,790,113
	Leaves	0.91	2,122,268	17,659	37,357,365	146,814,444
Ethanol Tops	Ethanol (energy)	33.5	158,933	29,677	4,716,654	18,536,452
	Ethanol (AUD 1/L)	33.5	158,933	—	—	201,435,995
	dry solid	131.5	624,859	17,659	11,034,385	43,365,133
Ethanol Leaves	Ethanol (energy)	95.1	222,274	29,677	6,596,425	25,923,952
	Ethanol (AUD 1/L)	95.1	222,274	—	—	281,716,096
	dry Solid	356	832,014	17,659	14,692,535	57,741,663
ABE tops	ABE (energy)	22.6	107,391	31,377	3,369,607	13,242,557
	ABE (AUD 1/L)	22.9	107,391	—	—	132,581,481
	dry solid	94	470,000	17,659	8,299,730	32,617,938
ABE leaves	ABE (energy)	66.9	156,592	31,377	4,913,387	19,309,611
	ABE (AUD 1/L)	66.9	156,592	—	—	193,323,456
	dry solid	240	1,200,000	17,659	21,190,800	83,279,844

Table A1. *Cont.*

Utilisation Method		Amount, kg/t Material	Total Amount, t/Year	Energy Content, kJ/kg	Total Energy, GJ	Value, AUD
Furfural + acetic acid Tops	Furfural (AUD 1200/t)	3.1	14,609	—	—	17,530,800
	Acetic acid (AUD 550/t)	1.4	6669	—	—	3,667,950
	Dry solid	0.163	774,149	17,659	13,670,697	53,725,839
Furfural + acetic acid Leaves	Furfural	6.4	15,034	—	—	18,040,800
	Acetic acid	2.9	6863	—	—	3,774,650
	Dry solid	0.472	1,103,579	17,659	19,388,779	76,197,903
Hydrogen, Trash—Tops, water extraction	H2(energy)	3.06	8440	142,000	1,198,480	4,710,026
	H2 (AUD 2.7/kg)	3.06	8440	—	—	22,788,000
	Dry solid	260	969,040	17,659	17,112,277	67,251,250
Hydrogen, Trash—Leaves, water extraction	H2 (energy)	3.06	9920	142	1,408,640	5,535,955
	H2 (AUD 2.7/kg)	3.06	9920	—	—	26,784,000
	Dry solid	260	1,090,845	17,659	19,263,231	75,704,501
Biogas + Ethanol (tops + leaves)	Biogas (energy)	86.02	337,784	22,000	7,431,248	29,204,804
	Biogas (Electricity)	86.02	337,784	—	—	61,682,362
	Ethanol (energy)	201.2	790,054	29,677	23,446,432	92,144,479
	Ethanol (AUD 1/L)	201.2	790,054	—	—	1,001,337,060
Pyrolysis at 900 °C for 1 h	Tops	—	—	—	—	191,281,664
	Leaves	—	—	—	—	149,777,152
Gasification at 900 °C	Tops	—	—	—	—	224,960,408
	Leaves	—	—	—	—	176,148,244
Pulp Tops	Pulp (AUD 1132/t)	0.088	158,799	—	—	179,761,455
	Remain dry solid	0.88	1,587,998	17,659	28,042,456	110,206,854
	Total					289,968,309

References

1. De Albuquerque Wanderley, M.C.; Martín, C.; de Moraes Rocha, G.J.; Gouveia, E.R. Increase in ethanol production from sugarcane bagasse based on combined pretreatments and fed-batch enzymatic hydrolysis. *Bioresour. Technol.* **2013**, *128*, 448–453. [CrossRef]
2. Kent, A.; Mercer, D. Australia's mandatory renewable energy target (MRET): An assessment. *Energy Policy* **2006**, *34*, 1046–1062. [CrossRef]
3. Pillai, I.; Banerjee, R. Renewable Energy in India: Status and Potential. *Energy* **2009**, *34*, 970–980. [CrossRef]
4. Fischer, R.A.; Byerlee, D.; Edmeades, G.O. *Crop Yields and Global Food Security: Will Yield Increase Continue to Feed the World?* Australian Centre for International Agricultural Research: Canberra, Australia, 2014.
5. Chisti, Y. Biodiesel from microalgae beats bioethanol. *Trends Biotechnol.* **2008**, *26*, 126–131. [CrossRef] [PubMed]
6. O'Hara, I. Cellulosic Ethanol from Sugarcane Bagasse in Australia: Exploring Industry Feasibility through Systems Analysis, Techno-Economic Assessment and Pilot Plant Development. Ph.D. Thesis, Queensland University of Technology, Brisbane, Australia, 2011.
7. Covey, G.; Thomas, R.; Dennis, S. A New Opportunity to Pulp Bagasse in Australia? In *Proceedings of the Australian Society of Sugar Cane Technologists*; Hogarth, D., Ed.; QUT ePrints: Mackay, Australia, 2006.
8. O'Hara, I.M. The potential for ethanol production from sugarcane in Australia. In *Proceedings of the Australian Society of Sugar Cane Technologists 2010*; Australian Society of Sugar Cane Technologists: Bundaberg, Australia, 2010; Volume 32, pp. 600–609.
9. Hay, R.K.M. Harvest index: A review of its use in plant breeding and crop physiology. *Ann. Appl. Biol.* **1995**, *126*, 197–216. [CrossRef]
10. Laj, P.; Klausen, J.; Bilde, M.; Plaß-Duelmer, C.; Pappalardo, G.; Clerbaux, C.; Baltensperger, U.; Hjorth, J.; Simpson, D.; Reimann, S.; et al. Measuring atmospheric composition change. *Atmos. Environ.* **2009**, *43*, 5351–5414. [CrossRef]

11. Zhu, C.; Xu, X.; Wang, D.; Zhu, J.; Liu, G.; Seneweera, S. Elevated atmospheric [CO$_2$] stimulates sugar accumulation and cellulose degradation rates of rice straw. *GCB Bioenergy* **2015**, *8*, 579–587. [CrossRef]
12. Furtado, A.; Lupoi, J.S.; Hoang, N.V.; Healey, A.; Singh, S.; Simmons, B.A.; Henry, R.J. Modifying plants for biofuel and biomaterial production. *Plant Biotechnol. J.* **2014**, *12*, 1246–1258. [CrossRef] [PubMed]
13. Lee, W.C.; Kuan, W.C. 2015. Miscanthus as cellulosic biomass for bioethanol production. *Biotechnol. J.* **2015**, *10*, 840–854. [CrossRef]
14. Vohra, M.; Manwar, J.; Manmode, R.; Padgilwar, S.; Patil, S. Bioethanol production: Feedstock and current technologies. *J. Environ. Chem. Eng.* **2014**, *2*, 573–584. [CrossRef]
15. Sugar Research Australia (SRA). Strategic Plan 2013/2014—2017/2018. Available online: www.sugarresearch.com.au (accessed on 15 August 2021).
16. Australian Sugar (AS) & Milling Council (MC). Australian Sugarcane Industry Overview. Available online: http://asmc.com.au/industry-overview/statistics/ (accessed on 15 August 2021).
17. Pippo, W.A.; Luengo, C.A. Sugarcane energy use: Accounting of feedstock energy considering current agro-industrial trends and their feasibility. *Int. J. Energy Environ. Eng.* **2013**, *4*, 10. [CrossRef]
18. The Agricultural Production Indices FAOSTAT. Food and Agricultural Organisation of the United Nations (FAO). Available online: http://faostat.fao.org/ (accessed on 15 August 2021).
19. Qureshi, M.E.; Wegener, M.K.; Mallawaarachchi, T. The economics of sugar mill waste management in the Australian Sugar Industry: Mill mud case study. In Proceedings of the 45th Annual Conference of the Australian Agricultural and Resource Economics Society 2001, Adelaide, Australia, 23–25 January 2001.
20. Bortolussi, G.; O'Neill, C.J. Variation in molasses composition from eastern Australian sugar mills. *Aust. J. Exp. Agric.* **2006**, *46*, 1455–1463. [CrossRef]
21. Meat and Livestock Australia (MLA). Tips & Tools: Expanded Use of Molasses for Intensive Beef Cattle Feeding. Fact Sheet Published in 2008. Available online: www.mla.com.au/publications (accessed on 15 August 2021).
22. Selina Wamucci. Australia Molasses Prices. Available online: https://www.selinawamucii.com/insights/prices/australia/molasses/ (accessed on 15 August 2021).
23. Bundaberg Molasses (BM). Published in 2015. Available online: https://www.bundabergmolasses.com.au/molasses.html (accessed on 15 August 2021).
24. Meier, E.A.; Thorburn, P.J.; Wegener, M.K.; Basford, K.E. The availability of nitrogen from sugarcane trash on contrasting tropics of North Queensland. *Nutr. Cycl. Agroecosyst.* **2006**, *75*, 101–114. [CrossRef]
25. Franco, H.C.J.; Pimenta, M.T.B.; Carvalho, J.L.N.; Magalhães, P.S.G.; Rossell, C.E.V.; Braunbeck, O.A.; Vitti, A.C.; Kölln, O.T.; Neto, J.R. Assessment of sugarcane trash for agronomic and energy purposes in Brazil. *Sci. Agric.* **2013**, *70*, 305–312. [CrossRef]
26. CANEGROWERS. Green Versus Burnt Cane Harvesting. Cited in 2015. Available online: http://www.bdbcanegrowers.com.au/images/images/fact_sheets/Green_Versus_Burnt_Cane_Harvesting.pdf (accessed on 15 August 2021).
27. OECD/FAO. OECD-FAO Agricultural Outlook. OECD Agriculture Statistics (Database) 2021. doi:10.1787/agr-outldata-en. Available online: https://www.agri-outlook.org/commodities/oecd-fao-agricultural-outlook-meat.pdf (accessed on 15 August 2021).
28. Biofuel Association of Australia (BAA). Ethanol in Australia. 2015. Available online: https://www.epw.qld.gov.au/__data/assets/pdf_file/0020/15950/biofuels-association-australia-biofuel-submission.pdf (accessed on 15 August 2021).
29. Lavarack, B.P. Estimates of ethanol production from sugar cane feedstocks. In Proceedings of the 2003 Conference of the Australian Society of Sugar Cane Technologists, Townsville, Australia, 6–9 May 2003; p. 25.
30. Qureshi, N.; Saha, B.C.; Hector, R.E.; Hughes, S.R.; Cotta, M.A. Butanol production from wheat straw by simultaneous saccharification and fermentation using Clostridium beijerinckii: Part I -Batch fermentation. *Biomass Bioenergy* **2008**, *32*, 168–175. [CrossRef]
31. Qureshi, N.; Saha, B.C.; Cotta, M.A. Butanol production from wheat straw by simultaneous saccharification and fermentation using Clostridium beijerinckii: Part II -Batch fermentation. *Biomass Bioenergy* **2008**, *32*, 176–183. [CrossRef]
32. Jonglertjunya, W.; Makkhanon, W.; Siwanta, T.; Prayoonyong, P. Dilute Acid Hydrolysis of Sugarcane Bagasse for Butanol Fermentation. *Chiang Mai J. Sci.* **2014**, *41*, 60–70.
33. Congcong, L.M.S. Butanol Production from Lignocellulosic Feedstocks by Acetone-Butanol-Ethanol Fermentation with Integrated Product Recovery. Ph.D. Thesis, The Ohio State University, Columbus, OH, USA, 2011.
34. Uppal, S.K.; Gupta, R.; Dhillon, R.S.; Bhatia, S. Potential of sugarcane bagasse for production of furfural and its derivatives. *Sugar Tech.* **2008**, *10*, 298–301. [CrossRef]
35. Wondu Business and Technology Services (WBTS). *Furfural Chemicals and Biofuels from Agriculture*; Publication No. 06/127. Project No. WBT-2A; Rural Industries Research and Development Corporation: Sydney, Australia, 2006.
36. Tanisho, S. Feasibility study of biological hydrogen production from sugarcane by fermentation. *Hydrog. Energy* **1995**, *6*, 2.
37. Hamawand, I.; Sandell, G.; Pittaway, P.; Chakrabarty, S.; Yusaf, T.; Chen, G.; Seneweera, S.; Al-Lwayzy, S.; Bennet, J.; Hopf, J. Bioenergy from Cotton Industry Wastes: A review and potential. *Renew. Sustain. Energy Rev.* **2016**, *66*, 435–448. [CrossRef]
38. Hallenbeck, P.C.; Benemann, J.R. Biological hydrogen production; fundamentals and limiting processes. *Int. J. Hydrog. Energy* **2002**, *27*, 1185–1193. [CrossRef]

39. Wang, X.; Jin, B. Process optimization of biological hydrogen production from molassesby a newly isolated *Clostridium butyricum* W5. *J. Biosci. Bioeng.* **2009**, *107*, 138–144. [CrossRef]
40. Janke, L.; Leite, A.; Nikolausz, M.; Schmidt, T.; Liebetrau, J.; Nelles, M.; Stinner, W. Biogas Production from Sugarcane Waste: Assessment on Kinetic Challenges for Process Designing. *Int. J. Mol. Sci.* **2015**, *16*, 20685–20703. [CrossRef]
41. Paturau, J.M.; Alternative Uses of Sugarcane and Its by-Products in Agro-Industries. FAO Corporation Document. 2014. Available online: http://www.fao.org/docrep/003/s8850e/s8850e03.htm (accessed on 15 August 2021).
42. Beeby, C. Drying in Superheated Steam-Fluidized Bed. Ph.D. Dissertation, University of Monash, Melbourne, Australia, 1984.
43. Trommelen, A.M.; Crosby, E.J. Evaporation and drying of drops in superheated vapours. *AIChE J.* **1969**, *16*, 857–867. [CrossRef]
44. Deventer, H.C.; Heijmans, R.M.H. Drying with superheated steam. *Dry. Technol.* **2001**, *19*, 2033–2045. [CrossRef]
45. Pronyk, C.; Cenkowski, S.; Muir, W.E. Drying foodstuff with superheated steam. *Dry. Technol.* **2004**, *22*, 899–916. [CrossRef]
46. Tang, Z.; Cenkowski, S. Dehydration dynamics of potatoes in superheated steam and hot air. *Can. Agric. Eng.* **2000**, *42*, 6.
47. Wimmerstedt, R.; Hager, J. Steam drying—Modelling and applications. *Dry. Technol.* **1996**, *14*, 1099–1119. [CrossRef]
48. Hamawand, I.; Silva, W.P.; Eberhard, F.; Antille, D.L. Issues related to waste sewage sludge drying under superheated steam. *Pol. J. Chem. Technol.* **2015**, *17*, 5–14. [CrossRef]
49. Figueroa, J.E.J.; Ardila, Y.C.; Lunelli, B.H.; Filho, R.M.; Maciel, M.R.W. Evaluation of Pyrolysis and Steam Gasification Processes of Sugarcane Bagasse in a Fixed Bed Reactor. *Chem. Eng. Trans.* **2013**, *32*, 925–930.
50. Marshall, A.J. Commercial Application of Pyrolysis Technology in Agriculture. 2013. Available online: https://ofa.on.ca/wp-content/uploads/2017/11/Pyrolysis-Report-Final.pdf (accessed on 15 August 2021).
51. IISD & SSI. Global Sustainable Sugarcane Production Shows Significant Growth. Available online: https://www.iisd.org/ssi/announcements/global-sustainable-sugarcane-production-shows-significant-growth-report/ (accessed on 30 January 2020).
52. Lucas, F.; Philip, M.; Fearnside. Sugarcane Threatens Amazon Forest and World Climate; Brazilian Ethanol Is Not Clean. 2019. Available online: https://news.mongabay.com/2019/11/sugarcane-threatens-amazon-forest-and-world-climate-brazilian-ethanol-is-not-clean-commentary/ (accessed on 30 January 2020).
53. Ihsan, H.; Saman, S.; Pubudu, K.; Jochen, B. Nanoparticle technology for separation of cellulose, hemicellulose and lignin nanoparticles from lignocellulose biomass: A short review. *Nano-Struct. Nano-Objects* **2020**, *24*, 100601.
54. Atlas, B. Countries by Sugarcane Production. Available online: https://www.atlasbig.com/en-au/countries-by-sugarcane-production (accessed on 15 August 2021).
55. Shahbandeh, M. Sugar Production Worldwide in 2020/2021, by Leading Country (in Million Metric Tons). 2021. Available online: https://www.statista.com/statistics/495973/sugar-production-worldwide/ (accessed on 15 August 2021).
56. Strain, K.V.; Akshaya, G.M. Conversion of Lignocellulosic Sugarcane Bagasse Waste into Bioethanol using Indigenous Yeast. *Iosci. Biotechnol. Res. Asia* **2020**, *17*, 559–565.
57. Yogita, L.; Rohit, R.; Ashish, A.; Prabhu, P.M.; Meenu, H.; Vinod, K.; Sachin, K.; Anuj, K.; Chandel, R.S. Recent Advances in Bioethanol Production from Lignocelluloses: A comprehensive focus on enzyme engineering and designer biocatalysts. *Biofuel Res. J.* **2020**, *7*, 1267–1295.
58. Nerijus, P.; Marius, P.; Justinas, P.; Eugenija, F.D. Analysis of Wood Chip Characteristics for Energy Production in Lithuania. *Energies* **2021**, *14*, 13. [CrossRef]
59. Shweta, J.; Keerthi Prabhu, M.K.; Mascarenhas, R.J.; Shetti, N.P.; Tejraj, M.A. Recent advances and viability in biofuel production. *Energy Convers. Manag.* **2021**, *10*, 100070.
60. Zhang, X.; Zhou, Y.; Jia, X.; Feng, Y.; Dang, Q. Multi-Criteria Optimization of a Biomass-Based Hydrogen Production System Integrated With Organic Rankine Cycle. *Front. Energy Res.* **2020**, *8*, 584215. [CrossRef]
61. Rana, A.A.; Al-Zuhairi, F.K. Biofuels (Bioethanol, Biodiesel, and Biogas) from Lignocellulosic Biomass: A Review. *J. Univ. Babylon Eng. Sci.* **2020**, *28*, 1.
62. Deodatus, K.; Justin, N.; Godlisten, K. A Review of Intermediate Pyrolysis as a Technology of Biomass Conversion for Coproduction of Biooil and Adsorption Biochar. *J. Renew. Energy* **2021**, *2021*, 5533780.
63. Bikash, K.; Nisha, B.; Komal, A.; Venkatesh, C.; Pradeep, V. Current perspective on pretreatment technologies using lignocellulosic biomass: An emerging biorefinery concept. *Fuel Process. Technol.* **2020**, *199*, 106244.
64. Bastos Lima, M.G. Corporate Power in the Bioeconomy Transition: The Policies and Politics of Conservative Ecological Modernization in Brazil. *Sustainability* **2021**, *13*, 12. [CrossRef]
65. D'Adamo, I.; Falcone, P.M.; Morone, P. A New Socio-economic Indicator to Measure the Performance of Bioeconomy Sectors in Europe. *Ecol. Econ.* **2020**, *176*, 106724. [CrossRef]
66. D'Adamo, I.; Falcone, P.M.; Morone, P. A circular economy model based on biomethane: What are the opportunities for the municipality of Rome and beyond? *Renew. Energy* **2021**, *163*, 1660–1672. [CrossRef]

Article

Experimental Research on the Macroscopic and Microscopic Spray Characteristics of Diesel-PODE$_{3-4}$ Blends

Yulin Chen [1], Songtao Liu [1], Xiaoyu Guo [1], Chaojie Jia [1], Xiaodong Huang [1,*], Yaodong Wang [2,*] and Haozhong Huang [1,*]

[1] College of Mechanical Engineering, Guangxi University, Nanning 530004, China; 1911391008@st.gxu.edu.cn (Y.C.); 2011301033@st.gxu.edu.cn (S.L.); 2011401013@st.gxu.edu.cn (X.G.); 1611392002@alu.gxu.edu.cn (C.J.)

[2] Department of Engineering, Durham University, Durham DH1 3LE, UK

* Correspondence: xdhuang@gxu.edu.cn (X.H.); yaodong.wang@durham.ac.uk (Y.W.); hhz421@gxu.edu.cn (H.H.)

Abstract: Polyoxymethylene dimethyl ether (PODE) is a low-viscosity oxygenated fuel that can improve the volatility of blended fuels. In this work, the macroscopic and microscopic spray characteristics of diesel-PODE$_{3-4}$ under different ambient temperatures and injection pressures (IP) are studied. The studied blends consisted of pure diesel (P0), two diesel blend fuels of 20% (P20) and 50% (P50) by volume fraction of PODE$_{3-4}$. The Mie scattering and Schlieren imaging techniques are used in the experiment. The results show that with the increase in IP, the vapor phase penetration distance and the average cone angle of the three fuels increased, and the Sauter mean diameter (SMD) of the three fuels decreased. When the ambient temperature increased, the vapor phase projection area and the average vapor phase cone angle of P20 and P50 increased, and the SMD decreased, but the vapor phase projection area of pure diesel did not change significantly. The results indicate that the blended fuel with PODE$_{3-4}$ has better spray characteristics than P0 at low temperature, and the SMD hierarchy between the three fuels is P0 > P20 > P50. Through the visualization experiment, it is helpful to further understand the evaporation characteristics of different fuel properties and develop appropriate alternative diesel fuel.

Keywords: macroscopic spray; microscopic spray; Mie scattering; schlieren images; PODE$_{3-4}$

Citation: Chen, Y.; Liu, S.; Guo, X.; Jia, C.; Huang, X.; Wang, Y.; Huang, H. Experimental Research on the Macroscopic and Microscopic Spray Characteristics of Diesel-PODE$_{3-4}$ Blends. *Energies* **2021**, *14*, 5559. https://doi.org/10.3390/en14175559

Academic Editor: Idiano D'Adamo

Received: 30 July 2021
Accepted: 1 September 2021
Published: 6 September 2021

Publisher's Note: MDPI stays neutral with regard to jurisdictional claims in published maps and institutional affiliations.

1. Introduction

Diesel engines are widely used in ships, generators and heavy trucks due to their high power output and high thermal efficiency [1–3]. However, due to the related environmental problems, the shortage of oil resources and the requirements of national laws and regulations in recent years, the internal combustion engines, which depend on traditional oil as the power source, have been severely challenged, and the quest to find a novel fuel mixed with diesel fuel is seen as a solution to these problems [4,5]. At present, the most common method is to add biomass oxygenated fuel to diesel to form a suitable blend.

Biodiesel, alcohols and ethers are the most suitable additives for compression ignition internal combustion engines [6,7]. Many countries are committed to developing biofuels. In Europe, bioeconomy has been well developed especially in Ireland, Denmark, Portugal and Austria [8]. D'Adamo et al. [9] researched a circular economy model and found that applying biomethane in the transport system of Rome leads to a reduction of emissions. Compared with diesel, biodiesel is renewable, non-toxic and has higher hexadecane value. Biodiesel contains no aromatic hydrocarbons, which is capable for reduction of unburned hydrocarbons (HC), particulate matter (PM) and carbon monoxide (CO) emissions. Thus, Colombia's government has implemented laws about promoting biodiesel production industry [10]. Hassaan et al. [11] also suggested that the Egyptian government should pay more attention to the construction of biogas and biomethane production plants in the future.

The high density and viscosity of biodiesel, which is not conducive to atomization, results in engine combustion performance decline. Meanwhile, due to the high oxygen content of biodiesel, it produces higher nitrogen oxide (NOx), especially since most engines are turbocharged [12–14]. Alcoholic fuels have great advantages for combustion and resources, but they have high latent heat of vaporization, low viscosity and low cetane number, resulting in poor ignition and lubricity of alcohols. Although alcohol fuel and diesel oil have good solubility, their stability is easily affected by moisture. If the engine is not changed, alcoholic fuels are difficult to apply directly in diesel engines [15,16]. Due to this reason, ether-based fuels are used more often in diesel engines. $PODE_n$ (polyoxymethylene dimethyl ethers) has the structural formula of $H_3CO(CH_2O)_nCH_3$ (where 'n' represents the degree of polymerization) and is regarded as the most promising alternative fuel for developing diesel engine fuels due to its high cetane number, high oxygen content, good solubility with diesel fuel and no need to modify the engine after blending with diesel [17–19]. When $n < 2$, it is volatile, has a low boiling point and its safety for use in transportation cannot be guaranteed. When $n > 6$, it is easy to precipitate after mixing with diesel oil, and therefore, n is generally between 3 and 5. Burger et al. [20] used methanol and trimeric formalin as reactants in a reacting furnace using distillation. They proposed a new process for the preparation of $PODE_n$ and investigated the physical and chemical properties, synthesis and purification of $PODE_n$. The same authors reported that more than a million tons of $PODE_n$ could be produced using their proposed process. They also observed that $PODE_n$ has lower saturated vapor pressure, stable solubility when mixed with diesel oil and can reduce the formation of soot after combustion. In addition, many experts and scholars have studied the combustion performance and emission characteristics of diesel oil mixed with a certain proportion of $PODE_n$. Lumpp et al. [21] produced a $PODE_n$-diesel blend with 20% $PODE_n$ through a transient and steady-state engine cycle and used it in a heavy-duty diesel engine, which met Euro V emission standards and achieved simultaneous reduction of PM and soot emissions. They also reported that the use of a diesel oil blend with 10% $PODE_n$ in cylinder diesel engines reduces the PM emissions by 40% and has the potential to reduce carbon dioxide emissions. Pellegrini et al. [22] investigated the diesel blended $PODE_{3-5}$ in single-cylinder and multi-cylinder diesel engines and found that the formation of soot emissions can be significantly suppressed when the blend is used in single-cylinder diesel engines, whereas the PM and NO_x emissions can be synchronously reduced after mixing the diesel with 50% $PODE_{3-5}$ in multi-cylinder diesel engines. Furthermore, they also reported that the noise of the multi-cylinder diesel engine was also optimized. Later, Pellegrini et al. [23] studied the diesel blended $PODE_{3-5}$ in a Euro II diesel engine, and found that PM emissions were reduced by 18% when diesel was blended with 10% $PODE_{3-5}$ and by 77% when pure $PODE_{3-5}$ was burned, which are far lower than Euro IV emission standards. However, it is worth noticing that the combustion of pure $PODE_{3-4}$ leads to an increase in NOx and CO emissions. Iannuzzi et al. [24] first tested the burning process of diesel blended $PODE_n$ in a constant volume device and found that, with the increase in the $PODE_n$ blending ratio, soot product was greatly reduced. When pure $PODE_n$ burnt, the smoke was almost zero. Subsequently, Iannuzzi et al. [25] investigated the emission and performance of different proportions of $PODE_n$ in diesel blends in a single-cylinder heavy-duty diesel engine, and found that, compared to pure diesel, when the $PODE_n$'s blending ratio reached 10%, soot emissions were reduced by 34% and the thermal efficiency was guaranteed, although NOx emissions did not change much. Similar conclusions have been reported by Huang et al. [26] and Liu et al. [27]. Liu et al. [28] investigated the combustion performance and emission characteristics of diesel blended $PODE_n$ on a four-cylinder supercharged diesel engine and found that, with the increase in $PODE_n$'s blending ratio, the ignition delay was shortened. When the proportion of $PODE_n$ blended diesel successively reached 10%, 20% and 30% at full load, compared with the pure diesel fuel, the carbon smoke decreased by 27.6%, 41.5% and 47.6%, respectively, while the HC and CO emissions decreased significantly, although the NOx emissions showed a slight increase. Song et al. [18] compared the combustion characteristics of dual fuels and found that the

PODE$_n$/natural gas blend resulted in fewer hydrocarbon, CO and soot emissions compared to the diesel/natural gas blend, and significantly improved the thermal efficiency.

These works have mainly focused on the in-cylinder combustion process of the engine. In fact, the fuel injection, atomization and mixing with air are critical to the entire combustion process, which is of great significance to energy saving and emissions reduction. Li et al. [29] researched the macroscopic and microscopic spray characteristics of diesel/PODE blended fuels and found that, with the increase in the proportion of PODE in diesel-PODE blend, the spray penetration distance decreased, though the average spray cone angle increased. In addition, both the characteristic diameters and the Sauter mean diameter decreased with the increase in the proportion of PODE in the diesel/PODE blend, though the range of relative size of droplets showed little change. However, Li et al. did not study the evaporation characteristics of PODE, whereas the test temperature was only the room temperature. Currently, the techniques of Mie scattering and diffuse back-illumination (DBI) images are widely used methods for measuring liquid spray characteristics, while schlieren image technology is suitable for studying spray evaporation characteristics [30–32]. Huang et al. [33] researched the spray, evaporation and combustion characteristics of ethanol/diesel blends under low temperature combustion (LTC) using DBI and schlieren image techniques. The results showed that the evaporation characteristics of the fuel blend increased the spray spreading angle and projected area, though they had little influence on the spray penetration distance, whereas the combustion phenomenon spread from the periphery behind the spray tip to the forward of the spray and then to the nozzle. Ma et al. [34] investigated the evaporation and spray characteristics of n-pentanol/diesel blends using DBI and schlieren image techniques, and found that pure diesel has a longer spray penetration distance and a smaller spray cone angle than n-pentanol in the absence of evaporation. When the ambient temperature exceeds 800 K under evaporation conditions, the spray penetration distance of pure diesel is reduced, although the spray penetration distance of n-pentanol increases and this trend is particularly remarkable for higher proportions of n-pentanol in diesel. Payri et al. [35] used Mie scattering and schlieren images techniques to study the effects of cylindrical and conical nozzles on the liquid and vapor phase spray characteristics of n-heptane, n-hexane and diesel alternative fuels (composed of n-tetradecane, n-decane and methylnaphthalene), and found that, for the same fuel under the same working conditions, the cylindrical nozzle has a smaller vapor phase penetration distance and liquid phase length than the conical nozzle. However, the spray spreading angle shows the opposite trend. Subsequently, Payri et al. [32] compared the experimental and calculated values of the vapor and liquid phase spray penetration distances of pure diesel oil at different IPs, ambient density and ambient temperature, and found that the experimental and calculated values can be in good agreement. Therefore, it is concluded that the IP affects the vapor phase length and the ambient temperature affects the liquid phase length.

However, the research on the spray (liquid penetration distance, liquid phase cone angle and liquid phase projection area) and evaporation characteristics (vapor phase cone angle, vapor phase penetration distance and vapor phase projection area) of a diesel/PODE$_n$ blend (chain length of PODE$_n$ is: $n = 3$–4) is still lacking, especially regarding the droplet size distribution characteristics of blended fuel. Many studies [20,36] have shown that the mixing of diesel and PODE$_{3-4}$ is the most suitable for application in diesel engines. Therefore, in this work, the effect of different ambient temperatures and IPs on the macroscopic and microscopic spray characteristics of diesel/PODE$_{3-4}$ blend is analyzed. From the results, the evaporation characteristics of different fuel properties and appropriate alternative diesel fuels can be further understood.

2. Experimental

2.1. Experimental Setup

Figure 1 shows the schematic of the experimental setup, which mainly consists of a constant volume combustion bomb (CVCB), fuel supply system and image acquisition system. The specific parameters of CVCB are described in detail in a previous work [37]. In order to ensure the optical path and camera shooting, three quartz windows with the diameter of 110 mm were located on the side of the CVCB. A single hole electromagnetic valve injector was used for the experiments, and the fuel supply system was procured from Bosch's third-generation high-pressure common rail test rig. Each set of experiments was repeated three times and the average of the three test data was taken. The uncertainties of the apparatus are shown in Table 1.

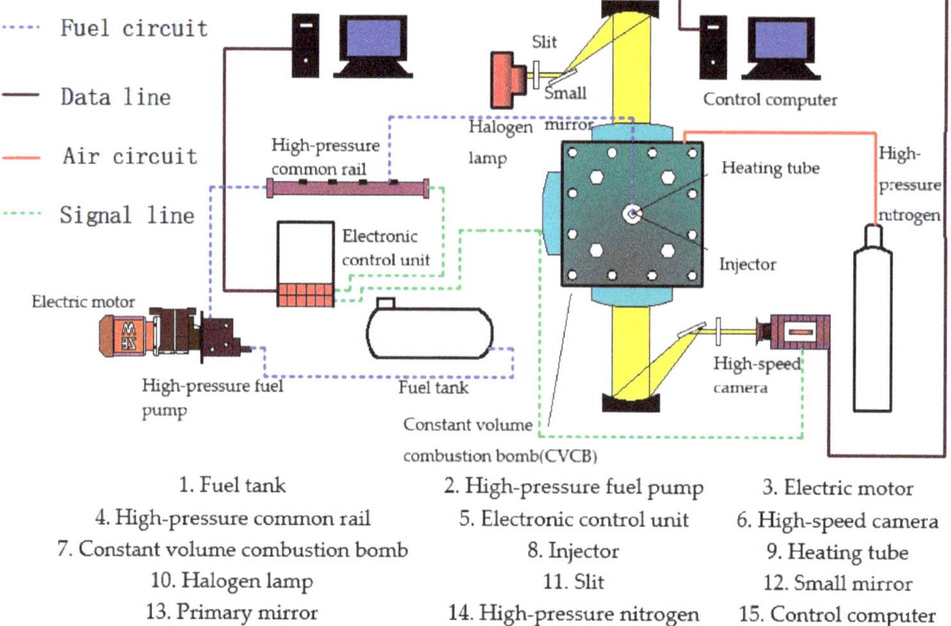

1. Fuel tank	2. High-pressure fuel pump	3. Electric motor
4. High-pressure common rail	5. Electronic control unit	6. High-speed camera
7. Constant volume combustion bomb	8. Injector	9. Heating tube
10. Halogen lamp	11. Slit	12. Small mirror
13. Primary mirror	14. High-pressure nitrogen	15. Control computer

Figure 1. Schematic of the constant volume combustion chamber system.

Table 1. Uncertainties and experimental measurement techniques/instruments.

Measurement	% Uncertainty	Measurement Technique
Pressure pickup	±0.1	Magnetic pickup principle
Temperature	±0.15	Thermocouple
Diesel fuel measurement	±1	Volumetric measurement
$PODE_{3-4}$ fuel measurement	±1	Volumetric measurement

The images were taken using a FASTCAM-SA7 high-speed camera, which was manufactured by PhotronCorp, Tokyo, Japan. During the experiments, the external computer issued an injection command, and then the electronic control unit (ECU) in a high-pressure common rail system drove the injector and high-speed camera to work synchronously according to the detected fuel injection signal. The schlieren device adopted a Z-shaped arrangement, which is mainly composed of a light source slit system and a knife-edge camera system.

Figure 2 shows the schematic of the Mie scattering device, which is mainly used to measure the liquid spray characteristics. During the experiments, two tungsten halogen lamps were placed in each of the two quartz windows. During operation, the tungsten halogen lamp emitted a light to illuminate the spray. Then, the scattered light, reflected by the spray, was received by the high-speed camera. Finally, the liquid phase spray boundary image was displayed on the computer screen. A micro-nano particle size analyzer (Winner 318A) was used to measure the microscopic characteristics of the spray and the micro experimental test structure diagram as shown in Figure 3a. The working principle of Winner 318A is shown in Figure 3b, which obtains particles' size by using a laser beam to test the intensity of scattering spectrum of particles.

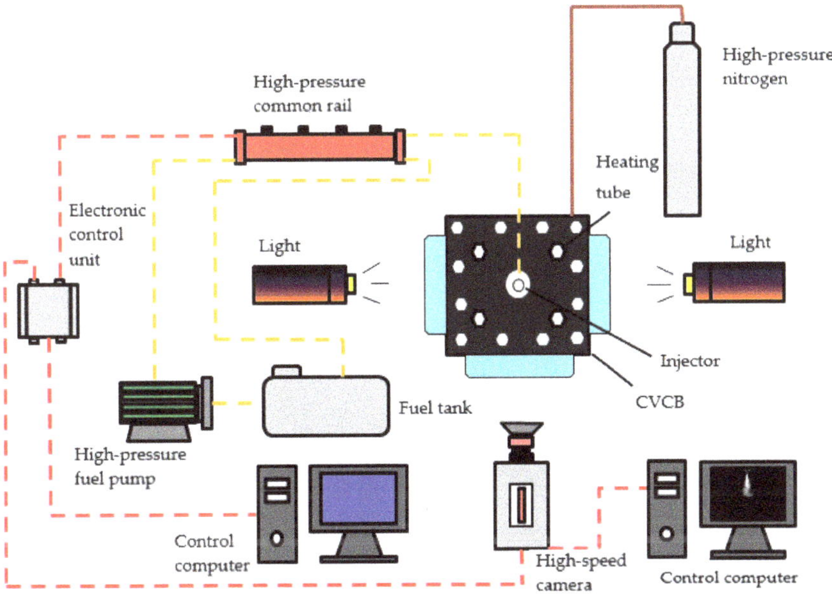

Figure 2. Optical setup for Mie scattering.

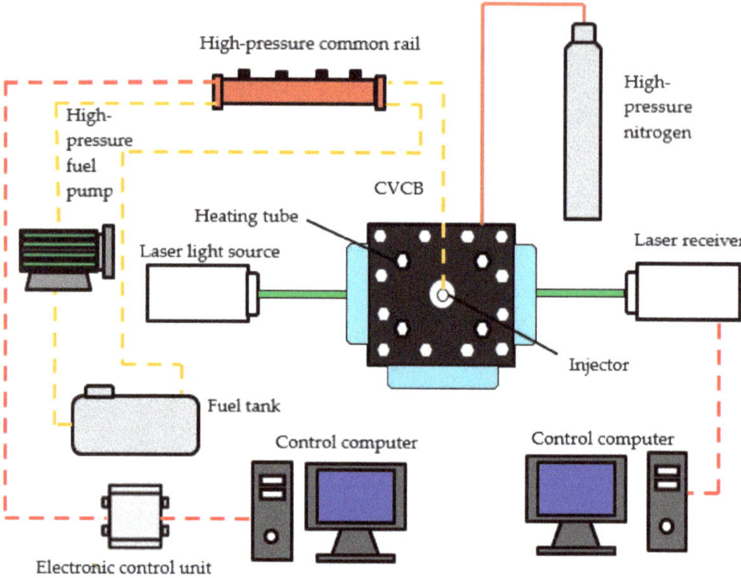

1. Laser light source   2. Control computer   3. Electronic control unit
4. High-pressure common rail   5. Fuel tank   6. Constant volume combustion bomb
7. Injector   8. Heating tube   9. Laser receiver
10. High-pressure nitrogen

(**a**)

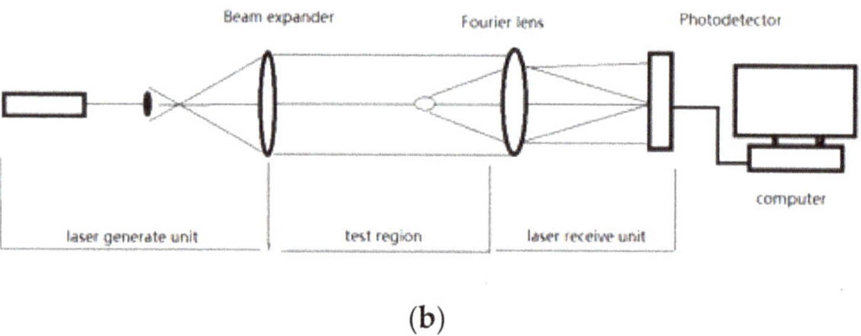

(**b**)

Figure 3. (**a**) Micro experimental test structure diagram. (**b**) Working principle of Winner 318A.

2.2. Image Processing

The Mie scattering method can capture the liquid phase spray boundary image, while the schlieren method can capture the spray boundary image of the vapor phase. Both the methods use MATLAB's self-programming process for image processing. The steps of processing the image are similar for both methods. Firstly, the background image and the spray image are subtracted by using the MATLAB program to achieve the removal of background. Then, the edge pixels of the spray are detected using two thresholds of the Sobel and the cany operator, and the boundary curve of the spray is determined to obtain the liquid phase spray characteristic parameters. The Sobel operator and cany operator are used to edge the spray image detection, obtaining the coordinate values for the edge point of the image. Finally, the spray characteristic parameters of the liquid or vapor phase are obtained. Figure 4 shows the processed images from Mie scattering and schlieren methods.

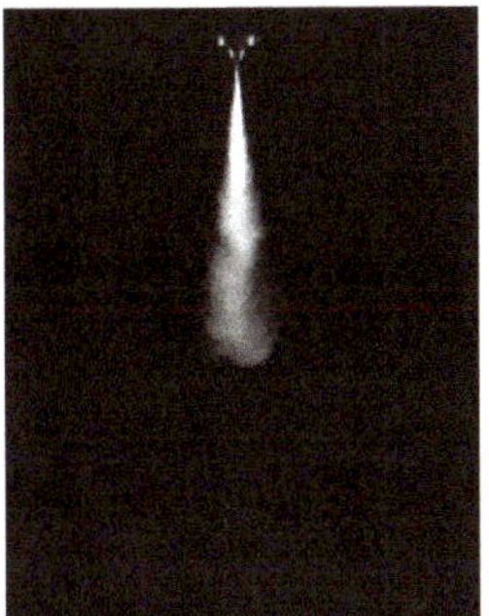

(**a**) Mie scattering method original image.

(**b**) Mie scattering method processed image.

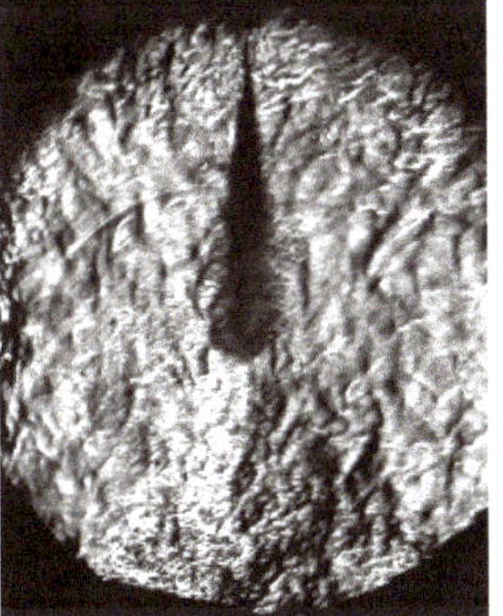

(**c**) Schlieren method original image.

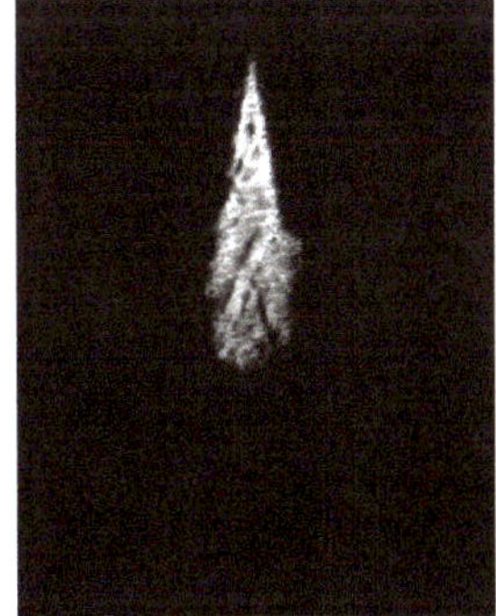

(**d**) Schlieren method processed image.

Figure 4. Comparison of processed and raw images using Mie scattering and schlieren methods.

2.3. Experimental Procedure

The experimental conditions are presented in Table 2. The fuel sample is based on No. 0 diesel fuel sold in China's petroleum market. The blended $PODE_n$ is obtained from Qingdao Tong Chuan Petrochemical Engineering Company, Qingdao, China, and is mainly composed of $PODE_2$, $PODE_3$ and $PODE_4$ with the mass fractions of 2.6%, 88.9% and 8.5%, respectively. Since the main components are $PODE_3$ and $PODE_4$, this article uses $PODE_{3-4}$ to represent $PODE_n$. There were three different volume ratios of diesel/$PODE_{3-4}$ blends used in the experiments, 0%, 20% and 50% with respect to $PODE_{3-4}$, and were abbreviated as P0, P20 and P50, respectively. The relevant physical and chemical properties of the blends are presented in Tables 3 and 4.

Table 2. Experimental parameters used in the current work.

Parameter	Numerical Value
Fuels	P0, P20, P50
Injection pressure/MPa	80, 120, 160
Ambient temperature/K	573,623,673 (macroscopic) 303 363 (microscopic)
Ambient Pressure/MPa	5 (macroscopic) 0.1 (microscopic)
Filming speed/fps	20,000
Injection pulse width/ms	1.0

Table 3. Physical and chemical parameters of PODE$_n$ (n = 2–4) [29,38–41].

Parameter	PODE$_2$	PODE$_3$	PODE$_4$
Density (298 K)/g·cm^{-3}	0.96	1.02	1.06
Viscosity (298 K)/(mm^2·s^{-1})	0.64	1.05	1.75
Oxygen content/(%)	45.3	47.1	48.2
Cetane value	63	70	90
Flash point/(T·K^{-1})	289.1	293.1	350.1
Low calorific value/(MJ·kg^{-1})	22.44	19.14	18.39
Sulfur content/(%)	0	0	0

Table 4. Physical and chemical parameters of various fuel blends used in the experiments.

Parameter	PODE$_{3-4}$	P0
Density (298 K)/g·cm^{-3}	1.02 *	0.86
Viscosity (298 K)/(mm^2·s^{-1})	1.05 **	3.44
Surface tension/(10^{-3}N·m^{-1})	35.67 *	27.74
Oxygen content/(%)	46.98 *	0
Cetane value	78.6 *	56.5
Flash point/(T·K^{-1})	297.64 *	>328.1
Low calorific value/(MJ·kg^{-1})	19.2 *	42.80
Sulfur content/(%)	0	0

* Calculated from Ref. [41]. ** Calculated from Ref. [42].

3. Results and Analysis

3.1. Definition of Spray Parameters

The evaporation characteristics of the spray influence the droplet size, whereas the size of the vapor penetration distance determines the rate of fuel impingement on the combustion chamber. The definition of the projected area is the calculated area inside the spray boundary. The projected area of vapor and liquid phase reflects the range of fuel spray diffusion, which is the combined effect of cone angle and penetration distance, while its size can reflect the quality of mixing of fuel and the surrounding environment gas [43]. The Mie scattering method is used to obtain the liquid phase spray boundary image, while the schlieren method is used to obtain the spray boundary image of the vapor phase. Therefore, it is necessary to define and distinguish the spray image parameters obtained by the two shooting methods. Figure 5 illustrates the way the spray characteristic parameters are defined in this work. The subscripts L and V represent the liquid phase and vapor phase spray characteristic parameters, respectively. The penetration distance for each phase is defined as the axial distance, which the fuel can reach farthest from the nozzle to the spray front, and is denoted by S_V and S_L for vapor and liquid phase, respectively. The spray cone angle for each phase is defined based on the method of Naber et al. [44], and is defined as the angle between the line from fuel injection from the nozzle to the half of the spray penetration distance and the tangent line along the spray contour. In Figure 5, the spray cone angle is represented by θ_V and θ_L for vapor and liquid phase, respectively. The sum of the values of the pixel area included in the entire spray image is the spray projection area, whereas the vapor and liquid phase projected areas are represented by A_V and A_L, respectively.

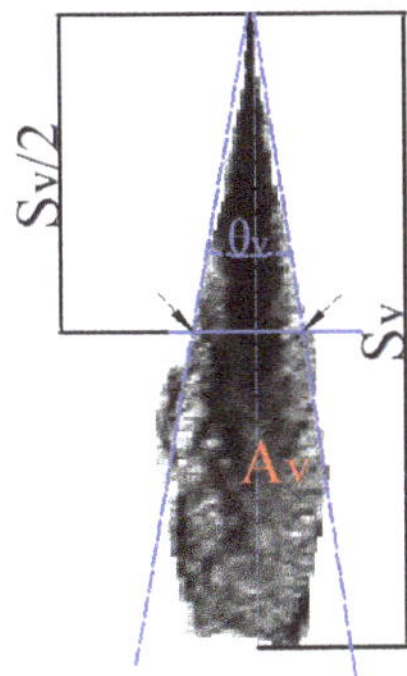

(**a**) Vapor phase spray image.

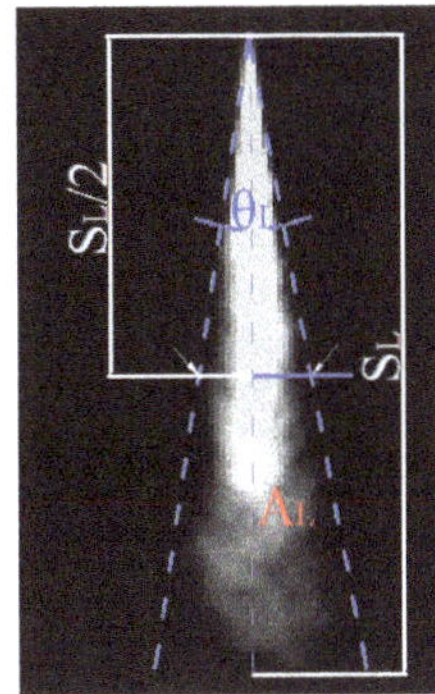

(**b**) Liquid phase spray image.

Figure 5. Definition of various spray parameters.

3.2. Effect of Ambient Temperature on Fuel Spray Characteristics

Figure 6 shows the evolution of vapor and liquid spray patterns for different fuel blends over time at different ambient temperatures (573 K, 623 K and 673 K) and the IP of 160 MPa. It can be known from Figure 6a,c,e that the vapor phase spray for each fuel blend at different ambient temperatures gradually increases with time. Meanwhile, the spray gradually changes from "dark black" (liquid phase) to "transparent color" (vapor phase) with the increase in temperature. This trend is particularly pronounced as the proportion of $PODE_{3-4}$ in the blend increases. It can be seen from Figure 6b,d,f that the liquid phase spray of the three fuels gradually increases with time at 573 K. However, when the ambient temperature is 623 K, and both the P0 and P20 are in the middle and late spray (after 0.7 ms), the front end of the liquid phase spray gradually becomes blurred, producing a "mist", which is composed of a lot of small droplets. On the other hand, the sample P50 shows a rapid reduction of spray at the same temperature (623 K). As the ambient temperature increases to 673 K, the mist area of the three blends decreases. This is because the rate of evaporation of droplets becomes faster due to an increase in the ambient temperature, which results in rapid evaporation of small droplets formed at the edge of the liquid core and the front end of spray during the spraying process.

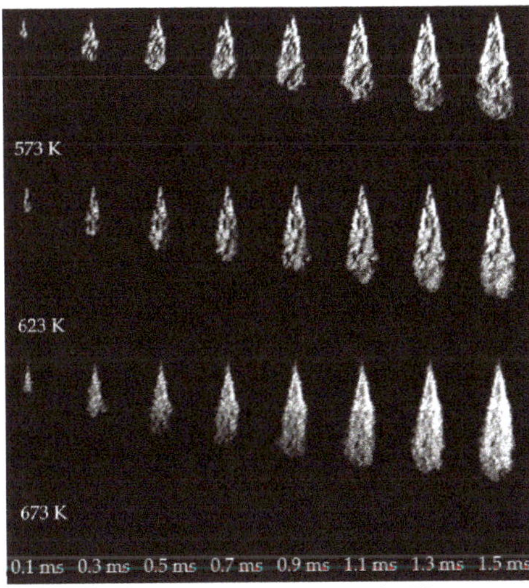

(**a**) Vapor phase spray image for P0.

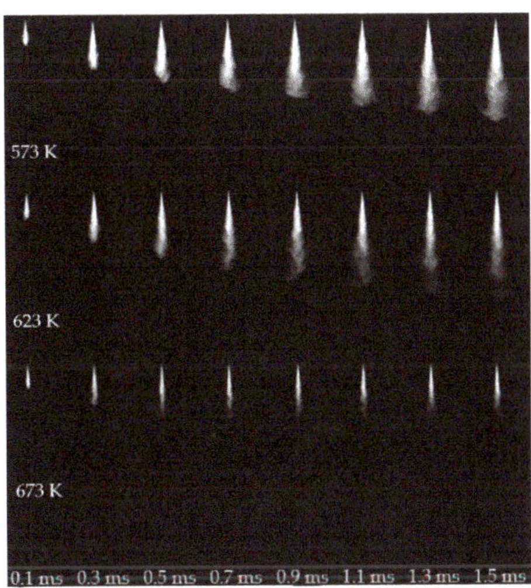

(**b**) Liquid phase oil beam image for P0.

Figure 6. *Conts.*

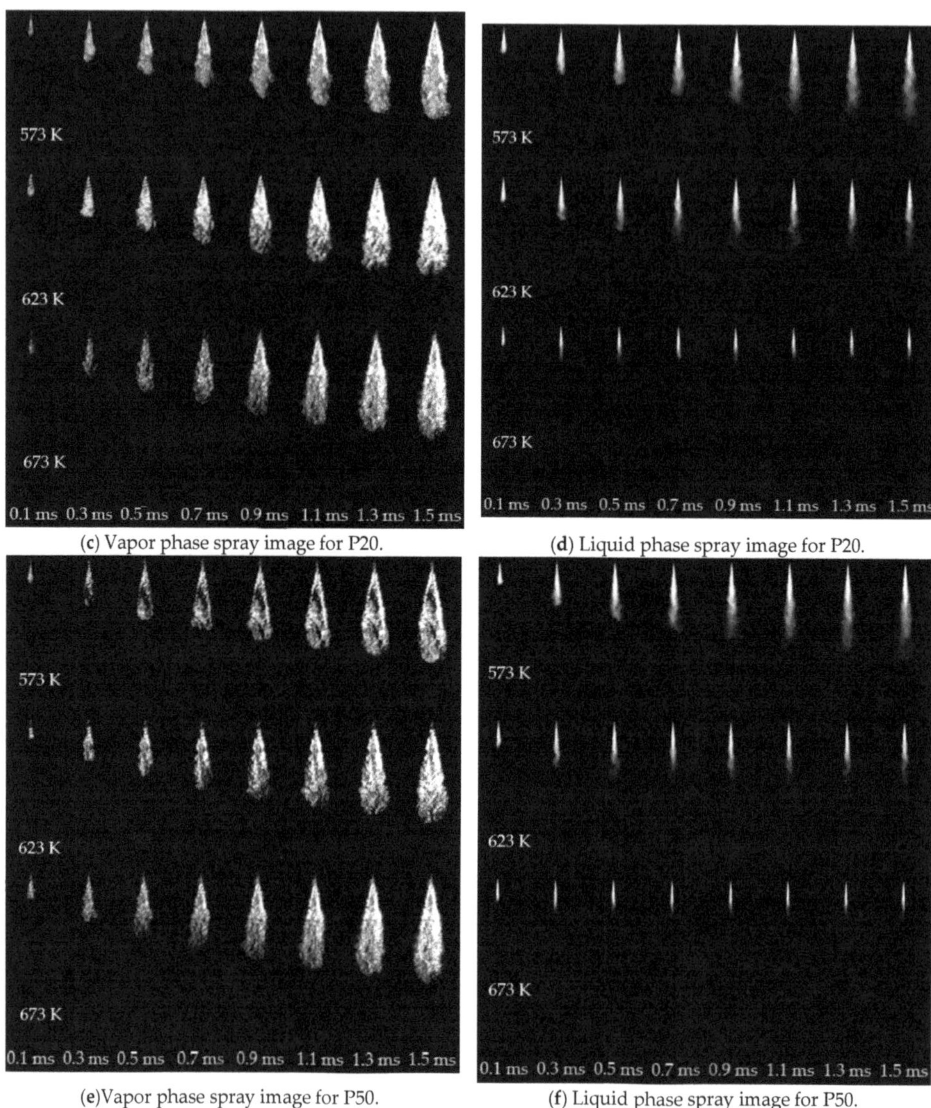

(c) Vapor phase spray image for P20.

(d) Liquid phase spray image for P20.

(e) Vapor phase spray image for P50.

(f) Liquid phase spray image for P50.

Figure 6. Spray development for different blends at different ambient temperatures.

Figure 7 shows the vapor and liquid phase penetration distances for the three blends at different ambient temperatures. It can be clearly seen from the picture that the vapor phase penetration distances of P0, P20 and P50 increase with time. However, when the ambient temperature is 673 K, the vapor phase penetration distance of the three fuels is significantly lower than those for 573 K and 623 K. The reason is that, at lower temperatures, the spray mainly develops in liquid form, while at a high temperature, the fuel evaporates in a large amount, causing the spray mainly to develop in vapor form. Furthermore, the vapor spray develops at a lower rate than that in liquid phase. Therefore, when the ambient temperature is 673 K, the vapor phase penetration distances of the three blends are smaller than the vapor phase penetration distances at low temperatures. For the liquid phase penetration distance, the fuel did not undergo significant evaporation at the temperatures of 573 K and 623 K.

When the temperature was increased to 673 K, the liquid phase penetration distance for P0 was substantially similar to the vapor phase penetration distance during the initial stage of the spray. However, the passage of time gradually separated both the values. The liquid penetration distance for each of P20 and P50 is expressed as the situation, in which the initial value of the spray quickly reaches a certain value and keeps fluctuating around this value. Therefore, it is indicated that diesel will slowly evaporate after the background temperature reaches a certain value until a certain period of time. However, when the diesel blend contains a large proportion of PODE$_{3-4}$, the liquid phase and vapor phase spray are separated at the initial moment of the spray. The liquid phase penetration distance reaches the maximum value and becomes stable. Meanwhile, the amount of evaporation and the amount of fuel injection achieve certain equilibrium. In addition, compared with the ambient temperature of 573 K and 623 K, when the ambient temperature is 673 K, the vapor and liquid phase penetration distances of the three fuels are all lower. This is consistent with the results reported by Gimeno et al. [30], who stated that "the ambient temperature rises, and the fuel vapor and liquid phase penetration distances decrease".

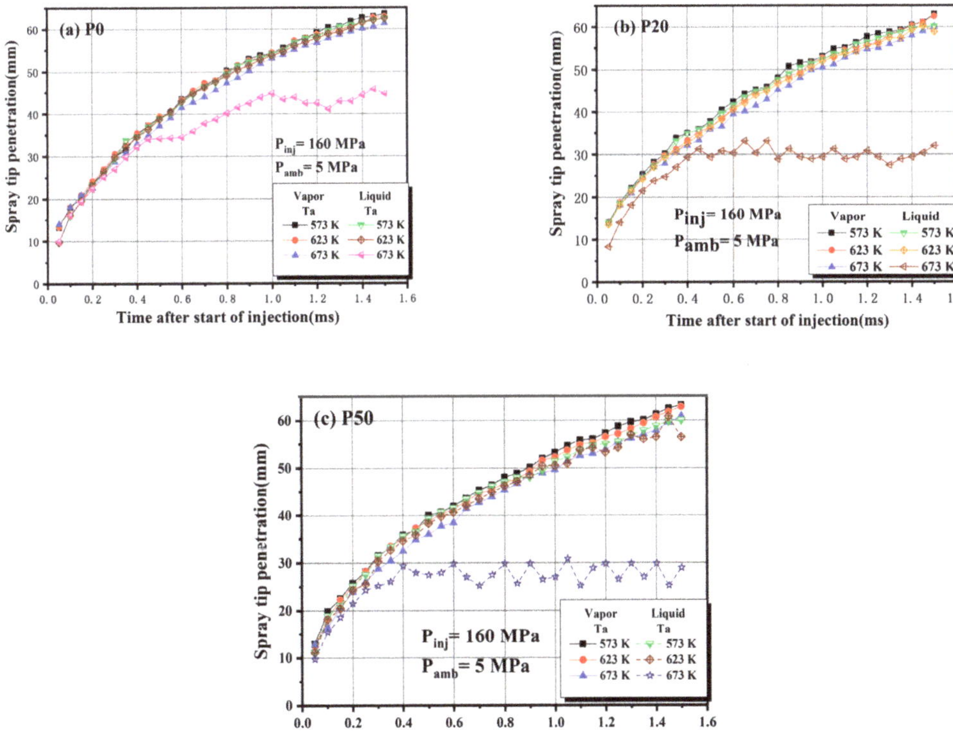

Figure 7. Penetration distances of blends at different temperatures: (**a**) P0, (**b**) P20, (**c**) P50.

Figure 8 shows the vapor and liquid cone angles of the three fuels at different ambient temperatures. It can be known that the vapor phase cone angles of P0, P20 and P50 increase with the increase in ambient temperature. However, the change in vapor phase cone angle does not agree with the increase in vapor phase conge angle of P0 before 0.6 ms, whereas the temperature increases gradually after 0.6 ms. The reason is that the viscosity of the diesel oil is high, which initially causes the droplets to stick together after the fuel is sprayed from the nozzle. Due to this, it is not easy for the droplets to diffuse in the radial direction. As time and temperature increase, the spray sharply evaporates, and the situation improves. In addition, when the temperature is 673 K, compared with the pure diesel, the average vapor cone angles of P20 and P50 increased by 1.9° and 3.3°, respectively, and the average liquid cone angle reduced by 1° and 2.2°, respectively, indicating that the addition of PODE$_{3-4}$ to diesel in a certain proportion can improve the evaporation and diffusion of the spray. It can also be seen that the liquid phase cone angles of P0, P20 and P50 decrease with the ambient temperature increase. The reason is that the ambient temperature increases, causing the heat exchange between the liquid phase fuel in the spray front and the edge region and the high-temperature ambient gas to increase. Therefore, the rate of evaporation of liquid fuel greatly increases, resulting in the decrease in liquid phase spray cone angle.

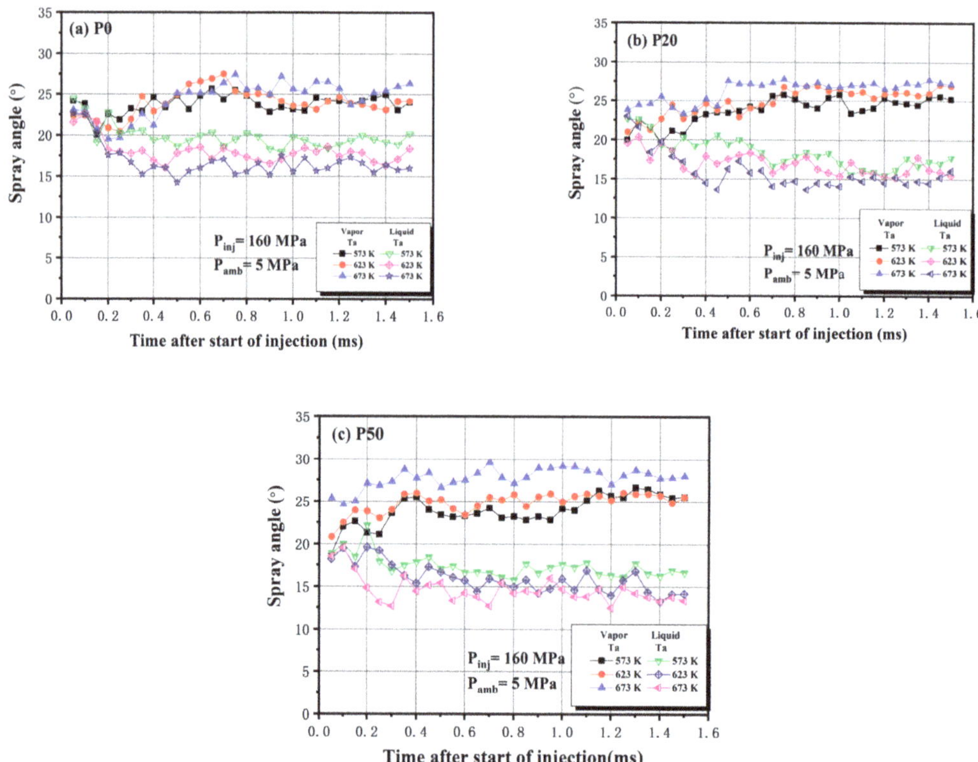

Figure 8. Cone angle of the three blends at different temperatures: (**a**) P0, (**b**) P20, (**c**) P50.

Figure 9 shows the vapor and liquid phase projected areas of the three tested fuels at different ambient temperatures. It can be seen that the vapor and liquid phase projection areas of P0 are not much different for the ambient temperatures of 573 K and 623 K and before 0.9 ms. However, the vapor phase projection area of P0 after 0.5 ms at ambient temperature of 673 K is larger than the other two temperatures. The results for P20 and P50 show that the higher the temperature, the larger the vapor phase projection area and smaller the liquid phase projection area. It shows that the increase in pure diesel in the low temperature range (within 623 K) has little influence on the evaporation characteristics and spray diffusion, while the fuel blend with PODE$_{3-4}$ still has good diffusion capability and evaporability at low temperatures. For the liquid fuel projected area of the mixed fuel, with the ambient temperature, the projected area decreases. When the temperature is 673 K, the liquid projection area of the three blends tends to flatten with time. The reason is that the higher the ambient temperature, the more evaporation on both sides of the liquid fuel spray and the faster the evaporation rate, which results in a smaller projected area. When the evaporation speed of the droplets is equal to the diffusion speed, the liquid projected area reaches a stable state.

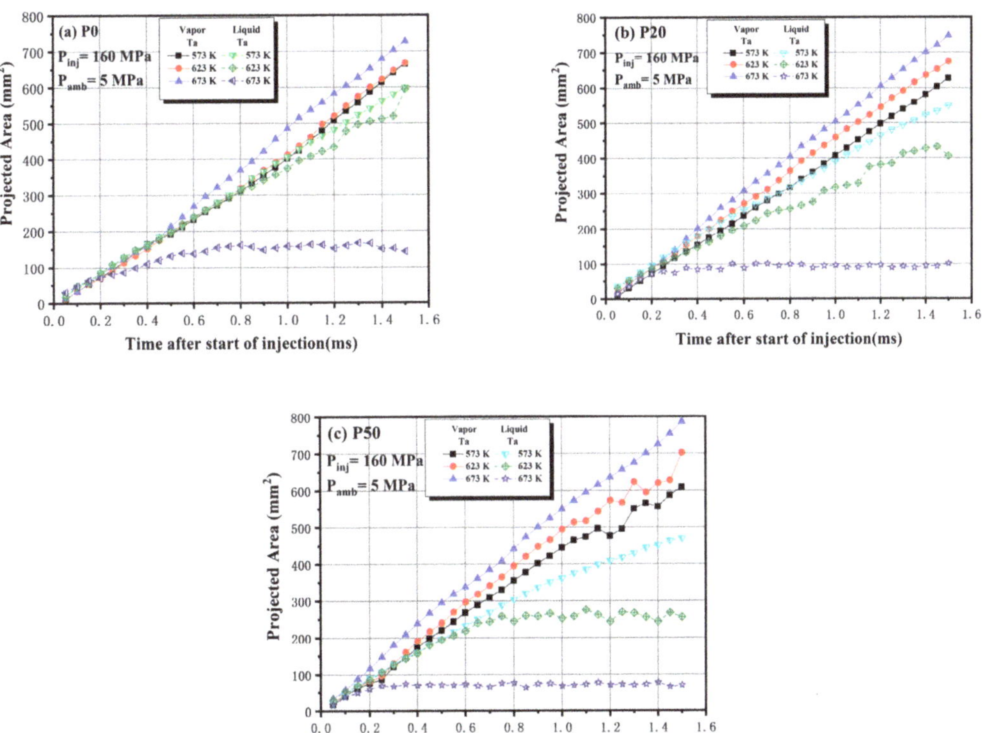

Figure 9. Projection areas of the three fuels at different ambient temperatures: (**a**) P0, (**b**) P20, (**c**) P50.

3.3. Effect of Injection Pressure on Spray Characteristics

Figure 10 shows the development of vapor and liquid spray patterns at different temperatures for different IPs (80, 120 and 160 MPa) at the ambient temperature of 673 K. It can be seen that the vapor phase and the liquid phase spray lengths for P0, P20 and P50 increase with the IP increase. For the vapor phase oil bundle, the vapor phase spray profile grows evenly larger under high IPs. Before 0.3 ms, the vapor phase fuel spray regions of the three fuels are dark black, whereas the vapor phase and liquid phase spray regions substantially coincide. However, the liquid phase of the fuel is relatively large. After 0.3 ms, the vapor phase oil jets of the three blends showed obvious vapor and liquid phase separation with time. This situation is particularly remarkable under the condition of high IP and high mixing ratio of PODE$_{3-4}$. For liquid phase spray, the liquid phase spray length of pure diesel oil under the same IP is significantly larger than the liquid phase spray lengths of P20 and P50. In addition, at a higher IP, the front end of the liquid fuel spray of the three fuels produces a "mist", which is composed of broken atomized droplets, indicating that increasing the IP promotes the atomization of the fuel.

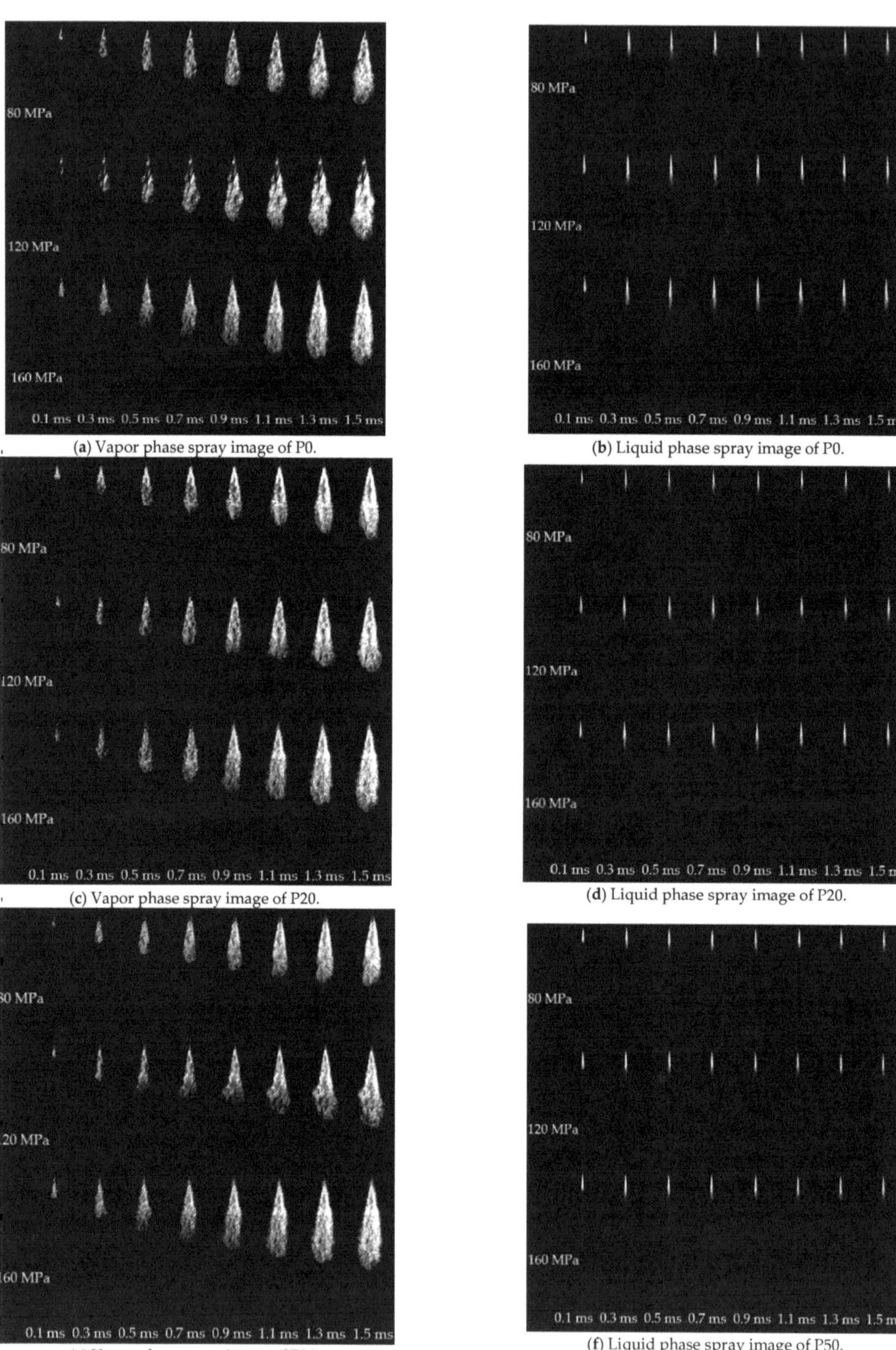

(**a**) Vapor phase spray image of P0.

(**b**) Liquid phase spray image of P0.

(**c**) Vapor phase spray image of P20.

(**d**) Liquid phase spray image of P20.

(**e**) Vapor phase spray image of P50.

(**f**) Liquid phase spray image of P50.

Figure 10. Spray image of three fuels under different IPs.

Figure 11 shows the penetration distance for P0, P20 and P50 over time for different IPs. It can be seen that the vapor and liquid phase penetration distances of the three fuels increase with the increase in IP. This is because when the IP increases, the pressure difference between the inside and outside of the nozzle increases, and the initial kinetic energy of the oil droplets becomes higher [45]. This increased kinetic energy makes the spray penetrate to a longer distance. In addition, the liquid phase penetration distances of P0 and P20 change significantly with the increase in IP. However, the liquid phase penetration distance of P50 does not show much change with the increase in injection pressure. This is because the P50 blend has a lower viscosity relative to P0 and P20, and the lower viscosity will make the fuel break into small droplets more easily after being sprayed from the nozzle [46]. Additionally, at high temperatures, small droplets at the liquid spray front are rapidly evaporated, which cause the liquid penetration distance of P50 to change little with the increase in IP. This process is beneficial to reduce the proportion of liquid phase fuel in the flame region after ignition and improve the combustion efficiency in the cylinder.

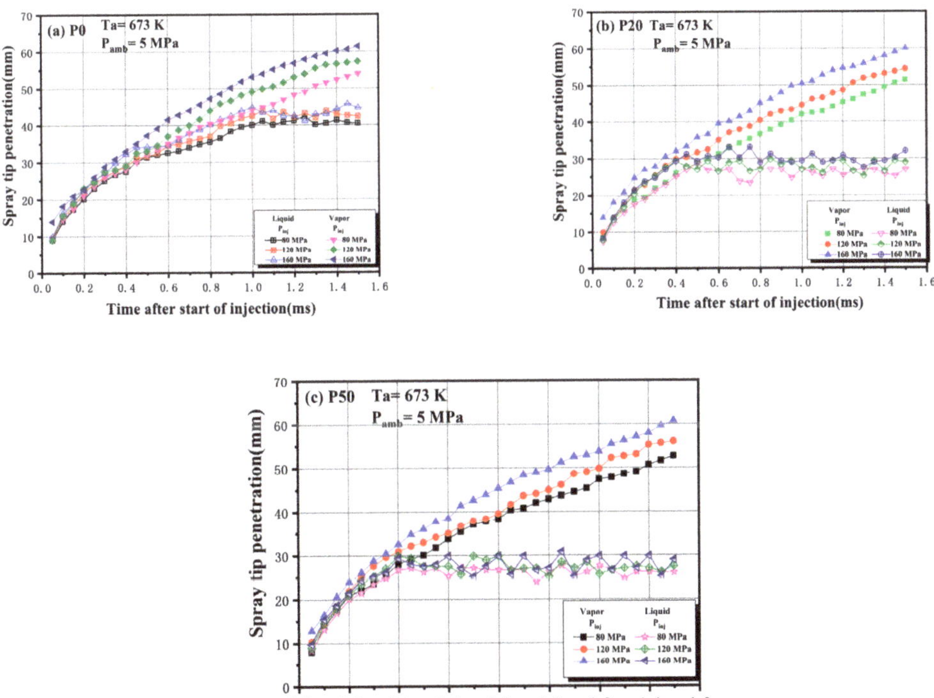

Figure 11. Penetration distances of the three fuels under different IPs: (**a**) P0, (**b**) P20, (**c**) P50.

Figure 12 shows the vapor and liquid cone angles for P0, P20 and P50 under different IPs. The results show that it is not intuitive to see the regularity of the vapor and liquid phase cone angles of the three fuels as a function of IP. Therefore, the average values of the vapor and liquid phase cone angles are shown in Figure 13. It can be seen that the average vapor phase cone angles of the three fuels increase with the increase in IP. When the IP is increased from 80 MPa to 160 MPa, the average vapor phase cone angles for P0, P20 and P50 increase by 1.7°, 1.8° and 2.5°, respectively. This is because when the IP is increased, the kinetic energy of the fuel from the nozzle outlet increases, which in turn leads to enhanced mixing of the spray with the surrounding gas [47], thus resulting in an increase in the vapor cone angle. In addition, regardless of the IP, the average vapor phase cone angles for the three samples are found in the following ascending order: P0 < P20 < P50. This is because as the proportion of PODE$_{3-4}$ in the blend increases, the viscosity of the mixed fuel decreases compared to pure diesel. This leads to an increase in the resistance of spray to ambient gas [30], causing the average vapor cone angle to increase. These results are consistent with those reported by Valentino et al. [48], who concluded that "reducing the viscosity of the fuel will increase the spray cone angle". Interestingly, the average liquid phase cone angle decreases as the proportion of PODE$_{3-4}$ in the blend increases. This is because the viscosity of the mixed fuel is lowered, which causes the fuel to break more easily and hence, atomize after being ejected from the nozzle. Therefore, it becomes easier to evaporate under high temperature conditions, and results in a decrease in average liquid phase cone angle. It can also be seen from Figure 13b that the average liquid phase cone angle of the three fuels does not differ much under different IPs. This because although increasing the IP will cause the spray to spread to both sides, at the same time, the higher IP will cause the fuel to break more severely. Due to this reason, the spray evaporation speed will increase, and therefore, the change in average liquid cone angle will not be obvious.

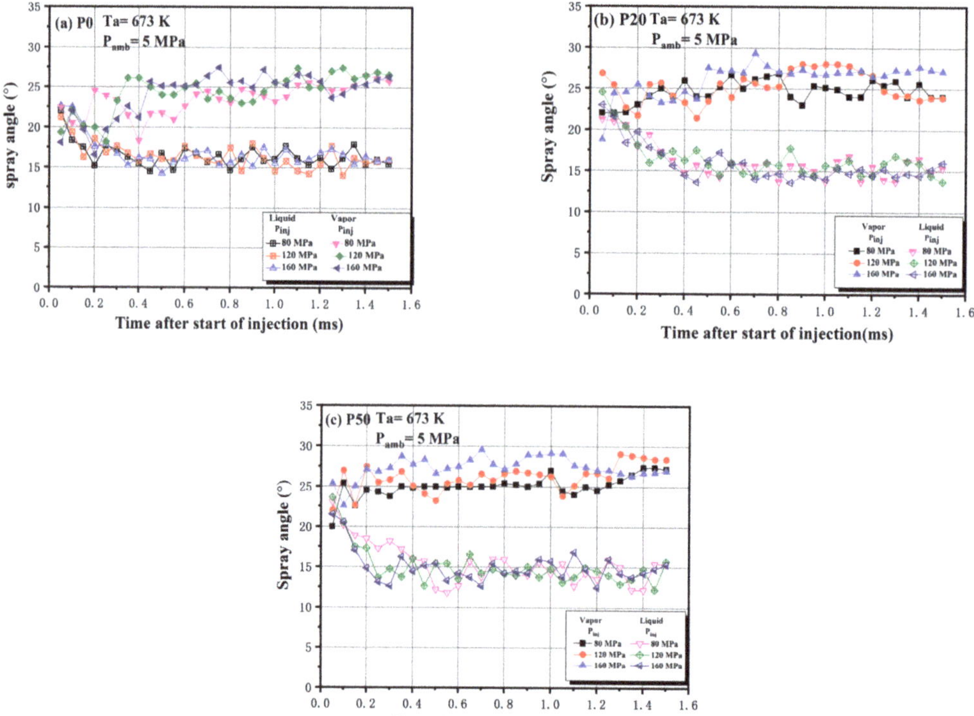

Figure 12. Cone angle of three fuels under different IPs: (**a**) P0, (**b**) P20, (**c**) P50.

Figure 14 shows the vapor and liquid phase projection areas of P0, P20 and P50 under different IPs. It can be seen that the vapor phase projected areas of P0, P20 and P50 increase with the IP increase, indicating that increasing the IP will increase the propagation speed of the spray. This will also improve the spread of spray to surrounding [49], thereby improving the utilization of air in the cylinder. It can be seen from the liquid projection area that the liquid projection area increases slightly with the increase in injection pressure. This is because with higher injection pressure, the axial velocity and radial momentum of the fuel are greater after it is ejected from the nozzle. Due to this, the range of oil beam space diffusion becomes wider. However, higher IP will aggravate the degree of breakage of droplets, which leads to the speed of partial evaporation of oil beam outline, due to which the overall liquid projection area does not increase much. In addition, it can be seen that the liquid phase projected area curve of P0 is substantially coincidental with the vapor phase projected area before 0.5 ms and for the IPs of 80 MPa and 120 MPa. However, for P20 and P50, these timings are 0.3 ms and before 0.2 ms, respectively, and after these time intervals, curves for both the samples gradually start to separate from each other. It shows that, compared with the pure diesel, the mixed fuel can produce diffusion and evaporation earlier, which is very important for the fuel and gas mixing in the cylinder when the engine is actually working.

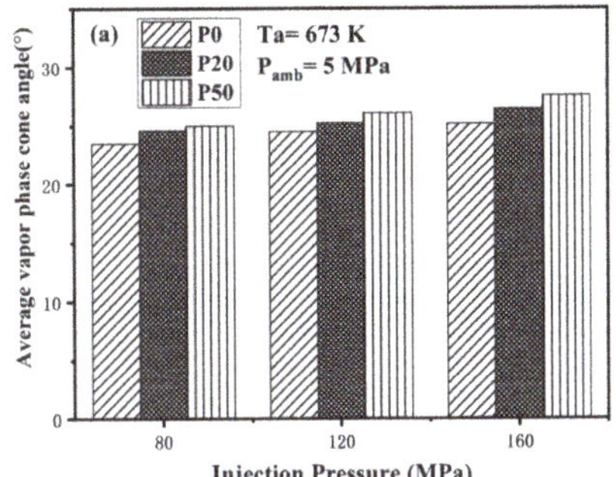

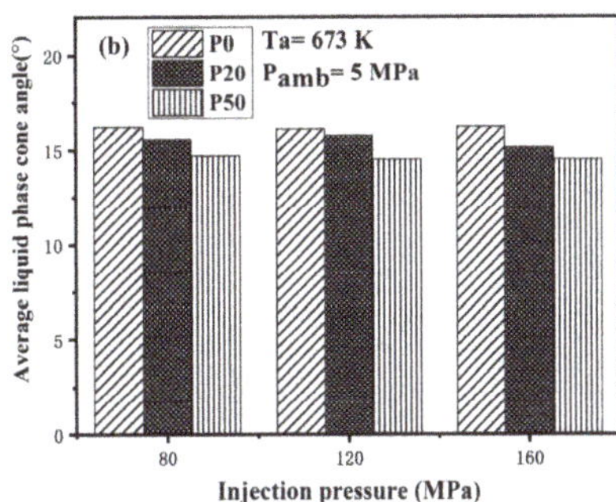

Figure 13. Average cone angles of the three fuels under different fuel Ips: (**a**) average vapor phase cone angle, (**b**) average liquid phase cone angle.

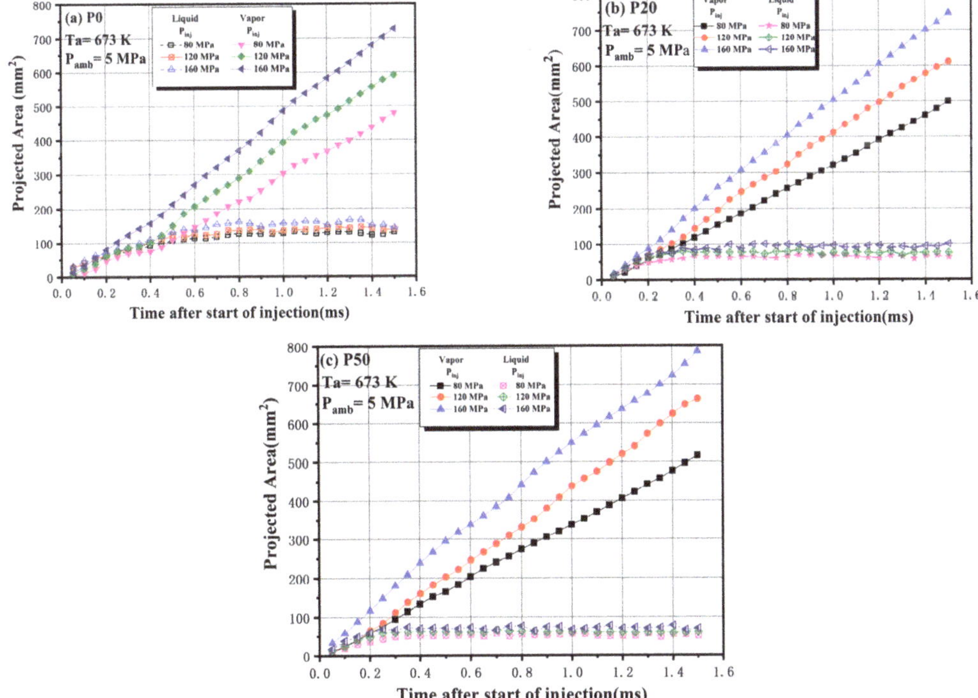

Figure 14. Projection area of the three fuels under different fuel IPs: (**a**) P0, (**b**) P20, (**c**) P50.

3.4. Microscopic Spray Characteristics

To further understand the atomization characteristics of blended fuel, in this section, the microscopic spray characteristics of blended fuels at different temperatures (303 K, 363 K) and injection pressures are studied.

Figure 15 shows the influence of different temperature and IPs on the SMD of test fuels' droplets. As can been seen from the figure, with the IP and temperature increases, the SMD of the test fuels' droplets are reduced. This is due to the increase in IP, which increases the degree of oil bundle breakage and tends to produce more small droplets; the higher of temperature, the more volatile the fuel is, and more tiny droplets can be added, so the SMD of the test fuel is reduced. It can also be seen from the figure that as the proportion of $PODE_{3-4}$ added increases, the fuel's SMD decreases. This is because of the low viscosity of $PODE_{3-4}$ (see Table 4), as the proportion of $PODE_{3-4}$ increases, the viscosity of blended fuel decreases, which is conducive to the improvement of volatility and the SMD reduced.

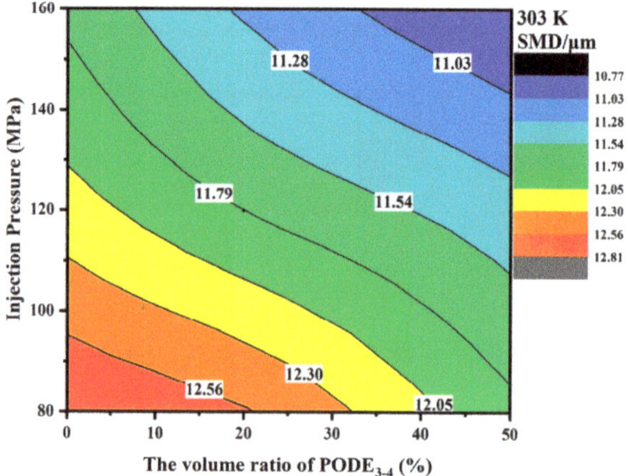

(a) 303 K

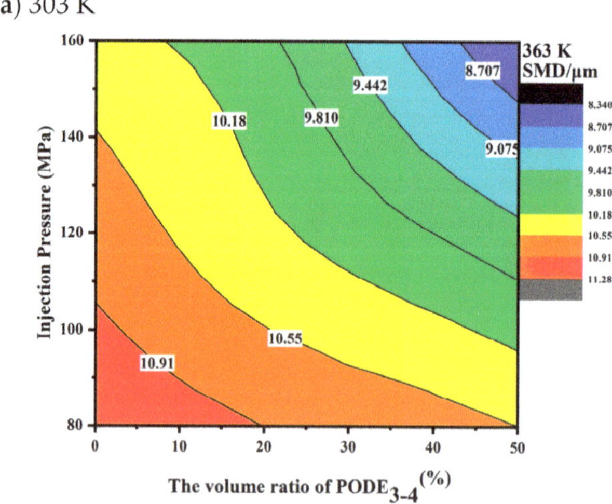

(b) 363 K

Figure 15. The influence of different temperature and IPs on the SMD of test fuels' droplets.

Figure 16 shows the effects of different temperature and IPs on the droplet size distribution. It can be seen from the figure, when the IP is 80 MPa, the P20 and P50 particle size curves shift to the left with the temperature rises, while P0 shifts to the right. This is because as the temperature rises, the volatility of the blended fuel increases, and a large number of small diameter droplets are generated. As the viscosity of P0 is relatively high, the oil bundle is easy to be broken and form large diameter droplets, while the kinetic energy of the larger droplets is relatively small and easy to be absorbed and merged into larger droplets, so more large diameter droplets are generated. With the IP increases, the droplets' size distribution curve of the fuel shifts to the left at 303 K. This is due to the higher IP, which improves the atomization and generates more small droplets, especially P20 and P50, which are highly volatile, generating more droplets of small size than P0. When the IPs is 160 MPa, with the temperature rises and the P0 curve shifts to the left, while the P20 and P50 curves shift to the right and with a bimodal distribution. This is because under high temperature and high IP, P0 can have a higher degree of atomization and smaller droplet size, but for P20 and P50, under high IP, it is easier to generate more smaller droplets (as shown in Figure 16d). However, as the temperature rises, the movement speed of the droplets increases and the probability of mutual adsorption between droplets increases, so the probability of large-size droplets appearing increases.

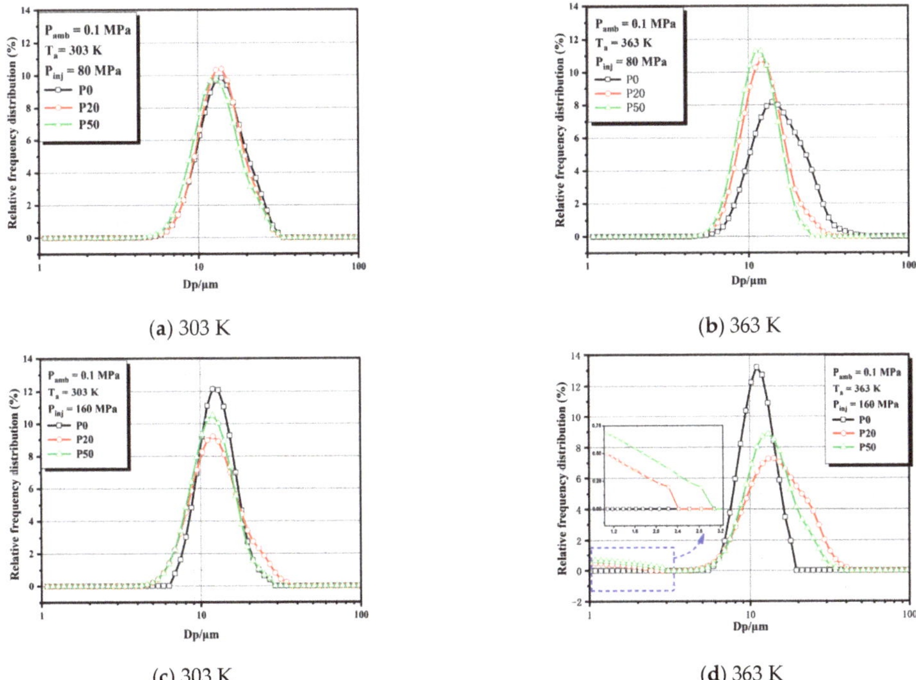

(**a**) 303 K (**b**) 363 K

(**c**) 303 K (**d**) 363 K

Figure 16. The effects of different temperature and IPs on the droplet size distribution.

4. Conclusions

In the CVCB premixed combustion device, the macroscopic and microscopic spray characteristics of the three different diesel and PODE blends (P0, P20 and P50) were studied under different temperatures and IPs. Based on the results, the following conclusions are drawn.

1. When the ambient temperature increases, the liquid cone angles decrease and the vapor cone angles increase for test fuels. When the ambient temperature is 673 K, P0 has the smallest average vapor phase cone angles, but the average liquid cone angle is largest.
2. For the ambient temperatures of 573 K and 623 K, the P0, P20 and P50 vapor and liquid phase penetrations do not change much. When the temperature increases to 673 K, the vapor and liquid phase penetration distances of the three fuels decrease, while the liquid phase penetration distance decreases the most.
3. The vapor phase projection areas of P20 and P50 show that the higher the ambient temperature, the larger the projected area, whereas the liquid phase projected area shows the opposite trend. The vapor and liquid phase projections of P0 at ambient temperatures of 573 K and 623 K do not change much, and the situation improved when the temperature increased to 673 K.
4. With the IP increase, the vapor phase penetration distance and the vapor phase cone angles of P0, P20 and P50 increase. Meanwhile, the average liquid phase cone angles of the three fuels will decrease. The liquid phase penetration distances increase in P0 and P20 with the IP increase, although the change in the liquid phase penetration distance of P50 is not obvious.
5. As the IP increases, the vapor and liquid phase projection areas of the test fuels increase. Compared with the P0, the vapor and liquid phase projection area curves of the blends can be distinguished at an early stage, indicating that the blended fuel can undergo earlier diffusion and evaporation, which are critical to the mixing of oil and gas in the cylinder during actual engine operation.
6. For the vapor phase, both the smallest cone angle and penetration distance is at 673 K and 80 MPa, while the largest projected area is at 673 K and 160 MPa of P50. For the liquid phase, it has the largest cone angle at 673 K and 160 MPa, and smallest penetration at 673 K and 80 MPa. At 573 K, 160 MPa, the liquid phase projected area of P0 is the largest of all tested points.
7. When the temperature and IPs increase, the SMD of the three fuels decrease, and the SMD hierarchy between the three fuels is P0 > P20 > P50. As the IPs and temperature increase, the droplet size decreases, especially when the IP is 160 MPa and the temperature is 363 K, the droplet size distribution of P50 and P20 is bimodal, and the droplet size is smaller.

This article accomplished research on the characteristics of the three different diesel and PODE blends, but more situations should be considered in the next experiments. In the future, the microscopic spray characteristics in high ambient temperature or ambient pressure should become the key point of discussion.

Author Contributions: Methodology, X.H. and H.H.; investigation, X.G. and Y.W.; resources, X.H. and H.H.; data curation, S.L.; writing—original draft preparation, Y.C.; writing—review and editing, C.J.; project administration, X.H. and H.H.; funding acquisition, Y.W. and H.H. All authors have read and agreed to the published version of the manuscript.

Funding: This work was supported by the National Natural Science Foundation of China (51966001) and the projects of EPSRC funded projects (EP/R041970/2 and EP/S032134/1).

Institutional Review Board Statement: Not applicable.

Informed Consent Statement: Not applicable.

Data Availability Statement: Data presented in this study can be made available upon request from the corresponding authors.

Acknowledgments: The authors gratefully acknowledge the support from the Engineering and Physical Sciences Research Council of the UK.

Conflicts of Interest: The authors declare no conflict of interest.

Nomenclature

V	the vapor phase	ECU	electronic control unit
L	the liquid phase	HC	unburned hydrocarbon
S_V	the vapor phase spray tip penetration	IP	injection pressure
S_L	the liquid phase spray tip penetration	LTC	low temperature combustion
θ_V	the vapor phase spray cone angle	NO_X	nitrogen oxide
θ_L	the liquid phase spray cone angle	P0	pure diesel
A_V	the vapor phase projected areas	PODE	polyoxymethylene dimethyl ether
A_L	the liquid phase projected areas	PM	particulate matter
CVCB	constant volume combustion bomb	P20	80% diesel + 20% $PODE_{3-4}$
CO	carbon monoxide	P50	50% diesel + 50% $PODE_{3-4}$
DBI	diffuse back-illumination	SMD	Sauter mean diameter

References

1. Yannopoulos, M.; EL-Seesy, A.; Abdel-Rahman, A.; Bady, M. Influence of adding aluminum oxide nanoparticles to diesterol blends on the combustion and exhaust emission characteristics of a diesel engine. *Exp. Therm. Fluid Sci.* **2018**, *98*, 634–644.
2. Valipour, M. Future of agricultural water management in Africa. *Arch. Agron. Soil Sci* **2014**, *61*, 907–927. [CrossRef]
3. Ma, Y.; Huang, S.; Huang, R.; Zhang, Y.; Xu, S. Ignition and combustion characteristics of n-pentanol–diesel blends in a constant volume chamber. *Appl. Energy* **2017**, *185*, 519–530. [CrossRef]
4. Han, K.; Yang, B.; Zhao, C.; Fu, G.; Ma, X.; Song, G. Experimental study on evaporation characteristics of ethanol–diesel blend fuel droplet. *Exp. Therm. Fluid Sci.* **2016**, *70*, 381–388. [CrossRef]
5. Guo, H.; Liu, S.; He, J. Performances and emissions of new glycol ether blends in diesel fuel used as oxygenated fuel for diesel engines. *J. Energy Eng.* **2016**, *142*, 04015003. [CrossRef]
6. Awad, O.; Mamat, R.; Ali, O.; Sidik, N.; Yusaf, T.; Kadirgama, K.; Kettner, M. Alcohol and ether as alternative fuels in spark ignition engine: A review. *Renew. Sustain. Energy Rev.* **2018**, *82*, 2586–2605. [CrossRef]
7. Jamrozik, A. The effect of the alcohol content in the fuel mixture on the performance and emissions of a direct injection diesel engine fueled with diesel-methanol and diesel-ethanol blends. *Energy Convers. Manag.* **2017**, *148*, 461–476. [CrossRef]
8. D'Adamo, I.; Falcone, P.; Morone, P. A New Socio-economic Indicator to Measure the Performance of Bioeconomy Sectors in Europe. *Ecol. Econ.* **2020**, *176*, 106724. [CrossRef]
9. D'Adamo, I.; Falcone, P.; Huisingh, D.; Morone, P. A circular economy model based on biomethane: What are the opportunities for the municipality of Rome and beyond? *Renew. Energy* **2021**, *163*, 1660–1672. [CrossRef]
10. Restrepo, J.B.; Paternina-Arboleda, C.D.; Bula, A.J. 1,2—Propanediol Production from Glycerol Derived from Biodiesel's Production: Technical and Economic Study. *Energies* **2021**, *14*, 5081. [CrossRef]
11. Hassaan, M.A.; El Nemr, A.; Elkatory, M.R.; Ragab, S.; El-Nemr, M.A.; Pantaleo, A. Synthesis, Characterization, and Synergistic Effects of Modified Biochar in Combination with α-Fe_2O_3 NPs on Biogas Production from Red Algae Pterocladia capillacea. *Sustainability* **2021**, *13*, 9275. [CrossRef]
12. Lapuerta, M.; Armas, O.; Rodriguez-Fernandez, J. Effect of biodiesel fuels on diesel engine emissions. *Prog. Energy Combust.* **2008**, *34*, 198–223. [CrossRef]
13. Kent-Hoekman, S.; Robbins, C. Review of the effects of biodiesel on NOx emissions. *Fuel Process. Technol.* **2012**, *96*, 237–249. [CrossRef]
14. Szybist, J.; Song, J.; Alam, M.; Boehman, A. Biodiesel combustion, emissions and emission control. *Fuel Process. Technol.* **2007**, *88*, 679–691. [CrossRef]
15. Geng, P.; Cao, E.; Tan, Q.; Wei, L. Effects of alternative fuels on the combustion characteristics and emission products from diesel engines: A review. *Renew. Sustain. Energy Rev.* **2016**, *71*, 523–534. [CrossRef]
16. Li, F.; Fu, W.; Yi, B.; Song, L.; Liu, T.; Wang, X.; Wang, C.; Lei, Y.; Lin, Q. Comparison of macroscopic spray characteristics between biodiesel-pentanol blends and diesel. *Exp. Therm. Fluid Sci.* **2018**, *98*, 523–533. [CrossRef]
17. Suh, H.; Yoon, S.; Chang, S. Effect of Multiple Injection Strategies on the Spray Atomization and Reduction of Exhaust Emissions in a Compression Ignition Engine Fueled with Dimethyl Ether (DME). *Energy Fuel* **2010**, *24*, 1323–1332. [CrossRef]
18. Song, H.; Liu, C.; Li, F.; Wang, Z.; He, X.; Shuai, S.; Wang, J. A comparative study of using diesel and PODEn as pilot fuels for natural gas dual-fuel combustion. *Fuel* **2017**, *188*, 418–426. [CrossRef]
19. Li, B.; Li, Y.; Liu, H.; Liu, F.; Wang, Z.; Wang, J. Combustion and emission characteristics of diesel engine fueled with biodiesel/PODE blends. *Appl. Energy* **2017**, *206*, 425–431. [CrossRef]

20. Burger, J.; Siegert, M.; Ströfer, E.; Hasse, H. Poly(oxymethylene) dimethyl ethers as components of tailored diesel fuel: Properties, synthesis and purification concepts. *Fuel* **2010**, *89*, 3315–3319. [CrossRef]
21. Lumpp, B.; Rothe, D.; Pastötter, C.; Lämmermann, R.; Jacob, E. Oxymethylene ethers as diesel fuel additives of the future. *MTZ World Emaga.* **2011**, *72*, 34–38. [CrossRef]
22. Pellegrini, L.; Marchionna, M.; Patrini, R.; Beatrice, C.; Del Giacomo, N.; Guido, C. Combustion behaviour and emission performance of neat and blended polyoxymethylene dimethyl ethers in a light-duty diesel engine. *Microelectron Eng.* **2012**, *87*, 778–781.
23. Pellegrini, L.; Marchionna, M.; Patrini, R.; Florio, S. Emission performance of neat and blended polyoxymethylene dimethyl ethers in an old light-duty diesel car. In Proceedings of the SAE International SAE 2013 World Congress& Exhibition, Detroit, MI, USA, 16–18 April 2013.
24. Iannuzzi, S.E.; Barro, C.; Boulouchos, K.; Burger, J. Combustion behavior and soot formation/oxidation of oxygenated fuels in a cylindrical constant volume chamber. *Fuel* **2016**, *167*, 49–59. [CrossRef]
25. Iannuzzi, S.E.; Barro, C.; Boulouchos, K.; Burger, J. POMDME-diesel blends: Evaluation of performance and exhaust emissions in a single cylinder heavy-duty diesel engine. *Fuel* **2017**, *203*, 57–67. [CrossRef]
26. Huang, H.; Liu, Q.; Teng, W.; Pan, M.; Liu, C.; Wang, Q. Improvement of combustion performance and emissions in diesel engines by fueling n-butanol/diesel/PODE3–4 mixtures. *Appl. Energy* **2017**, *227*, 38–48. [CrossRef]
27. Liu, H.; Wang, Z.; Zhang, J.; Wang, J.; Shuai, S. Study on combustion and emission characteristics of Polyoxymethylene dimethyl Ethers/diesel blends in light-duty and heavy-duty diesel engines. *Appl. Energy* **2017**, *185*, 1393–1402. [CrossRef]
28. Liu, H.; Wang, Z.; Wang, J.; He, X.; Zheng, Y.; Tang, Q.; Wang, J. Performance, combustion and emission characteristics of a diesel engine fueled with polyoxymethylene dimethyl ethers (PODE3-4)/diesel blends. *Energy* **2015**, *88*, 793–800. [CrossRef]
29. Li, D.; Gao, Y.; Liu, S.; Ma, Z.; Wei, Y. Effect of polyoxymethylene dimethyl ethers addition on spray and atomization characteristics using a common rail diesel injection system. *Fuel* **2016**, *186*, 235–247. [CrossRef]
30. Gimeno, J.; Bracho, G.; Martí-Aldaraví, P.; Peraza, J.E. Experimental study of the injection conditions influence over n-dodecane and diesel sprays with two ECN single-hole nozzles. Part I: Inert atmosphere. *Energy Convers. Manag.* **2016**, *126*, 1146–1156. [CrossRef]
31. Manin, J.; Bardi, M.; Pickett, L. Evaluation of the liquid length via diffused back-illumination imaging in vaporizing diesel sprays. In Proceedings of the Proposed for Presentation at the COMODIA 2012, Fukuoka, Japan, 23–26 July 2012; p. 2012.
32. Payri, R.; Giraldo, J.; Ayyapureddi, S.; Versey, Z. Experimental and analytical study on vapor phase and liquid penetration for a high pressure diesel injecto. *Appl. Therm. Eng.* **2018**, *137*, 721–728. [CrossRef]
33. Huang, S.; Deng, P.; Huang, R.; Wang, Z.; Ma, Y.; Dai, H. Visualization research on spray atomization, evaporation and combustion processes of ethanol–diesel blend under LTC conditions. *Energy Convers. Manag.* **2015**, *106*, 911–920. [CrossRef]
34. Ma, Y.; Huang, S.; Huang, R.; Zhang, Y.; Xu, S. Spray and evaporation characteristics of *n*-pentanol–diesel blends in a constant volume chamber. *Energy Convers. Manag.* **2016**, *130*, 240–251. [CrossRef]
35. Payri, R.; Viera, J.P.; Gopalakrishnan, V.; Szymkowicz, P.G. The effect of nozzle geometry over the evaporative spray formation for three different fuels. *Fuel* **2017**, *188*, 645–660. [CrossRef]
36. Tan, Y.; Botero, M.; Sheng, Y.; Dreyer, J.A.; Xu, R.; Yang, W.; Kraft, M. Sooting characteristics of polyoxymethylene dimethyl ether blends with diesel in a diffusion flame. *Fuel* **2018**, *224*, 499–506. [CrossRef]
37. Huang, H.; Liu, Q.; Shi, C.; Wang, Q.; Zhou, C. Experimental study on spray, combustion and emission characteristics of pine oil/diesel blends in a multi-cylinder diesel engine. *Fuel Process. Technol.* **2016**, *153*, 137–148. [CrossRef]
38. Liu, J.; Wang, H.; Li, Y.; Zheng, Z.; Xue, Z.; Shang, H.; Yao, M. Effects of diesel/PODE (polyoxymethylene dimethyl ethers) blends on combustion and emission characteristics in a heavy duty diesel engine. *Fuel* **2016**, *177*, 206–216. [CrossRef]
39. Kang, M.; Song, H.; Jin, F.; Chen, J. Synthesis and physicochemical characterization of polyoxymethylene dimethyl ethers. *J. Fuel Chem. Technol.* **2017**, *45*, 837–845. [CrossRef]
40. Xiong, Y.; Li, Z.; Jiang, Z.; Huang, R. Test of the Influence of Injection Pressure on the Combustion of Mixed Fuel. *Equip. Manuf. Technol.* **2018**, *5*, 142–144.
41. Wang, J.; Wu, F.; Xiao, J.; Shuai, S. Oxygenated blend design and its effects on reducing diesel particulate emissions. *Fuel* **2009**, *88*, 2037–2045. [CrossRef]
42. Ramirez-Verduzco, L. Density and viscosity of biodiesel as a function of temperature: Empirical models. *Renew. Sustain. Energy Rev.* **2013**, *19*, 652–665. [CrossRef]
43. Sohrabiasl, I.; Gorji-Bandpy, M.; Hajialimohammadi, A.; Mirsalim, M.A. Effect of open cell metal porous media on evolution of high pressure diesel fuel spray. *Fuel* **2017**, *206*, 133–144. [CrossRef]
44. Naber, J.; Siebers, D. Effects of gas density and vaporization on penetration and dispersion of diesel Sprays. *SAE Tech. Pap.* **1996**, *105*, 82–111.
45. Kim, W.; Lee, K.; Chang, S. Spray and atomization characteristics of isobutene blended DME fuels. *J. Nat. Gas Sci. Eng.* **2015**, *22*, 98–106. [CrossRef]
46. Dahmen, N.; Arnold, U.; Djordjevic, N.; Henrich, T.; Kolb, T.; Leibold, H.; Sauer, J. High pressure in synthetic fuels production. *J. Supercrit. Fluid* **2015**, *96*, 124–132. [CrossRef]

47. Yu, S.; Yin, B.; Jia, H.; Wen, S.; Li, X.; Yu, J. Theoretical and experimental comparison of internal flow and spray characteristics between diesel and biodiesel. *Fuel* **2017**, *2018*, 20–29. [CrossRef]
48. Valentino, G.; Allocca, L.; Iannuzzi, S.; Montanaro, A. Biodiesel/mineral diesel fuel mixtures: Spray evolution and engine performance and emissions characterization. *Energy* **2011**, *36*, 3924–3932. [CrossRef]
49. Wang, Z.; Xu, H.; Jiang, C.; Wyszynski, M.L. Experimental study on microscopic and macroscopic characteristics of diesel spray with split injection. *Fuel* **2016**, *174*, 140–152. [CrossRef]

 *sustainability*

Article

Lignocellulosic Corn Stover Biomass Pre-Treatment by Deep Eutectic Solvents (DES) for Biomethane Production Process by Bioresource Anaerobic Digestion

Akinola David Olugbemide [1,†], Ana Oberlintner [2,3,†], Uroš Novak [2] and Blaž Likozar [2,*]

[1] Department of Basic Sciences, Auchi Polytechnic, Auchi 312101, Nigeria; akinoladave@gmail.com
[2] Department of Catalysis and Chemical Reaction Engineering, National Institute of Chemistry, Hajdrihova 19, 1001 Ljubljana, Slovenia; ana.oberlintner@ki.si (A.O.); uros.novak@ki.si (U.N.)
[3] Jožef Stefan International Postgraduate School, Jamova Cesta 39, 1000 Ljubljana, Slovenia
* Correspondence: blaz.likozar@ki.si
† These authors contributed equally to the work.

Abstract: The valorization study of the largely available corn stover waste biomass after pretreatment with deep eutectic solvent (DES) for biomethane production in one-liter glass bioreactors by anaerobic digestion for 21 days was presented. Ammonium thiocyanate and urea deep eutectic solvent pretreatments under different conditions in terms of the components ratio and temperature were examined on corn stover waste biomass. The lignocellulose biomass was characterized in detail for its chemistry and morphology to determine the effect of the pretreatment on the natural biocomposite. Furthermore, the implications on biomethane production through anaerobic digestion with different loadings of corn stover biomass at 35 g/L and 50 g/L were tested. The results showed an increase of 48% for a cumulative biomethane production for a DES-pretreated biomass, using a solid-to-liquid ratio of 1:2 at 100 °C for 60 min, which is a strong indication that DES-pretreatment significantly enhanced biomethane production.

Keywords: deep eutectic solvents; lignocellulose biomass pretreatment; biomethane production; anaerobic digestion; corn stover waste valorization

Citation: Olugbemide, A.D.; Oberlintner, A.; Novak, U.; Likozar, B. Lignocellulosic Corn Stover Biomass Pre-Treatment by Deep Eutectic Solvents (DES) for Biomethane Production Process by Bioresource Anaerobic Digestion. *Sustainability* **2021**, *13*, 10504. https://doi.org/10.3390/su131910504

Academic Editor: Idiano D'Adamo

Received: 3 August 2021
Accepted: 15 September 2021
Published: 22 September 2021

Publisher's Note: MDPI stays neutral with regard to jurisdictional claims in published maps and institutional affiliations.

1. Introduction

The bio-based economy, an emerging concept that advances new uses of bioresources, is one of the main future drivers of sustainable economic growth, a big contributor to the 2030 UN Sustainable development goals agenda, and the focus in the transition to a fossil-free society [1]. In December 2018, a new renewable energy (RE) directive (Directive, 2018/2001/EU) entered into force, setting a target minimum 32% share of RE by 2030 [2]. Agricultural waste and forest residues, as well as municipal waste, are the drivers of biorefinery technology due to their wide availability and accessibility. Still, only in Europe, 60 million tons of organic solid waste is mismanaged and displaced in landfills, which presents a loss in potential energy [3,4]. Anaerobic digestion (AD) is highly preferred for energy generation and waste disposal because of its adaptability and sustainability [5]. After AD of the biomass, two major products are obtained: biogas, consisting mainly of CH_4 and CO_2 (60–40%), and digestates. In order to obtain biomethane, biogas upgrading—a process of removing CO_2—is required. Currently, six main upgrading techniques are used: water scrubbing, physical scrubbing, chemical scrubbing, pressure swing adsorption, membrane technology, and cryogenic separation, all well-analyzed by Carranza-Abid et al. [6]. Further, when CO_2 is separated from biomethane, various carbon capture and storage techniques and more sustainable carbon capture and utilization techniques have been proposed. Recently, Baena-Moreno et al. searched for an added-value product from CO_2 and obtained $CaCO_3$ through the precipitation of FGD gypsum from a power station [7]. The second product of AD is also known to be good fertilizers for agriculture or horticulture [8]. Rezaee et al. 2020 suggested that these two products would play some significant roles in shaping the future of the circular economy/bioeconomy through resource recovery enhancement and nutrient recycling [9]. The circular economy, which at present is in its infancy, is confronted with some challenges in reaching its full potentials. AD could be a veritable tool in actualizing this goal and could serve as an alternative to the linear or take-make-waste economy approach, which is both exploitative and environmentally unfriendly [10]. Biomethane production via AD has the capability of reducing greenhouse emissions through the utilization of wastes that would otherwise have ended up in the landfill or be burned, which, under uncontrolled natural anaerobic fermentation, could have led to the generation of methane and CO_2 (nearly 150 million tons in India) that would be released directly into the atmosphere [11,12]. Furthermore, the use of biofertilizers from the AD process can help in mitigating the adverse effects of the mineral fertilizers on the environment, which includes contamination of groundwater, surface water, and the atmosphere [13].

Various analyses on the economical feasibility of biomethane production, taking into account marketing biogas, as well as digestates, showed that currently, governmental incentives and subsidies play key roles in expanding to the wider market and paving the way to a fossil-free society [2,8,14].

Lignocellulosic materials are recalcitrant, and a barrier to their optimization for bio-based solutions is in the material preparation. Pretreatment is therefore imperative to improve biodegradability, thus making the lignocellulosic materials more amenable for microbial digestion [15]. Many pretreatment methods on raw biomass have been researched, employing mechanical pretreatment, physico-chemical pretreatment, biological pretreatment, chemical pretreatment, and a combination of these methods [16–18]. The chemical pretreatment method, in comparison to other methods, has the advantage of easiness, fastness, and efficiency [19]. The physical structures and chemical compositions of lignocellulosic materials could be altered or modified through various chemical pretreatments, making the compositions in lignocellulosic materials more accessible and more readily biodegradable to anaerobic microorganisms, thus increasing digestion efficiency and biogas production [20]. The chemical pretreatment of biomass, recently reported also together with sonification, has been reported to have a positive significant effect on the native structure of lignocellulosic biomass and, thus, their influence on biogas production [17,21]. However, some of these commonly used chemicals, such as acids, alkalis, and organic solvents, have their drawbacks such as the formation of inhibitors, high pretreatment severity, safety concerns, and high energy requirements. The need to mitigate these knotty issues has encouraged research endeavors to develop or produce solvents that would be more environmentally friendly, less toxic or nontoxic, and significantly improve the biomass structure for downstream processes [22]. Ionic liquids (ILs) were preferred to the conventional solvents because of their low melting point, high thermal stability, high reaction rates, and low volatility. The use of ionic liquids for the pretreatment of biomass prior to anaerobic digestion (AD) has been reported to notably increase the production of biogas [23,24]. Some of the commonly used ionic liquids are N-methylmorphine-N-oxide (NMMO), 1-ethyl-3-methylimidazolium acetate [EMIM]-[OAc], and 1-butyl-3-methylimidazolium acetate [BMIM]-[OAc]. Karp et al. [25] noted that the ability of ILs to solubilize polysaccharides in biomass has given them a prominent place among researchers. Padrino et al. [26] reported 28% and 80% increases in methane yield when [BMIM]-[OAc]-pretreated barley was anaerobically digested at mesophilic and thermophilic temperatures, respectively, for 35 days. However, toxicity and high prices have been identified as some of the drawbacks in using ILs for biomass pretreatment [27,28]. As a new class of ILs, deep eutectic solvents (DES) are the preferred alternative for the pretreatment of lignocellulosic biomass in recent times [29]. These green solvents are formed by the selection of appropriate hydrogen bond donors and hydrogen bond acceptors, which are both typically natural compounds, making this type of solvent biocompatible and biodegradable among the most noticeable green factors [30]. Furthermore, its ease of synthesis without the need for purification, as well as its lower cost, recyclability, and environmental benignity, are the most pronounced benefits compared to ILs [31]. DESs have been used in the pretreatment of lignocellulosic biomass with varying degrees of impact on the chemical and physical compositions of the biomass [32,33]. Procentese et al. [22] reported the processing of lettuce leaves with choline chloride–glycerol as an efficient pretreatment in biobutanol production, which is energetically more economical than alkaline treatment. Very recently, Lima et al. [34] studied the pretreatment of lignocellulosic biomass with choline chloride–oxalic acid DES, and demonstrated the positive effect on biogas production at lower concentrations, while concentrations of DES as high as 19.8 g/L showed strong toxic effects. However, to the best of the authors' knowledge, there are no reports on the use of ammonium thiocyanate:urea DES-pretreated biomass for biomethane production via AD. Therefore, the objectives of this work were to evaluate the physico-chemical properties of ammonium thiocyanate:urea DES-pretreated corn stover, and to determine its biomethane potentials at different loading rates through AD.

2. Materials and Methods

2.1. Materials

Analytical-grade ammonium thiocyanate (NH_4SCN) and urea (NH_2CONH_2) were purchased from Sigma-Aldrich (Darmstadt, Germany). H_2SO_4 (95–97%) was bought from Merck, and $CaCO_3$ for neutralization from Riedel-de Haën. The chemicals were used as received without further purification.

2.2. Preparation of Deep Eutectic Solvent

Deep eutectic solvent was prepared from ammonium thiocyanate (HBA) with urea (HBD) in molar ratios 1:1 and 1:2, respectively. The components were measured in the desired ratio and placed in a sealed beaker heated on a hotplate with magnetic stirring at the rate of 500 rpm at 80 °C for one hour, when the clear homogeneous solvent was obtained.

2.3. Sample Collection and Preparation

Corn stover (CS) was obtained from Irrua, Esan Central, Edo State, Nigeria. The sample was cleaned of adhering soil, ground, and passed through a 600 micron sieve. The sample was kept in an air-tight plastic bottle at room temperature before pretreatment.

2.4. Deep Eutectic Solvent Pretreatment of Corn Stover

Corn stover was pretreated with DES at 80 and 100 °C for one hour in an oven at solid-to-liquid ratios of 1:2 and 1:4. Then, 5 g and 10 g of the biomass were thoroughly mixed with 20 g of DES. The pretreated samples were cooled to room temperature and washed with water to remove all the DES components, until the pH approached neutral. The washed biomass was then dried at 70 °C for 3 h. The effect of varying temperature, DES molar ratios, and solid-to-liquid ratios on the samples was analyzed to evaluate the structural and chemical changes of the lignocellulosic biomass. In all eight corn stover biomass samples labeled from A to H, with the raw sample used as the control, labeled I, the biomass recovery was calculated as described by Procentese and Rehmann [35].

2.5. Biomass Characterization

2.5.1. Determination of Lignin, Organic Carbon, Ash, and Moisture Content

Acid-soluble lignin (ASL), acid-insoluble lignin (AIL), ash content, and organic carbon content (CHO) were determined according to the standard procedure for the determination of structural carbohydrates and lignin in biomass [36]. Moisture content (MC) was determined with a Moisture Analyzer (Mettler Toledo, Greifensee, Switzerland).

2.5.2. Fourier-Transform Infrared Spectroscopy Analysis

Fourier-transform infrared spectroscopy (FT-IR) was performed with FT-IR Spectrum Two (PerkinElmer, Waltham, Massachusetts, USA). The spectra were recorded at room temperature, with wavenumbers ranging from 4000 cm^{-1} to 400 cm^{-1}, a resolution of 4 cm^{-1}, and 32 scans being performed for each measurement.

2.5.3. X-ray Powder Diffraction Analysis

Crystallinity was assessed through XRD analysis with the PANanalytical XPert PRO (Malvern Panalytical, UK) high-resolution diffractometer using a Cu-κ1 radiation source (1.5406 Å). The analysis was performed at an energy of 40 kV and electric current of 40 mA, in the 2θ range from 10° to 80° (100 s per step at step size 0.034°). The crystallinity index (CrI) was calculated through the Segal method [37] according to Equation (1):

$$CrI = (I_{200} - I_{am})/I_{200} \times 100 \tag{1}$$

I200 is the maximum intensity of the crystalline plane (200) at 2θ = 22.8° and Iam is the minimum intensity of the amorphous region that is measured at 2θ = 18°.

2.5.4. Morphology of the Corn Stover Biomass

The samples for SEM imaging were taped onto carbon tape and observed by the scanning electron microscope SUPRA 35VP (Carl Zeiss, Jena, Germany).

2.5.5. Elemental Analysis

The carbon, nitrogen, hydrogen, and sulfur (CHNS) content was determined. Approximately 15 mg of sample was weighted and analyzed using the Elemental Analyzer vario EL Cube (Elementar, Langenselbold, Hesse, Germany). All samples were measured at least three times.

2.6. Calculation of Higher and Lower Heating Values

Higher heating values (HHV) and lower heating values (LHV) of raw and DES-pretreated corn stover were calculated through Equations (2) and (3), as described by [38].

$$HHV = 337\,C + 1428\,((H - O)/8) + 95\,S \tag{2}$$

$$LHV = HHV - 2465\,M_w \tag{3}$$

where HHV and LHV are expressed in kJ·kg^{-1}; C, H, O, and S present the weight percentages on a dry basis of carbon, hydrogen, oxygen, and sulfur, respectively, while Mw is a product of the fraction of hydrogen in the sample in 9 kg.

2.7. Anaerobic Digestion Process

Based on the physico-chemical properties and morphology of the samples, sample G was used in the production of biomethane in a batch-mode digester at an average ambient temperature of 28 ± 2 °C for a retention time of 21 days. Digesters were one-liter glass reactors labeled accordingly, i.e., I35 and I50 represented raw samples at the feeding regimes of 35 and 50 g/L, respectively, while G35 and G50 represented pretreated samples at the same feeding regimes. Inoculum was obtained from the effluent from a laboratory digester run on plantain peels for 21 days under mesophilic conditions. The inoculum was kept for about two months under anaerobic conditions prior to the time it was used in the anaerobic digestion (AD) process. Each of the digesters was seeded with 30 mL of inoculum. Biomethane productions were determined by the method described by Fernández-Cegrí, Ángeles De la Rubia, Raposo, and Borja [39].

2.8. Statistical Analysis

Statistical analysis was carried out using one-way ANOVA with the confidence level of 95% ($p < 0.05$) in conjunction with Tukey's honestly significant difference post hoc test. All experiments were performed in duplicate or triplicate and the results were expressed as the mean $\pm$ standard deviation.

3. Results and Discussion

3.1. Effect of Deep Eutectic Solvent Pretreatment on Yield Recovery

The yield recoveries of the DES-treated corn stover samples are shown in Table 1, which ranged from 81.8% to 86.3% with sample H having the lowest value (81.8%) while sample A had the highest (86.3%) followed by sample B (85.6%). The range in our study is higher than the 50% reported by Procentese and Rehmann [35] for coffee silverskin (CS) pretreated with choline chloride/glycerol DES at 150 °C for three hours.

Table 1. Pretreatment conditions and the obtained yield of the recovered DES-treated corn stover.

Sample	DES/Molar Ratio	Solid-to-Liquid Ratio (w:w)	Pretreatment Temperature (°C)	Pretreatment Time (min)	Yield (%)
A	AU 1:1	1:2	80	60	86.3
B	AU 1:1	1:4	80	60	85.6
C	AU 1:1	1:2	100	60	83.1
D	AU 1:1	1:4	100	60	82.0
E	AU 1:2	1:2	80	60	82.8
F	AU 1:2	1:4	80	60	83.0
G	AU 1:2	1:2	100	60	84.3

3.2. Phsicochemical Properties of Raw and Deep Eutectic Solvent Pretreated Samples

The physicochemical properties of raw (I) and pretreated samples (A–G) are shown in Table 2.

Table 2. Pretreatment conditions and obtained yield of the recovered DES-treated corn stover.

Sample	MC [1] (%)	Ash (%)	C/N	TS [2] (%)	VS [3] (%)	AIL [4] (%)	ASL [5] (%)	CHO (%)
A	8.1 ± 0.8	1.00 ± 0.03	46.4	92 ± 1 *	91 ± 1 *	17.3 ± 0.9 *	1.00 ± 0.03 *	73.56
B	9.0 ± 0.3	1.2 ± 0.7	46.4	91 ± 0.4	90 ± 2	14.7 ± 0.1 *	1.3 ± 0.2 *	75.08
C	8 ± 2	1.1 ± 0.5	37.3	92 ± 2 *	91.0 ± 2	16.6 ± 0.6 *	1.5 ± 0.2 *	74.09
D	10 ± 1	1.4 ± 0.8	37.8	90 ± 2	88 ± 3 *	16 ± 1 *	1.60 ± 0.06	71.94
E	8 ± 1	3.3 ± 0.9	41.9	92 ± 2	89 ± 3 *	19 ± 1 *	1.30 ± 0.06	72.44
F	7 ± 1	2.3 ± 0.9	45.9	93 ± 1 *	91 ± 3 *	15 ± 4 *	1.4 ± 0.1	76.86
G	7.5 ± 0.1	0.26 ± 0.05	38.5	92.5 ± 0.1	92.4 * ± 0.1	13 ± 3 *	1.60 ± 0.09 *	78.10
H	8.1 ± 0.2	1.9 ± 0.2	39.7	91.9 ± 0.3	90.0 ± 0.6	14 ± 3 *	1.5 ± 0.1 *	76.43
I	8.1 ± 0.8	4.8 ± 0.3	49.9	92 ± 1	87 ± 1	17 ± 4	1.4 ± 0.2	72.85

[1] Moisture Content (MC); [2] Total Solubility (TS); [3] Volatile Solids (VS); [4] Acid-Insoluble Lignin (AIL); [5] Acid-Soluble Lignin (ASL); * The samples have significantly different mean values ($p < 0.05$).

The ash content of pretreated samples was lower than that of the raw sample, which could be an indication of the high solubility of inorganic compounds in ammonium thiocyanate:urea deep eutectic solvents [40]. The organic carbon contents of all the pretreated samples were all significantly higher than those of the raw samples. The AIL ranged from 12.84% to 17.31% with sample A having the highest while sample G had the lowest value. These values are lower than the 21.7% reported for corn stover by Zhu et al. [23]. For the ASL, sample A recorded the lowest value of 1.00%, while samples D and G had the highest values of 1.63% and 1.62%, respectively. The values for ASL from samples D and G are higher than the 1.5% reported for wheat straw by [41]. The AIL degradation ranged from 2.35% to 23.53%. Sample G had the highest lignin fraction reduction while sample C had the lowest (Table 3). On the other hand, samples A and E had an increase in lignin fraction after pretreatment; this observed phenomenon has been attributed to the formation of pseudo-lignin during pretreatment [42]. A decrease in AIL could improve the biogas production capability of the samples as the lignin-carbohydrates complex would have been broken, making it more amenable to enzymatic attack [43]. Regarding MC, TS, and VC, only small and inconsistent changes were detected; therefore, it is impossible to make a reliable conclusion about the effect of DES pretreatment on these parameters. The statistical analysis using one-way ANOVA showed statistical significance of the results in Table 2 ($p < 0.05$). Further analysis with Tukey's honestly significant difference post hoc test revealed significant differences of the treated samples compared to the initial sample I shown in Table 2.

Table 3. Degradation of acid-insoluble and acid-soluble lignin after DES pretreatment.

Sample	AIL Degradation (%)	ASL Degradation (%)
A	NA	NA
B	13.53	7.14
C	2.35	NA
D	5.88	NA
E	NA	7.14
F	11.77	NA
G	23.53	NA
H	17.65	NA

NA: Not Applicable; AIL: acid-insoluble lignin; ASL: acid-soluble lignin.

The ASL degradation for samples A, B, and C ranged from 5.15% to 26.47% with sample A having the highest followed by sample B with 7.35%. On the other hand, samples C, D, F, G, and H had an increase in ASL after pretreatment. Table 2 revealed that there was an increase in carbohydrate contents of the samples after pretreatment except for samples D and E.

3.2.1. Elemental Analysis

Elemental analysis of the samples revealed that carbon, hydrogen, and sulfur contents increased with treatment, unrelated to the treatment conditions, while the nitrogen content decreased (Figure 1a–d). The decrease in nitrogen could be attributed to the removal of proteins, where treatments with an A:U ratio of 1:1 and a temperature of 80 °C were most efficient. Moreover, the removal of nitrogen can influence the composition ratios between other elements as well.

The nitrogen content values ranged between 0.88% and 1.20% with the raw sample having the lowest value while sample C had the highest value. There was a significant increase in nitrogen content after DES pretreatment, due to the selective removal of the nitrogen-rich materials (proteins) form the corn stover biomass. The C/N ratio was calculated, which ranged from 37.39% to 49.95%. This range was higher by 15–30%, which was recommended for the AD [44].

3.2.2. Fourier-Transform Infrared Spectroscopy Analysis

FT-IR analysis revealed the same structural characteristics in all samples. In Figure 2, the spectra of the untreated sample (I) and sample (G) are presented. The broadbands observed at the wavelength of 3330 cm^{-1} are characteristic for the stretching vibration of the hydroxyl group in polysaccharides. The band at 2900 cm^{-1} is attributed to the stretching vibration of the C–H bond. The adsorption bands between 1756 cm^{-1} and 1546 cm^{-1} are associated with the hemicellulose complex [45]. The stretching in the range from 1700 cm^{-1} to 1550 cm^{-1} may be correlated to the water adsorption. The bands at 1418 cm^{-1}, 1369 cm^{-1}, and 1319 cm^{-1} are attributed to CH$_2$, in-plane CH deformation, and CH$_2$ wagging, respectively. The bands at wavelengths 1156 cm^{-1} and 1029 cm^{-1} are characteristic for cellulose and are associated with the asymmetric C–O–C bridge stretching of the anhydroglucose ring and C–O–C pyranose ring skeletal vibration that are present in cellulose [46,47].

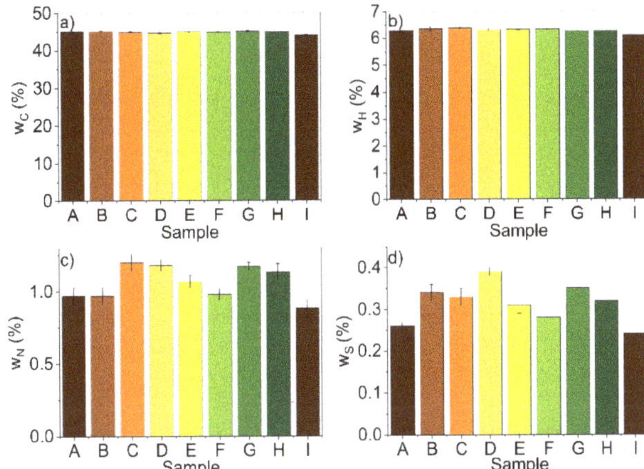

Figure 1. Elemental analysis of the pretreated (A–H) and untreated (I) samples for (**a**) carbon (C); (**b**) hydrogen (H); (**c**) nitrogen (N); (**d**) sulfur (S).

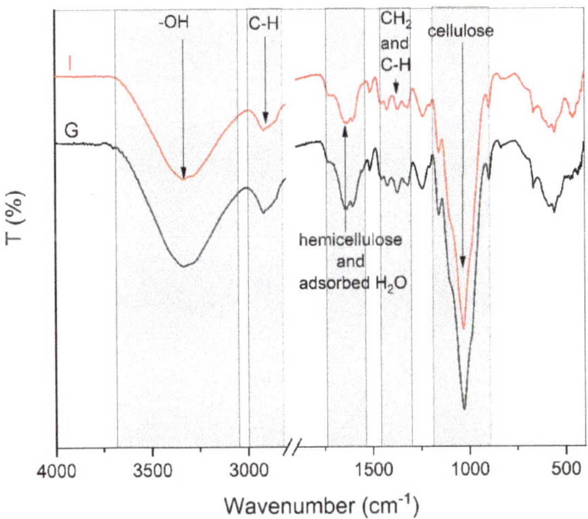

Figure 2. FT-IR spectra of pretreated (A–H) and raw (I) samples, grouped according to the pretreatment temperature.

3.2.3. X-ray Powder Diffraction Analysis

In XRD spectra, peaks for different crystal planes characteristic of cellulose I were observed. The peaks at 14.9° and 16.7°, which represent (110) and (110) planes, respectively, are overlapped in all spectra, which would point to a diamond-shaped cross-section of the crystallites [48]. There were no major differences in the characteristics of the XRD spectra of samples, which point to the fact that cellulose does not undergo structural changes during pretreatment. No shift in peaks was detected, which means that cellulose I did not transition to cellulose II during the dissolution [49]. The crystallinity indexes (CrI) of the treated and untreated samples are presented in Table 3. The CrI of the untreated material was 54.9%. It was observed that samples treated with DES had some increase in crystallinity except samples A and C. The highest CrI was reached by sample F with 62.1%

(Table 4). The increase in CrI is in agreement with the results reported for cotton stalk pretreated with different pretreatment methods by Zhang et al. [50]. They proposed that the increase was a result of the removal of the noncrystalline portions (amorphous regions) of the samples by the DES pretreatments and may not necessarily mean an increase in cellulose crystallinity.

Table 4. Crystallinity indexes of treated and untreated samples.

	I	A	B	C	D	E	F	G	H
CrI (%)	54.9	54.1	55.9	52.9	55.1	59.3	62.1	58.4	57.1

3.2.4. Morphology

The samples were observed with a scanning electron microscope. Regarding morphology, no differences were observed between samples. In Figure 3, untreated sample (I) and sample B are presented. Both samples appeared rod-shaped with a rough surface and were structured out of thinner, fiber-like particles.

Figure 3. SEM images of: (**a,b**) surface of untreated sample I, (**c**) treated sample B; (**d–g**) surface and morphology of sample G, which was used for further production of biomethane.

Furthermore, sample G was more thoroughly investigated as it was used for the production of biogas. The micrographs of sample G are presented in Figure 3d–g. It is possible to observe that the samples consisted of fiber-like particles, but their surface was smoother than in sample B. The surface seemed more cracked than in the untreated sample.

3.3. Determination of Higher and Lower Heating Values

Using Equations (2) and (3), the calculation of HHV and LHV for corn stover biomass samples was performed. HHV values contain the latent heat of the water vapor products of combustion, while the LHV represents the correction to HHV due to moisture in the fuel (biomass) or water vapor formed during the combustion of hydrogen in the fuel [51]. The calculated values for higher and lower heating values are presented in Figure 4. The higher heating values ranged from 15,659 to 17,676 kJ kg^{-1}, while the lower heating values ranged from 14,308 to 16,282 kJ kg^{-1}. The two samples (A and E) with the highest lignin contents had the highest values for both HHV and LHV (17,676 kJ kg^{-1} and 16,282 kJ kg^{-1}; 16,321 kJ kg^{-1} and 14,917 kJ kg^{-1}). A direct correlation between lignin content and HHV has been already reported in the literature [52]. The HHV for sample A was comparable to 17,530 kJ kg^{-1} reported for palm oil mill effluent (POME), while its LHV is higher than 13,872 kJ kg^{-1} reported for the same sample of POME by Jekayinfa and Omisakin [53].

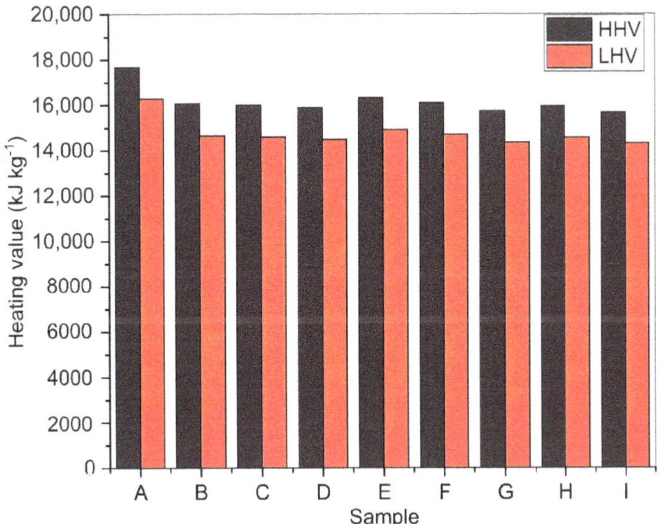

Figure 4. Calculated higher and lower heating values of raw and DES-pretreated corn stover.

3.4. Daily and Cumulative Biomethane Production

Daily biomethane production is shown in Figure 5a for raw samples I and DES-pretreated samples G in digesters with 35 g/L and 50 g/L of biomass loadings. The digester I35 production commenced on day 1, while in digester I50, there was no production recorded until day 10, when the production reached 30 mL. For the DES-pretreated digesters, G35 had started its production one day after I35, while digester G50 experienced a delay in production for two days compared to I35. The latter had its peak value of 90 mL on day 1; digester G35 attained its peak value of 110 mL on day 3, while G50 recorded 80 mL on day 3.

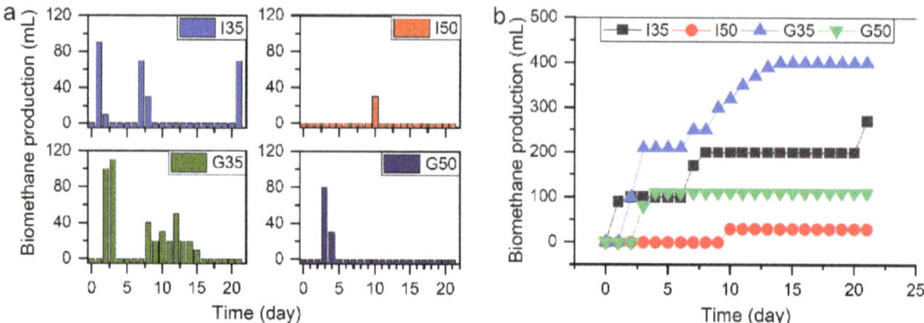

Figure 5. (**a**) Daily biomethane production from raw and DES-pretreated corn stover and (**b**) the cumulative biomethane production from raw and DES-pretreated corn stover.

Generally, there was a fast acclimation of microorganisms in digesters to the substrates, as evident in the lag phase day(s), except digesters G50 with a lag phase of two days. Digester I35 had no lag phase, while digester G35 had a lag phase of one day. Lag phase is an important factor in AD because it shows the rapidity with which the microorganisms acclimatize to the substrates, and from an economic point of view, the less lag phase that is present, the more beneficial it is for the AD process. The cumulative biomethane production is shown in Figure 5b, which revealed that a lower feeding regime produced more biogas than the higher feeding regime. It is imperative to determine the appropriate feeding regime for any biomass to be used for biogas production as there is no fit-it-all feeding regime for potential feedstocks; therefore, each has to be treated on its own merit.

The biomethane production of digester G35 was 48% higher than its counterpart (I35), a strong indication that DES-pretreatment significantly enhanced biomethane production at this feeding regime. This was due largely to the changes in the physical and chemical composition of the substrate after treatment with ammonium thiocyanate-urea-based DES at 100 °C (G). The percentage increase of 48% recorded in this study is higher than the previously reported 15.5% increase for wheat plant pretreated with dilute sulfuric acid by Taherdanak, Zilouei, and Karimi [54].

Digester G50 had a lower biomethane yield than I35 did, which showed that the 35 g L^{-1} feeding regime was better for biomethane production for the DES-pretreated sample. The lower feeding regime performed better in both the raw and DES-pretreated digesters. Digester I50 produced only 30 mL of biomethane and, afterward, there was no production until the end of the experiment (day 21). It has been reported previously that overloading a digester could lead to low methane yield or digester failure due to the accumulation of volatile fatty acid [55]. Although the complete study of the DES pretreatment effects on the production of biomethane is not shown, the collected experimental data showed a high potential for further exploration on the topic of waste corn stover biomethane production employing AD.

The values obtained by different DES treatments of corn stover in Table 2 are statistically significant ($p < 0.005$); however, due to the single measurement of the anaerobic corn stover, digestion cannot be claimed the same with a high degree of certainty; therefore, further studies on that topic also including the techno economics should be considered.

4. Conclusions

The pretreatment of corn stover with ammonium thiocyanate:urea DES under varying conditions revealed changes in physicochemical and structural properties compared to the untreated sample. The implications of these changes on biomethane production from the samples were discussed. The most promising of the samples was used in a novel attempt to determine the effect of ammonium thiocyanate:urea-based DES pretreatment on the biomethane potential of corn stover at two feeding regimes of 35 g L^{-1} and 50 g L^{-1}. DES pretreatment significantly enhanced methane production at a 35 g L^{-1} loading rate with a 48% increase over the untreated sample.

In the future, a larger study of the DES pretreatment effects on the production of biomethane should be carried out, as the collected experimental data showed a high potential for further exploration on the topic of waste corn stover biomethane production employing AD, in addition to the study of digestate composition, its utilization, and its role in resources circulation. Furthermore, there is a need to explore other DESs for the pretreatment of different substrates in order to evaluate their biomethane potentials and optimize the process efficiency for an improved economy of AD.

Author Contributions: Conceptualization, A.D.O. and B.L.; methodology, A.O. and U.N.; formal analysis, A.D.O., A.O. and U.N.; investigation, A.D.O., A.O. and U.N.; resources, A.D.O., U.N. and B.L.; data curation, A.D.O., A.O. and U.N.; writing—original draft preparation, A.D.O. and A.O.; writing—review and editing, A.D.O., U.N. and B.L.; visualization, A.O.; supervision, U.N. and B.L.; funding acquisition, A.D.O. and B.L. All authors have read and agreed to the published version of the manuscript.

Funding: This research was funded by the Ph.D. research grant (Ana Oberlintner) and Slovenian Research Agency (Program P2-0152).

Acknowledgments: The authors wish to thank Joshua Isaac Oludiran and his colleagues for their technical support. Anže Prašnikar is acknowledged for preparation of the SEM images.

Conflicts of Interest: The authors declare no conflict of interest.

References

1. Bennich, T.; Belyazid, S.; Stjernquist, I.; Diemer, A.; Seifollahi-Aghmiuni, S.; Kalantari, Z. The bio-based economy, 2030 Agenda, and strong sustainability—A regional-scale assessment of sustainability goal interactions. *J. Clean. Prod.* **2021**, *283*, 125174. [CrossRef]
2. D'Adamo, I.; Falcone, P.M.; Gastaldi, M.; Morone, P. RES-T trajectories and an integrated SWOT-AHP analysis for biomethane. Policy implications to support a green revolution in European transport. *Energy Policy* **2020**, *138*. [CrossRef]
3. Hladnik, L.; Vicente, F.A.; Novak, U.; Grilc, M.; Likozar, B. Solubility assessment of lignin monomeric compounds and organosolv lignin in deep eutectic solvents using in situ Fourier-transform infrared spectroscopy. *Ind. Crop. Prod.* **2021**, *164*, 113359. [CrossRef]
4. Kohli, K.; Prajapati, R.; Sharma, B.K. Bio-based chemicals from renewable biomass for integrated biorefineries. *Energies* **2019**, *12*, 233. [CrossRef]
5. Wainaina, S.; Awasthi, M.K.; Sarsaiya, S.; Chen, H.; Singh, E.; Kumar, A.; Ravindran, B.; Awasthi, S.K.; Liu, T.; Duan, Y.; et al. Resource recovery and circular economy from organic solid waste using aerobic and anaerobic digestion technologies. *Bioresour. Technol.* **2020**, *301*, 122778. [CrossRef] [PubMed]
6. Carranza-Abaid, A.; Wanderley, R.R.; Knuutila, H.K.; Jakobsen, J.P. Analysis and selection of optimal solvent-based technologies for biogas upgrading. *Fuel* **2021**, *303*, 121327. [CrossRef]
7. Baena-Moreno, F.M.; le Saché, E.; Price, C.A.H.; Reina, T.R.; Navarrete, B. From biogas upgrading to CO$_2$ utilization and waste recycling: A novel circular economy approach. *J. CO2 Util.* **2021**, *47*, 101496. [CrossRef]
8. D'Adamo, I.; Falcone, P.M.; Huisingh, D.; Morone, P. A circular economy model based on biomethane: What are the opportunities for the municipality of Rome and beyond? *Renew. Energy* **2021**, *163*, 1660–1672. [CrossRef]
9. Rezaee, M.; Gitipour, S.; Sarrafzadeh, M.-H. Different pathways to integrate anaerobic digestion and thermochemical processes: Moving toward the circular economy concept. *Environ. Energy Econ. Res.* **2020**, *4*, 57–67. [CrossRef]
10. Rodríguez, R.W.; Pomponi, F.; Webster, K.; D'Amico, B. The future of the circular economy and the circular economy of the future. *Built Environ. Proj. Asset Manag.* **2020**, *10*, 529–546. [CrossRef]
11. Selvaggi, R.; Valenti, F.; Pecorino, B.; Porto, S.M.C. Assessment of tomato peels suitable for producing biomethane within the context of circular economy: A GIS-based model analysis. *Sustainability* **2021**, *13*, 5559. [CrossRef]
12. Porichha, G.K.; Hu, Y.; Rao, K.T.; Xu, C.C. Crop residue management in India: Stubble burning vs. other utilizations including bioenergy. *Energies* **2021**, *14*, 4281. [CrossRef]
13. Suhartini, S.; Heaven, S.; Banks, C.J. Can anaerobic digestion of sugar beet pulp support the circular economy? A study of biogas and nutrient potential. *IOP Conf. Series Earth Environ. Sci.* **2018**, *131*, 012048. [CrossRef]
14. Baena-Moreno, F.M.; Malico, I.; Marques, I.P. Promoting sustainability: Wastewater treatment plants as a source of biomethane in regions far from a high-pressure grid. A real Portuguese case study. *Sustainability* **2021**, *13*, 8933. [CrossRef]
15. Mirmohamadsadeghi, S.; Karimi, K.; Azarbaijani, R.; Yeganeh, L.P.; Angelidaki, I.; Nizami, A.-S.; Bhat, R.; Dashora, K.; Vijay, V.K.; Aghbashlo, M.; et al. Pretreatment of lignocelluloses for enhanced biogas production: A review on influencing mechanisms and the importance of microbial diversity. *Renew. Sustain. Energy Rev.* **2021**, *135*, 110173. [CrossRef]

16. Thompson, D.N.; Campbell, T.; Bals, B.; Runge, T.; Teymouri, F.; Ovard, L.P. Chemical preconversion: Application of low-severity pretreatment chemistries for commoditization of lignocellulosic feedstock. *Biofuels* **2013**, *4*, 323–340. [CrossRef]

17. Agbor, V.B.; Cicek, N.; Sparling, R.; Berlin, A.; Levin, D.B. Biomass pretreatment: Fundamentals toward application. *Biotechnol. Adv.* **2011**, *29*, 675–685. [CrossRef]

18. Isikgor, F.H.; Becer, C.R. Lignocellulosic biomass: A sustainable platform for the production of bio-based chemicals and polymers. *Polym. Chem.* **2015**, *6*, 4497–4559. [CrossRef]

19. Hernández-Beltrán, J.U.; Hernández-De Lira, I.; Cruz-Santos, M.; Saucedo-Luevanos, A.; Hernández-Terán, F.; Balagurusamy, N. Insight into Pretreatment Methods of Lignocellulosic Biomass to Increase Biogas Yield: Current State, Challenges, and Opportunities. *Appl. Sci.* **2019**, *9*, 3721. [CrossRef]

20. Ariunbaatar, J.; Panico, A.; Esposito, G.; Pirozzi, F.; Lens, P.N.L. Pretreatment methods to enhance anaerobic digestion of organic solid waste. *Appl. Energy* **2014**, *123*, 143–156. [CrossRef]

21. Ouahabi, Y.R.; Bensadok, K.; Ouahabi, A. Optimization of the biomethane production process by anaerobic digestion of wheat straw using chemical pretreatments coupled with ultrasonic disintegration. *Sustainability* **2021**, *13*, 7202. [CrossRef]

22. Procentese, A.; Raganati, F.; Olivieri, G.; Russo, M.E.; Rehmann, L.; Marzocchella, A. Low-energy biomass pretreatment with deep eutectic solvents for bio-butanol production. *Bioresour. Technol.* **2017**, *243*, 464–473. [CrossRef]

23. Zhu, J.; Wan, C.; Li, Y. Enhanced solid-state anaerobic digestion of corn stover by alkaline pretreatment. *Bioresour. Technol.* **2010**, *101*, 7523–7528. [CrossRef]

24. Gao, J.; Chen, L.; Yuan, K.; Huang, H.; Yan, Z. Ionic liquid pretreatment to enhance the anaerobic digestion of lignocellulosic biomass. *Bioresour. Technol.* **2013**, *150*, 352–358. [CrossRef] [PubMed]

25. Karp, E.M.; Donohoe, B.S.; O'Brien, M.H.; Ciesielski, P.N.; Mittal, A.; Biddy, M.J.; Beckham, G.T. Alkaline pretreatment of corn stover: Bench-scale fractionation and stream characterization. *ACS Sustain. Chem. Eng.* **2014**, *2*, 1481–1491. [CrossRef]

26. Padrino, B.; Serrano, M.L.; Morales-Delarosa, S.; Campos-Martín, J.M.; Fierro, J.L.G.; Martínez, F.; Melero, J.A.; Puyol, D. resource recovery potential from lignocellulosic feedstock upon lysis with ionic liquids. *Front. Bioeng. Biotechnol.* **2018**, *6*, 119. [CrossRef] [PubMed]

27. Grilc, M.; Likozar, B.; Levec, J. Kinetic model of homogeneous lignocellulosic biomass solvolysis in glycerol and imidazolium-based ionic liquids with subsequent heterogeneous hydrodeoxygenation over NiMo/Al$_2$O$_3$ catalyst. *Catal. Today* **2015**, *256*, 302–314. [CrossRef]

28. Weerachanchai, P.; Lee, J.-M. Effect of organic solvent in ionic liquid on biomass pretreatment. *ACS Sustain. Chem. Eng.* **2013**, *1*, 894–902. [CrossRef]

29. Bjelić, A.; Hočevar, B.; Grilc, M.; Novak, U.; Likozar, B. A review of sustainable lignocellulose biorefining applying (natural) deep eutectic solvents (DESs) for separations, catalysis and enzymatic biotransformation processes. *Rev. Chem. Eng.* **2020**, 2019077. [CrossRef]

30. Elgharbawy, A.A.M.; Hayyan, M.; Hayyan, A.; Basirun, W.J.; Salleh, H.M.; Mirghani, M.E.S. A grand avenue to integrate deep eutectic solvents into biomass processing. *Biomass Bioenergy* **2020**, *137*, 105550. [CrossRef]

31. Smith, E.L.; Abbott, A.P.; Ryder, K.S. Deep eutectic solvents (DESs) and their applications. *Chem. Rev.* **2014**, *114*, 11060–11082. [CrossRef] [PubMed]

32. Roy, R.; Rahman, S.; Raynie, D.E. Recent advances of greener pretreatment technologies of lignocellulose. *Curr. Res. Green Sustain. Chem.* **2020**, *3*, 100035. [CrossRef]

33. Xue, B.; Yang, Y.; Tang, R.; Xue, D.; Sun, Y.; Li, X. Efficient dissolution of lignin in novel ternary deep eutectic solvents and its application in polyurethane. *Int. J. Biol. Macromol.* **2020**, *164*, 480–488. [CrossRef] [PubMed]

34. Lima, F.; Branco, L.C.; Lapa, N.; Marrucho, I.M. Beneficial and detrimental effects of choline chloride–oxalic acid deep eutectic solvent on biogas production. *Waste Manag.* **2021**, *131*, 368–375. [CrossRef]

35. Procentese, A.; Rehmann, L. Fermentable sugar production from a coffee processing by-product after deep eutectic solvent pretreatment. *Bioresour. Technol. Rep.* **2018**, *4*, 174–180. [CrossRef]

36. Sluiter, A.; Hames, B.; Ruiz, R.; Scarlata, C.; Sluiter, J.; Templeton, D.; Crocker, D. *Determination of Structural Carbohydrates and Lignin in Biomass*; NREL/TP-510-42618; National Renewable Energy Laboratory: Golden, CO, USA, 2004.

37. Segal, L.; Creely, J.J.; Martin, A.E.; Conrad, C.M. An empirical method for estimating the degree of crystallinity of native cellulose using the X-ray diffractometer. *Text. Res. J.* **1959**, *29*, 786–794. [CrossRef]

38. Podolski, W.F.; Conrad, V.; Lowenhaupt, D.E.; Winschel, R.A. Materials of construction. In *Perry's Chemical Engineers' Handbook*; Perry, R.H., Green, D.W., Eds.; The McGraw-Hill Companies, Inc.: New York, NY, USA, 2008; pp. 24–25. [CrossRef]

39. Fernández-Cegrí, V.; de la Rubia, M.A.; Raposo, F.; Borja, R. Effect of hydrothermal pretreatment of sunflower oil cake on biomethane potential focusing on fibre composition. *Bioresour. Technol.* **2012**, *123*, 424–429. [CrossRef] [PubMed]

40. Suopajärvi, T.; Ricci, P.; Karvonen, V.; Ottolina, G.; Liimatainen, H. Acidic and alkaline deep eutectic solvents in delignification and nanofibrillation of corn stalk, wheat straw, and rapeseed stem residues. *Ind. Crop. Prod.* **2020**, *145*, 111956. [CrossRef]

41. Zhang, X.; Nghiem, N.P. Pretreatment and fractionation of wheat straw for production of fuel ethanol and value-added co-products in a biorefinery. *AIMS Bioeng.* **2014**, *1*, 40–52. [CrossRef]

42. Mulat, D.G.; Dibdiakova, J.; Horn, S.J. Microbial biogas production from hydrolysis lignin: Insight into lignin structural changes. *Biotechnol. Biofuels* **2018**, *11*, 61. [CrossRef]

43. Sasmal, S.; Mohanty, K. Pretreatment of lignocellulosic biomass toward biofuel production. In *Biorefining of Biomass to Biofuels*; Springer: Cham, Switzerland, 2018; pp. 203–221.
44. Petravić-Tominac, V.; Nastav, N.; Buljubašić, M.; Šantek, B. Current state of biogas production in Croatia. *Energy Sustain. Soc.* **2020**, *10*. [CrossRef]
45. Kubovský, I.; Kačíková, D.; Kačík, F. Structural changes of oak wood main components caused by thermal modification. *Polymers* **2020**, *12*, 485. [CrossRef] [PubMed]
46. Poletto, M.; Pistor, V.; Zeni, M.; Zattera, A.J. Crystalline properties and decomposition kinetics of cellulose fibers in wood pulp obtained by two pulping processes. *Polym. Degrad. Stab.* **2011**, *96*, 679–685. [CrossRef]
47. Lao, W.; Li, G.; Zhou, Q.; Qin, T. Quantitative analysis of biomass in three types of wood-plastic composites by FTIR spectroscopy. *Bioresources* **2014**, *9*, 6073–6086. [CrossRef]
48. Duchemin, B. Size, shape, orientation and crystallinity of cellulose Iβ by X-ray powder diffraction using a free spreadsheet program. *Cellulose* **2017**, *24*. [CrossRef]
49. Moyer, P.; Kim, K.; Abdoulmoumine, N.; Chmely, S.; Long, B.K.; Carrier, D.J.; Labbé, N. Structural changes in lignocellulosic biomass during activation with ionic liquids comprising 3-methylimidazolium cations and carboxylate anions. *Biotechnol. Biofuels* **2018**, *11*, 265. [CrossRef] [PubMed]
50. Zhang, H.; Ning, Z.; Khalid, H.; Zhang, R.; Liu, G.; Chen, C. Enhancement of methane production from Cotton Stalk using different pretreatment techniques. *Sci. Rep.* **2018**, *8*, 3463. [CrossRef]
51. Demirbas, B. Biomass business and operating. *Energy Educ. Sci. Technol. Part A* **2010**, *26*, 37–47.
52. Sujan, S.; Kashem, M.; Fakhruddin, A. Lignin: A valuable feedstock for biomass pellet. *Bangladesh J. Sci. Ind. Res.* **2020**, *55*, 83–88. [CrossRef]
53. Jekayinfa, S.O.; Omisakin, O.S. The energy potentials of some agricultural wastes as local fuel materials in Nigeria. *Agric. Eng. Int. CIGR EJ.* **2005**, *7*, 1–10.
54. Taherdanak, M.; Zilouei, H.; Karimi, K. The influence of dilute sulfuric acid pretreatment on biogas production form wheat plant. *Int. J. Green Energy* **2016**, *13*, 1129–1134. [CrossRef]
55. Mirmohamadsadeghi, S.; Karimi, K.; Horváth, I.S. Improvement of solid-state biogas production from wood by concentrated phosphoric acid pretreatment. *Bioresources* **2016**, *11*, 3230–3243. [CrossRef]

Article

Development and Validation of Mass Reduction Prediction Model and Analysis of Fuel Properties for Agro-Byproduct Torrefaction

Seok-Jun Kim [1], Kwang-Cheol Oh [2], Sun-Yong Park [1], Young-Min Ju [3], La-Hoon Cho [1], Chung-Geon Lee [1], Min-Jun Kim [1], In-Seon Jeong [1] and Dae-Hyun Kim [1,*]

[1] Department of Interdisciplinary Program in Smart Agriculture, Kangwon National University, Hyoja 2 Dong 192-1, Chuncheon-si 24341, Korea; ksj91@kangwon.ac.kr (S.-J.K.); psy0712@kangwon.ac.kr (S.-Y.P.); jjola1991@kangwon.ac.kr (L.-H.C.); cndrjs9605@kangwon.ac.kr (C.-G.L.); alswns3262@naver.com (M.-J.K.); jis0714@kangwon.ac.kr (I.-S.J.)

[2] Green Materials & Processes R&D Group, Korea Institute of Industrial Technology, 55, Jongga-ro, Jung-gu, Ulsan 44413, Korea; okc@kitech.re.kr

[3] Department of Forest Products, Division of Wood Chemistry, National Institute of Forest Science, 57, Hoegi-ro, Dongdaemun-gu, Seoul 02455, Korea; joo889@naver.com

* Correspondence: daekim@kangwon.ac.kr; Tel.: +82-33-250-6496; Fax: +82-33-259-5561

Abstract: Global warming is accelerating due to the increase in greenhouse gas emissions. Accordingly, research on the use of biomass as energy sources, is being actively conducted worldwide to reduce CO_2 emissions. Although the production of agro-byproducts is vast, their utilization for energy production has not been fully investigated. This study suggests an optimal torrefaction process condition for agro-byproducts, such as grape branch and perilla, that have moisture content but low calorific values. To determine whether these agro-byproducts can be used for energy sources as substituents of fossil fuels, a mass reduction model was established and validated via experimental results. Thermogravimetric analysis was conducted for different heating rates, and the activation energy and frequency factor were derived through the analysis. The model was developed by changes in rate constants, moisture content, ash content, and lignocellulose content in biomass. To ascertain the optimal torrefaction conditions, fuel characteristic analysis and changes in energy yield of torrefied grape branch and perilla were investigated. The optimal torrefaction conditions for grape branch and perilla were 200 °C for 40 min and 230 °C for 30 min, respectively. The comparison result of the experiment and simulation at the optimum conditions of mass reduction were 1.42%p and 1.51%p, and 15 °C/min and 7.5 °C/min at heating rate, respectively.

Keywords: agro-byproducts; torrefaction; mass reduction model; mass yield; heating rate

Citation: Kim, S.-J.; Oh, K.-C.; Park, S.-Y.; Ju, Y.-M.; Cho, L.-H.; Lee, C.-G.; Kim, M.-J.; Jeong, I.-S.; Kim, D.-H. Development and Validation of Mass Reduction Prediction Model and Analysis of Fuel Properties for Agro-Byproduct Torrefaction. *Energies* **2021**, *14*, 6125. https://doi.org/10.3390/en14196125

Academic Editors: Idiano D'Adamo and Antonio Zuorro

Received: 26 July 2021
Accepted: 22 September 2021
Published: 26 September 2021

Publisher's Note: MDPI stays neutral with regard to jurisdictional claims in published maps and institutional affiliations.

1. Introduction

Owing to the global increase in energy demands and the phasing-out of nuclear energy, the Republic of Korea has established plans for increasing their portion of renewable energy to 20% by 2030 [1,2]. Hence, alternative facilities and energy sources that can generate electricity have increased as replacements of nuclear power plants. Thermal power plants are one such facility. In terms of energy source, interest in biomass as a carbon-neutral fuel source that emits only 1/12th the CO_2 compared to fossil fuels has increased [2]. Research on reducing carbon emissions and generating profits by using biomass and waste as energy sources is actively underway. There are several studies that suggest conversion technologies from sugarcane waste into biooil and biogas [3]. In order to use biowaste or byproducts as energy sources, a model to predict productivity of byproduct was developed [4,5] and the economic sustainability of renewable energy potential from agriculture, forestry, and other biomass was evaluated [6]. Rice straw and chaff are the most widely investigated biomass sources (Figure 1), having been used as compost, animal feed for livestock, and other purposes. However, other biomass sources, such as perilla and pepper stem, have a relatively smaller number of uses. In addition, these sources have disadvantages such as low calorific value, high moisture content, and difficulty of storage [7,8]. Torrefaction, a thermochemical conversion process, is a thermal pretreatment of biomass with a temperature range of 200–300 °C under lean or anoxic conditions within 1 h [9–13]. After torrefaction, fixed carbon in biomass increases, along with its calorific value. This property can be advantageous for storage and transportation of the torrefied product owing to improvement in its waster resistance [8,9]. However, as the torrefied sample can show different physical, chemical, and fuel properties depending on the process time and temperature, determining the optimal conditions to maximize energy efficiency is time consuming and costly.

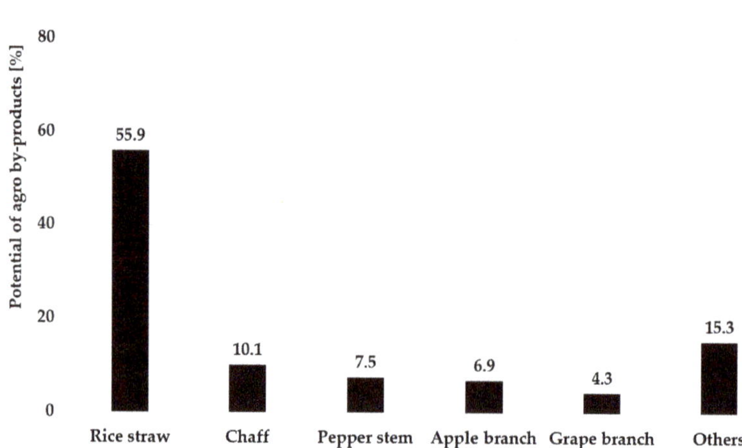

Figure 1. Estimation of geographical and technical potential of biomass resources [1].

To determine the optimum conditions, several studies have explained the biomass torrefaction prediction model by employing various process conditions and thermochemical and physical changes during torrefaction [2,14–23]. However, studies pertaining to the utilization of unused agro-biomass as a replacement energy source for fossil fuels are lacking. Accordingly, this study proposes a mass reduction model based on moisture content, ash content, and temperature duration. This process has been developed and validated for utilizing grape branch and perilla as energy sources. However, mass reduction using the torrefaction process is associated with energy loss, therefore, there were accuracy limits regarding the optimal torrefaction process based on mass yield only. Energy yield, a parameter for relating mass yield and fuel properties, was calculated by elemental analysis and fuel characteristic analysis, such as calorific value. An optimal torrefaction process was suggested, and the model accuracy was evaluated.

2. Materials and Methods

2.1. Sample

Each 20 kg of grape branch and perilla, naturally dried, was collected in Chuncheon, Gangwon Province. The collected samples are pulverized in powder form, and the change in fuel characteristics and the possibility of use as fuel are confirmed through the torrefaction process. Their bulk density and calorific value were measured, and proximate and elemental analyses were conducted on wet basis.

2.2. Experimental Method

A sample less than 2.36 mm size and 3 g in weight was placed into a prototype capsule (metal: carbon steel (AISI 304, $\varnothing$ 28 × 85H × 2T)) (Figure 2). To prevent rapid reaction with external air, the capsule was sealed and put into an electrical furnace. The change in mass reduction rate was measured after 3 process repetitions (including measuring heating value, mass yield, and energy yield). Experiment was conducted interval of 10, 20, and 40 min after reaching 200, 230, and 270 °C, respectively [24], followed by cooling for 30 min at room temperature; since ignition occurs when a sample with a high temperature above the ignition point reaction with oxygen, cooling is required.

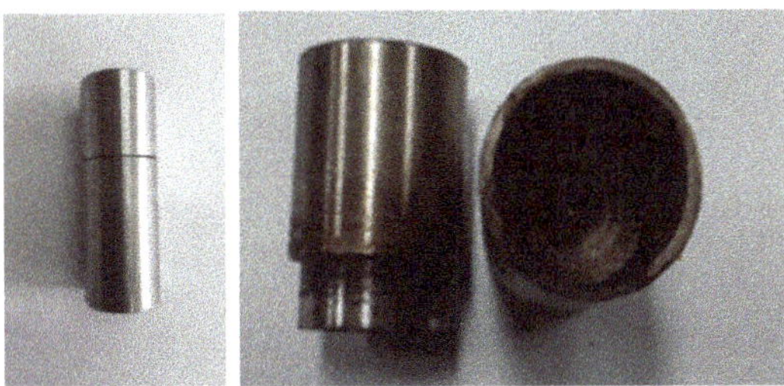

Figure 2. Prototype capsule.

2.3. Analysis Method

2.3.1. Bulk Density Analysis

The bulk density was measured as follows: particles smaller than 2.36 mm were poured from a height of approximately 200–300 mm from a 5-L container. The full container is dropped three times vertically from a height of approximately 150 mm from a flat, hard surface. The particles remaining on the container are removed with a flat object and then weighed. This process was repeated twice. The bulk density, in kg/m^3, was calculated using the following Equation (1), and the resulting value was rounded off to the first digit.

$$BD = \frac{M_p - M_c}{V} \tag{1}$$

where BD is the bulk density (kg/m^3); M_p is the weight of container with particles (kg); M_c is the weight of empty container (kg); and v is the volume of empty container (m^3).

2.3.2. Thermal Analysis

Woody biomass mainly comprises cellulose, hemicellulose, and lignin. Water evaporation occurs at 100 °C. Degradation, discoloration, and carbonization occur in the order of hemicellulose, cellulose, and lignin. This occurs in the temperature range of 150–350, 275–350, and 250–500 °C, respectively [6]. Condensed and non-condensed gases are generated at 110–300 °C due to evaporated volatile matter [9,25]. To determine mass changes caused by degradation, evaporation, and gasification in terms of temperature changes, thermogravimetric analysis was conducted using a thermogravimetric analyzer (DSA Q2000/SDT Q600, TA Instruments, New Castle, DE, USA). The temperature was increased at different heating rates (7.5, 15, and 22.5 °C/min).

2.3.3. Fuel Property Analysis

Torrefied biomass was dried for 3 h in oven dryer at 105 °C, and its calorific value was measured using a calorimeter (6400, Parr, Moline, IL, USA). Measurements were conducted 3 times. According to ISO 18122:2015, an element analyzer (EA-3000, Eurovector, Cuzio, Pavia, Italy) was used to determine the changes in elemental composition and plotted using a Van Krevelen diagram [26,27]. There exists a direct relationship between mass yield, energy yield, and fuel properties. Mass, energy yield, and energy density were calculated using Equations (2)–(4) on wet basis.

$$Y_M = \frac{M_{torr}}{M_{BO,I}} \times 100 \tag{2}$$

where Y_M is the mass yield (%); M_{torr} is the biomass mass after torrefaction (g); $M_{BO,I}$ is the initial biomass mass on wet basis (g).

$$EY = Y_M \times \frac{HHV_{torr}}{HHV_{raw}} \tag{3}$$

where EY is the energy yield (%); HHV_{torr} is the torrefied biomass higher heating value [MJ/kg]; and HHV_{raw} is the raw biomass higher heating value on wet basis [MJ/kg].

$$ED = \frac{EY}{Y_M} \tag{4}$$

where ED is the energy density (-); Y_M is mass yield (%); and EY is energy yield (%).

3. Simulation Analysis

3.1. 1-D Mass Reduction Prediction Model

In this study, the finite element method (FEM) was used to predict the mass reduction due to temperature change of biomass during the torrefaction. In order to use the finite element method, the steel layer and the biomass layer were divided, and the biomass layer was further divided into a total of 25 nodes (Figure 3). T_{m-1}, T_m, and T_{m+1} are arbitrarily positions divided according to FEM.

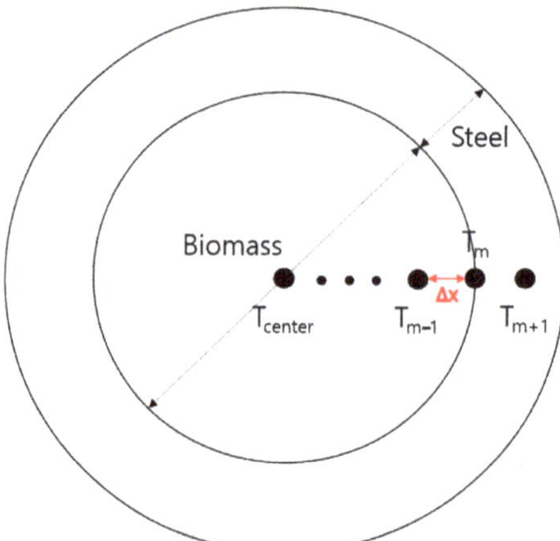

Figure 3. Finite element analysis to predict each node temperature in prototype capsule.

To calculate temperature changes according to thermal diffusivity (α) of the mass reduction prediction simulation model, absolute temperature (T), thermal diffusivity using specific heat (C_p) and heat conduction coefficient (K) were used following Equations (5)–(8) [28–32].

$$\begin{aligned} (K_{Steel} = 15 \, [W/m{\cdot}K], \; C_{p,Steel} = 477[J/g{\cdot}K]) \\ K_{Grape} = 0.13 + 0.0003 \times (T - 273.15) \, [W/m{\cdot}K] \end{aligned} \tag{5}$$

$$C_{p,Grape} = \left(-9.12 \times 10^{-2} + 4.4 \times 10^{-3} \times T\right) \times 1000 \, [J/g{\cdot}K] \tag{6}$$

$$K_{Perilla} = 0.00249 + 0.0000145 \times B.d_{perilla} + 0.000184 \times (T - 273.15) \ [W/m \cdot K] \quad (7)$$

$$C_{p, Perilla} = \left(-9.12 \times 10^{-2} + 4.4 \times 10^{-3} \times T\right) \times 1000 \ [J/g \cdot K] \quad (8)$$

$$\alpha = \frac{K}{\rho \times C_p} \ \left[m^2/s\right] \quad (9)$$

where the $B.d_{perilla}$ is the bulk density, Equation (10) represents the rate of change of the energy contents of the nodes; the first and second terms on the right-hand side are the rat of heat conduction on the right and left surfaces, respectively. Equation (11) was represented using Equations (9) and (10) [2].

$$\rho A \Delta x C_p \frac{dT_m}{dt} = kA \frac{T_{m+1} - T_m}{\Delta x} + kA \frac{T_{m-1} - T_m}{\Delta x} \ [W] \quad (10)$$

$$dT_m = \frac{\alpha}{\Delta x^2} \times (T_{m-1} - 2 \times T_m + T_{m+1}) \times dt \ [k] \quad (11)$$

3.2. Thermal Change Analysis of Biomass

During torrefaction, absolute temperature (T) changes may occur differently based on the density of biomass, moisture content, thermal permeability, and thermal property. The absolute temperature (T) changes were calculated and used in Equation (12) to derive the rate constant. To calculate the degree of reaction through thermal changes, the rate constant was used. Rate constant (k) was derived following Arrhenius' empirical equation (Equation (12)) [33–37].

$$k(T) = A \times exp^{\frac{-E_a}{R \times T}} \ [1/s] \quad (12)$$

After taking natural logs of both sides of Equation (13):

$$\ln(k) = \left(\frac{-E_a}{R}\right) \times \left(\frac{1}{T}\right) + \ln A \ [1/s] \quad (13)$$

In this study, frequency factor (A) and activation energy (E_a) were derived by thermogravimetric analysis (TGA). The frequency factor means the number of frequency of intermolecular collisions (1/s). The activation energy refers to the minimum energy required for a chemical reaction to proceed (kJ/mol). R is ideal gas constant (J/mol $\cdot$ k). As shown in Equation (13), rate constant (k) can change the frequency factor (A), activation energy (E_a), and also absolute temperature (T).

3.3. Mass Reduction Model

For predicting mass reduction during torrefaction, thermal conversion of biomass was studied via two pseudoelementary reactions (Figure 4). Elementary reaction 1 comprises moisture evaporation of biomass by drying and elementary reaction 2 involves thermal changes to the lignocellulosic components and volatile matter [2,37–44]. Mass reduction was calculated by considering the moisture content of raw sample (grape branch, perilla) (W), lignocellulosic (including volatile and fixed carbon) components (L), and separately (Table 1). Total mass reduction was calculated as the summation of measured quantities of these components (Equation (16)). Change in mass with time was multiplied by lignocellulosic components and rate constant. Decrease in lignocellulosic components and moisture content was derived from Equations (14) and (15), and k_t and k_d were rate constants of lignocellulose and water, respectively. Mass after torrefaction was derived by subtracting the sum of decrease in total mass (that is, the mass of lignocellulose, moisture content) from the initial mass (Equation (16)). Here, t is time(s); total mass$_{initial}$ is the entire biomass composition components before the torrefaction process; mass$_{final}$ is the biomass composition component, having removed the water and lignocellulosic components after torrefaction process.

$$\dot{L} = \frac{dL}{dt} = -k_t \times L \quad (14)$$

$$\dot{W} = \frac{dw}{dt} = -k_d \times W \tag{15}$$

$$Mass_{final} = Total\ mass_{initial} - \dot{W} - \dot{L} \tag{16}$$

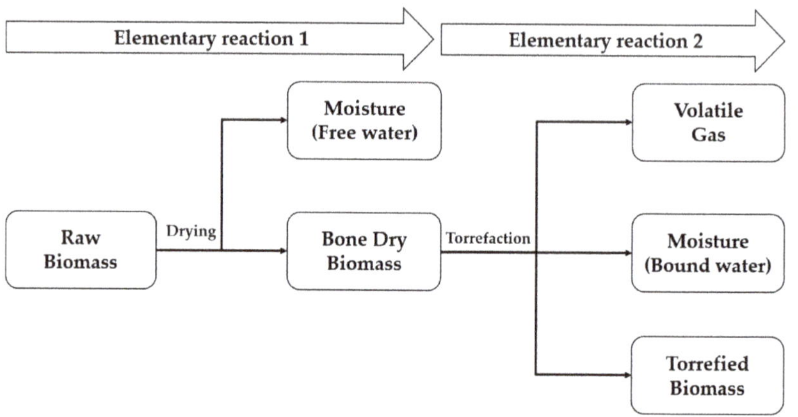

Figure 4. Thermochemical conversion of biomass [2].

Table 1. Raw sample properties of Grape and Perilla.

	Bulk Density (kg/m^3)	Higher Heating Value (MJ/kg)	Proximate Analysis (%)				Element Analysis (%)			
			Moisture	Ash	Volatile	Fixed Carbon	C	H	N	O
Grape	290	19.1	11.4	3.55	76	9.05	44.41	6.07	0.97	40.2
Perilla	150	18.9	7.77	5.83	71.2	15.21	41.19	5.68	1.15	31.49

4. Results

4.1. Fuel Properties

Figure 5 shows the Van Krevelen diagram of the elemental analysis result of torrefied grape branch and perilla. With an increase in process temperature and time, the composition ratio of carbon was increased, while that of oxygen and hydrogen was decreased. For the raw and torrefied grape branch comparison result, the composition ratio of carbon increased by approximately 2–25%p, but the ratio of oxygen and hydrogen were decreased by 22–40%p and 5–10%p, respectively. Comparing the raw and torrefied perilla, the composition ratio of carbon was increased by 14–60%p, but the ratio of oxygen and hydrogen decreased by 19–44%p and 2–15%p, respectively. Calorific values are summarized in Table 2. The calorific value of grape branch was 19.46–22.77 MJ/kg, and for perilla, it was 19.06–21.77 MJ/kg. There was a 2–19%p and 1.3–15%p increase compared with the raw sample of grape branch and perilla, respectively. The energy yield was derived using Equation (3) and summarized in Table 3. The energy yield of grape branch and perilla are 74.1–87.8% and 76–90.5%, respectively. Energy yield was also decreased with an increase in process time and temperature.

Table 2. Heating value of torrefied grape and perilla.

	Grape (MJ/kg)			Perilla (MJ/kg)		
	200 °C	230 °C	270 °C	200 °C	230 °C	270 °C
20 min	19.46	19.56	21.27	19.06	19.27	20.95
30 min	19.46	20.36	21.90	19.10	19.65	21.59
40 min	19.51	20.75	22.77	19.14	19.95	21.77

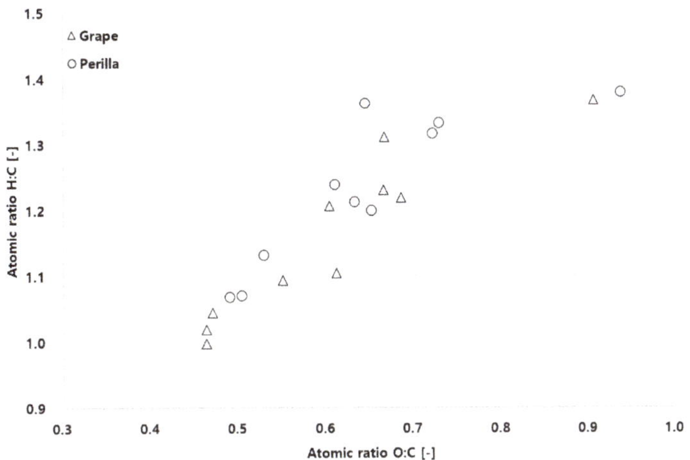

Figure 5. Van Krevelen diagram of torrefied grape and perilla.

Table 3. Energy yield of torrefied grape and perilla.

	Grape (%)			Perilla (%)		
	200 °C	230 °C	270 °C	200 °C	230 °C	270 °C
20 min	87.8	86.3	76.5	90.5	87.5	80.8
30 min	87.2	85.0	74.1	89.5	86.5	79.3
40 min	86.7	84.0	75.1	88.2	83.3	76.0

Figures 6 and 7 depict mass yield, energy yield, and energy density following torrefaction for each material. The higher the process temperature and longer time, the larger the mass reduction. In the case of grape branch, the initial moisture content and volatile content were higher than those of perilla; hence, mass reduction was larger than perilla according to the process condition. Energy yield was decreased even though heating values were increased with longer time and higher temperature, due to the larger mass loss according to Equation (3). The energy density increased with higher process temperature and longer process times. Thus, considering mass yield, energy yield (Figures 6 and 7) and calorific value, the optimal conditions of grape branch and perilla were 200 °C for 40 min and 230 °C for 30 min, respectively.

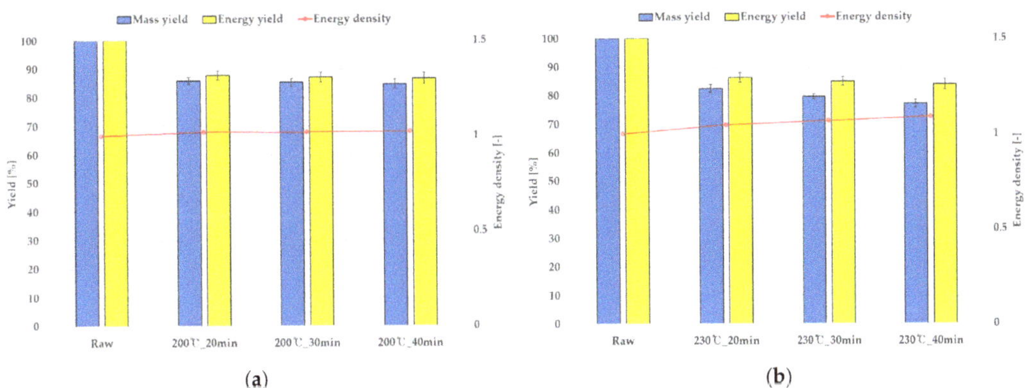

Figure 6. *Cont.*

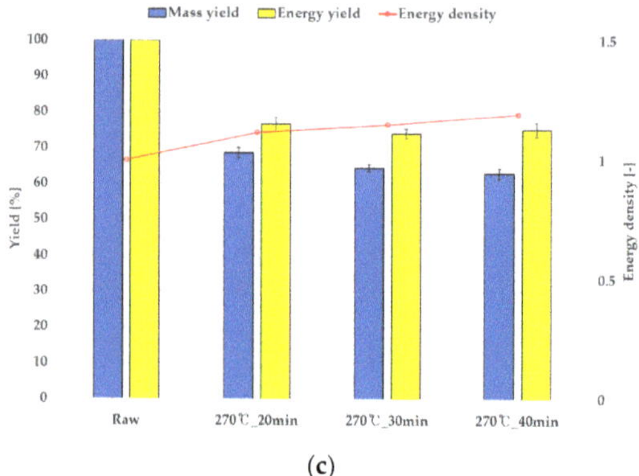

(c)

Figure 6. Grape mass and energy yields and energy density according to torrefaction process time and temperature. (**a**) At 200 °C, (**b**) at 230 °C, and (**c**) at 270 °C.

(a)   (b)

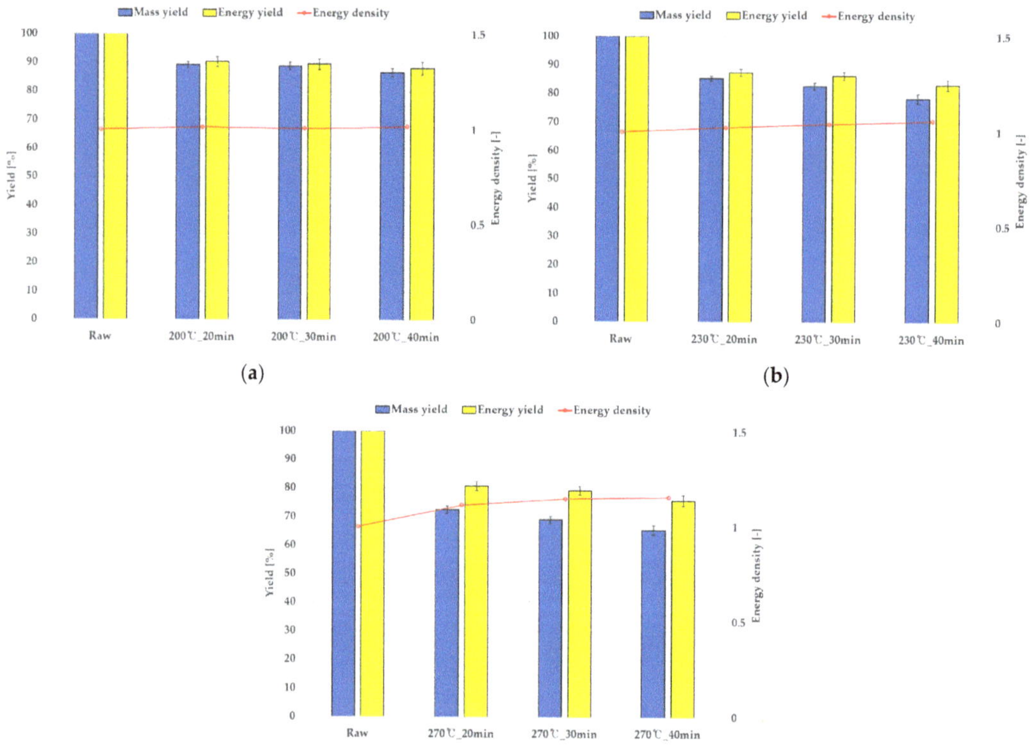

(c)

Figure 7. Perilla mass and energy yields and energy density according to torrefaction process time and temperature. (**a**) At 200 °C, (**b**) at 230 °C, and (**c**) at 270 °C.

4.2. Thermogravimetric Analysis (TGA)

Table 4 shows TGA results. The peak temperature varies depending on the heating rate, and the mass reduction occurs at the peak temperature. The burnout temperature is defined as the temperature where the rate of mass loss falls below 1%/min [2,45,46].

Results of thermogravimetric analysis for grape branch and perilla are shown in Figure 8. The figures show different mass reductions at all heating rates below 350 °C. Biomass was exposed to more heat at the low heating rate since the attainment of the target temperature required more time. Consequently, more mass loss was observed. Figure 9 shows the Arrhenius plot of torrefaction process temperature based on TGA results. The observed scattering in the TGA graph might have been attributed to the fine movement of the scale in the analyzer during TGA analysis. In addition, since the sample was not in a uniform shape, the weight might have been affected by the movement of the center of the sample according to the temperature change. Rate constants differed owing to the differences in the heating rate at a constant temperature. To derive activation energy and frequency factor, temperature range was divided into the water-evaporating and the composition degradation parts, as shown in Figure 10. Tables 5 and 6 list the activation energy, coefficient of determination (r^2), and frequency factor based on the temperature range for grape branch and perilla, respectively. Based on the TGA results, the 0–200 °C temperature range is where initial mass reduction occurs. Here, water-evaporation for the grape branch occurred within 130 °C, while water evaporation of perilla occurred below 140 °C, absorbing heat for the phase change between 140–200 °C with less mass reduction. Tables 5 and 6 show that the activation energy and frequency factor of the samples increased in the first temperature range, which was divided into the temperature range, in which the coefficient of determination of activation energy and frequency factor was 0.9 or higher. This showed significant water reduction but changed to negative values under the second temperature range, in which the reaction rate decreased after water evaporation. The reason for the negative activation energy might have been attributed to the exothermic reaction of the constituents of the biomass appears more prominent than the endothermic reaction. Thereafter, the activation energy and frequency factor increased under the third temperature range, during which the reaction rate increased.

Table 4. Thermogravimetric analysis results.

Material	Heating Rate (°C/min)	Experiment Time (min)	Peak Temperature (°C)	Burnout Temperature (°C)	Purge Gas (mL/min)	Particle Size (mm)
	7.5	102.9	337.0	370.0		
Perilla	15	52.2	350.4	390.0		
	22.5	34.7	357.0	432.0	N_2 (100)	<0.154
	7.5	102.9	331.5	363.3		
Grape	15	52.2	341.0	380.5		
	22.5	34.7	349.5	402.5		

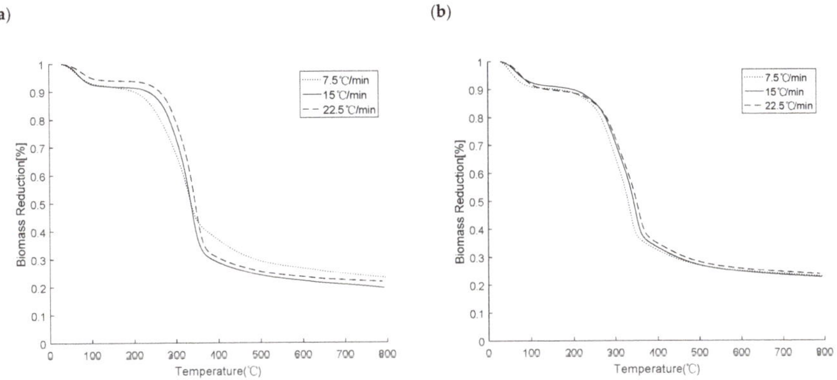

Figure 8. Biomass reduction curves for grape branch (**a**) and perilla (**b**) at various heating rates.

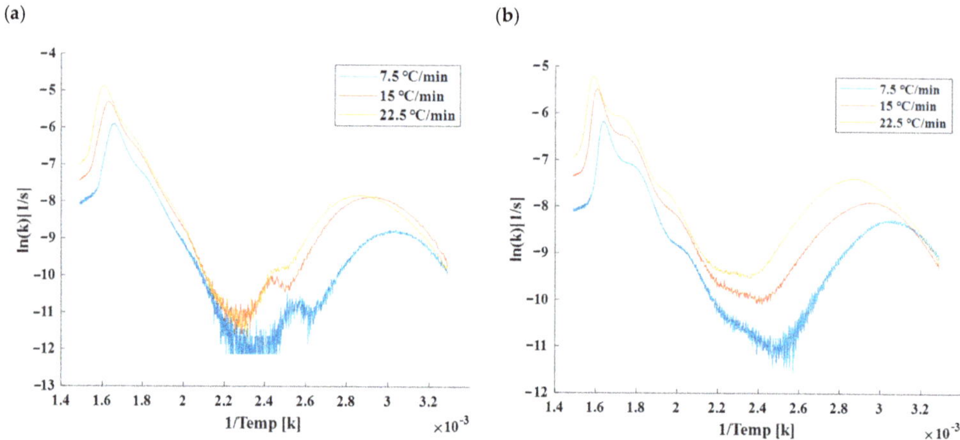

Figure 9. Arrhenius plot of TGA data measured at different heating rates of grape branch (**a**) and perilla (**b**).

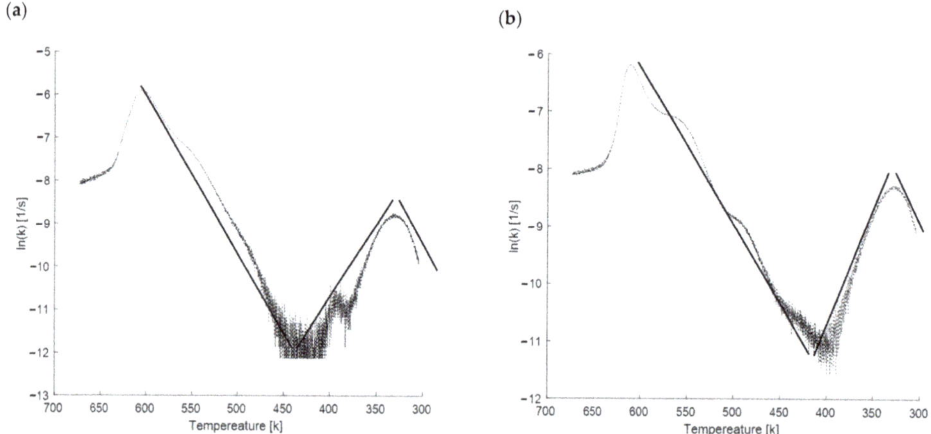

Figure 10. Regression lines of characteristic temperature for grape branch (**a**) and perilla (**b**) at a heating rate of 7.5 °C/min.

Table 5. Frequency factor and activation energy of grape branch from TGA results.

	Temperature range (°C)	30–55	65–130	130–350
7.5 °C/min	A (L/s)	1.99×10^2	2.14×10^{-11}	5.72×10^3
	E_a (L/Jmol)	3.8×10^4	-4.39×10^4	7.38×10^4
	r^2	0.944	0.918	0.972
	Temperature range (°C)	30–70	70–140	140–350
15 °C/min	A (L/s)	412.08	1.88×10^{-11}	3.01×10^4
	E_a (L/Jmol)	3.8892×10^4	-4.89×10^4	7.95×10^4
	r^2	0.924	0.955	0.993
	Temperature range (°C)	30–75	75–165	165–350
22.5 °C/min	A (L/s)	1.16×10^3	2.19×10^{-11}	1.57×10^3
	E_a (L/Jmol)	4.22×10^4	-4.96×10^4	6.55×10^4
	r^2	0.956	0.944	0.903

Table 6. Frequency factor and activation energy of perilla from TGA results.

	Temperature range (°C)	30–55	55–120	130–350
7.5 °C/min	A (L/s)	8.156	1.07×10^{-11}	2.30×10
	E_a (L/Jmol)	2.801×10^4	-4.746×10^4	4.89×10^4
	r^2	0.934	0.950	0.973
	Temperature range (°C)	30–60	60–140	140–350
15 °C/min	A (L/s)	9.63×10^2	2.15×10^{-9}	4.83×10^2
	E_a (L/Jmol)	4.060×10^4	-3.457×10^4	5.00×10^4
	r^2	0.975	0.950	0.978
	Temperature range (°C)	30–75	75–150	150–350
22.5 °C/min	A (L/s)	2.958×10^2	1.020×10^{-9}	8.790×10
	E_a (L/Jmol)	3.714×10^3	-3.952×10^4	5.17×10^4
	r^2	0.967	0.972	0.978

4.3. Torrefaction Mass Reduction and Comparison with Simulations

The simulation was derived through the summation of each node through Equation (11), and experimental results were compared, and the values for the grape branch and perilla are presented in Table 8, respectively. Under experimental conditions and different temperatures, mass reduction of grape branch and perilla ranged from 13.98% to 37.03% and from 10.69% to 34.36%, respectively. Meanwhile, simulated mass reductions ranged, from between 4.07% and 69.37% and from 5.93 to the maximum of 69.87%, respectively. Comparing the mass reduction values between the simulation and the experiment, the highest accuracy for the grape branch was obtained at the heating rate of 15 °C/min and a process temperature of 200 °C. For perilla, the heating rate was 7.5 °C/min, and the process temperature was 200 °C. Under these conditions, the root mean square error (RMSE) for the grape branch and perilla was 0.0356 and 0.0285, respectively. When Equation (16) was derived from Equations (14) and (15), in the case of grapes, the simulation result of 200 °C through 15 °C/min of heating rate showed higher accuracy than other heating rates. In addition, the error with the mass reduction amount of the 200 °C for 40 min process, which is the optimal process condition derived through the experiment, was 1.42%p. A large amount of heat transfer was required since the bulk density of grape branch was higher than that of perilla. In the case of perilla, the optimal condition through the experiment was derived at 230 °C for 30 min, with an error of approximately 1.5%p (at heating rate 7.5 °C/min) when compared with the simulation. Perilla has low bulk density and relatively lower moisture content than grape branch. Therefore, the mass loss was sufficiently reduced even with a low heat-transfer rate at a low heating rate. These results were judged to fall within the mass yield error range for each optimal condition (Figures 6a and 7a). This also confirmed that an error occurred as a result of the simulation mass reduction derivation. An error occurred since the mass reduction was derived using the frequency factor and activation energy derived from the temperature range along the trend line (Figure 10), deriving the frequency factor and activation energy by dividing the temperature range in detail and then applying it to the simulation to reduce the error. In addition, through the developed model, it is possible to reduce the amount of waste by converting unused agricultural byproducts into an energy source.

Table 7. Grape mass reduction comparison.

Temp. (°C)	Time (min)	Experiment Mass Reduction (%)	Simulation 7.5 °C/min Mass Reduction (%)	r^2	RMSE	Simulation 15 °C/min Mass Reduction (%)	r^2	RMSE	Simulation 22.5 °C/min Mass Reduction (%)	r^2	RMSE
200	20	13.98	4.07			8.30			8.44		
	30	14.57	6.25	0.996	0.0847	12.61	0.996	0.0356	12.93	0.996	0.0604
	40	15.26	8.35			16.68			17.17		
230	20	17.52	11.30			19.26			20.48		
	30	20.41	16.96	0.997	0.0411	28.34	0.993	0.0929	30.25	0.995	0.1109
	40	22.84	22.70			36.73			39.06		
270	20	31.28	28.0			39.48			43.12		
	30	35.48	41.50	0.964	0.0956	54.98	0.976	0.2057	59.01	0.981	0.2408
	40	37.03	52.10			65.70			69.37		

Table 8. Perilla mass reduction comparison.

Temp. (°C)	Time (min)	Experiment Mass Reduction (%)	Simulation 7.5 °C/min Mass Reduction (%)	r^2	RMSE	Simulation 15 °C/min Mass Reduction (%)	r^2	RMSE	Simulation 22.5 °C/min Mass Reduction (%)	r^2	RMSE
200	20	10.69	5.93			7.60			9.86		
	30	11.07	9.95	0.858	0.0285	13.08	0.854	0.0368	16.95	0.851	0.0964
	40	13.31	14.02			18.51			23.87		
230	20	14.61	10.80			14.46			18.95		
	30	17.16	18.67	0.979	0.0379	25.0	0.975	0.0910	32.27	0.969	0.1604
	40	21.46	26.59			35.14			44.36		
270	20	27.43	20.12			27.29			35.42		
	30	30.48	34.64	0.985	0.0866	45.52	0.977	0.1670	56.36	0.963	0.2578
	40	34.36	46.78			59.07			69.87		

5. Conclusions

In this study, to investigate the possibility of using agro-byproducts, specifically grape branch and perilla, as energy sources, a torrefaction process was used. Based on the experimental results, mass reduction of grape branch and perilla was 13.98–37.03% and 10.69–34.36%, respectively. Results of fuel properties analysis showed that range of the calorific value of grape branch and perilla was 19.46–22.77 and 10.69–21.77 MJ/kg, respectively. Although calorific value increased, energy yield decreased due to higher mass loss. Considering mass yield, energy yield and calorific value, the optimal condition of grape branch and perilla was 200 °C for 40 min and 230 °C for 30 min, respectively. Based on TGA, a mass reduction model during torrefaction was established and validated with experimental data. Fuel properties were observed using proximate analysis, elemental analysis, and measuring calorific value. Based on different heating rates, the rate constant was derived and applied to the mass reduction model, and experimental and simulation results were compared. RMSE of grape branch was 0.0356 under a heating rate of 15 °C/min at 200 °C and RMSE of perilla was 0.0285 under a heating rate 7.5 °C/min at 200 °C. Mass reduction differences between the simulation and experiment with grape branch under the heating rate 15 °C/min to 200 °C in 40 min and perilla at 7.5 °C/min to 230 °C in 30 min were 1.42%p and 1.51%p, respectively. The frequency factor and activation energy derived from the TGA results for each heating rate had an effect on the mass reduction due to the temperature change of grape and perilla. Studies using TGA at various heating rates and applying specific heat and thermal conductivity coefficient equations through various references to improve the accuracy of the simulation model are warranted in the future. Further research could be performed on not only forestry byproducts but also industrial wastes into fuel.

Author Contributions: Conception and design of study: S.-J.K., S.-Y.P. and D.-H.K.; acquisition of data: S.-J.K., S.-Y.P., L.-H.C. and C.-G.L.; analysis and/or interpretation of data: S.-J.K., S.-Y.P., L.-H.C., Y.-M.J. and K.-C.O.; drafting the manuscript: S.-J.K., S.-Y.P., K.-C.O., M.-J.K. and I.-S.J.; revising the manuscript critically for important intellectual content: S.-J.K., S.-Y.P., K.-C.O., Y.-M.J., L.-H.C., C.-G.L. and D.-H.K.; approval of the version of the manuscript to be published: S.-J.K., K.-C.O., S.-Y.P., Y.-M.J., L.-H.C., C.-G.L., M.-J.K., I.-S.J and D.-H.K. All authors have read and agreed to the published version of the manuscript.

Funding: This work was supported by the National Research Foundation of Korea [grant number NRF-2016R1D1A3B04935457].

Institutional Review Board Statement: Not applicable.

Informed Consent Statement: Not applicable.

Data Availability Statement: Not applicable.

Conflicts of Interest: The authors declare no conflict of interest.

References

1. Lee, J.P.; Park, S.C. Estimation of geographical and technical potential for biomass resources. *New Renew. Energy* **2016**, *12*, 53–58. [CrossRef]
2. Oh, K.C.; Park, S.Y.; Kim, S.J.; Choi, Y.S.; Lee, C.G.; Cho, L.H.; Kim, D.H. Development and validation of mass reduction model to optimize torrefaction for agricultural byproduct biomass. *Renew. Energy* **2019**, *139*, 988–999. [CrossRef]
3. Hamawand, I.; da Silva, W.; Seneweera, S.; Bundschuh, J. Value Proposition of Different Methods for Utilisation of Sugarcane Wastes. *Energies* **2021**, *14*, 5483. [CrossRef]
4. D'Adamo, I.; Falcone, P.M.; Huisingh, D.; Morone, P. A circular economy model based on biomethane: What are the opportunities for the municipality of Rome and beyond? *Renew. Energy* **2021**, *163*, 1660–1672. [CrossRef]
5. D'Adamo, I.; Falcone, P.M.; Morone, P. A New Socio-economic Indicator to Measure the Performance of Bioeconomy Sectors in Europe. *Ecol. Econ.* **2020**, *176*, 1–12. [CrossRef]
6. De Toro, A.; Gunnarsson, C.; Jonsson, N.; Sundberg, M. Effects of Variable Weather Conditions on Baled Proportion of Varied Amounts of Harvestable Cereal Straw, Based on Simulations. *Sustainability* **2021**, *13*, 9449. [CrossRef]
7. Holmgren, P.; Wagner, D.R.; Strandberg, A.; Molinder, R.; Wiinikka, H.; Umeki, K.; Broström, M. Size, shape, and density changes of biomass particles during rapid devolatilization. *Fuel* **2017**, *206*, 342–351. [CrossRef]
8. Nonaka, H. Change of chemical composition during wood pelletization: Whole wood pellet as a raw material for biorefinery. *J. Jpn. Inst. Energy* **2014**, *93*, 1005–1009. [CrossRef]
9. Basu, P. *Biomass Gasification, Pyrolysis and Torrefaction: Practical Design and Theory*; Academic Press: Cambridge, MA, USA, 2018.
10. Granados, D.A.; Chejne, F.; Basu, P. A two dimensional model for torrefaction of large biomass particles. *J. Anal. Appl. Pyrol.* **2016**, *120*, 1–14. [CrossRef]
11. Park, S.W.; Yang, J.K.; Baek, K.R. Fuel ratio and combustion characteristics of torrefied biomass. *J. Korea Soc. Waste Manag.* **2013**, *30*, 376–382. [CrossRef]
12. Chansaem, P. Modeling and Optimal Design of Biomass Torrefaction Process. Ph.D. Thesis, The Graduate School of Seoul National University, Seoul, Korea, 2014. Available online: http://hdl.handle.net/10371/119714 (accessed on 30 August 2014).
13. Tumuluru, J.S.; Sokhansanj, S.; Hess, J.R.; Wright, C.T.; Boardman, R.D. A review on biomass torrefaction process and product properties for energy applications. *Ind. Biotechnol.* **2011**, *7*, 384–401. [CrossRef]
14. Pentananunt, R.; Rahman, A.M.; Bhattacharya, S.C. Upgrading of biomass by means of torrefaction. *Energy* **1990**, *15*, 1175–1179. [CrossRef]
15. Li, J.; Brzdekiewicz, A.; Yang, W.; Blasiak, W. Co-firing based on biomass torrefaction in a pulverized coal boiler with aim of 100% fuel switching. *Appl. Energy* **2012**, *99*, 344–354. [CrossRef]
16. Fisher, E.M.; Dupont, C.; Darvell, L.I.; Commandré, J.M.; Saddawi, A.; Jones, J.M.; Grateau, M.; Nocquet, T.; Salvador, S. Combustion and gasification characteristics of chars from raw and torrefied biomass. *Bioresour. Technol.* **2012**, *119*, 157–165. [CrossRef] [PubMed]
17. Broström, M.; Nordin, A.; Pommer, L.; Branca, C.; Di Blasi, C. Influence of torrefaction on the devolatilization and oxidation kinetics of wood. *J. Anal. Appl. Pyrol.* **2012**, *96*, 100–109. [CrossRef]
18. Berrueco, C.; Recari, J.; Güell, B.M.; Del Alamo, G. Pressurized gasification of torrefied woody biomass in a lab scale fluidized bed. *Energy* **2014**, *70*, 68–78. [CrossRef]
19. Deng, J.; Wang, G.J.; Kuang, J.H.; Zhang, Y.L.; Luo, Y.H. Pretreatment of agricultural residues for co-gasification via torrefaction. *J. Anal. Appl. Pyrol.* **2009**, *86*, 331–337. [CrossRef]
20. Pittman, C.U., Jr.; Mohan, D.; Eseyin, A.; Li, Q.; Ingram, L.; Hassan, E.B.; Mitchell, B.; Guo, H.; Steele, P.H. Characterization of bio-oils produced from fast pyrolysis of corn stalks in an auger reactor. *Energy Fuels* **2012**, *26*, 3816–3825. [CrossRef]
21. Zheng, A.; Zhao, Z.; Chang, S.; Huang, Z.; Wang, X.; He, F.; Li, H. Effect of torrefaction on structure and fast pyrolysis behavior of corncobs. *Bioresour. Technol.* **2013**, *128*, 370–377. [CrossRef] [PubMed]

22. Granados, D.A.; Velásquez, H.I.; Chejne, F. Energetic and exergetic evaluation of residual biomass in a torrefaction process. *Energy* **2014**, *74*, 181–189. [CrossRef]
23. Bates, R.B.; Ghoniem, A.F. Biomass torrefaction: Modeling of volatile and solid product evolution kinetics. *Bioresour. Technol.* **2012**, *124*, 46–49. [CrossRef]
24. Bates, R.B.; Ghoniem, A.F. Modeling kinetics-transport interactions during biomass torrefaction: The effects of temperature, particle size, and moisture content. *Fuel* **2014**, *137*, 216–229. [CrossRef]
25. Sullivan, A.L.; Ball, R. Thermal decomposition and combustion chemistry of cellulosic biomass. *Atmos. Environ.* **2012**, *47*, 133–141. [CrossRef]
26. Van der Stelt, M.J.C.; Gerhauser, H.; Kiel, J.H.A.; Ptasinski, K.J. Biomass upgrading by torrefaction for the production of biofuels: A review. *Biomass Bioenergy* **2011**, *35*, 3748–3762. [CrossRef]
27. Sarvaramini, A.; Assima, G.P.; Larachi, F. Dry torrefaction of biomass–Torrefied products and torrefaction kinetics using the distributed activation energy model. *Chem. Eng. J.* **2013**, *229*, 498–507. [CrossRef]
28. Hankalin, V.; Ahonen, T.; Raiko, R. On thermal properties of a pyrolysing wood particle. *Finn.-Swed. Flame Days* **2009**, *16*, 1–16.
29. Grønli, M.G. A Theoretical and Experimental Study of the Thermal Degradation of Biomass. 1996. Available online: https://www.osti.gov/etdeweb/servlets/purl/642467 (accessed on 10 August 2021).
30. Koch, P. Specific heat of ovendry spruce pine wood and bark. *Wood Sci.* **1968**, *1*, 203–214.
31. Gupta, M.; Yang, J.; Roy, C. Specific heat and thermal conductivity of softwood bark and softwood char particles. *Fuel* **2003**, *82*, 919–927. [CrossRef]
32. Spearpoint, M.J.; Quintiere, J.G. Predicting the burning of wood using an integral model. *Combust. Flame* **2000**, *123*, 308–325. [CrossRef]
33. Park, S.W.; Jang, C.H.; Baek, K.R.; Yang, J.K. Torrefaction and low-temperature carbonization of woody biomass: Evaluation of fuel characteristics of the products. *Energy* **2012**, *45*, 676–685. [CrossRef]
34. Kök, M.V.; Pamir, M.R. Comparative pyrolysis and combustion kinetics of oil shales. *J. Anal. Appl. Pyrol.* **2000**, *55*, 185–194. [CrossRef]
35. Martín-Lara, M.A.; Blázquez, G.; Zamora, M.C.; Calero, M. Kinetic modelling of torrefaction of olive tree pruning. *Appl. Therm. Eng.* **2017**, *113*, 1410–1418. [CrossRef]
36. Kang, S.H.; Ryu, J.H.; Park, S.N.; Byun, Y.S.; Seo, S.J.; Yun, Y.S.; Lee, J.W.; Kim, Y.J.; Kim, J.H.; Park, S.R. Kinetic studies of pyrolysis and Char-CO_2 gasification on low rank coals. *Korean Chem Eng. Res.* **2011**, *49*, 114–119. [CrossRef]
37. Tzou, D.Y. On the wave theory in heat conduction. *J. Heat Trans.* **1994**, *116*, 526–535.
38. Prins, M.J.; Ptasinski, K.J.; Janssen, F.J. Torrefaction of wood: Part 1. Weight loss kinetics. *J. Anal. Appl. Pyrol.* **2006**, *77*, 28–34. [CrossRef]
39. Prins, M.J.; Ptasinski, K.J.; Janssen, F.J. Torrefaction of wood: Part 2. Analysis of products. *J. Anal. Appl. Pyrol.* **2006**, *77*, 35–40. [CrossRef]
40. Bach, Q.V.; Chen, W.H.; Chu, Y.S.; Skreiberg, Ø. Predictions of biochar yield and elemental composition during torrefaction of forest residues. *Bioresour. Technol.* **2016**, *215*, 239–246. [CrossRef]
41. Ozisik, M.N. *Heat Transfer: A Basic Approach*; McGraw-Hill: New York, NY, USA, 1985.
42. Kreith, F.; Black, W.Z. *Basic Heat Transfer*; Harper & Row: New York, NY, USA, 1980.
43. Kaviany, M. *Principles of Heat Transfer in Porous Media*; Springer Science & Business Media: Berlin/Heidelberg, Germany, 2012.
44. Grieco, E.; Baldi, G. Analysis and modelling of wood pyrolysis. *Chem. Eng. Sci.* **2011**, *66*, 650–660. [CrossRef]
45. El-Sayed, S.A.; Mostafa, M.E. Pyrolysis characteristics and kinetic parameters determination of biomass fuel powders by differential thermal gravimetric analysis (TGA/DTG). *Energy Convers. Manag.* **2014**, *85*, 1–8. [CrossRef]
46. Lu, J.J.; Chen, W.H. Investigation on the ignition and burnout temperatures of bamboo and sugarcane bagasse by thermogravimetric analysis. *Appl. Energy* **2015**, *160*, 49–57. [CrossRef]

 *sustainability*

Review

A Critical Overview of the State-of-the-Art Methods for Biogas Purification and Utilization Processes

Muhamed Rasit Atelge [1,*], Halil Senol [2], Mohammed Djaafri [3], Tulin Avci Hansu [4], David Krisa [5], Abdulaziz Atabani [6,*], Cigdem Eskicioglu [5], Hamdi Muratçobanoğlu [7], Sebahattin Unalan [6], Slimane Kalloum [8], Nuri Azbar [9] and Hilal Demir Kıvrak [10]

[1] Department of Mechanical Engineering, Faculty of Engineering, Siirt University, Siirt 56100, Turkey
[2] Department of Electrical and Electronics Engineering, Giresun University, Giresun 28200, Turkey; halilsenol@yahoo.com
[3] Unité de Recherche en Energie Renouvelables en Milieu Saharien (URERMS), Centre de Développement des Energies Renouvelables (CDER), Adrar 01000, Algeria; djaafrimoh@yahoo.fr
[4] Department of Chemical Engineering, Faculty of Engineering, Siirt University, Siirt 56100, Turkey; tulin.hansu@siirt.edu.tr
[5] UBC Bioreactor Technology Group, School of Engineering, Okanagan Campus, The University of British Columbia, 3333 University Way, Kelowna, BC V1V 1V7, Canada; david.krisa@gmail.com (D.K.); cigdem.eskicioglu@ubc.ca (C.E.)
[6] Alternative Fuels Research Laboratory (AFRL), Energy Division, Department of Mechanical Engineering, Faculty of Engineering, Erciyes University, Kayseri 38039, Turkey; sebahattinunalan@gmail.com
[7] Department of Environmental Engineering, Faculty of Engineering, Niğde Ömer Halisdemir University, Niğde 51240, Turkey; hamdimuratcobanoglu@gmail.com
[8] Laboratory of Energy, Environment and Information System, Adrar University, Adrar 01000, Algeria; kalloum_sli@yahoo.fr
[9] Bioengineering Department, Faculty of Engineering, Ege University, Bornova, Izmir 35100, Turkey; nuri.azbar@ege.edu.tr
[10] Department of Chemical Engineering, Faculty of Engineering and Architectural Sciences, Eskisehir Osmangazi University, Eskisehir 26040, Turkey; hilaldemir.kivrak@ogu.edu.tr
[*] Correspondence: rasitatelge@gmail.com (M.R.A.); aeatabani@gmail.com or a.atabani@erciyes.edu.tr (A.A.)

Citation: Atelge, M.R.; Senol, H.; Djaafri, M.; Hansu, T.A.; Krisa, D.; Atabani, A.; Eskicioglu, C.; Muratçobanoğlu, H.; Unalan, S.; Kalloum, S.; et al. A Critical Overview of the State-of-the-Art Methods for Biogas Purification and Utilization Processes. *Sustainability* **2021**, *13*, 11515. https://doi.org/10.3390/su132011515

Academic Editor: Idiano D'Adamo

Received: 10 September 2021
Accepted: 11 October 2021
Published: 18 October 2021

Publisher's Note: MDPI stays neutral with regard to jurisdictional claims in published maps and institutional affiliations.

Abstract: Biogas is one of the most attractive renewable resources due to its ability to convert waste into energy. Biogas is produced during an anaerobic digestion process from different organic waste resources with a combination of mainly CH_4 (~50 mol/mol), CO_2 (~15 mol/mol), and some trace gasses. The percentage of these trace gases is related to operating conditions and feedstocks. Due to the impurities of the trace gases, raw biogas has to be cleaned before use for many applications. Therefore, the cleaning, upgrading, and utilization of biogas has become an important topic that has been widely studied in recent years. In this review, raw biogas components are investigated in relation to feedstock resources. Then, using recent developments, it describes the cleaning methods that have been used to eliminate unwanted components in biogas. Additionally, the upgrading processes are systematically reviewed according to their technology, recovery range, and state of the art methods in this area, regarding obtaining biomethane from biogas. Furthermore, these upgrading methods have been comprehensively reviewed and compared with each other in terms of electricity consumption and methane losses. This comparison revealed that amine scrubbing is one the most promising methods in terms of methane losses and the energy demand of the system. In the section on biogas utilization, raw biogas and biomethane have been assessed with recently available data from the literature according to their usage areas and methods. It seems that biogas can be used as a biofuel to produce energy via CHP and fuel cells with high efficiency. Moreover, it is able to be utilized in an internal combustion engine which reduces exhaust emissions by using biofuels. Lastly, chemical production such as biomethanol, bioethanol, and higher alcohols are in the development stage for utilization of biogas and are discussed in depth. This review reveals that most biogas utilization approaches are in their early stages. The gaps that require further investigations in the field have been identified and highlighted for future research.

Keywords: biofuels; biogas components; purification; utilization of biogas

1. Introduction

The bioeconomy is a new approach economy model that produces energy, food, and materials from renewable biological resources [1]. Bioenergy, which is produced from biomass, may be used in any energy industry, such as electricity and transport. The circular bioeconomy has become more competitive with conventional production pathways every day. The productivity of resources, especially bioenergy, has been shown to affect economic growth [2]. As a result, it is clear that the expansion of the bioenergy business aids the reduction in pollution and unemployment [3]. Several authors have concluded that bioenergy has a significant role to play in the decarbonization of society. In this context, it is critical to widen and deepen investigations of bioenergy materials that could be employed to aid in the attainment of sustainability objectives [4]. D'Adamo et al. [1] has explained and highlighted very well the future trend of renewable energy resources and especially their sustainability aspect with new concepts definitions. Renewable energies are critical components of the energy revolution, which aims to replace fossil fuel production with renewable energy [1]. Biomasses, when compared to other green sources such as wind, photovoltaic, and hydropower, can have the greatest influence in this context. Biomasses are suitable for use in local energy supply and consumption systems. This means that energy production from locally sourced biomass is far more sustainable than energy production from biomass sourced from other areas, sometimes even across national borders [1]. This practice must be closely monitored, and it must be accompanied by sustainability evaluations that justify its use. The goal is to promote the use of renewable energy sources in order to achieve circularity while maintaining an appropriate tradeoff between food production and biobased energy resource generation [1]. It is important to remember that the existing transportation industry generates a substantial amount of global emissions, and the percentage of renewable energy in this sector has not yet reached acceptable levels [1]. Among the renewable energy production pathways, biogas produced through the anaerobic digestion process is under the spotlight due to being able to answer those concerns mentioned above.

Anaerobic digestion is a series of successive biochemical reactions, namely hydrolysis, acidogenesis, acetogenesis, and methanogenesis, performed under strict anaerobic conditions [5–7]. These reactions lead to a gas mixture production known as digestion gas or biogas. The primary gas of economic value among these components is methane. For this reason, the term "biogas" is an inaccurate and imprecise term, because the carbon dioxide (CO_2) gas produced by aerobic decomposition is also "biogas" in a sense—just like other biogasses, it is the result of the biodegradation. However, the term "biogas" is specifically used to refer to the CH_4–CO_2 combustible mixture produced by the anaerobic decomposition of organic matter [8]. This biogas is made up of 45–75% methane (CH_4), the remainder being mainly CO_2 between 20–55%, with traces of other gaseous compounds (impurities) such as hydrogen sulfide (H_2S), nitrogen (N_2), hydrogen (H_2), oxygen (O_2), and others, which are explained further in detail below. This biogas becomes flammable at methane levels greater than 45% [9–11].

Impurities appear for various reasons in raw biogas. Among them, it was found that the feedstock that is introduced into the reactor contains some impurities; these were later found in the generated biogas after the evaporation of impurities in the digester. Siloxanes are an example of such compounds. Similarly, during anaerobic digestion, impurities can likewise be formed. Ammonia and hydrogen are an example of such impurities. Additionally, the temperature inside the reactor and the volatility of the compound influence the quantity that evaporates. In raw biogas, water is also found [12,13]. A good number of these impurities are malodorous, among them, there are H_2S, HCl, HF, H_2, CO, O_2, N_2, NH_3, and volatile organic compounds (VOCs). They are divided into organic and inorganic compounds. Organic compounds, in addition to containing methane, also include VOCs like siloxanes, iodomethane, toluene, xylenes, ethers, benzene, ketones, naphthalene, alcohols, esters, furans, and undecane. These VOCs also contain nitrogen compounds (due to degradation of protein waste), volatile fatty acids (VFAs), and volatile sulfur compounds (VSC). Inorganic gases produced in the anaerobic digester by anoxic respiration (denitrification) are nitrous oxide (N_2O) and molecular nitrogen (N_2). It is also possible to produce these inorganic gases by adding some compounds to increase the alkalinity of digestion, which contain nitrates such as sodium nitrate ($NaNO_3$) or by nonoxidative respiration during the transfer of sludge to the digester with the transfer of nitrate ions (NO_3^-). Among the inorganic and the most undesirable gases, which may lead to damage to the digester equipment and are produced during anaerobic digestion, include hydrogen sulfide (H_2S) along with dichlorine monoxide (Cl_2O), chlorine (Cl_2), chloric acid ($HClO_3$), and hypochlorous acid ($HClO$). The production of H_2S is due to the proteinaceous compounds containing sulfur that are transferred to the digester with the organic waste feed [6,14,15].

Depending on the subsequent use of the biogas, these impurities linked to the trace compounds must be removed. For example, some applications do not need high-quality energy, such as cooking and lighting and biofuel as biomethane for transportation. In such cases, carbon dioxide removal from biogas (upgrading) becomes unnecessary [16]. However, carbon dioxide removal when using biogas as a vehicle fuel is very necessary. Conversely, the presence of the other impurities in small quantities in the biogas damages equipment, engines, metal parts, etc., as is the case with H_2S and water vapor which generate highly corrosive compounds such as H_2SO_4, and their presence reduces the equipment's lifespan. Therefore, these impurity types require a deep elimination from biogas before any use [17,18].

Biogas treatment usually aims to be a purifying and upgrading process. The biogas cleaning process (purification) includes firstly drying it by dewatering, then removing hydrogen sulfide, and finally removing other impurities. The upgrading process is precisely the separation of methane from carbon dioxide in the biogas to obtain higher purity methane as a biomethane. [19–21]. There are several methods for purifying and upgrading biogas. Despite its high requirements of energy and chemicals, biological, physical, and chemical methods are the most commonly used. Among these methods, there is chemical or physical adsorption with a high surface area, gas absorption, condensation, washing or scrubbing with specific liquid solvents, catalytic conversion, and membrane separation. Due to the growing demand for energy and chemicals from these technologies, biogas upgrading by biotechnological processes has experienced rapid development in recent years. The biological techniques are considered a promising alternative because of their economic competitiveness and superior environmental sustainability. This technology is based on the use of microbial consortia capable of efficient application even on a small scale. Biofiltration is one of the most important methods used in this field [22,23]. The final product is biomethane typically containing 95–99% of CH_4, 1–5% of CO_2, and a significantly low level of H_2S [24].

After biogas cleaning and upgrading, natural gas can be replaced by the final biomethane obtained and become a direct alternative when $CH_4 > 96\%$, which is the same percentage as natural gas [25]. However, this methane level standard varies from one country to another in the European Union since it is technically a mandatory requirement according to some countries. To illustrate, if the methane content of biogas is higher than 85%, it can be injected into the natural gas grids in the Netherlands, while this percentage must reach 96% and 97% in Switzerland and Sweden, respectively [25]. It can be compressed to be used as compressed renewable natural gas (CNG), or else liquefied to be used as liquefied renewable natural gas (LNG) [26]. Biomethane can also be used in a wide range of applications including as fuel for engines and gas turbines to generate electricity, as a conditioner for the storage and preservation of fruits and vegetables, as fuel for fuel cells, as raw materials for modern industry, for disinfection and storage of seeds, and many other uses [27–29].

The global biogas industry market has seen an accelerated increase over the last decade (2009–2019) estimated at more than 126% as reported by the International Renewable Energy Agency (IRENA). Where the overall potential of the industry has increased from

46 terawatt hours (TWh) in 2009 to 91.8 TWh in 2019 [30,31]. The European Union produces 69% of this amount with 63.3 TWh. Germany produces approximately half of the biogas in the entire EU with more than 32.9 TWh followed by Italy and UK with more than 8.2 and 7.5 TWh, respectively. Moreover, there are currently 93 agricultural biogas plants in Poland and their energy production potential is 131 PJ/year [32]. Outside of the European Union, the United States produces 12.6 TWh, China produces 3.8 TWh, Thailand produces 2.6 TWh, and Turkey produces more than 2.5 TWh. The remaining countries outside of the European Union produce less than 2 TWh. The above statistics are for 2019 and were published by IRENA in 2021 [31]. In recent years, the motivation for the choice of biogas-upgrading technologies is determined by local markets based on the number of existing biomethane projects [33]. The biomethane sector is mostly developed in the European Union, with Germany holding a strong position due to its decision to utilize biomethane in combined heat and power plants. Sweden, on the other hand, employs biomethane as a biofuel for public transportation [34]. Moreover, a study for biomethane feasibility reveals that when the capacity of biomethane production plants is higher than 500 m^3/h, the plants show significant economic improvements in comparison to small scale plants. Additionally, the economic losses may be between EUR 370,000 and EUR 2.9 million for each year the construction or conversion of biomethane plants is postponed [34]. In the EU, there are 367 operational biomethane plants, and 178 of them which produce 8.5 TWh of biomethane annually, are located in Germany [35]. Moreover, 25 operational biomethane plants are located in the United States, six in Japan, and five in South Korea [35].

The aim of this review is to study biogas in depth, starting with the properties and components of biogas. Moreover, the study investigates impurities in biogas and the possible reasons why they exist in biogas. Afterward, the next section is focused on the elimination of these impurities as a cleaning process with recent developments. In addition, to obtain biomethane from biogas, the upgrading procedures are systematically assessed according to their state of the art technology and recovery range of methane in this area. Furthermore, various upgrading methods have been thoroughly examined and compared in terms of electricity usage of methods and their methane losses. For methane losses and system energy demand, the most promising strategy of the upgrading process was determined. Moreover, raw biogas and biomethane have been assessed with recently available data in the literature in the utilization of the biogas section. Biogas utilization pathways were investigated regarding their electrical production efficiency. Furthermore, the review reveals that biogas can be used as a biofuel for vehicles in an internal combustion engine. The usage of biogas in vehicles as a biofuel was discussed with its advantages and disadvantages. Finally, gaps in the field that needs to be investigated further have been highlighted for future research in this field.

2. Biogas Properties and Components

Biogas is produced by organic matter degradation through the anaerobic digestion process using anaerobic digesters or directly from landfills and ponds. The main biogas components are CH_4 and CO_2. Biogas burns easily due to the presence of methane, while the second noncombustible compound, CO_2, lowers the biogas calorific value [16]. However, the biogas becomes flammable as soon as the CH_4 content is greater than 45%, and a mixture of 60% CH_4 and 40% CO_2 is capable of keeping a steady flame with a calorific value of approximately 5340 kcal/m^3 at 15 °C, compared to 9000 kcal/m^3 for pure CH_4 [36]. The typical biogas properties are presented in Table 1.

Adding to its two main components (CH_4 and CO_2), raw biogas also contains some impurities in small amounts. These impurities appear because the production and raw biogas composition are affected by several factors including fermentation technology, operating conditions (such as pH, temperature, organic loading rate), substrate type and its organic matter concentration, collection method, etc. [29,37]. Among these impurities there is H_2S (0.005–2%), NH_3 (<1%), N_2 (0–2%), H_2 (0–4%), H_2O (5–10%), VOCs (< 0.6%), CO (<0.6%), siloxanes (0–0.02%), and O_2 (0–1%) [12,19,21,38,39].

Table 1. The properties of typical biogas from an anaerobic digestion process [40–43].

Properties	Values
Composition	55–70% methane (CH_4), 30–45% carbon dioxide (CO_2), traces of other gases
Energy content	6.0–6.5 kWh/m^3
Wobble index	19.5 MJ / m^3
Fuel equivalent	0.60–0.65 l oil/m^3 biogas
Explosion limits	6–12% biogas in air
Ignition temperature	650–750 °C (with the above-mentioned methane content)
Critical pressure	75–89 bar
Critical temperature	−82.5 °C
Normal density	1.2 kg /m^3
Heat of vaporization	0.5 MJ/kg
Smell	Rotten egg (the smell of desulfurized biogas is hardly noticeable)
Molar mass	16.043 kg/kmol
Methane number	124–150
Flame speed	25 cm/s
Lower heating value	17 MJ/kg

The proportion of these impurities in raw biogas is governed by several parameters. Therefore, the percentages differ from one substrate to another and differ within the same substrate if the anaerobic digestion conditions differ. For example, the percentage of H_2S in raw biogas varies depending on the percentage of the proteinaceous and other sulfur compounds in the substrate [13]. Additionally, the percentage of O_2 and N_2 in the raw biogas varies according to the percentage of air introduced into the digester to remove the hydrogen sulfide by oxidation in certain cases, or the air entering the landfill by reducing the gas pressure in order to extract the gas in other cases. There are also other reasons for certain impurities in the the substrate, which can evaporate during the anaerobic digestion process in the reactor due to the process temperature and the component's volatile nature, such as siloxanes and H_2O found in raw biogas compounds [12,15]. The typical composition of biogas generated from different origins versus natural gas is summarized in Table 2. Hereinafter, each of these impurities will be discussed separately and in depth.

2.1. Carbon Dioxide

CO_2 is the second main component of biogas. During the complex degradation processes of organic materials, different types of bacteria are involved to produce biogas. CO_2 is produced during some of these steps and acts as an electron acceptor for methanogenic microorganisms [13]. The percentage of this component in the biogas is influenced by temperature, pressure, and the liquid content in the digester [42]. The concentration of CO_2 dissolved in water decreases with increasing temperature during the fermentation process which causes an increase in the CO_2 level in the produced biogas. Unlike the other two parameters (the pressure and the liquid content in the digester), when they are higher, they lead to an increase in the concentration of CO_2 dissolved in the water, which causes a decrease in the CO_2 level in the produced biogas. This is considered to be beneficial for biogas [42].

Table 2. Composition of biogas generated from different origins versus standard biogas (EBA) and natural gas [12,44–46].

Compound	Agricultural Waste	Landfills	Industrial Waste	Wastewater	Standard Biogas (EAB *)	Natural Gas
CH_4 %	50–80	35–80	50–70	60–70	50–75	81–97
CO_2 %	19–50	15–50	30–50	19–40	24–45	0.2–1.5
H_2O %	≤ 6	1–5	1–5	1–5	1–2	-
N_2 %	0–1	0–3	0–3	0–1	1–5	0.28–14
O_2 %	0–1	0–1	0–5	<0.5	Traces	-
H_2 %	0–2	0–5	0–3	0	0-3	-
H_2S ppm	2160–10,000	0.1	0.8	0-4,000	0.1–0.5%	1.1–5.9
NH_3 ppm	50–144	~5	-	100	-	-
CO %	0–1	0–1	0–1	-	0–0.3	-
Total Cl mg/m^3	-	5	-	100	-	-
Siloxanes %	Traces	Traces	Traces	-	-	-

* EBA: European biomass association.

The volumetric energy content of biogas decreases with increasing CO_2 percentage. When biogas is used as fuel for vehicles, CO_2 is considered an impurity; therefore, it should be removed from raw biogas. However, it can be tolerated for power and heat generation. When CO_2 is mixed with water, CO_2 will be transformed into carbonic acid. This formation will cause some damage to the equipment [20].

2.2. Water Vapor

The anaerobic digestion process is categorized based on humidity in the digester content, which is linked to the moisture content of the feedstock. There are two main anaerobic digestion types: wet (liquid) anaerobic digestion with a moisture content from 85% to 99.5% and dry (high solids) anaerobic digestion with humidity from 60% to 85% [47]. Accordingly, during the anaerobic digestion process, a small amount of this water evaporates to become one of the biogas components. Its proportion in the biogas is dependent on the internal digester temperature and pressure [20].

The water in raw biogas can create some problems such as corrosion of equipment and reaction with other components like CO_2, NH_3, and H_2S to produce an acidic solution. Further, it can cause blockage of pipelines, flow meters, valves, compressors, etc., during subsequent cooling. The two phases coexist; liquid and gas cause flow oscillations which can interfere with the operations control. Moreover, water decreases the energy content and affects biogas heat value [6,13,48,49]. Briefly, water creates a negative impact on biogas utilization, so it is necessary to dry biogas before use.

2.3. Hydrogen Sulfide

Hydrogen sulfide is considered one of the most highly toxic gases as it can lead to serious risks to human health and can kill quickly (from 30 minutes to one hour) if its concentration reaches 0.05% of inhaled air [50]. It can also be used as a powerful nerve poison. When this gas combines with alkaline substances in tissues, it can form sodium sulfide and cause damage to the respiratory system and eyes. When it arrives in the bloodstream, it associates with hemoglobin which leads to nonreducible hemoglobin formation that causes toxic symptoms. Similarly, it excites human mucous in a strong way. It is quickly taken by the lungs and stomach and can cause conjunctivitis [20,50].

This gas occurs in small amounts (ppm levels) during the protein's degradation throughout the anaerobic digestion process from organically bound sulfur (S-bearing proteins). This gas production rate differs with the protein rate in the substrate entering the digester. Low rates are recorded with vegetable waste. While the highest rate comes from protein-rich materials like molasses, which produce more than 3% by volume of H_2S. However, the average rates are produced by animal waste such as poultry droppings and cattle and pig manure, with an H_2S level of 0.5% and 0.3%, respectively. Usually, this gas in biogas from wastewater treatment plants is higher compared with biogas from landfills. Another source of H_2S gas comes from the biochemical transformation of mineral sulfur (sulfates) to sulfide by sulfate-reducing bacteria. Sulfides are inhibitory to methanogens and can decrease methane production. Another source of H_2S gas comes from the biochemical transformation. During the AD process, when the degradation of sulfur (S) compounds and the desulfurization of sulfates (SO_4^{2-}) occur, H_2S is produced. Microorganisms need SO_4^{2-} because it is not only converted into cellular materials with enzymes but also behaves like an electron acceptor while organic matter is oxidized. If SO_4^{2-} exists inside the AD reactor, H_2S is produced from H_2 and SO_4^{2-}. Sulfate-reducing bacteria and methanogens compete with each other to obtain H_2 in the AD reactor. In this circumstance, sulfate-reducing bacteria can dominate over methanogens due to their higher microbial growth rate, higher demand of H_2, and higher energy yield during reduction [50–53]. Therefore, if sulfate is present during anaerobic digestion, the sulfate-reducing bacteria always produce H_2S.

H_2S is the most problematic and common impurity in raw biogas as shown in Table 3. It is a colorless gas with a strong rotten-egg smell that appears even at very low concentrations (0.05 to 500 ppm). It is inflammable. When mixed with oxygen, it forms an explosive mixture. On combustion, when it reacts with water it forms sulfur dioxide (SO_2) which is a toxic material and causes the flue gas to appear corrosive due to the formation of sulfuric acid. The latter causes acid rain to precipitate which is harmful to the environment. The airborne emissions from SO_2 are limited in several countries. H_2S also produces weak acids when dissolved in water. At room temperature, it can produce metal sulfides by reaction with certain metal oxides such as zinc oxide and ferric oxide. It can also produce hydrosulfide or metal sulfide by reaction with alkalis. Similarly, it can produce low solubility sulfides by reaction with metal ions in liquid form except for alkali metals and ammonium. For copper compounds, this gas type is caustic. Moreover, it also damages many engines components [13,20,49,50].

Table 3. Impurities in biogas and their negatives impacts [13,21,22,24,25,42,53–58].

Compounds	Negatives Impacts
CO_2	Reduces the overall calorific value; Promotes metal part corrosion due to the formation of low carbonic acid; Affect and damages alkaline fuel cells; Alteration of combustion properties, which affects the efficiency and safety of end-user equipment.
H_2O	A major contributor to corrosion in aggregate compressors, gas storage tanks, engines and pipelines by forming acid with other compounds such as H_2S, NH_3, and CO_2; Condensation formation resulting in damage to instruments; Freezing of accumulated water under high pressure and low-temperature conditions; Corrosion, rust, lubrication washes, clogging of pipes; Absorption/accumulation of other contaminants; Reduce combined heat and power production in cogeneration engine efficiency; Damage gas compressors; Condensation due to high pressure; accumulation in pipes.
H_2S	Acts as a corrosive in pipelines; Causes SO_2 emissions after combustion or H_2S emissions in case of incomplete combustion. Poisons catalytic converter; Corrosive to steel reactors, compressors, gas storage tanks, engines and instruments; Toxic at 450 ppm or (>5 mL/m^3); Due to furious combustion, there is the formation of SO_2, SO_3, which is more highly toxic than H_2S and can form H_2SO_4 causing more severe corrosion; Adsorbs irreversibly on the adsorbent and poisons it; Poison of the catalytic converter; Inhibits adsorption of fuel molecules and thus affects fuel oxidation; Affects fuel reforming, causes resistance to mass transport across electrodes caused by sulfur blocking sites; Unpleasant odor; Toxic to PSA adsorbents.
Siloxanes	Formation of SiO_2 and microcrystalline quartz during the combustion process and deposition on engine surfaces, valves, spark plugs, and cylinder heads, abrading the surface and causing grinding and malfunctioning of the engine part; Abrasion of engines, insulators, and spark plugs.

Table 3. *Cont.*

Compounds	Negatives Impacts
NH_3, O_2	NH_3: Toxic to anaerobic bacteria; Leads to an increase in the anti-knock properties of engines; Increased NOx emissions after combustion; Toxic compound: health problems; Corrosion in equipment due to reaction with H_2O to form a base; Forms nitrogen oxides during combustion in gas engines, foul odor. O_2: Explosive at high O_2 concentration in biogas.
VOCs	Toxic, and forms polyhalogenated dioxins and furans; Corrosive to combustion engines; Effect on elastomers and plastics: system integrity problems; Carcinogenic and toxic: health problems; Soot formation during the combustion of PAHs; Impact on safety and performance of end-user equipment.
Particulates (dust)	Damages vents and exhaust by clogging; Clogging equipment and engine because of deposition in compressors and gas storage tanks.

2.4. Siloxanes

Siloxanes are organic compounds based on a combination and repitition of silicon and oxygen atoms encircled by methyl groups. Usually they are classified as VOCs because they are degraded into volatile methyl siloxanes (VMS). The latter have a linear or cyclic structure and their degradation makes them a low molecular-weight species [42,59,60].

Siloxanes are mostly added to health care consumer products to improve lubricity. They are used in a wide range of products including cosmetics, shampoos, soaps, detergents, in industrial uses as precursors in polymeric silicone products, paints, etc. [59–61]. Most of these compounds are drained to wastewater treatment plants (WWTP) after rinsing, where they are absorbed by sludge. Since these siloxanes are resistant to biological and chemical degradation and have low solubility, they accumulate inside the sludge, making it a reservoir of siloxanes. Another reservoir for siloxanes is landfill, where most of the residues of the product containing siloxanes are disposed of [57,59,61].

While treating sludge by an anaerobic digestion process or waste in landfills, the siloxanes are present in the raw biogas as impurities. Therefore, it is expected that the amount of the siloxanes in the biogas produced from digesters that use food waste, agricultural product residues, or animal manure are much less than those produced by landfills or digesters treating sludge. Previous studies showed that the maximum amount of siloxanes expected to be found in biogas produced from landfills is from 4 to 9 ppm, while this percentage exceeds 41 ppm in biogas from digesters treating sludge [27,53].

As mentioned above, the siloxanes present in raw biogas are problematic impurities. As this affects the biogas performance as an energy source, it also affects the equipment used during energy production. Whereas during the biogas combustion process, oxygen reacts with the VMS under high temperature and pressure, the white SiO_2 deposits are formed in different morphological forms (white, glassy, crystal, microcrystalline, and amorphous deposits) [62,63]. In some cases, these deposits are associated with other components like calcium or sulfur [64]. These deposits can generate abrasive and thermal insulating properties, that can cause damage to turbines, reduce engine life and efficiency, and cause damage to their accessories such as valves, engine heads, and spark plugs. Additionally, these deposits may cause explosions in the combustion chambers, poisoning of catalysts, and may blocks pipes. For that, in raw biogas, the recommended siloxanes limit is 0.2 mg m^{-3}, and in some industries, the strict rule is <0.1 mg/m^3 [42,57,59,65].

2.5. Nitrogen, Ammonia Nitrogen, and Oxygen

Nitrogen and oxygen are generally absent in the reactor because of anaerobic conditions. If nitrogen is detected in raw biogas, it is a strong sign of denitrification or air leakage in the reactor. Additionally, among the existing causes of nitrogen compounds in biogas, the main cause is the release of nitrogen compounds after the bacterial reduction of proteins in the reducing medium which are subsequently transformed into ammonia [20,49]. Biogas produced from organic or agricultural waste digesters generally contains a low nitrogen proportion (usually 0.1%) compared to that produced from landfills (mostly from 5% to

15%) [66]. On the contrary, aqueous ammonia could be used for biogas purification and upgrading by absorbing H_2S and CO_2 [67]. A recent study demonstrates a significant positive effect of ammonia nitrogen on biogas upgrading. With 5500 mg/L of ammonia nitrogen, the CH_4 content in biogas reached 94.1% [68].

A small amount of oxygen is found in raw biogas produced from anaerobic digestion, because oxygen will react with H_2S or be consumed by facultative anaerobic bacteria at the start of the anaerobic digestion process, and oxygen itself will not be detected. Similarly, landfill gas extraction leads to low internal pressure, which causes some air absorption [20,42,69]. The oxygen could cause an explosion if the CH_4 content is 60% in biogas and the air reaches a range between 8.5% and 20.7% at 25 °C [42]. Additionally, oxygen can lead to flammable mixture formation [20]. Therefore, the ratio of air should be carefully adjusted. An oxygen amount of 1% to 2% is cited as ideal [70]. Likewise, a 4:1 nitrogen to oxygen ratio in raw biogas is mentioned as ideal [8].

2.6. Volatile Organic Compounds

Volatile organic compounds (VOCs) exist in trace amounts in biogas. They are divided into different chemical families such as aromatics, alkanes, alcohols, halogens, sulfur compounds (excluding hydrogen sulfide), carbonyls, and siloxanes [9,53]. These compounds' type and concentrations in biogas depend on the substrate's origin. For example, a very low concentration of these compounds was recorded in farm digester biogas. On the contrary, this concentration increased in the gases produced by wastewater, household waste, and landfills [71]. Materials that produce these components include cleaning compounds, cosmetic products, silicon compounds, lacquers, foaming agents, pharmaceuticals, aggregates, adhesives, solvents, pesticides, synthetic plastics, propellants, textiles, coatings, etc. [66,72]. The concentration increase of these compounds in biogas is attributed to direct emission from the substrate or to the volatilization after the compounds' degradation (complex molecules) and their transformation into low molecular mass. Consequently, the VOCs measured in biogas at the start of digestion come essentially from direct volatilization, while the concentration of those measured after a certain duration of digestion in biogas are dictated by the substrate biodegradation rate [9].

Despite their low concentration in raw biogas (usually 1% by volume), just like hydrogen sulfide, VOCs lead to equipment problems and negative environmental effects such as greenhouse gases, groundwater contaminants, and disagreeable odor, and they can affect human health [9,73]. For example, acids can be formed during halogenated hydrocarbons combustion (in the presence of water) such as hydrofluoric acid (HF) or hydrochloric acid (HCl) which can cause acidification, premature equipment degradation, and corrosion of materials and catalytic surfaces [20,72]. Similarly, organochloride in biogas can lead to combustion-engine corrosion, and silicon and chloride compounds make landfill gas as vehicle fuel much more expensive and often too complicated [52,66]. Generally, VOC accumulation affects the proper functioning of systems for converting biogas into energy; for this reason, their contents in biogas must be carefully controlled.

2.7. Particulates

Several researchers have reported the presence of particulates (dust) in raw biogas produced from digestion or landfills. This can lead to mechanical damage to gas turbines and gas engines due to their abrasive properties, or cause blockages when they deposit in the gas storage tank and compressor. They can also form condensation nuclei of water droplets [6,20,45,74,75]. For these reasons, the total content of particles and aerosols in the biogas must be maintained under 0.01 mg/m^3 [56,69].

3. Current Biogas Purification Technologies

In this section, the latest technologies for the improvement and upgrade of biogas and the resulting biological methanation processes are summarized. This paper examines the main principles of various upgrade methods, the technical and scientific features/consequences for biomethanation efficiency, the challenges that need to be addressed for further improvement, and the applicability of upgrade concepts.

As mentioned in the previous section, the ratio of CH_4 and CO_2 in the biogas is mostly in the range of 50–70% and 30–50%, respectively, depending partly on the organic content and pH of the substrate. [19]. In addition to CH_4 and CO_2 gases, biogas may contain N_2 gas in the concentration range of 0–3% depending on the volume of the head space at the start of gas production in the reactors [76]. The most important minor components are carbon monoxide (CO), O_2, hydrogen sulfide, H_2, and NH_3 [29]. Depending on the substrate source in the reactors, biogas may contain other contaminants such as siloxanes (0–41 mg Si/m^3), volatile hydrocarbons (alcohols, fatty acids, terpenes) or fluorinated hydrocarbons, chlorinated, heavy metal vapors, and aromates [77].

All gases except CH_4 contained in biogas are undesirable gases and are known as biogas pollutants. For the biogas purification process, the first step is biogas cleaning and the second step is biogas upgrading. Biogas cleaning is performed to remove harmful or toxic components such as H_2S, Si, CO, siloxanes, VOCs, and NH_3 [19]. The second step, biogas upgrading, aims to increase the calorific value of biogas and convert it to a fuel standard. In addition, biogas upgrading is a multistage gas-separation process that involves the separation of CO_2 gas, drying the gas to remove moisture content, and extracting and compressing other small components [77]. In order for biogas to be used in different applications, its methane content must be at least 90% (v/v) [17]. Upgraded biogas that has a 95% (v/v) methane content is called biomethane [78].

3.1. Carbon Dioxide Removal

Almost all of the noncombustible portion of biogas is CO_2 gas, and this reduces the calorific value per unit volume of biogas. This limits biogas for uses such as biofuel for transportation, cooking, and lighting, etc., where direct combustion technology is needed. CO_2 creates disadvantages such as the extra space occupied in the biogas storage area and the use of extra energy in the compression of the biogas. Dry ice is formed as a result of the compression of biogas with CO_2 and this creates the problems of lump formation and freezing in valves or measurement points. This makes it difficult to store biogas in containers for transportation and limits its use. For such reasons, the removal of CO_2 in biogas gains importance in terms of its use in larger scales [17].

Various commercial technologies are available to separate CO_2 from biogas. The higher solubility of CO_2 in water than CH_4 gas enables them to be separated from each other by taking advantage of the difference in their solubility in water. At 25 °C, the solubility of CO_2 in water is 26 times higher than the water solubility of CH_4. Biomethane containing 95–99% (v/v) CH_4 can be obtained as a result of the separation process in this way [21]. Water scrubbing is one of the most common methods used for biogas cleaning [19]. Figure 1 shows a schematic of a process flow diagram of a recirculating water scrubber. Biogas is sent to the absorption column by pressurizing from the bottom (6–10 bar) and water is supplied from the top of the column at the same time. The absorption column used is filled with random packaging material in order to work more effectively [21]. The saturated water is transferred to the flash tank where the pressure is dropped to around three bars to minimalize methane loss. The water leaves the flash tank and goes to the desorption column. Air is taken into the desorption tank due to increasing the driving force for CO_2 desorption by decreasing its partial pressure. A 1000 Nm^3/h raw biogas water scrubber upgrading system needs to circulate 200 m^3/h water at 8 bars of pressure at 20 °C [79]. The benefit of the water scrubber is that H_2S can be eliminated with the removal of CO_2 because the solubility of H_2S in water is higher than CO_2. If the H_2S is removed with CO_2 at the same time, the water quality can decrease rapidly, therefore, using fresh water is recommended. The upgrading system has some advantages such as being economical, having high efficiency, not requiring additional chemicals, and gaining a high level of methane recovery above 97% [75]. However, the high initial investment and high energy demand during water regeneration are the primary drawbacks of the system.

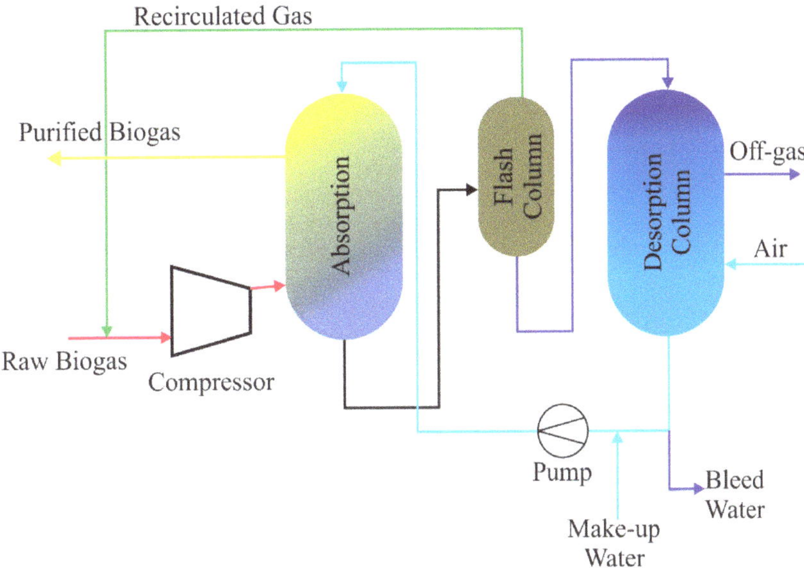

Figure 1. The schematic flow chart of a recirculating water scrubber (adapted from [79]).

By choosing a washing liquid that is more effective than water in the separation of CO_2 and CH_4 in biogas, further improvement of absorption can be achieved. In this context, the most commonly used chemicals are aqueous solutions of amines such as diethanolamine, methyldiethanolamine, monoethanolamine, and diglycolamine [21]. Such biogas-boosting methods are also called amine scrubbing. Figure 2 shows a schematic of biogas upgrading by amine scrubbing of CO_2. The raw biogas enters the absorption tank from the bottom, and amine solvent enters through the top of the tank. During the counter-current flow, CO_2 reacts with the amine solvent. During the reaction, the temperature increases from between 20–40 °C to between 45–65 °C due to an exothermic reaction [80]. Generally, an increase in the temperature decreases the solubility of materials but the solubility of CO_2 in amine solvent increases with an increase in temperature [75]. After the reaction, the liquid goes to the heat exchanger to increase its temperature and is boiled at 120–150 °C. This is the regeneration step of amine solvent, as CO_2 is released from the amine solution. Recovery of CH_4 is higher than 99% due to the sensitivity of the reaction with CO_2 but methane loss may increase up to 4% because of dissolution in amine [81]. While the system has high sensitivity and low operational costs, the high initial investment and significant energy demand for regeneration of amine are some of its drawbacks [75]. By using these chemicals as solvents in the liquid phase, an extremely low CH_4 adsorption is provided and CH_4 recovery is achieved at a rate of approximately 99.95% [77]. In addition, the need for a lower operating pressure in the absorption processes with these chemicals compared to all other biogas-upgrading technologies makes these technologies advantageous. In many amine washing plants, only one blower is used instead of a compressor, and this significantly reduces the electrical energy requirement. However, a disadvantage of amine scrubbing is the possible oxidative or thermal degradation of amine solutions. This increases the chemical consumption, corrosion potential, and emission potential of hazardous decomposition products [82].

Another method of separation of CO_2 and CH_4 is the adsorption of gas molecules to a solid surface. The adsorbent solid surfaces used for this process are porous materials with highly specific surface areas. The other name of this technique is pressure swing adsorption (PSA). This technique enables the selective separation of biogas by using different adsorption equilibria that adsorb larger amounts of CO_2 or different adsorption kinetics that adsorb CH_4 faster than CO_2. The adsorbent materials used for this technique are titanosilicates, zeolites, silica gels, activated carbon, and carbon molecular sieves [24,83]. In the adsorption process to the solid surface, the water vapour in the biogas must be preseparated in order to prevent potential poisoning of the adsorbent material [29]. In addition, the pressure of the biogas fed must be 10 bar in order to provide sufficient driving force in this adsorption process. This method is able to adsorb N_2 and O_2 simultaneously with CO_2. The technology is well developed and commercially available in the market with various range capacities between 10 and 10,000 m^3/h [75]. The flow chart is shown in Figure 3. In this process, there are four vertical columns that are used for adsorption, depressurization, desorption, and pressurization sequences. CH_4 recovery can be around 98–99% as a result of solid surface adsorption processes. The biggest advantage of this technology is that it enables N_2, O_2 and CO_2 to be separated from raw biogas at the same time [29]. However, it is a disadvantage that the active sites on the adsorbent material used in adsorption separation technologies are blocked by NH_3 and H_2S gases [58]. Additionally, the process efficiency can be influenced by impurities of raw biogas. Additionally, 2–4% of CH_4 is lost during the process and reduction in this loss should recirculate the output of gases into the PSA system [24].

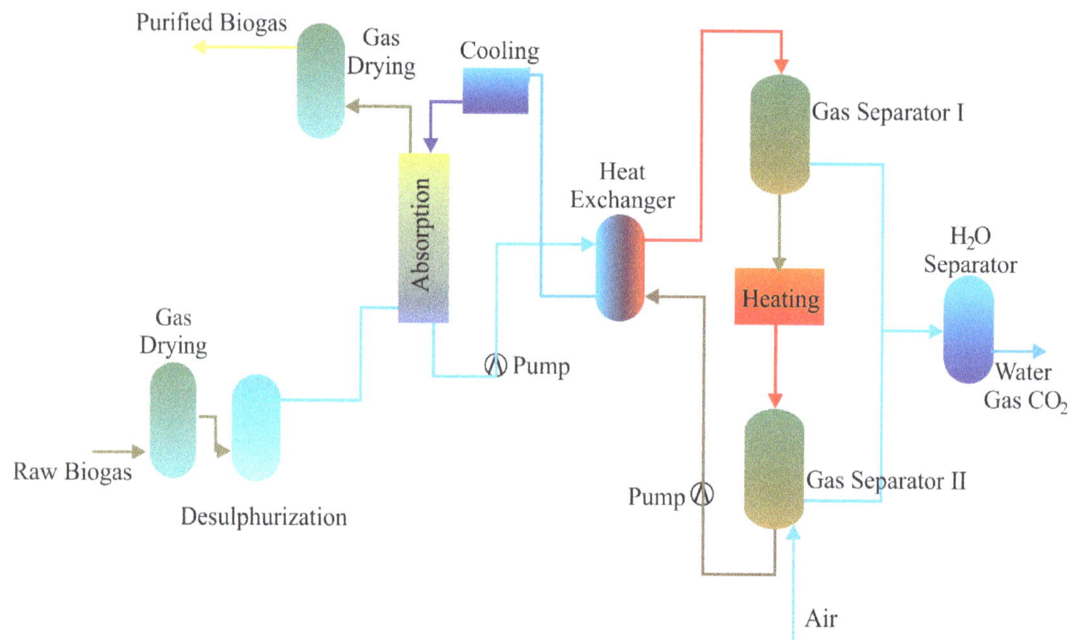

Figure 2. The schematic flow chart of chemical amine scrubbing (adapted from [24]).

Electrical swing adsorption (ESA) [84] and temperature swing adsorption (TSA) [85] are other swing adsorption methods. Temperature increases at constant pressure in the TSA process; therefore, thermal energy is required for the regeneration of the adsorbent material [86]. In an ESA process, the current passes through the saturated adsorbents for regeneration, so that CO_2 is released from adsorbents as a result of heat generation by the Joule-heating phenomena. Even though the operating cost of ESA is lower than TSA and PSA, conductive adsorbent is needed. Activated carbon is a promising adsorbent due to its semiconductor properties with large surface area and porous structure [87].

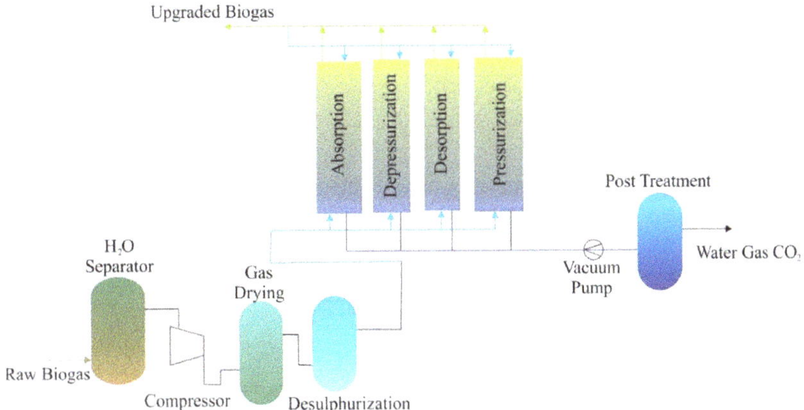

Figure 3. The schematic flow chart of pressure swing adsorption (PSA) (adapted from [24]).

Another biogas-upgrading technique is membrane-based gas permeability, and this technology has gained increasing importance in recent years [88]. This technique is based on different solubilities of gas types in particular membrane materials [89]. Figure 4 shows a schematic of process design for a membrane upgrading process. Biogas is fed into the membrane module at pressures of 5–30 bar [90]. Gas types with higher permeability in biogas preferably pass the membrane to the low-pressure permeable side, while gas types with lower permeability accumulate on the high-pressure side and leave the membrane module as a retentate. One of the challenges at membrane systems is energy demand to maintain pressure for separation, further studies should be conducted to reduce pressure demand. The most commonly used membrane materials in this technique are polysulphone, polyimide, and cellulose acetate [88]. It is a cheap membrane because of its cellulose-based material with good CO_2–CH_4 sensitivity. However, a cellulose acetate (CA) membrane has low plasticization pressure around 8 bars because CO_2 can be dissolved in the matrix of the membrane due to its OH^--rich structure [91]. The membrane separation technique provides up to 99.5% CH_4 recovery and provides biomethane quality containing at least 99% CH_4, especially when membrane modules are used sequentially [92]. As a result, polymeric material-based membranes are heavily used in raw biogas separation. This method is advantageous in terms of its robustness, ease of use, lack of need for chemicals, flexibility of scaling up, low operating costs, low initial investment costs, low energy demand, and compact design [58]. Moreover, it should be considered that while CH_4 is separated from biogas by membrane systems, another valuable and biologically produced product, CO_2, is also separated from biogas for further usage. The main challenge is the relationship between permeability and selectivity. Therefore, a low permeable membrane is used together with a low selectivity one [75].

Inorganic membranes have excellent thermal stability, mechanical strength, and resistance against chemicals compared to organic membranes. The membranes are made from different materials such as zeolite, silica, activated carbon, metal-organic framework, and carbon nanotubes [75]. The most difficult part is to produce a defect-free structure for an inorganic membrane because the fabrication process requires continuous monitoring and a small mistake can create a defect due to their very fragile structure [93]. The defects in the structure are closely related to the sensitivity of membranes.

Considering that biogas components liquefy or sublimate at different temperatures, they can be used to remove impurities from biogas by the cryogenic separation method. Cryogenic separation is a technique that relies on the liquefaction of gases under different temperatures and pressures. Biogas is cooled down to $-55\,^\circ C$, and moisture, NH_3, and H_2S are removed and CO_2 is separated as a liquid in the last step [21]. CO_2 can be removed in solid form through sublimation when the temperature is decreased at $-85\,^{\circ}\mathrm{C}$. With this technique, the CH_4 content of the gas phase at equilibrium is higher than 97% (v/v). Since cooling for purification is synergistic with cooling for liquefaction, this technique is particularly valuable in the case of producing liquefied biomethane. The process is run at a temperature of $-170\,^{\circ}\mathrm{C}$ and under a pressure of 80 bars, which requires several compressors and heat exchangers [75]. In addition, other gasses such as N_2, O_2 and siloxanes in raw biogas can be separated. After this process, the final product is liquid biomethane which is free of N_2 and O_2 and is equivalent to liquid natural gas. The cryogenic separation method can produce not only over 97% pure biomethane in liquid form, but also the marketable liquid form of CO_2 [76,94]. The schematic flow chart of the cryogenic separation of biogas is shown in Figure 5.

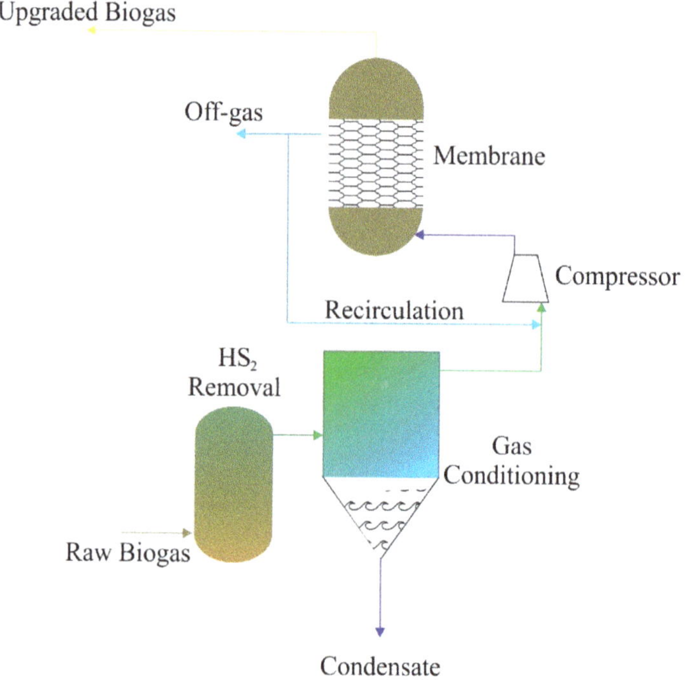

Figure 4. The schematic flow chart of a membrane separation process (adapted from [79]).

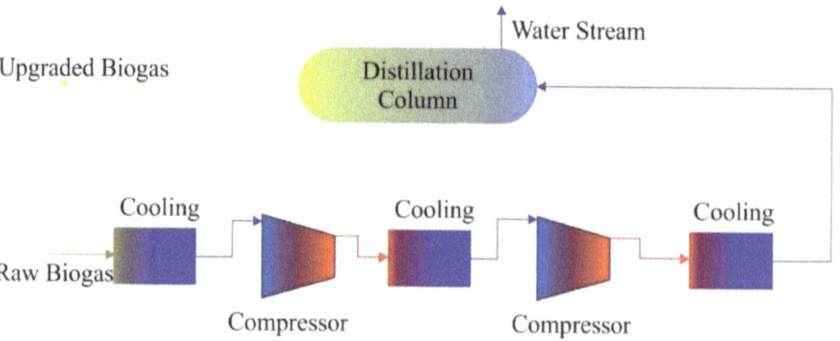

Figure 5. The schematic flow chart of the cryogenic separation of biogas (adapted from [24]).

Although the cryogenic separation method has promising results, it is still in development and operates on a commercial scale for only a few facilities. High investment and operating costs, and CH_4 losses and blockage due to solid CO_2 are a few of the problems that limit the use of this technique [19].

3.2. Biological Techniques for CO_2 Removal

Biological upgrade techniques can be classified as photosynthetic and chemoautotrophic. Most of these techniques have proven themselves in the experimental stage for being the early stage of full-scale applications. These technologies contribute to sustainable cyclic energy recovery by converting CO_2 into other energy-containing products (i.e., CH_4, H_2, etc.). CO_2 in biogas is converted to CH_4 using H_2, according to the chemoautotrophic upgrade method (Reaction (1)).

$$4H_2 + CO_2 \rightarrow CH_4 + H_2O \quad \Delta G^0 = -130.7 \text{ KJ/mol} \tag{1}$$

H_2 gas should be obtained from renewable energy sources. For this, water is hydrolyzed and the remaining electrical energy can be provided by windmills or solar panels for this hydrolysis [95]. With the electrolysis of water made using this energy, water is divided into O_2 and H_2. However, the low volumetric energy density is a disadvantage that makes it difficult to store the H_2 gas produced by this method [95]. The low initial investment costs are due to the fact that CH_4 has an energy of 36 MJ/m^3, H_2 has an energy greater than 10.88 MJ/m^3, and the existing possibilities of the biogas plant are used in the improvement processes which make this technology advantageous [96].

H_2 gas-assisted biogas upgrade methods exist in three different forms: ex situ, in situ, and hybrid design techniques (Figure 6).

In the in situ concept, H_2, externally added to anaerobic digester is converted to CH_4 by combining with CO_2 in the digester and by the action of autochthonous methanogenic archaea [97]. The biggest disadvantage of this technology is that it reduces the activity of the methanogenesis stage by increasing the pH value to above 8.5 in the biogas process. CO_2 gas is decomposed into H^+ and HCO_3^- as shown in Reaction (2). Thus, the use of CO_2 causes a reduction in H^+ gas and causes an increase in the pH of the environment.

$$H_2O + CO_2 \rightarrow H^+ + HCO_3^- \tag{2}$$

In order to overcome this difficulty, it has been proposed that codigestion with acidic wastes can stop the pH increase [98]. Luo and Angelida [98] used hollow fiber membranes to inject H_2 into an anaerobic reactor that processes cattle manure and whey in a continuously fed reactor and obtained methane gas containing 96% CH_4. In another study, a hollow fiber membrane in an upflow anaerobic sludge blanket reactor was placed in an external degassing unit, and biomethane containing 94% CH_4 was obtained from the in situ biogas upgrade process [99].

The ex situ biogas upgrade method is the process of transforming products into CH_4 by supplying H_2 and CO_2 externally to a secondary vessel containing hydrogenotrophic culture outside the main anaerobic digester, as shown in Figure 6 [97]. This method has some advantages over in situ methods. These advantages include the benefit of the secondary vessel which makes the biochemical process simpler due to the absence of organic substrate degradation and makes the process independent of biomass. With this method, biomethane efficiency can be between 79–98%. This yield may vary according to digester types. For example, 98–99% purity biomethane can be obtained due to the formation of a biofilm of a mixed anaerobic consortium, which acts as a good biocatalyst of a trickle bed reactor [100]. As a result, ex situ and in situ processes have taken their place among the hydrogen gas-assisted biogas upgrade techniques in the literature with their experimental results, but the hybrid design technique is currently in the developmental stage and the results of this technology will soon begin to emerge.

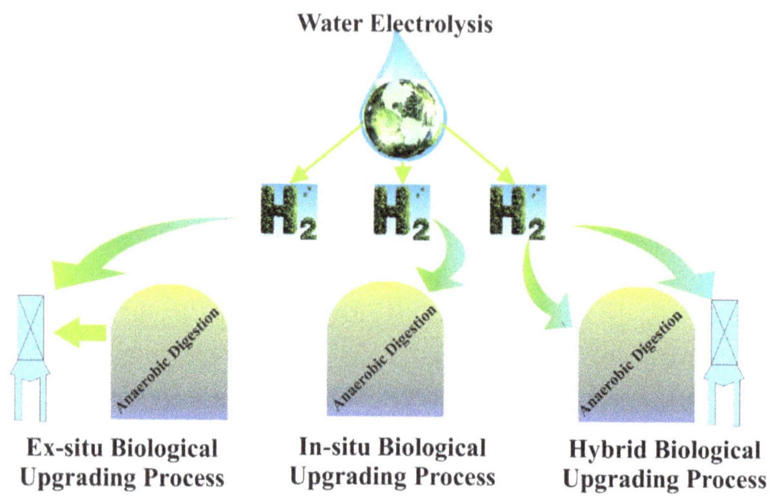

Figure 6. Hybrid, in situ, and ex situ biological upgrading technologies (adapted from [19]).

Another alternative method of CO_2 separation is the photosynthetic biogas uptake technique. In addition to CO_2, H_2S is also removed with this method and >54% CO_2 is removed. The biomethane production efficiency of photoautotrophic techniques can increase up to 97%, depending on the digester type and substrate. Closed systems are advantageous due to their low land and water requirements and high photosynthetic performance. However, open photobioreactors differ from closed systems in that they require fewer resources for construction and processing, have low photosynthetic CO_2 removal, and have high natural-resource requirements. In the biogas-upgrading process, biogas is directly injected into photobioreactors, and then photoautotrophic microorganisms can efficiently take up O_2 and CO_2 to generate heat. Thus, up to 2–6% CO_2 can be found in the upgraded biogas and the CH_4 ratio increases [101].

Various microalgae or cyanobacteria with high photosynthetic activity have been used to raise biogas quality. The most common of these are *Spirulina, Chlorella, and Arthrospira* [76]. Using microalgae to reduce the CO_2 component in raw biogas has taken place successfully and achieved up to 97.07% of CO_2 removal [102]. Another approach is indirectly upgrading biogas with microalgae as claimed by Chi et al. [103], which is a novel technique and was suggested as a two-step process to upgrade biogas while avoiding the risk of explosion. In the first step, biogas was passed through a carbonate solution to capture CO_2 as a bicarbonate form. The second step is the regeneration of the carbonate solution. Figure 7 demonstrates direct and indirect biogas upgrading with microalgae.

3.3. Removal of Hydrogen Sulfide

H_2S is another common component of raw biogas. Its quantity is between 100 and 10,000 ppm and the variation depends on the substrate combination [24]. If the substrate is rich in protein content, H_2S production will be high. Due to its negative effects such as causing the corrosion of pipes, pumps, and engines, it has to be cleaned before upgrading to biomethane or any other use of biogas. H_2S can be transformed into SO_2 and sulphuric acid (H_2SO_4) which increase environmental concerns and corrosion risk.

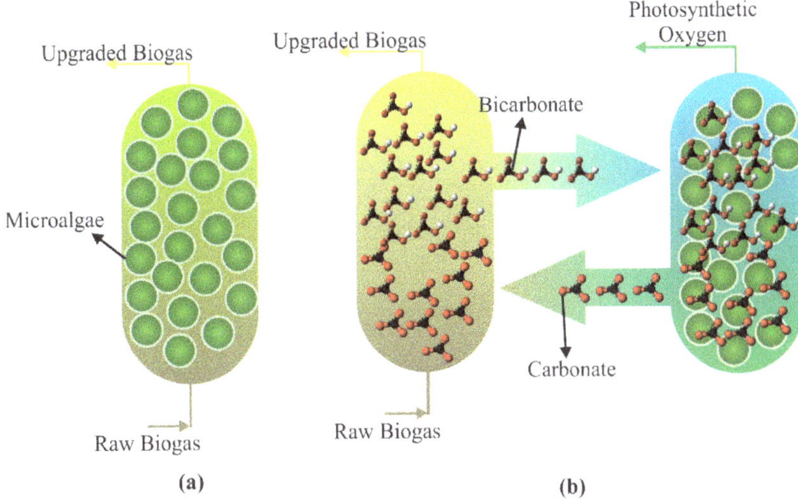

Figure 7. The schematic flow chart of (**a**) direct and (**b**) indirect upgrading biogas using microalgae (adapted from [104]).

The simplest desulphurisation methods are accomplished by introducing pure oxygen or air into the gas phase or adding iron hydroxides or chlorides to the liquid phase [21]. The addition of iron hydroxides or chlorides to the liquid phase causes ferrous sulfide formation and precipitation and can significantly reduce the concentration of H_2S in the gas phase. O_2 added to the gas phase is consumed by *Thiobacillus* bacteria along with the oxidation of H_2S to elemental sulphur deposited on surfaces or absorbed by the slurry [105]. These techniques create minimum additional operating and investment costs but can provide removal of 80–99% H_2S concentrations. However, it is known to be less efficient in obtaining stable and low sulphur contents required for biomethane production. The disadvantages of the method are that oxygen can affect the anaerobic digestion process negatively if too much air is injected. Additionally, the remaining nitrogen could be a problem for future processes. For example, it is difficult to upgrade by separating nitrogen and methane [106]. Application to filter reactors is commonly used on the output of the biogas pipe. With this method, combinations of H_2S in raw biogas can achieve a range of 50–100 ppm [107].

The oldest technique used in the removal of H_2S from biogas is scrubbing with chemically active liquids. The most used scrubbing chemicals are sodium hydroxide (NaOH), calcium hydroxide ($Ca(OH)_2$), iron (II) chloride, iron (III) hydroxide, ethylenediaminetetraacetic acid, and monoethanolamine [108–110]. $Ca(OH)_2$ and NaOH allow for the formation of sulphurous salts. It has been reported that with these techniques, 90–100% cleansing is normally achieved. The biggest disadvantages of chemical upgrading techniques are the necessity and high costs of using chemicals [108]. In this method, Fe^{3+}/EDTA catalyst is used where the reaction product is elemental sulphur. During the reaction, Fe^{3+} is reduced to Fe^{2+} as shown in Reaction (3). The regeneration process is represented in Reactions (4) and (5) [24]. During the H_2S cleaning process, a small amount of Fe^{3+}/EDTA solution is used for the regeneration step. H_2S removal of 90-100% is achieved when raw biogas and solution flow rates are 1 dm^3/min and 83.6 cm^3/min, respectively [21].

$$2S^{2-} + 2F^{3+} \rightarrow S + 2Fe^{2+} \tag{3}$$

$$O_2 + 2H_2O \rightarrow O_2 \tag{4}$$

$$0.5O_2 + 2Fe^{2+} \rightarrow 2Fe^{3+} + 2OH \tag{5}$$

H_2S can be separated from biogas by adsorption in hydroxides or metal oxides, and the most commonly used chemicals are iron, zinc and copper oxides [21]. At the end of this process, sulphur binds as metal sulphur and is released into the gas thanks to a mild endothermic reaction.

The membrane separation technique can also be applied to separate H_2S from biogas [111]. However, it should be noted that the efficiency of H_2S removal will be limited when glassy polymeric material is used only for bulk CO_2 removal. High CH_4/H_2S separation can be achieved by selecting a special, rubbery, polymeric membrane material. This technique is very attractive for biogas with H_2S content higher than 2% [112].

High-pressure water washing (HPWS) is a biogas-boosting technology that is frequently used industrially [113]. Water is used for single-pass absorption; however, without a regeneration step, the process has high water consumption [24]. Therefore, it prefers a small absorption tank volume. This method is more efficient for smaller concentrations of H_2S or combinations of CO_2 removal. As shown in Figure 8, the adsorption process occurs in a high pressure (6–10 bar) column in which the biogas is washed with an HPWS and biomethane exiting the top of the column is obtained. A disadvantage of this technique is that it is affected by higher water consumption than other adsorbing techniques [114].

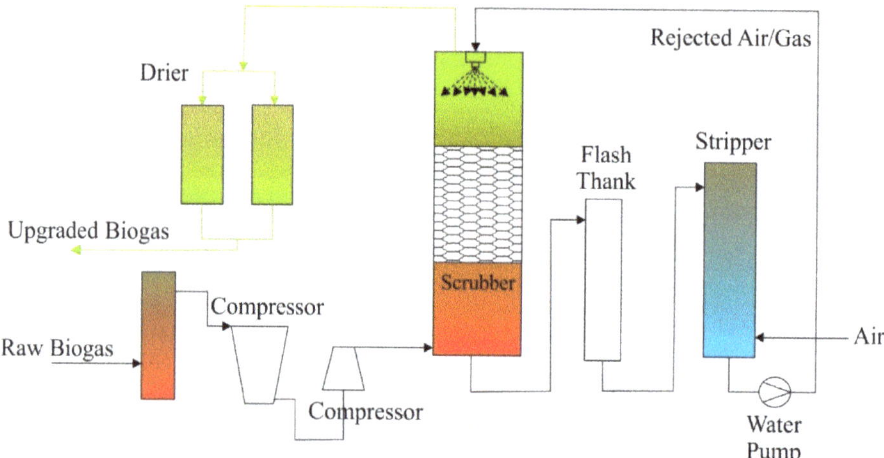

Figure 8. The schematic flow chart of water scrubbing of H_2S (adopted from [24]).

H_2S removal from biogas can be achieved using adsorption to activated carbons impregnated with potassium iodide or H_2SO_4 or not impregnated (untreated) [24]. For this, 4–6% O_2 must be added to the biogas to reduce H_2S to elemental sulphur (Reaction 6) [21].

$$2H_2S + O_2 \rightarrow 2S + 2H_2O \tag{6}$$

The temperature required for this reaction to occur is 50–70 °C and the required pressure is 7–8 bar. As seen in Figure 9, the elemental sulphur obtained is adsorbed by activated carbon. To improve the reaction rate, adsorption is done by impregnating the activated carbon with an oxide or alkali such as sodium potassium iodine, potassium hydroxide and potassium permanganate. This process can increase the H_2S removal from 10–20 kg H_2S/m^3 to 120–140 kg H_2S/m^3 [115].

While this technology has advantages such as high H_2S cleaning efficiency and high purity, it has disadvantages such as the need for replacing activated carbon instead of regeneration when the solid is saturated with sulphur, increasing environmental concerns about proper disposal methods (such as creating waste oxygen), and removing dust and water prior to activated carbon treatment [76].

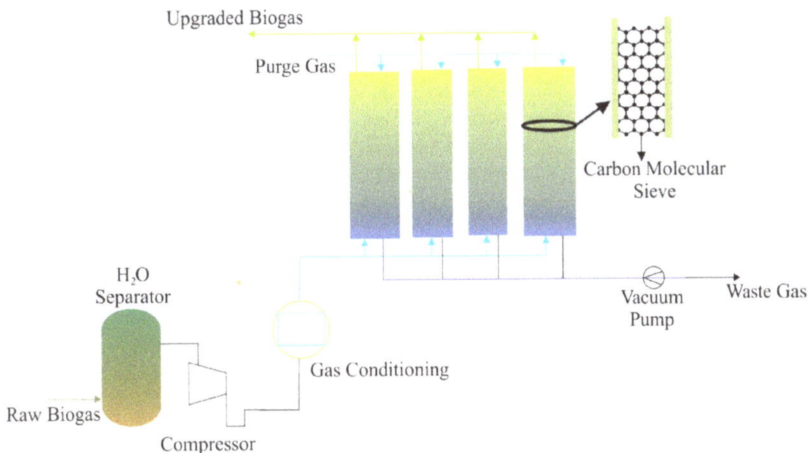

Figure 9. PSA technique flow chart for H_2S removal (adapted from [24]).

3.4. Biological Techniques for Removal of H_2S

Biological oxidation can be performed ex situ in a distinct apparatus with packed bed scrubbers or trickling beds [21]. H_2S is absorbed in a liquid film and bio-oxidized to sulfur or sulfate. Washing liquid is drained and replaced if the pH falls below the threshold level. O_2 should be dosed as pure air or as pure O_2 (Reactions 7 and 8). Although the separation performance of this technique is effective, the fact that H_2S levels of less than 100 ppm cannot be reached continuously for the remaining O_2 content prevents the continuous application of this technique in biomethane production [109].

$$H_2S + 0.5O_2 \rightarrow S + H_2O \tag{7}$$

$$H_2S + 2O_2 \rightarrow SO_4^{2-} + 2H^+ \tag{8}$$

In addition to this technique, it has been reported in the literature that NO_3 can be used as an electron acceptor instead of O_2 for H_2S oxidation in the biofiltration unit (Reactions 9 and 10) [116].

$$3H_2S + NO_3^- \rightarrow 3S + 0.5N_2 + 3H_2O \tag{9}$$

$$3H_2S + 4NO_3^- \rightarrow 3SO_4^{2-} + 2N_2 + 6H^+ \tag{10}$$

The sulphur-oxidizing microorganisms used here belong to the *Thiomonas, Thiobacillus, Acidithiobacillus, Paracoccus, Halothiobacillus* or *Sulfurimonas* genera [117]. This technique has been mainly applied in a biotrickling filter due to its low cost and nutrient requirement.

With this technology, the H_2S concentration can be reduced from 3000–5000 ppm to 50–100 ppm, and at the same time, NH_3 is removed. In the biotic filter, HD-QPAC, pall rings or polyurethane foam are used as filled bed column materials for H_2S removal [24,76].

3.5. Removal of Other Compounds

3.5.1. Removal of Siloxanes

SiO_2 deposits in the spark plug, cylinder head, and combustion chamber which can cause damage to the engine. Maximum siloxane concentrations in biogas should be between 0.03 and 28 mg/m^3 according to engine manufactures [21]. A study showed that while most siloxanes can be eliminated with long carbon-chain organic solvents, they can be completely removed with chemical absorption [118]. Removal of 95% of the siloxane concentration at 60 °C was achieved using a combination of sulphuric acid and nitric acid at 480 and 650 L/m^3, respectively [118]. In addition, selexol can be used to remove at least 98% of siloxane in biogas plants [119].

Using silica gel is another method and experimental reports show that its removal capacity is higher than activated carbon; however, using silica gel may only be effective in large-scale uses [21].

The activated carbon adsorption method can be used for siloxane removal from biogas. The siloxane adsorption efficiency of this method depends on the adsorption capacity, physicochemical properties, and microporous volumes of the activated carbon [120]. Schweigkofler and Niessner [118] have reported on using different types of adsorbents such as activated carbon, polymer beads, silica gel, and molecular sieves for siloxane removal, and found that they exhibit good adsorption capacity. The cryogenic separation method is another advantageous technology. With the cryogenic separation method, siloxane concentrations can be removed from raw biogas at 12%, 25.9%, 90%, and 99.3% at temperatures of 5 °C, −25 °C, −30 °C, and −70 °C, respectively. The initial investment and operating costs of any technique involving cryogenation are higher than other adsorption methods such as activated carbon [121].

The membrane separation technique can alternatively be used to remove siloxanes from biogas. For this purpose, membranes with high methane/siloxane content can be used [122]. Although membrane separation techniques are very well established for mostly CO_2 removal from biogas in industrial applications, there are only a few published results applying the membrane technique for more than 80% siloxane removal. Recently, a new method has been developed to remove siloxanes using a cryogenic temperature condensation system by Piechota [62]. In the study, the temperature of the system ranged between +40 to −50 °C with different biogas flow rates, and 99.87% of siloxane removal was achieved; moreover, the obtained biogas met biomethane requirements from the European Union [62]. The same author has published another method for the removal of siloxanes and the improvement of biogas quality [123]. A specially designed adsorptive packed column system was developed, and the remarkable results obtained show that siloxanes were removed entirely and nonsilica impurities were eliminated by 99.76% in comparison to the intake biogas [123].

3.5.2. Removal of Water

Biogas is mostly saturated with water vapour after leaving the digesters operating at mesophilic temperatures. Water in the form of vapour in biogas can condense in gas pipelines or cause corrosion with sulphur compounds. Raw biogas, which has 5% water at 35 °C as saturated water vapour, can be removed by changing the pressure and temperature. When the pressure is increased or the temperature is decreased, water vapour will condense as a physical separation method. Additionally, absorption and adsorption methods can be applied to remove the water as a chemical drying method. Adsorption separation techniques include activated charcoal, molecular sieves, or silica. Other impurities such as siloxanes and particles can dissolve in water [106]. When removing water vapour in raw biogas, these impurities will be simultaneously removed.

Physical separation methods can be divided into three main groups which are namely cyclone separators which use centrifugal force for separation of water droplets, moisture traps which use a pressure differential to create a low temperature and condense water, and water traps which use a design of pipes to collect water [24,76,107]. These methods have a chronic operating problem where water freezes on the surface of a heat exchanger [76]. The dew point of biogas can be decreased to −40 °C using water adsorption under 6–10 bar pressure in the columns. Silica, alumina, magnesium oxide, or activated carbon is placed inside the pressurized columns. For continuous operation, two columns are needed. When one of them runs until saturation, the other regenerates at low pressure simultaneously. A benefit is that this system has a low operating cost. However, high initial investment costs and the required precleaning of dust and oil particles are drawbacks of the system [76]. When using the water absorption method with glycols, the biogas dew point can be decreased to −15 °C. One of the benefits of this system is that precleaning of dust and oil particles is not necessary. However, the solvent regenerates at 200 °C which causes high operating costs. Ryckebosch et al. [21] reported that the biogas flow rate should be at least 500 m^3/h for the economic benefits of the glycol-based system.

3.5.3. Removal of Ammonia

Ammonia is another common impurity. If the substrate contains proteins, ammonia is produced during hydrolysis. NH_3 is removed during the recovery or primary drying of the biogas and during the adsorption process.

Guo et al. [124] designed a packed bed N-TRAP system impregnated with H_2SO_4 for NH_3 removal from biogas. The adsorbent in the N-TRAP system was prepared with waste wood-shaving sand and anaerobic digestion biowaste which had a higher adsorption capacity than activated carbon when under the same conditions. In addition, this system is advantageous in that it can be applied in the presence of water, unlike activated carbon, and the adsorbent saturated with NH_3 can be used as fertilizer.

Co-removal processes involving biological techniques for the removal of H_2S and NH_3 as a mixture have also been reported. Jiang et al. [125] suggested a horizontal biotrickling filter packed with exhausted activated carbon for the co-removal of H_2S and NH_3 from biogas. The biofilter used was inoculated with nitrifying bacteria that oxidize sulphur. The removal efficiency for H_2S was 95% for a 20–100 ppm operating range and 8 s gas residence time. For gas residence time longer than 4 s, NH_3 removal efficiency reached 98%. However, NH_3 degradation resulting from high H_2S concentration has been observed [126]. These effects may be due to the accumulation of ammonium sulfate $((NH_4)SO_4)$ in the system.

Alternatively, NH_3 can be extracted from the biogas along with other unwanted components. For example, as mentioned in the previous sections, sulphur compounds and halogenated compounds in biogas can be removed in addition to ammonia by using a water-scrubbing method [75]. In addition, NH_3 and water vapour can also be removed together using a PSA system.

3.5.4. Removal of Oxygen and Nitrogen

N_2 and O_2 are generally absent in the reactor because of anaerobic conditions. If N_2 is detected in raw biogas, it is a strong sign of air leakage in the reactor because O_2 will react with H_2S and oxygen itself will not be detected. Its presence in biogas is undesirable due to the diluent effect that may cause a decrease in methane content. O_2 can be tolerated up to some amount in biogas, but high amounts are undesirable as the risk of biogas explosion increases.

O_2 and N_2 can be removed from biogas by the adsorption method using a molecular sieve or activated carbon. PSA or membranes used for sulphur and CO_2 removal can remove some of the O_2 and N_2. As a result, both components are difficult and expensive to extract from biogas [109]. Therefore, it is necessary to develop and apply new biogas-boosting technologies to remove N_2 and O_2.

3.5.5. Comparison of Different Technologies to Upgrade Biogas

The most common upgrade technologies today are PSA, amine wash, and water scrubbing. PSA can actually be improved with newer and better-performing adsorbents [79]. Organic solvent washing is a highly effective upgrade technique, but there is no clear trend for this technology to increase or decrease within the biogas upgrade market. The development of this technology mostly depends on studies such as the optimization of pressure and temperature adsorption amounts. Additionally, organic solvent washing technology requires annual compensation of solvent lost to the atmosphere, which will increase the cost [79].

New process cycles are being developed to optimize methane yields, removal of unwanted gases from biogas, and energy demand [89]. The least-mature technology among these is membrane separation technology, but it has recently begun to develop rapidly in industrial applications [19]. The membrane separation technologies are attractive in terms of lower investment and operating costs compared to other techniques. Membrane separation technologies have proven to be suitable to replace other upgrading technologies. Polymeric membranes used in membrane separation technologies are preferred because of their low cost and flexibility [127]. However, the high cost of this technique and the impermeability of membranes to foreign matter limit the use of this technique. It was also stated that the membrane separation technique is relatively more expensive than the PSA technique [128].

If biogas-upgrading technologies are evaluated in terms of biomethane purity, the most important parameter is to remove CO_2. Many upgrade technologies can obtain biomethane with 97% methane content, but high energy demand and increased costs occur for biomethane higher than 97%. For example, in the case of substantial N_2 in biogas, a complex adsorbent bed configuration can be used to remove this gas. This will result in both little refinement and a relatively high cost.

Table 4 compares the effects of different biogas uptake techniques in terms of methane recovery, biomethane purity, and electrical energy demands. Among the biogas-boosting techniques, the amine-scrubbing method appears to have relatively less electricity consumption, CH_4 losses, and purer biomethane compared to other methods. However, amine scrubbers may need antifoaming agents. Additionally, since the amine washing technique is chemical adsorption, it should not be forgotten that the efficiency and effects of this technique will vary depending on the chemical used.

Table 4. Comparison of the effects of some biogas-upgrading technologies.

Biogas Upgrading Methods	CH_4 Losses (% *v/v* CH_4)	Biomethane Purity (% *v/v* CH_4)	Electrical Energy Consumption (kWh/Nm^3)	Refs.
Water Scrubbing	1	97	0.20–0.30	[129]
	-	>97	-	[130]
	<2	80–99	-	[29]
	<2	95–98	0.2–0.5	[131]
	<2	96–98	0.25–0.3	[19]
	2	95–97	0.25	[12]
	1	96.5	0.26–0.27	[114]
	0.5–2	96	0.25–0.30	[132]
	1–2	96–98	0.20–0.32	[29]
	1	96–98	0.20–0.30	[133]
	1-3	95–98	0.20–0.50	[17]
Pressure swing adsorption	4	96–98	0.23–0.30	[19]
	1–5	90–98.5	0.20–0.25	[134]
	2	98	0.21–0.46	[78]
	4	95–99	0.24	[75]
	1–3	96	0.20–0.25	[129]
	2	96	0.18	[114]
	4	95–99	-	[130]
	<3	>96–98	0.16–0.43	[131]
	2–4	95–99	0.25	[12]
	1–3.5	96–98	0.16–0.43	[17]
	2	96	0.22	[114]

Table 4. *Cont.*

Biogas Upgrading Methods	CH$_4$ Losses (% *v/v* CH$_4$)	Biomethane Purity (% *v/v* CH$_4$)	Electrical Energy Consumption (kWh/Nm^3)	Refs.
Membrane separation	4	96	0.19–0.77	[94]
	-	>96	-	[130]
	0.6	96–98	0.18–0.20	[19]
	<5	90–99	0.18–0.35	[131]
	0.5	98	0.20–0.30	[79]
	0.5	98	0.20–0.30	[133]
	1	97	0.30	[114]
	0.3–0.5	95	0.25	[129]
	0.8	99.5	0.26	[135]
	0.7	97.5	0.29	[136]
Amine scrubbing	0.1	96–99	0.05–0.15	[19]
	<1	>99.0	0.22	[130]
	0.1	99.9	0.12–0.14	[79]
	0.1	99	0.15	[129]
	0.04–0.1	98	0.05–0.25	[17]
	<0.5	98	0.05–0.18	[131]
	0.1	99	0.12–0.14	[133]
	0.1	99	0.12–0.14	[12]
	0.1	99	0.05–0.25	[134]

Another parameter effective in comparing the upgrade techniques is the size of the plant. For example, although the water-scrubbing technique in small plants is relatively less costly compared to other techniques, it has shown a higher cost compared to the PSA technique. Considering the external costs, the choice of upgrade technologies for each system is not affected much by the other. Considering this situation, chemical adsorption technology is preferred as it causes less CH$_4$ loss compared to other techniques [114]. Water-scrubbing and PSA techniques suffer from higher methane loss and lower biomethane yield than other technologies. However, these techniques are quite suitable for small installations [128].

Converting biogas to biomethane is a strategic goal in many countries. Nonbiological (physicochemical) upgrade techniques have taken place in the literature with many applications and are at the level of high technology preparation. Biological techniques, on the other hand, are quite new technologies compared to other techniques and are advantageous in terms of feasibility and technological facilities. The development of biological upgrade technologies will be rapid and advantageous as the identified challenges are overcome.

4. Utilization of Biogas

As a renewable energy resource, biogas has gained enormous attention and the use of biogas is expected to increase considerably in the near future. According to estimations, the utilization of biogas will be 29.5 GW globally in 2022 while it was only 14.5 GW in 2012 [137]. Biogas can be used in a raw or upgraded form with different applications such as in a boiler to produce heat, combined heat and power (CHP) to produce heat and power, fuel cells to produce electricity, production of chemicals, production of vehicle fuels, and injection to the gas grid [26,138]. In the EU, electricity generated from biogas was 22 TW h/year in 2020 [139]. Moreover, the expectation is that it will reach 640 TW h/year, a 30-fold increase [140]. After 2017, greenhouse gas emissions were reduced by 60% due to biogas utilization [75]. Additionally, the estimation of biogas utilization for vehicle fuel shows that it may increase from 2% in 2017 to 27% in 2050 [75]. In the US, there were 283 manure-based biogas production plants in 2021 and Figure 10 represents the end-used biogas utilization systems according to their technologies between 2000 and 2021 [141]. As of 2017, there was a divergent increasing trend in numbers of pipeline and CNG biogas projects in the US. This may be due to legal regulations in 2016 and the widespread use of biogas as a fuel for agricultural purposes.

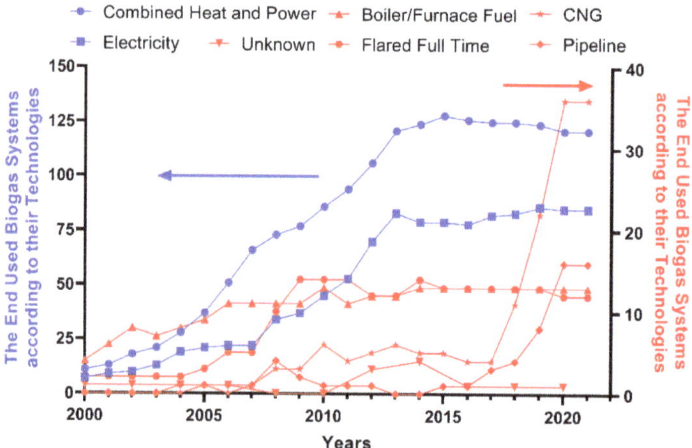

Figure 10. The end-used biogas utilization system according to their technologies in the US from 2000 to 2021 [141].

When producing electricity from biogas, different methods have different impacts on emission rates due to their combustion processes. For example, gas engines produce low emissions, while diesel engines which use ignition oil have high gas emissions in their exhaust fumes. Electricity production in engines of less than 100 kW has lower efficiency and a higher rate of exhaust gas emissions when compared with electricity production at higher than 300 kW [142]. It is clearly implied in the literature that biogas utilization for the production of electricity is much more environmentally friendly compared to electricity generation from fossil fuels.

Biogas can also be used as a hybrid with other renewable energy sources to stabilize energy systems such as alongside wind and solar [143]. Additionally, it has been utilized for cooking and lighting purposes for a long time in developing countries. In this section, the utilization of biogas and associated technologies will be reviewed. Biogas utilization pathways are represented in Figure 11.

4.1. Boilers, Gas Engines and Gas Turbines

Boilers are the most common way to produce heat from biogas, and internal combustion engines are the most commercialized process for power production from biogas at biogas power plants. Boiler efficiencies are between 75 and 85% [144] for heat production from biogas. Typical boilers can be made suitable for biogas utilization with small modifications such as gas-airflow rates. The advantage of these boilers is that they are able to burn low-quality biogas [145]. Internal combustion engines, which are developed for natural gas, can be used for biogas without any modification. The four-stroke engines' capacities are between a few kW and 10 MW and their lifetime is approximately 60,000 hours; however, their electrical efficiency is low, between 35 and 40% [42]. Their efficiency and antiknock properties are increased with a reduction in CO_2 concentration in biogas [145]. The most common system for the production of power and heat simultaneously is known as combined heat and power (CHP) plants. In the EU, half of the biogas plants have CHP units, which are run with four-stroke engines, and half of these engines are diesel engines [42]. A modern CHP plant's heat and power efficiency is 85–90% [145]; however, the electrical efficiency is under 40% and only 2.4 kWh of electricity can be produced when 1 m^3 biogas is combusted [42]. In addition, gas turbines are used to produce electricity from biogas. Various capacities of gas turbines are available in the market, from 500 kW to 250 MW, and single-cycle gas turbine efficiencies are between 20–45% and their efficiency increases with the size [145]. The distinguishing feature of a gas turbine is low NO$_x$ emissions, which are less than 25 ppm when the cleaned biogas is used [42].

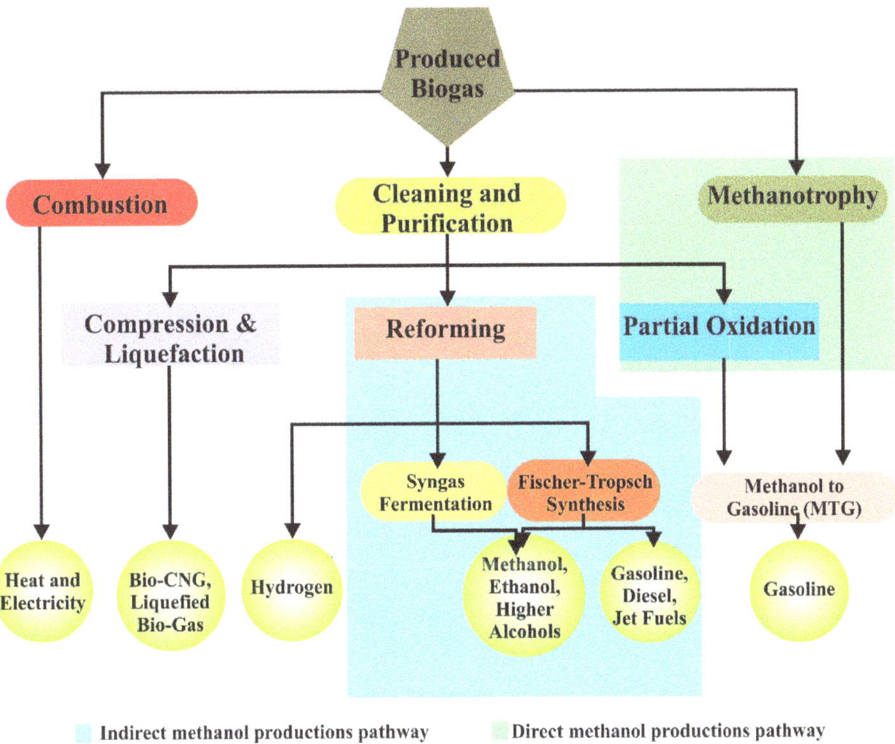

Figure 11. The schematic representation of biogas utilization pathways (adopted from [58]).

Using small-sized engines for the utilization of biogas is still problematic due to low efficiency [146]. Currently, the homogeneous charge compression ignition (HCCI) method has been heavily researched and has shown to be a promising technique. Saxane et al. [147] and Reitx et al. [148] made comprehensive reviews about HCCI technology and compared it with spark-ignited and diesel engines. HCCI uses low-temperature combustion which produces low NO$_x$ emissions. A study tested the HCCI thermal efficiency and found it to be around 50%, which is close to diesel engine performance when diesel and biogas mixtures were used [149]. In fact, additional biogas in the fuel decreases thermal efficiency due to its low calorific value. Bedoya et al.'s [150] experimental study clearly showed that HCCI is a promising method to reach high thermal efficiencies of 44% and low NO$_x$ emissions of 0.11 g/kWh. Additionally, Blizman et al. [151] reported that NO$_x$ emissions were at the same level for landfill gases with the HCCI method. Another method for using biogas is flameless combustion. This method is able to increase biogas efficiency and also decreases particles and NO$_x$ emissions. The study investigated biogas utilization using flameless combustion, and the results revealed that electrical efficiency was 53% and CHP efficiency could reach 82% using exhausted gases for energy recovery with low NO_x emissions which was around 2% of the volume of exhausted gases [152].

Innovative methods for the utilization of biogas include dual-fuel engines, stirling engines, and micro gas turbines. Dual-fuel engines use biogas by cofiring with very small amounts of biodiesel, bioethanol, or diesel [153]. Stirling engines convert thermal energy into mechanical energy; thus, they are theoretically perfect for producing electricity from waste thermal energy. The initial investment cost is still very high and their electrical efficiency is around 20% [154]. Micro gas turbines are compact, operate at high speed, are low noise, and are vibration-free where gas turbines and their power range is between 25 and 500 kW. If the methane content in biogas is higher than 30%, the biogas can be used with a micro gas turbine [145]. Their electrical efficiency is higher than 45% [143].

In the literature, biogas and diesel in the dual-fuel mode in combustion ignition (CI) have been studied since 1980. Mathur et al. [155] used biogas as a biofuel in CI and Prakash et al. [156] improved a mathematical equation to predict the ignition delay of dual-fuel mode CI. Mustafi et al. [157] investigated dual-fuel mode CI with different CH_4 and CO_2 ratios of biogas and diesel. The results revealed that 70% of particulate matter (PM) emissions and 37% of NO_x emissions were reduced compared with diesel fuel when the biogas contained 70% of CH_4 [157]. Cacue et al. [158] investigated the effect of enriched oxygen in the air and dual-fuel mode in CI. They reported that increased O_2 concentration in the air gave a positive impact on biogas and diesel dual-fuel mode for emission and brake thermal efficiency (BTE) according to the control group. BTE increased by 28% when the O_2 concentration in the air reached 27% [158]. Another comprehensive study focused on combustion characteristics and emissions in dual-fuel mode with different feeding ratios of biogas [159]. They used diesel and four different biogas feeding rates of 0.3, 0.6, 0.9, and 1.2 kg/h. The results represented that 14.8, 26.6, 36.9, and 43.9% of energy could be met from biogas in dual-fuel mode at the full load when biogas flow rates were 0.3, 0.6, 0.9, and 1.2 kg/h, respectively. Moreover, with a 1.2 kg/h biogas feeding rate, the exhaust temperature and NO_x were decreased by 14.2 and 42.8%, respectively when compared with control fuel at full load [159]. Another study was focused on the exergy analysis of biogas and diesel fuel under the dual-fuel mode and the result revealed that exergy efficiency does not drop as energy efficiency does when biogas was used as a primary fuel [160]. Recently, Wang et al. [161] studied a reactivity-controlled compression ignition mode based on dual-fuel combustion. In their report, the BTE could reach 40% in the optimal condition. Another promising development has occurred in fuel in recent years. This development is known as modified fuel, in which liquid fuels are mixed with nanoparticles. The nanoparticles act as combustion catalysts, especially with CI engines [162]. Diesel or biodiesel have been mixed with different nanoparticles, which are classified as metal, metal oxide, nanofluid, and carbon nanotubes [163]. However, additive nanoparticles give several advantages to fuels, and researchers have been generally focused on diesel and blending diesel and biodiesel fuels. In the literature, there is a huge gap to be filled regarding other biofuel resources such as modified fuels. Modified fuels can be used with biogas in dual-fuel mode. To the authors' best knowledge, only one group has been working on this topic. Feroskhan et al. [164] investigated a dual-fuel CI engine with fueled biogas and modified diesel. They modified diesel with 15, 25, and 35 ppm CeO_2 as an additive material and they used this fuel with different biogas feeding rates in dual-fuel CI [164]. In their first study, modified diesel with 15 and 25 ppm CeO_2 and 4 L/min biogas feeding rate showed higher BTE than both diesel and modified diesel [165]. In their second study, they worked on the optimization of CeO_2 concentration in dual-fuel mode CI. Their results revealed that biogas and modified fuel with 25 ppm CeO_2 reduced all emissions and increased BTE [164]. Table 5 shows the current studies with their key results.

Table 5. Biogas utilization in the compression ignition engine with dual-fuel mode.

Type of Fuel	CH$_4$ (%)	Compression Rate	Control Fuel	NO$_x$ (%)	HC (%)	CO (%)	Smoke (%)	BTE (%)	Refs.
Biogas-Biodiesel-DEE	90.6	16.5–18.5	Biodiesel	↑7.6	↓39.5	↓42.2	↓42.8	↑7	[166]
Biodiesel-Biogas Dual Fuel	73	17.5	Biodiesel	↑5.5	↓18.2	↓17.1	↓2.1	↑6.6	[167]
Simulated Biogas-Biodiesel	30–40–50	19.5	Biodiesel	↓80 *	↑96 *	↑92 *	-	↓32.6	[153]
Diesel, WCOME + Biogas	71	17.5	Diesel	↑6.1–8.9	↓14.8–44.4	↑21.4–29.8	↑2.3–3.4	↓9.7	[168]
55–85 °C Preheated Biogas-Air Mixture	70	17.2	non-preheated mixture	↑11.73	↑150–250 *	↑6–371 *	-	↑51.4	[169]
Diesel-Purified Biogas	70.4–93.8	17.2	Diesel	-	-	↓10–50 *	-	↓10–15 *	[170]
Diesel-Biogas-10-20%H$_2$	70	18	Diesel	↑6	↓25	↓30	-	↑32	[171]
Diesel-Biogas-H$_2$	48-80	20	Direct and indirect injection condition	↑150		↓20	No change	↑6	[172]

* was calculated from the paper data. ↑ was increase, ↓ was decrease.

Biogas can be easily used in spark ignition (SI) engines due to mixing with gasoline, after which it can be fed into the cylinder, while dual-fuel mode has to be used for compression ignition with mixing air. A study reported that when the biogas was used with a spark engine with a compression ratio of 10 and a BSFC increased by 66% while BTE, CO and NO$_x$, emissions were reduced by 12%, 40%, and 81.5%, respectively, compared to fueling with gasoline [173]. Another study investigated the compression ratios of SI in the range of 6 to 9 [174]. The results revealed that the power obtained from the test engine fueled with biogas was increased by 13.4% and BSFC was decreased by 18.18% when the compression ratio was 8 [174]. A similar study reported the same finding for the compression ratio of SI. When ethanol and biogas were used, HC, CO, and NO$_x$ emissions were decreased while BTE was increased [175]. Table 6 represents the findings of the recent studies when biogas was used as a fuel in an SI engine.

Table 6. Biogas utilization in the spark ignition engine.

Type of Fuel	CH$_4$ (%)	Compression Rate (CR)	Control Fuel	NO$_x$ (%)	HC (%)	CO (%)	BTE (%)	Refs.
Gasoline+Biogas	55.6	10	Gasoline	↓81.5	↑6.8	↓40	↓12	[173]
Simulated Biogas	60	12	100% CH$_4$	↓71.74	-	-	-	[176]
Biogas-Ethanol	64.96	13.6	Ethanol	↓30	↓60	↓50	↑20	[175]
Syngas-Biogas	4–65	12.9	Biogas	↓5	-	-	↑3	[177]
Biogas	55.6	10–12	Biogas from 10 to 12 CR	↑10.17	↑15.6	↑0.01–0.258	↑26.69–30.32	[178]
Biogas	55.6	10	Biogas from 33 to 47 °C A ignition advance	↓88	↓40	↑50 *	↓13 *	[179]
Gasoline–Biogas-LPG	55.6	8	Gasoline	↓50.09	↓23	↓15	↓16.04	[180]
Different ratio of Hydroxy and Biogas	70	8.5	Ratio of Hydroxy and Biogas	1192 ppm	-	0.74%	204 J/cycle	[181]

* was calculated from the paper data. ↑ was increase, ↓ was decrease.

4.2. Fuel Cells

One of the most efficient ways to generate electricity using biogas is through a fuel cell. The method uses an electrochemical reaction where oxygen and hydrogen atoms react to produce energy and water. The fuel cell has an anode and cathode that are separated from each other with a nonconductive solid or liquid electrolyte. Fuel flows through the anode and oxygen goes through the cathode. The ions, which are produced at the anode, are able to pass through the electrolyte; however, the electrons pass through an external circuit and generate electricity [145]. Fuel cells are a relatively new technology with high efficiency and low emissions as reviewed by Alves et al. [182] and Bocci et al. [183]. Biogas utilization with fuel cells requires fuel purification and a cleaning process depending on fuel cell types. Therefore, biogas is more challenging than natural gas to use as a fuel. Additionally, another drawback of the fuel cells is their relatively high operating temperature, which depends on the type of fuel cell used. There are generally three different fuel cell types which utilize biogas.

4.2.1. Solid Oxide Fuel Cells

Solid Oxide Fuel Cells (SOFCs) are made of solid oxide or ceramic materials. This method has recently gained much attention. Ni-based material is generally used as an anode without deterioration of the fuel cell because it is cheap [184]. However, the fuel cell performance has a close relationship with impurities in biogas, especially H_2S. Directly feeding biogas to SOFC is a heavily studied subject both experimentally [184–189], and with modeling [190,191]. Shiratori et al. [189] reported that Ni–ScSZ cermet was used as an anode material at 1000 °C operating temperature, and they achieved a cell voltage above 0.9V after cleaning the H_2S in biogas. Moreover, they demonstrated that SOFC can tolerate H_2S impurity up to 1 ppm. Another study claimed that there is not any significant degradation after H_2S removal and the biogas can be used as a fuel for electricity generation with SOFC [188]. To decrease impurities in biogas, air can be added into biogas with no significant changes in the output voltage [143].

The system can work under either intermediate (600–800 °C) [184,187] or high (800–1000 °C) [189] temperature conditions. The temperature is a significant parameter for SOFC because electrochemical reactions are fast above 600 °C [46]. In addition, Ma et al. [187] reported that biogas is more suitable than H_2 due to carbon degradation.

SOFC-based power-plant efficiency was estimated to be 74% when the method of partial oxidation reforming of the biogas was used [192]. Additionally, the efficiency can reach 80.5% with a heat recovery system. SOFC thermoeconomic analysis showed that pressurized SOFC can be more economical and can have a 25% lower electricity cost at 20 bars of pressure [193].

4.2.2. Proton Exchange Membrane Fuel Cells

Proton Exchange Membrane Fuel Cells (PEMFCs) run at a relatively low temperature and pressure compared to SOFCs. Their temperature and pressure range is between 50 and 100 °C, and 1 and 2 bars, respectively [194]. This fuel cell requires clean, hydrogen-rich fuel; therefore, it is not suitable for biogas utilization. Gual et al. [195] modeled biogas utilization with PEMFC. According to their results, electrical efficiency was 40% and the whole system efficiency reached 82% when the fuel cell was combined with CHP. Schmersahl et al. [194] studied PEMPC experimentally. They claimed that its performance was lower with biogas rather than natural gas. However, its efficiency was still higher than internal combustion engines. Another study confirmed their result with their model [196]. This study calculated that the electrical efficiency of PEMFC is around 40%.

4.2.3. Molten Carbonate Fuel Cells

Molten Carbonate Fuel Cells (MCFC) are made of porous nickel electrodes and molten carbonate salt. This fuel cell has several advantages such as using different fuel sources, high efficiency, and low environmental impacts. The operating temperature is 600–700 °C and CO can be used as a fuel; however, H_2S has to be cleaned [197]. The first MCFC plant was built in the US in 1996 and was running with natural gas. It produced 2 MW of electricity with 43.6% efficiency [198].

Trogisch et al. [199] studied MCFC with different biogas sources and they achieved 50% electrical efficiency. Another study focused on the H_2S toleration level of MCFC [200]. They suggested that sulphur-tolerant materials should be used in MCFC. Lanzini et al. [201] reviewed SOFC and MCFC plants and the effects of biogas containing significant impurities. All in all, MCFC needs further investigation to be commercialized.

4.3. Biomethane

Biomethane is almost equivalent to natural gas; therefore, it could be used for the transportation sector and for direct injection to gas grids. As mentioned in the previous section, raw biogas can be upgraded to biomethane by separating out CO_2 and other impurities. If raw biogas is upgraded to higher than ~95% CH_4, it can be directly injected into the gas grid and used as a vehicle fuel. In comparison to other countries, Sweden has best-utilized biogas as vehicle fuel. They have more than 36,000 vehicles on the road in 2011 that are running on biogas which can be refueled at 500 filling stations all over the country [202,203]. Another efficient way to use biomethane is to directly inject it into gas grids, because utilization of biogas at or near to production facilities is not feasible. Biomethane can be transferred through a large energy-demanding area with gas grids. Additionally, the separation of CO_2 and CH_4 can convert biogas into hydrogen-rich biogas via synthesis technology from Fischer-Tropsch [58]. Additionally, CO_2 can be transformed into CH_4, known as methanation through the Sabatier pathway [46]. Methanol production is able to pressurize CO_2 with hydrogen under 20 bars of pressure. In the literature, hydrogen is obtained from water through electrolysis using renewable energy resources such as wind or solar power for obtaining CH_4-rich biogas [95,204].

4.4. Biosyngas and Biohydrogen

Biogas can be used for the production of biosyngas with a reforming process. Catalytic reforming including partial oxidation reforming with a metal catalyst such as platinum (Pt), palladium (Pd), nickel (Ni), cobalt (Co) and iron (Fe) has been studied [205,206]. Although the catalytic reforming process is possible, biogas is not well suitable for this process because of the poisoning of the catalysts and a significantly increased production price. A gliding arc reactor was used to produce biosyngas. Tu et al.'s [207] experimental study showed that CH_4 and CO_2 were reformed into 31% H_2-rich gases, and they claimed that a lower feeding gas flow allowed more hydrogen and carbon monoxide production and less hydrocarbon production. In addition, a tri-reforming process was studied for biogas utilization and their results showed high hydrogen yields, which were between 59.5% and 73.5% [208].

4.5. Biomethanol, Bioethanol and Higher Alcohols

Methanol can be used as a vehicle fuel. It can be converted to gasoline via a methanol-to-gasoline process by catalytic conversion of syngas to methanol. Methanol has recently been produced from biomass using a biomass gasification process. In 2012, 200,000 tons of biomethanol was produced from biomass [209]. Biogas is also able to convert into methanol through processes such as oxidation of methane, photocatalytic conversion, biological conversion, and an indirect pathway, which reforms biogas and then methanol synthesis occurs. Methanol production has been reviewed by Park et al. [210]. Partial oxidation of methane is a well-established and the most widely used pathway to convert biogas to methanol. The process requires high pressure, which is between 0.5–15 MPa, or high temperature, which is between 700 and 750 °C, to reach high yield [58]. Hydroxyl radicals are used to produce methanol with a photocatalytic conversion pathway. Hydroxyl radicals react to CH_4, and methyl radicals are produced. When these radicals and water molecules react, methanol and hydrogen are produced as a final product [211]. With a biological pathway, the bacteria use methane and ammonia to produce methanol [211]. Recently, Su et al. [212] achieved 0.33 g/L of methanol production with a biological methanol production pathway and a methanol conversion ratio of 0.47 mol methanol/mol methane.

Bioethanol and higher alcohols produced from biogas is a relatively new topic and has gained attention due to its energy density and low carbon and particulate matter emissions. Ethanol blend fuels have been used in the US; however, direct bioethanol and higher alcohols produced from biogas are still under research and development. With the catalytic approach patented by Exxon, CH_4 is able to convert into ethanol and methanol directly [58]. For this method, $Fe_2(SO_4)_3$ or $Fe(ClO_4)_3$ were used to produce ethanol and methanol and different noble materials have been investigated for this method [58]. However, an availability date is not contained in the literature.

Another method to produce ethanol and higher alcohols is a two-step approach. The first one is to reform biogas to syngas, and the next step is to convert syngas into ethanol and higher alcohols. The second step can take place in biological or catalytic conversion. With biological conversion, bacteria use the acetyl-CoA path. The reaction depends on the H_2 to CH_4 ratio. This method is a promising way to produce ethanol and higher alcohols in an ambient environment with no potential metal poisoning [58]. The yield of ethanol production from syngas was reported at 48 g/L [213]. With the catalytic conversion method, the reaction is an alternative way to produce ethanol and higher alcohols. The operating conditions require pressure and temperature of around 20 MPa and 300–400 °C, respectively [58].

5. Conclusion and Future Trend of Biogas

Biogas production from anaerobic digestion processes has gained great attention during the last two decades because of its many positive benefits, which include it being a low energy-demand process, it generating relatively low sludge production for disposal, and it being a form of renewable energy. Additionally, this process is key to make the best use of wastes and to obtain energy. The production of biogas with a well-optimized system is not the only difficulty, but upgrading the biogas quality is also necessary for it to meet certain requirements so it can be used instead of fossil fuels. Cleaning and upgrading technologies are currently being heavily studied. However, there is a lack of information available such as for commercial-scale applications including economic and environmental information. The cleaning and upgrading processes require high energy use and the input of additional chemicals. According to the results of this survey, the amine-scrubbing method appears to be more beneficial than other methods because of lower methane losses, lower energy demand, and higher final methane content.

Additionally, the removal of CO_2, which is the second main product of biogas, and the removal of H_2S, has been investigated with a novel approach using microalgae. Although it has some issues that need to be overcome such as O_2 concentration during upgrading, it could be commonly used soon in commercial applications. However, biological methods are relatively new compared to other techniques but appear to be advantageous in terms of overall feasibility and advantages. The development of biological upgrade technologies may occur quickly once the identified challenges are overcome.

Another promising upgrading method is the membrane technique, due to economic and environmental concerns. In the literature, researchers have generally focused on reducing methane losses; however, the membrane material compatibility and improvement of resistance with different impurities in raw biogas should be be investigated in future research. New membrane materials and theories are still under development and are being tested in labs. The pilot systems should be applied and would fill the gap between lab-scale and large-scale systems. Additionally, in situ production and upgrading reactor design should be studied more to gain additional information and to be able to apply them to large-scale systems.

Biogas can be used as an energy source in different kinds of applications such as heat, vehicle fuels, gas-grid injection, biochemical production, and higher alcohols; however, the cleaning and upgrading process is necessary for methane purification. The maximum energy could be obtained from biogas using a combined heat and power system. Moreover, there is a notable finding for internal combustion engines when adding nanoparticles which both reduces the harmful exhaust emissions and improves the engine performance when the engine is fueled with biogas under dual-fuel mode. Researchers should focus on this area to reveal the benefits and drawbacks.

Overall, biogas production and utilization are a great and promising pathway to reduce mankind's negative impacts on the environment and may reduce dependence on fossil fuels. Policymakers have to think about the underutilized significant potential of biogas production and increase their support for biogas utilization. With their support and researchers' efforts, biogas can be a significant solution towards a reduction in greenhouse gas emissions, the production of renewable energy, and the management of waste disposal.

Author Contributions: Writing—original draft preparation, M.R.A., H.S., M.D., T.A.H. and D.K.; writing—review and editing A.A., M.R.A., C.E., H.M., S.U., S.K., N.A., H.D.K.; supervision A.A., C.E., N.A., H.D.K. All authors have read and agreed to the published version of the manuscript.

Funding: This research received no external funding.

Conflicts of Interest: All the authors declare no conflict of interest.

Nomenclature

BTE	Brake thermal efficiency		$HClO_3$	Chloric acid
CH_4	Methane		LNG	Liquefied renewable natural gas
CHP	Combined heat and power		TSA	Temperature swing adsorption
CI	Combustion ignition		H_2SO_4	Sulphuric acid
Cl_2O	Dichlorine monoxide		PSA	Pressure swing adsorption
CO_2	Carbon dioxide		Cl_2	Chlorine
H_2	Hydrogen		EBA	European biomass association
H_2S	Hydrogen sulfide		HClO	Hypochlorous acid
HCCI	Homogeneous charge compression ignition		CA	Cellulose acetate
HPWS	High-pressure water washing		ESA	Electrical swing adsorption
N_2	Nitrogen		CNG	Compressed renewable natural gas
N_2O	Nitrous oxide		HF	Hydrofluoric acid
$NaNO_3$	Sodium nitrate		HCl	Hydrochloric acid
NO_3^-	Nitrate ions		CO	Carbon monoxide
O_2	Oxygen		SO_2	Sulfur dioxide
PM	Particulate matter		NaOH	Sodium hydroxide
SI	Spark ignition		PEMFCs	Proton exchange membrane fuel cells
SOFCs	Solid oxide fuel cells		MCFCs	Molten carbonate fuel cells
VFAs	Volatile fatty acids		WWTP	Wastewater treatment plants
VOCs	Volatile organic compounds		VMS	Volatile methyl siloxanes
VSC	Volatile sulfur compounds		VOCs	Volatile organic compounds
$Ca(OH)_2$	Calcium hydroxide			

References

1. D'Adamo, I.; Morone, P.; Huisingh, D. Bioenergy: A Sustainable Shift. *Energies* **2021**, *14*, 5661. [CrossRef]
2. Busu, M. Assessment of the Impact of Bioenergy on Sustainable Economic Development. *Energies* **2019**, *12*, 578. [CrossRef]
3. Alsaleh, M.; Abdulwakil, M.; Abdul-Rahim, A. Does Social Businesses Development Affect Bioenergy Industry Growth under the Pathway of Sustainable Development? *Sustainability* **2021**, *13*, 1989. [CrossRef]
4. Pehlken, A.; Wulf, K.; Grecksch, K.; Klenke, T.; Tsydenova, N. More Sustainable Bioenergy by Making Use of Regional Alternative Biomass? *Sustainability* **2020**, *12*, 7849. [CrossRef]
5. Fu, S.; Angelidaki, I.; Zhang, Y. In situ Biogas Upgrading by CO_2 -to- CH_4 Bioconversion. *Trends Biotechnol.* **2020**, *39*, 336–347. [CrossRef] [PubMed]
6. Buffiere, P.; Germain, R.B.P. Freins et Développements de la Filière Biogaz: Les Besoins en Recherche et Développement. Available online: https://inis.iaea.org/collection/NCLCollectionStore/_Public/47/106/47106366.pdf (accessed on 12 October 2021).

7. Atelge, M.R.; Krisa, D.; Kumar, G.; Eskicioglu, C.; Nguyen, D.D.; Chang, S.W.; Atabani, A.E.; Al-Muhtaseb, A.H.; Unalan, S. Biogas Production from Organic Waste: Recent Progress and Perspectives. *Waste Biomass Valorization* **2020**, *11*, 1019–1040. [CrossRef]

8. Deublein, D.; Steinhauser, A. *Biogas from Waste and Renewable Resources*; Wiley: Hoboken, NJ, USA, 2008.

9. Boulinguiez, B.; Le cloirec, P. Purification de biogaz. Élimination des COV et des siloxanes. *Tech. L'ingénieur* **2011**, *33*, 1–26.

10. Demirbas, A.; Taylan, O.; Kaya, D. Biogas production from municipal sewage sludge (MSS). *Energy Sources Part A Recovery Util. Environ. Eff.* **2016**, *38*, 3027–3033. [CrossRef]

11. Atelge, M.R.; Atabani, A.E.; Banu, J.R.; Krisa, D.; Kaya, M.; Eskicioglu, C.; Kumar, G.; Lee, C.; Yildiz, Y.Ş.; Unalan, S.; et al. A critical review of pretreatment technologies to enhance anaerobic digestion and energy recovery. *Fuel* **2020**, *270*, 117494. [CrossRef]

12. Baena-Moreno, F.M.; Rodríguez-Galán, M.; Vega, F.; Vilches, L.F.; Navarrete, B. Review: Recent advances in biogas purifying technologies. *Int. J. Green Energy* **2019**, *16*, 401–412. [CrossRef]

13. Wellinger, A.; Jerry, M.; Baxter, D. *The Biogas Handbook: Science, Production and Applications*; Elsevier: Amsterdam, The Netherlands, 2013; pp. 1–476.

14. Gerardi, M.H. The microbiology of anaerobic digesters. In *Wastewater Microbiology Series A*; John Wiley & Sons. Inc.: Piscataway, NJ, USA, 2003.

15. Paolini, V.; Petracchini, F.; Carnevale, M.; Gallucci, F.; Perilli, M.; Esposito, G.; Segreto, M.; Occulti, L.G.; Scaglione, D.; Ianniello, A.; et al. Characterisation and cleaning of biogas from sewage sludge for biomethane production. *J. Environ. Manag.* **2018**, *217*, 288–296. [CrossRef]

16. Kapoor, R.; Ghosh, P.; Tyagi, B.; Vijay, V.K.; Vijay, V.; Thakur, I.S.; Kamyab, H.; Nguyen, D.D.; Kumar, A. Advances in biogas valorization and utilization systems: A comprehensive review. *J. Clean. Prod.* **2020**, *273*, 123052. [CrossRef]

17. Kapoor, R.; Ghosh, P.; Kumar, M.; Vijay, V.K. Evaluation of biogas upgrading technologies and future perspectives: A review. *Environ. Sci. Pollut. Res.* **2019**, *26*, 11631–11661. [CrossRef]

18. Zheng, L.; Cheng, S.; Han, Y.; Wang, M.; Xiang, Y.; Guo, J.; Cai, D.; Mang, H.P.; Dong, T.; Li, Z.; et al. Bio-natural gas industry in China: Current status and development. *Renew. Sustain. Energy Rev.* **2020**, *128*, 109925. [CrossRef]

19. Angelidaki, I.; Treu, L.; Tsapekos, P.; Luo, G.; Campanaro, S.; Wenzel, H.; Kougias, P.G. Biogas upgrading and utilization: Current status and perspectives. *Biotechnol. Adv.* **2018**, *36*, 452–466. [CrossRef] [PubMed]

20. Deng, L.; Liu, Y.; Wang, W. Biogas Cleaning and Upgrading. In *Biogas Technology*; Deng, L., Liu, Y., Wang,, W., Eds.; Springer: Singapore, 2020; pp. 201–243.

21. Ryckebosch, E.; Drouillon, M.; Vervaeren, H. Techniques for transformation of biogas to biomethane. *Biomass Bioenergy* **2011**, *35*, 1633–1645. [CrossRef]

22. Garnaik, P.P.; Pradhan, R.R.; Chiang, Y.W.; Dutta, A. Biomass-Based CO_2 Adsorbents for Biogas Upgradation with Pressure Swing Adsorption. In *Climate Change and Green Chemistry of CO_2 Sequestration, Green Energy and Technology*; Goel, M., Satyanarayana, T., Sudhakar, M., Agrawal, D.P., Eds.; Springer: Singapore, 2021.

23. Struk, M.; Kushkevych, I.; Vítězová, M. Biogas upgrading methods: Recent advancements and emerging technologies. *Rev. Environ. Sci. Biotechnol.* **2020**, *19*, 651–671. [CrossRef]

24. Awe, O.W.; Zhao, Y.; Nzihou, A.; Minh, D.P.; Lyczko, N. A Review of Biogas Utilisation, Purification and Upgrading Technologies Review. *Waste Biomass Valorization* **2017**, *8*, 267–283. [CrossRef]

25. Savickis, J.; Zemite, L.; Zeltins, N.; Bode, I.; Jansons, L.; Dzelzitis, E.; Koposovs, A.; Selickis, A.; Ansone, A. The biomethane injection into the natural gas networks: The EU's gas synergy path. *Latv. J. Phys. Tech. Sci.* **2020**, *57*, 34–50. [CrossRef]

26. Hakawati, R.; Smyth, B.M.; Rosa, F.D.; Rooney, D. What is the most energy efficient route for biogas utilization: Heat, electricity or transport ? *Appl. Energy* **2017**, *206*, 1076–1087. [CrossRef]

27. Chang, I.S.; Zhao, J.; Yin, X.; Wu, J.; Jia, Z.; Wang, L. Comprehensive utilizations of biogas in Inner Mongolia, China. *Renew. Sustain. Energy Rev.* **2011**, *15*, 1442–1453. [CrossRef]

28. Deng, L.; Liu, Y.; Wang, W. Biogas Utilization. In *Biogas Technology*; Springer: Singapore, 2020; Volume 8, pp. 269–318.

29. Sun, Q.; Li, H.; Yan, J.; Liu, L.; Yu, Z.; Yu, X. Selection of appropriate biogas upgrading technology-a review of biogas cleaning, upgrading and utilisation. *Renew. Sustain. Energy Rev.* **2015**, *51*, 521–532. [CrossRef]

30. IRENA. *Renewable Energy Statistics 2019*; The International Renewable Energy Agency: Abu Dhabi, United Arab Emirates, 2019.

31. IRENA. *Renewable Energy Statistics 2021*; The International Renewable Energy Agency: Abu Dhabi, United Arab Emirates, 2021.

32. Igliński, B.; Piechota, G.; Iwański, P.; Skarzatek, M.; Pilarski, G. 15 Years of the Polish agricultural biogas plants: Their history, current status, biogas potential and perspectives. *Clean Technol. Environ. Policy* **2020**, *22*, 281–307. [CrossRef]

33. Schmid, C.; Horschig, T.; Pfeiffer, A.; Szarka, N.; Thrän, D. Schmid; Horschig; Pfeiffer; Szarka; Thrän, Biogas Upgrading: A Review of National Biomethane Strategies and Support Policies in Selected Countries. *Energies* **2019**, *12*, 3803. [CrossRef]

34. D'Adamo, I.; Falcone, P.M.; Huisingh, D.; Morone, P. A circular economy model based on biomethane: What are the opportunities for the municipality of Rome and beyond? *Renew. Energy* **2021**, *163*, 1660–1672. [CrossRef]

35. Herbes, C. *Marketing of Biomethane*; Springer: Cham, Switzerland, 2017; pp. 151–171.

36. Rodenhizer, J.S.; Boardman, G.D. Collection, Analysis, and Utilization of Biogas from Anaerobic Treatment of Crab Processing Waters. *J. Aquat. Food Prod. Technol.* **1999**, *8*, 59–75. [CrossRef]

37. Chandra, R.; Isha, A.; Kumar, S.; Khan, S.A.; Subbarao, P.M.V.; Vijay, V.K.; Chandel, A.K.; Chaudhary, V.P. Potentials and Challenges of Biogas Upgradation as Liquid Biomethane. In *Biogas Production*; Balagurusamy, N., Chandel, A.K., Eds.; Springer: Cham, Switzerland, 2020; pp. 307–328.

38. Atelge, M.R.; Atabani, A.E.; Abut, S.; Kaya, M.; Eskicioglu, C.; Semaan, G.; Lee, C.; Yildiz, Y.Ş.; Unalan, S.; Mohanasundaram, R.; et al. Anaerobic co-digestion of oil-extracted spent coffee grounds with various wastes: Experimental and kinetic modeling studies. *Bioresour. Technol.* **2021**, *322*, 124470. [CrossRef] [PubMed]

39. Lahbab, A.; Djaafri, M.; Kalloum, S.; Benatiallah, A.; Atelge, M.R.; Atabani, A.E. Co-digestion of vegetable peel with cow dung without external inoculum for biogas production: Experimental and a new modelling test in a batch mode. *Fuel* **2021**, *306*, 121627. [CrossRef]

40. Bora, D.; Barbora, L.; Borah, A.J.; Mahanta, P. A Comparative Assessment of Biogas Upgradation Techniques and Its Utilization as an Alternative Fuel in Internal Combustion Engines. In *Alternative Fuels and Advanced Combustion Techniques as Sustainable Solutions for Internal Combustion Engines. Energy, Environment, and Sustainability*; Singh, A.P., Kumar, D., Agarwal, A.K., Eds.; Springer: Singapore, 2021.

41. Braun, R. *Anaerobic digestion: A multi-faceted process for energy, environmental management and rural development. Improvement of Crop Plants for Industrial End Uses*; Springer: Dordrecht, The Netherlands, 2007; pp. 335–416.

42. Deublein, D.; Steinhauser, A. *Biogas from Waste and Renewable Resources: An Introduction*, 2nd ed.; John Wiley & Sons: Piscataway, NJ, USA, 2011.

43. Hassaan, M.A.; El Nemr, A.; Elkatory, M.R.; Ragab, S.; El-Nemr, M.A.; Pantaleo, A. Synthesis, Characterization, and Synergistic Effects of Modified Biochar in Combination with α-Fe2O3 NPs on Biogas Production from Red Algae Pterocladia capillacea. *Sustainability* **2021**, *13*, 9275. [CrossRef]

44. Felipe, I.; Barros, R.M.; Lucio, G.; Filho, T. Electricity generation from biogas of anaerobic wastewater treatment plants in Brazil: An assessment of feasibility and potential. *J. Clean. Prod.* **2016**, *126*, 504–514.

45. Kumara Behera, B.; Varma, A. Chapter 2: Biomethanization. In *Microbial Resources for Sustainable Energy*; Springer: Cham, Switzerland, 2016; pp. 35–122.

46. Yentekakis, I.V.; Goula, G. Biogas Management: Advanced Utilization for Production of Renewable Energy and Added-value Chemicals. *Front. Environ. Sci.* **2017**, *5*, 7. [CrossRef]

47. Chen, X.; Yan, W.; Sheng, K.; Sanati, M. Comparison of high-solids to liquid anaerobic co-digestion of food waste and green waste. *Bioresour. Technol.* **2014**, *154*, 215–221. [CrossRef] [PubMed]

48. Devia, C.R. Biogaz en Vue de son Utilisation en Production D'energie: Séparation des Siloxanes et du Sulfure D'hydrogene. Ph.D. Thesis, Nantes School of Mines, Nantes, France, 2013; pp. 1–202.

49. Timmerberg, S.; Dieckmann, C.; Mackenthun, R.; Kaltschmitt, M. Biomethane in Transportation Sector. In *Encyclopedia of Sustainability Science and Technology*; Meyers, R.A., Ed.; Springer: New York, NY, USA, 2018.

50. Marchaim, U. *Les Procédés de Production de Biogaz Pour le Développement de Technologies Durables*; FAO: Rome, Italy, 1994.

51. Wang, H.; Larson, R.A.; Runge, T. Impacts to hydrogen sulfide concentrations in biogas when poplar wood chips, steam treated wood chips, and biochar are added to manure-based anaerobic digestion systems. *Bioresour. Technol. Rep.* **2019**, *7*, 100232. [CrossRef]

52. Ali, M.; Hassan, W.; Haj, M.E.; Allagui, A.; Cha, S.W. On the technical challenges affecting the performance of direct internal reforming biogas solid oxide fuel cells. *Renew. Sustain. Energy Rev.* **2019**, *101*, 361–375.

53. Wasajja, H.; Lindeboom, R.E.F.; Lier, J.B.V.; Aravind, P.V. Techno-economic review of biogas cleaning technologies for small scale off grid solid oxide fuel cell applications. *Fuel Process. Technol.* **2020**, *197*, 106215. [CrossRef]

54. Abd, A.A.; Othman, M.R.; Naji, S.Z.; Hashim, A.S. Methane enrichment in biogas mixture using pressure swing adsorption: Process fundamental and design parameters. *Mater. Today Sustain.* **2021**, *11*, 100063. [CrossRef]

55. Atelge, R. Türkiye'de Sığır Gübresinden Biyoyakıt Olarak Biyogaz Üretiminin Potansiyeli ve 2030 ve 2053 Yıllarında Karbon Emisyonlarının Azaltılmasına Öngörülen Etkisi. *Int. J. Innov. Eng. Appl.* **2021**, *5*, 56–64.

56. Rafiee, A.; Khalilpour, K.R.; Prest, J.; Skryabin, I. Biogas as an energy vector. *Biomass and Bioenergy* **2021**, *144*, 105935. [CrossRef]

57. Shen, M.; Zhang, Y.; Hu, D.; Fan, J.; Zeng, G. A review on removal of siloxanes from biogas: With a special focus on volatile methylsiloxanes. *Environ. Sci. Pollut. Res.* **2018**, *25*, 30847–30862. [CrossRef]

58. Yang, L.; Ge, X.; Wan, C.; Yu, F.; Li, Y. Progress and perspectives in converting biogas to transportation fuels. *Renew. Sustain. Energy Rev.* **2014**, *40*, 1133–1152. [CrossRef]

59. Ghidotti, M.; Fabbri, D.; Torri, C. Determination of linear and cyclic volatile methyl siloxanes in biogas and biomethane by solid-phase microextraction and gas chromatography-mass spectrometry. *Talanta* **2019**, *195*, 258–264. [CrossRef]

60. Wang, N.; Tan, L.; Xie, L.; Wang, Y.; Ellis, T. Investigation of volatile methyl siloxanes in biogas and the ambient environment in a landfill. *J. Environ. Sci.* **2020**, *91*, 54–61. [CrossRef] [PubMed]

61. Eichler, C.M.A.; Wu, Y.; Cox, S.S.; Klaus, S.; Boardman, G.D. Evaluation of sampling techniques for gas-phase siloxanes in biogas. *Biomass Bioenergy* **2018**, *108*, 1–6. [CrossRef]

62. Piechota, G. Removal of siloxanes from biogas upgraded to biomethane by Cryogenic Temperature Condensation System. *J. Clean. Prod.* **2021**, *308*, 127404. [CrossRef]

63. Tansel, B.; Surita, S.C. Managing siloxanes in biogas-to-energy facilities: Economic comparison of pre- vs post-combustion practices. *Waste Manag.* **2019**, *96*, 121–127. [CrossRef]

64. Piechota, G. Siloxanes in Biogas: Approaches of Sampling Procedure and GC-MS Method Determination. *Molecules* **2021**, *26*, 1953. [CrossRef]
65. Liu, Y.H.; You, Z.; Ji, M.; Wang, Y.; Fei, Y.; Zi, D.; Ma, C. Removal of siloxanes from biogas using acetylated silica gel as adsorbent. *Pet. Sci.* **2019**, *16*, 920–928. [CrossRef]
66. Rasi, S.Ã.; Veijanen, A.; Rintala, J. Trace compounds of biogas from different biogas production plants. *Energy* **2007**, *32*, 1375–1380. [CrossRef]
67. He, Q.; Tu, T.; Yan, S.; Yang, X.; Duke, M.; Zhang, Y. Relating water vapor transfer to ammonia recovery from biogas slurry by vacuum membrane distillation. *Sep. Purif. Technol.* **2018**, *191*, 182–191. [CrossRef]
68. Li, Z.; Wachemo, A.C.; Yuan, H.; Mustafa, R.; Li, X. High levels of ammonia nitrogen for biological biogas upgrading. *Int. J. Hydrog. Energy* **2020**, *45*, 28488–28498. [CrossRef]
69. Koonaphapdeelertm, S.; Aggarangsi, P.; Moran, J. Biogas Cleaning and Pretreatment. In *Biomethane, Green Energy and Technology*; Springer: Singapore, 2020.
70. Porpatham, E.; Ramesh, A.; Nagalingam, B. Experimental studies on the effects of enhancing the concentration of oxygen in the inducted charge of a biogas fuelled spark ignition engine. *Energy* **2018**, *142*, 303–312. [CrossRef]
71. Lohmann, H.; Krassowski, J.; Salazar, J.I.G. Determination of volatile organic compounds from biowaste and co- fermentation biogas plants by single-sorbent adsorption. *Chemosphere* **2016**, *153*, 48–57.
72. Papadias, D.D.; Ahmed, S.; Kumar, R. Fuel quality issues with biogas energy—An economic analysis for a stationary fuel cell system. *Energy* **2012**, *44*, 257–277. [CrossRef]
73. Randazzo, A.; Asensio-ramos, M.; Melián, G.V.; Venturi, S.; Padrón, E.; Hernández, P.A. Volatile organic compounds (VOCs) in solid waste land fill cover soil: Chemical and isotopic composition vs. degradation processes. *Sci. Total Environ.* **2020**, *726*, 138326. [CrossRef] [PubMed]
74. Mahmoodi, R.; Yari, M.; Ghafouri, J.; Poorghasemi, K. Effect of reformed biogas as a low reactivity fuel on performance and emissions of a RCCI engine with reformed biogas/diesel dual-fuel combustion. *Int. J. Hydrog. Energy* **2021**, *46*, 16494–16512. [CrossRef]
75. Ullah, I.; Ha, M.; Othman, D.; Hashim, H.; Matsuura, T.; Ismail, A.F.; Rezaei-dashtarzhandi, M.; Azelee, I.W. Biogas as a renewable energy fuel – A review of biogas upgrading, utilisation and storage. *Energy Convers. Manag.* **2017**, *150*, 277–294. [CrossRef]
76. Muñoz, R.; Meier, L.; Diaz, I.; Jeison, D. A review on the state-of-the-art of physical/chemical and biological technologies for biogas upgrading. *Rev. Environ. Sci. Bio Technol.* **2015**, *14*, 727–759. [CrossRef]
77. Miltner, M.; Makaruk, A.; Harasek, M. Review on available biogas upgrading technologies and innovations towards advanced solutions. *J. Clean. Prod.* **2017**, *161*, 1329–1337. [CrossRef]
78. Niesner, J.; Jecha, D.; Stehlík, P. Biogas upgrading technologies: State of art review in European region. *Chem. Eng. Trans.* **2013**, *35*, 517–522.
79. Bauer, F.; Persson, T.; Hulteberg, C.; Tamm, D. Biogas upgrading – technology overview, comparison and perspectives for the future. *Biofuels Bioprod. Biorefining* **2013**, *7*, 499–511. [CrossRef]
80. Privalova, E.; Rasi, S.; Mäki-Arvela, P.; Eränen, K.; Rintala, J.; Murzin, D.Y.; Mikkola, J.P. CO_2 capture from biogas: Absorbent selection. *RSC Adv.* **2013**, *3*, 2979–2994. [CrossRef]
81. Lasocki, J.; Kołodziejczyk, K.; Matuszewska, A. Laboratory-Scale Investigation of Biogas Treatment by Removal of Hydrogen Sulfide and Carbon Dioxide. *Pol. J. Environ. Stud.* **2015**, *24*, 1427–1434. [CrossRef]
82. Voice, A.K.; Closmann, F.; Rochelle, G.T. Oxidative Degradation of Amines With High-Temperature Cycling. *Energy Procedia* **2013**, *37*, 2118–2132. [CrossRef]
83. Kaya, M.; Atelge, M.R.; Bekirogullari, M.; Eskicioglu, C.; Atabani, A.E.; Kumar, G.; Yildiz, Y.S.; Unalan, S. Carbon molecular sieve production from defatted spent coffee ground using $ZnCl_2$ and benzene for gas purification. *Fuel* **2020**, *277*, 118183. [CrossRef]
84. Grande, C.A.; Ribeiro, R.P.L.; Oliveira, E.L.G.; Rodrigues, A.E. Electric swing adsorption as emerging CO_2 capture technique. *Energy Procedia* **2009**, *1*, 1219–1225. [CrossRef]
85. Mason, J.A.; Sumida, K.; Herm, Z.R.; Krishna, R.; Long, J.R. Evaluating metal–organic frameworks for post-combustion carbon dioxide capture via temperature swing adsorption. *Energy Environ. Sci.* **2011**, *4*, 3030–3040. [CrossRef]
86. Yamamoto, T.; Endo, A.; Ohmori, T.; Nakaiwa, M. Porous properties of carbon gel microspheres as adsorbents for gas separation. *Carbon* **2004**, *42*, 1671–1676. [CrossRef]
87. Zhao, Q.; Wu, F.; He, Y.; Xiao, P.; Webley, P.A. Impact of operating parameters on CO_2 capture using carbon monolith by Electrical Swing Adsorption technology (ESA). *Chem. Eng. J.* **2017**, *327*, 441–453. [CrossRef]
88. Scholz, M.; Melin, T.; Wessling, M. Transforming biogas into biomethane using membrane technology. *Renew. Sustain. Energy Rev.* **2013**, *17*, 199–212. [CrossRef]
89. Makaruk, A.; Miltner, M.; Harasek, M. Membrane biogas upgrading processes for the production of natural gas substitute. *Sep. Purif. Technol.* **2010**, *74*, 83–92. [CrossRef]
90. Haider, S.; Lindbråthen, A.; Hägg, M.-B. Techno-economical evaluation of membrane based biogas upgrading system: A comparison between polymeric membrane and carbon membrane technology. *Green Energy Environ.* **2016**, *1*, 222–234. [CrossRef]
91. Bos, A.; Pünt, I.G.M.; Wessling, M.; Strathmann, H. CO_2-induced plasticization phenomena in glassy polymers. *J. Membr. Sci.* **1999**, *155*, 67–78. [CrossRef]

92. Miltner, M.; Makaruk, A.; Harasek, M. Selected Methods of Advanced Biogas Upgrading. *Chem. Eng. Trans.* **2016**, *52*, 463–468.

93. Barboiu, C.; Sala, B.; Bec, S.; Pavan, S.; Petit, E.; Colomban, P.; Sanchez, J.; de Perthuis, S.; Hittner, D. Structural and mechanical characterizations of microporous silica–boron membranes for gas separation. *J. Membr. Sci.* **2009**, *326*, 514–525. [CrossRef]

94. Andriani, D.; Wresta, A.; Atmaja, T.D.; Saepudin, A. A Review on Optimization Production and Upgrading Biogas Through CO_2 Removal Using Various Techniques. *Appl. Biochem. Biotechnol.* **2014**, *172*, 1909–1928. [CrossRef] [PubMed]

95. Jürgensen, L.; Ehimen, E.A.; Born, J.; Holm-Nielsen, J.B. Utilization of surplus electricity from wind power for dynamic biogas upgrading: Northern Germany case study. *Biomass Bioenergy* **2014**, *66*, 126–132. [CrossRef]

96. Luo, G.; Johansson, S.; Boe, K.; Xie, L.; Zhou, Q.; Angelidaki, I. Simultaneous hydrogen utilization and in situ biogas upgrading in an anaerobic reactor. *Biotechnol. Bioeng.* **2012**, *109*, 1088–1094. [CrossRef]

97. Kougias, P.G.; Treu, L.; Benavente, D.P.; Boe, K.; Campanaro, S.; Angelidaki, I. Ex-situ biogas upgrading and enhancement in different reactor systems. *Bioresour. Technol.* **2017**, *225*, 429–437. [CrossRef]

98. Luo, G.; Angelidaki, I. Co-digestion of manure and whey for in situ biogas upgrading by the addition of H_2: Process performance and microbial insights. *Appl. Microbiol. Biotechnol.* **2013**, *97*, 1373–1381. [CrossRef]

99. Bassani, I.; Kougias, P.G.; Angelidaki, I. In-situ biogas upgrading in thermophilic granular UASB reactor: Key factors affecting the hydrogen mass transfer rate. *Bioresour. Technol.* **2016**, *221*, 485–491. [CrossRef]

100. Savvas, S.; Donnelly, J.; Patterson, T.; Chong, Z.S.; Esteves, S.R. Biological methanation of CO_2 in a novel biofilm plug-flow reactor: A high rate and low parasitic energy process. *Appl. Energy* **2017**, *202*, 238–247. [CrossRef]

101. Meier, L.; Pérez, R.; Azócar, L.; Rivas, M.; Jeison, D. Photosynthetic CO_2 uptake by microalgae: An attractive tool for biogas upgrading. *Biomass Bioenergy* **2015**, *73*, 102–109. [CrossRef]

102. Mann, G.; Schlegel, M.; Schumann, R.; Sakalauskas, A. Biogas-conditioning with microalgae. *Agron. Res.* **2009**, *7*, 33–38.

103. Chi, Z.; O'Fallon, J.V.; Chen, S. Bicarbonate produced from carbon capture for algae culture. *Trends Biotechnol.* **2011**, *29*, 537–541. [CrossRef] [PubMed]

104. Xia, A.; Herrmann, C.; Murphy, J.D. How do we optimize third-generation algal biofuels? *Biofuels Bioprod. Biorefining* **2015**, *9*, 358–367. [CrossRef]

105. Miltner, M.; Makaruk, A.; Krischan, J.; Harasek, M. Chemical-oxidative scrubbing for the removal of hydrogen sulphide from raw biogas: Potentials and economics. *Water Sci. Technol.* **2012**, *66*, 1354–1360. [CrossRef] [PubMed]

106. Petersson, A. *Biogas Cleaning*; Elsevier: Amsterdam, The Netherlands, 2013; pp. 329–341.

107. Persson, M.; Jonsson, O.; Wellinger, A. Biogas Upgrading To Vehicle Fuel Standards and Grid. *IEA Bioenergy* **2006**, *37*, 1–32.

108. Maia, D.S.; Cardozo, F.; Frare, L.; Gimenes, M.; Pereira, N. Purification of biogas for energy use. *Chem. Eng. Trans.* **2014**, *37*, 643–648.

109. Petersson, A.; Wellinger, A. Biogas upgrading technologies–developments and innovations. *IEA Bioenergy* **2009**, *20*, 1–19.

110. Tippayawong, N.; Thanompongchart, P. Biogas quality upgrade by simultaneous removal of CO_2 and H_2S in a packed column reactor. *Energy* **2010**, *35*, 4531–4535. [CrossRef]

111. Harasimowicz, M.; Orluk, P.; Zakrzewska-Trznadel, G.; Chmielewski, A. Application of polyimide membranes for biogas purification and enrichment. *J. Hazard. Mater.* **2007**, *144*, 698–702. [CrossRef]

112. Makaruk, A.; Miltner, M.; Harasek, M. Biogas desulfurization and biogas upgrading using a hybrid membrane system – modeling study. *Water Sci. Technol.* **2013**, *67*, 326–332. [CrossRef]

113. Khoshnevisan, B.; Tsapekos, P.; Alvarado-Morales, M.; Rafiee, S.; Tabatabaei, M.; Angelidaki, I. Life cycle assessment of different strategies for energy and nutrient recovery from source sorted organic fraction of household waste. *J. Clean. Prod.* **2018**, *180*, 360–374. [CrossRef]

114. Lombardi, L.; Francini, G. Techno-economic and environmental assessment of the main biogas upgrading technologies. *Renew. Energy* **2020**, *156*, 440–458. [CrossRef]

115. Allegue, L.B.; Hinge, J.; Allé, K. *Biogas and Bio-Syngas Upgrading*; Danish Technological Institute: Taastrup, Denmark, 2012; pp. 5–97.

116. Allegue, L.B.; Hinge, J. *Biogas Upgrading Evaluation of Methods for H_2S Removal*; Danish Technological Institute: Taastrup, Denmark, 2014; p. 31.

117. Montebello, A.M. Aerobic biotrickling filtration for Andrea Monzón Montebello. *J. Hazard. Mater* **2013**, *280*, 200–208. [CrossRef] [PubMed]

118. Schweigkofler, M.; Niessner, R. Removal of siloxanes in biogases. *J. Hazard. Mater.* **2001**, *83*, 183–196. [CrossRef]

119. Wheless, E.; Pierce, J. Siloxanes in Landfill and Digester Gas Update. In *Proceedings of the 27th SWANA Landfill Gas Symposium*; Silver Springs: San Antonio, TX, USA, 2004.

120. López, M.E.; Rene, E.R.; Veiga, M.C.; Kennes, C. Biogas Technologies and Cleaning Techniques. In *Environmental Chemistry for a Sustainable World: Volume 2: Remediation of Air and Water Pollution*; Lichtfouse, E., Schwarzbauer, J., Robert, D., Eds.; Springer: Dordrecht, The Netherlands, 2012; pp. 347–377.

121. Ruiling, G.; Shikun, C.; Zifu, L. Research progress of siloxane removal from biogas. *Int. J. Agric. Biol. Eng.* **2017**, *10*, 30–39.

122. Favre, E.; Bounaceur, R.; Roizard, D. Biogas, membranes and carbon dioxide capture. *J. Membr. Sci.* **2009**, *328*, 11–14. [CrossRef]

123. Piechota, G. Multi-step biogas quality improving by adsorptive packed column system as application to biomethane upgrading. *J. Environ. Chem. Eng.* **2021**, *9*, 105944. [CrossRef]

124. Guo, X.; Tak, J.K.; Johnson, R.L. Ammonia removal from air stream and biogas by a H_2SO_4 impregnated adsorbent originating from waste wood-shavings and biosolids. *J. Hazard. Mater.* **2009**, *166*, 372–376.

125. Jiang, X.; Yan, R.; Tay, J.H. Simultaneous autotrophic biodegradation of H_2S and NH_3 in a biotrickling filter. *Chemosphere* **2009**, *75*, 1350–1355. [CrossRef] [PubMed]

126. Malhautier, L.; Gracian, C.; Roux, J.-C.; Fanlo, J.-L.; Le Cloirec, P. Biological treatment process of air loaded with an ammonia and hydrogen sulfide mixture. *Chemosphere* **2003**, *50*, 145–153. [CrossRef]

127. Baena-Moreno, F.M.; le Saché, E.; Pastor-Pérez, L.; Reina, T. Membrane-based technologies for biogas upgrading: A review. *Environ. Chem. Lett.* **2020**, *18*, 1649–1658. [CrossRef]

128. Norja, J. Value Creation by Automation in Biogas Upgrading. Master's Thesis, Tampere University, Tampere, Finland, 2020.

129. Collet, P.; Flottes, E.; Favre, A.; Raynal, L.; Pierre, H.; Capela, S.; Peregrina, C. Techno-economic and Life Cycle Assessment of methane production via biogas upgrading and power to gas technology. *Appl. Energy* **2017**, *192*, 282–295. [CrossRef]

130. Chen, X.Y.; Vinh-Thang, H.; Ramirez, A.A.; Rodrigue, D.; Kaliaguine, S. Membrane gas separation technologies for biogas upgrading. *RSC Adv.* **2015**, *5*, 24399–24448. [CrossRef]

131. Sahota, S.; Shah, G.; Ghosh, P.; Kapoor, R.; Sengupta, S.; Singh, P.; Vijay, V.; Sahay, A.; Vijay, V.K.; Thakur, I.S. Review of trends in biogas upgradation technologies and future perspectives. *Bioresour. Technol. Rep.* **2018**, *1*, 79–88. [CrossRef]

132. Hosseinipour, S.A.; Mehrpooya, M. Comparison of the biogas upgrading methods as a transportation fuel. *Renew. Energy* **2019**, *130*, 641–655. [CrossRef]

133. Singhal, S.; Agarwal, S.; Arora, S.; Sharma, P.; Singhal, N. Upgrading techniques for transformation of biogas to bio-CNG: A review. *Int. J. Energy Res.* **2017**, *41*, 1657–1669. [CrossRef]

134. Hahn, H.; Krautkremer, B.; Hartmann, K.; Wachendorf, M. Review of concepts for a demand-driven biogas supply for flexible power generation. *Renew. Sustain. Energy Rev.* **2014**, *29*, 383–393. [CrossRef]

135. Harasek, M. Biogas, biomethane and power-to-gas–new options to increase biomethane injection. In Proceedings of the EGATEC2015-European Gas Technology Conference, Vienna, Austria, 25–26 November 2015.

136. Ardolino, F.; Parrillo, F.; Arena, U. Biowaste-to-biomethane or biowaste-to-energy? An LCA study on anaerobic digestion of organic waste. *J. Clean. Prod.* **2018**, *174*, 462–476. [CrossRef]

137. Raboni, M.; Urbini, G. Production and use of biogas in Europe: A survey of current status and perspectives. *Ambiente E Agua Interdiscip. J. Appl. Sci.* **2014**, *9*, 445–458.

138. Holm-Nielsen, J.B.; Al Seadi, T.; Oleskowicz-Popiel, P. The future of anaerobic digestion and biogas utilization. *Bioresour. Technol.* **2009**, *100*, 5478–5484. [CrossRef] [PubMed]

139. O'Connor, S.; Ehimen, E.; Pillai, S.C.; Black, A.; Tormey, D.; Bartlett, J. Biogas production from small-scale anaerobic digestion plants on European farms. *Renew. Sustain. Energy Rev.* **2021**, *139*, 110580. [CrossRef]

140. Brémond, U.; Bertrandias, A.; Steyer, J.-P.; Bernet, N.; Carrere, H. A vision of European biogas sector development towards 2030: Trends and challenges. *J. Clean. Prod.* **2021**, *287*, 125065. [CrossRef]

141. EPA AgSTAR Data and Trends. Available online: https://www.epa.gov/agstar/agstar-data-and-trends#adfacts (accessed on 9 July 2021).

142. Hijazi, O.; Munro, S.; Zerhusen, B.; Effenberger, M. Review of life cycle assessment for biogas production in Europe. *Renew. Sustain. Energy Rev.* **2016**, *54*, 1291–1300. [CrossRef]

143. Budzianowski, W.M. A review of potential innovations for production, conditioning and utilization of biogas with multiple-criteria assessment. *Renew. Sustain. Energy Rev.* **2016**, *54*, 1148–1171. [CrossRef]

144. Krich, K.; Augenstein, D.; Batmale, J.; Benemann, J.; Rutledge, B.; Salour, D. *Biomethane from Dairy Waste: A Sourcebook for the Production and use of Renewable Natural Gas in California*; USDA Rural Development: Washington, DC, USA, 2005.

145. Kaparaju, P.; Rintala, J. Generation of heat and power from biogas for stationary applications: Boilers, gas engines and turbines, combined heat and power (CHP) plants and fuel cells. In *The Biogas Handbook Science, Production and Applications*; Elsevier: Amsterdam, The Netherlands, 2013; pp. 404–427.

146. Jatana, G.S.; Himabindu, M.; Thakur, H.S.; Ravikrishna, R.V. Strategies for high efficiency and stability in biogas-fuelled small engines. *Exp. Therm. Fluid Sci.* **2014**, *54*, 189–195. [CrossRef]

147. Saxena, S.; Bedoya, I.D. Fundamental phenomena affecting low temperature combustion and HCCI engines, high load limits and strategies for extending these limits. *Prog. Energy Combust. Sci.* **2013**, *39*, 457–488. [CrossRef]

148. Reitz, R.D.; Duraisamy, G. Review of high efficiency and clean reactivity controlled compression ignition (RCCI) combustion in internal combustion engines. *Prog. Energy Combust. Sci.* **2015**, *46*, 12–71. [CrossRef]

149. Swami Nathan, S.; Mallikarjuna, J.M.; Ramesh, A. An experimental study of the biogas–diesel HCCI mode of engine operation. *Energy Convers. Manag.* **2010**, *51*, 1347–1353. [CrossRef]

150. Bedoya, I.D.; Saxena, S.; Cadavid, F.J.; Dibble, R.W.; Wissink, M. Experimental study of biogas combustion in an HCCI engine for power generation with high indicated efficiency and ultra-low NOx emissions. *Energy Convers. Manag.* **2012**, *53*, 154–162. [CrossRef]

151. Blizman, B.J.; Makel, D.B.; Mack, J.H.; Dibble, R.W. Landfill Gas Fueled HCCI Demonstration System. In Proceedings of the ASME 2006 Internal Combustion Engine Division Fall Technical Conference, Sacramento, CA, USA, 5–8 November 2006; ASME: New York, NY, USA, 2006; pp. 327–347.

152. Hosseini, S.E.; Wahid, M.A. Biogas utilization: Experimental investigation on biogas flameless combustion in lab-scale furnace. *Energy Convers. Manag.* **2013**, *74*, 426–432. [CrossRef]

153. Kalsi, S.S.; Subramanian, K.A. Effect of simulated biogas on performance, combustion and emissions characteristics of a bio-diesel fueled diesel engine. *Renew. Energy* **2017**, *106*, 78–90. [CrossRef]

154. Renzi, M.; Brandoni, C. Study and application of a regenerative Stirling cogeneration device based on biomass combustion. *Appl. Therm. Eng.* **2014**, *67*, 341–351. [CrossRef]

155. Mathur, H.B.; Babu, M.K.G.; Prasad, Y.N. *A Thermodynamic Simulation Model for a Dual Fuel Open Combustion Chamber Compression Ignition Engine*; SAE International: Warrendale, PA, USA, 1986.

156. Prakash, G.; Ramesh, A.; Shaik, A.B. An Approach for Estimation of Ignition Delay in a Dual Fuel Engine. *SAE Trans.* **1999**, *108*, 399–405.

157. Mustafi, N.N.; Raine, R.R.; Verhelst, S. Combustion and emissions characteristics of a dual fuel engine operated on alternative gaseous fuels. *Fuel* **2013**, *109*, 669–678. [CrossRef]

158. Cacua, K.; Amell, A.; Cadavid, F. Effects of oxygen enriched air on the operation and performance of a diesel-biogas dual fuel engine. *Biomass and Bioenergy* **2012**, *45*, 159–167. [CrossRef]

159. Barik, D.; Murugan, S. Investigation on combustion performance and emission characteristics of a DI (direct injection) diesel engine fueled with biogas–diesel in dual fuel mode. *Energy* **2014**, *72*, 760–771. [CrossRef]

160. Atelge, R. Kısmi Yük Koşullarında Dizel-Biyogaz Kullanılarak Çift Yakıtlı Dizel Motorun Enerji ve Ekserji Analizi. *Eur. J. Sci. Technol.* **2021**, *27*, 334–346.

161. Wang, X.; Qian, Y.; Zhou, Q.; Lu, X. Modulated diesel fuel injection strategy for efficient-clean utilization of low-grade biogas. *Appl. Therm. Eng.* **2016**, *107*, 844–852. [CrossRef]

162. Mujtaba, M.A.; Kalam, M.A.; Masjuki, H.H.; Gul, M.; Soudagar, M.E.M.; Ong, H.C.; Ahmed, W.; Atabani, A.E.; Razzaq, L.; Yusoff, M. Comparative study of nanoparticles and alcoholic fuel additives-biodiesel-diesel blend for performance and emission improvements. *Fuel* **2020**, *279*, 118434. [CrossRef]

163. Kegl, T.; Kovač Kralj, A.; Kegl, B.; Kegl, M. Nanomaterials as fuel additives in diesel engines: A review of current state, opportunities, and challenges. *Prog. Energy Combust. Sci.* **2021**, *83*, 100897. [CrossRef]

164. Feroskhan, M.; Ismail, S.; Gosavi, S.; Tankhiwale, P.; Khan, Y. Optimization of performance and emissions in a biogas–diesel dual fuel engine with cerium oxide nanoparticle addition. *Proc. Inst. Mech. Eng. Part D J. Automob. Eng.* **2018**, *233*, 1178–1193. [CrossRef]

165. Feroskhan, M.; Ismail, S.; Kumar, A.; Kumar, V.; Aftab, S.K. Investigation of the effects of biogas flow rate and cerium oxide addition on the performance of a dual fuel CI engine. *Biofuels* **2017**, *8*, 197–205. [CrossRef]

166. Barik, D.; Murugan, S.; Samal, S.; Sivaram, N.M. Combined effect of compression ratio and diethyl ether (DEE) port injection on performance and emission characteristics of a DI diesel engine fueled with upgraded biogas (UBG)-biodiesel dual fuel. *Fuel* **2017**, *209*, 339–349. [CrossRef]

167. Barik, D.; Murugan, S.; Sivaram, N.M.; Baburaj, E.; Shanmuga Sundaram, P. Experimental investigation on the behavior of a direct injection diesel engine fueled with Karanja methyl ester-biogas dual fuel at different injection timings. *Energy* **2017**, *118*, 127–138. [CrossRef]

168. Khayum, N.; Anbarasu, S.; Murugan, S. Combined effect of fuel injecting timing and nozzle opening pressure of a biogas-biodiesel fuelled diesel engine. *Fuel* **2020**, *262*, 116505. [CrossRef]

169. Prabhu, A.V.; Avinash, A.; Brindhadevi, K.; Pugazhendhi, A. Performance and emission evaluation of dual fuel CI engine using preheated biogas-air mixture. *Sci. Total Environ.* **2021**, *754*, 142389. [CrossRef]

170. Vijin Prabhu, A.; Manimaran, R.; Jeba, P.; Babu, R. Effect of methane enrichment on the performance of a dual fuel CI engine. *Int. J. Ambient Energy* **2021**, *42*, 325–330. [CrossRef]

171. Bouguessa, R.; Tarabet, L.; Loubar, K.; Belmrabet, T.; Tazerout, M. Experimental investigation on biogas enrichment with hydrogen for improving the combustion in diesel engine operating under dual fuel mode. *Int. J. Hydrog. Energy* **2020**, *45*, 9052–9063. [CrossRef]

172. Bui, V.G.; Bui, T.M.T.; Hoang, A.T.; Nižetić, S.; Nguyen Thi, T.X.; Vo, A.V. Hydrogen-Enriched Biogas Premixed Charge Combustion and Emissions in Direct Injection and Indirect Injection Diesel Dual Fueled Engines: A Comparative Study. *J. Energy Resour. Technol.* **2021**, *143*, 1–13. [CrossRef]

173. Hotta, S.K.; Sahoo, N.; Mohanty, K. Comparative assessment of a spark ignition engine fueled with gasoline and raw biogas. *Renew. Energy* **2019**, *134*, 1307–1319. [CrossRef]

174. Sadiq Y, R.; Iyer, R.C. Experimental investigations on the influence of compression ratio and piston crown geometry on the performance of biogas fuelled small spark ignition engine. *Renew. Energy* **2020**, *146*, 997–1009. [CrossRef]

175. da Costa, R.B.R.; Valle, R.M.; Hernández, J.J.; Malaquias, A.C.T.; Coronado, C.J.R.; Pujatti, F.J.P. Experimental investigation on the potential of biogas/ethanol dual-fuel spark-ignition engine for power generation: Combustion, performance and pollutant emission analysis. *Appl. Energy* **2020**, *261*, 114438. [CrossRef]

176. Kim, Y.; Kawahara, N.; Tsuboi, K.; Tomita, E. Combustion characteristics and NO_X emissions of biogas fuels with various CO_2 contents in a micro co-generation spark-ignition engine. *Appl. Energy* **2016**, *182*, 539–547. [CrossRef]

177. Kan, X.; Zhou, D.; Yang, W.; Zhai, X.; Wang, C.-H. An investigation on utilization of biogas and syngas produced from biomass waste in premixed spark ignition engine. *Appl. Energy* **2018**, *212*, 210–222. [CrossRef]

178. Hotta, S.K.; Sahoo, N.; Mohanty, K.; Kulkarni, V. Ignition timing and compression ratio as effective means for the improvement in the operating characteristics of a biogas fueled spark ignition engine. *Renew. Energy* **2020**, *150*, 854–867. [CrossRef]

179. Hotta, S.K.; Sahoo, N.; Mohanty, K. Ignition advancement study for optimized characteristics of a raw biogas operated spark ignition engine. *Int. J. Green Energy* **2019**, *16*, 101–113. [CrossRef]

180. Simsek, S.; Uslu, S. Investigation of the impacts of gasoline, biogas and LPG fuels on engine performance and exhaust emissions in different throttle positions on SI engine. *Fuel* **2020**, *279*, 118528. [CrossRef]

181. Buí, V.G.; Tran, V.N.; Hoang, A.T.; Bui, T.M.T.; Vo, A.V. A simulation study on a port-injection SI engine fueled with hydroxy-enriched biogas. *Energy Sources Part A Recovery Util. Environ. Eff.* **2020**, 1–17. [CrossRef]

182. Alves, H.J.; Bley Junior, C.; Niklevicz, R.R.; Frigo, E.P.; Frigo, M.S.; Coimbra-Araújo, C.H. Overview of hydrogen production technologies from biogas and the applications in fuel cells. *Int. J. Hydrog. Energy* **2013**, *38*, 5215–5225. [CrossRef]

183. Bocci, E.; Di Carlo, A.; McPhail, S.J.; Gallucci, K.; Foscolo, P.U.; Moneti, M.; Villarini, M.; Carlini, M. Biomass to fuel cells state of the art: A review of the most innovative technology solutions. *Int. J. Hydrog. Energy* **2014**, *39*, 21876–21895. [CrossRef]

184. Papadam, T.; Goula, G.; Yentekakis, I.V. Long-term operation stability tests of intermediate and high temperature Ni-based anodes' SOFCs directly fueled with simulated biogas mixtures. *Int. J. Hydrog. Energy* **2012**, *37*, 16680–16685. [CrossRef]

185. Lackey, J.; Champagne, P.; Peppley, B. Use of wastewater treatment plant biogas for the operation of Solid Oxide Fuel Cells (SOFCs). *J. Environ. Manag.* **2017**, *203*, 753–759. [CrossRef] [PubMed]

186. Lanzini, A.; Leone, P.; Guerra, C.; Smeacetto, F.; Brandon, N.P.; Santarelli, M. Durability of anode supported Solid Oxides Fuel Cells (SOFC) under direct dry-reforming of methane. *Chem. Eng. J.* **2013**, *220*, 254–263. [CrossRef]

187. Ma, J.; Jiang, C.; Connor, P.A.; Cassidy, M.; Irvine, J.T.S. Highly efficient, coking-resistant SOFCs for energy conversion using biogas fuels. *J. Mater. Chem. A* **2015**, *3*, 19068–19076. [CrossRef]

188. Papurello, D.; Borchiellini, R.; Bareschino, P.; Chiodo, V.; Freni, S.; Lanzini, A.; Pepe, F.; Ortigoza, G.A.; Santarelli, M. Performance of a Solid Oxide Fuel Cell short-stack with biogas feeding. *Appl. Energy* **2014**, *125*, 254–263. [CrossRef]

189. Shiratori, Y.; Oshima, T.; Sasaki, K. Feasibility of direct-biogas SOFC. *Int. J. Hydrog. Energy* **2008**, *33*, 6316–6321. [CrossRef]

190. Janardhanan, V.M. Multi-scale Modeling of Biogas Fueled SOFC. *ECS Trans.* **2015**, *68*, 3051–3058. [CrossRef]

191. Ni, M. Modeling and parametric simulations of solid oxide fuel cells with methane carbon dioxide reforming. *Energy Convers. Manag.* **2013**, *70*, 116–129. [CrossRef]

192. Tjaden, B.; Gandiglio, M.; Lanzini, A.; Santarelli, M.; Järvinen, M. Small-Scale Biogas-SOFC Plant: Technical Analysis and Assessment of Different Fuel Reforming Options. *Energy Fuels* **2014**, *28*, 4216–4232. [CrossRef]

193. Gandiglio, M.; Lanzini, A.; Leone, P.; Santarelli, M.; Borchiellini, R. Thermoeconomic analysis of large solid oxide fuel cell plants: Atmospheric vs. pressurized performance. *Energy* **2013**, *55*, 142–155. [CrossRef]

194. Schmersahl, R.; Mumme, J.; Scholz, V. Farm-Based Biogas Production, Processing, and Use in Polymer Electrolyte Membrane (PEM) Fuel Cells. *Ind. Eng. Chem. Res.* **2007**, *46*, 8946–8950. [CrossRef]

195. Guan, T.; Alvfors, P.; Lindbergh, G. Investigation of the prospect of energy self-sufficiency and technical performance of an integrated PEMFC (proton exchange membrane fuel cell), dairy farm and biogas plant system. *Appl. Energy* **2014**, *130*, 685–691. [CrossRef]

196. Birth, T.; Heineken, W.; He, L. Preliminary design of a small-scale system for the conversion of biogas to electricity by HT-PEM fuel cell. *Biomass Bioenergy* **2014**, *65*, 20–27. [CrossRef]

197. Verotti, M.; Servadio, P.; Bergonzoli, S. Biogas upgrading and utilization from ICEs towards stationary molten carbonate fuel cell systems. *Int. J. Green Energy* **2016**, *13*, 655–664. [CrossRef]

198. Eichenberger, P.H. The 2 MW Santa Clara Project. *J. Power Sources* **1998**, *71*, 95–99. [CrossRef]

199. Trogisch, S.; Hoffmann, J.; Daza Bertrand, L. Operation of molten carbonate fuel cells with different biogas sources: A challenging approach for field trials. *J. Power Sources* **2005**, *145*, 632–638. [CrossRef]

200. Ciccoli, R.; Cigolotti, V.; Lo Presti, R.; Massi, E.; McPhail, S.J.; Monteleone, G.; Moreno, A.; Naticchioni, V.; Paoletti, C.; Simonetti, E.; et al. Molten carbonate fuel cells fed with biogas: Combating H_2S. *Waste Manag.* **2010**, *30*, 1018–1024. [CrossRef]

201. Lanzini, A.; Madi, H.; Chiodo, V.; Papurello, D.; Maisano, S.; Santarelli, M.; Van herle, J. Dealing with fuel contaminants in biogas-fed solid oxide fuel cell (SOFC) and molten carbonate fuel cell (MCFC) plants: Degradation of catalytic and electro-catalytic active surfaces and related gas purification methods. *Prog. Energy Combust. Sci.* **2017**, *61*, 150–188. [CrossRef]

202. Månsson, A.; Sanches-Pereira, A.; Hermann, S. Biofuels for road transport: Analysing evolving supply chains in Sweden from an energy security perspective. *Appl. Energy* **2014**, *123*, 349–357. [CrossRef]

203. Olsson, L.; Fallde, M. Waste(d) potential: A socio-technical analysis of biogas production and use in Sweden. *J. Clean. Prod.* **2015**, *98*, 107–115. [CrossRef]

204. Monnerie, N.; Roeb, M.; Houaijia, A.; Sattler, C. Coupling of Wind Energy and Biogas with a High Temperature Steam Electrolyser for Hydrogen and Methane Production. *Green Sustain. Chem.* **2014**, *4*, 60–69. [CrossRef]

205. Kharton, V.V.; Yaremchenko, A.A.; Valente, A.A.; Sobyanin, V.A.; Belyaev, V.D.; Semin, G.L.; Veniaminov, S.A.; Tsipis, E.V.; Shaula, A.L.; Frade, J.R.; et al. Methane oxidation over Fe-, Co-, Ni- and V-containing mixed conductors. *Solid State Ion.* **2005**, *176*, 781–791. [CrossRef]

206. Yang, S.; Kondo, J.N.; Hayashi, K.; Hirano, M.; Domen, K.; Hosono, H. Partial oxidation of methane to syngas over promoted C12A7. *Appl. Catal. A Gen.* **2004**, *277*, 239–246. [CrossRef]

207. Tu, X.; Whitehead, J.C. Plasma dry reforming of methane in an atmospheric pressure AC gliding arc discharge: Co-generation of syngas and carbon nanomaterials. *Int. J. Hydrog. Energy* **2014**, *39*, 9658–9669. [CrossRef]
208. Izquierdo, U.; Barrio, V.L.; Requies, J.; Cambra, J.F.; Güemez, M.B.; Arias, P.L. Tri-reforming: A new biogas process for synthesis gas and hydrogen production. *Int. J. Hydrog. Energy* **2013**, *38*, 7623–7631. [CrossRef]
209. IRENA. *Production of Bio-Methanol Technology Brief*; IRENA: Masdar City, Abu Dhabi, United Arab Emirates, 2013; pp. 1–28.
210. Park, D.; Lee, J. Biological conversion of methane to methanol. *Korean J. Chem. Eng.* **2013**, *30*, 977–987. [CrossRef]
211. Shamsul, N.S.; Kamarudin, S.K.; Rahman, N.A.; Kofli, N.T. An overview on the production of bio-methanol as potential renewable energy. *Renew. Sustain. Energy Rev.* **2014**, *33*, 578–588. [CrossRef]
212. Su, Z.; Ge, X.; Zhang, W.; Wang, L.; Yu, Z.; Li, Y. Methanol Production from Biogas with a Thermotolerant Methanotrophic Consortium Isolated from an Anaerobic Digestion System. *Energy Fuels* **2017**, *31*, 2970–2975. [CrossRef]
213. Munasinghe, P.C.; Khanal, S.K. Biomass-derived syngas fermentation into biofuels: Opportunities and challenges. *Bioresour. Technol.* **2010**, *101*, 5013–5022. [CrossRef] [PubMed]

Article

Energy Stored in Above-Ground Biomass Fractions and Model Trees of the Main Coniferous Woody Plants

Rudolf Petráš [1], Julian Mecko [1], Ján Kukla [2], Margita Kuklová [2,*], Danica Krupová [1], Michal Pástor [1], Marcel Raček [3] and Ivica Pivková [2]

[1] Forest Research Institute, National Forest Centre, 960 01 Zvolen, Slovakia; petras@nlcsk.org (R.P.); mecko@nlcsk.org (J.M.); danica.krupova@nlcsk.org (D.K.); pastor@nlcsk.org (M.P.)

[2] Institute of Forest Ecology, Slovak Academy of Sciences, 960 53 Zvolen, Slovakia; kukla@ife.sk (J.K.); ivica.pivkova@ife.sk (I.P.)

[3] Institute of Landscape Architecture, Slovak University of Agriculture in Nitra, 949 76 Nitra, Slovakia; marcel.racek@uniag.sk

* Correspondence: kuklova@ife.sk

Abstract: The paper considers energy stored in above-ground biomass fractions and in model trees of the main coniferous woody plants (*Picea abies* (L.) H. Karst., *Abies alba* Mill., *Pinus sylvestris* (L.), *Larix decidua* Mill.), sampled in 22 forest stands selected in different parts of Slovakia. A total of 43 trees were felled, of which there were 12 spruces, 11 firs, 10 pines, and 10 larches. Gross and net calorific values were determined in samples of wood, bark, small-wood, twigs, and needles. Our results show that these values significantly depend on the tree species, biomass fractions, and sampling point on the tree. The energy stored in the model trees calculated on the basis of volume production taken from yield tables increases as follows: spruce < fir < pine < larch. Combustion of tree biomass releases an aliquot amount of a greenhouse gas—CO_2, as well as an important plant nutrient, nitrogen—into the atmosphere. The obtained data must be taken into account in the case of the economic utilization of energy stored in the fractions of above-ground tree biomass and in whole trees. The achieved data can be used to assess forest ecosystems in terms of the flow of solar energy, its accumulation in the various components of tree biomass, and the risk of biomass combustion in relation to the release of greenhouse gases.

Keywords: coniferous trees; biomass fractions; calorific values; energy reserves

Citation: Petráš, R.; Mecko, J.; Kukla, J.; Kuklová, M.; Krupová, D.; Pástor, M.; Raček, M.; Pivková, I. Energy Stored in Above-Ground Biomass Fractions and Model Trees of the Main Coniferous Woody Plants. *Sustainability* **2021**, *13*, 12686. https://doi.org/10.3390/su132212686

Academic Editors: Idiano D'Adamo and Piergiuseppe Morone

Received: 5 October 2021
Accepted: 12 November 2021
Published: 16 November 2021

Publisher's Note: MDPI stays neutral with regard to jurisdictional claims in published maps and institutional affiliations.

1. Introduction

The country and its soil are considered highly important but limited resources both in Europe and worldwide, and are currently facing pressure from anthropogenic increases in greenhouse gases in the atmosphere of terrestrial ecosystems, climate change, and a loss of biodiversity. Landscape and soil management should therefore include environmental monitoring and measures aimed at reducing the release of industrial pollutants into the air, the gradual interconnection of green spaces to reduce landscape fragmentation, or expanding protected areas to preserve natural diversity.

A lack of multi-dimensional data is one of the major gaps which limits the knowledge and assessment possibilities of European forests. Nowadays, the most extensive and complete data on European forest statuses are provided by the National Forest Inventories, which provide information about the extent of forest resources and their composition and structure [1]. Forest inventory methods are the primary tools used to assess the current state and development of forests over time [2]. On the other hand, long-term experimental plots provide information on forest stand dynamics, which cannot be derived from forest inventories or small temporary plots [3]. By measuring the remaining as well as the removed stand, the survey of long-term experiments provides the total production at a given site, which is most relevant for examining the relationship between site conditions and stand productivity on one hand and between stand density and productivity on the other.

The assessment of above-ground biomass stocks in the coniferous forests of inland northwest USA is important both for the inventory of wood, bioenergy, and carbon, as well as for wildfire risk determination [4]. The use of bioenergy is increasing rapidly due to the need to reduce greenhouse gas emissions [5]. According to the above authors, tree biomass is characterized not only by its mechanical, physical, and chemical properties, but also by its energy content, which accumulates in the process of photosynthetic assimilation, and which can be released later. In the case of its wider use for energy purposes, it is necessary to know the energy content of both whole trees and their individual parts. The energy of the trees can be determined directly by the destruction method, by means of which the fresh and dry mass of the biomass fractions of individual trees is determined first, and then their calorific value.

The calorific value of plants is an important parameter for evaluating and indexing material cycles and energy conversion in forest ecosystems [6]. The effective heating value of wood correlates best with the lignin content, of inner bark with carbohydrates, and outer bark with carbohydrates and the extractives soluble in alkaline solvents [7]. A similar determination through the content of chemical elements C, H, N and S is reported by [8]. The determination of the heating value might be used thus as an indicator of the cellulose content of coniferous wood.

The most abundant data on biomass properties are found achieved in research on wood density which is usually associated with wood mechanical properties. Based on many literary sources, such a correlation has been derived for 103 tree species [9]. The lowest density is of soft deciduous species, followed by conifers and hard deciduous species. Some authors report that the density of wood varies not only with tree species but also with the vertical or radial position of the wood on the tree trunk and in the tree crown [10,11]. Wood density also depends on the width of tree rings [12,13], the proportion of spring and summer wood, and the tree age [14,15]. The density of wood of branches with bark is significantly higher than that of stem wood [16]. For more accurate calculations of the weight of the whole tree, however, in addition to its volume, it is necessary to know the density of all its components, specifically the density of the round-wood, its bark, and branches [17].

Partial data can also be found for the calorific value of individual biomass fractions. Data have been published on the calorific values of the stem wood, branches, and roots of the bark of spruce and beech trees [18], the calorific value of wood and cones of four coniferous trees [19], as well as on the relationship of the calorific value of fir wood and the width of the annual rings [20]. A very detailed study of the heating values of seven tree species was conducted in Finland [21].

Due to the high laboriousness of the method of destruction, it is not possible to process larger and more representative experimental material when researching the energy accumulated in the tree species of forest stands. It is therefore more accessible and efficient to build on existing knowledge on the amount of tree biomass, which is expressed in volume units in forestry. The generally known models of forest tree volume tables exist in the form of mathematical functions and simulate the volume of above-ground biomass of not only whole trees, but also their main parts, such as wood, bark, and branches with bark [22]. Therefore, they can be used effectively in the conversion about biomass volume to dry weight and subsequently to the energy reserve of the tree. For this purpose, it is necessary to know not only the calorific value of tree biomass, preferably according to its basic fractions, but also the density of tree biomass fractions.

Renewable energies are essential parts of the energy revolution in which the goal is to replace energy production from fossil fuels with those from renewable sources. There is a view that many of the available resources are not fully utilized (for example, the correct use of forest biomass, organic residues from agriculture, forestry or landscaping or residues from the animal breeding sector). In that sense, biomaterials are essential renewables which require changes in attitudes, visions, strategies, and activities, based on the principles of resource sharing [23].

However, studies pertaining to the utilization of forest biomass as a replacement energy source for fossil fuels are lacking. Forest biomass meets sustainability criteria and has significant potential for CO_2 sequestration. Therefore, the effort to increase the forest cover of the country and of hectare wood stocks significantly contributes to increasing carbon stocks and sustainability of terrestrial ecosystems. This approach is essential if, by a concerted effort, we want to contribute to stop the process of destabilizing natural ecosystems and, consequently, of society due to climate change.

The aim of this study is to obtain data usable in the economic use of solar energy stored in the fractions of aboveground tree biomass, as well as in whole trees and forest stands of four main coniferous woody plants (spruce, fir, pine and larch) based on gross calorific values and basic densities of aboveground fractions of wood, bark, small-wood, twigs, needles, and volume production of forest stands. We hypothesize that the energy content of the biomass of coniferous woody species depends mainly on: (1) tree species, (2) the biomass fraction, (3) the density of the biomass fraction, (4) the sampling point on the tree, and (5) the tree developmental stage.

The obtained knowledge can be useful in assessing the flow of solar energy in forest ecosystems, the risks of accidental and deliberate forest fires, the burning of fossil fuels, climate change, and the possibility to achieve local and global carbon neutrality.

2. Materials and Methods

2.1. Data Collection and Sampling

The experimental material was obtained in 22 forest stands located in the Slovak territory (Figure 1), which covers a large part of the Western Carpathians. Most of the forest stands are located in central Slovakia, with others in western and eastern Slovakia. The growth conditions of sampled coniferous stands are provided in Table 1.

The forest stands situated at an altitude of 165–1070 m a.s.l. consist of the Norway spruce (*Picea abies* (L.) H.Karst.), Silver fir (*Abies alba* Mill.), Scots pine (*Pinus sylvestris* (L.)) and European larch (*Larix decidua* Mill.) (Table 2). The stands are on slopes of 25–60%, with the exception of most pine forests located on the plain. The production level of these stands is expressed by the site index, which is in the range of 24–42. This index represents the mean height of trees (m) that the stand would reach at the age of 100 years.

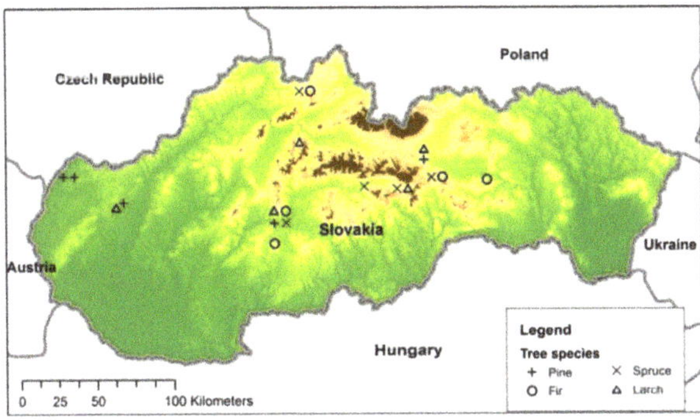

Figure 1. Location of examined coniferous stands.

Table 1. Geobiocoenological classification of stands of examined woody plants (in the sense of [24,25]).

Woody Plant	Vegetation Grade	Edaphic-Hydric Order	Edaphic-Trophic Order	Group of Geobiocoene Types
Norway spruce (*Picea abies* (L.) H.Karst.)	3rd, Oak-Beech	wetted	Mesotrophic	*Querceto-Fagetum*
	5th, Fir-Beech	normal		*Abieto-Fagetum inf.*[a]
			hemioligotrophic	*Fageto-Abietum inf.*
	6th, Spruce-Beech Fir			*Fageto-Abietum sup.*[b]
Silver fir (*Abies alba* Mill.)	3rd, Oak-Beech	wetted	mesotrophic	*Querceto-Fagetum*
	4th, Beech	normal		*Fagetum pauper Fagetum typicum*
		little restricted	heminitrophilous	*Fagetum tiliosum*
	5th, Fir-Beech 6th, Spruce-Beech Fir	normal	mesotrophic hemioligotrophic	*Abieto-Fagetum inf. Fageto-Abietum sup.*
Scots pine (*Pinus sylvestris* L.)	1st, Oak			*Carpineto-Quercetum*
	2nd, Beech-Oak	normal	mesotrophic	*Fageto-Quercetum*
European larch (*Larix decidua* Mill.)	3rd, Oak-Beech	wetted		*Querceto-Fagetum*
	5th Fir-Beech	normal		*Abieto-Fagetum inf.*
			hemioligotrophic	*Fageto-Abietum inf.*

[a] inf.—inferiora; [b] sup.—superiora.

Table 2. Characteristics of trees from which biomass samples were taken.

Tree Species	Number of Sampled Trees	DBH (cm)	h (m)	Age	Site Index	Altitude (m)
Picea abies	12	20–62	23–38	35–105	26–42	435–1070
Abies alba	11	23–75	22–39	35–153	24–40	390–950
Pinus sylvestris	10	25–51	24–30	75–108	24–30	165–940
Larix decidua	10	26–56	24–35	40–100	28–40	275–1070
Overview	43	20–75	22–39	35–153	24–42	165–1070

A total of 43 trees were felled in these stands, of which there were 12 spruces, 11 firs, 10 pines, and 10 larches (Table 2). A higher amount of spruce and fir trees were felled due to their higher representation in the forests of Slovakia. Based on their diameter, height and age, the majority of trees had the parameters of mature trees. The stands in which the trees were cut down are located at various exposures with a slope of 25–60%. Only in the case of pines did most plots have zero inclination.

The following samples were taken from each tree: three circular cut-outs (one from the trunk foot, another from the middle of the trunk, and a third from the middle part of the tree crown) divided into a wood and bark fraction (as there was not enough bark volume on the circular cut-outs, another bark was sampled from neighbouring places on the trunk); about 20–25 cm long cuttings thinner than 7 cm of small-wood (with bark) from the central part of the crown and twigs overgrown with green needles divided into needles and twigs after drying. The small-wood fraction was taken from the main (primary) branches that grow up directly from the trunk of the tree.

A total of 9 samples were taken from each tree (3 of wood, 3 of bark and another 3 of small-wood, twigs, and needles). The only exception was larch, from which no twigs and needles were taken. A total of 367 biomass samples were taken from 43 coniferous trees (129 from trunk bark, 129 from trunk wood, 43 from small-wood, 33 from twigs and 33 from needles).

2.2. Data Processing and Analysis

For calorimetric determination, bark, wood, small-wood, twigs, and needle samples were taken from 5 trees of each woody plant (spruce, fir, pine and larch). Samples smaller than 30 mm were dried at $103 \pm 2\ °C$ and ground using an SM 100 cutting mill (Retsch) with bottom sieves with 2 mm square holes and a circular filter to a size of approximately <1 mm at 1450 rpm for 5 min.

Gross calorific value was determined using an IKA C-4000 calorimeter (program C-402). Two determinations were performed on each sample with an accuracy of up to 120 J [26]. The elements C, H, N, and S necessary to calculate the net calorific value have been determined using CNS Flash EA 1112 from Thermo Finnigan. Two determinations per sample were performed to the C, H, N and to the S [27,28]. Oxygen content was obtained by subtracting the sum of the percentages of C, H, N, S, and ash from 100% [29]. Ash content was determined gravimetrically by combustion of samples in an electric muffle furnace at 500 °C in triplicate [30].

2.3. Statistical Analysis

The variability of the calorific value of basic biomass fractions within each of the examined species of coniferous woody plants and within each of the basic biomass fractions taken from different coniferous woody plants was evaluated using the program Statistica 9 (StatSoft, 2008). A one-way ANOVA test followed by a Fisher-LSD test was used to detect significant differences between observed characteristics. In the analysis of variance of the calorific values the tree species (spruce, fir, pine, larch) and the biomass fractions (bark, wood, small-wood, twigs, and needles) were used as factors. Results were expressed as mean $\pm$ standard deviation (SD). Differences between means were considered significant when they occurred at $p < 0.05$.

2.4. Calculation Procedures

The energy stored in the above-ground biomass of the mean trees of the studied coniferous species was calculated on the basis of the measured calorific values, tables of model volumes of trees [22], as well as the model density values for bark, wood, and small-wood [31]. Mathematical models of classical tree volume tables show the volume of whole trees v (m^3) in relation to their diameters DBH (cm), heights h (m) and main components (round-wood with or without bark, bark, and small-wood). The volume tables do not contain twigs and needles. Therefore, their dry weight values, which depend on tree diameter and height, were taken from [32]. According to this source, for example, the fresh weight of needles for trees with a DBH of 40 cm and a height of 30 m is approximately 90 kg in the case of spruce, and 35 kg in the case of pine. The dry weight is approximately 42% of the fresh weight of spruce and pine. Spruce models were also taken for fir. The energy reserves of the trees at different stages of their development were then calculated according to the following formula:

$$CH(DBH, h) = \left[\sum v_i \cdot \rho_i \cdot CH_i + m_0 \cdot CH_{twne} \right] \cdot 10^{-3} \tag{1}$$

where: CH(DBH, h)—the energy reserve of the tree (GJ tree^{-1});

v_i—the volume of i-th fraction (stem wood, stem bark, small-wood) taken over from the volume tables [22] (m^3);

ρ_i—basic density of the i-th fraction taken from [31] (kg m^{-3});

CH_i—the calorific value of the i-th fraction (MJ kg^{-1});

m_0—the dry matter weight of the twigs and needles taken from [32] (kg);

CH_{twne}—the calorific value of the twigs and needles (MJ kg^{-1}).

The dry weight of the above-ground biomass of the model trees at different stages of their development was calculated using equation (1), from which the items CH_i and CH_{twne} representing the combustion heat were omitted. The same equation was also used for the calculation of dry weight of aboveground biomass of adult model trees (DBH 60 cm, h. 30 m).

3. Results

3.1. Gross Calorific Values of Biomass Fractions of Examined Woody Plants

The variability of the gross calorific values of the biomass fractions within individual coniferous woody plants is shown in Figure 2a. In the case of spruce, the lowest values were found in bark and wood taken from the middle part of the trunk and the highest in bark and wood taken from the base of the trunk. In general, however, spruce bark and wood have similar gross calorific values. The same can be said for the bark and wood fractions of fir. The gross calorific values of wood fractions taken from different parts of the tree trunk do not differ significantly. Larch wood has the lowest values. The values found for small-wood are slightly higher, but they are also very similar to each other. There is a significantly higher concentration of energy in the bark of pine and larch taken from the foot of the trunk, as well as in the twigs and needles of all examined woody plants. In contrast, the energy values of bark taken from the middle and crown part of pine trunk were the lowest of all woody plants.

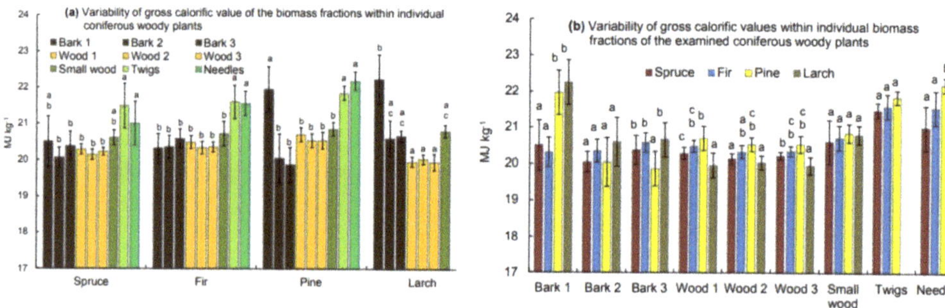

Figure 2. Variability of gross calorific values (**a**) ANOVA, Fischer LSD test; spruce: $F_{(8, 36)} = 6.1397$, $p = 0.0001$; fir: $F_{(8, 36)} = 13.950$, $p = 0.0001$; pine: $F_{(8, 36)} = 23.228$, $p = 0.0001$; larch: $F_{(6, 28)} = 27.058$, $p = 0.0001$. (**b**) 1—on the tree foot, 2—at the middle of the stem, 3—at the middle of the crown; ANOVA, Fischer LSD test; bark 1: $F_{(3, 16)} = 13.11$, $p = 0.0001$; bark 2: $F_{(3, 16)} = 1.65$, $p = 0.2171$; bark 3: $F_{(3, 16)} = 5.33$, $p = 0.0098$; wood 1: $F_{(3, 16)} = 10.79$, $p = 0.0004$; wood 2: $F_{(3, 16)} = 9.29$, $p = 0.0009$; wood 3: $F_{(3, 16)} = 9.10$, $p = 0.0010$; small-wood: $F_{(3, 16)} = 1.09$, $p = 0.4107$; twigs: $F_{(2, 12)} = 0.81$, $p = 0.4696$; needles: $F_{(2, 12)} = 9.91$, $p = 0.0029$. Significantly different mean values ($p < 0.05$) are indicated by different letters.

The coefficients of variation of the gross calorific value of spruce and fir wood were in the range 0.5–0.9%, and in the range of 0.7–1.6% for pine and larch. The coefficients of variation of the gross calorific value of small-wood (0.8–1.7%) are very close to the trunk wood. In the case of bark (0.8–3.4%), the highest coefficients are approximately twice as high, especially in the bark of pine and larch. The coefficients of variations of twigs and needles range from 0.9 to 2.9%, these values being closer to the values found in the bark than in the wood.

The variability of gross calorific values within individual biomass fractions of the examined coniferous woody plants is shown in Figure 2b. The gross calorific values of bark taken from the foot of pine and larch trunks are significantly higher compared to the values of the equivalent spruce and fir bark. On the other hand, in the case of bark taken from the middle of the trunks and crowns of trees, the differences in the values of different woody plants were relatively small. The only exception was the significantly lower value of bark taken from the pine crown. Larch wood has the lowest gross calorific value regardless of the sampling point on the tree. Spruce wood has a slightly higher value, and the highest value was found for pine wood. Significant differences were found between the gross calorific values of pine wood compared to larch wood and spruce wood 1 and 2, and larch wood compared to spruce wood and fir wood 1 and 3. The gross calorific values of the small-wood of the examined woody plants did not differ much and were only slightly higher compared to the values found in the wood fractions. The gross calorific value of the needles increased markedly from spruce to pine, but the differences were not significant.

The variability of gross calorific values of the biomass fractions within individual woody plants is given in Table 3a. It can be seen that the gross calorific values of the spruce bark and wood, and the bark, wood and small-wood of fir and pine differ significantly from the values of twigs and needles. On the other hand, the gross calorific value of spruce small-wood is not significantly different from the values of the other spruce fractions. In the case of larch, the values of the bark and small-wood fractions are similar and significantly different from the values of the wood fraction.

Table 3. Variability of gross calorific values (arithmetic mean $\pm$ SD in MJ kg^{-1}): (**a**—ANOVA, Fischer LSD test; spruce: $F_{(4, 40)} = 11.577$, $p = 0.0001$; fir: $F_{(4, 40)} = 27.931$, $p = 0.0001$; pine: $F_{(4, 40)} = 8.027$, $p = 0.0008$; larch: $F_{(2,32)} = 14.22$, $p = 0.0004$; **b**—ANOVA, Fischer LSD test; wood: $F_{(3, 56)} = 29.322$, $p = 0.0001$; bark: $F_{(3, 56)} = 3.568$, $p = 0.019$; small-wood: $F_{(3, 16)} = 1.019$, $p = 0.410$; twigs: $F_{(2, 12)} = 0.806$, $p = 0.469$; needles: $F_{(2, 12)} = 9.903$, $p = 0.000$).

Tree Species	Bark	Wood	Small-Wood	Twigs	Needles
a—Variability within each of the examined woody plants					
Picea abies	20.314 $\pm$ 0.495 [a]	20.222 $\pm$ 0.142 [a]	20.613 $\pm$ 0.208 [a,b]	21.495 $\pm$0.621 [b]	21.005 $\pm$ 0.600 [b]
Abies alba	20.422 $\pm$0.330 [a]	20.395 $\pm$ 0.164 [a]	20.727 $\pm$ 0.342 [a]	21.602 $\pm$ 0.468 [b]	21.563 $\pm$ 0.338 [b]
Pinus sylvestris	20.627 $\pm$0.112 [a]	20.596 $\pm$ 0.253 [a]	20.856 $\pm$ 0.180 [a]	21.857 $\pm$ 0.198 [b]	22.196 $\pm$ 0.253 [b]
Larix decidua	21.171 $\pm$ 0.910 [b]	19.988 $\pm$ 0.159 [a]	20.811 $\pm$ 0.166 [b]	-	-
b—Variability within each of the basic biomass fractions of the examined woody plants					
Picea abies	20.314 $\pm$ 0.495 [a]	20.222 $\pm$ 0.142 [b]	20.613 $\pm$ 0.208 [c]	21.495 $\pm$ 0.621 [c]	21.005 $\pm$ 0.600 [a]
Abies alba	20.422 $\pm$ 0.330 [a]	20.395 $\pm$ 0.164 [b]	20.727 $\pm$ 0.342 [c]	21.602 $\pm$ 0.468 [c]	21.563 $\pm$ 0.338 [a]
Pinus sylvestris	20.627 $\pm$ 0.112 [a,c]	20.596 $\pm$ 0.253 [a]	20.856 $\pm$ 0.188 [c]	21.857 $\pm$ 0.198 [c]	22.196 $\pm$ 0.253 [c]
Larix decidua	21.171 $\pm$ 0.910 [c]	19.988 $\pm$ 0.159 [c]	20.811 $\pm$ 0.166 [c]	–	–

Note: significantly different mean values ($p < 0.05$) are indicated by different letters ([a,b,c]).

The variability of gross calorific values within individual biomass fractions of investigated woody plants is provided in Table 3b. Twigs and needles have the highest values, and the lowest were found for wood. The values of spruce and fir bark are similar and significantly lower compared to the values of pine and especially larch bark. The same can be said for the wood fraction, with the exception of larch wood, which, on the other hand, has the absolute lowest energy value. The differences in values found for the small-wood and twig fractions of examined woody plants are not very large. The gross calorific values of spruce and fir needles are also similar, but significantly lower compared to the value of pine needles.

The coefficients of variation are in the range of 1–4%. Twigs, needles and small-wood have lower variability, and bark of larch and pine trees a higher one. It can be seen that wood with bark, and the twigs with needles of spruce and fir trees belong to the same set of the average calorific values. Pine wood and bark and the small-wood of all four species of conifers also have similar calorific values. The variability of calorific values is relatively small. The coefficients of variation are in the range of 1–4%. Twigs, needles and small-wood have lower variability and the bark of larch and pine a higher one. The significantly different gross calorific value was in the bark of larch trees.

3.2. Net Calorific Values of Biomass Fractions of Examined Woody Plants

An important indicator of biomass energy content is the net calorific value, which depends on the elemental composition, moisture content and ash content. The ultimate analysis of soft-wood species is generally 51% carbon, 42% oxygen, 6.3% hydrogen, 0.1% nitrogen and 0.02% sulphur. In hardwood, the C content is 49%, O 44%, H 6.2%, N 0.1%, and S 0.02% [29]. The differences are mainly due to different carbon content (main energy source) and different ash content (not combustible material). The net calorific values of basic biomass fractions of examined coniferous woody plants are in provided in Table 4.

Table 4. Net calorific values of biomass fractions of examined tree species (n = 5).

Tree Species	Biomass Fraction		C	O	H	N	S	Ash	Min	Max	Average ± SD
						(%)				$(MJ\ kg^{-1})$	
Picea abies	bark	1	49.99 ± 0.2	42.64	5.91 ± 0.1	0.57 ± 0.0	0.05 ± 0.1	0.32 ± 0.1	18.388	19.828	19.217 ± 0.7 [b]
		2							18.445	19.118	18.764 ± 0.3 [b]
		3							18.449	19.498	19.094 ± 0.4 [b]
	wood	1	50.90 ± 0.0	39.93	5.99 ± 0.0	0.10 ± 0.0	0.05 ± 0.0	3.55 ± 0.3	18.722	19.163	18.985 ± 0.2 [b]
		2							18.705	19.056	18.843 ± 0.1 [b]
		3							18.801	19.074	18.929 ± 0.1 [b]
	small-wood		50.55 ± 0.2	41.79	6.16 ± 0.0	0.31 ± 0.0	0.03 ± 0.0	1.16 ± 0.1	18.964	19.545	19.273 ± 0.2 [a,b]
	twigs		50.50 ± 0.6	40.19	6.04 ± 0.0	1.01 ± 0.0	0.11 ± 0.0	2.15 ± 0.0	19.089	20.646	20.180 ± 0.6 [a]
	needles		49.45 ± 0.5	38.35	6.15 ± 0.1	1.50 ± 0.0	0.13 ± 0.0	4.42 ± 0.1	18.699	20.289	19.675 ± 0.6 [a]
Abies alba	bark	1	51.95 ± 0.2	43.75	6.20 ± 0.0	0.46 ± 0.0	0.02 ± 0.0	0.28 ± 0.1	18.482	20.627	19.229 ± 0.9 [b]
		2							18.766	19.953	19.273 ± 0.4 [b]
		3							19.009	20.319	19.499 ± 0.5 [b]
	wood	1	49.65 ± 0.3	39.30	6.21 ± 0.0	0.09 ± 0.0	0.04 ± 0.0	2.05 ± 0.1	18.872	19.281	19.141 ± 0.2 [b]
		2							18.737	19.227	18.989 ± 0.2 [b]
		3							18.801	19.144	19.009 ± 0.1 [b]
	small-wood		51.80 ± 0.1	40.10	6.24 ± 0.0	0.31 ± 0.0	0.04 ± 0.0	1.51 ± 0.0	18.989	19.759	19.371 ± 0.3 [a]
	twigs		52.50 ± 0.1	37.37	6.21 ± 0.0	1.04 ± 0.0	0.11 ± 0.0	2.77 ± 0.0	19.656	20.799	20.254 ± 0.5 [a]
	needles		51.25 ± 0.2	37.94	6.03 ± 0.0	1.17 ± 0.0	0.11 ± 0.0	3.50 ± 0.1	19.742	20.663	20.252 ± 0.3 [a]

Table 4. Cont.

Tree Species	Biomass Fraction		C	O	H	N	S	Ash	Min	Max	Average ± SD
					(%)					(MJ kg⁻¹)	
Pinus sylvestris	bark	1	51.35 ± 0.2	42.58	5.99 ± 0.0	0.41 ± 0.0	0.02 ± 0.0	0.40 ± 0.1	18.586	21.700	20.368 ± 0.1 [a]
		2							18.315	19.929	18.745 ± 0.7 [b]
		3							18.078	19.270	18.566 ± 0.5 [b]
	wood	1	50.78 ± 0.2	39.23	6.10 ± 0.0	0.12 ± 0.0	0.08 ± 0.0	2.94 ± 0.1	19.022	19.806	19.380 ± 0.3 [b]
		2							19.027	19.527	19.214 ± 0.2 [b]
		3							18.988	19.589	19.218 ± 0.2 [b]
	small-wood		51.25 ± 0.1	41.06	6.23 ± 0.0	0.32 ± 0.0	0.03 ± 0.0	1.11 ± 0.1	19.194	20.780	19.765 ± 0.6 [a,b]
	twigs		53.05 ± 0.2	36.83	6.34 ± 0.0	0.88 ± 0.0	0.07 ± 0.0	2.83 ± 0.0	20.253	20.713	20.482 ± 0.2 [a]
	needles		53.3 ± 0.1	35.96	6.31 ± 0.0	1.55 ± 0.0	0.13 ± 0.0	2.70 ± 0.1	20.527	21.209	20.828 ± 0.3 [a]
Larix decidua	bark	1	52.90 ± 0.1	46.42	6.20 ± 0.0	0.24 ± 0.0	0.03 ± 0.0	0.12 ± 0.0	19.895	22.207	21.157 ± 0.9 [a]
		2							18.438	19.773	19.252 ± 0.5 [b,c]
		3							19.074	19.506	19.308 ± 0.2 [b,c]
	wood	1	47.48 ± 0.4	39.21	5.87 ± 0.0	0.08 ± 0.0	0.03 ± 0.0	1.42 ± 0.1	18.478	18.848	18.692 ± 0.1 [b]
		2							18.581	18.922	18.762 ± 0.1 [b]
		3							18.306	19.978	18.924 ± 0.6 [b,c]
	small-wood		51.49 ± 0.4	41.20	5.97 ± 0.0	0.26 ± 0.0	0.01 ± 0.0	1.07 ± 0.1	19.354	19.736	19.511 ± 0.2 [c]

Note: significantly different mean values (p < 0.05) are indicated by different letters ([a,b,c]).

The net calorific values depend on the content of biogenic elements (C, O, H, N) as well as S and ash in tree biomass. The most represented element was C (from 47.48% for larch wood to 53.35% for pine needles) followed by O (from 35.96% for pine needles to 46.42% for larch bark), H (from 5.87% for larch wood to 6.34% for pine twigs), N (from 0.08% for larch wood to 1.55% for pine needles) and S (from 0.02% for fir and pine bark to 0.13% for spruce and pine needles). The highest content of O, N and S was in twigs and needles, H in small-wood, twigs, or needles, while the content of C did not show any regularity. The highest ash content was in the needles of fir and spruce (3.50–4.42%), the lowest was always in the bark of examined woody plants (0.12–0.40%).

The net calorific values of examined woody plants range from a minimum of 18.078–20.527 MJ kg^{-1} to a maximum of 18.848–22.207 MJ kg^{-1} (Table 4). The highest values were found in twigs and needles. Average calorific values ranged from 18.843 to 21.157 MJ kg^{-1} and their variability was relatively low. The coefficient of variation ranged between 0.70–4.45% and reached the maximum values in the case of the bark fraction of the examined woody plants. The calorific values of the bark and wood of spruce, fir and pine trees are very similar. Significant differences occurred only in the case of larch. However, these values differed significantly from those found in small-wood, twigs, and needle fractions. Calorific values of small-wood fractions of spruce and pine significantly differ from all other fractions.

3.3. Dry Weight of Above-Ground Biomass Fractions of Model Trees

The dry weight and percentage of biomass fractions of the model trees of examined woody plants are given in Table 5. The proportion of the bark fraction reaches 4.8–16%, of wood 76–84%, of small-wood 5.5–8.2%, of twigs 0.2–2.5%, and of needles 2.6–5.5%. Larch and fir bark have the highest share, pine bark the lowest. Pine, on the other hand, has the largest share of the wood and small-wood fractions. Spruce, in turn, has the largest proportion of twigs and needles. The dry weight of larch and pine is the same and at the same time the absolute highest, and the weight of spruce is the lowest.

Table 5. Dry weight and percentage of above-ground biomass fractions of the model trees (DBH 60 cm, h 30 m) of examined woody plants.

Tree Species	Bark		Wood		Small-Wood		Twigs		Needles		Model Tree
	(kg)	(wt %)	(kg)	(wt %)	(kg)	(wt %)	(kg)	(wt %)	(kg)	(wt %)	(kg tree^{-1})
Picea abies	101	6.9	1169	79.7	81	5.5	36	2.4	80	5.5	1467
Abies alba	172	10.7	1220	76.2	94	5.9	36	2.2	80	5.0	1602
Pinus sylvestris	82	4.8	1434	84.2	139	8.2	4	0.2	44	2.6	1703
Larix decidua	272	16.0	1322	77.5	111	6.5	-	-	-	-	1705
Average	157	9.6	1286	79.4	106	6.5	19	1.2	51	3.3	1619

The dry weight of the above-ground biomass of the model trees of examined woody plants shown in Figure 3 is the sum of the dry weights of all fractions of their biomass, excluding larch twigs and needles.

The dry weight of larch biomass is highest, and of spruce lowest. This fact is mainly due to a higher basic density of larch wood by approximately 35% compared to spruce wood. The energy reserves of trees growing at medium-quality sites are significantly lower and the lowest are in the case of trees growing at the worst sites. At these sites, pine has the highest dry weight, while the order of weight of other species does not change.

The reliability of the determination of the energy stored in the tree depends on the variability of gross calorific values, basic densities, and volume of tree biomass fractions. However, it must be said that the relative proportion of tree biomass fractions will be the decisive factor in this case. Small trees have a significant share of small-wood, but large trees have a significant share of trunk wood. This is of crucial importance, as the wood of the trunk has the largest share, but also the lowest variability of calorific value.

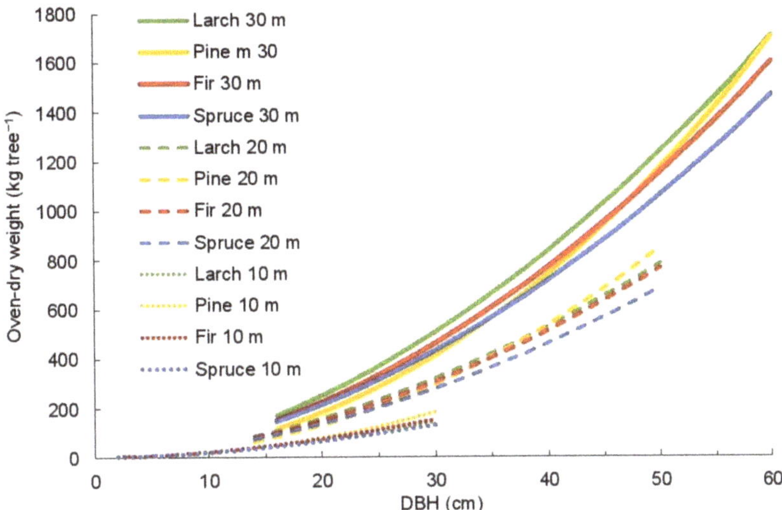

Figure 3. Oven-dry weight of the above-ground biomass of the model trees of examined woody plants depending on their diameter (DBH) and height (10, 20, 30 m).

3.4. Energy Density of Above-Ground Biomass Fractions and of Model Trees

The energy density of the biomass fractions depends on their basic density and gross calorific value. Figure 4 shows the energy density of the basic biomass fractions and of the model trees of examined woody plants.

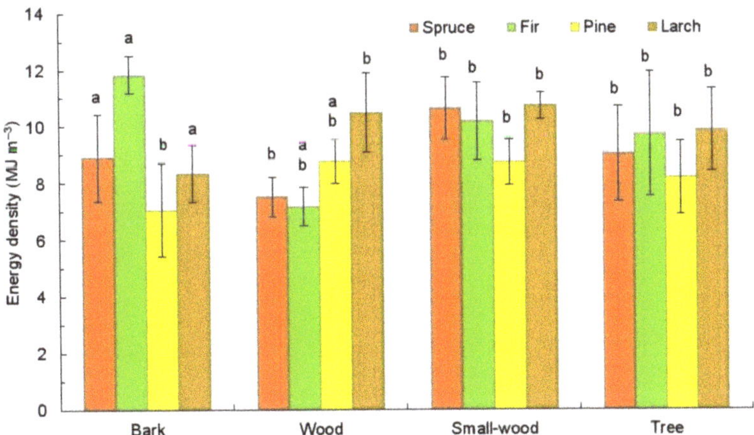

Figure 4. Variability of energy density of above-ground biomass fractions and of model trees. Significantly different mean values ($p < 0.05$) are indicated by different letters.

The average energy density of the above-ground biomass of pine is 8.18 GJ m^{-3}, spruce 9.01 GJ m^{-3}, fir and larch 9.73 and 9.84 GJ m^{-3}, and for all examined woody plants is approximately 9.19 GJ m^{-3}. The highest energy density was found in fir bark (almost 11.84 GJ m^{-3}), the lowest in pine bark (7.05 GJ m^{-3}) and fir wood (7.18 GJ m^{-3}).

On the other hand, the energy density of larch wood and small-wood of spruce, fir and larch reaches 10.16–10.72 GJ m^{-3}. Higher energy density found in whole trees of fir and larch was due to the higher energy density of bark and small-wood (fir), and wood and small-wood (larch).

The energy reserves of above-ground biomass fractions and of model trees of examined woody plants are shown in Table 6. The energy reserve of pine bark is the lowest and that of larch bark the highest. The most energy is stored in wood fraction and small-wood fraction of pine, the least in the wood fraction and the small-wood fraction of spruce, and in twigs and needles of pine. The energy reserve of the model trees of examined woody plants ranges from about 30 GJ for spruce to 34–35 GJ for larch and pine.

Table 6. Energy reserves of above-ground biomass fractions and of model trees (DBH 60 cm, h 30 m) of examined woody plants (GJ tree^{-1}).

Tree Species	Biomass Fraction					Model Tree
	Bark	**Wood**	**Small-Wood**	**Twigs**	**Needles**	
Picea abies	2.05	23.64	1.67	0.77	1.68	29.82
Abies alba	3.51	24.88	1.95	0.78	1.73	32.85
Pinus sylvestris	1.69	29.53	2.90	0.09	0.98	35.19
Larix decidua	5.76	26.42	2.31	-	-	34.49
Average	3.25	26.12	2.21	0.55	1.461	33.09

The energy reserves of the examined woody plants are shown in Figure 5. The highest energy reserve is again in larch (similar to dry weight) and the lowest in fir and spruce. Since the differences in the calorific values of the basic biomass fractions of studied woody plants are not large, we can conclude that the higher energy content of larch trees may be caused by their higher oven-dry weight compared to other tree species.

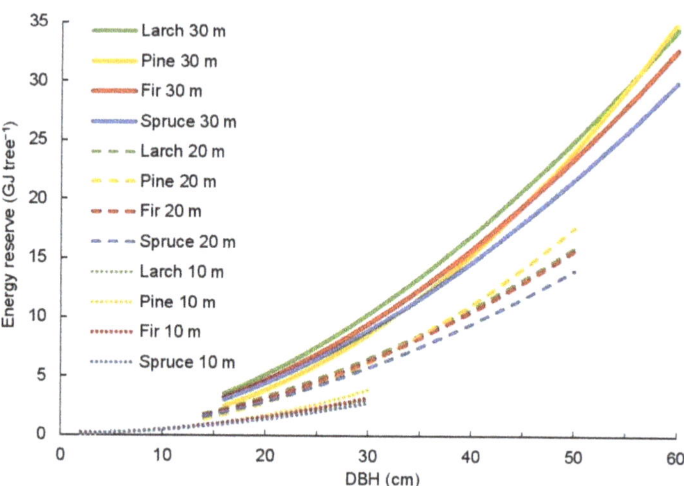

Figure 5. The energy reserves of above-ground biomass of the model trees of the examined woody plants depending on their diameter (DBH) and height (10, 20, 30 m).

The energy reserves of the examined woody plants are shown in Figure 5. The highest energy reserve is again in larch (similar to dry weight) and the lowest in fir and spruce. Since the differences in the calorific values of the basic biomass fractions of studied woody plants are not large, we can conclude that higher energy content of larch trees may be caused by their higher oven-dry weight compared to other tree species.

The carbon accumulated during photosynthesis forms a substantial part of the dry weight of the tree (Table 7). At C 12 and O 16 atomic weights, CO_2 has a molecular weight of 44. It is thus 3.67 times heavier than carbon. With an average C content of 50.62–51.96%, the biomass of model trees (DBH 60 cm, h 30 m) of the studied woody plants contains from 744 kg C (spruce) to 867 kg C (pine).

Table 7. Amount of CO_2 greenhouse gas released after burning of model trees (DBH 60 cm, h 30 m).

Tree Species	*Picea abies*			*Abies alba*			*Pinus sylvestris*			*Larix decidua*		
Biomass Fraction	C		CO_2	C		CO_2	C		CO_2	C		CO_2
	(%)	(kg)	(kg)	(%)	(kg)	(kg)	(%)	(kg)	(kg)	(%)	(kg)	(kg)
Bark	49.99	50.49	185.3	51.95	89.35	327.9	51.35	42.11	154.5	52.90	143.89	528.1
Wood	50.90	595.02	2183.7	49.65	605.73	2223.0	50.78	728.19	2672.5	47.48	627.69	2303.6
Small-wood	50.55	40.95	150.3	51.80	41.96	154.0	51.25	71.24	261.5	51.49	57.15	209.7
Twigs	50.50	18.18	66.7	52.50	18.90	69.4	53.05	2.12	7.8	–	–	–
Needles	49.45	39.56	145.2	51.25	41.00	150.5	53.35	23.47	86.1	–	–	–
Model tree	50.28	744.20	2731.2	51.43	796.94	2924.8	51.96	867.13	3182.46	50.62	828.73	3041.4

Combustion of this biomass releases 2731–3182 kg of CO_2 into the atmosphere, which is generally considered to be a significant greenhouse gas. At a density of 300 trees per 1 ha, 223–260 t C is accumulated in the tree biomass, the combustion of which releases 819–955 t CO_2 into the air. At the same time, 21 to 47 kg of ash is produced. The highest proportion of ash is in spruce and pine wood (40.5–42.2 kg), the lowest in larch wood (18.8 kg). Nitrogen and sulfur are equally important for a living tree. By the burning of model trees there would be released from 1.99 kg N (larch) to 3.56 kg N (spruce, fir) into the air. Most N would be released from wood and needles. There is also most S in wood and needles. Its content in model trees varies from 0.49 kg (larch) to 1.26 kg (pine).

4. Discussion

The combustion heat values obtained in the present study are similar to those reported by other authors. The lowest combustion heat was found in wood and bark of all tree species, except for larch bark. The highest values were accumulated in the bark of larch and in needles of spruce, fir, and pine. Published calorific values for spruce biomass fractions are as follows (MJ kg^{-1}): stem wood 19.048–19.083, stem bark 18.803–19.621, whole stem 19.022–19.161, wood of branches 19.432–20.052, bark of branches 19.870–20.390, crown (branches) 19.772–20.108, foliage 19.224–19.298, and whole tree 19.286–19.478 [7]. The calorific values determined by the author for the pine biomass fractions were slightly higher (MJ kg^{-1}): stem wood 19.308–19.392, stem bark 19.529–19.981, whole stem 19.333–19.479, wood of branches 19.796–20.839, bark of branches 20.668–21.629, crown (branches) 20.234–20.873, foliage 20.800–20.950, and whole tree 19.525–19.763. The differences in the calorific values between different parts of the tree can be greater than the differences between species. Bark generally has a higher calorific value due to the high concentration of extractives and lignin. There is a large difference in the basic density between tree species and tree parts, and this difference results in differences in the heating value per unit volume. For example, the Norway spruce branches have considerably higher densities than the stem wood, whereas the bark in general has a lower density compared to the stem wood of Norway spruce and Scots pine [33].

The values that we found for spruce wood and bark (20.338 ± 318 kJ g^{-1}) are near the lower end of the range for wood in stems, branches and roots (20.36–20.79 kJ g^{-1}) and for bark (20.34–21.14 kJ g^{-1}), as reported in [34]. On the other hand, the values reported by this author for needles and fine roots (20.74–20.79 kJ g^{-1}) were slightly lower compared to our results (21.416 ± 537 kJ g^{-1} for needles and twigs). Other authors [35–38] determined calorific value for spruce wood in the range of 18.8–20.5 MJ kg^{-1}. In the middle of this range are the average calorific values of the various stumps and root fractions of Norway spruce (19.0 and 19.3 kJ g^{-1}) as reported by [38].

The gross calorific values of 20.08 ± 0.87 MJ kg^{-1} for Norway spruce, 20.79 ± 0.61 MJ kg^{-1} for the silver fir, 19.04 ± 0.70 MJ kg^{-1} for Scots pine, and 20.37 ± 0.48 MJ kg^{-1} for European larch are reported by [19]. The values reported for spruce and larch are only slightly higher, and those for fir and pine lower compared to our results. The calorific value of dry matter for European conifer species is 20.45 kJ g^{-1} [38].

As noted above, the gross calorific value of 19.04 ± 0.70 MJ kg^{-1} is reported for Scots pine [19]. Several authors determined calorific value for pine wood in the range of 19.2–21.2 MJ kg^{-1} [35–39]. These values are consistent with our results (20.612 ± 801 kJ g^{-1}). The average calorific value of a pine cone is in the range of 17.81–19.86 MJ kg^{-1}, but empty cones have a significantly higher calorific value and heat of combustion than the wood of spruce, larch and fir [40]. In the case of pine, cones and wood did not differ significantly. The calorific value of a pine cone was 18.78 MJ kg^{-1} and of pine needles 20.14 MJ kg^{-1} [41]. The last of the values is considerably lower compared to our findings (22.196 ± 253 kJ g^{-1} pine needles). For samples from the *Pinus taeda* stem the average combustion temperature was 20.0 MJ kg^{-1} and the variation range was 19.3 to 21.7 MJ kg^{-1} [42].

Closer to our results are the values for *Pinus massoniana* needles found by [43]. This author found the following ash-free calorific values in organs of Masson pines in order from the largest to the smallest: foliage (23.55 kJ g^{-1}), branches (22.25 kJ g^{-1}), stem bark (21.71 kJ g^{-1}), stem wood (21.35 kJ g^{-1}), and root (21.52 kJ g^{-1}). All of these values are slightly higher than our results. The mean ash-free calorific value of a whole tree of Masson pine was 21.74 kJ g^{-1} [6]. The calorific value decreased in the following order: foliage > branch > stem bark > root > stem wood. This value increased from the top to the lower sections of the trunk. Mean calorific values of above-ground parts were significantly higher than those of belowground parts (roots).

The determined values of combustion heat are comparable also with the values found for conifers in North America [44]. The bark of nine principal commercial timber species of the Northern Rocky Mountains (*Thuja plicata, Abies grandis, Larix occidentalis, Pinta monticola, Picea engelmannii, Pinus contorta, Tsuga heterophylla, Pseudotauga mensiesii,* and *Pinus ponderosa*) had the highest average heating value and the lowest foliage level. The principal exception was *Thuja plicata*, wherein foliage had the highest average heating values and bark had the lowest. For bark, the average heating values ranged from the low of 20.16 MJ kg^{-1} for *Thuja plicata* to a high of 25.23 MJ kg^{-1} for *Pseudotauga mensiesii* (for all species, including all determinations, it was 22.01MJ kg^{-1}). The range of average values for the twigs was from 20.26 MJ kg^{-1} for *Thuja plicata* to 23.32 MJ kg^{-1} for *Pinus ponderosa* (overall the average value was 21.48 MJ kg^{-1}). For foliage, the average values ranged from 20.24 MJ kg^{-1} for *Larix occidentalis* to 22.40 MJ kg^{-1} for *Thuja plicata* (overall the average value was 21.67 MJ kg^{-1}). These values do not differ much from our findings.

In the case of spruce, the energy reserve was affected by the density of the wood, if appropriate also the resin content. The calorific value of the resin is considerable and reaches a value of up to 36.87 kJ g^{-1} [34]. The heating value of liquid resin collected from mechanically injured southern pine trees reaches up to 34.0–37.8 MJ kg^{-1} [42]. This author found unusually high calorific values in the resin wood from the mature stumps of the ancient *Pinus palustris*. The average heat of combustion for samples from *Pinus taeda* stem wood was 20.0 MJ kg^{-1}, and the range of variation from 19.3 to 21.7 MJ kg^{-1}. Since conifers usually contain more lignin and resin than hardwoods, they tend to have slightly higher heating values per unit mass of stem wood.

For the same reason, in conifers the heating values are higher in foliage and branches than in stem wood, whereas in hardwoods the lignin content and heating value are slightly lower in branch wood [45]. The outer bark of birch wood also has an unusually high calorific value, which significantly exceeds, for example, the calorific value of the inner bark and needles [42]. In this case, however, this was due to the high suberin content.

The above data explain the higher content of the calorific value in spruce, fir, pine, and larch biomass compared to other, mainly deciduous, tree species. Trees have a stronger assimilation ability than shrubs and herbs, due to the greater leaf density and access to light [6]. Tree species are richer in energy than herbaceous species, and, in general, the energy content depends directly on the carbon content in each substance [46]. The higher fixed carbon content of a biomass feedstock, the higher the heating value [47]. Among the plant substances, the highest energy content is lignin (26.4 kJ g^{-1}), lipids (38.9 kJ g^{-1}) and terpenes (up to 46.9 kJ g^{-1}).

The calorific value of tree biomass also depends on the content of other non-woody substances. These special structures accumulate high calorific components, such as fats and proteins, which increase calorific content of biomass to some extent [48,49]. Conifer species contain wax which is even higher in calories than in fats and proteins [50].

5. Conclusions

Forest biomass is currently one of the important sources of renewable energy. The indicator of its amount is the gross and net calorific value, which depends mainly on the content of biogenic elements, moisture, and ash. The stated values, together with the values of the basic density and biomass volume taken from the yield tables, make it possible to determine both the energy density of the above-ground biomass fractions and the energy reserves of model trees and whole stands.

From the woody plants examined in this work, larch wood has the absolute lowest gross calorific value, with slightly higher values in spruce, fir and pine wood, and the highest values in larch bark and in twigs and needles of other tree species. The dry weights of larch and pine are the same and the absolute highest, and of spruce the lowest. The average energy density of the above-ground biomass of the examined woody plants decreases as follows: pine > spruce > fir > larch. The highest value was found for fir bark and the lowest for pine bark. With the same dimensions of model trees, spruce has the lowest energy reserve and larch and pine have the highest, similar to dry weight.

In this work, we focused mainly on the study of energy accumulated in above-ground fractions of four main woody plants forming the basis of coniferous stands in Slovakia. The data obtained can be useful both in planning the economic use of energy stored in the above-ground fractions of tree biomass, as well as in whole trees and stands. They can also be used in the evaluation of forest ecosystems in terms of solar energy flow, its accumulation in the individual components of tree biomass, and the risks of forest fires. Belowground biomass has not been studied in this work, so in the future it will be necessary to also focus on this issue in order to evaluate the whole cycle of storage and release of solar energy in coniferous forest ecosystems.

Author Contributions: Conceptualization, J.K. and R.P.; methodology, J.K., M.K., J.M. and R.P.; software, R.P. and M.K.; validation, J.K., R.P. and M.P.; formal analysis, M.P., D.K. and M.R.; investigation, R.P., J.M. and M.K.; resources, R.P., M.K., and M.R.; data curation, J.K. and I.P.; writing—original draft preparation, R.P.; writing—review and editing, J.K. and I.P.; visualization, J.K., M.K., M.P. and I.P.; supervision, J.K.; project administration, D.K., I.P. and J.M.; funding acquisition, R.P., M.R., M.P. and M.K. All authors have read and agreed to the published version of the manuscript.

Funding: This research was funded by the Slovak Research and Development Agency (under the contract no. APVV-16-0344 and APVV-20-0326), the Scientific Grant Agency of the Ministry of Education of Slovak Republic and Slovak Academy of Sciences (project VEGA no. 2/0009/21) and the Cultural and Educational Grant Agency of the Ministry of Education, Science, Research and Sport of the Slovak Republic (project KEGA 021SPU-4/2019).

Institutional Review Board Statement: Not applicable.

Informed Consent Statement: Not applicable.

Data Availability Statement: Not applicable.

Conflicts of Interest: The authors declare no conflict of interest.

References

1. Galluzzi, M.; Giannetti, F.; Puletti, N.; Canullo, R.; Rocchini, D.; Bastrup-Birk, A.; Chirici, G. A plot-level exploratory analysis of European forest based on the results from the BioSoil Forest Biodiversity project. *Eur. J. For. Res.* **2019**, *138*, 831–845. [CrossRef]
2. Hill, A.; Buddenbaum, H.; Mandallaz, D. Combining canopy height and tree species map information for large-scale timber volume estimations under strong heterogeneity of auxiliary data and variable sample plot sizes. *Eur. J. For. Res.* **2018**, *137*, 489–505. [CrossRef]
3. Pretzsch, H.; Del Río, M.; Biber, P.; Arcangeli, C.; Bielak, K.; Brang, P.; Dudzińska, M.; Forrester, D.I.; Klädtke, J.; Kohnle, U.; et al. Maintenance of long-term experiments for unique insights into forest growth dynamics and trends: Review and perspectives. *Eur. J. For. Res.* **2019**, *138*, 165–185. [CrossRef]
4. Affleck, D.L.R. Above-ground biomass equations for the predominant conifer species of the Inland Northwest USA. *For. Ecol. Manag.* **2019**, *432*, 179–188. [CrossRef]
5. Hytönen, J. Biomass, nutrient content and energy yield of short-rotation hybrid aspen (*P. tremula* × *P. tremuloides*) coppice. *For. Ecol. Manag.* **2018**, *413*, 21–31. [CrossRef]
6. Zeng, W.S.; Tang, S.Z.; Xiao, Q.H. Calorific values and ash contents of different parts of Masson pine trees in southern China. *J. For. Res.* **2014**, *25*, 779–786. [CrossRef]
7. Nurmi, J. Heating values of the above ground biomass of small-sized trees. *Acta For. Fenn.* **1993**, *236*, 1–30. [CrossRef]
8. Dzurenda, L.; Pňakovič, Ľ. The influence of the combustion temperature of the non-volatile combustible wood matter of de-ciduous trees upon ash production and its properties. *Acta Fac. Xylol.* **2016**, *58*, 95–104.
9. Niemz, P.; Sonderegger, W. Untersuchungen zur Korrelation ausgewählter Holzeigenschaften untereinander und mit der Rohdichte unter Verwendung von 103 Holzarten [Investigations on the correlation of selected wood properties with each other and with the bulk density using 103 species of woody plants]. *Schweiz. Z. Forstwes.* **2003**, *154*, 489–493.
10. Husch, B.; Beers, T.W.; Kershaw, J.A. *Forest Mensuration*, 4th ed.; John Wiley & Sons: New York, NY, USA, 2003; 447p.
11. Repola, J. Models for vertical wood density of Scots pine, Norway spruce and birch stems, and their application to determine average wood density. *Silva Fenn.* **2006**, *40*, 673–685. [CrossRef]
12. Petty, J.A.; Macmillan, D.C.; Steward, C.M. Variation of density and Growth Ring Width in Stems of Sitka and Norway Spruce. *Forestry* **1990**, *63*, 39–49. [CrossRef]
13. Vavrčík, H.; Gryc, V. Analysis of the annual ring structure and wood density relations in English oak and Sessile oak. *Wood Res* **2012**, *57*, 573–580.
14. Wimmer, R. Beziehungen zwischen Jahrringparametern und Rohdichte von Kiefernholz [Relationships between tree ring parameters and density of pine wood]. *Holzforsch. Holzverwert.* **1991**, *43*, 79–82.
15. Fabisiak, E.; Drogoszewski, B.; Kocjan, H.; Marcinkowska, A.; Molinski, W.; Roszyk, E. Selected physical properties of larch wood (Larix decidua Mill.) from plantation. *For. Wood Technol.* **2003**, *53*, 90–95.
16. Gryc, V.; Horáček, P.; Šlezingerová, J.; Vavrčík, H. Basic density of spruce wood with bark. and bark of branches in locations in the Czech Republic. *Wood Res* **2011**, *56*, 23–32.
17. Petráš, R.; Mecko, J.; Neuschlová, E. Density of basic components of above-ground biomass of poplar clones. *Wood Res* **2010**, *55*, 113–122.
18. Pretzsch, H. *Forest Dynamics Growth and Yield. From Measurement to Model*; Springer: Berlin/Heidelberg, Germany, 2009; 664p.
19. Aniszewska, M.; Gendek, A.; Zychowicz, W. Analysis of selected physical properties of conifer cones with relevance to energy production efficiency. *Forests* **2018**, *9*, 405. [CrossRef]
20. Wasik, R.; Michalec, K. Comparative analysis of the calorific value of giant fir timber (*Abies grandis* Lindl.) from various stands in southern Poland. *Acta Sci. Pol.—Silvarum Colendarum Ratio Ind. Lignaria* **2012**, *11*, 65–76.
21. Nurmi, J. Heating values of mature trees. *Acta For. Fenn.* **1997**, *256*, 1–28. [CrossRef]
22. Petráš, R.; Pajtík, J. Sústava česko-slovenských objemových tabuliek drevín [System of Czech-Slovak volume tables of woody plants]. *For. J.* **1991**, *37*, 49–56.
23. D'Adamo, I.; Morone, P.; Huisingh, D. Bioenergy: A Sustainable Shift. *Energies* **2021**, *14*, 5661. [CrossRef]
24. Zlatník, A. *Přehled Slovenských Lesů Podle Skupin Lesních Typů [Overview of Slovak Forests by Groups of Forest Types]*; VŠZ: Brno, Czechoslovakia, 1959; 195p.
25. Zlatník, A. Přehled skupin typů geobiocénů původně lesních a křovinných [Overview of groups of types of geobiocoenes originally forest and shrubby in the C.S.S.R.]. *Zprávy Geogr. úst. ČSAV Brno* **1976**, *13*, 55–64.
26. STN EN ISO 18125. *Solid Biofuels—Determination of Calorific Value. Office for Standardization*; Vestník ÚNMS SR 10/17; Metrology and Testing of the Slovak Republic: Bratislava, Slovakia, 2017; 56p.
27. STN EN ISO 16948. *Solid Biofuels—Determination of Total Content of Carbon, Hydrogen and Nitrogen*; Vestník ÚNMS SR 10/15; Office for Standardization, Metrology and Testing of the Slovak Republic: Bratislava, Slovakia, 2016; 9p.
28. STN EN ISO 16994. *Solid Biofuels—Determination of Total Content of Sulfur and Chlorine*; Vestník ÚNMS SR. č. 11/16; Office for Standardization, Metrology and Testing of the Slovak Republic: Bratislava, Slovakia, 2016; 11p.
29. Telmo, C.; Lousada, J.; Moreira, N. Proximate analysis, backwards stepwise regression between gross calorific value, ultimate and chemical analysis of wood. *Bioresour. Technol.* **2010**, *101*, 3808–3815. [CrossRef] [PubMed]
30. STN EN ISO 18122. *Solid Biofuels— Determination of Ash Content*; Vestník ÚNMS SR11/16; Office for Standardization, Metrology and Testing of the Slovak Republic: Bratislava, Slovakia, 2016; 6p.

31. Petráš, R.; Mecko, J.; Krupová, D.; Slamka, M.; Pažitný, A. Above-ground biomass basic density of softwoods tree species. *Wood Res.* **2019**, *64*, 205–212.

32. Petráš, R.; Košút, M.; Oszlányi, J. Listová biomasa stromov smreka, borovice a buka [Leaf biomass of spruce, pine and beech trees]. *For. J.* **1985**, *31*, 121–136.

33. Hakkila, P. *Utilization of Residual Forest Biomass*; Springer Series in Wood Science; Springer: Berlin/Heidelberg, Germany, 1989; 568p.

34. Ellenberg, H. *Vegetation Ecology of Central Europe*, 4th ed.; Strutt, G.K., Ed.; Cambridge University Press: Cambridge, UK, 1973; 756p.

35. Günther, B.; Gebauer, K.; Barkowski, R.; Rosenthal, M.; Bues, C.-T. Calorific value of selected wood species and wood products. *Eur. J. Wood Wood Prod.* **2012**, *70*, 755–757. [CrossRef]

36. So, C.L.; Eberhardt, T. A mid-IR multivariate analysis study on the gross calorific value in longleaf pine: Impact on correlations with lignin and extractive contents. *Wood Sci. Technol.* **2013**, *47*, 993–1003. [CrossRef]

37. Stolarski, M.; Szczukowski, S.; Tworkowski, J.; Krzyżaniak, M.; Gulczyński, P.; Mleczek, M. Comparison of quality and production cost of briquettes made from agricultural and forest origin biomass. *Renew. Energy* **2013**, *57*, 20–26. [CrossRef]

38. Uri, V.; Aosaar, J.; Varik, M.; Becker, H.; Kukumägi, M.; Ligi, K.; Pärn, L.; Kanal, A. Biomass resource and environmental effects of Norway spruce (Picea abies) stump harvesting: An Estonian case study. *For. Ecol. Manag.* **2015**, *335*, 207–215. [CrossRef]

39. Runge, M. *Energieumsätze in den Biozönosen terrestrischer Ökosysteme. Untersuchungen im "Sollingprojekt" [Energy Turnover in the Biocenoses of Terrestrial Ecosystems. Investigations in the "Solling Project"]*; Scripta Geobotanica 4; Verlag Erich Goltze KG: Göttingen, Germany, 1973; 74p.

40. Aniszewska, M.; Gendek, A. Comparison of heat of combustion and calorific value of the cones and wood of selected forest trees species. *For. Res. Pap.* **2014**, *75*, 231–236. [CrossRef]

41. Font, R.; Conesa, J.A.; Moltó, J.; Muñoz, M. Kinetics of pyrolysis and combustion of pine needles and cones. *J. Anal. Appl. Pyrolysis* **2009**, *85*, 276–286. [CrossRef]

42. Howard, E.T. Heat of combustion of various southern pine materials. *Wood. Sci.* **1973**, *5*, 194–197.

43. Guan, L.L.; Zhou, X.Y.; Luo, Y. A review on the study of plant caloric value in China. *Chin. J. Ecol.* **2005**, *24*, 452–457.

44. Kelsey, R.G.; Shafizadeh, F.; Lowery, D.P. Heat Content of Bark, Twigs, and Foliage of Nine Species of Western Conifers. Forestry, Paper 69, 1979. USDA Forest Service, Research Note INT-261. 1979. Available online: https://digitalcommons.usu.edu/govdocs_forest/69 (accessed on 20 August 2020).

45. Harris, R.A. Fuel values of stems and branches in white oak, yellow-poplar and sweetgum. *For. Prod. J.* **1984**, *34*, 25–26.

46. Larcher, W. *Physiological Plant Ecology: Ecophysiology and Stress Physiology of Functional Groups*; Springer: Berlin/Heidelberg, Germany, 2003.

47. Nelson, N.; Darkwa, J.; Calautit, J.; Worall, M.; Mokaya, R.; Adjei, E.; Kemausuor, F.; Ahiekpor, J. Potential of Bioenergy in Rural Ghana. *Sustainability* **2021**, *13*, 381. [CrossRef]

48. Sinclair, T.R.; Vadez, V. Physiological traits for crop yield improvement in low N and P environments. *Plant Soil* **2002**, *245*, 1–15. [CrossRef]

49. Qu, G.; Wen, M.; Guo, J. [Energy accumulation and allocation of main plant populations in Aneurolepidium chinense grassland in Songnen Plain]. *Ying Yong Sheng Tai Xue Bao = J. Appl. Ecol.* **2003**, *14*, 685–689.

50. Jin, Z.X.; Li, J.M.; Ma, J.E. Ash content and caloric value in the leaves of Sinocalycanthus chinensis and its accompanying spe-cies. *Acta. Ecol. Sin.* **2011**, *31*, 5246–5254.

Article

Potential Use of Plant Biomass from Treatment Wetland Systems for Producing Biofuels through a Biocrude Green-Biorefining Platform

Marco Antonio Rodriguez-Dominguez [1,2,*], **Patrick Biller** [3], **Pedro N. Carvalho** [2,4], **Hans Brix** [1,2] **and Carlos Alberto Arias** [1,2]

[1]   Department of Biology, Aarhus University, Ole Worms Allé 1, Building 1135, 8000 Aarhus, Denmark; hans.brix@bio.au.dk (H.B.); carlos.arias@bio.au.dk (C.A.A.)

[2]   Aarhus University Centre for Water Technology WATEC, Aarhus University, Ny Munkegade 120, Building 1521, 8000 Aarhus, Denmark; pedro.carvalho@envs.au.dk

[3]   Department of Biological and Chemical Engineering—Process and Materials Engineering, Aarhus University, Hangøvej 2, 8200 Aarhus, Denmark; pbiller@bce.au.dk

[4]   Department of Environmental Science, Aarhus University, Frederiksborgvej 399, 4000 Roskilde, Denmark

*   Correspondence: mard@bio.au.dk; Tel.: +45-521-722-560-7063

Citation: Rodriguez-Dominguez, M.A.; Biller, P.; Carvalho, P.N.; Brix, H.; Arias, C.A. Potential Use of Plant Biomass from Treatment Wetland Systems for Producing Biofuels through a Biocrude Green-Biorefining Platform. *Energies* **2021**, *14*, 8157. https://doi.org/10.3390/en14238157

Academic Editors: Idiano D'Adamo and Piergiuseppe Morone

Received: 26 October 2021
Accepted: 3 December 2021
Published: 5 December 2021

Abstract: The potential of using the biomass of four wetland plant species (*Iris pseudacorus, Juncus effusus, Phragmites australis* and *Typha latifolia*) grown in treatment wetland systems and under natural conditions were tested to produce high-value materials using hydro-thermal liquefaction (HTL). The results show that the wetland plants biomass is suitable for biocrude and biochar production regardless of the origin. The hydrothermal liquefaction products' (biocrude, biochar, aqueous and gaseous phase) yields vary according with the specific biomass composition of the species. Furthermore, the results show that the biomass composition can be affected by the growing condition (treatment wetland or natural unpolluted conditions) of the plants. None of the single components seems to have a determinant effect on the biocrude yields, which reached around 30% for all the analyzed plants. On the contrary, the biochar yields seem to be affected by the composition of the biomass, obtaining different yields for the different plant species, with biochar yields values from around 12% to 22%, being that *Phragmites australis* is the one with the highest average yield. The obtained aqueous phase from the different plant species produces homogeneous compounds for each plant species and each growing environment. The study shows that biomass from treatment wetlands is suitable for biocrude production. The environmental value of this biomass lies on the fact that it is considered a residual product with no aggregated value. The treatment wetland biomass is a potential sustainable source for biofuel production since these plants do not need extra land or nutrients for growing, and the biomass does not compete with other uses, offering new sources for enhancing the bioeconomy concepts.

Keywords: treatment wetlands; biocrude; hydrothermal liquefaction; biomass; biorefinery; biofuels; wastewater treatment; biochar; aqueous phase

1. Introduction

Currently, the world faces diverse and serious environmental challenges. The transport sector, responsible for a significant share of CO_2 emissions, and other sectors such as the agriculture, forestry, or manufacturing ones, demand for sustainable, resilient, affordable, and fair sources of energy that can ease the pressure exerted to nature.

To face those challenges, the bioeconomy model proposes the use of renewable biological resources from the land and sea (e.g., animals, crops, fish, forests and microorganisms) to produce energy, food and materials [1] to reduce the environmental impact of human activity.

In this context, the combined use of green biorefining (GBR) technologies and nature-based solutions (NBS) brings the opportunity of controlling pollution as well as recovering neglected biological resources for use as a sustainable source of biomass for biofuel and later bioenergy production. GBR is a complex and full-integrated technology system that aims to protect the environment and natural resources, making comprehensive use of materials and energy available in green biomasses [2]. NBS are technologies focused in establishing ecosystem services to address societal challenges, such as climate change, food and water security, to tackle natural disasters, and to enhance well-being and biodiversity benefits [3].

Treatment wetlands (TW) are one of the most attractive NBS for wastewater treatment; TW technology mimics processes present in nature, but optimizes them through engineering designs [4]. TW are systems where wastewater, local climate, and energy (mainly coming from the sun) are integrated into an engineered ecosystem formed by plants, bacteria, and depending on the type of TW, filling media to remove pollutants from waters aiming at meeting local discharge standards. The technology is convenient due to the relatively low operative and maintenance cost [5,6], and other ancillary benefits, such as sustaining biodiversity [7], climate change mitigation [8], carbon sequestration [9], hydrological flow regime regulation, public use, education, and habitat conservation [10].

Until now, TW were considered a sustainable technology for wastewater treatment effective for pollution control of domestic wastewater [11–14], rainwater run-off [15–17], slaughterhouses [18,19], industrial [20–22] and urban sewage [8,23], and many other types of polluted waters. However, TW seem to offer a novel and sustainable opportunity to recover neglected resources coming from the produced biomass which can be used to produce plant-based materials and fuels.

To use the TW biomass is attractive because of (1) the high primary productivity reported for plants grown in CW [24], (2) the enhanced capacity for removing pollutants reported from harvested TW [25,26], and (3) the promising results from previous studies regarding the potential for producing high-value products from residual biomass through green biorefining (GBR) processes [2,27,28].

GBR technologies are based on the utilization of plant or other organic materials to produce different high-value materials that can substitute the petroleum-based materials such as plastics, fuels, textiles, and chemicals [29,30]. One of the most promising GBR processes seems to be hydrothermal liquefaction (HTL) technology, which produces biocrude. The HTL process transforms biomass, using high pressure and high temperature under subcritical conditions to obtain a combination of biocrude that can produce biofuel, biochar, gases (mainly CO_2), and an aqueous phase (AqP), rich in organic compounds. The hydrothermal process is essentially similar to how nature has produced fossil fuels over millions of years, heating aqueous organic slurries at elevated pressures to produce an energy carrier with increased energy density [31].

The biomass composition used for HTL processes determines the composition and the yields of the products of the reaction, named biocrude, biochar and AqP. In previous studies, microalgae (*Chlorella*) slurry was processed, obtaining 40 wt% biocrude yields [32]. *Miscanthus*, *Spirulina* and sewage sludge produced biocrudes with average yields of 26 wt%, 33 wt% and 25 wt%, respectively [33]. Maize, oats, and ryegrass had yields of 27.6 wt%, 25.1 wt%, and 22 wt%, respectively [34]. These observations confirm the fact that the predictability of biocrude characteristics is still a topic that needs attention [35].

The HTL process was previously evaluated for the valorization of wastewater-treatment wastes, offering an alternative solution for sludge management. Sludge has been used for biocrude production, reporting yields from 37 wt% to 43 wt% [36]. Other studies have evaluated the biocrude production from willow biomass irrigated with wastewaters at supercritical water conditions (400 °C), reporting a yield of 40 wt% [37].

However, none of the previous studies have evaluated the differences in the capacity of the TW plants for producing biocrude compared with the same plants grown in natural conditions (NC). This study aims to assess the potential of four of the most used plants in

TW for producing biocrude through the HTL technology. The study includes a comparison between the biocrude and biochar yields, as well as the AqP composition, to evaluate how the TW conditions affect the potential of the plant biomass to be used in a HTL platform. The energy balance from the biocrude and biochar is also determined to estimate the potential to produce environmentally friendly renewable biofuels. Additionally, this research evaluates the AqP composition, including a screening of the content of emergent pollutants after the HTL reaction, highlighting the potential environmental problems that can be faced due the complex composition of it. Lastly, a mass balance study is performed to track the fate of carbon (C) and nitrogen (N) in the HTL process, to assess where in the environment those elements could potentially end up. When the C and N of the feed biomass are converted to biocrude, biochar, AqP or a gaseous state, they can later either be emitted to the atmosphere as a gas if the biochar or biocrude are used as fuels, be discharged in the water if the AqP is not used in any other process, or be added to the soil if the biochar is used as a soil amendment acting also as a carbon sequestrator [38–40].

This article is a micro-approach to the development of biofuels from raw materials. The study assesses the potential of TW biomass to be considered as a source of new, renewable biomass to produce biofuels. The biomass from TW is considered sustainable because it reduces the environmental impact, being grown in sustainably way (in the TW), being considered until now a residue, and having not conflicted with land use for agricultural products [1]. TW biomass is considered a by-product with no or low value, but due to the operational characteristics, is an unlimited source of biomass. Up to our knowledge, this approach has not been studied. The combined use of TW for wastewater treatment and biorefinery processes to produce biofuels from the TW biomass can contribute to reaching the UN sustainable goals dealing with the water management and the recovering of resource in a circular bioeconomy. The use of biomass that otherwise would be disposed of can contribute to reaching some of the UN goals as follows: "6, clean water and sanitization", "7, Affordable and clean energy", "11, Sustainable cities and communities", "12, Responsible consumption and production", and "13, climate action." [41].

2. Materials and Methods

2.1. The Plants

During the summer of 2020, four plant species, from two different locations, were selected and harvested to compare their potential for being used in the HTL process. Triplicates of individual samples of *Iris pseudacorus*, *Juncus effusus*, *Phragmites australis*, and *Typha latifolia* were harvested from two different locations. The first place was a local TW system located in the Tilst neighborhood in Aarhus, Denmark, (56°10′20.5″ N, 10°06′21.9″ E). The same plant species were harvested during the same period from Påskehøjgaard growth facilities of Aarhus University (56°13′48.7″ N, 10°07′38.3″ E), where the plants were grown under natural conditions (not exposed to pollutants), referred to herein as NC (natural conditions).

2.2. Ash Content and Fixed Carbon

The plants were dried at 65 °C for 72 h to drive off water. Samples of the residue were cooled, weighed, and combusted at 550 °C for 3 h to drive off volatile solids. The total solids, volatile solids, and ash content were determined by comparing the mass of the sample before and after each step. The method followed was that of Ref. [42].

A TGA Mettler Toledo SDTA851 was used to analyze the raw materials and biochar samples. The TGA was operated using a constant heating rate of 10 K min^{-1} from 50 °C to 900 °C under nitrogen followed by 10 min under air at constant temperature. A minimum of 5 mg of the mass sample was placed in the TGA ceramic crucibles. The fixed carbon was obtained by calculating the mass difference in the sample, between the weight at the air injection and the final weight in the test.

2.3. Elemental Analysis

For the CHNS content, the solid and biocrude samples were determined using an Elementar vario Macro Cube elemental analyzer (Langenselbold, Germany). The protein content was estimated using a nitrogen Jone's factor of 6.25 [43].

The higher heating value (HHV) was calculated according to the the Channiwala–Parikh [44] correlation

$$\mathrm{HHV}\left[\frac{\mathrm{MJ}}{\mathrm{kg}}\right] = 0.3491C + 1.1783H + 0.1005S - 0.1034O - 0.0151N - 0.0211A \quad (1)$$

The equation used to calculate the energy recovery of the biocrude, and biochar was:

$$\text{Energy yield } (\eta_{th}) = \frac{\mathrm{HHV}_{\mathrm{oil}}\left[\frac{\mathrm{MJ}}{\mathrm{kg_{oil}}}\right] \cdot \mathrm{yield}_{\mathrm{oil}}\left[\frac{\mathrm{kg_{oil}}}{\mathrm{kg_{feed}}}\right]}{\mathrm{HHV}_{\mathrm{feed}}\left[\frac{\mathrm{MJ}}{\mathrm{kg_{feed}}}\right]} \times 100 \quad (2)$$

$$\text{Energy yield } (\eta_{th}) = \frac{\mathrm{HHV}_{\mathrm{char}}\left[\frac{\mathrm{MJ}}{\mathrm{kg_{char}}}\right] \cdot \mathrm{yield}_{\mathrm{char}}\left[\frac{\mathrm{kg_{char}}}{\mathrm{kg_{feed}}}\right]}{\mathrm{HHV}_{\mathrm{feed}}\left[\frac{\mathrm{MJ}}{\mathrm{kg_{feed}}}\right]} \times 100 \quad (3)$$

2.4. Compositional Analysis

The structural carbohydrates and lignin content were determined according to [45]. The value of the hemicellulose is from the addition of the obtained values of xylan, arabinan, and galactan. The values of lignin were obtained by the addition of the Klason lignin and the soluble lignin content.

2.5. HTL Reaction

To perform hydrothermal liquefaction of the biomass, and determine the product yields of bio-crude, gas, solid residue (biochar), and water-soluble products (aqueous phase), the reactions were carried out at 340 °C for 15 min residence time in small bomb-type 20 mL batch reactors, and no catalyst was used. Biomass slurries were prepared by mixing 20 wt% and 80 wt% demineralized water. The feedstock consisted of the 8 biomasses described previously. Reactors were sealed and lowered into a preheated fluidized sand bath at 340 °C, then 20 min reaction time was applied. Subsequently, the reactor were cooled quickly to ambient temperature in a water bath. The reactors were vented, and the aqueous phase (AqP) was decanted into a centrifuge tube and centrifuged for 5 min before the AqP was transferred with a glass pipette to a preparative glass. The AqP was then stored at 5 °C for further analysis. The centrifuge tube was washed with 2 mL of dichloromethane and the reactor was extracted with around 4 mL of dichloromethane, which were combined. The dichloromethane phase was vacuum filtered, and the residue washed with dichloromethane until the filtrate appeared clear. Dichloromethane was evaporated under a stream of nitrogen until a constant weight. Each experiment was performed in triplicate and the average values are reported.

2.6. COD, TOC, and TN in Aqueous Phase

AqP samples were analyzed for chemical oxygen demand (COD) content, using Merck Spectroquant cell. The total organic carbon (TOC) and total nitrogen (TN) of the AqP samples were analyzed using a scalar FORMACS HT-I TOC/TN analyzer.

2.7. High-Resolution Mass Spectrometry (HRMS) Analysis of the Aqueous Phase

For a broader screening of the chemical composition of the AqP, a liquid chromatograph (LC) coupled to a 6600 quadrupole-time-of-flight (QTOF) instrument (SCIEX) was used. An Acquity UPLC BEH Shield RP18 1.7 μm column (2.1 mm × 30 mm) (Waters) was used for chromatographic separation. Eluents A and B were water and methanol, both with 0.1% formic acid, respectively. The chromatographic conditions, as well as detailed parameters of the instrument, were similar to those used before [46]. The samples were injected directly (10 µL) to the LC-QTOF and measured in a data-dependent acquisition mode (DDA), with a TOF mass range 100–1000 Da, 250 ms accumulation time, and successive fragmentation and recording of product ion spectra of the 10 highest peaks for 40 ms each. The samples were measured in positive and negative ionization modes with electrospray ionization (with ±5300 V ion spray voltage).

The high-resolution MS data were processed with Marker View (SCIEX) for the molecular feature extraction (feature is defined by exact m/z and retention time), according to [46]. Briefly, one feature was only considered if present in 3 out of the 3 injected replicate samples with a mass tolerance of 20 ppm and a retention time tolerance of 0.1 min. All features present in control blank samples were filtered.

2.8. Statistics

The differences of the analyzed parameter between the two different selected sites (TW and NC) were analyzed using an ANOVA test. Prior to statistical analysis, all data were tested for homogeneity of variance by Levene's test. For clarity, all data are presented as untransformed values. Post hoc Tukey HSD tests were applied to identify significant differences between samples. All statistical analyses were conducted in R studio at a significance level of 0.05 and all figures were prepared in GraphPad Prism 7.00.

For the AqP analysis, a principal component analysis was performed with no weighting and Pareto scaling to analyze the triplicate samples from the 4 plant species, both TW and NC.

3. Results

3.1. Characterization of Harvested Biomass

The characterization of the biomass is relevant since the composition and yield of HTL products depend strongly on the type and composition of the biomass used as feed [39]. A first characterization of the biomass, shown in Table 1, reports the content of cellulose, hemicellulose, lignin, protein, and ash found in the analyzed biomass. A second characterization of the biomass, shown in Table 2 shows the results of the elemental analysis of the feed biomass, reported in C, H, O, N, and S.

From Table 1, it is possible to observe that the cellulose content in the plants did not differ between the two growing environments, having no significative difference for the *Iris pseudacorus*, *Phragmites australis* and *Typha latifolia*, and being lower for around 4.5% for the *Juncus effusus*, grown in TW. The hemicellulose content had a lower concentration in the *Phragmites australis* and *Typha latifolia*, being lower for around 6.6% and 5% respectively. Regarding lignin content, the plants showed significative differences between both growing environments, having a lower value in around 0.3% for *Iris pseudacorus* and *Juncus effusus*, grown in TW, and higher concentration in around 0.3% for *Phragmites australis* and *Typha latifolia* grown in TW. The protein content was significantly higher for all the plant species grown in TW, with differences between 0.5% and 1.4%. Lastly, the ash content was also higher for all the plant species grown in TW conditions, reporting differences of 1.6, 1.3, 0.3, and 2.7% for the *Iris pseudacorus*, *Juncus effusus*, *Phragmites australis* and *Typha latifolia*, respectively.

Regarding the elemental analysis of the biomass, only the N content was higher for all the TW plants. The rest of the elements in the plants did not present significant differences or showed lower concentration in the plants grown in TW. For *Juncus effusus*, *Phragmites australis*, and *Typha latifolia*, the C content was lower in the TW plants. For *Phragmites australis* and *Typha latifolia*, the H content was lower in the TW, and the O was lower in the TW for *Iris pseudacorus*, being that the S in the *Iris pseudacorus* was the only element and plant that showed a higher content for the TW.

This confirms that the composition of the plants grown in TW is not negatively affected due to the presence of wastewater, comparable to other plants grown as a biological source for the production of biofuels grown under natural conditions.

Table 1. Cellulose, hemicellulose, lignin, and protein content in the selected plants for the different growing environments.

Plant Description	Cellulose (%)			Hemicellulose (%)			Lignin (%)			Protein (%)			Ash (%)		
	TW		NC	TW		NC	TW		NC	TW		NC	TW		NC
1 Iris pseudacorus	37.6 ± 2.6	=	38.9 ± 0.4	17.8 ± 4.5	=	14.6 ± 1.7	5.6 ± 0.4	↓	5.8 ± 0.6	4.8 ± 0.5	↑	3.5 ± 0.3	8.6 ± 0.0	↑	7.0 ± 0.0
2 Juncus effusus	36.9 ± 2.6	↓	41.5 ± 4.9	32.8 ± 2.1	=	32.3 ± 1.0	5.8 ± 0.6	↓	6.2 ± 0.7	6.7 ± 0.7	↑	4.5 ± 0.3	5.0 ± 0.2	↑	3.7 ± 0.0
3 Phragmites australis	35.0 ± 4.3	=	37.2 ± 1.1	21.0 ± 4.6	↓	27.6 ± 2.9	5.9 ± 0.5	↑	5.6 ± 1.0	7.2 ± 0.2	↑	6.4 ± 0.4	5.0 ± 0.0	↑	4.7 ± 0.0
4 Typha latifolia	41.4 ± 4.0	=	37.9 ± 3.8	17.2 ± 4.7	↓	22.2 ± 5.3	7.1 ± 2.4	↑	5.8 ± 1.9	5.8 ± 0.6	↑	4.7 ± 0.2	8.5 ± 0.1	↑	5.8 ± 0.1

↑ Indicates a positive effect of the TW in the concentration of the reported element. ↓ Indicates a negative effect of the TW in the concentration of the reported element. = Indicates a not significative effect of the TW in the concentration of the reported element.

Table 2. C, H, O, N, S content in the selected plants for the different growing environments.

Plant Description	C [%]			H [%]			O [%]			N [%]			S [%]		
	TW		NC	TW		NC	TW		NC	TW		NC	TW		NC
1 Iris pseudacorus	42.4 ± 0.2	=	41.4 ± 0.1	6.4 ± 0.1	=	6.5 ± 0.0	41.8 ± 0.2	↓	44.6 ± 0.2	0.7 ± 0.0	↑	0.5 ± 0.0	0.2 ± 0.1	↑	0.0 ± 0.0
2 Juncus effusus	43.9 ± 0.1	↓	45.5 ± 0.2	6.5 ± 0.1	=	6.7 ± 0.1	43.4 ± 0.3	=	43.4 ± 0.2	1.1 ± 0.1	↑	0.7 ± 0.0	0.1 ± 0.0	=	0.1 ± 0.0
3 Phragmites australis	44.9 ± 0.0	↓	46.1 ± 0.1	6.4 ± 0.1	↓	6.7 ± 0.0	42.4 ± 0.0	=	41.2 ± 0.1	1.2 ± 0.0	↑	1.1 ± 0.0	0.2 ± 0.1	=	0.2 ± 0.0
4 Typha latifolia	41.9 ± 0.2	↓	45.4 ± 3.5	6.1 ± 0.1	↓	6.6 ± 0.5	42.6 ± 0.2	=	41.4 ± 4.2	0.9 ± 0.0	↑	0.8 ± 0.0	0.1 ± 0.0	=	0.0 ± 0.0

↑ Indicates a positive effect of the TW in the concentration of the reported element. ↓ Indicates a negative effect of the TW in the concentration of the reported element. = Indicates a not significative effect of the TW in the concentration of the reported element.

3.2. HTL Yields

Table 3 presents the biocrude yields obtained from the different reactions and biomass. Biocrude yields presented a significative difference only for the *Typha latifolia*, which reported a lower biocrude yield (-6%) in the TW plants. For the biochar, the differences were significant for almost all the plant species, decreasing the yields in the case of *Iris pseudacorus* and *Juncus effusus*, and increasing it for the *Phragmites australis*, grown in TW, without any difference for *Typha latifolia*. The results suggest that the yields of biochar are more sensible than the yields of biocrude to the plant species and composition.

Table 3. Fraction yields for all the selected plant for the different growing environments after HTL reaction.

	Plant Description	Biocrude (%)			Biochar (%)			AqP (%)		Gas (%)	
		TW		NC	TW		NC	TW	NC	TW	NC
1	*Iris pseudacorus*	25.6 ±1.2	=	29.6 ± 6.5	12.4 ± 2.4	↓	22.0 ± 3.3	16.5 ± 8.4	16.7 ± 3.5	26.6 ± 3.7	32.0 ± 5.8
2	*Juncus effusus*	31.7 ± 2.6	=	33.9 ± 3.1	14.5 ± 0.7	↓	17.4 ± 1.4	31.2 ± 18.7	22.7 ± 18.8	22.6 ± 18.4	27.1 ± 13.7
3	*Phragmites australis*	26.3 ± 6.2	=	28.6 ± 0.5	23.3 ± 1.7	↑	19.3 ± 3.0	21.6 ± 6.4	20.8 ± 6.2	26.8 ± 3.1	31.4 ± 5.4
4	*Typha latifolia*	31.6 ± 8.9	↓	37.6 ± 7.9	18.7 ± 0.2	=	19.3 ± 6.2	20.1 ± 6.5	15.5 ± 6.9	30.3 ± 5.1	47.9 ± 7.7

↑ Indicates a positive effect of the TW in the concentration of the reported element. ↓ Indicates a negative effect of the TW in the concentration of the reported element. = Indicates a not significative effect of the TW in the concentration of the reported element.

In general, it is expected that the biocrude formation follows the trend lipids/fats > proteins > carbohydrates [40]; nonetheless, it is possible to observe from the results of this study that this trend does not fit with the results obtained from the lignocellulosic analyzed biomass. From Figure 1, it is possible to observe that the biomass with the highest content of protein is not the ones with the highest biocrude or biochar yields, e.g., meanwhile, the plant 3A (*Phragmites australis* grown in TW) is the one with the highest content of protein, the biocrude yield is one of the lowest, and the biochar yield one of the highest. It is important to highlight that the differences in protein content in the biomass, do not represent an increase larger than 3.5% between the protein content of the plant with the highest content and the lowest. Additionally, the highest protein content of all species is 7%, meaning that protein has a relatively small contribution to the overall bio-crude yield; hence, the slight change in protein contents amongst samples does not follow the general trend of biocrude lipids/fats > proteins > carbohydrates.

It was described that the C5 and C6 carbohydrates (from cellulose and hemicellulose) tend to produce mainly biochar in the HTL reactions [31], but according with the obtained results, the plants with the highest content of cellulose (plant 2B and 4A) did not report the highest yield of biochar. In other case, the plant with the highest biochar yield (3A) reported the highest content of protein but the lowest content of cellulose.

From the particular plant analysis, it is not possible to observe any pattern, e.g., in the case of *Phragmites australis*, which reported a significative lower concentration of hemicellulose in the TW plants, the biochar yield was significantly higher, and for *Typha latifolia*, even the lower concentration of hemicellulose in the TW plants, the biochar yield did not show significative differences.

Nevertheless, the results do not contradict the results from other authors since the feed biomass is not a uniform mass with a completely known content. On the contrary, lignocellulosic biomasses are complex systems, with some unknown components, and even though is possible to know the generality of them, when the biomass is forced to react under HTL conditions, all the present elements have a role in the reaction and interfere with the final biocrude and biochar yields and composition.

The interaction in the HTL between the protein, the saccharide and the lignin were studied previously. A study mixed soya with cellulose, xylan, and alkaline lignin to model the interactions of the protein with the other elements, reaching models with an accuracy of around 94% [47]. Some others have studied synergistic effects between the lignocellulosic biomass compounds, reporting that mixtures of protein and cellulose, protein and xylose, cellulose and lignin, and xylose and lignin seem to have synergistic effects on biocrude yield, and mixtures of soybean oil and lignin showed an antagonistic effect [48]. The results seem to confirm the synergistic effect between the protein and cellulose, lignin, and hemicellulose. Figure 1 shows that plants with the higher lignin and protein and low ashes, such as 2A and 2B, report biocrude yields over 30%.

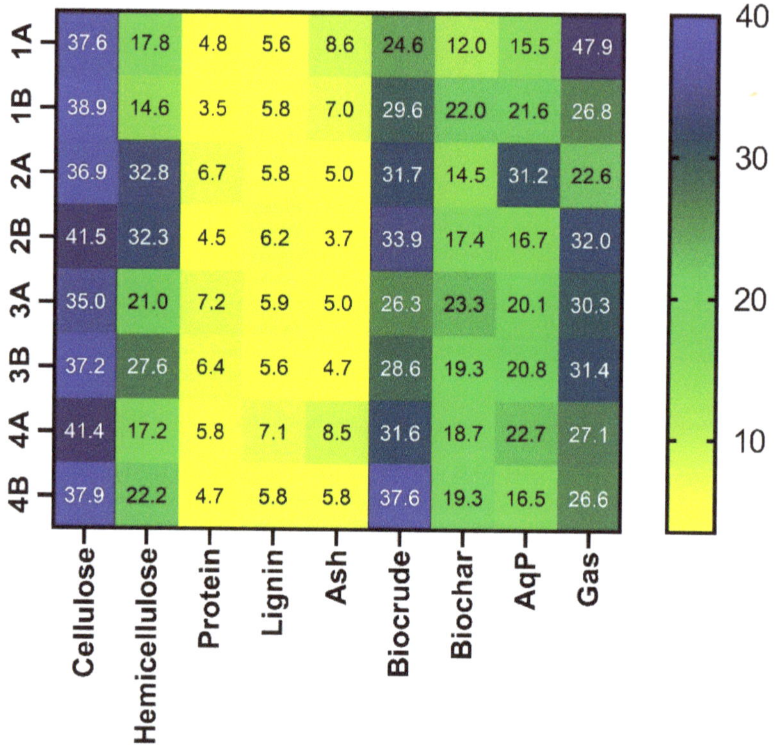

Figure 1. Heat map comparing feed composition and HTL fractions yields. The values in each square are the mean content of each parameter named in the columns. The content is reported in mass fraction (%). In the rows, the numbers refer to each plant species named: 1, *Iris pseudacorus*; 2, *Juncus effusus*; 3, *Phragmites australis*; and 4, *Typha latifolia*. The letters A and B refer to the growing environments: A = TW, and B = NC environment.

The predictably of the HTL fraction yields still needs to be studied further. This study can elucidate how the lignocellulosic biomass, even with no significative differences regarding its composition, can produce different yields. Nonetheless, and despite the differences, the results are consistent with previous studies regarding biocrude, and biochar yields produced from lignocellulosic biomass. The yields, in all the cases, range from 25.6 ± 1.2% to 37.6 ± 7.9% for biocrude, and from 12.4 ± 2.4% to 23.3 ± 1.7% for the biochar, which correspond with previous studies about cellulosic biomass used in the HTL reaction, such as biocrude yields of 26% obtained from *Miscanthus* [33], 22% reported for ryegrass [34], or the yield of 39.7% reported for willow biomass at supercritical water conditions (400 °C) [37].

These results are important because they provide new information from the simultaneous evaluation of different plant species of different composition for producing biocrude and the different HTL subproducts (biochar, AqP, and gases), showing that even with plants with different composition, the biocrude yields are consistent and comparable. However, the results also indicate that the predictions about biocrude yields can be more effective if they are made based on the plant species more than the specific composition. There is no significative difference between the biocrude yield from plants of the same species grown in TW or NC, but a difference between plant species was found, being that *Juncus effusus* and *Typha latifolia* are the most attractive plants for the production of biocrude.

3.3. Characterization of the HTL Products

3.3.1. Biocrude and Biochar

From Tables 4 and 5 is observed that the N, C, and S content in the plants is densified in the biocrude and biochar, the H is densified only in the biocrude, and O decreases its density in both, biocrude and biochar, this pattern being consistent with previous studies. The N content in the produced biocrude from TW plants seems to have a lower concentration than biocrudes previously studied, e.g., biocrude obtained from *Miscanthus* [33] reported N content from 1.0% to 1.6%, which is a higher concentration than that obtained in this study, from 0.5% to 1.2%. Additionally, the O reported for the biocrude in this study is also higher than the 17.7% reported for the biocrude obtained from *Miscanthus* in the same reference. These differences in the elemental analysis influence the HHV of the biocrudes, and together with the product yield can define the global value or potential of the studied biocrudes.

Table 4. C, H, O, N, and S content in the biocrude produced by the selected plants for the different growing environments.

	Plant Description	C [%]		H [%]		O [%]		N [%]		S [%]	
		TW	NC	TW	NC	TW	NC	TW	NC	TW	NC
1	*Iris pseudacorus*	66.9 ± 11.4	58.3 ± 24.0	7.37 ± 0.6	9.4 ± 3.1	22.9 ± 12.0	30.8 ± 27.8	2.6 ± 0.0	1.5 ± 0.6	0.3 ± 0.0	0.1 ± 0.1
2	*Juncus effusus*	61.4 ± 3.8	49.1 ± 27.4	8.61 ± 0.5	5.1 ± 2.8	27.6 ± 3.9	44.0 ± 31.2	1.6 ± 0.1	1.6 ± 0.9	0.8 ± 0.4	0.2 ± 0.2
3	*Phragmites australis*	69.4 ± 4.3	63.2 ± 6.2	8.47 ± 0.9	7.6 ± 0.7	18.7 ± 4.2	26.9 ± 6.7	3.2 ± 0.1	2.1 ± 0.4	0.3 ± 0.2	0.2 ± 0.0
4	*Typha latifolia*	64.9 ± 5.3	63.5 ± 10.9	8.38 ± 0.4	8.4 ± 0.7	24.0 ± 6.2	26.3 ± 12.1	2.5 ± 0.9	1.6 ± 0.5	0.2 ± 0.1	0.2 ± 0.1

The results offer new information regarding the distribution of elements in the different plant species used in TW. The higher amount of N from the TW plants results in higher concentration of N in the resultant biocrude and the biochar. This is relevant for future studies as well as for the refining of the biocrude since excess N can affect the quality and can potentially generate N emissions.

3.3.2. Aqueous Phase

The elements that constitute this HTL fraction are multiple and with an undefined number of chemicals. In this study, a first approach to the chemical content of the AqP was done using referential parameters as TN, TOC, IC, and COD to determine total nitrogen, the organic, and inorganic carbon content, and the COD as a reference of the polluting potential of the AqP. Then, a second evaluation was performed, using LC-HRMS analysis to compare how different or similar the different samples were, regarding the chemical content, using a full in-depth analysis of the full dataset and characterization of the features considered outside the scope of the present work.

Table 6 shows the obtained values for the COD, TOC, TN, and IC in the AqP obtained from each plant and site. It is possible to observe that IC was not present in any of the cases, being that organic carbon was the only kind of detected carbon. The highest value for COD was reported for *Iris pseudacorus* grown in NC (69 g/L) and the lowest was obtained for *Iris pseudacorus* grown in TW (47 g/L). The highest TOC was obtained for *Phragmites australis* grown in TW (21 g/L), and the lowest was reported for *Juncus effusus* grown in TW (17.4 g/L). The highest TN concentration was obtained for *Iris pseudacorus* grown in TW (1.1 g/L), and the lowest was reported for *Phragmites australis* grown in NC (0.7 g/L).

Table 5. C, H, O, N, S and ash content in the biochar produced by the selected plants for the different growing environments.

Plant Description		C [%]		H [%]		O [%]		N [%]		S [%]		Ash [%]	
		TW	NC	TW	NC	TW	NC	TW	NC	TW	NC	TW	NC
1	*Iris pseudacorus*	61.9 ± 1.6	63.8 ± 1.4	4.4 ± 0.2	4.7 ± 0.2	11.0 ± 5.5	21.3 ± 0.6	2.7 ± 0.2	1.8 ± 0.1	1.6 ± 0.2	0.2 ± 0.1	17.8 ± 3.5	11.9 ± 2.4
2	*Juncus effusus*	70.4 ± 0.1	70.5 ± 0.1	4.6 ± 0.2	4.5 ± 0.1	13.0 ± 0.4	13.3 ± 0.6	2.2 ± 0.1	2.7 ± 0.1	0.3 ± 0.1	0.7 ± 0.1	8.2 ± 1.0	11.1 ± 6.2
3	*Phragmites australis*	67.7 ± 1.2	70.5 ± 1.2	4.6 ± 0.1	4.7 ± 0.4	15.6 ± 1.4	21.6 ± 1.3	3.6 ± 0.1	2.8 ± 0.4	0.2 ± 0.0	0.4 ± 0.1	9.5 ± 0.9	7.5 ± 0.1
4	*Typha latifolia*	69.0 ± 1.1	69.0 ± 1.5	4.6 ± 0.2	4.9 ± 0.3	23.8 ± 1.6	24.0 ± 1.4	2.3 ± 0.5	1.9 ± 0.6	0.2 ± 0.1	0.2 ± 0.1	8.3 ± 0.4	7.7 ± 1.8

Table 6. Pollutant analysis in the AqP produced after the HTL by the selected plants for the different growing environments.

Plant Description		COD (g/L)			TOC (g/L)			TN (g/L)			IC (g/L)	
		TW		NC	TW		NC	TW		NC	TW	NC
1	*Iris pseudacorus*	47.0 ± 4	↓	69.0 ± 3	18.8 ± 0	↑	17.6 ± 1	1.1 ± 0.1	=	1.0 ± 0.1	BLD	BLD
2	*Juncus effusus*	57.0 ± 2	↑	49.5 ± 2	17.4 ± 1	↓	19.4 ± 1	0.9 ± 0.1	↓	1.1 ± 0.4	BLD	BLD
3	*Phragmites australis*	54,5 ± 2	↑	48.7 ± 4	21.0 ± 2	↑	18.1 ± 0	1.3 ± 0.8	↑	0.7 ± 0.0	BLD	BLD
4	*Typha latifolia*	57.7 ± 2	=	55.2 ± 7	18.3 ± 1	↑	17.5 ± 1	1.3 ± 0.2	↑	0.9 ± 0.4	BLD	BLD

↑ indicates a positive effect of the TW in the concentration of the reported element. ↓ Indicates a negative effect of the TW in the concentration of the reported element. = Indicates a not significative effect of the TW in the concentration of the reported element. BLD: below limit of detection.

It is possible to observe from all the comparisons a high heterogeneity between all the cases. The COD concentration for *Typha latifolia* did not show a significative difference between the TW and NC environments. On the contrary, the COD concentration for *Juncus effusus* and *Phragmites australis* was higher in the TW with a corresponding negative effect on the water quality and eventually on the environment. Lastly, it was observed for *Iris pseudacorus*, and *Salix viminalis* a lower COD concentration in the TW AqP compared with the concentration of the NC one.

The statistical analysis shows significative differences between all the cases regarding TOC concentration. Only two comparisons, 1B–4B and 2A–4B, did not show a significative difference, observing no tendency in the TOC concentrations.

The high values of TOC, TN, and COD in the AqP are important and demand special attention since the reported values are high when compared with regular wastewater. For example, while the AqP of the present experiment reports COD values between 47 g/L and 69 g/L, the municipal wastewater COD is in the range of 0.3 g/L to 1 g/L, and after biological treatment, the COD drops to 0.02–1.0 g/L [49]. It means that the concentration of pollutants in the AqP is up to 70 times higher than regular wastewater. The AqP results with transformation products of aromatic and other organic compounds. The study presents an initial screening of these compounds to the composition of this AqP, showing the relation between the plants used, and the compounds that are present in the liquid. The AqP should receive special attention since is the phase that is the least known regarding the compounds that constitute it, which can be evaluated for further use.

3.3.3. LC-HRMS Analysis of the Aqueous Phase

The complexity of the LC-QTOF-MS dataset was reduced by the application of a PCA analysis shown in Figure 2 and Supplementary Materials. The PCA results showed that the three replicates of each sample type are in general grouped, stressing the reproducibility of the different processed reactions in the HTL. The major differences between samples are due to plant species (e.g., 1 *Iris pseudacorus* vs. 3 *Phragmites australis*), meaning that the type of biomass fed in the HTL reactors produced an aqueous phase of different chemical composition. The *Iris pseudacorus* (1) samples seemed more different than all others, while those from *Typha latifolia* (4), as well as *Juncus effusus* (2) and *Phragmites australis* (3), tended to be more similar. Moreover, it is also clear that for the same plant type, plants sourced in treatment wetlands generated a different aqueous phase than that from natural systems (e.g., 1A vs. 1B or 3A vs. 3B). These difference between TW and NC is especially marked for *Iris pseudacorus* (1), *Juncus effusus* (2) and *Phragmites australis* (3).

The high heterogeneity in the AqP points to the fact that each plant species, depending on the growing environment, have the potential to produce certain components in the AqP. This is confirmed by the LC-HRMS analysis, which shows that not only the type of plant, but also the growing environment conditions the composition of the AqP.

This characterization is useful also because HTL could have the potential to treat biomass polluted with a specific kind of persistent components. It would be interesting for further studies to evaluate how the HTL process deals with biomass that have already uptaken persistent pollutants. Previous studies have evaluated the potential of HTL to degrade pesticides and pharmaceutical compounds, due to the reactions at supercritical conditions [50] and this kind of component being present in treatment sludges [44] or in wetland biomass [45], with promising results.

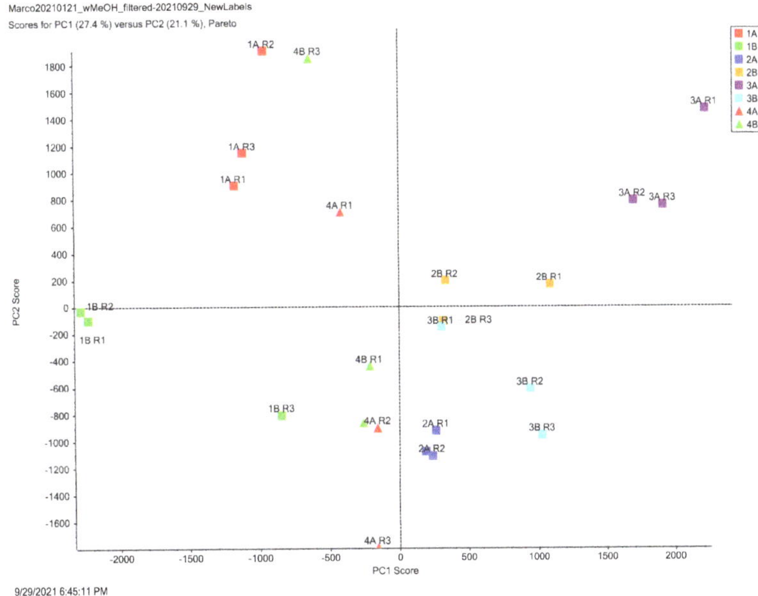

Figure 2. PCA plot from the LC-QTOF-MS dataset.

3.4. Energy Balance

The relation between the different components in the biomass, named C, H, N, S, O, and ash, determine the HHV of the analyzed biomass, and the relation between these compounds and the obtained yields of the different HTL fractions determine their energy yields (η_{th}). The energy yield, then, is a reference parameter about the potential of a feedstock to be used for producing biocrude and biochar through HTL reactions. From the obtained results (shown in Table 7), it is possible to observe that there is no significant difference between the biocrude energy yields for the same species grown in the two different environments, TW and NC. Conversely, the biochar energy yield seems to be more sensible to the biomass composition, it being possible to observe that two of the TW plants, *Iris pseudacorus* and *Phragmites australis*, produced biochar with a significative difference of energy yield compared with the plants grown in NC. However, these differences point to two different directions: while the biochar obtained from *Iris pseudacorus* in the TW had a lower energy yield, the one obtained from *Phragmites australis* produced a higher one.

Table 7. Calculated HHV for the feed biomass, biocrude and biochar for the selected plants and the different growing environments.

Plant Description		Feed Biomass		Biocrude		Biochar		Biocrude Energy Yield			Biochar Energy Yield		
		HHV (MJ/kg)		HHV (MJ/kg)		HHV (MJ/kg)		η_{th}			η_{th}		
		TW	TW	TW	TW	TW	NC	TW		NC	TW		NC
1	*Iris pseudacorus*	17.8 ± 0.1	17.3 ± 0.1	29.7 ± 5.9	28.2 ± 14.9	25.4 ± 1.2	25.3 ± 0.6	42.8 ± 10.0	=	33.4 ± 7.4	17.9 ± 4.4	↓	32.2 ± 5.3
2	*Juncus effusus*	18.3 ± 0.2	19.2 ± 0.1	28.8 ± 2.2	27.8 ± 2.9	28.5 ± 0.5	28.4 ± 0.2	49.7 ± 3.4	=	49.1 ± 4.5	22.4 ± 1.5	=	25.8 ± 2.1
3	*Phragmites australis*	18.8 ± 0.1	19.7 ± 0.1	32.3 ± 1.7	28.3 ± 3.0	27.2 ± 0.6	27.5 ± 1.3	44.9 ± 8.6	=	41.1 ± 4.8	33.8 ± 3.0	↑	27.0 ± 5.1
4	*Typha latifolia*	17.2 ± 0.2	19.2 ± 2.3	30.1 ± 3.0	29.4 ± 5.8	26.9 ± 0.7	27.1 ± 1.0	42.8 ± 10.0	=	33.4 ± 7.4	29.2 ± 1.0	=	27.9 ± 11.5

↑ indicates a positive effect of the TW in the concentration of the reported element. ↓ indicates a negative effect of the TW in the concentration of the reported element. = indicates a not significative effect of the TW in the concentration of the reported element.

Figure 3 shows that since the HHV is more homogeneous for all the studied plants, the element which generates the difference in the energy yields is the biocrude or biochar yield. It points to the fact that the combination between all the big chemical structures in the biomass, named structural sugars, lignin, and protein, is more relevant for the global potential of a feedstock than the molecular composition (amount of C, H, O, N, and S). It can be a direct consequence of the chemical reactions that take place in the HTL process because the elements do not interact directly but are associated with different molecules, which dictate the chemical pathway. Additionally, the high molecular weight of the organic compounds contains a lot of energy because of the many chemical bonds between the elements, particularly carbon, in the molecules, where other elements, such as N and S, are more loosely bound, which may even occur in inorganic forms.

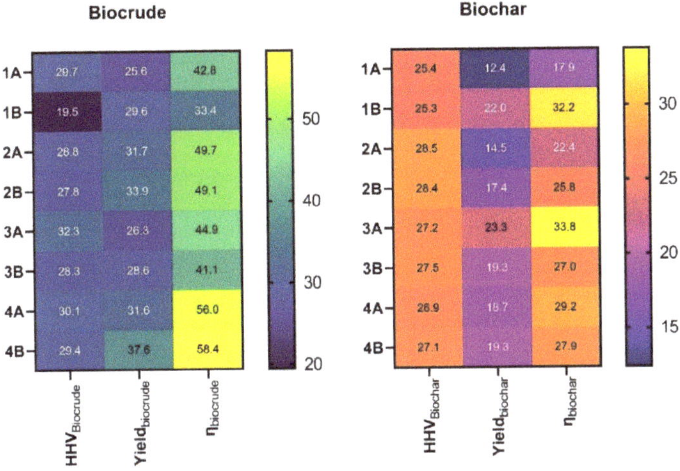

Figure 3. Heat map comparing the energy yields of the obtained biocrude and biochar. The HHV is reported in MJ/kg and the yields in %. In the rows, the numbers refer to each plant species: 1, *Iris pseudacorus*; 2, *Juncus effusus*; 3, *Phragmites australis*; and 4, *Typha latifolia*. The letters A and B refer to the growing environments: A = TW, and B = NC environment.

TW conditions do not seem to affect significantly the biocrude energy yield. For the same plant species in different environments, the biocrude energy yield does not present significative differences. However, Figure 2 shows significative differences between the energy yield reported by different plant species. The results suggest that for the biocrude energy yield, it is more important to select a species with higher biocrude yield (associated with the content of specific molecules) than the influence of the TW environment and its effect on the presence of the different elements. On the other hand, it seems that some plant species tend to produce different energy yields for the biochar, being relevant, in some cases, even if grown in the same place, to the conditions where they grow.

From the biofuel and bioeconomy perspective, it is relevant to make an assertive selection of the plants used in the TW in order to obtain higher energy yields in possible later biocrude production from these biomasses. This study presents for the first time an energy balance to approach which plant species, commonly used in TW, are more attractive for biocrude production, with *Typha latifolia* and *Juncus effusus* being the most attractive for this purpose.

3.5. Mass Balance

Carbon Balance

Table 8 shows how C is distributed in the different HTL products, biocrude being the fraction accumulating the highest portion, reporting values from 35% for the *Juncus effusus* grown in NC, to values of around 59% for *Typha latifolia* grown in NC, with a global average of 43.4%. The biochar reported values from around 18% to around 35% for *Iris pseudacorus* in TW and *Phragmites australis* in TW, respectively, with a global average of 28.5%. The AqP is the third phase that accumulates the most C, reporting values from around 17% for *Juncus effusus* in TW to around 20% for *Iris pseudacorus* in TW, having a global average of 18.5%. Lastly, the gas phase reported values from around 3.5% to 21.2%, and an average of 10.1%.

Table 8. Carbon balance [1] for the different HTL fractions obtained from the selected plants and the different growing environment.

Plant Description	Biocrude			Biochar			AqP [3]			Gas [2]	
	TW		NC	TW		NC	TW		NC	TW	NC
Iris pseudacorus	40.5 ± 8.5	=	44.1 ± 28.5	18.2 ± 4.0	↓	33.9 ± 5.7	20.1 ± 0.4	↑	18.4 ± 0.7	21.2 ± 4.7	3.5 ± 24.0
Volatile C				10.2 ± 1.7		20.2 ± 3.7					
Fixed C				7.6 ± 2.3	↓	13.0 ± 2.0					
Juncus effusus	44.3 ± 2.4	=	35.4 ± 18.2	23.2 ± 1.3	=	27.1 ± 2.2	16.7 ± 1.4	↓	18.3 ± 0.5	15.8 ± 4.0	19.2 ± 17.6
Volatile C				12.2 ± 1.2		13.2 ± 1.3					
Fixed C				10.7 ± 2.5	↓	13.6 ± 0.9					
Phragmites australis	40.2 ± 7.0	=	39.1 ± 4.4	35.2 ± 3.0	↑	29.5 ± 5.0	19.5 ± 1.7	↑	16.8 ± 1.3	5.1 ± 6.3	14.6 ± 0.9
Volatile C				18.3 ± 2.0		15.2 ± 2.1					
Fixed C				16.5 ± 0.9	↑	13.8 ± 2.8					
Typha latifolia	49.6 ± 17.4	=	53.4 ± 16.6	30.8 ± 0.6	=	29.7 ± 11.4	18.5 ± 0.2	↓	19.5 ± 5.9	1.1 ± 17.1	0.6 ± 21.8
Volatile C				15.9 ± 0.5		15.1 ± 4.9					
Fixed C				14.4 ± 0.2	=	14.1 ± 6.1					
Mean		43.4			28.5			18.5			10.1
SD		13.9			6.9			2.3			14.7

[1] The values are reported in % in reference to the amount of carbon present in the biomass feed. [2] The value of gas is the difference between 100% and the addition of the biocrude, biochar and AqP fractions; it means gas = 100 − (biocrude + biochar + AqP). [3] The value is obtained from the TOC concentration in the AqP and the amount of water in the reaction. ↑ indicates a positive effect of the TW in the concentration of the reported element. ↓ indicates a negative effect of the TW in the concentration of the reported element. = indicates a not significative effect of the TW in the concentration of the reported element

Regarding the fixed carbon, it is possible to observe that the highest value of fixed carbon was found for *Phragmites australis* grown in TW (16.5%) and the lowest was reported for *Iris pseudacorus* in the NC (7.6%). The statistical analysis for the comparison of fixed carbon in the different biochar shows for *Iris pseudacorus*, *Juncus effusus*, and *Phragmites australis* a significant difference in the fixed carbon content between the plants grown in TW and the ones grown in NC. The TW showed a positive effect in the fixed carbon content for the *Phragmites australis*, and a negative effect on the *Iris pseudacorus* and *Juncus effusus*. Typha latifolia did not show significant differences between the two growing environments.

4. Conclusions

The results of this study show that biomass from TW is suitable for biocrude production. In general, the TW conditions did not affect the biocrude yields of the HTL reaction; however, the biochar yield seems to be more sensitive to the biomass composition, showing yield differences between both environments, TW and NC, and those differences are more evident between plant species. The amount of fixed carbon in the biochar also seems to depend on the plant species, showing also differences between the growing environments. Lastly, the AqP composition showed to be different according with the growing environment and the plant species. This was confirmed by LC-HRMS analysis performed.

The biomass produced in TW can produce biochar and biocrude in similar amounts to the plants grown in NC, being a suitable source of sustainable biomass to be used in biorefinery processes, which demand, based on the bioeconomy concepts, sustainable and sources of biomass. The environmental value of the TW plants lies in the fact that the biomass is considered a residual product and up to now, no aggregated value has been found from this wastewater treatment; the plants capture the CO_2 that later can be emitted by the combustion of the biofuels, and to produce these TW plants, no extra land or nutrients are needed for growing the plants.

However, further studies dealing with the primary production of the plants, HTL pathways, uses of the AqP, and alternatives to treat the AqP are needed. Additionally, and to increase the global impact of the study, it would be relevant to select other species used in TW, especially for systems running in warm and tropical countries where TW are becoming widely used. The approach from this study can transform the wastewater treatment process into a productive system, simultaneously improving water quality and generating resources.

Lastly, the use of the HTL technology represents new environmental concerns, such as the management of the AqP, which has up to 70 times the pollutant concentration found in typical municipal wastewater. In this context, it seems that it is necessary to explore more and new biorefining process to produce from the AqP new products that take advantage of the high concentration of N and C.

This study is evidence of the possibilities of the circular economy concepts, where the biomass that already has been used for treating wastewater in a previous and sustainable way can be used and re-used not only once, but many more times. Moreover, it shows that the use of sustainable sources of biomass for producing biofuel and later bioenergy allows to bring to the present the future externalities caused by fossil fuels, which are even cheaper in present value and have a high cost because of the future environmental externalities.

Lastly, the study supports the bioeconomy values, where environmental externalities are controlled and managed in the present time. This model can create an invisible self-regulated system, compared to and readapted from the concept of the invisible hand of Adam Smith [1], where the development is invisibly controlled by the capacity of the systems to keep and control in the present, using sustainable resources for energy and goods production, the environmental externalities for their activities, avoiding the inheritance of environmental problems by the future generations because of present actions.

Supplementary Materials: The following are available online at https://www.mdpi.com/article/10.3390/en14238157/s1, Figure S1: Marco20210121_wMeOH_filtered-20210929_NewLabels Loadings for PC1 (27.4%) vs. PC2 (21.1%), Pareto.

Author Contributions: Conceptualization, M.A.R.-D., C.A.A. and P.B.; methodology, M.A.R.-D. and P.B.; validation, M.A.R.-D. and P.B.; formal analysis, M.A.R.-D., P.N.C. and P.B.; investigation, M.A.R.-D.; resources, C.A.A., P.N.C. and P.B.; data curation, M.A.R.-D., P.N.C. and P.B.; writing—original draft preparation, M.A.R.-D.; writing—review and editing, M.A.R.-D., P.B., P.N.C., H.B. and C.A.A.; visualization, M.A.R.-D., P.B. and C.A.A.; supervision, H.B. and C.A.A.; project administration, C.A.A.; funding acquisition, P.B. and C.A.A. All authors have read and agreed to the published version of the manuscript.

Funding: This research was funded by PAVITR project (grant agreement No 821410) a EU Horizon 2020 research and innovation program, the CONACYT-CENER escrow in Mexico (CVU 301254), the Aarhus Center for Water Technology (WATEC) co-founded the LC-QTOF-MS, and the company Green Growth Group Mexico SA de CV.

Data Availability Statement: Data is contained within the article.

Acknowledgments: The author acknowledges the bioengineering colleges, Juliano Souza dos Passos and Lars Bjørn Silva Thomsen, lab technician Cecilie Nielsen, biology lab technician Lone Jytte Lund Ottosen, and Birgitte Kretzschmar Tagesen in Aarhus University for the access to their facilities and support. Additionally, the author acknowledges the access to the advanced mass spectrometry facilities from Kai Bester's group at the Department of Environmental Sciences.

Conflicts of Interest: The authors declare no conflict of interest. The funders had no role in the design of the study; in the collection, analyses, or interpretation of data; in the writing of the manuscript, or in the decision to publish the results.

References

1. D'Adamo, I.; Morone, P.; Huisingh, D. Bioenergy: A Sustainable Shift. *Energies* **2021**, *14*, 5661. [CrossRef]
2. Birgit, K.; Michel, K.; Patrick, R.G., Stefan, K. Biorefinery Systems-An Overview. In *Biorefineries—Industrial Processes and Products. Status Quo and Future Directions*; Elsevier: Amsterdam, The Netherlands, 2006; Volume 1, pp. 3–39. ISBN 3527310274.

3. Walters, G.; Janzen, C.; Maginnis, S. *Nature-Based Solutions to Address Global Societal Challenges*; Cohen-Shacham, E., Walters, G., Janzen, C., Maginnis, S., Eds.; IUCN International Union for Conservation of Nature: Gland, Switzerland, 2016; ISBN 9782831718125.

4. Rodriguez-Dominguez, M.A.; Konnerup, D.; Brix, H.; Arias, C.A. Constructed Wetlands in Latin America and the Caribbean: A Review of Experiences during the Last Decade. *Water* **2020**, *12*, 1744. [CrossRef]

5. Uggetti, E.; Ferrer, I.; Arias, C.; Brix, H.; García, J. Carbon footprint of sludge treatment reed beds. *Ecol. Eng.* **2012**, *44*, 298–302. [CrossRef]

6. Mannino, I.; Franco, D.; Piccioni, E.; Favero, L.; Mattiuzzo, E.; Zanetto, G. A Cost-Effectiveness Analysis of Seminatural Wetlands and Activated Sludge Wastewater-Treatment Systems. *Environ. Manag.* **2007**, *41*, 118–129. [CrossRef]

7. McKinney, R.A.; Raposa, K.B.; Cournoyer, R.M. Wetlands as habitat in urbanizing landscapes: Patterns of bird abundance and occupancy. *Landsc. Urban Plan.* **2011**, *100*, 144–152. [CrossRef]

8. Stefanakis, A. The Role of Constructed Wetlands as Green Infrastructure for Sustainable Urban Water Management. *Sustainability* **2019**, *11*, 6981. [CrossRef]

9. Brix, H.; Sorrell, B.; Lorenzen, B. Are Phragmites-dominated wetlands a net source or net sink of greenhouse gases? *Aquat. Bot.* **2001**, *69*, 313–324. [CrossRef]

10. Brix, H. How "green" are aquaculture, constructed wetlands and conventional wastewater treatment systems? *Water Sci. Technol.* **1999**, *40*, 45–50. [CrossRef]

11. Konnerup, D.; Koottatep, T.; Brix, H. Treatment of domestic wastewater in tropical, subsurface flow constructed wetlands planted with Canna and Heliconia. *Ecol. Eng.* **2009**, *35*, 248–257. [CrossRef]

12. Vymazal, J. Constructed Wetlands for Wastewater Treatment: Five Decades of Experience. *Environ. Sci. Technol.* **2011**, *45*, 61–69. [CrossRef]

13. Wallace, S. Feasibility, Design Criteria, and O & M Requirements for Small Scale Constructed Wetland Wastewater Treatment Systems. *Water Intell. Online* **2006**, *5*, 304. [CrossRef]

14. Brix, H.; Arias, C.A. The use of vertical flow constructed wetlands for on-site treatment of domestic wastewater: New Danish guidelines. *Ecol. Eng.* **2005**, *25*, 491–500. [CrossRef]

15. Malaviya, P.; Singh, A. Constructed Wetlands for Management of Urban Stormwater Runoff. *Crit. Rev. Environ. Sci. Technol.* **2012**, *42*, 2153–2214. [CrossRef]

16. Istenič, D.; Arias, C.A.; Matamoros, V.; Vollertsen, J.; Brix, H. Elimination and accumulation of polycyclic aromatic hydrocarbons in urban stormwater wet detention ponds. *Water Sci. Technol.* **2011**, *64*, 818–825. [CrossRef]

17. Istenič, D.; Arias, C.A.; Vollertsen, J.; Nielsen, A.H.; Wium-Andersen, T.; Hvitved-Jacobsen, T.; Brix, H. Improved urban stormwater treatment and pollutant removal pathways in amended wet detention ponds. *J. Environ. Sci. Health Part A* **2012**, *47*, 1466–1477. [CrossRef]

18. Gutiérrez-Sarabia, A.; Fernandez-Villagómez, G.; Martínez-Pereda, P.; Rinderknecht-Seijas, N.; Poggi-Varaldo, H.M. Slaughterhouse wastewater treatment in a full-scale system with constructed wetlands. *Water Environ. Res.* **2004**, *76*, 334–343. [CrossRef]

19. Soroko, M. Treatment of wastewater from small slaughterhouse in hybrid constructed wetlands systems. *Ecohydrol. Hydrobiol.* **2007**, *7*, 339–343. [CrossRef]

20. Vymazal, J. Constructed wetlands for treatment of industrial wastewaters: A review. *Ecol. Eng.* **2014**, *73*, 724–751. [CrossRef]

21. Hadad, H.R.; Mufarrege, M.M.; Pinciroli, M.; Di Luca, G.A.; Maine, M.A. Morphological Response of Typha domingensis to an Industrial Effluent Containing Heavy Metals in a Constructed Wetland. *Arch. Environ. Contam. Toxicol.* **2010**, *58*, 666–675. [CrossRef]

22. Maine, M.; Sanchez, G.; Hadad, H.; Caffaratti, S.; Pedro, M.; Mufarrege, M.; Di Luca, G. Hybrid constructed wetlands for the treatment of wastewater from a fertilizer manufacturing plant: Microcosms and field scale experiments. *Sci. Total. Environ.* **2019**, *650*, 297–302. [CrossRef]

23. Pozo-Morales, L.; Moron, C.; Garvi, D. Ecologically Designed Sanitary Sewer Based on Constructed Wetlands Technology—Case Study in Managua (Nicaragua). *J. Green Eng.* **2017**, *7*, 421–450. [CrossRef]

24. Vymazal, J. Emergent plants used in free water surface constructed wetlands: A review. *Ecol. Eng.* **2013**, *61*, 582–592. [CrossRef]

25. Zhu, N.; An, P.; Krishnakumar, B.; Zhao, L.; Sun, L.; Mizuochi, M.; Inamori, Y. Effect of plant harvest on methane emission from two constructed wetlands designed for the treatment of wastewater. *J. Environ. Manag.* **2007**, *85*, 936–943. [CrossRef]

26. Zheng, Y.; Wang, X.; Ge, Y.; Dzakpasu, M.; Zhao, Y.; Xiong, J. Effects of annual harvesting on plants growth and nutrients removal in surface-flow constructed wetlands in northwestern China. *Ecol. Eng.* **2015**, *83*, 268–275. [CrossRef]

27. Camelo, A.; Genuino, D.A.; Maglinao, A.L.; Capareda, S.C.; Paes, J.L.; Owkusumsirisakul, J. Pyrolysis of Pearl Millet and Napier Grass Hybrid (PMN10TX15): Feasibility, byproducts, and comprehensive characterization. *Int. J. Renew. Energy Res.* **2018**, *8*, 682–691.

28. Karmee, S.K. Liquid biofuels from food waste: Current trends, prospect and limitation. *Renew. Sustain. Energy Rev.* **2016**, *53*, 945–953. [CrossRef]

29. Kamm, B.; Hille, C.; Schönicke, P.; Dautzenberg, G. Green biorefinery demonstration plant in Havelland (Germany). *Biofuels Bioprod. Biorefining* **2010**, *4*, 253–262. [CrossRef]

30. Balan, V.; Kumar, S.; Bals, B.; Chundawat, S.; Jin, M.; Dale, B.E. Biochemical and Thermochemical Conversion of Switchgrass to Biofuels. In *Appraisal: From Theory to Practice*; Springer: Singapore, 2012; pp. 153–185.

31. Biller, P.; Ross, A. Production of biofuels via hydrothermal conversion. In *Handbook of Biofuels Production*; Elsevier: Amsterdam, The Netherlands, 2016; pp. 509–547.

32. Biller, P.; Sharma, B.K.; Kunwar, B.; Ross, A.B. Hydroprocessing of bio-crude from continuous hydrothermal liquefaction of microalgae. *Fuel* **2015**, *159*, 197–205. [CrossRef]

33. Anastasakis, K.; Biller, P.; Madsen, R.B.; Glasius, M.; Johannsen, I. Continuous Hydrothermal Liquefaction of Biomass in a Novel Pilot Plant with Heat Recovery and Hydraulic Oscillation. *Energies* **2018**, *11*, 2695. [CrossRef]

34. Biller, P.; Lawson, D.; Madsen, R.B.; Becker, J.; Iversen, B.B.; Glasius, M. Assessment of agricultural crops and natural vegetation in Scotland for energy production by anaerobic digestion and hydrothermal liquefaction. *Biomass Convers. Biorefinery* **2017**, *7*, 467–477. [CrossRef]

35. Pedersen, T.H.; Rosendahl, L. Production of fuel range oxygenates by supercritical hydrothermal liquefaction of lignocellulosic model systems. *Biomass Bioenergy* **2015**, *83*, 206–215. [CrossRef]

36. Biller, P.; Johannsen, I.; dos Passos, J.S.; Ottosen, L.D.M. Primary sewage sludge filtration using biomass filter aids and subsequent hydrothermal co-liquefaction. *Water Res.* **2018**, *130*, 58–68. [CrossRef]

37. Conti, F.; Toor, S.S.; Pedersen, T.H.; Nielsen, A.H.; Rosendahl, L.A. Biocrude production and nutrients recovery through hydrothermal liquefaction of wastewater irrigated willow. *Biomass Bioenergy* **2018**, *118*, 24–31. [CrossRef]

38. Tenenbaum, D.J. Biochar: Carbon Mitigation from the Ground Up. *Environ. Health Perspect.* **2009**, *117*, A70–A73. [CrossRef]

39. Kwapinski, W.; Byrne, C.M.P.; Kryachko, E.; Wolfram, P.; Adley, C.; Leahy, J.J.; Novotny, E.H.; Hayes, M.H.B. Biochar from Biomass and Waste. *Waste Biomass Valorization* **2010**, *1*, 177–189. [CrossRef]

40. Qambrani, N.A.; Rahman, M.; Won, S.; Shim, S.; Ra, C. Biochar properties and eco-friendly applications for climate change mitigation, waste management, and wastewater treatment: A review. *Renew. Sustain. Energy Rev.* **2017**, *79*, 255–273. [CrossRef]

41. United Nations Sustainable Development Goals. Available online: https://sdgs.un.org/goals (accessed on 4 October 2021).

42. Sluiter, A.; Hames, B.; Hyman, D.; Payne, C.; Ruiz, R.; Scarlata, C.; Sluiter, J.; Templeton, D.; Nrel, J.W. Determination of total solids in biomass and total dissolved solids in liquid process samples. *Natl. Renew. Energy Lab.* **2008**, *9*, 3–5.

43. Jones, D.B. *Factors for Converting Percentages of Nitrogen in Foods and Feeds into Percentages of Proteins*; US Department of Agriculture: Washington, DC, USA, 1941; Volume 183, p. 22.

44. Channiwala, S.; Parikh, P. A unified correlation for estimating HHV of solid, liquid and gaseous fuels. *Fuel* **2002**, *81*, 1051–1063. [CrossRef]

45. Sluiter, A.; Hames, B.; Ruiz, R.; Scarlata, C.; Sluiter, J.; Templeton, D.; Crocker, D. Determination of Structural Carbohydrates and Lignin in Biomass—NREL/TP-510-42618. Laboratory Analytical Procedure (LAP). 2012. Available online: nrel.gov/docs/gen/fy13/42618.pdf (accessed on 10 September 2021).

46. Tisler, S.; Liang, C.; Carvalho, P.N.; Bester, K. Identification of more than 100 new compounds in the wastewater: Fate of polyethylene/polypropylene oxide copolymers and their metabolites in the aquatic environment. *Sci. Total. Environ.* **2021**, *761*, 143228. [CrossRef] [PubMed]

47. Yang, J.; He, Q.; Corscadden, K.; Niu, H.; Lin, J.; Astatkie, T. Advanced models for the prediction of product yield in hydrothermal liquefaction via a mixture design of biomass model components coupled with process variables. *Appl. Energy* **2019**, *233–234*, 906–915. [CrossRef]

48. Lu, J.; Liu, Z.; Zhang, Y.; Savage, P.E. Synergistic and Antagonistic Interactions during Hydrothermal Liquefaction of Soybean Oil, Soy Protein, Cellulose, Xylose, and Lignin. *ACS Sustain. Chem. Eng.* **2018**, *6*, 14501–14509. [CrossRef]

49. Frimmel, F.; Abbt-Braun, G. Sum Parameters: Potential and Limitations. In *Treatise on Water Science*; Elsevier: Amsterdam, The Netherlands, 2011; pp. 3–29.

50. Thomsen, L.B.S.; Carvalho, P.; dos Passos, J.S.; Anastasakis, K.; Bester, K.; Biller, P. Hydrothermal liquefaction of sewage sludge; energy considerations and fate of micropollutants during pilot scale processing. *Water Res.* **2020**, *183*, 116101. [CrossRef]

sustainability

MDPI

Article

Biomass Valorization to Bioenergy: Assessment of Biomass Residues' Availability and Bioenergy Potential in Nigeria

Uchechukwu Stella Ezealigo [1,*], Blessing Nonye Ezealigo [2], Francis Kemausuor [3], Luke Ekem Kweku Achenie [4] and Azikiwe Peter Onwualu [1]

[1] Department of Materials Science and Engineering, African University of Science and Technology, Abuja 900107, Nigeria; aonwualu@aust.edu.ng
[2] Department of Physics and Astronomy, University of Nigeria, Nsukka 410001, Nigeria; blessing.ezealigo@gmail.com
[3] Department of Agricultural and Biosystems Engineering, Kwame Nkrumah University of Science and Technology, Kumasi AK 4448-6434, Ghana; kemausuor@gmail.com
[4] Department of Chemical Engineering, Virginia Polytechnic Institute and State University, Blacksburg, VA 24061, USA; luke.achenie@gmail.com
* Correspondence: uezealigo@aust.edu.ng

Abstract: The bioenergy sector in Nigeria currently lacks a proper assessment of resource availability. In this study, we investigated the bioenergy potential of agricultural residues and municipal solid and liquid waste using data from 2008 to 2018, and we applied a computational and analytical approach with mild assumptions. The technical potential for the production of cellulosic ethanol and biogas was estimated from the available biomass. It was discovered that higher energy was generated from biogas than cellulosic ethanol for the same type of residue. The available crop residue technical potential of 84 Mt yielded cellulosic ethanol and biogas of 14,766 ML/yr (8 Mtoe) and 15,014 Mm^3/yr (13 Mtoe), respectively. Biogas has diverse applications ranging from heat to electric power generation and therefore holds great potential in solving the current electricity crisis in Nigeria. It will also position the nation towards achieving the 7th sustainable development goal (SDG 7) on clean and affordable energy.

Keywords: biomass; residues; cellulosic ethanol; bioenergy potential; biogas

Citation: Ezealigo, U.S.; Ezealigo, B.N.; Kemausuor, F.; Achenie, L.E.K.; Onwualu, A.P. Biomass Valorization to Bioenergy: Assessment of Biomass Residues' Availability and Bioenergy Potential in Nigeria. *Sustainability* **2021**, *13*, 13806. https://doi.org/10.3390/su132413806

Academic Editors: Idiano D'Adamo and Piergiuseppe Morone

Received: 25 September 2021
Accepted: 23 November 2021
Published: 14 December 2021

Publisher's Note: MDPI stays neutral with regard to jurisdictional claims in published maps and institutional affiliations.

1. Introduction

Biomass from agricultural products is abundant, and it has a strong potential for sustainable renewable energy generation [1]. Currently, biomass is responsible for about 14% of the primary energy consumed globally [2]. Agricultural residues from crops and forestry can be converted to energy carriers (solid fuel, biogas, and cellulosic ethanol) through several techniques. They have found applications in transport fuels, electricity, and heat generation [3].

Nigeria depends principally on fossil fuels (about 86%) and hydropower plants for electricity generation [4]. The overdependence on fossil fuels has negative implications for environmental sustainability [5,6]. The lack of diversity and the high power demand are factors leading to inconsistency in the electricity supply in the country. Therefore, there is a need to adopt green energy sources with less environmental impact that will complement the hydro-plants, thereby decreasing pollution arising from the combustion of fossil fuels. Although Nigeria has a high population (over 200 million) and agricultural production, due to economic problems and lack of proper assessment of available biomass [7], there has not been significant progress in transitioning to renewable energy sources.

Jekayinfa and Scholz [8] estimated residues generated from nine crops in Nigeria for 2000–2004. Their findings were restricted to only crop residues and for five years. In the same vein, Simonyan and Fasina [9] estimated the bioenergy potential of residues

311

from crops, perennial plantation, forestry, animal waste, and urban municipal waste in Nigeria using data for 2010 only. However, their study did not relate the estimated energy potential to a specific energy carrier. Alhassan et al. [10] used five crop residues obtained in Kwara State, Nigeria, to estimate the energy potentials for power solutions. In their assessment, they used theoretical potential values rather than the technical potential for these residues. The challenge is the limitation imposed by the use of the latter potential due to its unreliability for energy application [11]. Therefore, there exists a knowledge gap in adequately quantifying the bioenergy potential.

The present work aimed at estimating the total energy obtainable from agricultural residues (crops, forests, and livestock) and municipal waste for biofuel application. We investigated an 11-year (2008–2018) span to arrive at a holistic perspective and meaningful conclusions. Specifically, we adopted a computational/analytical approach to determine the bioenergy potential from cellulosic ethanol and biogas. In conclusion, we highlighted some possible challenges to the generation of bioenergy and implications on the bio-economy of Nigeria, and we made recommendations. Our findings are relevant to stakeholders, investors, and organisations in the sustainable environment and renewable energy sector for the government to adopt best practices towards the diversification of electric power generation in Nigeria.

2. Materials and Methods

2.1. Case Study

In this study, biomass resources in Nigeria were evaluated. These resources include crop residues, forest residues, livestock dung, and municipal waste generated in the country. The residue availability and bioenergy potential were assessed based on a resource-focused computational and analytical approach, using the technical potential generated from residue produced in year the 2008–2018. Data were sourced from the Food and Agriculture organization of the United Nations statistics (FAOSTAT) database [12]. The bioenergy potential of residues was estimated statistically. Although this method is simple, reproducible, low cost, and transparent, it is deficient in accounting for the economic dimensions required for evaluating the availability of land for energy crop production, the impact of bioenergy production on the environment, as well as social constraints for some key factors that elucidate the influence on soil, biodiversity, climate, cost, and other macro-economic factors on bioenergy potential.

The conceptual framework for the research is shown in Figure 1. The biomass residues are classified as agricultural residues and municipal waste. The various agricultural residues considered included crops (soya beans, seed cotton, sugar cane, sorghum, plantain, groundnut, coconut, rice, cocoa, millet, cowpea, cassava, yam, sweet potatoes, cocoyam, maize, and oil palm), forests (round wood processing such as logging, sawing, and timber processing), and livestock (dung from cattle, chicken, goats, pigs, and sheep). Solid and liquid municipal waste generated was evaluated from the estimated population of 16 major cities, which represents the four geographical regions in Nigeria. Suitable conversion technologies were computationally implemented to transform these residues and wastes into energy carriers, which include solid fuel (from crude crop residues), cellulosic ethanol (from forest and crop residues), and biogas (from the forest, crop residues, livestock, and municipal solid and liquid waste). It is worth noting that in this work, primary biomass (wood fuel and staple crops) was not considered because their conversion to energy carriers is detrimental to the environment (soil status, biodiversity, climate change) and food security. Additionally, certain energy crops (such as *Jatropha curcas*), grasses (e.g., switchgrass and seaweeds), and microfauna (such as algae) were excluded due to the limitation of certified or reliable data. Table 1 shows the categories of residues considered in this assessment.

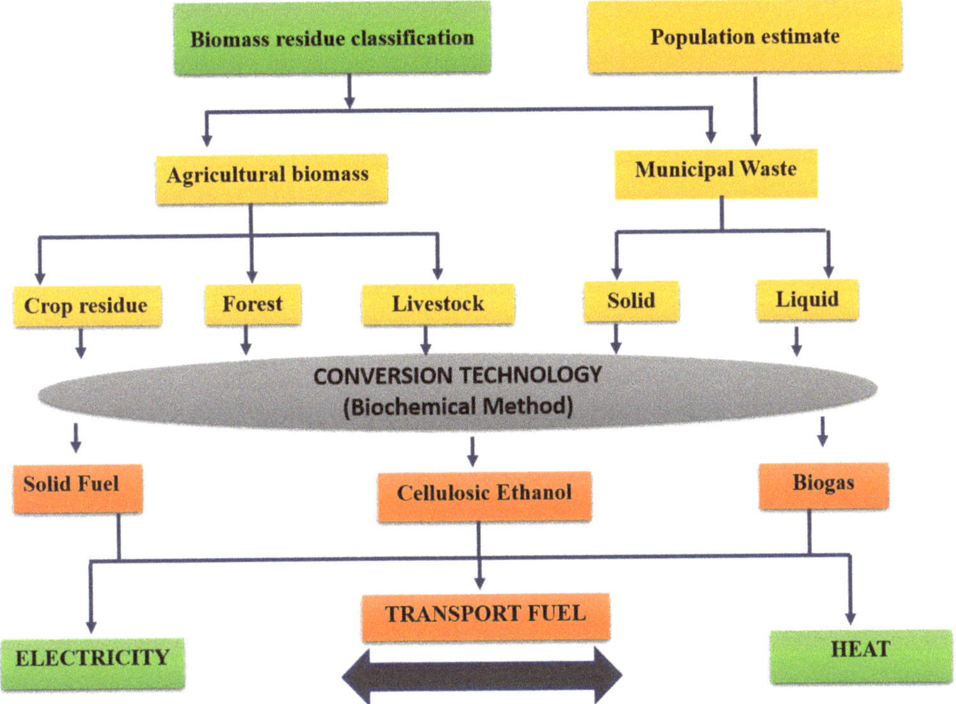

Figure 1. Analytical framework for estimating cellulosic ethanol and biogas from residues.

Table 1. Categories of biomass resources used for the bioenergy potential assessment.

S/N	Class of Residues	Category	Examples
1	Agricultural residues	Primary by-product	All residues from crops (Table 2), during harvesting
		Secondary residues	All crop residues during processing (Table 2)
		Tertiary residues	Municipal solid waste (*MSW*) and municipal liquid waste (*MLW*)
2	Forest residues	Primary by-product	Wood bark and wood slab
		Secondary residues	Sawdust
3	Livestock	Primary by-product	Manure

2.2. Crop Residues

The crop residues investigated were resources from existing farmlands. However, some assumptions (Section 2.2.1) were made to account for the key parameters for sustainability. Table 2 shows the annual crop production in Nigeria; data were obtained from the FAOSTAT database [12]. The total crop production was highest in 2016, as 164.695, 158.807, and 159.947 million tonnes (Mt) were generated in 2016, 2017, and 2018, respectively (Table 2). Fluctuations were observed in the production of these crops across the 11 years. Furthermore, a total of 27 residues (Table 3) from 17 crops were considered.

2.2.1. Sustainability Assumptions

Some assumptions that were considered are:

- Land availability: The primary energy crop (PEC) was not considered, hence, there was no land competition for animal husbandry or crop cultivation. There are no certified data regarding the annual production of the PEC, yield, and cultivated land. Therefore, the land-use competition was not taken into account. Only cultivable land was used for the estimation, and no expansion on arable land was included. There was no future projection on PEC.

- Land use: Since crops are given priority (more lands are allocated to food and fibers), the efficient use of land produces biomass that accounts for a large extent of the available residue for bioenergy assessment. In addition to land availability and use, farm management practices such as the use of improved seed, fertilizer, pest, and weed control with better technology (research and development (R&D)) are the norm for farmers. It, therefore, supports residue availability. These agricultural practices ensure sustainable residue supply from existing farmlands.
- Soil quality: Soil quality is also an important factor. Lands with rich soil quality will yield more harvest (more residues) than those with poor soil nutrients. Hence, double cropping, alternate crop rotation, appropriate mineral fertilizer, and the use of compost on farmland may increase residue production [13,14].
- Biodiversity: Biodiversity is limited as there is negligible forest encroachment since only farmlands already in use were considered in this assessment. Additionally, the use of technical residue potential preserves biodiversity, because they are utilized for other purposes.
- Climate change: The right crop management system on farmlands can reduce climate change.
- Water: The rain-fed condition was assumed as Nigeria has suitable agro-climatic conditions.
- Farm practice (animal husbandry): Regarding the livestock manure production, improved feeds with large pasture land support livestock production. With the use of the technical residue potential, the pasture for livestock and manure for soil nutrient renewal is guaranteed.

Table 2. Crop production in Nigeria.

Crop Type	Crop Production (Mt)										
	2008	2009	2010	2011	2012	2013	2014	2015	2016	2017	2018
Soya beans	0.591	0.427	0.365	0.493	0.650	0.518	0.624	0.589	0.615	0.730	0.758
Seed cotton	0.492	0.364	0.602	0.538	0.288	0.270	0.290	0.278	0.279	0.291	0.271
Sugar cane	1.410	1.400	0.850	0.756	1.090	1.270	1.410	1.450	1.490	1.490	1.420
Sorghum	9.320	5.280	7.140	5.690	5.840	5.300	6.880	7.010	7.560	6.940	6.860
Plantain	2.730	2.700	2.680	2.680	2.950	2.960	3.010	3.080	3.030	3.060	3.090
Groundnut	2.870	2.980	3.800	2.960	3.310	2.470	3.400	3.470	3.580	2.420	2.890
Coconut	0.234	0.243	0.264	0.265	0.265	0.266	0.268	0.269	0.283	0.282	0.285
Rice	4.180	3.550	4.470	4.610	5.430	4.820	6.000	6.260	7.560	6.610	6.810
Cocoa	0.367	0.364	0.399	0.391	0.383	0.367	0.330	0.302	0.298	0.324	0.333
Millet	9.060	4.930	5.170	1.270	1.280	0.910	1.400	1.490	1.550	1.500	2.240
Cowpea	2.920	2.370	3.370	1.640	5.150	4.630	2.140	2.310	3.020	2.490	2.610
Cassava	44.60	36.80	42.50	46.20	51.00	47.40	56.30	57.60	59.60	59.40	59.50
Yam	35.00	29.10	37.30	33.10	32.30	35.60	45.20	45.70	49.40	47.90	47.50
Sweet potatoes	3.320	3.300	3.470	3.520	3.590	3.680	3.670	3.820	3.890	3.960	4.030
Cocoyam	5.390	3.030	2.960	3.010	3.200	2.930	3.270	3.280	3.230	3.270	3.300
Maize	7.530	7.360	7.680	8.880	8.690	8.420	10.10	10.60	11.50	10.40	10.20
Oil palm	8.500	8.500	8.000	8.000	8.100	8.000	7.970	7.890	7.810	7.740	7.850
TOTAL	138.514	112.698	131.02	124.003	133.516	129.811	152.262	155.398	164.695	158.807	159.947

Source: FAOSTAT [12].

2.2.2. Theoretical and Technical Crop Residue Potentials

The theoretical residue potential, for each crop, was obtained from the product of the total specific crop available for a given year and the residue-to-product ratio (*RPR*). *RPR* is an index that indicates the weight of residue a particular crop generates, based on the produced amount [15]. Taking into account the variability of the *RPR* values due to several factors identified by Simonyan and Fasina [9], the mean *RPR* was used. The theoretical potential of the crop residues was estimated using Equation (1):

$$P_{th} = P_{crop} \times RPR \tag{1}$$

where P_{th} = theoretical residue potential; P_{crop} = crop production; and *RPR* = the residue-to-product ratio.

The use of theoretical residue potential was not realistic because other forms of crop residue utilization may compete with its availability for bioenergy production. Hence, we considered only the recoverable residue fraction for each crop, referred to as the technical residue potential. The latter is defined as the surplus residue after considering the competition among other uses and spatial restrictions. It is estimated using Equation (2). The obtained value gives the quantitative amount of the excess residues available for energy purposes.

$$P_{tech} = P_{th} \times Rf \tag{2}$$

P_{tech} = technical residue potential; Rf = recoverable fraction

The technical residue potential was used to estimate the energy potential of cellulosic ethanol and biogas.

2.2.3. Solid Fuel Energy Potential

The bioenergy potential in dried crop residues in their crude forms was calculated using Equation (3). The estimated solid fuel made from crop residues was obtained by multiplying the total annual technical crop residue potential and the lower heating values (Table 3).

$$P_{SFE} = P_{tech} \times LHV \tag{3}$$

P_{SFE} = solid fuel energy potential; LHV = lower heating value (MJ/kg).

Table 3. Parameters used in estimating bioenergy potentials from crop residues.

Crop Residues	RPR	Rf (%)	LHV [b] (MJ/kg)
Soya beans straw	2.50 [a]	100	12.38
Soya beans pods	1.00 [a]	100	12.38
Seed cotton stalk	2.88	80	18.61
Sugar cane tops/leaves	0.11	80	15.81
Sugar cane bagasse	0.18	100	18.10
Sorghum straw	1.99	80	12.38
Plantain trunks and leaves	0.50	80	15.48 [c]
Groundnut straw	1.25	100	17.58
Groundnut shell	0.37	100	15.66
Coconut husk	0.42	100	18.63
Coconut shell	0.25	100	18.09
Rice husk	0.26	100	19.33
Rice straw	1.66	80	16.02
Cocoa bean pods	0.93	80	15.12
Millet straw	1.83	80	12.38
Cowpea shell	1.75	100	19.44
Cassava stalk	0.06	80	17.50
Cassava peeling	0.25	20	10.61
Yam straw	0.50	80	14.24
Sweet potatoes straw	0.50	80	14.24

Table 3. *Cont.*

Crop Residues	RPR	Rf (%)	LHV [b] (MJ/kg)
Cocoyam straw	0.50	80	14.24
Maize stalk	1.59	80	19.66
Maize husk	0.20	100	15.56
Maize cobs	0.29	100	16.28
Oil palm EFB	0.17	100	8.16
Oil palm kernel shell	0.07	100	18.83
Oil palm fibre	0.14	100	11.34

The mean values of *RPR* and *Rf* were obtained from Kemausuor et al. [16]. Other values with alphabetic superscripts were sourced as indicated [a] [17]; [b] [9]; [c] [8].

2.2.4. Cellulosic Ethanol Potential

To estimate the bioenergy potential and cellulosic ethanol conversion of the crop residues by anaerobic digestion, some pre-treatment processes such as hydrolysis, enzymatic activities, and microbial fermentation were taken into account. The cellulosic ethanol production from crop residues was estimated using Equation (4):

$$Y_{CE} = P_{tech} \cdot C_{glu} \cdot y_{hyd} \cdot y_{eth} \cdot \eta_{pre} \cdot \eta_{enz} \tag{4}$$

where:

Y_{CE} = yield of cellulosic ethanol;
P_{tech} = technical potential;
C_{glu} = concentration of glucan;
y_{hyd} = yield of enzymatically hydrolyzed glucan;
y_{eth} = stoichiometric yield from glucose;
η_{pre} = efficiency of pretreatment;
η_{enz} = efficiency enzymatic cellulose conversion

In estimating the cellulosic ethanol production, we assumed fermentation and distillation processes to be 100%, as no loss was considered. The assumed values used for the estimation of cellulosic ethanol production are shown in Table 4.

Table 4. Summary of indices for cellulosic ethanol production from crop and forest residues.

Conditions	Y_{eth}	Y_{hyd}	η_{Pre} (%)	η_{enz} (%)	ρ_{Distil} (%)	ρ_{Ferm} (%)	η_{Scale} (%)
No pre-treatment	0.51	1.11	-	30	100	100	50
With pre-treatment	0.51	1.11	80	90	100	100	80

Where ρ_{Distil} = distillation efficiency; ρ_{Ferm} = fermentation efficiency. Values were sourced from Kemausuor et al. [16].

During the hydrolysis of crop residues for cellulosic ethanol production, two scenarios were considered: no pre-treatment and pre-treatment. In the no pre-treatment case, the enzymatic activity was assumed to be minimal (about 30%) with a production of cellulosic ethanol scale-up (η_{Scale}) of about 50%. In the pre-treatment scheme, the enzymatic efficiency was assumed to be 90%, to yield cellulosic ethanol of 80%. The bioenergy potential of cellulosic ethanol was estimated from the lower heating value (*LHV*) of 28.9 MJ/kg and an ethanol density of 0.789 kg/L.

2.2.5. Biogas Potential

The estimation of biogas was performed using the technical residue potential generated for the crop residues. To obtain the biomethane potential (*BMP*), the Buswell *BMP* equivalent (Equation (5)) was first determined.

$$Y_{BMP\ Buswell} = \left(Y_{Buswell,\ glu} \times C_{glu} \right) + \left(Y_{Buswell,\ hem} \times C_{hem} \right) \tag{5}$$

BMP is defined as the theoretical estimate based on the experimental evaluation of a given feedstock for the determination of the maximum volume of methane generated. It is the optimal methane volume per gram of volume solid (VS) of a substrate (i.e., the biodegradable fraction).

$Y_{BMP\ Buswell}$ = estimated biodegradable fraction in specific crop residue (feedstock) for biogas production using Buswell formula;

$Y_{Buswell.glu}$ = estimated glucan in specific residue using Buswell formula;

$Y_{Buswell.hem}$ = estimated hemicellulose using Buswell formula;

C_{glu} = concentration of glucan;

C_{hem} = concentration of hemicellulose.

The maximum biogas estimate/potential was determined using Equation (6):

$$Y_{Biogas} = P_{tech} \times Y_{BMP\ Buswell} \times \eta_{Scale} \tag{6}$$

where; Y_{Biogas} = biogas yield; η_{scale} = average efficiency for continuous biogas production.

For the energy potential of biogas, calculations were based on the following assumptions: 1 m^3 biomethane has a calorific value of 10 kWh STP; the energy potential of CH$_4$ conversion and the conversion factor of TJ to Mtoe is 0.278 GWh/yr and 24, respectively.

2.3. Forest Residues

From the FAOSTAT database [18], we obtained data on the average industrial round wood harvested yearly in Nigeria. The residues generated from the logging, sawing, and timber processing activities of round wood were determined using the assumption proposed by Koopmans and Koppejan [19]. These residues were classified into three: wood slab, wood bark, and sawdust. Wood slabs were taken to be 40% and 38% for logging and sawmilling processes, respectively, while, for sawdust, the values were 12% and 20%, in the same processes. In addition, the sawdust from the particleboard was 10%, while the residue from the wood bark during sawmilling was 12%. These values were adopted following Simonyan and Fasina [9] and Koopmans and Koppejan [19].

2.3.1. Cellulosic Ethanol from Forest Residues

Similar to the ethanol estimation from crop residues, the cellulosic ethanol potential from wood residues was determined using Equation (7).

$$Y_{CE}\ (forest\ residues) = P_{FR} \times C_{glu} \times Y_{hyd} \times Y_{eth} \times \eta_{Pre} \times \eta_{enz} \tag{7}$$

P_{FR} = annual production of forest residue.

2.3.2. Biogas Potential from Forest Residues

The maximum biomethane (biogas) production from forest residues was determined based on Buswell's formula using an expression similar to Equation (5). However, an industrial-scale efficiency of 40% was assumed for biogas production from forest residues. Hence, the biogas estimated at the industrial scale was obtained from Equation (8).

$$Y_{Biogas}(Forest) = P_{FR} \times Rf \times \left(Y_{Buswell,\ glu} \times C_{glu}\right) + (Y_{Buswell,\ hem} \times C_{hem}) \times \eta_{Scale} \tag{8}$$

2.4. Livestock Residues

The data for the livestock population from 2008–2018 was obtained from FAOSTAT [20]. The residue considered was excreta (dung) estimated for each livestock following Equations (9) and (10).

$$Y_{man}(theoretical\ potential) = P_{livestock} \times EMP \tag{9}$$

Y_{man} = manure produced; EMP = estimated manure produced per day.

$$Y_{man}(technical\ potential) = Y_{man}(theoretical\ potential) \times Rf \tag{10}$$

Biogas Potential from Livestock Residue

The biogas potential from manure was estimated from Equation (11), with the biomethane potential (Y_{BMP}) = 0.26111 m^3 CH_4/kg VS.

$$LMM = Y_{man}(technical\ potential) \times C_{TS} \times VS \times Y_{BMP} \tag{11}$$

LMM = livestock manure methane;
VS = volume solid;
C_{TS} = total solid concentration.

2.5. Municipal Waste

2.5.1. Municipal Solid Waste (*MSW*)

The quantity of municipal solid waste (*MSW*) was calculated from the population of major cities like Lagos [21] using Equation (12). Sixteen (16) cities were considered. The organic fraction concentration (C_{OF}) of the *MSW* was obtained from the literature on the various cities.

$$P_{MSW} = EP \times WG \times O_{wc} \tag{12}$$

where P_{MSW} = total waste production; EP = estimated population per city; WG = waste generated (kg/person/day); O_{wc} = organic waste content (%).

The estimate of biogas potential from municipal solid waste was determined using Equation (13).

$$Y_{biogas}(MSW) = P_{MSW} \times C_{OF} \times C_{TS} \times Y_{BMP} \tag{13}$$

2.5.2. Municipal Liquid Waste (*MLW*)

The potential biogas from municipal liquid waste (*MLW*) is a function of the product of the quantity of liquid waste from the estimated population, the concentration of total solids, and the biomethane potential, as shown in Equations (14) and (15):

$$P_{MLW} = EP \times AWE \tag{14}$$

EP = estimated population per city; AWE = average weight excreta per person per day (250 g) as derived by Feachem et al. [22].

$$Y_{biogas}(MLW) = P_{MLW} \times C_{TS} \times VS \tag{15}$$

P_{MLW} = municipal liquid waste production.

For municipal liquid waste, the concentration of total solids (TS) was assumed to be 8.9275 g TS/100 g [23]. Other factors used for the conversion are shown in Table 5.

Table 5. Indices for estimating the biogas potential of residues and wastes.

Factors	Unit Value	Reference
Volatile solid (VS)	64.7%	[24]
Lower calorific value of CH_4	10 kWh/m^3 STP	[25]
Methane yield VS reduction	0.24 m^3/kg (24%)	[26]
CH_4 yield	0.525 m^3 CH_4/kg VS	[23]
Energy potential of CH_4 conversion	0.278 GWh/yr	
TJ to Mtoe conversion factor	24	[16]

2.6. Data Analysis

The data collected were analyzed using Microsoft Office Excel version 2016. Originlab 9 was used to plot the graphs.

3. Results

3.1. Crop Production and Residue Potentials

The residues from the crops considered included the straws, stalks, cobs, pods, shells, peels, and husks from the harvesting (field-based residues) and processing (process-based residues) activities.

The annual theoretical residues from a total of 27 sources (17 crops) showed total values of 126, 116, and 119 Mt for 2016, 2017, and 2018, respectively (Table 6). The technical residues were also found to be 97, 89, and 91 Mt for 2016, 2017, and 2018, respectively (Table 6). However, in 2009, both residue potentials (i.e., theoretical and technical) had the least values.

Table 6. Estimated crop potential residues and bioenergy potentials.

Year	Theoretical (Mt)	Technical (Mt)	Cellulosic Ethanol		Biogas	
			ML/yr	Mtoe	Mm^3 CH$_4$/yr	Mtoe
2008	115.82	90.53	15,578.77	8.52	15,859.26	13.69
2009	90.45	71.10	12,237.41	6.69	12,405.56	10.71
2010	106.94	84.18	14,429.09	7.89	14,534.41	12.55
2011	93.88	72.37	13,023.50	7.12	13,198.36	11.39
2012	103.69	81.06	13,835.72	7.56	14,024.91	12.11
2013	97.14	75.60	12,800.56	7.00	13,095.00	11.31
2014	112.97	86.81	15,529.04	8.49	15,747.22	13.59
2015	115.95	89.13	15,926.22	8.71	16,173.43	13.96
2016	125.79	97.20	17,226.24	9.42	17,571.09	15.17
2017	116.14	88.66	15,754.78	8.61	16,144.15	13.94
2018	118.54	90.84	16,088.55	8.79	16,404.53	14.16
Average	108.85	84.32	14,766.35	8.07	15,014.36	12.96

Mt = million tonnes; Mm^3 = mega cubic meter (volume); Toe: tonne of oil equivalent is a unit of energy defined as the amount of energy released by burning one tonne of crude oil. Mtoe = one million toe.

The average crop production and theoretical and technical residues across the investigated period were 142, 109, and 84 Mt, respectively. These values differ from the lowest and highest obtained data. Therefore, it is inferred that crop production and technical residues can sustain biofuel production.

3.2. Bioenergy Potential from Crop Residues

3.2.1. Solid Biofuel Potential

Wood biomass is still used for energy purposes (in the form of wood fuel) in Nigeria. The production of wood fuel showed an increasing trend from 2008–2018 (Figure 2). This trend can escalate due to high demand with respect to the population. Further increases in the use of wood fuel contribute to climate change. However, maximizing the energy potential in crude crop residues can drastically reduce the direct combustion of wood. The solid fuel energy available in these crop residues was highest in 2016, followed by 2018 and 2017 (Figure 3).

3.2.2. Cellulosic Ethanol and Biogas Production from Crop Residue

The estimated cellulosic ethanol production was highest from 2016 to 2018 (Table 6). Similarly, the energy from cellulosic ethanol followed the same trend. Since the volume of ethanol produced is greatly influenced by the quantity of residues, the particle size and enzymatic digestion are very important.

3.3. Residue and Bioenergy Potential from Forestry

3.3.1. Estimated Residue from Forestry

The estimated residues (sawdust, wood bark, and wood slab in volume) generated during the harvest and processing of round wood for industrial use are given in Table 7. The variation in the generated residues from 2008–2013 and 2014–2018 was mainly due to the significant increase in the volume of industrial round wood harvested and processed in 2014. It is worth noting that the two groups (2008–2013 and 2014–2018) emerged due to a significant increase in wood production in 2014 (Table S2, Supplementary Materials). Hence, we adopted such a classification for better comparison and discussion.

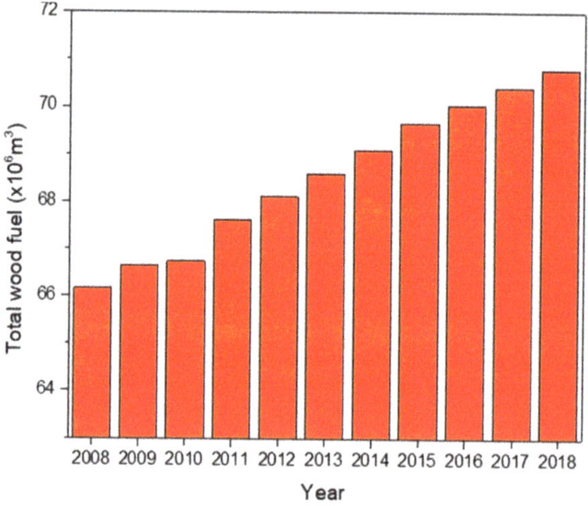

Figure 2. Total annual wood fuel production. (Total wood fuel = wood fuel + charcoal; Table S1, Supplementary Materials).

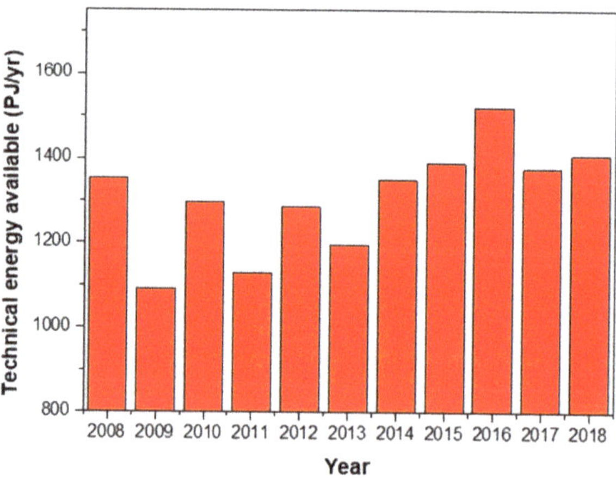

Figure 3. Solid fuel potential showing the technical energy available (TEA) in crop residues generated annually in Nigeria.

Table 7. Estimated residues generated from forestry.

Residues	Estimated Average Residues Generated (m^3)	
	2008–2013	**2014–2018**
Saw dust	360,408	379,249
Wood bark	71,621	75,424
Wood slab	1,352,490	1,422,960
Total	1,784,519	1,877,633

3.3.2. Cellulosic Ethanol Production from Forest Residues

Cellulosic ethanol production from forest residues (wood slabs, wood bark, and sawdust) was also higher in 2014–2018 compared to 2008–2013 (Figure 4). The treatment conditions were selected for estimating and assessing the maximum quantity of cellulosic ethanol, given the recalcitrant nature of the cell walls of forest trees. In both the 2008–2013 and 2014–2018 groups, a higher cellulosic ethanol yield was obtained when compared with the no pre-treatment scenario (Figure 4). The pre-treatment condition is an important factor for maximum cellulosic ethanol yield from forest residues.

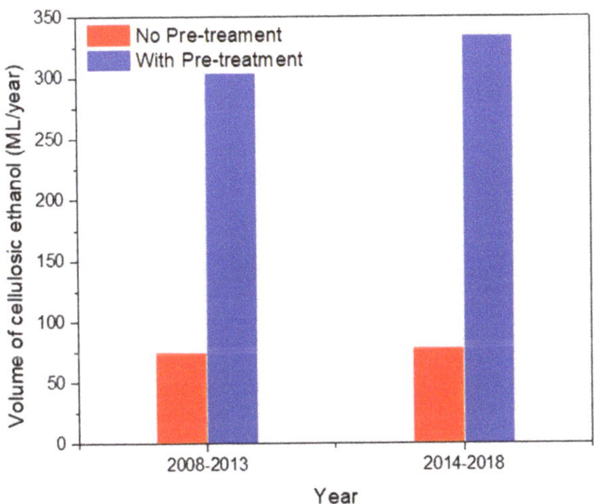

Figure 4. Volume of cellulosic ethanol produced from forest residue with and without pre-treatment.

3.3.3. Biogas Potential from Forest Residue

The biogas production from forest residue was relatively higher for the 2014–2018 period compared with that estimated for the 2008–2013 period (Figure 5; Table S3, Supplementary Materials).

3.4. Livestock

3.4.1. Livestock Production

The total livestock production varied from 272 million (in 2014) to 308 million livestock (in 2011). Although, in 2011, individual livestock such as chicken and pigs experienced a significant drop in production. However, pig production, unlike chicken production, showed a substantial increase and exceeded that of 2010. Despite these changes, the total annual livestock production showed a rising trend in the later years (i.e., 2014–2018). This can be attributed to the growing population (Figure 6; Table S4, Supplementary Materials).

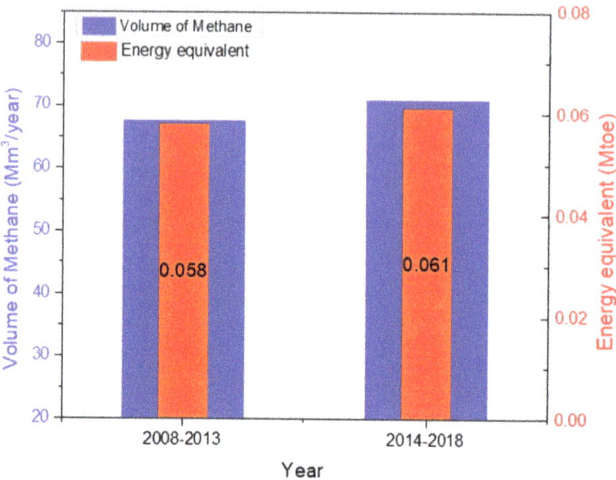

Figure 5. Estimated volume of methane (biogas) and the energy equivalent.

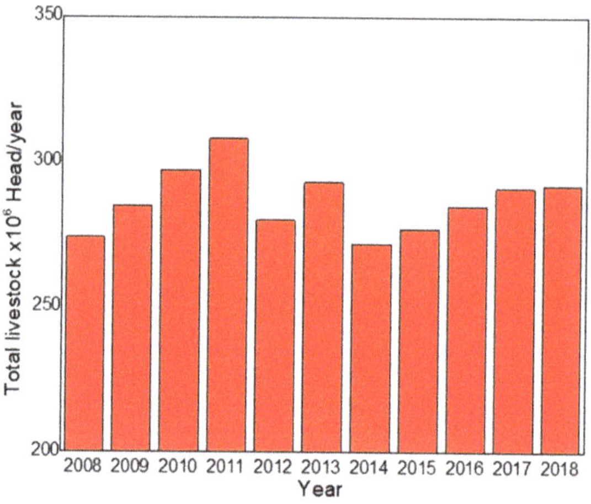

Figure 6. Total annual livestock production.

3.4.2. Biogas Potential from Livestock Manure

The bioenergy potential measured from recoverable livestock dung in the form of biogas was determined (Figure 7). The result recorded the highest and least recoverable dung in 2018 and 2008, respectively (Table 8). Additionally, the biogas produced within the investigated 11-year period showed an increasing trend. A remarkable increase in bioenergy was observed in 2011, which may be due to the high production of cattle, goats, and sheep recorded in that year (Table S4, Supplementary Materials). From Figure 7, a linear relationship was observed in the methane potential and the estimated energy equivalent.

Table 8. Details of estimated livestock dung generated.

Year	Dung Produced (million kg)	Recoverable Dung (million kg)	Dung *VS per day (10^6)	Dung VS per yr (10^9)
2008	345	77.6	10.7	3.90
2009	351	79.1	10.9	3.98
2010	357	80.6	11.1	4.07
2011	401	87.9	12.3	4.49
2012	408	89.6	12.5	4.58
2013	414	91.0	12.7	4.65
2014	422	92.9	13.0	4.75
2015	430	94.9	13.2	4.83
2016	438	96.6	13.5	4.92
2017	449	98.8	13.8	5.05
2018	457	100	14.0	5.13

* VS = volume solid.

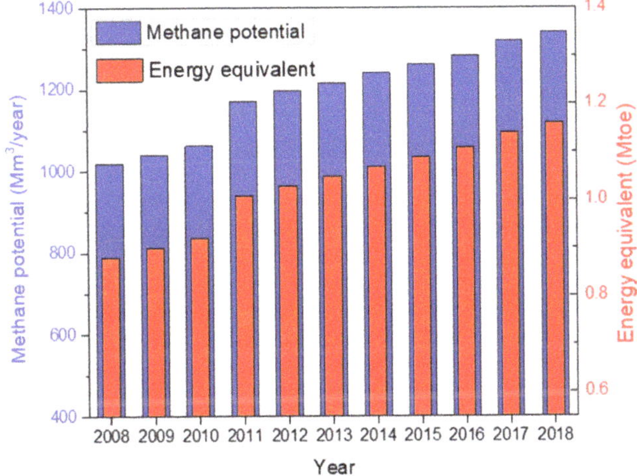

Figure 7. Estimated energy potential of biogas from livestock.

3.5. Municipal Wastes

3.5.1. Municipal Solid Wastes

The waste generated by the population of 16 major cities (representing all four geographical regions in Nigeria) was evaluated for its biogas potential. An increase in population gave a corresponding rise in the waste generated from food and other biodegradable materials (Figure 8; Table S5, Supplementary Materials). These cities were: north (Abuja, Kano, Makurdi, Maiduguri, and Kaduna), south (Benin City, Port Harcourt), east (Onitsha and Enugu), and west (Ife, Ilorin, Akure, Ado-Ekiti, Abeokuta, Lagos, and Ibadan).

3.5.2. Energy Potential from Municipal Liquid Wastes (*MLW*)

The municipal liquid waste of the 16 major cities was estimated based on the assumption that a person produces an average of 250 g fecal waste daily [22,27]. The estimated liquid waste increases per year with population growth, which subsequently leads to a rise in the biogas potential (Figure 9; Table S6, Supplementary Materials).

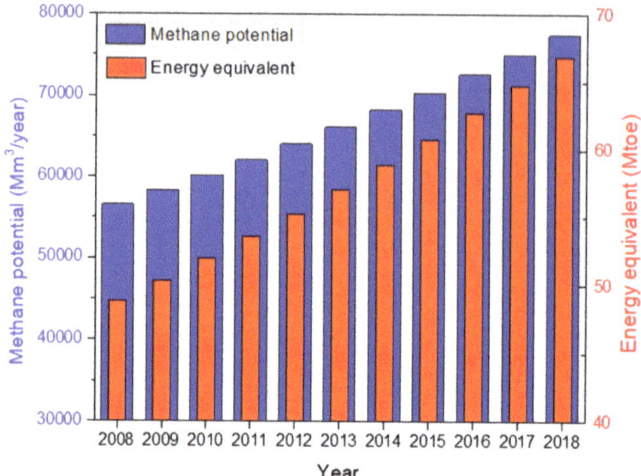

Figure 8. Methane potential and energy equivalent generated from municipal solid waste.

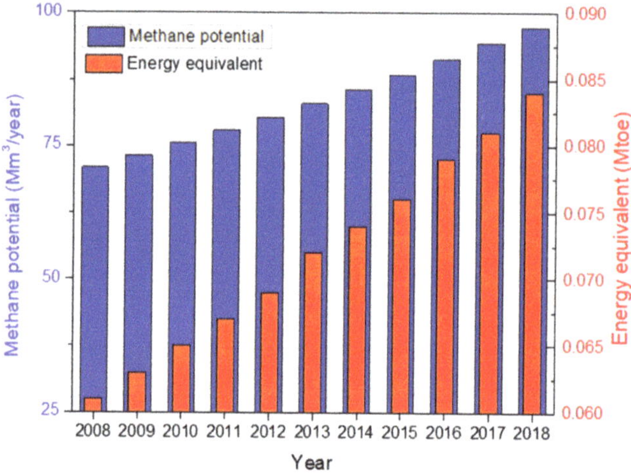

Figure 9. Methane potential and energy equivalent generated from municipal liquid waste.

4. Discussion

The present study on crop production in the last 11 years does not follow the increasing crop yield as reported by Jekayinfa and Schloz [8]. High quantities of technical potentials were recorded, and different forms of energy carriers with increased energy efficiency were estimated. However, when other potentials (such as environmental, socio-economic, and sustainable potentials) are taken into account, the overall generated residue potential may reduce. On the other hand, both the theoretical and technical residue potentials fluctuated within the investigated period. In the agricultural sector, in particular, farmers need to be enlightened on the importance of residues for energy generation. This will enable better collection and storage practices. Additionally, the awareness can potentially increase the number of agricultural residues. Crop residues can be processed by various techniques, which include gasification, pyrolysis, and combustion (for biogas, bio-oil, and biochar); fermentation (for cellulosic ethanol); and briquettes (as solid fuel) [28]. Solid biofuels (in the form of pellets and briquettes) made from residues of forest and crops are good alternatives to wood fuel and charcoal, as they potentially reduce the felling of trees and deforestation. Residual biomass from the enzymatic or fermentation process for cellulosic ethanol may further be processed into pellets [29] for combustion purposes. There is a market for these in Nigeria because the use of wood fuel is high [30]. The bioenergy produced from solid fuel depends on the generated technical crop residues. Similarly, the potential energy from crop residues follows the crop production trend. Briquettes and pellets made from crop residues can serve as wood fuel, thus reducing the demand for conventional wood fuel and charcoal. Cellulosic ethanol is a liquid fuel obtained from the digestion of lignocellulose components of crop residues, which can be used in place of petrol [16]. On the one hand, the quantity of cellulosic ethanol produced was high from 2016 to 2018. However, the conversion processes of crop residues to biofuel, as well as the cost, must be considered. Moreover, the selection of suitable techniques is necessary for optimal ethanol yield. Although the estimated cellulosic ethanol has a huge potential as transport fuel with high-performance efficiency (in vehicles including racing cars), their optimal production is limited due to the recalcitrant structure of the cell wall [31,32]. On the other hand, biogas production is more efficient compared to cellulosic ethanol, as indicated by the inherent potential energy measured in the fossil fuel equivalent (Mtoe) in Table 6.

The increase in the use of wood fuel (Figure 3) is primarily a result of the rise in population and poverty. Correspondingly, high wood fuel demand leads to deforestation. The felling of trees for energy purposes plagues Nigeria with the tragedy of climate change, soil infertility (due to erosion), and forest area depletion. Secondary biomass, which includes forest residues, serves as an alternative to wood fuel, for diverse energy forms. These residues are from fallen branches and wood barks during sawmilling and logging processes. Cellulosic ethanol and biogas can be obtained from forest residues. The energy efficiency for biogas implies that biogas is suitable for electricity generation and can positively influence the power condition in Nigeria if properly appropriated. These power sources can serve the inhabitants of the rural areas where bioenergy plants are likely to be situated. Biofuel will not only reduce the adverse effect of smoke from the direct combustion of wood fuel during cooking on the health of the rural dwellers but will also provide an alternative clean cooking energy source [33]. The degree of the recalcitrant varies with the age and maturity of forest residues. For optimizing cellulosic ethanol production, the type of pre-treatment selected should ensure a very high estimate and resulting biofuel. Additionally, reducing the particle size of the residue enhances the surface area for effective hydrolysis. Moreover, a smaller particle size promotes the solubility and biodegradability of organic matter, leading to a significant increase in the cellulosic ethanol yield (Figure 4).

The animals produced in large quantities in Nigeria include chickens, goats, sheep, cattle, and pigs (Table S4, Supplementary Materials). There is a direct relationship between the amount of manure generated and the quality of food intake when considering the weight of the animal. As shown in Table 8, the estimated dung generated and the amount recovered for biogas production was rising monotonously per year (Figure 7) despite the fluctuating livestock production (Figure 6). This result agrees with the work of Suberu et al. [34] and also confirms that Nigeria has a high potential of generating an enormous amount of biogas from animal dung. The present study does not include data from domestic livestock farmers from rural households in Nigeria due to the lack of certified data. The recoverability of the manure from livestock is quite a challenge except in the case of large and mechanized farms that utilize intensive farm practice for commercial purposes. Cattle have the potential of producing higher manure, but most farmers in Nigeria use the nomadic approach. The latter limits the amount of cattle dung for energy purposes. Hence, the quantity of manure recovered is about 50%. Better farm practices and management can enhance the recoverability of animal dung. Nigeria may have to impose mandatory intensive cattle rearing practices. Moreover, intensive farm practices are also economical in food management as the cattle eat more and burn fewer calories; as a result, a higher quantity of manure can be generated.

The high volume of biogas from both *MSW* and *MLW* (Figures 8 and 9) may be ascribed to the high population, which is a consequence of migration to these major cities. This migration is mostly an indirect effect of social factors such as the job search, a quest for improved living standards, industrialization, urbanization, and insurgency. The quantity of feces and urine excreted per day is a function of the climate, diet, volume of water consumed, and the occupation of an individual.

In our assessment, among the various energy carriers, biogas presents the highest potential and capacity for the development of both integrated and flexible bioenergy strategies in Nigeria. According to World Bank data and world info, Nigeria consumed an average of about 2.2 Mtoe (24.72 bn KWh) of electric power per year [35,36], of which the average estimated energy equivalent of biogas from crop residues and municipal solid waste combined can yield over 30% increase in energy for consumption. Therefore, biomass has a significantly high potential to improve the available electric energy supply, thereby providing a solution to the power outage problem currently experienced in the country. Our findings are in agreement with Sobamowo and Ojolo [37]. Although there is a linear relationship between the methane potential and the energy equivalent of biogas, the estimated energy was lower than the volume of methane (Figures 8 and 9). This result may be ascribed to the thermodynamic factors involved in the conversion of biogas to heat energy.

From an economic point of view, waste is a resource in the production process, which reduces the extraction of fresh materials and the related energy consumption. The circular economy is a regenerative system that supports the optimal use of resources and waste, thus leading to an economic and ecological resource closed-loop [38–40]. In the context of the present study, the circular economy approach prevents resource depletion (resulting from improper waste incineration or decomposition) and a high carbon footprint and ensures production–consumption operations that promote sustainable growth along with the social well-being of Nigerians.

5. Biofuel Potentials and Challenges

5.1. Cellulosic Ethanol and Biogas Potentials

The potential for energy generation from waste, as well as its ability to control waste management, is of great benefit to the rapidly growing population. Nigeria can leverage the latter and the vast arable land for the production of crops and residue generation for energy purposes.

Biomass gasification technology produces relatively clean energy that consists of methane and hydrogen gas from the carbon-based feedstock. The effluent from anaerobic digestion can be used as fertilizer to enhance the soil nutrients and maintain high crop production [23]. The lignocellulose nature of crop and forestry residues possess high biogas energy potential due to its rich methane content.

The conversion technology employed to transform biomass to biofuel depends on the quality of the feedstock. Poor feedstocks with 60–65% moisture content are preferably processed into other forms of biofuel. This diversification ensures an optimum biofuel recovery. The application of pre-treatment conditions (such as drying the biomass) improves its quality for gasification. Nigeria has high solar radiation capable of drying feedstock at a low cost. Besides, solar resources are abundant in regions where sufficient cereal residues are produced. The benefits and challenges of producing biogas or cellulosic ethanol from biomass residues are presented in Table 9.

Table 9. Potential benefits and challenges in cellulosic ethanol and biogas production in Nigeria.

	Factors	Biogas	Cellulosic Ethanol
1	Bio-digester	Simple.	Complex to handle due to multiple purification processes.
2	Feedstock type	Relatively dry and low moisture biomass are preferred for biogas production.	All types of feedstock type are suitable as water is required.
3	Energy cost	No drying is required.	High energy is needed for drying, grinding, and purification of ethanol.
4	Technology	Low technical know-how is needed at a low or medium scale.	Advance technology is essential both in the design and installation of hardware for industrial ethanol production.
5	Research	Little research and development in the area of inoculation for constituent biogas production.	To overcome the recalcitrant nature of the biomass, constant R&D is necessary, even in the area of genetic modification of cellulose.
6	Products	Methane, CO_2, H_2, etc.	Cellulosic ethanol, water, fertilizer, and other recyclable products.
7	Cost	Relatively low-cost compared to ethanol production.	Enzymes and microbes for hydrolysis and fermentation; equipment are capital intensive.
8	Engine modification	Needs regular adjustment.	No intensive adjustment is required.

Source: [41,42].

The comparison between biogas and cellulosic ethanol production (Table 9) has shown that the process of biogas production is simple, feasible, and less expensive [43]. Therefore, it is more appropriate to start with biogas production.

5.2. Challenges

The production of either biogas or cellulosic ethanol is feasible in principle, considering the availability of different types of residues and the high demand for a steady power supply. However, some challenges could potentially limit its viability in Nigeria, as discussed subsequently.

First, the assessment of biomass residues, as well as the estimation of total bioenergy potentials, involves many uncertainties. The latter can affect the available residue potential. Secondly, the technical residue potential is usually lower than the theoretical one. This reduction emanates from the various value chains of the residues. The competition makes it expedient to source biomass residues solely for energy production. In this regard, there is a need to identify other crop residues that have little or no competitive use. These crops include energy crops, grasses, algae, and other aquatic plants. Furthermore, poor mechanization may limit the collection as well as the conversion method involved in processing the residues [44]. The lack of data on some biomasses (e.g., grass) with high bioenergy potential has contributed to insufficient information on the total residue estimate available in Nigeria. A more comprehensive residue valuation should include energy crops such as *Jatropha curcas* and aquatic weeds (water hyacinth, water lettuce, and bracken fern), which are abundant in swampy regions. There is also a need to regularly update the national biomass database.

The estimates for solid and liquid waste produced in Nigeria focused on the major cities and are shown in Tables S5 and S6 (Supplementary Materials). Although these cities account for the large and diverse forms of waste estimated due to the high population, it represents only a fraction of the total population (16 major cities out of 36 states in Nigeria). Nonetheless, it is difficult to assess the data for major cities, and it is needless to consider the rural areas. This barrier hinders the detailed assessment of municipal waste generated in Nigeria. Currently, only the city of Abuja practices a central sewage system, while others practice a system where a few households are connected to a septic tank. Regarding *MSW*, the nation needs to adopt a solid waste disposal practice, properly sorting waste into different categories. This will ensure better processing of *MSW* into energy carriers.

Another challenge in the realization of biofuel production hinges on infrastructure. This includes investment in bio-digesting systems, structural facilities, and technologies required for an efficient biofuel yield.

5.3. Implications on the Bio-Economy of Nigeria

An essential focus of the bio-economy is the production and processing of biomass wastes into value-added products [45]. The valorization of biomass residue is connected to the sustainable utilization of renewable biological resources (which includes food, bio-based products, and bioenergy) leading to the restoration and preservation of biodiversity. Therefore, the bio-economic perspective provides a balance to the social, environmental, and economic benefits that promote the use of renewable resources, allowing an optimal trade-off between food and bioenergy production.

The implication of our assessment on the bio-economy of Nigeria includes the following:

1. Prompts the implementation of good farm practices that will increase crop production, food security, and residue generation and, consequently, will create jobs for the unemployed. Additionally, it leads to a sustainable ecosystem.
2. Provides business opportunities for innovative start-ups that will attract foreign investment in value-based products for a global market. This could position Nigeria at the forefront of the bioenergy market in Africa.
3. Diversification into bioenergy generation will enable a healthy environment by reducing greenhouse gas emissions from fossil fuels.
4. Decrease our overdependence on foreign nations, thereby making Nigeria's economy tend towards self-reliance (reducing external debits).
5. Enforce collaboration among researchers of various fields as well as the cooperation between Nigeria and other countries towards the establishment of functional bioenergy plants.
6. Facilitates the transition from a circular economy to a bio-economy, as information on the residues generated, their availability, and the bioenergy potential are valuable for policy-making.

5.4. Recommendations

The energy equivalent from crop residues is higher for biogas production than for cellulosic ethanol. Moreover, livestock manure, *MSW*, and *MLW* can be preferably processed into biogas, hence leading to a higher volume of biogas compared to cellulosic ethanol. Since biogas can easily be converted to electricity, Nigeria can partly deal with its electricity challenge by focusing on biogas production. Furthermore, the assessment and estimation of the bioenergy potential from biomass residues in Nigeria are but one side of the coin. A more holistic approach that accounts for the cost of establishing a functional biogas plant for residue conversion should also be taken into consideration. The concept of bioenergy from biomass resources involves a multi-dimensional study that includes raw material availability, assessment, and energy potential. It also covers various divisions from agriculture through the industrial, government, and power sectors. However, the socio-economic influence towards bioenergy establishment is another measure of its sustainability [46–48].

Finally, the implementation of proper biofuel policy is expedient; in this regard, the government plays a vital role in the exploitation of natural resources and the attainment of environmental sustainability [49]. However, sustained biofuel production requires the cooperation of other stakeholders [50,51], as illustrated in Figure 10. It is important to note that promoting the use of biogas in Nigeria may require the introduction of subsidies [52].

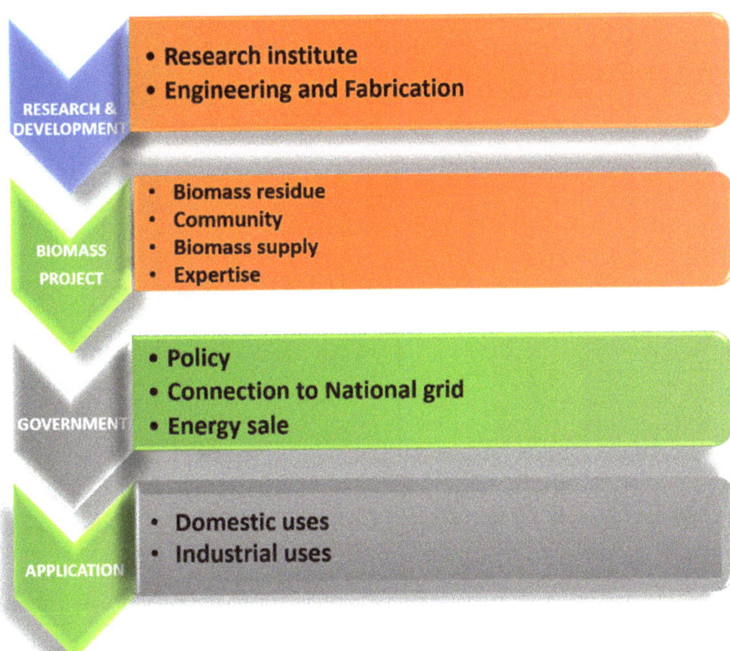

Figure 10. Structural framework showing the stakeholders in bioenergy from biomass resources.

6. Conclusions

The assessment of biomass residues and their bioenergy potential is often performed for either solid biofuel or biogas. However, in this work, we estimated the bioenergy potential from both solid biofuel and biogas perspectives. We discovered that 143 Mt of crop residues produces about 84 Mt of technical residue potential on average. Hence, only about 58% of the total residue is available for energy purposes. Our findings revealed that crop production is directly correlated with the quantity of biofuel produced. For the forest residues, enzyme pre-treatment led to higher cellulosic ethanol. Among the bioenergy carriers evaluated, biogas had the highest potential, with an average of 15,014 Mm^3 from crop residues. Therefore, it is a more promising energy carrier to be adopted in Nigeria. Although biogas production is favoured, there is a need to investigate its cost, feasibility, and the economic analysis of setting up the plant in Nigeria. Additionally, the pragmatic behaviour of the biomass residues during anaerobic activity (i.e., the breakdown of lignocellulose content) needs to be experimentally validated. Finally, the policies that will facilitate the optimum collection of these biomass residues are expedient.

Supplementary Materials: The following are available online https://www.mdpi.com/article/10.3390/su132413806/s1, Table S1: Annual wood production, Table S2: Annual industrial round wood production, Table S3: Forestry Residue, Table S4: Annual livestock production, Table S5: Municipal solid waste generated and bioenergy potential, Table S6: Municipal liquid waste generated and bioenergy potential.

Author Contributions: U.S.E.: manuscript writing, conceptualization, data curation, and formal analysis. B.N.E.: graph analysis, manuscript review, and editing. F.K.: statistical validation, manuscript review, and editing. L.E.K.A.: manuscript review, editing, and general supervision. A.P.O.: manuscript review, editing, and general supervision. All authors have read and agreed to the published version of the manuscript.

Funding: This research received no external funding.

Institutional Review Board Statement: Not applicable.

Informed Consent Statement: Not applicable.

Data Availability Statement: The data presented in this study are available in this manuscript and Supplementary Materials.

Acknowledgments: U.S. Ezealigo is grateful for the PAMI fellowship and the African development bank for the publication grant.

Conflicts of Interest: The authors declare no conflict of interest.

References

1. Hall, D.O.; Rao, K.K. *Photosynthesis*, 6th ed.; Cambridge University Press: Cambridge, UK, 1999.
2. Balat, M.; Ayar, G. Biomass Energy in the World, Use of Biomass and Potential Trends. *Energy Sources* **2005**, *27*, 931–940. [CrossRef]
3. Fernandes, U.; Costa, M. Potential of biomass residues for energy production and utilization in a region of Portugal. *Biomass Bioenergy* **2010**, *34*, 661–666. [CrossRef]
4. Oyewo, A.S.; Aghahosseini, A.; Bogdanov, D.; Breyer, C. Pathways to a fully sustainable electricity supply for Nigeria in the mid-term future. *Energy Convers. Manag.* **2018**, *178*, 44–64. [CrossRef]
5. Ikram, M.; Zhang, P.Q.; Sroufe, P.R.; Shah, P.S.Z.A. Towards a sustainable environment: The nexus between ISO 14001, renewable energy consumption, access to electricity, agriculture and CO_2 emissions in SAARC countries. *Sustain. Prod. Consum.* **2020**, *22*, 218–230. [CrossRef]
6. Ikram, M.; Ferasso, M.; Sroufe, R.; Zhang, Q. Assessing green technology indicators for cleaner production and sustainable investments in a developing country context. *J. Clean. Prod.* **2021**, *322*, 129090. [CrossRef]
7. Ezealigo, U.S.; Otoijamun, I.; Onwualu, A.P. Electricity and biofuel production from biomass in Nigeria: Prospects, challenges and way forward. In *IOP Conference Series: Earth and Environmental Science*; IOP Publishing: Bristol, UK, 2021; Volume 730, p. 012035. [CrossRef]
8. Jekayinfa, S.O.; Scholz, V. Potential availability of energetically usable crop residues in Nigeria. *Energy Sources Part A Recover. Util. Environ. Eff.* **2009**, *31*, 687–697. [CrossRef]
9. Simonyan, K.J.; Fasina, O. Biomass resources and bioenergy potentials in Nigeria. *Afr. J. Agric. Res.* **2013**, *8*, 4975–4989.
10. Alhassan, A.E.; Olaoye, O.J.; Olayanju, A.; Okonkwo, E.C. An investigation into some crop residues generation from farming activities and inherent energy potentials in Kwara State, Nigeria. In Proceedings of the IOP Conference Series: Materials Science and Engineering, Ota, Nigeria, 24–28 June 2019; Volume 640, p. 012093.
11. Batidzirai, B.; Smeets, E.; Faaij, A. Harmonising bioenergy resource potentials—Methodological lessons from review of state of the art bioenergy potential assessments. *Renew. Sustain. Energy Rev.* **2012**, *16*, 6598–6630. [CrossRef]
12. FAOSTAT. Food and Agriculture Organization of The United Nations—Statistics Division. Crop Production. 2021. Available online: https://www.fao.org/faostat/en/#data/QCL (accessed on 8 October 2020).
13. Edmeades, D.C. The long-term effects of manures and fertilisers on soil productivity and quality: A review. *Nutr. Cycl. Agroecosyst.* **2003**, *66*, 165–180. [CrossRef]
14. Loveland, P.; Webb, J. Is there a critical level of organic matter in the agricultural soils of temperate regions: A review. *Soil Tillage Res.* **2003**, *70*, 1–18. [CrossRef]
15. Iye, E.L.; Bilsborrow, P.E. Assessment of the availability of agricultural residues on a zonal basis for medium- to large-scale bioenergy production in Nigeria. *Biomass Bioenergy* **2013**, *48*, 66–74. [CrossRef]
16. Kemausuor, F.; Kamp, A.; Thomsen, S.T.; Bensah, E.; Østergård, H. Assessment of biomass residue availability and bioenergy yields in Ghana. *Resour. Conserv. Recycl.* **2014**, *86*, 28–37. [CrossRef]
17. Bhattacharya, S.C.; Pham, H.L.; Shrestha, R.M.; Vu, Q.V. CO_2 emissions due to fossil and traditional fuels, residues and wastes in Asia. AIT, 1992; Unpublished Work.
18. FAOSTAT. Food and Agriculture Organization of The United Nations–Statistics Division. Forestry Production. 2021. Available online: https://www.fao.org/faostat/en/#data/FO (accessed on 8 October 2020).
19. Koopmans, A.; Koppejan, J. Agricultural and Forest Residues—Generation, Utilization and Availability. Available online: http://www.fao.org/DOCREP/006/AD576E/ad576e00.pdf (accessed on 25 March 2020).
20. FAOSTAT. Food and Agriculture Organization of The United Nations–Statistics Division. Live Animal. 2021. Available online: http://www.fao.org/faostat/en/#data/QA/visualize (accessed on 8 October 2020).
21. Macrotrends. Nigeria Metro Area Population 1950–2020. 2020. Available online: https://www.macrotrends.net/cities/22007/lagos/population (accessed on 2 April 2020).
22. Feachem, R.G.; Bradley, D.J.; Garelick, H.; Mara, D.D. *Sanitation and Disease: Health Aspects of Excreta and Wastewater Management*; John Wiley and Sons: New York, NY, USA, 1983.
23. Arthur, R.; Brew-Hammond, A. Potential biogas production from sewage sludge: A case study of the sewage treatment plant at Kwame Nkrumah University of science and technology, Ghana. *Int. J. Energy Environ.* **2010**, *1*, 1009–1016.
24. Arthur, R. Feasibility Study for Institutional Biogas Plant at KNUST Sewage Treatment Plant. Master's Thesis, Kwame Nkrumah University of Science and Technology, Kumasi, Ghana, 2009.

25. Suhartini, S.; Lestari, Y.P.; Nurika, I. Estimation of methane and electricity potential from canteen food waste. In *IOP Conference Series: Earth and Environmental Science*; IOP Publishing: Bristol, UK, 2019; Volume 230, p. 012075. [CrossRef]

26. Chynoweth, D.P.; Owens, J.M.; Legrand, R. Renewable methane from anaerobic digestion of biomass. *Renew. Energy* **2001**, *22*, 1–8. [CrossRef]

27. Rose, C.; Parker, A.; Jefferson, B.; Cartmell, E. The characterization of feces and urine: A review of the literature to inform advanced treatment technology. *Crit. Rev. Environ. Sci. Technol.* **2015**, *45*, 1827–1879. [CrossRef] [PubMed]

28. Clauser, N.M.; González, G.; Mendieta, C.M.; Kruyeniski, J.; Area, M.C.; Vallejos, M.E. Biomass waste as sustainable raw material for energy and fuels. *Sustainability* **2021**, *13*, 794. [CrossRef]

29. Moreira, B.R.D.A.; Viana, R.D.S.; Cruz, V.H.; Magalhães, A.C.; Miasaki, C.T.; De Figueiredo, P.A.M.; Lisboa, L.A.M.; Ramos, S.B.; Sánchez, D.E.J.; Filho, M.C.M.T.; et al. Second-generation Lignocellulosic supportive material improves atomic ratios of C:O and H:O and thermomechanical behavior of hybrid non-woody pellets. *Molecules* **2020**, *25*, 4219. [CrossRef]

30. Jekayinfa, S.; Orisaleye, J.; Pecenka, R. An assessment of potential resources for biomass energy in Nigeria. *Resources* **2020**, *9*, 92. [CrossRef]

31. Gonzalez-Estrella, J.; Asato, C.M.; Jerke, A.C.; Stone, J.J.; Gilcrease, P.C. Effect of structural carbohydrates and lignin content on the anaerobic digestion of paper and paper board materials by anaerobic granular sludge. *Biotechnol. Bioeng.* **2017**, *114*, 951–960. [CrossRef]

32. Patinvoh, R.J.; Osadolor, O.A.; Chandolias, K.; Horváth, I.S.; Taherzadeh, M.J. Innovative pretreatment strategies for biogas production. *Bioresour. Technol.* **2017**, *224*, 13–24. [CrossRef]

33. Sa'ad, S.; Bugaje, M.I. Biomass consumption in Nigeria: Trends and policy issues. *J. Agric. Sustain.* **2016**, *9*, 127–157.

34. Suberu, M.Y.; Bashir, N.; Mustafa, M.W. Biogenic waste methane emissions and methane optimization for bioelectricity in Nigeria. *Renew. Sustain. Energy Rev.* **2013**, *25*, 643–654. [CrossRef]

35. World Bank Data. Available online: https://data.worldbank.org/indicator/EG.USE.ELEC.KH.PC?end=2014&locations=NG&start=1971&view=chart (accessed on 24 December 2020).

36. World Data Info. Available online: https://www.worlddata.info/africa/nigeria/energy-consumption.php (accessed on 24 December 2020).

37. Sobamowo, G.M.; Ojolo, S.J. Techno-economic analysis of biomass energy utilization through gasification technology for sustainable energy production and economic development in Nigeria. *J. Energy* **2018**, *2018*, 1–16. [CrossRef]

38. Haas, W.; Krausmann, F.; Wiedenhofer, D.; Heinz, M. How circular is the global economy? An assessment of material flows, waste production, and recycling in the European Union and the world in 2005. *J. Ind. Ecol.* **2015**, *19*, 765–777. [CrossRef]

39. Sariatli, F. Linear economy versus circular economy: A comparative and analyzer study for optimization of economy for sustainability. *Visegr. J. Bioecon. Sustain. Dev.* **2017**, *6*, 31–34. [CrossRef]

40. Morone, P.; Imbert, E. Food waste and social acceptance of a circular bioeconomy: The role of stakeholders. *Curr. Opin. Green Sustain. Chem.* **2020**, *23*, 55–60. [CrossRef]

41. Achinas, S.; Achinas, V.; Euverink, G.J.W. A technological overview of biogas production from biowaste. *Engineering* **2017**, *3*, 299–307. [CrossRef]

42. Bušić, A.; Marđetko, N.; Kundas, S.; Morzak, G.; Belskaya, H.; Šantek, M.I.; Komes, D.; Novak, S.; Šantek, B. Bioethanol production from renewable raw materials and its separation and purification: A review. *Food Technol. Biotechnol.* **2018**, *56*, 289–311. [CrossRef]

43. Ishola, M.M.; Brandberg, T.; Sanni, S.A.; Taherzadeh, M.J. Biofuels in Nigeria: A critical and strategic evaluation. *Renew. Energy* **2013**, *55*, 554–560. [CrossRef]

44. Oyedepo, S.O.; Dunmade, I.S.; Adekeye, T.; Attabo, A.A.; Olawole, O.C.; Babalola, P.O.; Oyebanji, J.A.; Udo, M.O.; Kilanko, O.; Leramo, R.O. Bioenergy technology development in Nigeria—Pathway to sustainable energy development. *Int. J. Env. Sustain. Dev.* **2019**, *18*, 175–205. [CrossRef]

45. D'Adamo, I.; Morone, P.; Huisingh, D. Bioenergy: A sustainable shift. *Energies* **2021**, *14*, 5661. [CrossRef]

46. D'Adamo, I.; Falcone, P.M.; Morone, P. A new socio-economic indicator to measure the performance of bioeconomy sectors in Europe. *Ecol. Econ.* **2020**, *176*, 106724. [CrossRef]

47. Saracevic, E.; Koch, D.; Stuermer, B.; Mihalyi, B.; Miltner, A.; Friedl, A. Economic and global warming potential assessment of flexible power generation with biogas plants. *Sustainability* **2019**, *11*, 2530. [CrossRef]

48. Bacenetti, J. Economic and environmental impact assessment of renewable energy from biomass. *Sustainability* **2020**, *12*, 5619. [CrossRef]

49. Abid, N.; Ikram, M.; Wu, J.; Ferasso, M. Towards environmental sustainability: Exploring the nexus among ISO 14001, governance indicators and green economy in Pakistan. *Sustain. Prod. Consum.* **2021**, *27*, 653–666. [CrossRef]

50. Attard, J.; McMahon, H.; Doody, P.; Belfrage, J.; Clark, C.; Ugarte, J.A.; Pérez-Camacho, M.N.; Martín, M.D.S.C.; Morales, A.J.G.; Gaffey, J. Mapping and analysis of biomass supply chains in Andalusia and the Republic of Ireland. *Sustainability* **2020**, *12*, 4595. [CrossRef]

51. Mahmood, A.; Wang, X.; Shahzad, A.; Fiaz, S.; Ali, H.; Naqve, M.; Javaid, M.; Mumtaz, S.; Naseer, M.; Dong, R. Perspectives on bioenergy feedstock development in Pakistan: Challenges and opportunities. *Sustainability* **2021**, *13*, 8438. [CrossRef]

52. D'Adamo, I.; Falcone, P.M.; Huisingh, D.; Morone, P. A circular economy model based on biomethane: What are the opportunities for the municipality of Rome and beyond? *Renew. Energy* **2021**, *163*, 1660–1672. [CrossRef]

Article

The Development of a Model of Economic and Ecological Evaluation of Wooden Biomass Supply Chains

Bohdan Hrechyn [1,*], Yevhen Krykavskyy [1,2] and Jacek Binda [2]

[1] Department of Marketing and Logistics, Lviv Polytechnic National University, 79000 Lviv Oblast, Ukraine; yevhen.v.krykavskyi@lpnu.ua

[2] Department of Finance and Information Technologies, Bielsko-Biala School of Finance and Law, 43-382 Bielsko-Biała, Poland; jbinda@wsfip.edu.pl

* Correspondence: bohdan.d.hrechyn@lpnu.ua; Tel.: +48-604-775-216

Abstract: This scientific publication is dedicated to the development of scientific methodological and practical recommendations about the formation of ecologistics approaches towards usage of the energetical potential of wooden biomass as a promising trend of economic activity subject development. The hierarchy of ecological chain build-up is established, which will allow one to effectively organize the logistics of supply of biomass to the place of energy production. The methodological approaches to modeling of economic and ecological evaluation of wooden mass supply chain were improved. It is aimed to the calculation of expanses and harmful emissions that depend on specific logistics processes in implementation of perspective actions of collection and recycling of wooden biomass and substitution of non-renewable energy sources by it, which, on the one hand, analyzes the actual state of affairs of knowledge in the field of ecological processes evaluation, and on the other hand, however, identifies restrictions on the amounts of potential provision of biomass. Due to the proposed model of economic and ecological evaluation of the supply chain of wooden biomass and the development of software with a database that covers information on specific logistics processes, it will be possible to conduct economic and ecological evaluation on each step of the logistics chain, present specific processes in cash equivalents, depict ecological effectiveness, and identify the most vulnerable points of the logistics system, opening vast opportunities for improvement of other supply systems.

Keywords: ecologistics; supply chains; ecological evaluation; economic evaluation; energetic potential; wooden biomass; environment; logistics processes

Citation: Hrechyn, B.; Krykavskyy, Y.; Binda, J. The Development of a Model of Economic and Ecological Evaluation of Wooden Biomass Supply Chains. *Energies* **2021**, *14*, 8574. https://doi.org/10.3390/en14248574

Academic Editor: Idiano D'Adamo

Received: 28 June 2021
Accepted: 19 August 2021
Published: 20 December 2021

Publisher's Note: MDPI stays neutral with regard to jurisdictional claims in published maps and institutional affiliations.

1. Introduction

The actualization of ecological problems in the global environment stimulates the scientific and practical research of social phenomenon and processes that affect the quality of natural and production environments in different fields of life activities. Economical, logistical, organizational, and managerial processes and analytical technologies to realize the energetical potential of wooden biomass must become the means to reach the aim of energy independence of Ukraine, on the one hand, and the strategic vector of state politics development in the sphere of environmental protection and production environment on the other. Ecologization is one of the main innovative competitive advantages of logistics activity and has to cohere with it. According to global practice, it is quite important to combine logistics and ecologizational processes, as the logistics part itself is a key alternative for economic activities of manufacturers and their agents, while ecological is an innovative part, both for subjects of economic activity and socioeconomical systems of higher levels of management.

2. Scientific Novelty of the Obtained Results

From an ecological point of view, Forsberg analyzed an optimum of production processes and electrical energy distribution in a country supplied with biomass by another country [1]. In terms of energy and money expenditure, Suurs explored this topic in his study and evaluated general strategies for the importing of biomass as an energy carrier for Europe and Latin America [2].

Studies of logistics starting from biomass collection to biofuel, heat, or electrical energy recycling and production were carried out by Pfohl [3], Weber [4], and Gudehus [5]. Their approaches were based on classical logistics planning and optimization; however, analysis and evaluation of environmental impacts caused by shipping and other logistics processes have not been sufficiently studied.

This article includes references to works written by researchers who focused on economic and/or ecological biomass evaluation and supply and provision processes and who also took changes and tendencies in energy industry into account. Pistorius [6] and Plöchl [7] conducted energy estimation of different biomass types. A considerable number of works is dedicated to estimating the whole life cycle of biomass, from cultivation production of energy (Jungbluth [8], Krotschek [9], Spath [10], Marheineke [11], Scholwin [12]) to biofuel (Mardon [13], Soimakallio [14], Zah [15]). However, logistics and provision processes are not described in detail in their works, and only central tendencies are measured; hence, there is a high level of abstraction.

Scientists such as Kanzian et al. [16], Neff et al. [17], Scholz et al. [18], and Suurs [2] described models that enabled the analysis of one or a few processes of biomass provision, taking into account needs for personnel, time, and materials, and that also enabled estimation of costs and waste levels. They believed that the figures estimated in the analyses would serve as reference points for future economic and ecological evaluations.

Eberhardinger evaluated ecological balance of production processes of electrical energy obtained from wood waste. His analysis emphasized the impact of the whole supply chain and wood biomass provision on the biofuel energy balance. They reached the conclusion that a logistical component of biomass supply and provision have the potential for improvement in terms of expenditure, energy usage, and quality [19].

Given the extent of pollution caused by fossil fuels and human concerns about health and environmental protection, it is necessary to focus on strategies, practices, and actions that allow for the implementation of a sustainable energy system based on bioenergy production. It is known that the COVID-19 pandemic period has led to a socioeconomic crisis, which will have serious consequences for international logistics systems. Given that the transport sector is responsible for a significant share of CO_2 emissions, special emphasis should be placed on biofuels and sustainable development at the global level. This is particularly true in the case of Ukraine, as an agricultural country, where a concentrated large part of biomass is obliged to act as an active participant in this process in the following ways:

- Use raw materials that meet the criteria of sustainability;
- Analyze the supply chain for the actual sustainability of the product;
- Use industrial symbiosis and energy communities as an effective element of cooperation among energy consumers;
- Implement a policy that supports the development of bioenergy and biofuels;
- Popularize the tendency of the willingness to pay for green or circular products among business and individual consumers;
- Support those business models that implement eco-initiatives in comparison with business models that include the consumption of conventional products based on fossil materials [20,21].

This research is related to the improvement of methodical approaches to making a model of economic and ecological evaluation of the wood biomass supply chain, which is oriented towards providing a high level of business entity competitiveness based on ecologistics. In the research were used such methods as the methods of observation, factor analysis, and economic and mathematical modeling. In contrast to existing studies [1–19,22,23], it allows one to secure a steady pace of wood biomass provision and consumption growth and to apply a diversified approach to the attraction of investment resources and employment of new technologies. Further developments have occurred in essential components of the supply chain in the process of wood biomass provision and consumption. They, in addition to the existing ones [1–19,22,23], take ecological standards and energy usage efficiency level into account and make it possible to minimize negative impacts on ecology and to secure resource saving, stable development of business entities, and increases Ukrainian business effectiveness. Comparative calculations, e.g., on the use of coal, briquettes, peat, and gas as a fuel, are usually based on the estimation of direct costs without taking into account associated costs (for example, costs aimed at minimizing the negative impact on the environment, etc.) in the supply chain.

This research and the proposed model are cross-applicable and relevant for all countries with a significant wood stock.

3. Statement of Basic Materials

Protection of natural and industrial environments, regulatory and logistics processes, balance of greenhouse waste emissions, and also bioenergy industry development are closely interconnected. These mutual relations form the basis for future projects, research, and elaboration of a database in the sphere of economic and/or ecological evaluation applied to forestry, the energy industry, and biomass logistics. In term of raw material reserves, central and western regions of Ukraine are the most promising for production organization (Figure 1).

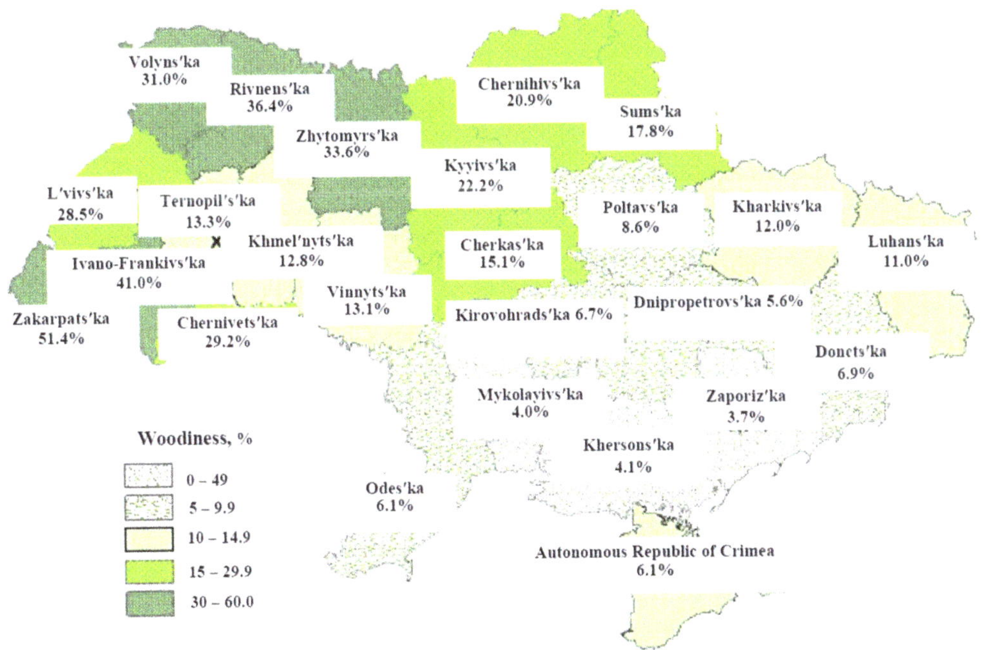

Figure 1. Raw material base for wood biomass provision in the territory of Ukraine [24].

The research suggests a typical wood biomass supply chain including gathering of wood raw materials (fallen trees, cutting standing trees) → loading raw material into a vehicle → transshipment of raw material (the biomass can be overloaded) → transporting of raw material to terminals (warehouses) → unloading of raw material → warehousing and storage of raw material → chopping raw material into woodchips → transporting of woodchips to a production department → production of biofuel (pellet) and delivery after production. With regard to internal and external technical and technological, infrastructural, financial, and organizational conditions, this chain transforms into an individual wood biomass supply chain of a business entity (Figure 2).

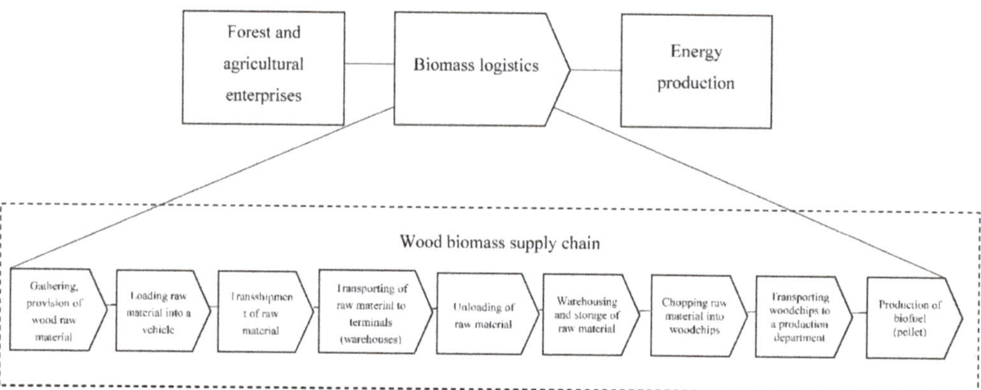

Figure 2. Biomass logistics: wood biomass supply chain.

According to the Ukrainian biofuel portal, woodchips production volumes are gradually increasing, although it is not a large-scale phenomenon. Woodchips are mainly produced by recycling waste obtained from industrial wood sawing; not more than 10% of logging waste is recycled [24,25]. In general, the situation with economic management by regions in Ukraine is stable. From year to year, economic processes become more efficient, wood stocks increase, log harvest decreases, and the amount of processing increases, leading to an increase in the amount of waste for energy fuel.

Miscanthus and willow are still underused, due to the alternative use of land for agricultural purposes being more profitable.

The application of methodic approaches, improved and developed in this research, will facilitate identification of "narrow" areas in modelling of wood biomass provision processes as well as economic and ecological evaluation of wood biomass supply chains, which will help to receive the information needed for making practical management solutions.

For example, if the task is to measure the impact of logistics processes on the natural environment during the whole process of biomass provision, then on the one hand, it should be clarified what types of impact on the environment will be analyzed, and, on the other hand, what methods will be used for estimation, to what extent the existing norms should be obeyed, and, finally, how will the outcomes be interpreted and presented [25,26].

Improvement of methodical approaches will be followed by elaboration of a model of economic and ecological evaluation of a wood biomass supply chain. The model will aim to calculate costs and measure the amount of hazardous waste emissions into the environment, which will enable evaluation of economic and ecological effectiveness of the whole process. Applying this model in practice requires a large number of parameters. The more detailed the data calculation description is, the more realistic the economic and ecological evaluation of wood biomass supply system's effectiveness of a business entity can be.

Economic evaluation of each process encompasses all the information on consumer spending and expenses on equipment (cars, vehicles, and other), personnel, and if needed on infrastructure.

Since railway and air transportation are usually considered as services provided by a third party and used according to demand, they will not be included into this calculation model. Types of road transport for freight shipping include common classic freight transport and shipping by agricultural machinery.

Transport costs include vehicle usage costs C_{veh}, personnel costs C_{per}, and vehicle repair and tuning up costs C_{tun}.

Fuel costs have considerable impacts on transportation cost calculations. Hazardous waste emissions into the environment, which are produced by road vehicles, should be taken into consideration. Information on emissions is provided for each type of vehicle (for example, a road train consisting of a semi-trailer truck and a trailer 34–40 t with the norm Euro-3) for a corresponding road category (for example, an agricultural road) in accordance with fuel costs on an empty or fully loaded vehicle, which corresponds to a load factor γ changing from 0% to 100%. Linear relations between fuel costs and a load factor are taken into account for fuel cost calculation. Figure 3 depicts dependency of diesel fuel costs F_{fuel} in grams/km on a vehicle load factor for different types of roads.

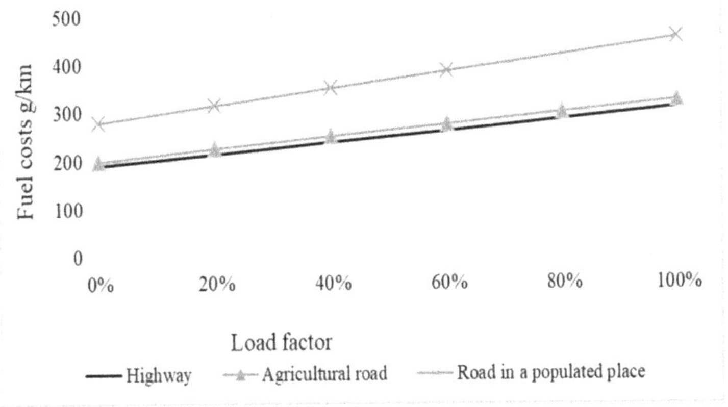

Figure 3. Dependency of fuel costs on a load factor [16].

The formula for calculating fuel costs C_{fuel} on freight transport m_{car}, which is shipped at a distance L, is as follows:

$$C_{fuel} = \frac{L \cdot m_{car}}{1000 \cdot m_{pload} \cdot \gamma} \cdot \left[\left(F_{fuel}^{100} - F_{fuel}^{0} \right) \cdot \gamma + k_{ret} \cdot F_{fuel}^{0} \right] \tag{1}$$

where F_{fuel}^{100} stands for fuel usage with a 100% loaded vehicle, and F_{fuel}^{0} stands for fuel usage with no load vehicle. In order to calculate fuel costs on a return road, a return coefficient is used, namely $k_{ret} = 2$ if return freight transport takes place and $k_{ret} = 1$ if a truck returns empty.

In order to calculate fuel costs of a partly loaded vehicle with m_{pload}, it is necessary to know its potential load $\left(m_{pload} \cdot \gamma \right)$. This index is presented to a client for space distribution in a vehicle. For example, a tank truck for dry bulk cargo used for transportation of pellet fuel can have enough room to transport the material for a few clients at a time.

In order to secure a full load for each transport change m_{pload}, it is necessary to calculate the difference between the maximum allowable weight of a vehicle and the weight of an empty vehicle. Simplified data applicable for all, without exception, Euro-standards are used to calculate the weight.

Semi-trailer trucks can be grouped according to engine power: vehicles with engine power P ≤ 210 kW, which belong to light semi-trailer trucks; and vehicles with engine power P > 210 kW, which belong to heavy semi-trailer trucks.

If a transport route consists of a few road types, this factor is included in square brackets (see (2)) in accordance with each road type k_r, for example, a controlled-access highway, or roads outside (tracks) or within populated areas. According to this, fuel consumption of one transport shift is calculated as follows:

$$C_{fuel}^1 = \sum k_r \cdot \left[\left(F_{fuel}^{100} - F_{fuel}^0 \right) \cdot \gamma + k_{ret} \cdot F_{fuel}^0 \right] \tag{2}$$

Total fuel consumption is calculated separately for each transport change:

$$C_{fuel}^{\Sigma} = \sum k_{tc} \cdot C_{fuel}^1 \tag{3}$$

Since 1 January 2016, all the cars imported from abroad have to meet the ecological standard Euro-5. As a result, there is a complete ban on importing cars produced earlier than 2010/2011 (or a compulsory engine redesign in accordance with the eco-standard), since, in fact, the ecological standard Euro-5 came into effect in EU-countries and the USA in 2009 and, consequently, car manufacturers commenced production of cars meeting Euro-5 standards after 2009.

Unfortunately, national business entities do not always monitor vehicles for Euro-5 standards. Thus, the suggestion is to introduce an average index of all Euro-standards Euroave, which are legislatively established (Euro 0–5). This will make it possible to estimate the level of hazardous emissions based on the statistical data of the vehicles' yearly kilometrage (km) and the volume of transported cargo (T-km).

It is recommendable to use the method suggested by a German scientist, Borken [16], for estimating fuel consumption of the vehicles used for wood biomass transportation. His method presupposes differential adjustment of fuel consumption based on five different applications of concentrated load to five different points, namely maximum load, normal load, minimum load, transportation (for example, transportation from a warehouse to a field), and idling (for example, during preparations or waiting), as shown in Figure 4.

Upon timing of concentrated load application points, it is possible to estimate vehicle fuel consumption according to the ratio

$$F_{burnt} = \sum k_{\%} \cdot C_{cons}, \tag{4}$$

where C_{cons} is fuel consumption for a specific application point and is applied in the research in order to estimate fuel consumption of agricultural machinery with the given rated powers for different concentrated load application points.

In addition to the fuel consumption C_{fuel}, there are costs of other materials employed C_{em} and road costs. In contrast to the vehicle usage costs C_{veh}, there are personnel costs C_{per} including vehicle repair C_{rv} and control costs C_{cv}. They are not constants and change with time, for example, lubricant costs.

Special software, "maKost", is applied in European practice. It has a database carefully selected for calculating all constant and changeable costs for each concrete agricultural vehicle model and evaluating its potential negative impacts on the natural environment. The program was elaborated by "KTBL", a German inspection board monitoring agricultural machinery and construction, and it is obviously multifunctional software. All parameters can be set individually, and the database can be used to calculate fuel costs and repair costs, to outline the needs, estimate economic factors, etc. [27–29]. In order to reduce time expenditure and make calculations as accurate as possible, it is recommendable that Ukrainian business entities use the aforementioned software. We also suggest that a national analogue of this program be elaborated, which would allow one to take into account the peculiarities of the vehicles used by our business entities.

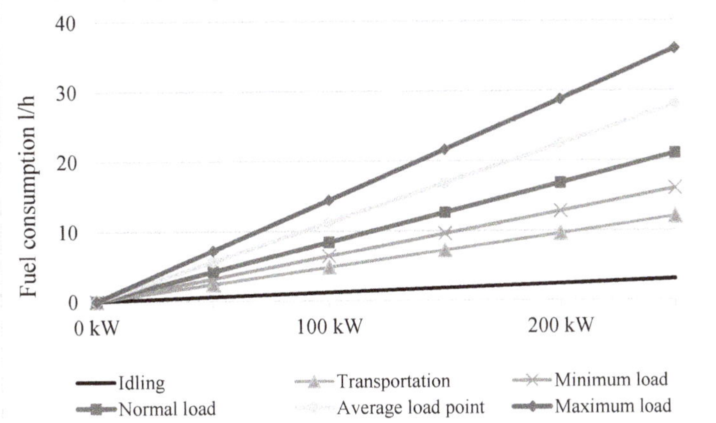

Figure 4. Correlation between fuel consumption and agricultural machinery power for different load application points [28].

In 2006, a Viennese scientist, Kanzian, introduced a model of calculating transportation time of short agricultural connections, in particular, for trucks making 4–12 km (from the forest to wood biomass processing department). The average speed of vehicles was assumed to be 17 km/h [16].

Wood biomass transshipment processes include cargo transshipment costs C_{tsh} where various machines and processes are involved. In general, the expenditure includes specific costs on infrastructure C_{tsh}^{inf}, with the area S_{tsh}^{a}, personnel C_{tsh}^{p}, machines C_{tsh}^{m}, and also energy expenditure C_{tsh}^{e} to get machinery ready for use:

$$C_{tsh} = t_{tsh} \cdot \left(S_{tsh}^{a} \cdot C_{tsh}^{inf} + C_{tsh}^{p} + C_{tsh}^{m} \right) + C_{tsh}^{e}, \tag{5}$$

where t_{tsh} is cargo transshipment time.

The observations of German scientist Dobers have shown that a flatbed truck with a crane can be loaded with branches in 10 min. Branches are aligned and piled up easily, which is quite productive as the time for unloading is one third quicker (while unloading branches are not to be aligned or piled) [30].

Evaluating the time needed for transshipment of woodchips using a pneumatic loader with shovel, it was estimated that loading of a container of 30 m^3 takes 15 min. Thus, the effectiveness of the process will be 120 m^3 per hour, which is 1 shovel lift in 1 min [31].

The process of unloading of woodchips off a vehicle is performed by turning it over; according to preliminary calculations, it is possible to unload 240 m^3 of round timber, firewood, or bales in an hour or 260 m^3 of woodchips or fuel pellets [2].

The storage expenses are divided into infrastructural expenses C_{infs} and extra expenses on container production and storage of cover materials C_{con}, and also the expenses on loading, transshipping, and unloading C_{load} are taken into account. Different types of wooden biomass require different types of storage options: in the warehouses with or without a roof, in silos, reservoirs, containers, bales, or large bags. All in all, for every option, the following formula to calculate the expenses on storage of cargo C_{stor} in a time t_{stor}, could be used:

$$C_{stor} = t_{stor} \cdot \left(C_{infs} + C_{con} \right) + C_{load} \tag{6}$$

Additional expenses C_{con} are estimated on the basis of the number of containers n_{con} or storage structures (wooden framework, hopper for bulk materials) and expenses on their production C_{prod}. In case the biomass is to be covered with waterproof material (area

S_{mat}) to protect from environmental conditions, then these expenses ($S_{mat} \cdot C_{mat}$) are added to the sum:

$$C_{con} = n_{con} \cdot C_{prod} + S_{mat} \cdot C_{mat} \tag{7}$$

If stored outdoors or covered, the expenses of the containers and the required amount of waterproof material required to cover a certain area as well as its assembling and expenses on storage space rental are determined.

Woodchips are loaded, unloaded, and transshipped by means of pneumatic loaders. Transshipment of the woodchips is also performed with ventilation measures and reduction of dry mass loss. If biomass is needed to be transshipped again during the storage, then the additional area S_{tshp} (e.g., larger hopper for bulk materials) is added to the number of times of loadings n_{tshp} in a time t_{tshp}. The loading, unloading, and transshipment expenses are calculated with the formula

$$C_{load} = (t_{load} + t_{unl}) \cdot \left(C_{load}^{p} + C_{load}^{m}\right) + n_{tshp} \cdot t_{tshp} \cdot \left(C_{sload}^{p} + C_{sload}^{m} + S_{tshp} \cdot C_{sload}^{inf}\right) \tag{8}$$

where t_{load} and t_{unl} are the time indicator of loading and unloading, respectively; C_{load}^{p} and C_{load}^{m} are the expenses of staff and machines for loading, transshipment, and unloading, respectively; C_{sload}^{p} and C_{sload}^{m} are the expenses of staff and machines for second loading, respectively; and C_{sload}^{inf} are specific expenses of infrastructure. The expenses of storage of wooden biomass are calculated with the formula

$$C_{stor} = t_{con} \cdot n_{con} \cdot \left(C_{load}^{p} + C_{load}^{m}\right) + \frac{t_{load} + t_{unl}}{3600} \cdot n_{con} \cdot \left(C_{stor}^{p} + C_{stor}^{m}\right) \tag{9}$$

where n_{con} is the number of containers; t_{con} is the time of the work of loader; and C_{stor}^{p} and C_{stor}^{m} are the staff expenses and machinery for storage, respectively.

Apart from this, we should take into account the loss of wooden biomass weight during the process of drying that is considerable in first weeks of storage. The loss of weight during the drying process happens in first six weeks and reaches from 80% to 100%.

The expenses on preliminary preparation and processing the wooden biomass C_{prep} are calculated similar to the expenses on transshipment of cargo, including the infrastructural expenses, staff and maintenance of technical means and energetic expenses in general.

$$C_{prep} = t_{prep} \cdot \left(S_{prep} \cdot C_{prep}^{inf} + C_{prep}^{p} + C_{prep}^{m}\right) + C_{prep}^{e}. \tag{10}$$

The cut branches are chopped with the wood chopping machine (no infrastructural expenses). The chopping expenses C_{ch} are calculated with the following formula:

$$C_{ch} = t_{ch} \cdot \left(n_{ch} \cdot C_{ch}^{p} + C_{ch}^{m}\right) + C_{ch}^{e}, \tag{11}$$

where n_{ch} is the amount of biomass for chopping.

If wood chips, during the chopping, are immediately blown into a vehicle or a container for storage, then one should take into account the readiness of the vehicle or the container, as well as the personnel, for example a driver, who is waiting at this time. In this case, another item of additional costs arises:

$$C_{con}^{ch} = t_{con}^{ch} \cdot \left(C_{con}^{p} + C_{con}^{m}\right), \tag{12}$$

where C_{con}^{ch} is costs of the servicing of the container during the chopping; t_{con}^{ch} is chopping time; C_{con}^{p} is personnel costs of the servicing of the container; C_{con}^{m} is machine costs of the servicing of the container.

Then the effectiveness of chopping should be verified. Since the readiness of the vehicle or the container, into which the chopped wood is loaded, is not established, redundant time is spent on waiting, and, consequently, redundant financial costs arise. Research has shown that waiting time and the time spent on the tuning of the wood chopping machine reduce the net chopping time to 60–70% [18,32].

During the filtration of wood chips, one should also take into consideration seasonal differences. The quota of small fractions during the usage of waste wood amounts to 50% in summer months and may plunge to 25% after the leaves fall down in winter months.

If the non-filtered wooden chips are loaded by pneumatic wheel loader, then the filtration costs C_{fil} are calculated in the following way:

$$C_{fil} = t_{fil} \cdot \left(S_{fil} \cdot C_{fil}^{inf} + C_{fil}^{p} + C_{fil}^{m} + C_{sload\ fil}^{p} + C_{sload\ fil}^{m} \right) \tag{13}$$

where C_{fil} is filtration costs; t_{fill} is filtration time; C_{fil}^{inf} is specific infrastructural costs of filtration; C_{fil}^{p} is personnel costs of filtration; C_{fil}^{m} is machine costs of filtration; $C_{sload\ fil}^{m}$ is personnel costs of transloading; $C_{sload\ fil}^{m}$ is machine costs of transloading.

Natural air drying of wooden biomass is a long-lasting process; however, the drying can be done using machinery, for example, using a drying cylinder or a drying container. In order to calculate drying costs C_{dr} in addition to heating energy costs C_{he}^{e}, we should consider the costs of the servicing of infrastructure, of drying machinery, and of appropriate containers (grate container)—C_{inf}^{teh}. Here one will also add personnel costs C_{dr}^{p} and costs of the servicing of drying equipment C_{de}^{m}, thus obtaining the following drying costs C_{dr}:

$$C_{dr} = t_{dr} \cdot C_{inf}^{tech} + C_{he}^{e} + \left(t_{dr} + t_{prep\ dr} \right) \cdot \left(C_{dr}^{p} + C_{dr}^{m} + C_{inf}^{tech} \right) \tag{14}$$

where t_{dr} is time for the preparation to drying; $t_{prep\ dr}$ is post-drying treatment time.

According to the research [2], having taken into account the heat and electricity consumption of the drying cylinder and the drier, we must calculate total costs of the drying of wood chips with water content from 40% to 10%. In case of six-month drying, financial costs that are needed to obtain dry mass amount from 6 Euro/t to more than 40 Euro/t, and in case of twelve-month drying, from 4 Euro/t to more than 24 Euro/t.

Ecological evaluation is carried out in the way analogous to the economical evaluation. For every process, data are calculated to define the level of harmful emissions into the environment during the usage of fuel, the preparation of equipment (machinery, means of transport, etc.), and the work of personnel, while ensuring the functioning of infrastructure.

In order to calculate the influence of any means of transport on the environment, one takes as a basis the factors of influence of greenhouse gases, such as carbon dioxide, methane, and nitrogen oxide emissions. The algorithm of their calculation is analogous to the calculation of fuel costs for every shift of transportation, type of road, and cargo type for an empty or fully loaded truck.

We take into consideration every separate type of road in the same way as in the process of economical evaluation:

$$V_{veh} = \frac{L \cdot m_{car}}{m_{pload} \cdot \gamma} \cdot \left[\left(F_{CO_2}^{100} - F_{CO_2}^{0} \right) \cdot \gamma + k_{ret} \cdot F_{CO_2}^{0} \right], \tag{15}$$

where V_{veh} is harmful emissions of a vehicle; $F_{CO_2}^{100}$ is emissions factor if the vehicle is loaded 100%; $F_{CO_2}^{0}$ is emissions factor if the vehicle is not loaded.

$$F_{CO_2} = \sum k_r \cdot \left[\left(F_{CO_2}^{100} - F_{CO_2}^{0} \right)^{veh} \cdot \gamma + \left(F_{CO_2}^{0} \right)^{veh} \cdot k_{ret} \right]. \tag{16}$$

We suggest that domestic enterprises calculate the decrease of the level of harmful emissions in a way that, when the biofuel burnt by the vehicles is estimated, the emissions of CO_2 and SO_x caused by it should be equal to 0. This means that, depending on the increase of the quota of biofuel in traditional fuel, the CO_2 and SO_x emissions will decrease in percentage, Table 1 is given as an example. The fact that the level of harmful emissions from 2011 till 2013 remained at 9% means that the quota content of biofuel in the vehicles of the enterprise remained unchanged during this period.

Table 1. Decrease of CO_2 emissions in percent in cases of year-to-year increases of biofuel quota in the vehicles of enterprises.

Years	Quota of CO_2 Emissions Decrease in Contrast to 2005 on Condition that Biofuel Quota Increases, %
2006	2
2006	4
2010	6
2012	8
2014	10
2016	12
2018	14
2020	16

[self-elaborated].

In order to calculate the direct emissions of road transport, with respect to CO_2-equivalent, we should use the derivatives of the emissions factors:

$$F_{CO_2} = F_{CO_2bio} + F_{CH_4} \cdot Q_{gg}^{CH4} + F_{N_2O} \cdot Q_{gg}^{N_2O}, \tag{17}$$

where F_{CO_2bio} is emissions factor of a vehicle that uses biofuel; F_{CH4} is greenhouse gases emissions factor; Q_{gg}^{CH4} is greenhouse gases emissions quota.

When we evaluate fuel consumption of the entire supply chain, during the ecological evaluation, we obtain the factors of the emissions of side mixtures (oils, etc.) F_{veh}^{sm}, infrastructure F_{veh}^{inf}, machinery F_{veh}^{m}, and other fuel consumption F_{veh}^{fuel} in the chain. This is why CO_2 emissions into the environment are added to direct F_{veh}^{de} and indirect emissions, and the harmful emissions of a vehicle are calculated as follows:

$$V_{veh} = F_{veh}^{de} + F_{veh}^{fuel} + F_{veh}^{sm} + F_{veh}^{m} + F_{veh}^{inf}, \tag{18}$$

In order to transship the biomass, trucks with crane bodies, pneumatic wheel loaders, and fork loaders are used. Since during the loading and unloading of the vehicles the emissions of harmful substances into the environment take place, then

$$V_{trs} = V^{de} + V_{trs}^{ee} + V_{trs}^{sm} + V_{trs}^{m} + F_{veh}^{m} + V_{trs}^{inf}, \tag{19}$$

where V_{trs} is emissions during the transshipment; V^{de} is direct emissions; V_{trs}^{ee} is engine emissions during the transshipment; V_{trs}^{sm} is emissions of side mixtures during the transshipment; V_{trs}^{m} is machinery emissions during the transshipment; V_{trs}^{inf} is infrastructure emissions during the transshipment.

Direct emissions of harmful substances are influenced by the usage of fuel during the transshipment of cargo V_{trs}^{de}, as well as while waiting, and by the idling of vehicles F_{veh}^{de}

$$V^{de} = V_{trs}^{de} + F_{veh}^{de}. \tag{20}$$

The first ones are influenced by the consumption of fuel and electric energy by the machinery (pneumatic wheel loaders and fork autoloaders), and the evaluation of indirect emissions (costs of infrastructure and consumables) are already included in the economic model.

The emissions of harmful substances during provision include chopping, filtration, and drying processes.

$$V_{prov} = V^{de} + V^{ee}_{prov} + V^{m}_{prov} + V^{inf}_{prov}. \qquad (21)$$

At this time, we obtain direct emissions caused by the usage of fuel for corresponding machines. We take into consideration the usage of fuel and electric energy, as well as machines (equipment) and corresponding infrastructure (maintenance of the place) during the entire logistics chain.

The composition of the model of the economic and ecological evaluation of a wooden biomass supply chain (Figure 5) is based on the evaluation of specific logistic processes of "provision–loading–transporting–transshipment–unloading–chopping–warehousing–storage", which depend on the functioning sequence, the evaluation of costs and harmful substances emission, as well as on the identification of the volumes of the potential wooden biomass provision.

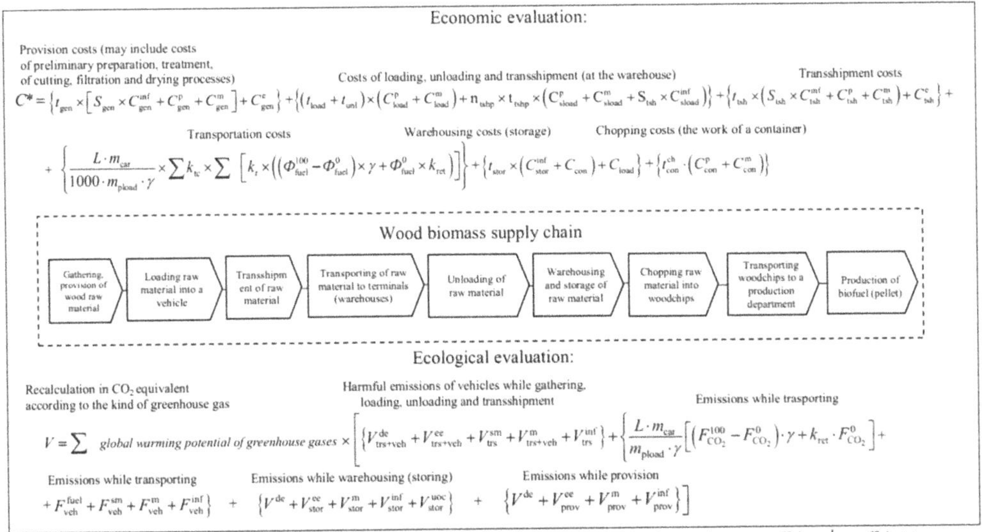

Note: * C – general costs; t – time; S – area; L – transportation distance; m – mass; n – quantity; γ – carrying capacity usage coefficient; k – coefficient; tc – transportation shift; ret – fuel costs of the way back; r – road type. $F^{100}_{fuel}, F^{0}_{fuel}, \Phi^{100}_{CO_2}, \Phi^{0}_{CO_2}$ – factors of fuel consumption and emissions if the truck is 100% loaded and without load; upper indices of consumption and emissions: p – personnel; m – machinery (equipment); inf – infrastructure; e – energy consumption; sm – side mixtures; ee – engine; de – direct emissions; fuel – other fuel consumption in the chain; uoc – consumables; V – general harmful emissions.

Figure 5. The model of the economic and ecologic evaluation of a wooden biomass supply chain. Source [self-elaborated].

A typical chain of chopped wood supply includes the processes of loading raw material (branches, trunks) by a mobile crane; road transportation by different trucks; unloading by turning over, emptying, or cutting; chopping and storing of wood chips at a rented warehouse, for example; as well as biofuel production.

However, not only physical processes (transportation, transshipment, warehousing, provision of raw material, etc.) are important in biomass supply, but it is also important to have a modern software product with a corresponding database, which encompasses information about the specific character of logistic processes (vehicle type, its carrying capacity, the level of the development of transportation and warehousing infrastructure, and its peculiarities in mountain regions, etc.). The following software products are used in global practice: the database "ecoinvent"; the library "Umberto"; the directory of emissions factors for traffic "HBEFA"; software for calculations "EcoTransiT World" and "MaKost"; the global model of emissions for integrated systems "GEMIS"; and basic data for environment management "PROBAS". The usage of the appropriate software products is possible on condition of their adaptation to the domestic databases.

Practical usage of the model of the economic and ecological evaluation of a wooden biomass supply chain will enable business entities not only to promote the increase of the quota of the usage of alternative energy sources in the general energy consumption of the country, but also to decrease general costs of gathering, transporting, warehousing, and processing wooden biomass and at the same time to decrease significantly the volumes of greenhouse gases emissions into the natural and industrial environment. A possible real application of the model can be divided into three areas:

- Business (formation of the business model of enterprises in the field of renewable energy sources);
- Fuel (the rationale for the choice of fuel for heating homes, infrastructure, hotels, recreation centers, and local authorities);
- Recycling (substantiation of directions of use of woodworking waste: chipboard, chemical processing, cardboard, paper, fuel) [33].

By using the elaborated model and on the basis of expert estimates of the scientists of Dortmund Technical University, it was established that business entities will be able to decrease the cargo turnover by about 15%, which will lead to the decrease of costs and harmful substance emissions by about 10% and will give possibility to analyze systematically their costs and the negative impacts on the natural and industrial environments in separate logistics processes or logistics chains [30]. The literature sources listed in the article testify to the significant potential of wood biomass in Ukraine. As well as for solar and wind energy there are state preferences for biomass, so a certain system of benefits can be offered to consumers themselves, including exemptions from VAT, export duties, etc.

It is necessary to conduct economic and ecological evaluations of wooden biomass supply chain at every stage with the help of analytical models, defining certain indices. Thanks to the suggested model of the economic and ecological evaluation of wooden biomass supply chains, concrete processes of business entities can be presented in monetary equivalents and can be evaluated in terms of greenhouse gases emissions, and the combination of these results will give possibility to reflect the ecological effectiveness of a logistics system. Moreover, the elaborated economic and ecological model enables the identification of the most "vulnerable spots" in the wooden biomass supply chain, which will become the basis for further optimization of the entire chain and will provide broad opportunities to improve the analogous evaluation of other supply systems with consideration of different types of transport. The proposed approach could be suitable for the supply chain of coal, briquettes, peat, and gas.

4. Conclusions

The first step towards a "green economy" in Ukraine (which is one of the key aims of energetic policy of the EU) will be a change to renewable sources of energy, considering that the main directions of energetical potential of biomass and biogas realization in Ukraine is the production of heat energy and electricity, along with the understanding that modern methods of energy production are environmentally harmful. Bioenergetics is one of the chief aims of renewable sources of energy sector development for Ukraine, taking the country's strong reliance on imported energy carriers into account, mainly natural gas, and the high biomass potential available for energy production. Modern technologies allow different types of biomass to be manufactured; as a result, even those economic entities are effective for which biofuel business is supplementary, based on their own raw materials and those that have active heat supply reorientation, for instance, using solid biofuel boilers that have proven their effectiveness and fast payoff.

Systematization of the results of scientific applied research afford grounds for the following conclusions:

Logistics research should be focused on new models and new solutions that will allow the manufacturers of ecological types of energy out of biomass to see concrete practical steps of the state and investors in encouraging new ideas, technologies, and investing solutions support.

The mechanism of constructing and structuring of logistics chains is provided with a description. The chains are oriented on ecologistics development and stable rates of wooden biomass provision and consumption growth, which will allow a high level of competitiveness of the subject economic activity and business efficiency to be ensured by raising investment resources and new technologies in Ukraine.

During the research on the interaction of elements in logistics in the resources supply sphere, the topical problems of formation and storage of necessary bio raw materials for their further recycling into biomass as one of the energy alternatives in regional industry that considers the specific activity of the subjects, market infrastructure, and other factors. The proposed model covers the sequence of functioning of such logistics processes as provision, loading, transporting, transshipment, unloading, grinding, warehousing, storage, and indexes, showing consumer spending and expenses on equipment (machinery, vehicles, etc.), staff, infrastructure and so on.

The method's approaches were improved to construct a model of economic ecological evaluation of supply chains of wooden biomass that will allow the toolkit of logistic management in supply processes management to be enriched, particularly when using biomass for production purposes, and to broaden the sphere of business interaction with biomass supplier and business entity that ensure its effective recycling. This will also allow concrete processes of business entities to be presented in monetary equivalents and evaluate them from the perspective of greenhouse gases emissions, and the resulting combination will allow one to show ecologistic effectiveness of the whole logistics system.

The paper presents and improves the core components of supply chains in the process of provision and consumption of wooden biomass, which will allow the negative impacts on the natural and production environment to be minimized in practice and the usage of non-renewable energy sources to be reduced, to ensure resource economy and to support the development of sustainable business entities.

The model of economic and ecological evaluation of logistics chains constructed by the authors indicates the perspective direction of further implementation of the ideas of integrated management of resources, expenses calculation, and harmful emissions in whole logistics chains and also channels further research to search for mechanisms of energy and resource consumption volume reduction due to the all-in-one vision of production and logistics processes.

Author Contributions: Conceptualization, Y.K. and B.H.; methodology, B.H.; software, J.B.; validation, Y.K., J.B., and B.H.; formal analysis, B.H.; investigation, B.H.; resources, B.H.; data curation, Y.K.; writing—original draft preparation, B.H.; writing—review and editing, B.H.; visualization, B.H.; supervision, Y.K.; project administration, J.B. All authors have read and agreed to the published version of the manuscript.

Funding: This research received no external funding.

Institutional Review Board Statement: Not applicable.

Informed Consent Statement: Not applicable.

Data Availability Statement: Exclude this statement, did not report any data.

Conflicts of Interest: The authors declare no conflict of interest.

References

1. Forsberg, G. Biomasse energz transport. Analysis of bioenergy transport chains using life cycle inventory method. *Biomass Bioenergy* **2000**, *19*, 17–30. [CrossRef]
2. Suurs, R. Long Distance Bioenergy Logistics. An Assessment of Costs and Energy Consumption for Various Biomass Energy Transport Chains. Master's Thesis, Universiteit Utrecht, Copernicus Institute, Department of Science, Technology and Society, Utrecht, The Netherlands, 2002; p. 78.
3. Pfohl, H.-C. *Ökologische Herausforderungen an Die Logistik in den 90er Jahren. Umweltschutz in der Logistikkette bei Ver-und Entsorgung;* Schmidt: Berlin, Germany, 1993; p. 10.

4.   Weber, J. Logistik-Controlling–Konzept und empirischer Stand. *Kostenrechn. Z. Für Control. Account. Syst.-Anwend.* **2001**, *45*, 275–282. [CrossRef]
5.   Gudehus, T. *Logistik. Bd 1: Grundlagen, Verfahren und Strategien*; Springer: Berlin, Germany, 2000; p. 660.
6.   Pistorius, T.; Zell, J.; Hartebrodt, C. *Untersuchungen zur Rolle des Waldes und der Forstwirtschaft im Kohlenstoffhaushalt des Landes Baden-Württemberg.* Forschungsbericht FZKA-BWPLUS.; Forstliche Versuchs-und Forschungsanstalt BadenWürttemberg, Institut für Forstökonomie: 2006. p. 224. Available online: https://www.irb.fraunhofer.de/bauforschung/baufolit/projekt/Untersuchungen-zur-Rolle-des-Waldes-und-der-Forstwirtschaft-im-Kohlenstoffhaushalt-des-Landes-Baden-W%C3%BCrttemberg/20100282/ (accessed on 23 August 2021).
7.   Plöchl, M.; Heiermann, M. *Ökologische Bewertung der Bereistellung Landwirtschaftlicher Kosubstrate zur Biogaserzeugung.* Bornimer Agrartechnischer Benchte. p. 32. Available online: https://opus4.kobv.de/opus4-slbp/files/4030/1_BAB_32_2.pdf (accessed on 23 August 2021).
8.   Jungbluth, N.; Frischknecht, R.; Faist, M. *Ökobilanz für die Stromerzeugung aus Holzbrennstoffen und Altholz*; Projekt 41458 Vertrag 81427; Herausgegeben von Bundesamt für Energie: Bern, Switzerland, 2002; p. 77.
9.   Krotschek, C.; König, F.; Obernberger, I. Ecological assessment of integrated bioenergy systems using the Sustainable Process Index. *Biomass Bioenergy* **2000**, *18*, 341–368. [CrossRef]
10. Mann, M.K.; Spath, P.L. A life cyde assessment of biomasscofiring in a coal-fired power plant. *Clean Prod. Process.* **2001**, *3*, 81–91. [CrossRef]
11. Marheineke, T.; Krewitt, W.; Neubarth, J.; Friedrich, R.; Voß, A. *Ganzheitliche Bilanzierung der Energie-und Stoffströme von Energieversorgungstechniken*; Forschungsbericht; Universität Stuttgart. Institut für Energiewirtschaft und rationelle Energieanwendung: Stuttgart, Germany, 2000; Volume 74, p. 107.
12. Scholwin, F. Ökologische Bewertung der Nutzung nachwachsender Rohstoffe zur Biogasgewinnung und-nutzung. Umwelteffekte und ihre Berücksichtigung im Anlagenbetrieb. In *Biogas. Energieträger der Zukunft; Tagung Osnabrück, 12. und 13. April 2005*; Gesellschaft Energietechnik, Hg.: Düsseldorf, Germany, 2005; pp. 77–87. Available online: https://www.ifeu.de/fileadmin/uploads/landwirtschaft/pdf/IFEU-TAB-Nawaro-stofflich.pdf (accessed on 23 August 2021).
13. Mardon, C. The feasibility of producing alcohol fuels from biomass in Australia. *Int. J. Glob. Energy Iss.* **2007**, *27*, 138–159. [CrossRef]
14. Soimakallio, S.; Mäkinen, T.; Ekholm, T.; Pahkala, K.; Mikkola, H.; Paappanen, T. Greenhouse gas balances of transportation biofuels, electricity and heat generation in Finland–Dealing with uncertainties. *Energy Policy* **2009**, *37*, 80–90. [CrossRef]
15. Zah, R. Umweltbilanz von Biotreibstoffen. *GWA* **2007**, *12*, 953–959.
16. Kanzian, C.; Fenz, B.; Holzleitner, F.; Stampfer, K. Waldhackguterzeugung aus Schlagrücklass. Fallbeispiele im Laub-und Nadelholz: Project Universität für Bodenkultur Wien. Department für Wald-und Bodenwissenschaften: Wien, Austria, 2006; p. 17.
17. Neff, A.; Nelles, M.; Loewen, A.; Hapla, F. Perspektiven für die energetische Verwertung von Landschaftspflegeholz. In *Jahrbuch der Baumpflege*; Dujesiefken, D., Kockerbeck, P., Eds.; Haymarket Media: Hamburg, Germany, 2006.
18. Scholz, V.; Idler, C.; Daries, W.; Egert, J. Lagerung von Feldholzhackgut. Verluste und Schimmelpilze. *Agrartech. Forsch.* **2005**, *11*, 100–113.
19. Eberhardinger, A.W.W.; Zormaier, F.; Schardt, M.; Huber, T.; Zimmer, B. *Prozessanalyse und Ökobilanzierung der Bereitstellug von Waldhackgut*; Schlussbericht for Lehrstuhl für Forstliche Arbeitswissenschaften und Angewandte Informatik; Technische Universität München: Freising, Germany, 2009; p. 166.
20. D'Adamo, I.; Falcone, P.M.; Huisingh, D.; Morone, P. A circular economy model based on biomethane: What are the opportunities for the municipality of Rome and beyond? *Renew. Energy* **2021**, *163*, 1660–1672. [CrossRef]
21. D'Adamo, I.; Falcone, P.M.; Morone, P. A new socio-economic indicator to measure the performance of bioeconomy sectors in Europe. *Ecol. Econ.* **2020**, *176*, 106724. [CrossRef]
22. Bretzke, W.R. *Nachhaltige Logistiksysteme–Anpassungsbedarfe in Einer Welt Steigend Energiekosten, Überlasteter Verkehrswege und Rigide Bekämpfter Schadstoffemissionen*; Tagungsband Magdeburger Logistiktagung: Magdeburg, Germany, 2009; p. 95.
23. Bringezu, S. Nutzungskonkurrenzen bei Biomasse–Auswirkungen der Verstärkten Nutzung von Biomasse im Energiebereich auf Die Stoffliche Nutzung in der Biomasse Verarbeitenden Industrie und Deren Wettbewerbsfähigkeit Durch Staatlich Induzierte Förderprogramme [Electronic Resourse]. 2008. Available online: http://www.rwiessen.de/media/content/pages/publikationen/rwi-projektberichte/PB_BiomasseNutzungskonkurrenz_Kurzfassung.pdf (accessed on 14 October 2013).
24. Orhanizatsiya Pererobky Lisosichnykh Vidkhodiv: Proekt/[L'vivs'ke Oblasne Upravlinnya Lisovoho ta Myslyvs'koho Hospodarstva]—L'viv, Ukraine 2016. p. 40. Available online: https://lvivlis.gov.ua/ (accessed on 23 August 2021).
25. Hrechyn, B.D. *Approaches of Ecologistics to the Usage of Wood Biomass Power Potential: Dys. kand. ek. Nauk: 08.00.06 Hrechyn Bohdan Dmytrovych*; University of Water Management and Environmental Engineering: Rivne, Ukraine, 2018; p. 265.
26. Hrechyn, B.D. Substantial Characterization of the Methods for Evaluating Negative Impact of Production and Logistics Activities of Enterprises in the Context of the Strategic Development of Eco-Logistics in Ukraine. *Business Inform.* **2016**, *4*, 169–176.
27. Borken, J. *Basisdaten für ökologische Bilanzierungen. Einsatz von Nutzfahrzeugen in Transport, Landwirtschaft und Bergbau*; Springer: Braunschweig, Germany, 1999; p. 223.
28. Eckel, H. *Energiepflanzen. Daten für die Planung des Energiepflanzenanbaus*; KTBL: Darmstadt, Germany, 2006; p. 372.
29. MaKost 2009b: Kuratorium für Technik und Bauwesen in der Landwirtschaft e. V. (KTBL) (Hg.): MaKost -Maschinenkosten und Reparaturkosten. [Electronic Resourse]. Available online: https://www.ktbl.de/ (accessed on 24 August 2021).

30. Dobers, K.; Clausen, U. *Bewertung der Ökoeffizienz von Logistiksystemen am Beispiel der Biomassebereitstellung (Logistik, Verkehr und Umwelt)*; Auflage: 1; Taschenbuch. Praxiswissen Service: Dortmund, Germany, 2011; p. 212.
31. Spickermann, C. *Hackschnitzellagerung und–Handling*; Westphalian University of Applied Sciences: Dorsten, Germany, 2007.
32. Wittkopf, S. Bereitstellung von Hackgut zur Thermischen Verwertung durch Forstbetriebe in Bayern. Ph.D. Thesis, Technische Universität München, Wissenschaftszentrum Weihenstephan für Ernährung, Landnutzung und Umwelt, Munich, Germany, 2005; p. 209.
33. Krykavskyy, Y. *Lohistyka Dlya Ekonomistiv: Pidruchnyk. Druhe Vydannya, Vypravlene ta Dopovnene*; Vydavnytstvo L'vivs'koyi Politekhniky: L'viv, Ukraine, 2014; p. 476.

Article

Bio-Char Characterization Produced from Walnut Shell Biomass through Slow Pyrolysis: Sustainable for Soil Amendment and an Alternate Bio-Fuel

Rami Alfattani [1], Mudasir Akbar Shah [2,*], Md Irfanul Haque Siddiqui [3], Masood Ashraf Ali [4] and Ibrahim A. Alnaser [3]

[1] Department of Mechanical Engineering, Umm Al-Qura University, P.O. Box 715, Makkah 24224, Saudi Arabia; rafattni@uqu.edu.sa

[2] Department of Chemical Engineering, Kombolcha Institute of Technology, Wollo University, Kombolcha 1145, Ethiopia

[3] Department of Mechanical Engineering, King Saud University, Riyadh 11421, Saudi Arabia; msiddiqui2.c@ksu.edu.sa (M.I.H.S.); ianaser@ksu.edu.sa (I.A.A.)

[4] Department of Industrial Engineering, College of Engineering, Prince Sattam Bin Abdulaziz University, Al-Kharj 16273, Saudi Arabia; mas.ali@psau.edu.sa

* Correspondence: shahmudasir22@nitsri.net

Abstract: Bio-char has the ability to isolate carbon in soils and concurrently improve plant growth and soil quality, high energy density and also it can be used as an adsorbent for water treatment. In the current work, the characteristics of four different types of bio-chars, obtained from slow pyrolysis at 375 °C, produced from hard-, medium-, thin- and paper-shelled walnut residues have been studied. Bio-char properties such as proximate, ultimate analysis, heating values, surface area, pH values, thermal degradation behavior, morphological and crystalline nature and functional characterization using FTIR were determined. The pyrolytic behavior of bio-char is studied using thermogravimetric analysis (TGA) in an oxidizing atmosphere. SEM analysis confirmed morphological change and showed heterogeneous and rough texture structure. Crystalline nature of the bio-chars is established by X-ray powder diffraction (XRD) analysis. The maximum higher heating values (HHV), high fixed carbon content and surface area obtained for walnut shells (WS) samples are found as ~ 18.4 MJ kg^{-1}, >80% and 58 m^2/g, respectively. Improvement in HHV and decrease of O/C and H/C ratios lead the bio-char samples to fall into the category of coal and confirmed their hydrophobic, carbonized and aromatized nature. From the Fourier transform infra-red spectroscopy (FTIR), it is observed that there is alteration in functional groups with increase in temperature, and illustrated higher aromaticity. This showed that bio-chars have high potential to be used as solid fuel either for direct combustion or for thermal conversion processes in boilers, kilns and furnace. Further, from surface area and pH analysis of bio-chars, it is found that WS bio-chars have similar characteristics of adsorbents used for water purifications, retention of essential elements in soil and carbon sequestration.

Keywords: walnut shells; pyrolysis; higher heating values; bio-char; surface area

Citation: Alfattani, R.; Shah, M.A.; Siddiqui, M.I.H.; Ali, M.A.; Alnaser, I.A. Bio-Char Characterization Produced from Walnut Shell Biomass through Slow Pyrolysis: Sustainable for Soil Amendment and an Alternate Bio-Fuel. *Energies* **2022**, *15*, 1. https://doi.org/10.3390/en15010001

Academic Editor: Idiano D'Adamo

Received: 30 September 2021
Accepted: 15 December 2021
Published: 21 December 2021

Publisher's Note: MDPI stays neutral with regard to jurisdictional claims in published maps and institutional affiliations.

1. Introduction

Continuous environmental issues, ascending prices of petroleum, energy crisis, exhaustion of fossil fuels, increasing application and need for energy are the serious motivations, due to which there is much insistence on substitute sustainable energy sources. Environmental friendliness, sustainability and biodegradability are the important characters which have made the biomass a primary candidate for the generation of bio-energy. The conversion technologies are the possible options to explore the economic potential of bio-resources. Biofuels and bio-chemicals are formed through thermo-chemical conversion, which includes pyrolysis, gasification, liquefaction and combustion [1]. Pyrolysis is the most striking process for converting biomass into bio-fuels [2]. Volatiles and semi-volatiles are discharged from the feedstock residues during pyrolysis of biomass and yields gases, bio-oil and chars. Further, bio-char may be formed with re-condensing vapors into the bio-char material depending on the residence time of vapor, which increases the bio-char products [3,4].

Bio-char is the solid yield of pyrolysis and has different properties in comparison to the corresponding feedstock's, and can deliver considerable and sustainable diversity in securing an upcoming resource of green energy [5]. It has got several commercial applications, such as fuel production [6,7], energy storage [8], soil improvement [9], soil conditioner [4,10] animal farming, building sector, drinking and wastewater treatments, biogas production, industrial materials (plastics, carbon fibers), exhaust filters, energy production (substitute for lignite, pellets), electronics semiconductors, batteries), paints and coloring (industrial paints, food colorants), cosmetics (therapeutic bath additives, skin-cream, soaps), medicines (detoxification, carrier for active pharmaceutical ingredients), etc. [10–14] as illustrated in Figure 1. Further it can be upgraded by using appropriate methods [15], to form activated bio-char and value-added yields. Due to these applications, bi-ochar may be used as soil amendment and solid fuel due its high porosity, specific surface area and heating value near to coal. In Taiwan, bio-char is largely used for soil alteration as a result of its high-water absorption and surface area and acting as an activated carbon [16].

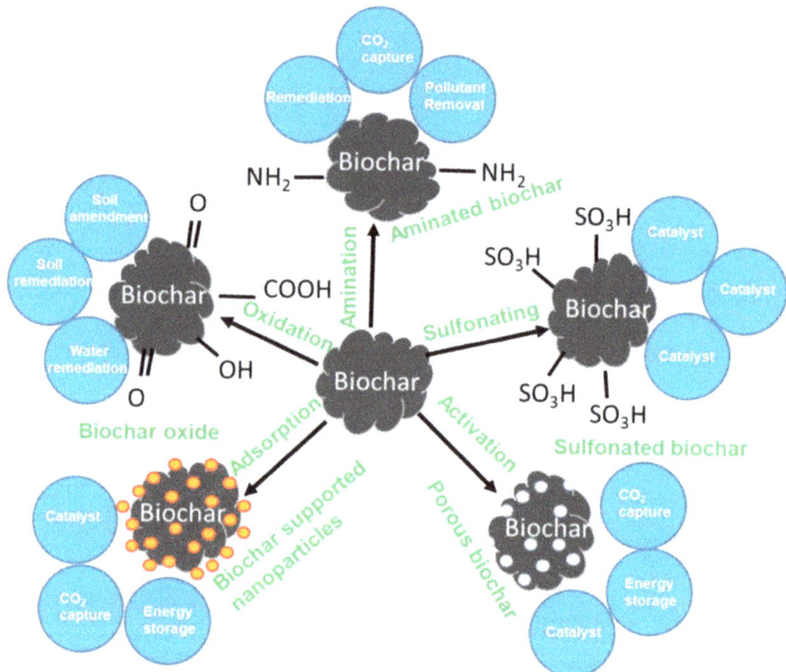

Figure 1. Bio-char potential applications of different sectors.

The origin of biomass source is an important parameter, which influences the characteristics of bio-char yield. Diversified potential biomass residues exist for bio-char formation including municipal wastes, animal manures, forestry and agricultural residues and another growing biomass. A large number of characters should be considered, however, when determining biomass feedstock suitability, like the sustainability requirements, possible toxicity of the bio-char, desired bio-char characteristics and end use [17]. The characteristics of the bio-char are affected by number of variables, such as pyrolysis temperature (maximum or minimum), feedstock size, retention time at the maximum temperature and the pyrolysis atmosphere [4,17,18]. A number of studied reported that the surface area of bio-char is higher as compared to their respective biomass [19] which makes bio-char is a suitable candidate to be either used as an adsorbent or as bio-fuel. Additionally, bio-char obtained from biomass is rich in minerals, therefore it can also be used to improve soil conditions [4,17]. Walnut shell bio-char may be a future and eco-friendly candidate for solid biofuel. Due to its high heating value and may replace the coal fuels (30 MJ kg^{-1}) in future. Jiang et al. [20], obtained chestnut bio-char by pyrolysis and the preparation were done by catalytic pre-oxidation with urea and sulfuric acid. The high heating (35.48 MJ kg^{-1}) value was recorded by this method.

Thermochemical conversion technologies such as pyrolysis are dominant to avert secondary pollution and beginning circular bioeconomy [21]. Environmentally sustainable and economically feasible technologies must be engaged to execute industrial-scale pyrolysis for manufacture of biochar, thus ease to commercialization and possible applications of biochar-based yields [4]. Pyrolysis of lignocellulosic biomass residues is an energy strategy and carbon-negative, needs profound investigation activities internationally [22] in the field of renewable energy substitutions [23], environmental pollution control [24], climate mitigation [25], sustainable towards food security and agriculture [26].

Ghodake et al. [27] reported an extensive work on pyrolysis mechanism and physicochemical properties of biochar. They discussed various aspects of in-management of biomass feedstocks supply chain, biomass feedstock composition and pyrolysis products. They discussed the possibility of a sustainable way of bio-char production and also how this could be a great material for soil amendment, agricultural and to achieve circular bioeconomy. Further, Lin et al. [28] have studied the torrefaction of fruit peel waste to yield environmentally friendly biofuel. In their work, it was reported that they used *Ananas comosus* peel and *Annona* squamosa peel samples to produce bio-char as a renewable energy source. Interestingly, it was found that the higher heating value of both bio-char was increased to 19.1–27.7 MJ/kg after torrefaction. Additionally, they reported a high energy return on investment for renewable energy. Moreover, it was emphasized that the application of bio-char for partial coal substitution can reduce CO_2 emissions by 83.7–94.3%. Further, Romanowska-Duda et al. [29] discussed the promotive effect of Cyanobacteria and Chlorella sp. foliar biofertilization to produce feedstock production, solid biofuel and biochar. It was reported that triple foliar plant spraying with non-sonicated monocultures of Cyanobacteria and Chlorella sp. exhibited a considerably progressive impact on metabolic activity and development of plants. Bio-char can be produced in all scales from individual, domestic as well as the industrial levels and is most prominent and leading industry at various socioeconomic settings. The opportunity of multi-functionality structures and sustainable bio-char production practices creates an increasing demand in the fields of cutting-edge materials, soil amendment, environmental protection, agricultural sustainability and to achieve mitigation of climate variation and circular bio economy. There is a necessity to understand the prediction of organic molecules, bioavailability, toxicity, concentration, surface functions, surface radicals, mobility and environmental fates about bio-char structures. The correlation between the structure, applications and mechanisms of bio-char is progressively developing to enhance their agronomic uses, to achieve precisely designed bio-char with a zero-waste dream [27–29].

In addition, circular bioeconomy focuses on the sustainable and resource-saving value of biomass in a broad and multi-generational production chain. Additionally utilizing residues and waste to optimize the value cascade of biomass in production. Recently, D'Adamo, Morone and Huisingh [30] discussed a sustainable shift towards bioenergy. In this work they reiterated that bioenergy should be included in the bioeconomy sector. In that case, it would also include the agriculture and forestry and new manufacturing sectors. Currently, several types of agrochemical and biochemical processes are adapted to convert lignocellulosic residues into value-added yields. The microbial delignification joined with hydrolysis to increase biofuel yields such as methane [31], butanol [32], ethanol [33], hydrogen production [34] and fuel briquettes [35]. None of the literature reported the comparative study of four different types of walnut shells. In addition to this, we have carried out research work on the biomass agriculture residues available in the State of Jammu and Kashmir, India, which has been again not studied earlier for this particular geographical area.

In the present study, walnut shells are considered as biomass residue for the production of bio-char. Walnut is cultivated mainly in Asian, European and American regions and is one of the important agricultural products for dry fruit industry. In the last decade, the production of walnut has increased by ~25% [36]. Walnut consists of oily material kernel (60%) and a hard covering shell (40%) [37]. The walnut shells, lignocellulosic biomass, have no utilization except being directly used for combustion in furnace, otherwise it is dumped in open areas. This study investigates the physio- and thermochemical characterization of bio-chars obtained from hard-shelled walnut (HSW), medium-shelled walnut (MSW), thin-shelled walnut (TSW) and paper-shelled walnut (PSW). HHV and molar ratios of hydrogen to carbon (H/C) and oxygen to carbon (O/C) are determined. The pH values and surface area of different bio-chars are determined to propose their specific applications. Three different heating rates are used for TG analysis. Furthermore, bio-chars are also investigated by scanning electron microscopy, X-ray diffraction and Fourier transform infrared spectrometry to understand the product profile of bio-char samples obtained from different walnut shell samples.

2. Materials and Methods

2.1. Sample Collection and Preparation

Walnut samples of HSW, MSW, TSW and PSW were collected from the walnut business unit Jammu and Kashmir, India, and sundried for three days at a temperature of 25 °C with less than 47% humidity. Thereon, it was crushed by high speed ball mill and passed through screens in order to obtain particle sizes of 2.5–3, 1.5–2.5 and 0.5–1.5 mm. The sample biomass residues (Figure 2) was packed in an airtight PVC jar and stored in desiccators for further experiments.

(a)   (b)   (c)

Figure 2. (**a**) The original shape of walnut (**b**) shell residues and (**c**) their respective bio-chars.

2.2. Pyrolysis Setup: Pyrolysis of Biomass in a Fixed-Bed Reactor

Pyrolysis was carried out in a fixed-bed reactor. The internal diameter and length of the reactor were 122.26 and 1200 mm, respectively. The design pressure and temperature were 12 bars and 950 °C, respectively. It was surrounded by an electric furnace with P&ID controller to supply power for heating. Ni-Cr thermocouple was used to sense the temperature inside the reactor. A batch of 300 g of the individual feedstock was charged for pyrolysis by increasing the temperature from ambient to 375, 450, 550, 650 and 750 °C, respectively, at a heating rate of 10, 20 and 50 °C/min^{-1}. Nitrogen at a flow rate of 50, 100 and 150 cm^3/min was used to keep the environment inert and oxygen-free, and also to carry over the condensable vapors produced during pyrolysis. These vapors were collected in a condensers I and II to collect oil and scrubbing tank to collect gas. The reaction was performed for 35 min, or till no further release of gas was observed. Further, the bio-char was collected after cooling down the pyrolizer to room temperature. The bio-char samples were stored in airtight PVC containers for further analysis.

2.3. Material Analysis

The proximate analysis of the bio-char samples was carried out as per ASTM standard procedures: E871–82 (2013), D1102-84 (2013) and E872-82(2013) for ash and volatile matter contents, respectively. Ash and volatile matters were determined at a temperature of 580 $^\circ$C for 30 min and at a temperature of 950 $^\circ$C for 7 min, respectively. Fixed carbon was calculated by subtracting the summation of the percentages of moisture, ash and volatile matters from 100. All percentages were on the same moisture reference basis (ASTM E871–82, 2013; (ASTM D1102–84, 2013; ASTM E872–82, 2013) [38–40].

Energy yield and densification of the bio-char yield produced at 375 $^\circ$C were also calculated, according to the method proposed by Chowdhury et al. (2017) [41]. The energy densification was obtained by high heating value (HHV) of bio-chars divided by the HHV of biomass residues and the energy yield was calculated as the energy densification multiplied by the bio-char yield [42]. Ultimate analysis was achieved by using a CHNS elemental analyzer (Euro EA3000, Euro vector, Pavia, Italy) as per ASTM procedure D5373-08(ASTM D5373–08, 2008) [43]. Higher heating values of the bio-char were determined by bomb calorimeter (CC01/M3; Toshniwal, New Delhi, India) using ASTM procedure D2015-85. 1.0 g bio-char samples was implanted in a calorimeter, and inflamed in the presence of oxygen. The heat of combustion was recorded for the calculation of HHV of bio-char samples (ASTM D–85, 2015) [44].

Surface area, pore volume and average pore size measurements of bio-char samples, obtained from different walnut shell residues, were investigated by nitrogen (N_2) adsorption/desorption isotherms at 77 K using Micromeritics ASAP 2060 V3.05 H Surface Area Analyzer (Brunauer–Emmett–Teller). About 3 mg of bio-char samples was degassed for 6 h at 200 $^\circ$C under vacuum. Pore volume and average pore size were observed by using Barrett-Joyner-Halenda (BJH) and surface area with BET method.

The pH bio-char samples was measured using the following procedure: A total of 20 mL of deionized water was mixed with 0.5 g of bio-char with the help of magnetic stirrer, for 24 h at 100 $^\circ$C at 150 rpm, in order to get homogeneous solution [43]. The suspension is filtered and equilibrium was reached after one hour. pH of the filtered sample was measured by using Orion pH meter (Thermo Scientific, Cambridge, MA, USA) for bio-char samples.

TGA was carried out by the thermogravimetric analyzer (SII 6300 Exstar; Hitachi, Tokyo, Japan). Runs were accomplished non-isothermally at three different heating rates (10, 20 and 50 $^\circ$C/min). Nitrogen was used as a carrier gas, at a flow rate of 100 mL min^{-1}. Bio-char samples were placed on an open platinum sample pans during the TG analysis. A bio-mass sample of 10 $\perp$ 0.26 mg was used in the experiments. The changes in mass of samples with temperature were recorded for further analysis.

The morphology of the bio-char samples of different walnut shell residues was determined using a SEM (S 3600; Hitachi, Japan) analysis. Images were taken at 15 kV with 10,000$\times$ magnification. The X-ray source was tungsten filament dazed with lanthanum hexaboride (LaB$_6$), which was equipped with a secondary electron detector (i.e., Evehart-Thornley detector (ETD)). The biomass samples were circulated on a carbon coated adhesive pursued by vapor-deposition with gold before investigation. XRD was performed on the bio-char samples using the diffractometer (D8-Advance; Bruker, MA, USA) (fitted with a Lynx eye high-speed strip detector) with Cu Ka radiation (λ = 0.15432 nm). One gram of each bio-char samples was granulated for powder diffraction using X-Ray source with 2.2 kW Cu anode (40 kV, 40 mA) under angular range 2θ (5−1200). For collection of data from 0.5° to 5° of 2θ a regular mode was employed at a scanning speed of 2°/min.

FTIR spectroscopy of different bio-char samples was carried out using a Nicolet 6700, Thermo scientific USA instrument. FTIR analysis the sample powder diluted in 1% potassium bromide (KBr). The FTIR spectrum in the range of 500–4000 cm^{-1} was measured with a resolution of 4 cm^{-1}.

The bio-char showed maximum yields at 375 °C and the characteristics of biochar, like proximate analysis, ultimate analysis, HHV and molar ratios of hydrogen to carbon (H/C) and oxygen to carbon (O/C), were determined at 375 °C. The pH values, surface area, TG analysis, scanning electron microscopy (SEM), X-ray diffraction (XRD) and Fourier transform infrared (FTIR) spectrometry were analyzed in the same temperature.

3. Results and Discussion

3.1. Proximate and Ultimate Analysis

The results obtained from proximate (on dry ash free (daf) basis) and ultimate analysis, and the corresponding H/C, O/C and HHV data, are shown in Tables 1 and 2. For the purpose of comparison, the earlier work reported on bio-chars obtained from different biomass residues, such as almond shell [45], palm shell [17], wheat straw [46] and also for coal [47], were included in Tables 1 and 2. For a bio-char sample to be considered as bio-fuel, moisture content is an important component. The higher moisture content lowers the heating value and hence affects physical properties and the quality of the yields, which in turns affects the behavior of fuel properties. The results show that moisture contents in the bio-chars are in the range of 0.2–0.8%, which are comparable with wheat straw [46] but much lower than the moisture contents reported for coconut shell bio-chars (7.1 wt%) [17], lignite (34 wt%) and bituminous coals (11 wt%) [47].

Fixed carbon (FC) and volatile matters (VM) contents of the bio-char are significant depending on the type of its utilization as an energy source. For bio-chars used in the present work, the VM contents fall in the range from 8.7–14.4%, which are comparable with palm shell [17] but higher than coconut shell [17] and wheat straw bio-chars [46]. However, the values are much lower than lignite (29%) and bituminous coals (35%) [47]. the highest and lowest VMs were observed in PSW and HSW bio-chars, respectively. The reduction of volatile matters was due to conversion of volatile matter into pyrolysis products.

FC contents varied from 78.4–85.6%, which are comparable with the values reported for almond shell [45] and palm shell chars [17]. However, they were found much higher than bituminous and lignite coals. The highest carbon content present in HSW bio-char (85.6%) implies that hard-shelled bio-char yield would be the largest among all types of bio-char samples.

Table 1. Proximate analysis, high heating values, energy density and energy yield of bio-char obtained at particle size of +1.5–2.5 mm, heating rate 20 °C/min at 375 °C and comparison with other bio-chars.

Types of Bio-Char	Proximate Analysis (wt% daf)							
	M	VM	A	FC	HHV (Mj/Kg)	ED	EY	References
PSW	0.8	14.4	5.5	78.5	14.8	1.08	72.6	
TSW	0.6	12.1	4.4	78.4	15.2	1.11	72.6	
MSW	0.3	11.7	3.2	84.2	16.6	1.15	69.7	Present work
HSW	0.2	8.7	2.02	85.6	18.4	1.26	71.9	
Almond shells char	-	21.2	1.9	76.9	28.2	-	-	[45]
Palm shell char	2.2	11.5	6.7	88.5	33.6	-	-	[17]
Coconut shell	7.1	8.1	4.1	91.9	33.7	-	-	
Wheat straw bio-char (WSB)	0.6 ± 0.01	7.3	8.2 ± 0.2	83.9	22.0±0.7	-	-	[46]
Coal (lignite)	34	43.93	9	46.96	9.3-19.3 *	-	-	[47]
Coal (bituminous)	11	39.93	10.11	50.56	27.9–34.89 **	-	-	

M = moisture; VM = volatile matter; A = ash content; FC = fixed carbon; HHV = high heating value; ED = energy densification; EY = energy yield.

Table 2. Elemental analysis and H/C and O/C values of bio-chars obtained at particle size of +1.5–2.5mm, heating rate 20 °C/min at 375 °C and comparison with other bio-chars and coals.

Type of Bio-Chars	Ultimate Analysis (wt %)						References
	C	H	N	O	H/C	O/C	
PSW	73.4	3.0	0.8	22.7	0.49	0.3	Present work
TSW	76.6	2.4	0.8	15.2	0.31	0.19	
MSW	81.8	3.0	0.6	19.6	0.36	0.23	
HSW	82.7	2.38	0.8	14.0	0.28	0.17	
Walnut shell	55.3	0.89	0.47	1.6			[35]
Apricot Kernel shell char	72.72	3.17	1.27	19.84	0.50	-	[19]
Barley straw	74.83	3.51	0.10	8.46			[48]
Wheat straw bio-char (WSB)	64.8	3.1	0.8	23.0	0.6	0.3	[46]
Coal (lignite)	56.4	4.2	1.6	18.4	-	-	[47]
Coal (bituminous)	73.1	5.5	1.4	8.7	-	-	

The chemical composition of the ash can create significant operational problems in a thermo-chemical conversion process, such as combustion processes due to formation of slag from ash at elevated temperatures. For the various bio-chars under study, the ash contents varied from 2.0–5.5%. Where, HSW contains the lowest value, 2.0%, and PSW bio-char exhibits the highest value, 5.5%. The ash contents of the walnut shell bio-chars are well within the range as reported for palm shell, wheat straw, coconut shell, almond shell bio-chars and coals (Table 1).

The ultimate analysis revealed that the carbon, hydrogen, nitrogen and oxygen contents are almost comparable with each other. The ranges may also be tallied with the values reported in the literature for apricot kernel shell [19] and wheat straw and other agricultural bio-chars [48–50]. However, some large variations have been observed in hydrogen and oxygen for walnut shell [50], and nitrogen and oxygen contents in barley straw [48]. They were highly carbonaceous, with carbon contents ranging from 73.4% to 82.7%, and much less nitrogen content ($n < 1\%$), which produces a lesser amount of NO_X during pyrolysis.

The energy densification and yield vary from 1.08 to 1.26 and from 69.7% to 72.6%, respectively. These values are almost comparable with each other for the bio-chars under investigation. However, the energy densification values are lower and yields are higher than the values reported by Rather et al. (2017) [51] for weeds bio-char.

H/C and O/C data plotted on the van Krevelen diagram (Figure 3) show that the energy quality of the bio-char is improved in comparison to the feedstock, and the bio-char may be compared with lignite coal and other walnut shell and apricot Kernel shell [19,47,50] bio-chars. The reduction in O/C and H/C ratios is due to the loss of oxygen and hydrogen during pyrolysis and may be attributed to decarboxylation and demethylation, respectively [50].

From Tables 1 and 2 it can be concluded that the fuel qualities of the different bio-char samples are improved tremendously, as compared to their respective biomass residues. On comparison of bio-char samples with other biomass residues, it was observed that volatile matters, oxygen and ash contents decreased and fixed carbon and carbon content increased. This shows that bio-char samples have high potential to be used as a solid fuel either for direct combustion or for thermal conversion processes.

HHVs vary from 14.8 to 18.4 MJ/kg. The values are comparable with the results for lignite coal as reported by McKendry (2002) [47]. The results obtained in the present investigation show that the fuel qualities in terms of energy value and fixed carbon contents of the bio-chars are improved in comparison to the respective biomass residues. Thus, the bio-chars have the potential to be used as solid fuels.

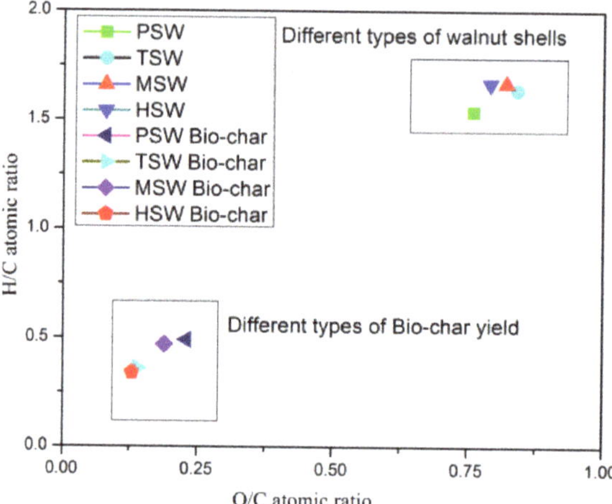

Figure 3. Van Krevelen diagram of different types of walnut shells and their respective bio-char yields.

3.2. TGA Analysis of Bio-Char

The results of TGA carried out in an oxidizing atmosphere at 10, 20, and 50 °C/min heating rates are revealed in Figure 4. The weight loss range can be classified into the three divisions. Every new slope indicates the beginning of a new stage. In the first division, about ≈2.5–7% mass loss was observed between the temperatures from 29 to 160 °C for all heating rates and for all types of bio-chars. In the first stage, the maximum mass loss seems to occur in TSW (7%) at 20 °C/min and the lowest in HSW bio-char (2.5%) at a heating rate of 50 °C/min (Figure 4b,d). The mass loss is due to the removal of moisture and sorbent water bounded by surface tension. The second zone starts at 330 °C and continued up to 475 °C where the mass loss from 75–81% at 10 °C/min, 78–89% at 20 °C/min, and 81–90% at 50 °C/min are recorded for all bio-chars. A huge weight loss of 90% for HSW at a heating rate of 50 °C/min and average weight loss of 82.5% were observed for rest of the bio-chars, see Figure 4d. In this zone the mass loss is due to the existence of cellulose and lignin contents which undergo a oxidation/devolatization reaction. The third zone starts at 475 °C and continues with almost negligible loss of mass. The decomposition of lignin takes place very slowly in the third zone as a result almost straight line is observed. The behavior of TGA analysis was found in agreement with the outcomes described in the literature [52–57].

The above statement is also endorsed by the research carried out many other researchers [58–60] with respect to the decomposition of cellulose, hemi-cellulose, and lignin within the given temperature ranges. The thermogravimetric analysis of various bio-char samples in an oxidizing atmosphere suggested that produced bio-char can be used as alternative solid fuel for various processes.

From the TG curves, a decomposition performance may be illuminated by the specific constituents of bio-chars, whereby the cellulose, hemi-cellulose and lignin are the main components and extractives are the minor components. It was also noticed that the decomposition of cellulose, hemi-cellulose and lignin was accomplished at temperature intervals of 310–400, 210–325, and 160–900 °C, respectively, which is comparable to other bio-chars [56–60]. Therefore, it can be concluded that the major and minor reactions, as detected, in the active pyrolysis zone may be credited to cellulose and hemi-cellulose decomposition. The final zone revealed much less mass loss due to slow degradation of lignin at 510 to 800 °C to produce bio-char as residue [61,62]. Similar observations have also been made by other researchers [61–63]. It was also observed that, at low heating rate, the pyrolysis above 550 °C was almost negligible.

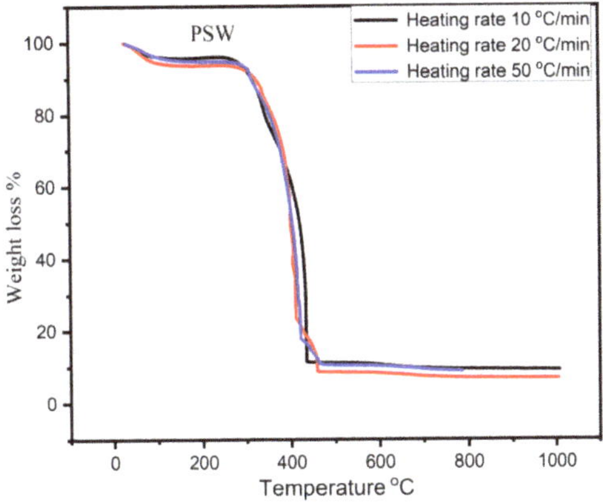

(**a**)

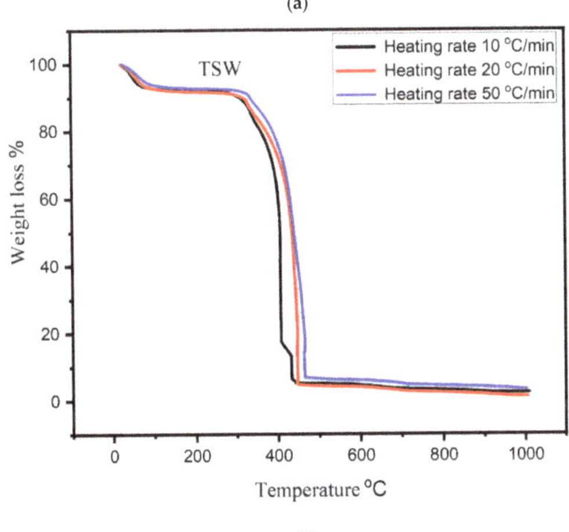

(**b**)

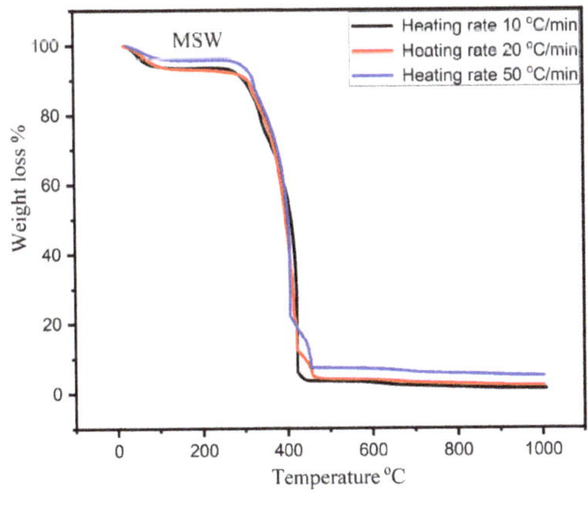

(**c**)

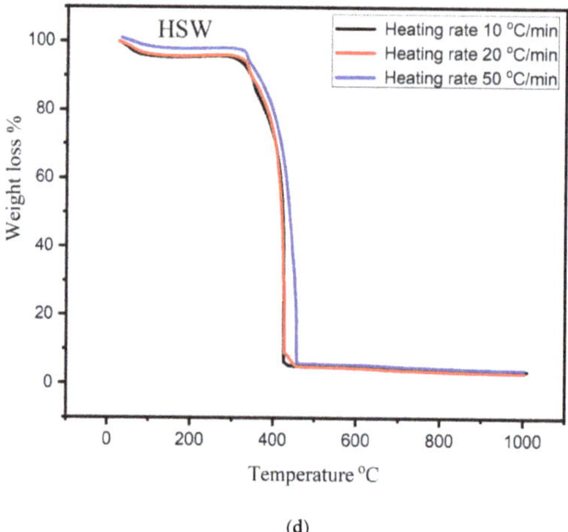

(**d**)

Figure 4. (**a–b**) TGA profiles of (**a**) PSW, (**b**) TSW, (**c**) MSW and (**d**) HSW bio-char in an oxidizing atmosphere at 10, 2, and 50 °C/min heating ranges.

3.3. SEM and XRD Analysis of Bio-Chars

The surface morphology of bio-chars, collected from SEM analysis, is exhibited in Figure 5. It is evident that PSW (a) shows porous cracks and HSW (d) shows the agglomerated rocky-like structure. However, TSW (b) and MSW (c) show planner sheet-like structures. The structures have rough textures and are heterogeneous in nature. The results are similar to those found by Guerrero et al. (2008) [64], where they mentioned melting followed by devolatization and finally vesicle formation responsible for the formation of such structures [64,65]. As temperature gets increased with high heating rate, there is release of various volatile components. Devolatilization results in morphological changes of bio-char, followed by the formation of high pore surface structure of bio-char samples [54].

The XRD patterns of the bio-chars at a temperature of 375 °C are shown in Figure 6. The peaks are in the range of 5–900 on the base line of the diffractograms. The peaks at 14° (d-space~5.96 Å), 15° (d-space~5.7 Å), 16° (d-space~5.3 Å), 22° (d-space~4.0Å), 26° (d-spa1ce~3.34 Å) and 35° (d-space~2.5 Å) were assigned to cellulose and hemi-cellulose, respectively. Different types of bio-char samples in the diffraction angle (2θ) have a wide halo in the 2θ range from 6 to 20°, showed that chain contains large number of carbons containing components substances. The band at 2θ = 22° showed disordered structure, which occurred due to presence of aliphatic and distorted arrangement of carbon chain. Broad peak at 2θ ≈ 26° of bio-char indicated presence of silica in the X-ray diffractogram. The peak at 15 and 16° were derived from cellulose constituent. Bio-char samples showed narrow and sharp bands over the examined 2θ, due to the presence of inorganic constituents in the carbon chain. The XRD graphs confirmed the aromatic and crystallinity nature of the bio-chars, which is in agreement with other bio-chars [65–67].

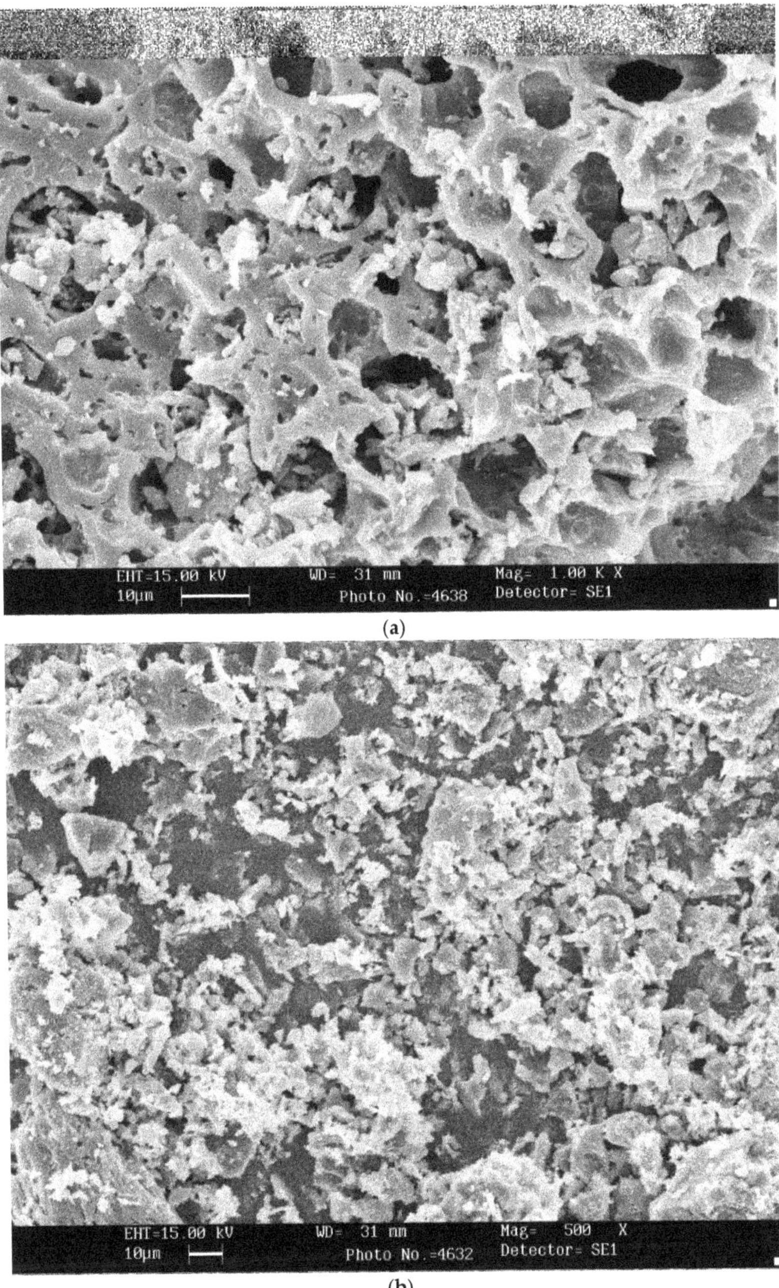

(**a**)

(**b**)

Figure 5. (**a–d**) Scanning electron micrographs (magnification 10,000×) of (**a**) PSW, (**b**) TSW, (**c**) MSW and (**d**) HSW biochars.

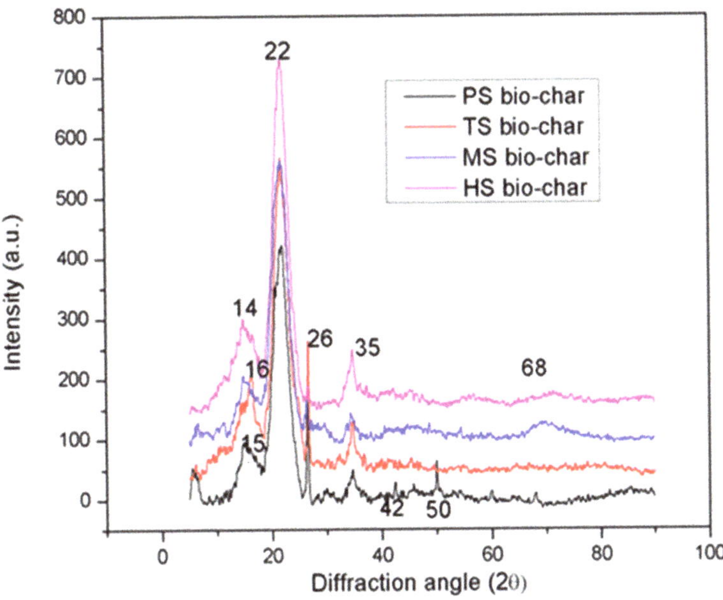

Figure 6. XRD of the different types of bio-char.

3.4. FTIR of Bio-Char

Functional group analysis of various bio-char products, obtained using FTIR spectroscopy, is shown in Figure 7, and the band assignment is discussed in Table 3. The FTIR spectrum in the range of 500–4500 cm^{-1} was measured with a resolution of 4 cm^{-1}. Major components of biomass are hemicellulose, cellulose and lignin. Lignin, unlike cellulose, possesses olefinic carbon-carbon ($-C = C-$) double bond in cyclic as well as side chains and is aromatic in nature [68]. The band peaks at the wave numbers of 3465, 3428, 3411 and 3442 cm^{-1} are for PSW, TSW, MSW and HSW bio-chars, respectively, which indicated the stretching vibration of $-OH$ hydroxyl groups of phenol. The second prominent peaks are at 3050 and 2849 cm^{-1}, and 3075, 2925 cm^{-1}, only shown by TSW and MSW bio-chars, which represent the $-CH$ stretching vibrations due to the presence of methyl/methylene group. The peak at 1587 cm^{-1} represents the aromatic C = C ring stretching vibration of lignin. The medium band intensity between 1398–1401 cm^{-1} may be assigned to aromatic skeleton vibrations combined with $C-H$ in plane deformations of bio-chars. The band peak at 1265 cm^{-1} of PSW bio-char confirmed the presence of aromatic CO$-$ and phenolic $-OH$ stretching due to the presence of cellulose, hemi-cellulose and lignin. The 750 cm^{-1} band peak showed 3–4 adjacent H deformation of all bio-char samples except TSW bio-char.

From Figure 7 it can be concluded that the different peaks for different bio-char samples are almost similar. A drift of wave number from a lower to higher value was due to an increase in temperature, which indicated more carbon content of the char. The wave number from 3400 to 3460 cm^{-1} indicated low frequency values between these peaks, suggested that hydroxyl groups are involved in hydrogen bonding. The non-involving OH bonds were above 3500 cm^{-1} for other groups (i.e., alcohols, phenols and carboxylic acid). Hemicelluloses and celluloses components are broken completely; it goes into either gases or liquid products. The peaks in the range of 1585–1127 cm^{-1} indicated the presence of hemi-cellulose components. The holo-cellulose (cellulose+hemi-cellulose) structure will collapse after wave number gets reduced. The band intensities were decreased at 3411 cm^{-1} (O$-$H stretching) and 1127 cm^{-1} (C$-$O stretching) due to the existence of a hydrogen bond, reduction of water and cellulose contents in bio-char. The band intensity of the absorbance of the $-OH$ decreased. Due to complete loss of alcoholic or phenolic groups, the oxygen:carbon ratio of the char decreased. Further, due to high rigid structure lignin was remain within the carbon chain at 1400 to 750 cm^{-1} in the bio-char, while unconverted lignin remains within the bio-char. Similar observations were reported for different bio-chars in the literature [64,68–70].

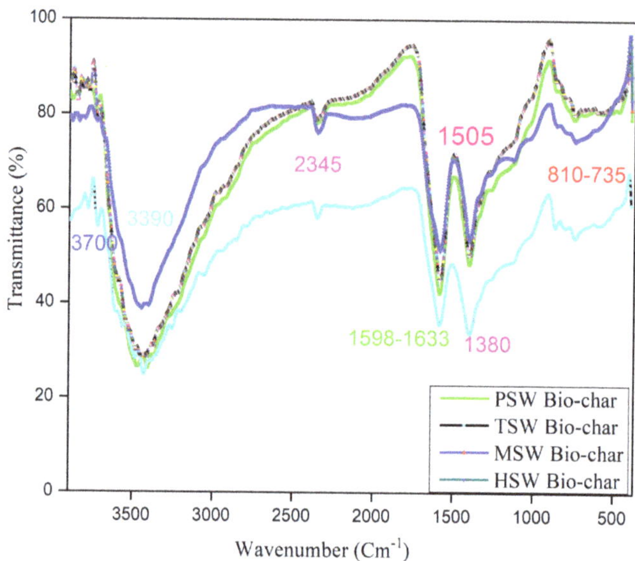

Figure 7. Infrared spectra of the different bio-chars.

Table 3. FTIR band assignments spectra of different types of bio-chars at 375 °C.

Band Assignment	Band Frequency (cm^{-1})			
	PSW	**TSW**	**MSW**	**HSW**
O–H stretching	3465	3428	3411	3442
C–H stretching	-	3050, 2849	3075, 2925	-
Aromatic vibrations of C–C/C = C stretching	1587	1585	1585	1586
C–O–H in-plane bending and aromatic vibrations	1401	1398	1400	1399
OH bending and CH deformation vibrations	1265	-	1265	1265
C–H in-plane deformation and C–OH stretch in syringyl	1121	1124	1124	1127
O–CH$_3$ and C–OH stretching	-	-	1021	1015
C–H out-of-plane stretching	868	870, 806	865, 803	868
	750	750	747	750

3.5. Surface Area, Total Pore Volume, Average Pore size Volume, pH and its Potential Applications

The surface area, total pore volume, average pore size volume and pH values of different bio-chars, shown in Table 4, are significant like other physical and chemical characteristics. It may strongly influence the combustion and reactive behavior of the bio-char. The bio-chars produced at pyrolysis temperature of 375 °C develop high porosity in the surface of bio-chars which emerged macro and micro porous particles. This occurs due to removal of volatile matters from different biomass residues [71]. Due to opportunity of high pores surface area and adsorption sites which contributes to adsorptive capacity and also provides spaces for nutrients/pollutants and water retention [53] in soil treatment applications.

For different bio-chars investigated in the present work, BET surface areas are found as 40–58 m^2/g, at particle size and heating rate +1.5–2.5, and 20 °C/min, respectively, for paper-, thin-, medium- and hard-shelled walnuts, at temperature of 375 °C. With an increasing temperature, reduction in the surface area values are predominantly detected, as shown in Table 4. The fusion of adjacent pores seems to predominate, leading to the decline in the surface area and thermal deactivation of the bio-chars. The bio-chars used in the present work have higher surface area and some others were comparable with the investigation reported by other examiners (Table 4) for different biomass residues [72–76]. The highest and lowest surface areas are in HSW and PSW bio-chars, respectively. Due to high BET surface area and quality of bio-chars, it could have high adsorption capacity. For the application of bio-char in wastewater treatment and soil remediation, the BET surface area and quality of bio-chars can be further enhanced by alkaline and acid treatment. Moreover, it can be transformed to activated carbon for water purification processes and in fuel utilities.

Pyrolysis of biomass involves eradication of organic/volatile matters, which enhances the alkali concentration [77]. The pH plays an important role for soil fertility, which effects the types of plants, availability of nutrients and microbes to be consumed [52]. The pH is found in between 8.1–8.3 for all types of bio-chars, respectively. All bio-char samples showed alkaline characteristics, and may be used for soil amendment to neutralize soil acidity, and also enhances the soil quality and improves the yield productivity [78]. Similar results have also been made by other investigators and are comparable with the obtained results, as reported in Table 4.

3.6. Circular Economy Models

The circular bio-economy is a concept for the transformation and management of land, food, health and industrial systems using renewable natural capital. It has the aim of achieving sustainable wellbeing in concord with nature. The prosperity of the recycling-based bioeconomy requires modern technology, innovation, traditional wisdom and biodiversity. That is ultimately the fundamental driving force for bioeconomy. Further, biodiversity affects the ability of biological systems to adapt to changing environments. Thus, it become important to ensure the resilience and sustainability of biological resources. It must be recognized its importance not only through proper nature maintenance policies, but also through locally adapted market-based means that encourage farmers, forest owners and bio-based companies to invest in biodiversity. Since the industrial revolution, human activity has been a major cause of global environmental change. Humans and the environment have a skewed connection, which has resulted in faced thresholds and turning points connected with planetary boundaries, such as biodiversity loss and the global climate catastrophe [84]. A sustainable bioeconomy also encompasses more than just the interchange of fossil and renewable resources. Low-carbon energy, sustainable supply chains, and promising disruptive conversion technologies are all required for the long-term conversion of renewable energy resources into high-quality bio-based goods, materials and fuels. The natural environment, human health and natural resources are one of the activities [85]. Circular economy means that it is fundamentally different from person to person. It has basically become an "essentially controversial concept". This is a phrase created by Gallie [86], and although there is consensus on the means and purpose of the concept, there is disagreement on its definition. Recently used to characterize the concept of the circular economy [87], the European Commission's bioeconomic strategy interprets the circulating bioeconomy as a framework for reducing dependence on natural resources. Manufacturing transformation: Promote sustainable production of renewable resources from land, fisheries and aquaculture. It will drive the transition to a variety of bio based products and bioenergy while creating new jobs and industries [88]. On the one hand, circular economy focuses on increasing efficiency and reducing speed, reducing and closing hardware loops to reduce resource consumption and system waste through reduced inputs, sustainable design, practice improvement, reuse and waste recycling [89,90]. In accordance with circular bioeconomy concepts, the bio-char was prepared form different walnut shells as a biomass residue at different temperatures, particle sizes and heating rates. The smaller particle size was considered as there may be higher temperature gradient in larger particles, which results into non-uniform heat distribution in the biomass particle. The different bio-chars showed high carbon (73.4–82.7) and lower nitrogen contents with high heating values (14.8 to 18.4 MJ/kg), which enhanced bio-char qualities and is comparable to high quality lignite coal, and, therefore, can be utilized as a renewable solid fuel.

Table 4. Surface area, total pore volume, average pore size volume and pH values of WS bio-chars with different temperatures, at 1.5–2.5 mm particle size and 20 °C/min heating value.

Bio-Char Types (Pyrolysis Temperature)	BET Surface Areas (m²/g)	Total Pore Volume cm³/g	Average Pore Size (Å)	pH	Applications	References
PSW	42	0.012	17	8.4	Fuel, energy storage, soil conditioner, building sector, drinking and wastewater treatments, biogas production, exhaust filters, industrial materials, electronics semiconductors, cosmetics, paints and coloring	Present work
TSW	44	0.013	18	8.2		
MSW	48	0.014	19	8.1		
HSW	58	0.016	19.5	8.3		
Sugarcane bagasse (500 °C)	10.85	0.011	43.7	8.1	Solid fuel, adsorbent, soil amendment	[53]
Apricot kernel shell (550 °C)	195	0.1124	-	-	Activated carbon, fuel applications, water purification, adsorption	[19]
Rice straw (600 °C)	4.76	0.0023	18.74	-	Adsorption Water purification, activated carbon, fuel applications	[74]
Coconut fiber (600 °C)	23.2	0.04		9.6	Sequester carbon in soils, improving soil quality and plant growth	[17]
Flax straw (550 °C)	28.7	0.009	37.5	-	Soil amendment, carbon sequestration, activated chars.	[72]
Wood (450 °C)	23			6.7	Soil reduced the C-mineralization rate compared against the control soil samples	[75]
Paddy Straw (500 °C)	45.8			10.5	Fertilizer consumption reduced, and sequestrate carbon	[76]
Durian wood sawdust (450 °C)	45.78	5.786	80.98	6.4	Provide suitable proportions for developing clusters of microorganisms, water retention capacity in soil and enhances soil fertility.	[79]
Spent *P. ostreatus* (500 °C)	18.05	0.061	136	10.37	Potential adsorbent for removing heavy metals from wastewater	[80]
Corn stalk (450 °C)	57.80	0.081	49.1		Adsorption characteristics and mechanism of bio-char on nonpolar pollutants	[81]
Rice husk (500 °C)	92.6	0.076	22.0		Adsorptive properties	[82]
Charcoal fines (500 °C)	43 ± 3	0.035			Water retention capacities and cation exchange capacity	[83]

4. Conclusions

Persistent environmental problems, rising oil prices, energy crisis, depletion of fossil fuels, and growing application and demand for energy are important reasons why people are resolute in demanding energy sources and alternative sustainable fuels. For the environment, sustainability and biodegradability are key characteristics that make biomass a prime candidate for bioenergy production. Bio-char has the ability to sequester carbon in the soil, while improving plant growth and soil quality, with high energy density. Further, it can also be used as an adsorbent for water treatment. In the present study, the characterization of four different bio-chars, obtained by slow pyrolysis at 375 °C, produced from hard, medium, thin seed residues and paper peels was investigated. The properties of biochar, such as proximate and ultimate analysis, heating value, surface area, pH value, thermal degradation behavior, morphological and crystalline nature and functional characterization using FTIR, were determined. The key outcome from the present work can be summarized as follows;

- The pyrolytic behavior of bio-char was studied using thermogravimetric analysis (TGA) in an oxidizing atmosphere. SEM analysis confirmed morphological change and showed heterogeneous and rough texture structure.
- Crystalline nature of the bio-chars was established by X-ray powder diffraction (XRD) analysis. The maximum higher heating values (HHV), high fixed carbon content and surface area obtained for walnut shells (WS) samples were found as ~18.4 MJ kg^{-1}, >80% and 58 m^2/g, respectively.
- Improvement in HHV and decrease of O/C and H/C ratios led the bio-char samples to fall into the category of coal and confirmed their hydrophobic, carbonized and aromatized nature.
- From the Fourier transform infra-red spectroscopy (FTIR), it was observed that there was alteration in functional groups with an increase in temperature, and illustrated higher aromaticity. Therefore, it could be concluded that the bio-char obtained from walnut shell has a high potential to be used as an efficient fuel in both industrial as well as domestic furnaces for energy production.
- Further, from surface area and pH analysis of bio-chars, it was found that WS bio-chars had similar characteristics to adsorbents used for water purifications, retention of essential elements in soil and carbon sequestration.
- The improvement in characteristics of different bio-chars, as compared to their respective biomass residues, showed that it may also be used as a good adsorbent for wastewater treatment as well as for enhancing soil fertility.

Author Contributions: Conceptualization, M.A.S. and R.A.; methodology, M.A.S. and M.I.H.S.; formal analysis, M.A.S., M.A.A. and I.A.A.; investigation, M.A.S., M.A.A. and I.A.A.; resources, M.A.S. and R.A.; writing—original draft preparation, M.A.S. and M.I.H.S.; writing—review and editing, M.A.S., R.A., M.A.A., I.A.A. and M.I.H.S. All authors have read and agreed to the published version of the manuscript.

Funding: The authors express their appreciation to the Deputyship for Research and Innovation, Ministry of education in Saudi Arabia for funding this research work through the project number 20-UQU-IF-P2-001.

Institutional Review Board Statement: Not applicable.

Informed Consent Statement: Not applicable.

Data Availability Statement: Not applicable.

Acknowledgments: We are thankful to the Deputyship for Research and Innovation, Ministry of education in Saudi Arabia for funding this research work through the project number 20-UQU-IF-P2-001.

Conflicts of Interest: The authors declare no conflict of interest.

References

1. Jia, Y.; Wang, Y.; Zhang, Q.; Rong, H.; Liu, Y.; Xiao, B.; Guo, D.; Laghari, M.; Ruan, R. Gas-carrying enhances the combustion temperature of the biomass particles. *Energy* **2021**, *239*, 121956. [CrossRef]
2. Hansson, A.; Haikola, S.; Fridahl, M.; Yanda, P.; Mabhuye, E.; Pauline, N. Biochar as multi-purpose sustainable technology: Experiences from projects in Tanzania. *Environ. Dev. Sustain.* **2021**, *23*, 5182–5214. [CrossRef]
3. Masek, O.; Brownsort, P.; Cross, A.; Sohi, S. Influence of production conditions on the yield and environmental stability of biochar. *Fuel* **2013**, *103*, 151–155. [CrossRef]
4. Manya, J.J. Pyrolysis for biochar purposes: A review to establish current knowledge gaps and research needs. *Environ. Sci. Technol.* **2012**, *46*, 7939–7954. [CrossRef] [PubMed]

5. Kane, S.N.; Mishra, A.; Dutta, A.K. Preface: International Conference on Recent Trends in Physics (ICRTP 2016). *J. Phys. Conf. Ser.* **2016**, *755*, 5. [CrossRef]

6. Ciolkosz, D.; Wallace, R. A review of torrefaction for bioenergy feedstock production. *Biofuels Bioprod. Biorefin.* **2011**, *5*, 317–329. [CrossRef]

7. Liu, Z.; Han, G. Production of solid fuel biochar from waste biomass by low temperature pyrolysis. *Fuel* **2015**, *158*, 159–165. [CrossRef]

8. Chen, W.-H.; Lin, B.-J.; Colin, B.; Chang, J.-S.; Pétrissans, A.; Bi, X.; Pétrissans, M. Hygroscopic transformation of woody biomass torrefaction for carbon storage. *Appl. Energy* **2018**, *231*, 768–776. [CrossRef]

9. Zhu, X. Effects and mechanisms of biochar-microbe interactions in soil improvement and pollution remediation: A review. *Environ. Pollut.* **2017**, *227*, 98–115. [CrossRef]

10. Oliveira, F.R.; Patel, A.K.; Jaisi, D.P.; Adhikari, S.; Lu, H.; Khanal, S.K. Environmental application of biochar: Current status and perspectives. *Bioresour. Technol.* **2017**, *246*, 110–122. [CrossRef]

11. Xu, G.; Lv, Y.; Sun, J.; Shao, H.; Wei, L. Recent Advances in Biochar Applications in Agricultural Soils: Benefits and Environmental Implications. *Clean Soil Air Water* **2012**, *40*, 1093–1098. [CrossRef]

12. Zhao, L.; Cao, X.; Masek, O.; Zimmerman, A. Heterogeneity of biochar properties as a function of feedstock sources and production temperatures. *J. Hazard. Mater.* **2013**, *256–257*, 1–9. [CrossRef]

13. Gupta, S.; Kua, H.W.; Low, C.Y. Low, Use of biochar as carbon sequestering additive in cement mortar. *Cem. Concr. Compos.* **2018**, *87*, 110–129. [CrossRef]

14. Gan, Y.Y. Torrefaction of microalgal biochar as potential coal fuel and application as bio-adsorbent. *Energy Convers. Manag.* **2018**, *165*, 152–162. [CrossRef]

15. Lee, H.W.; Kim, Y.M.; Kim, S.; Ryu, C.; Park, S.H.; Park, Y.-K. Review of the use of activated biochar for energy and environmental applications. *Carbon Lett.* **2018**, *26*, 1–10.

16. Chen, W.; Lee, K.T.; Chih, Y.K.; Eng, C.; Lin, H.P.; Chiou, Y.B.; Cheng, C.L.; Lin, Y.X.; Chang, J.S. Novel Renewable Double-Energy System for Activated Biochar Production and Thermoelectric Generation from Waste Heat. *Energy Fuels* **2020**, *34*, 3383–3393. [CrossRef]

17. Windeatt, J.H.; Ross, A.B.; Williams, P.T.; Forster, P.M.; Nahil, M.A.; Singh, S. Characteristics of biochars from crop residues: Potential for carbon sequestration and soil amendment. *J. Environ. Manag.* **2014**, *146*, 189–197. [CrossRef]

18. Moreira, R.; Vaz, J.M.; Spinace, E.V.; dos Reis Orsini, R.; Penteado, J.C. Production of Biochar, Bio-Oil and Synthesis Gas from Cashew Nut Shell by Slow Pyrolysis. *Waste Biomass Valorization* **2016**, *8*, 217–224. [CrossRef]

19. Demiral, I.; Kul, S.C. Pyrolysis of apricot kernel shell in a fixed-bed reactor: Characterization of bio-oil and char. *J. Anal. Appl. Pyrolysis* **2014**, *107*, 17–24. [CrossRef]

20. Jiang, K.-M.; Cheng, C.-G.; Ran, M.; Lu, Y.-G.; Wu, Q.-L. Preparation of a biochar with a high calorific value from chestnut shells. *New Carbon Mater.* **2018**, *33*, 183–187. [CrossRef]

21. Yaashikaa, P.R.; Kumar, P.S.; Varjani, S.; Saravanan, A. A critical review on the biochar production techniques, characterization, stability and applications for circular bioeconomy. *Biotechnol. Rep.* **2020**, *28*, e00570. [CrossRef] [PubMed]

22. Lu, H.R.; El Hanandeh, A. Life cycle perspective of bio-oil and biochar production from hardwood biomass; what is the optimum mix and what to do with it? *J. Clean. Prod.* **2019**, *212*, 173–189. [CrossRef]

23. Kim, M.; Park, J.; Yu, S.; Ryu, C.; Park, J. Clean and energy-efficient mass production of biochar by process integration: Evaluation of process concept. *Chem. Eng. J.* **2019**, *355*, 840–849. [CrossRef]

24. Wang, J.; Wang, S. Preparation, modification and environmental application of biochar: A review. *J. Clean. Prod.* **2019**, *227*, 1002–1022. [CrossRef]

25. Gurwick, N.P.; Moore, L.A.; Kelly, C.; Elias, P. A systematic review of biochar research, with a focus on its stability in situ and its promise as a climate mitigation strategy. *PLoS ONE* **2013**, *8*, e75932. [CrossRef]

26. Marmiroli, M.; Bonas, U.; Imperiale, D.; Lencioni, G.; Mussi, F.; Marmiroli, N.; Maestri, E. Structural and functional features of chars from different biomasses as potential plant amendments. *Front Plant Sci.* **2018**, *9*, 1119. [CrossRef]

27. Ghodake, G.S.; Shinde, S.K.; Kadam, A.A.; Saratale, R.G.; Saratale, G.D.; Kumar, M.; Palem, R.R.; AL-Shwaiman, H.A.; Elgorban, A.M.; Syed, A. Review on biomass feedstocks, pyrolysis mechanism and physicochemical properties of biochar: State-of-the-art framework to speed up vision of circular bioeconomy. *J. Clean. Prod.* **2020**, *297*, 126645. [CrossRef]

28. Lin, Y.L.; Zheng, N.Y.; Hsu, C.H. Torrefaction of fruit peel waste to produce environmentally friendly biofuel. *J. Clean. Prod.* **2020**, *284*, 124676. [CrossRef]

29. Romanowska-Duda, Z.; Szufa, S.; Grzesik, M.; Piotrowski, K.; Janas, R. The Promotive Effect of Cyanobacteria and Chlorella sp. Foliar Biofertilization on Growth and Metabolic Activities of Willow (Salix viminalis L.) Plants as Feedstock Production, Solid Biofuel and Biochar as C Carrier for Fertilizers via Torrefaction Process. *Energies* **2021**, *14*, 5262. [CrossRef]

30. D'Adamo, I.; Morone, P.; Huisingh, D. Bioenergy: A Sustainable Shift. *Energies* **2021**, *14*, 5661. [CrossRef]

31. Xu, T.; Bhattacharya, S. Mineral transformation and morphological change during pyrolysis and gasification of victorian brown coals in an entrained flow reactor. *Energy Fuels* **2019**, *33*, 6134–6147. [CrossRef]

32. Li, J.; Du, Y.; Bao, T.; Dong, J.; Lin, M.; Shim, H.; Yang, S.-T. n-Butanol production from lignocellulosic biomass hydrolysates without detoxification by Clostridium tyrobutyricum Dack-adhE2 in a fibrous-bed bioreactor. *Bioresour. Technol.* **2019**, *289*, 121749. [CrossRef]

33. Zabed, H.; Sahu, J.N.; Boyce, A.N.; Faruq, G. Fuel ethanol production from lignocellulosic biomass: An overview on feedstocks and technological approaches. *Renew. Sustain. Energy Rev.* **2016**, *66*, 751–774. [CrossRef]
34. Zhang, P.; Guo, Y.J.; Chen, J.; Zhao, Y.R.; Chang, J.; Junge, H.; Beller, M.; Li, Y. Streamlined hydrogen production from biomass. *Nat. Cat.* **2018**, *1*, 332–338. [CrossRef]
35. Sette, J.R.; Hansted, C.R.; Novaes, A.L.S.; Lima, E.; Rodrigues, P.A.F.; Santos, A.C.; Yamaji, D.R.S. Energy enhancement of the eucalyptus bark by briquette production. *Ind. Crops. Prod.* **2018**, *122*, 209–213. [CrossRef]
36. The International Nut and Dried Fruit Council Foundation, Stastical Year Book 2017–2018. Available online: https://www.nutfruit.org/files/tech/1524481168_INC_Statistical_Yearbook_2017-2018.pdf (accessed on 20 June 2021).
37. Shah, M.A.; Khan, M.N.S.; Kumar, V. Biomass residue characterization for their potential application as biofuels. *J. Therm. Anal. Calorim.* **2018**, *134*, 2137–2145. [CrossRef]
38. *ASTM E 871-82*; Moisture Analysis of Particulate Wood Fuels; ASTM: West Conshohocken, PA, USA, 2014. [CrossRef]
39. *ASTM D1102-84*; Standard Test Method for Ash in Wood; ASTM: West Conshohocken, PA, USA, 2013. [CrossRef]
40. *EN 14175-3:2003*; Fume Hoods—Type Test Methods; National Standards Association Ireland: Dublin, Ireland, 2003. [CrossRef]
41. Chowdhury, Z.Z.; Pal, K.; Johan, R.B.; Yehye, W.; Ali, M.E. Comparative Evaluation of Physiochemical Properties of a Solid Fuel Derived from Adansonia digitata Trunk Using Torrefaction. *Bioresources* **2017**, *12*, 3816–3833. [CrossRef]
42. Liu, Z.; Quek, A.; Kent, H.S.; Balasubramanian, R. Production of solid biochar fuel from waste biomass by hydrothermal carbonization. *Fuel* **2013**, *103*, 943–949. [CrossRef]
43. *ASTM D5373-08*; Standard Test Methods for Instrumental Determination of Carbon, Hydrogen, and Nitrogen in Laboratory Samples of Coal; ASTM: West Conshohocken, PA, USA, 2012. [CrossRef]
44. *ASTM D2015-85*; Standard Test Method for Gross Calorific Value of Coal and Coke by the Adiabatic Bomb Calorimeter; ASTM: West Conshohocken, PA, USA, 1975.
45. Gonzalez, J.F.; Ramiro, A.; Gonzalez-Garcia, C.M.; Ganan, J.; Encinar, J.M.; Sabio, E.; Rubiales, J. Pyrolysis of almond shells. Energy applications of fractions. *Ind. Eng. Chem. Res.* **2005**, *44*, 3003–3012. [CrossRef]
46. Nanda, S.; Mohanty, P.; Pant, K.K.; Naik, S.; Kozinski, J.A.; Dalai, A.K. Characterization of North American Lignocellulosic Biomass and Biochars in Terms of their Candidacy for Alternate Renewable Fuels. *Bioenergy Res.* **2013**, *6*, 663–677. [CrossRef]
47. McKendry, P. Energy production from biomass (part 1): Overview of biomass. *Bioresour. Technol.* **2002**, *83*, 37–46. [CrossRef]
48. Yang, Y.; Brammer, J.G.; Mahmood, A.S.N.; Hornung, A. Intermediate pyrolysis of biomass energy pellets for producing sustainable liquid, gaseous and solid fuels. *Bioresour. Technol.* **2014**, *169*, 794–799. [CrossRef]
49. Onay, O.; Beis, S.H.; Kockar, O.M. Pyrolysis of walnut shell in a well-swept fixed-bed reactor. *Energy Sources* **2004**, *26*, 771–782. [CrossRef]
50. Mukome, F.N.D.; Zhang, X.; Silva, L.C.R.; Six, J.; Parikh, S.J. Use of chemical and physical characteristics to investigate trends in biochar feedstocks. *J. Agric. Food Chem.* **2013**, *61*, 2196–2204. [CrossRef]
51. Rather, M.A.; Khan, N.S.; Gupta, R. Hydrothermal carbonization of macrophyte Potamogeton lucens for solid biofuel production: Production of solid biofuel from macrophyte Potamogeton lucens. *Eng. Sci. Technol. Int. J.* **2017**, *20*, 168–174.
52. Varma, A.K.; Mondal, P. Physicochemical characterization and kinetic study of pine needle for pyrolysis process. *J. Therm. Anal. Calorim.* **2016**, *124*, 487–497. [CrossRef]
53. Varma, A.K.; Mondal, P. Pyrolysis of sugarcane bagasse in semi batch reactor: Effects of process parameters on product yields and characterization of products. *Ind. Crops Prod.* **2017**, *95*, 704–717. [CrossRef]
54. Maiti, S.; Dey, S.; Purakayastha, S.; Ghosh, B. Physical and thermochemical characterization of rice husk char as a potential biomass energy source. *Bioresour. Technol.* **2006**, *97*, 2065–2070. [CrossRef] [PubMed]
55. Crombic, K.; Mašek, O.; Brownsort, P.; Sohi, S.P.; Cross, A. The effect of pyrolysis conditions on biochar stability as determined by three methods. *GCB Bioenergy* **2012**, *5*, 122–131. [CrossRef]
56. Souza, B.S.; Moreira, A.P.D.; Teixeira, A.M.R.F. TG-FTIR coupling to monitor the pyrolysis products from agricultural residues. *J. Therm. Anal. Calorim.* **2009**, *97*, 637–642. [CrossRef]
57. Bisht, A.S.; Kumar, S.R. Use of pine needle in energy generation. *Int. J. Res. Appl. Sci. Eng. Technol.* **2014**, *2*, 59–63.
58. Cai, J.M.; Bi, L.S. Kinetic analysis of wheat straw pyrolysis using isoconversional methods. *J. Therm. Anal. Calorim.* **2009**, *98*, 325–330. [CrossRef]
59. Chutia, R.S.; Kataki, R.; Bhaskar, T. Thermogravimetric and decomposition kinetic studies of Mesua ferrea L. deoiled cake. *Bioresour. Technol.* **2013**, *139*, 66–72. [CrossRef]
60. Mishra, G.; Bhaskar, T. Non isothermal model free kinetics for pyrolysis of rice straw. *Bioresour. Technol.* **2014**, *169*, 614–621. [CrossRef]
61. AcIkalln, K. Pyrolytic characteristics and kinetics of pistachio shell by thermogravimetric analysis. *J. Therm. Anal. Calorim.* **2012**, *109*, 227–235. [CrossRef]
62. Lopez-Velazquez, M.A.; Santes, V.; Balmaseda, J.; Torres-Garcia, E. Pyrolysis of orange waste: A thermo-kinetic study. *J. Anal. Appl. Pyrolysis* **2013**, *99*, 170–177. [CrossRef]
63. Lapuerta, M.; Hernández, J.J.; Rodríguez, J. Kinetics of devolatilisation of forestry wastes from thermogravimetric analysis. *Biomass Bioenergy* **2004**, *27*, 385–391. [CrossRef]
64. Guerrero, M.; Ruiz, M.P.; Millera, A.; Alzueta, M.U.; Bilbao, R. Characteriza-tion of biomass chars formed under different devolatilization conditions: Differences between rice husk and eucalyptus. *Energy Fuels* **2008**, *22*, 1275–1284. [CrossRef]

65. Morin, M.; Pécate, S.; Hémati, M.; Kara, Y. Pyrolysis of biomass in a batch fluidized bed reactor: Effect of the pyrolysis conditions and the nature of the biomass on the physicochemical properties and the reactivity of char. *J. Anal. Appl. Pyrolysis* **2016**, *122*, 511–523. [CrossRef]

66. Vassilev, S.V.; Baxter, D.; Andersen, L.K.; Vassileva, C.G.; Morgan, T.J. An overview of the organic and inorganic phase composition of biomass. *Fuel* **2012**, *94*, 1–33. [CrossRef]

67. Fu, P.; Hu, S.; Xiang, J.; Sun, L.; Li, P.; Zhang, J.; Zheng, C. Pyrolysis of maize stalk on the characterization of chars formed under different devolatilization conditions. *Energy Fuels* **2009**, *23*, 4605–4611. [CrossRef]

68. Sotoudehnia, F.; Baba, R.A.; Alayat, A.; McDonald, A.G. Characterization of bio-oil and biochar from pyrolysis of waste corrugated cardboard. *J. Anal. Appl. Pyrolysis* **2020**, *145*, 104722. [CrossRef]

69. Chen, L.; Wang, X.; Yang, H.; Lu, Q.; Li, D.; Yang, Q.; Chen, H. Study on pyrolysis behaviors of non-woody lignins with TG-FTIR and Py-GC / MS. *J. Anal. Appl. Pyrolysis* **2015**, *113*, 499–507. [CrossRef]

70. Xu, F.; Yu, J.; Wang, D.; Tesso, T. Qualitative and quantitative analysis of lignocellulosic biomass using infrared techniques: A mini-review. *Appl. Energy* **2013**, *104*, 801–809. [CrossRef]

71. Pradhan, D.; Singh, R.K.; Bendu, H.; Mund, R. Pyrolysis of Mahua seed (Madhuca indica)—Production of biofuel and its characterization. *Energy Convers. Manag.* **2016**, *108*, 529–538. [CrossRef]

72. Tushar, H.K.; Mahinpey, M.S.; Khan, N.A.; Ibrahim, H.; Kumar, P.; Idem, R. Production, characterization and reactivity studies of chars produced by the isothermal pyrolysis of flax straw. *Biomass Bioenergy* **2012**, *37*, 97–105. [CrossRef]

73. Angın, D. Effect of pyrolysis temperature and heating rate on biochar obtainedfrom pyrolysis of safflower seed press cake. *Bioresour. Technol.* **2013**, *128*, 593–597. [CrossRef]

74. Fu, P.; Hu, S.; Xiang, J.; Sun, L.; Su, S.; Wang, J. Evaluation of the porous structure development of chars from pyrolysis of rice straw: Effects of pyrolysis temperature and heating rate. *J. Anal. Appl. Pyrolysis* **2012**, *98*, 177–183. [CrossRef]

75. Ronsse, F.; van Hecke, S.; Dickinson, D.; Prins, W. Production and characterization of slow pyrolysis biochar: Influence of feedstock type and pyrolysis conditions. *GCB Bioenergy* **2013**, *5*, 104–115. [CrossRef]

76. Lee, Y.; Park, J.; Ryu, C.; Gang, K.S.; Yang, W.; Park, Y.; Jung, J.; Hyun, S. Comparison of biochar properties from biomass residues produced by slow pyrolysis at 500 °C. *Bioresour. Technol.* **2013**, *148*, 196–201. [CrossRef]

77. Nzihou, A.; Stanmore, B.; Lyczko, N.; Minh, D.P. The catalytic effect of inherent and adsorbed metals on the fast/flash pyrolysis of biomass: A review. *Energy* **2019**, *170*, 326–337. [CrossRef]

78. Bordoloi, N.; Narzari, R.; Chutia, R.S.; Bhaskar, T.; Kataki, R. Pyrolysis of Mesua ferrea and Pongamia glabra seed cover: Characterization of bio-oil and its sub-fractions. *Bioresour. Technol.* **2015**, *178*, 83–89. [CrossRef] [PubMed]

79. Chowdhury, Z.Z.; Karim, Z.M.; Ashraf, M.A.; Khalid, K. Influence of carbonization temperature on physicochemical properties of biochar derived from slow pyrolysis of durian wood (Durio zibethinus) sawdust. *BioResources* **2016**, *11*, 3356–3372. [CrossRef]

80. Wu, Q.; Xian, Y.; He, Z.; Zhang, Q.; Wu, J.; Yang, G. Adsorption characteristics of Pb(II) using biochar derived from spent mushroom substrate. *Sci. Rep.* **2019**, *9*, 1–11. [CrossRef]

81. Zhu, L.; Zhao, N.; Tong, L.; Lv, Y. Structural and adsorption characteristics of potassium carbonate activated biochar. *RSC Adv.* **2018**, *8*, 21012–21019. [CrossRef]

82. Kwiatkowski, M.; Kalderis, D. A complementary analysis of the porous structure of biochars obtained from biomass. *Carbon Lett.* **2020**, *30*, 325–329. [CrossRef]

83. Batista, E.M.C.C.; Shultz, J.; Matos, T.T.S.; Fornari, M.R.; Ferreira, T.M.; Szpoganicz, B. Effect of surface and porosity of biochar on water holding capacity aiming indirectly at preservation of the Amazon biome. *Sci. Rep.* **2018**, *8*, 1–9. [CrossRef]

84. Tan, E.C.D. Sustainable Biomass Conversion Process Assessment. In *Process Intensification and Integration for Sustainable Design*; Foo, D.C.Y., El-Halwagi, M.M., Eds.; John Wiley and Sons, Ltd: Weinheim, Germany, 2021; pp. 301–318.

85. Dewulf, J.; Benini, L.; Mancini, L.; Sala, S.; Blengini, G.A.; Ardente, F.; Recchioni, M.; Maes, J.; Pant, R.; Pennington, D. Rethinking the area of protection "natural resources" in life cycle assessment. *Environ. Sci. Technol.* **2015**, *49*, 5310–5317. [CrossRef]

86. Gallie, W.B. IX—Essentially contested concepts. *Proc. Arist. Soc.* **1956**, *56*, 167–198. [CrossRef]

87. Korhonen, J.; Nuur, C.; Feldmann, A.; Birkie, S.E. Circular economy as an essentially contested concept. *J. Clean. Prod.* **2018**, *175*, 544–552. [CrossRef]

88. European Commission. Bioeconomy. 2013. Available online: https://ec.europa.eu/programmes/horizon2020/en/h2020-section/bioeconomy (accessed on 20 November 2013).

89. Bocken, N.M.P.; de Pauw, I.; Bakker, C.; van der Grinten, B. Product design and business model strategies for a circular economy. *J. Ind. Prod. Eng.* **2016**, *33*, 308–320. [CrossRef]

90. Tan, E.C.D.; Lamers, P. Circular Bioeconomy Concepts—A Perspective. *Front. Sustain.* **2021**, *2*, 701509. [CrossRef]

Article

Quality Improvement and Cost Evaluation of Pellet Fuel Produced from Pruned Fruit Tree Branches

Yining Li [1,2]**, Kang Kang** [3,4,]*** and Wei Wang** [2,5]

1   School of Mechanical Engineering, Shaanxi Polytechnic Institute, Xianyang 712000, China; liyining@sxpi.edu.cn
2   Xianyang Key Laboratory of New Power and Intelligent Microgrid System, Xianyang 712000, China; wangwei05@sxpi.edu.cn
3   Institute for Chemicals and Fuels from Alternative Resources (ICFAR), Western University, London, ON N0M 2A0, Canada
4   College of Mechanical and Electronic Engineering, Northwest A&F University, Xianyang 712100, China
5   School of Aeronautical Engineering, Shaanxi Polytechnic Institute, Xianyang 712000, China
*   Correspondence: kknwafu@hotmail.com

Abstract: Biomass-based pellet is an important source of renewable energy. In this study, to obtain the high-quality fuel pellet via the densification of pruned branches of fruit trees, we investigated the optimization of blending ratios for different raw materials using branches from jujube (*Ziziphus jujuba* Mill.), which is a widely distributed waste biomass resource in China. Through the characterization of raw materials and pellets, the effects of different raw materials on the storage, transportation, and combustion performances of the pellets can be understood. The cost evaluation analysis showed that the two optimized, co-densified pellets had great cost advantages compared with the pure jujube branch pellets. This indicates the potential industrial value of optimized pellets. The results of this study can help to improve the application value of orchard residues and generate an additional profit for fruit plantations, simultaneously avoiding the environmental damage caused by its open combustion.

Keywords: orchard residues; densification; formula parameters; biomass; bioenergy

Citation: Li, Y.; Kang, K.; Wang, W. Quality Improvement and Cost Evaluation of Pellet Fuel Produced from Pruned Fruit Tree Branches. *Energies* **2022**, *15*, 113. https://doi.org/10.3390/en15010113

Academic Editors: Idiano D'Adamo and Piergiuseppe Morone

Received: 15 November 2021
Accepted: 23 December 2021
Published: 24 December 2021

Publisher's Note: MDPI stays neutral with regard to jurisdictional claims in published maps and institutional affiliations.

1. Introduction

Energy has always been crucial for human survival and development [1]. In recent years, the fast consumption of fossil energy has shown an increasingly negative impact on the global environment [2,3]. Hence, the development and utilization of alternative energy sources have become major concerns at the global scale [4,5]. In response to this, biomass energy received an increasing amount of attention as an environmentally friendly energy source [6,7].

Every type of organic matter produced directly or indirectly from the process of photosynthesis is considered biomass [8]. Orchard residues are an important type of biomass that is abundant, widely distributed, and renewable [9]. China is a major fruit-producing and exporting country, with a fruit cultivation area that reached 12,276,700 hm^2 by the end of 2019. The pruned branches of fruit trees produced during routine orchard management are one of the key producers of Chinese orchard residues [10,11]. However, due to their low bulk density and complicated geological distribution, pruned branches are often treated as solid waste. Additionally, their poor storage, transportation, and combustion performance makes economic benefits difficult, further resulting in excessive resource waste and environmental pollution through open combustion [12,13].

For this reason, the pruned branches of fruit trees and other orchard residues can be densified into pellets, and used for industrial power generation or home heating [14,15]. Compared to forestry residues, pruned branches of fruit trees have a higher ash content and lower energy density, so they are not a conventionally desirable raw material for producing pellets. Therefore, they can be blended with forestry residues such as pine wood, with the advantages of both combined [16]. However, this approach raises new issues. Specifically, the pruned branches of fruit trees and forestry residues, which can complement each other, are not always produced at the same geological location. Due to the far distance of the production source, the cost of biomass resources will increase dramatically with the increase in the transportation distance. Therefore, they are more suitable for local acquisition, processing, sales, and use. Hence, to improve pellet quality at an acceptable cost, it is necessary to consider other biomass fuels from the same origin as the pruned branches of fruit trees for co-densification. For example, by co-densifying the pruned branches and pomace of olive trees, the mechanical strength and bulk density of pellets can be significantly improved [17]. Another effective approach is to increase the fixed carbon content and higher heating value of biomass feedstock by further thermochemical processing, such as conversion into biochar [14]. However, biochar has a lower mechanical strength and often requires additives for binding during densification [18].

In order to solve the above problems, in this study, pruned branches of fruit trees and their biochar were mixed and co-densified in proportion, and an appropriate amount of biomass additives were used to obtain pellets with an excellent quality and controllable cost without increasing process complexity.

2. Materials and Methods

2.1. Raw Materials

The raw materials used for pellet production were categorized as main raw materials, secondary raw materials, and additives [19]. The jujube tree (*Ziziphus jujuba* Mill.) is native to China and often grows below 1700 m sea level in various landforms, mountains, hills, or plains. Currently, the jujube tree is widely planted in 21 provincial-level administrative regions (34 in total) in China and is one of the most distinctive local fruits in the country. China's jujube planting area covers about 3,250,000 hm^2, accounting for 26.47% of the country's total fruit tree planting area. Therefore, the use of pruned jujube tree branches (JB) as a primary feedstock, and their charcoals (JBC) as a secondary feedstock, to produce pellet could significantly reduce the quantity of the waste jujube biomass by converting it into clean fuel. In addition, considering the cost and potential pollution problems, it is best to use biomass-based raw materials, which are low-cost and available in large quantities, as additives. A preliminary experiment showed that coco coir (CC) and bone meal (BM) in garden flower fertilizer had better properties as additives. CC and BM meet the requirements for additives in the recommended standard of the Ministry of Agriculture of China (NY/T 1878-2010).

JB and JBC were purchased from Xuzhou Simaide Trading Co., Ltd. (Xuzhou, China). The JB originated from orchards of regularly pruned jujube trees during winter to save nutrient consumption and ensure the smooth overwintering of jujube trees. According to the producer, the JBC is produced by the pyrolysis and hardening of the JB in an SXGT-1000 rotary drum type carbonization furnace (Sanxiong Heavy Industry, Zhengzhou, China) at temperatures of 300–400 °C and a heating rate of approximately 2 °C/s. CC and BM were purchased from Dewoduo Fertilizer Co., Ltd. (Hengshui, China). The CC was produced by washing and crushing coconut shells, which were then dried and pressed into bricks. The BM was made by the steaming method, in which the animal bones were transferred into an autoclave and heated at 105–110 °C. Steam was continuously supplied to the autoclave, and the bones were dried and crushed after most of the grease and gum were removed. Ultra-pure water (UW) was prepared using a Molro 40 economic water purifier (Molecular, Shanghai, China).

Table 1 reports the physicochemical properties of the raw materials. The moisture contents of all raw materials did not exceed 6 wt%, which was within the optimal range of the densification requirements. The HHV (higher heating value) and bulk density of the JBC, obtained following pyrolysis, increased significantly due to the consumption of volatile components and the enhanced fixed carbon. This was reflected in the increased energy density by approximately 3.26 times compared to the JB. Additives are generally used to improve the physical stability of pellet that was blended and densified, as its lower energy density, and higher ash content may have a negative impact on the combustion performance of densified pellet [20], and thus the content of these additives in the pellet formulation need to be controlled to less than 10 wt% [21].

Table 1. Physicochemical properties of the raw materials used to produce pellets.

Sample Name	Proximate Analysis (ar [a], wt%)			
	MC [b]	VM	ASH	FC
JB [c]	5.45 ± 0.62	84.53 ± 0.59	1.84 ± 0.23	8.19 ± 0.62
JBC	2.45 ± 1.16	38.13 ± 0.97	4.37 ± 0.04	55.05 ± 1.16
CC	5.36 ± 0.88	61.28 ± 0.38	15.21 ± 0.81	18.15 ± 0.88
BM	0.72 ± 0.23	17.37 ± 0.10	81.90 ± 0.23	0.01 ± 0.23

Sample Name	Ultimate Analysis (ar, wt%)				
	N	C	H	O	S
JB	5.75	50.41	6.11	35.85	0.48
JBC	1.81	42.61	7.80	46.20	0.44
CC	2.15	44.74	6.08	44.40	0.16
BM	2.86	37.84	8.13	47.82	0.89

Sample Name	HHV [d] (MJ/kg)	Bulk Density (kg/m^3)	Energy Density (GJ/m^3)
JB	18.67	250	4.67
JBC	31.84	490	15.60
CC	17.45	140	2.44
BM	2.93	950	2.78

[a] Ar as received basis. [b] MC: Moisture content; VM: Volatile content; ASH: Ash content; FC: Fixed carbon. [c] JB: The pruned jujube tree branches; JBC: The charcoal of the pruned jujube tree branches; CC: Coco coir; BM: Bone meal. [d] HHV: Higher heating value.

2.2. Pre-Treatment of Raw Materials

Prior to the preparation of the densified pellet, the raw materials were pretreated by crushing and sieving. First, the raw materials were processed into lumps less than 10 mm in diameter using pruning shears and a hand hammer. Following this, 1.5 kg of raw materials were weighed and placed into an RS-FS1811 high-speed grinder (Royalstar, Hefei, China) for full pulverization. The crushed raw materials were then sieved using a ZDS-05 vibrating screening machine (OLAD, Quanzhou, China). The standard sieves were installed from top to bottom with a 10 (0.85–2.00 mm), 20 (0.60–0.85 mm), 30 (0.18–0.60 mm), and 80 (<0.18 mm) mesh. The sieved raw materials were weighed, and the volume was measured to calculate the bulk density.

2.3. Optimization of Pellet Formulation

2.3.1. Densification Process

The particles of the different raw materials (densification of some samples required the addition of UW) with a total weight of 1.00 g were initially mixed proportionally in a φ40 mm jar, and, subsequently, mixed thoroughly using an XH-C vortex mixer (AICE, Taizhou, China). The mixed raw materials were densified using a 769YP-30T manual powder tablet press (REOTAI, Guangzhou, China) with an alloy steel mold [22]. The inner diameter of the alloy steel mold was φ15 mm, with a loading zone height of 50 mm. The densification process was carried out at ambient temperature with a uniform manual pressure of 144 MPa [23]. After holding the pressure for 90 s, the pellet samples were removed using an ejector [23]. The as-prepared samples were stabilized for 15 min in air, then weighed using an FA2204C analytical balance (Techcomp, Shanghai, China), and their external dimensions were measured using a digital vernier caliper. Finally, the samples were labeled, the pellet density was calculated, and they were subsequently sorted and stored in sealed bags.

2.3.2. Experimental Design

The storage and transportation performance of a pellet are two crucial factors determining its industrialization potential; the combustion performance and cost effectiveness of the pellet are also essential for determining its suitability for industrialization. Hence, the experimental process was divided into three stages: (1) parameter range screening tests were employed to investigate the influence of individual factors on the storage and transportation performance of samples and to determine the factor level range required for the optimization test; (2) pellet formulation optimization tests were designed to evaluate the interaction effect of various factors on the storage and transportation performance of samples, and determined the optimized formula parameters in combination with their combustion performance; and (3) the pellet formulation comparison test was conducted to compare the storage and transportation characteristics, combustion characteristics and cost-effectiveness of different pellet samples to validate the superiority of pellet.

During the storage and transportation processes, the pellet may gradually break due to vibration, extrusion, etc., which can cause soil contamination and waste dispersion, and negatively impact its combustion efficiency. Hence, it is necessary to improve the physical stability of pellet [24]. In this study, drop resistance was employed as the primary evaluation index of the physical stability, and the primary indicator of storage and transportation performance. For the physical stability tests, the samples were subjected to free drop motion at a height of 1.85 m onto the steel plates on the ground, with each sample hitting the steel plate three times. The mass ratio percentage of the sample before and after the drop was used to indicate the physical stability of the pellet. Test results with values greater than or equal to 98% indicated that the pellet sample met the standard. The pellet density was employed as the secondary indicator of storage and transportation performance. According to the recommended standard of the Ministry of Agriculture of China (NY/T 1878-2010), the pellet density should be greater than or equal to 1000 kg/m^3 [25].

Table 2 reports the factors, levels, and formulation parameters for each stage of the drop test. In the parameter range screening test, JB particles with a particle size $\varphi < 0.18$ mm (80 mesh) were employed as the main raw materials in the first group of tests, and no additives were included. In the second group of tests, the JBC content in the secondary raw materials was maintained at 20 wt% without additives, while in the third group, 10 wt% additives were added based on the parameters of the previous two groups.

The pellet formulation optimization tests were performed using a four-factor, three-level orthogonal test. Orthogonal testing employs mathematical statistics to rationalize the test procedure [26]. This method can significantly reduce the number of tests without losing test information and is able to simultaneously analyze multiple factors and their interactions. The optimized formula parameters were obtained by the variance and range analysis of test results. The variance analysis decomposes the sum of squares of the total variance and subsequently performs statistical tests [27] to determine the influence of the controllable factors on the test index. Range refers to the maximum difference between the test results of each factor at different levels. By comparing the range of each factor, the order of magnitude can be used to determine the primary and secondary effects of each factor [28]. Based on the results of the range and the variance analysis, the JBC content was increased to the maximum value to obtain the optimal formula parameters of the highest JBC content.

The pellet formulation comparison tests were conducted to compare the characteristics of samples without JBC and additives according to the above optimal formulation and to analyze the differences and reasons through characterization. In Table 2, the samples are named in the form of JB + number 1 + (number 2) + C + number 3 + additive type. For example, JB56(80)-C37-CC means that the sample consists of 80 mesh JB with a mass fraction of 56 wt%, and JBC with a mass fraction of 37 wt%, with CC as the additive. For the three-stage experiment, all tests were repeated three times for each sample, and the mean and standard deviations were calculated.

Table 2. Factors, levels, and formula parameters of the three-stage experiment.

Parameter	Group Number	Factor	Level				
Range Screening Tests	1	JBC [a] content (wt%)	0	10	20	30	40
	2	JB particle size (mesh)	10	20	30	80	-
	3	Additive type	CC [b]	BP	UW	CG	-

Parameter	Total Number	Factor	Level		
Pellet Formulation Optimization Tests	27	JBC content (wt%)	25	30	35
		JB particle size (mesh)	20	30	80
		Additive type	CC	BM	UW
		Additive content (wt%)	10	7	4

Parameter	Formula Parameters				
	Sample Name	JBC Content (wt%)	JB ParticleSize (mm)	Additive Type	Additive Content (wt%)
Pellet Formulation Comparison Tests	JB100(80) [c]	-	<0.18	-	-
	JB63(80)-C37	37	<0.18	-	-
	JB56(80)-C37-CC	37	<0.18	CC	7
	JB100(30)	-	0.60–0.18	-	-
	JB69(30)-C31	31	0.60–0.18	-	-
	JB65(30)-C31-BP	31	0.60–0.18	BM	4

[a] JB: The pruned jujube tree branches; JBC: the charcoal of the pruned jujube tree branches. [b] CC: Coco coir; BM: Bone meal; UW: Ultra-pure water; CG: Control group. [c] The samples are named in the form of JB + number 1 + (number 2) + C + number 3 + additive type. For example, JB56(80)-C37-CC means that the sample consists of 80 mesh JB with a mass fraction of 56 wt%, and JBC with a mass fraction of 37 wt%, with CC as the additive.

2.4. Characterization

Proximate and elemental analyses were performed on the raw materials. The former adopted the analytical methods of the American Society for Testing and Materials (ASTM 1762-84 and 3173-87) [29], and the latter was conducted via a vario PYRO elemental analyzer (Elementar, Langenselbold, Germany). Higher Heating Value (HHV) measurements of raw materials and samples were obtained using an LC-VC-430 automatic calorimeter (UCHEN, Shanghai, China). The bulk density of raw materials was measured using a 500 mL density cup, while the pellet density of samples was calculated by the ratio of mass to volume [30]. The energy density of raw materials and samples was taken as the product of bulk density or pellet density and HHV.

The moisture absorption behavior of the samples was measured using an HWS-158 constant temperature and humidity incubator (Ningbo Southeast Instrument Co., Ltd., Ningbo, China). The samples were dried in a DHG-9070AS incubator (Ningbo Southeast Instrument Co. Ltd., Ningbo, China) at 105 °C for 12 h to remove the internal moisture, and then transferred into an incubator operated at the relative humidity of 70% at 30 °C [31,32]. The samples were weighed many times over a period of 32 h and tested for pellet. Each formulation was repeated three times. The moisture content was then calculated based on the changes in the sample weights.

Infrared spectra of the raw materials and samples were collected using an FT-IR spectrometer (Nicolet iS20, Thermo Scientific, Waltham, MA, USA). The materials were scanned in the range of 4000–400 cm^{-1} with a resolution of 4 cm^{-1} in transmission mode and 16 scan per spectrum. Thermogravimetric (TG) and differential thermogravimetric (DTG) analyses were performed on the raw materials and samples using a TG 209-F3 Tarsus thermal analyzer (NETZSCH, Selb, Germany). Approximately 15–20 mg of samples were heated from ambient temperature to 800 °C at a constant heating rate of 10 °C/min under an oxygen atmosphere with a 50 mL/min flow rate.

3. Results and Discussion

3.1. Analysis of Raw Materials

3.1.1. Particle Size and Density Distribution

The particle size and bulk density of raw materials have important impacts on the physical stability of the densified pellet [33]. As shown in Figure 1a, the relative contents of JB particles with different particle sizes were prevalent in all four intervals after full grinding, and the bulk density increased significantly with decreasing particle size. For the ease of densification, all materials, including the JBC and the additives, were pre-screened to unify the particles size. After the pre-screening, it was found that most of the materials reached the particle size of less than 80 mesh (≤0.18 mm). Specifically, 98.00 wt% for both the JBC and BM and 79.69 wt% for CC falls in the particle range size of less than 80 mesh, which is shown in Figure 1b. The bulk density of the JBC and BM within the 80 mesh was observed to be 1.17 and 2.29 times compared to that of the JB, while the bulk density of CC was only 32.71% of JB, respectively. This reveals the necessity to fully consider the effect of the JB with different particle sizes on the pellet performance during the densification process. The concentrated particle size distribution of the remaining raw materials can help simplify the formulation process.

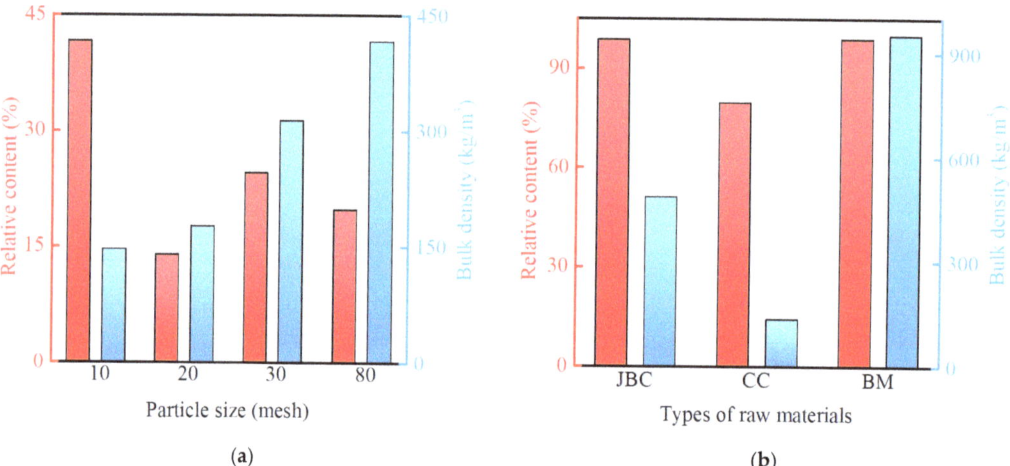

Figure 1. Particle size distributions and bulk densities of raw materials: (**a**) Data on JB; (**b**) Data on JBC, CC, and BM. JB: The pruned jujube tree branches; JBC: The charcoal of the pruned jujube tree branches; CC: Coco coir; BM: Bone meal.

3.1.2. Functional Groups

Figure 2 shows the FT-IR spectra of the raw materials. The hydroxyl-OH stretching vibration was observed around 3400 cm^{-1} [34], and the peak on the spectrum of JBC was significantly lower than that of the JB. This is attributed to the breakage of the hydrogen-bonded hydroxyl group following the carbonization of the JB, resulting in the detachment of bound water. Aliphatic C-H stretching vibrations [35] were observed around 2920 cm^{-1}. The peak at 1035 cm^{-1} corresponds to C-O-C stretching vibrations, where the characteristic JBC peak weakened due to the dehydrogenation and deoxygenation of the JB during the carbonization process [36]. The out-of-plane aromatic C-H bending vibration of the JBC was clear within the region 987–781 cm^{-1}, indicating an intensification in the dehydrogenation reaction during the carbonization process and an enhancement in the aromatization structure [37]. The bending vibration of aliphatic C-H around 1383 cm^{-1} indicated the formation of intermediate decomposition products and the polymerization of cellulose and lignin contained in the JBC [38]. The carbonyl functional groups C=O of esters were observed around 1739 cm^{-1}, and the stretching vibrations of aromatic rings in the lignin were present around 1509 cm^{-1} [39]. These peaks were absent in the JBC due to carbonization and decomposition. The 897 cm^{-1} and 781 cm^{-1} peaks of the JB and JBC, respectively, indicate a gradual shift of C-H in the aromatic ring structure to a lower wave number, resulting in the breakage of the lignin aromatic ring structure and the generation of more free radicals [40]. This is an overlap of the -OH stretching vibrations of surface-free water and calcium hydroxy-phosphate. Additional peaks were observed at 1442 and 870 cm^{-1}, 1030 cm^{-1}, and 694 and 536 cm^{-1}, corresponding to CO_3^{2-}, the stretching vibrations of PO_4^{3-}, and the bending vibrations of P-O, respectively.

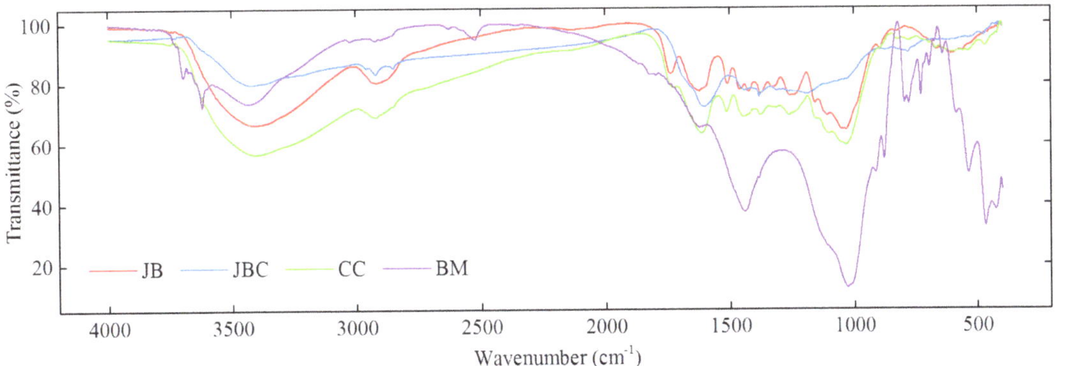

Figure 2. FT-IR spectra of the raw materials. JB: The pruned jujube tree branches; JBC: The charcoal of the pruned jujube tree branches; CC: Coco coir; BM: Bone meal.

3.1.3. Thermogravimetric Analysis

Figure 3 presents the combustion characteristics of raw materials investigated via thermogravimetric analysis. The TG curves of the JB and CC were essentially similar, indicating the similar combustion processes between the two, with the main difference being that the residue of the final non-combustible substances of the CC was about 2.00 times that of the JB (Figure 3a). The major mass loss temperature zone of JBC was significantly delayed compared to that of the JB, which indicated that the combustion of the JBC mainly occurred at higher combustion temperature. This is because the volatile component in the JBC was relatively low after the carbonization. The TG curve of the BM was significantly different to the other three biomass types in that there was no water evaporation phase, and slow changes were observed in the volatile component release and combustion phase. The residue from BM generally did not contain fixed carbon, with a residual mass of 84.05 wt%, the main components being calcium phosphate and calcium carbonate that had not reached their melting points. This is essentially consistent with the ash content of BM in Table 1 (81.90 $\pm$ 0.23 wt%).

Figure 3b depicts the DTG curves of the raw materials. By combining the TG curves with the ignition temperatures, burnout temperatures and burning times were calculated [41]. The ignition temperatures of the JB and JBC were determined as 271.6 °C and 379.3 °C, respectively, with the 100 °C difference indicating that JBC was more difficult to ignite due to the absence of volatile components. The ignition temperatures of CC and BM were 259.6 °C and 652.7 °C, respectively. The JB and CC exhibited two characteristic peaks, representing the volatile component release and combustion phase and fixed carbon combustion phase, respectively [42]. Unlike the CC, the volatile component release and combustion phase of the JB were highly reactive, while the fixed carbon combustion phase was moderate. This was consistent with the results of the proximate analysis. The JBC, which lacked volatile components due to carbonization, exhibited a single relatively smooth characteristic peak, indicating that the combustion process of the JBC was both gentle and highly persistent.

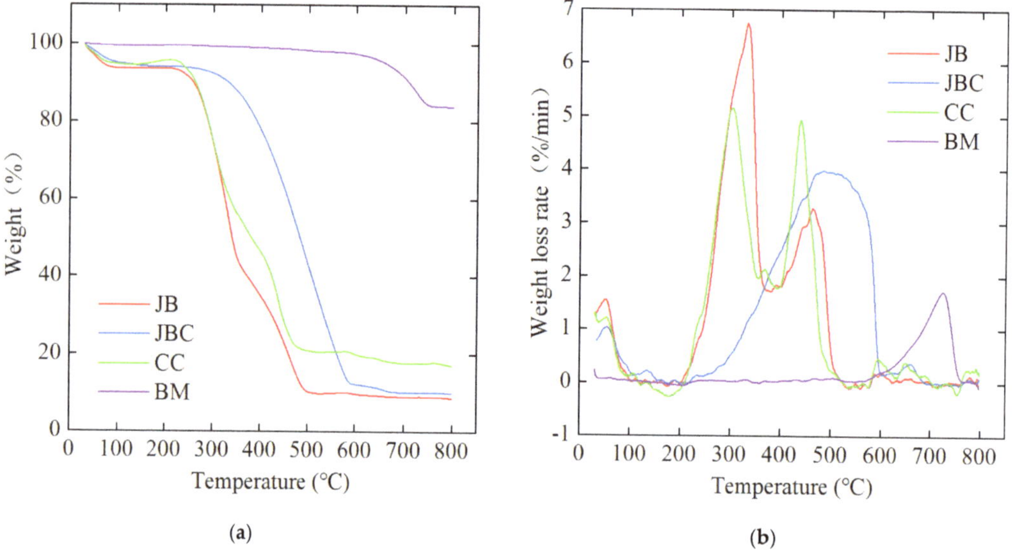

Figure 3. Thermogravimetric analysis of raw materials: (**a**) TG curve; (**b**) DTG curve. JB: The pruned jujube tree branches; JBC: The charcoal of the pruned jujube tree branches; CC: Coco coir; BM: Bone meal.

3.2. Optimization of Formula Parameters

3.2.1. Parameter Range Screening

Optimization of JBC Contents

The drop resistance and pellet density of samples were found to be negatively correlated with the JBC content (Figure 4). Specifically, as the JBC content increased, the drop resistance and pellet density of the samples decreased. When the JBC content was 20 wt%, the samples were in a critical state. Namely, the drop resistance was 98.14%, meeting the physical stability requirement. This corresponded to a pellet density of 994 kg/m^3, which was slightly lower than the standard of the Ministry of Agriculture of China (NY/T 1878-2010). As the JBC content increased, the standard deviation of the drop resistance and pellet density also expanded, and the storage and transportation performance of the samples gradually deteriorated. When the JBC content increased to 40 wt%, the drop resistance of the samples reduced significantly to 77.55 ± 10.18%. The difficulty of densifying between the JB particles was reduced by the lubricating effect of the moisture. In addition, the densifying was stabilized by the gradual increase in the lignocellulose surface viscosity via warming and softening under high pressure, followed by cooling to form a solid bridge [43]. The addition and mixing of smaller-sized JBC particles resulted in the brittleness and hydrophobicity of the JBC particles and the lack of viscosity after pyrolysis, preventing water flow and the mutual contact between JB particles with increasing JBC content. Thus, the particles could not adhere together to form sufficient solid bridges, resulting in the inability to form stable pellet. This was clear when the JBC content increased to 40 wt%, and the drop resistance of the samples was no longer able to meet the commercialization requirements (77.55 ± 10.18%).

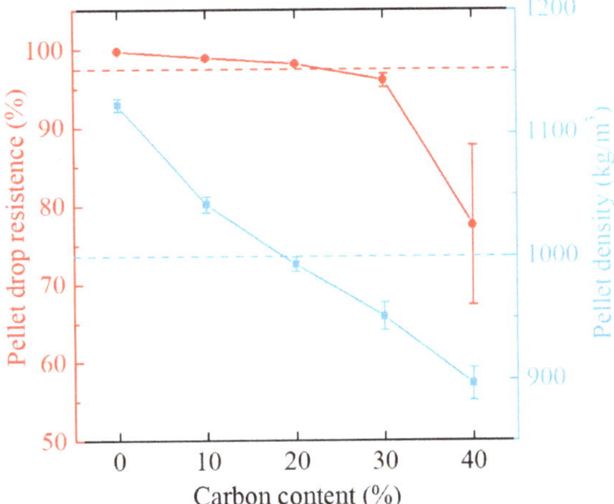

Figure 4. Effects of different JBC contents on the storage and transportation characteristics of the pellets. JBC: The charcoal of the pruned jujube tree branches.

Optimization of JB Particle Sizes

At the JBC content of 20 wt%, the drop resistance and pellet density of the samples varied with the JB particle size (Figure 5). The drop resistance and pellet density values peaked at the JB particle size of 30 mesh. The drop resistance of the JB particle sizes in the first two groups all exceeded 98%, while the pellet density was observed to be approximately 99.4% of the required standard value, meeting the physical stability requirements. During the densifying process, the JB particles were stacked on top of each other, and the surfaces were in full contact. For larger particle sizes, the drop resistance of the JB particles decreased. The possible reasons for this are twofold. First, the surfaces of larger JB particles were not in full contact with each other, with a greater number of voids in the center. This affected the flow of water and the formation of solid bridges. Second, owing to the elongated structure of lignocellulose, the larger the JB particle size, the higher the elastic modulus, and the greater the power consumption required to resist the internal elastic potential energy under the same compression parameters. Consequently, this complicated the densifying processes. When the JB particle size was 80 mesh, the drop resistance of the pellet was slightly lower than that of the pellet produced with the 30-mesh size JB particle. Despite the lower elastic potential energy of the 80-mesh size JB particles, the overlap of its substrate frame was affected by JBC particles of the same particle size, and rather than the JBC particles filling the frame gaps, they replaced the 80-mesh size JB particles as the framework. The JBC particles lacked the binding characteristics of the JB particles, thus reducing the overall strength of the pellet.

Effect of Additives

Figure 6 depicts the performance of the drop resistance and pellet density of the samples in different additive and control groups, with an 80 mesh JB particle size, 20 wt% JBC content, and 10 wt% additive content. The addition of UW, BM, and CC all increased the drop resistance of the pellet samples, with their effectiveness, from lowest to highest, in the order of UW (98.38%) < BM (99.22%) < CC (99.69%). The slight increase in strength via the addition of UW is related to the improved lubrication between the particles during densification following the external moisture replenishment, which reduced the energy consumption and resulted in tighter densifying. In the functional group analysis, the BM was observed to contain a low content of hydroxyl groups. Therefore, the BM exhibited a good hydrophilicity and facilitated the flow of internal water for lubrication and enhancing densification. Despite the similar particle sizes between the CC and JB, its longitudinal dimension was relatively longer. This improved the overall strength of the pellet by linking the polymeric agglomerates of particles that were close to each other but had insufficient surface contact when the blended particles were stacked on top of each other. A positive correlation was observed between the effect of the CC and BM on the pellet density (1020 kg/m^3 and 1070 kg/m^3), while UW was found to negatively impact the pellet density (900 kg/m^3). This negative correlation is attributed to the excessively large 10 wt% ratio. The excess water was squeezed out from the mold gap during the densifying process, resulting in an overall loss of mass. The highest density of pellet with BM additives was attributed to the bone meal density (1000 kg/m^3). Although the density of pellet with CC additives was lower; it increased the densifying strength and reduced the JBC loss.

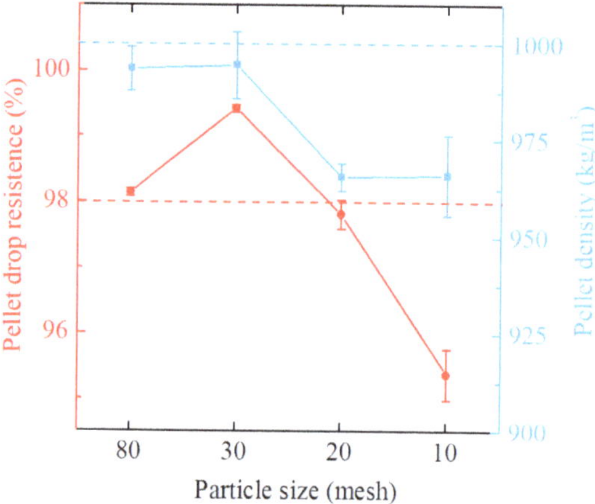

Figure 5. Effects of different JB particle sizes on the storage and transportation characteristics of the pellets. JB: The pruned jujube tree branches.

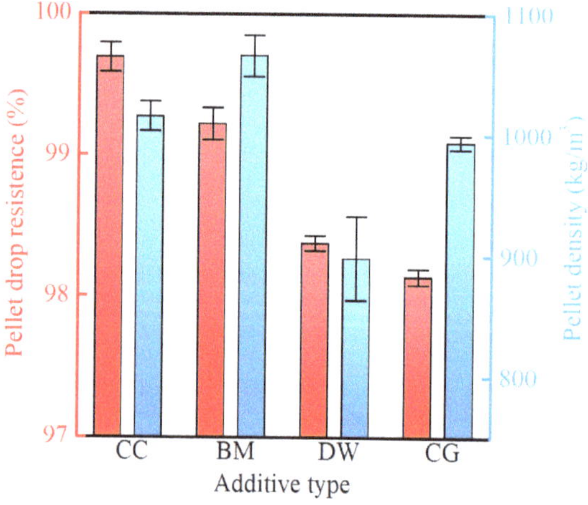

Figure 6. Effects of different additives on the storage and transportation characteristics of the pellets. CC: Coco coir; BM: Bone meal; UW: Ultra-pure water; CG: Control group.

3.2.2. Pellet Formulation Optimization

Experimental Results

Orthogonal tests were employed for the parameter optimization based on 27 samples (each sample was repeated three times) using four-factor, three-level interaction tests

(Figure 7). A total of 18 groups exhibited drop resistance values greater than or equal to 98%, and 7 groups with particle densities greater than or equal to 1000 kg/m^3. The results reveal the great influence of the different JBC contents, JB mesh numbers, additive types, and dosage combinations on the physical stability of pellets. In order to further analyze the relationships between these variables, variance and range analyses were performed. The drop resistance, which represents physical stability, was the primary control index of this study, and thus the focus of the analysis. Based on ensuring the drop resistance, the pellet formulation was optimized using the drop resistance and the pellet density.

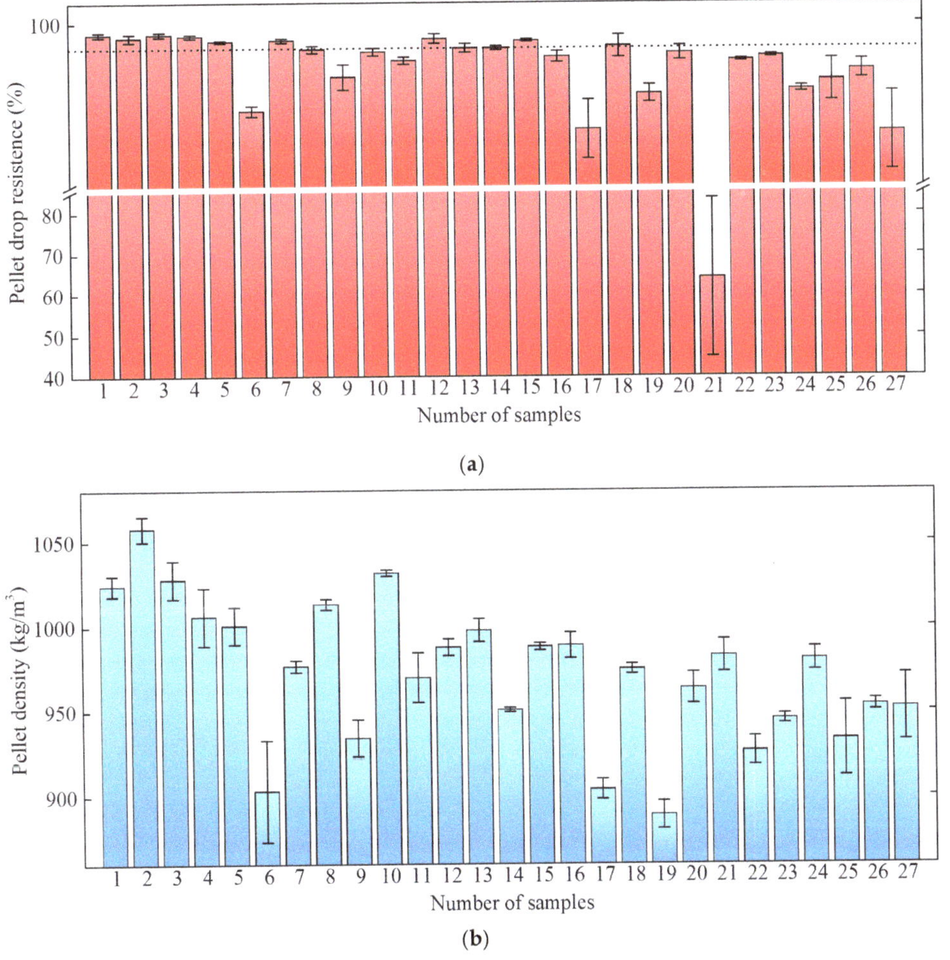

Figure 7. Results of the formulation optimization: (a) Pellet drop resistance; (b) Pellet density.

Variance Analysis

IBM SPSS Statistics 22 software was used to analyze the variance of the test results. Table 3 reports the variance analysis results of the drop resistance. The influences of each factor and their interactions on the drop resistance of samples were: A > AC > C > AD > BD > BC > CD > AB > B > D, and all were highly significant ($p < 0.01$). The JBC content (A) was the most important factor affecting the drop resistance of samples. During densification, the moisture and volatiles, as natural binders between particles, play a very important role [44,45]. The JBC particles, lacking moisture and volatiles, adhered to the JB and solid additive particles. This hindered the flow and lubrication of the water, as well as the softening and bonding with other particles. Hence, the higher the JBC content (A), the more clear the damage to the drop strength of the samples. The drop resistance of the samples was significantly improved by adding the appropriate additives (C) and controlling their content (D). The interaction between the JBC content and additive type and dosage (AC and AD) was also clear. The effect of the JB particle size (B) was not significant, neither was that of the additives (C and D) and the interaction between them (BC and BD), which also indicated the importance of the additives (C and D) in blending the densified pellet. This does not indicate that the JB particle size (B) was less influential, rather that it played a positive role in the underlying framework structure.

Table 3. Variance analysis results for the parameter optimization test.

Source	Sum of Squares III	DF	Mean Square	F	P
Modified model	3478.553 [a]	26	133.790	9.359	0.00000
Intercept	759194.670	1	759194.670	53105.501	0.00000
A: JBC [b] content (wt%)	437.407	2	218.704	15.298	0.00001
B: JB particle size (mm)	2.000	2	85.588	5.987	0.00448
C: Additive type	296.903	2	148.451	10.384	0.00015
D: Additive content (wt%)	163.104	2	81.552	5.705	0.00565
AB	208.997	2	104.498	7.310	0.00155
AC	344.141	2	172.070	12.036	0.00005
AD	245.629	2	122.814	8.591	0.00058
BC	221.672	2	110.836	7.753	0.00110
BD	222.472	2	111.236	7.781	0.00107
CD	212.987	2	106.493	7.449	0.00139
Error	771.982	54	14.296		
Total	763445.205	81			
Adjusted sum	4250.535	80			

[a] $R^2 = 0.818$ (Adjusted $R^2 = 0.731$); [b] JB: the pruned jujube tree branches; JBC: The charcoal of the pruned jujube tree branches.

Range Analysis

Figure 8 depicts the results of the range analysis of the parameter optimization test, where the ordinate represents the drop resistance of the sample; letters A, B, C, and D correspond to the four factors in the test, respectively; and the numbers 1, 2, and 3 after the letters are the levels of the corresponding factors, respectively. The optimization aimed to further improve combustion performance by increasing the JBC content while maintaining the storage and transportation performance. Combining the results of the variance analysis, two formulations were obtained: (a) A2C1D2B1: JBC content 30 wt%, JB particle size 80 mesh, and CC additive content 7 wt%; and (b) A2C2D3B2: JBC content 30 wt%, JB particle size 30 mesh, BM additive content 4 wt%. The drop resistance values of the formulations were 98.85% and 98.71%, respectively.

Optimized Formula Parameters

The formula parameters obtained by the variance and range analyses, which are based on the set parameters of the test group, do not reflect the maximum JBC content limit that can be achieved under these conditions. In order to further improve the combustion performance of pellet, the target drop resistance was set as $\geq 98\%$, and the JBC content was increased to the upper limit without changing the determined JB particle size, additive type, and content. A single-factor test was conducted with the initial value of JBC content set at 25 wt% and the addition of 2 wt% increments. The optimized parameters were as follows: (a) JB particle size—80 mesh, CC content—7 wt%, and maximum JBC content—37 wt%; and (b) JB particle size—30 mesh, BM content—4 wt%, and maximum JBC content—31 wt%.

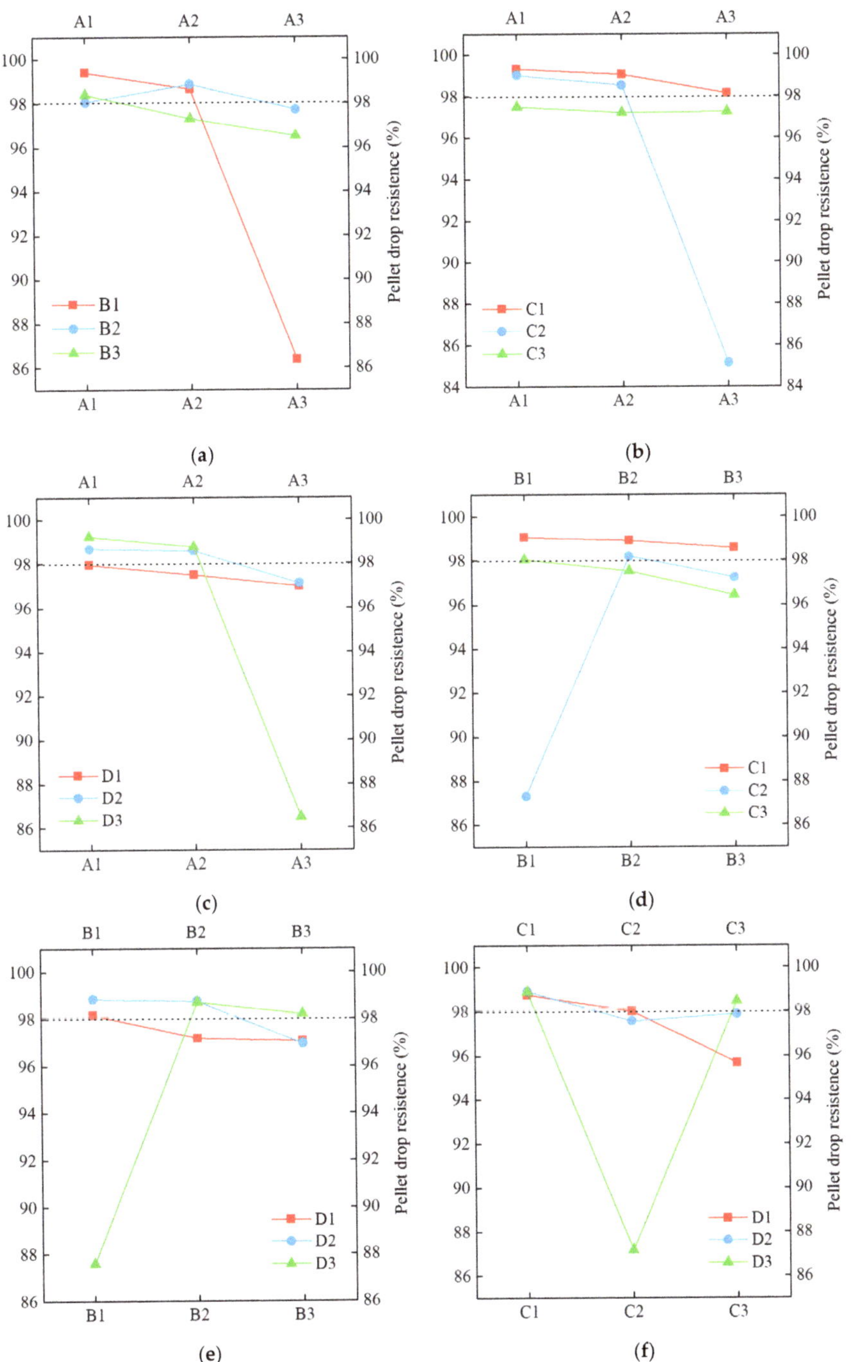

Figure 8. Range analysis results of the parameter optimization: (**a**) JBC content (A) and JB particle size (B); (**b**) JBC content (A) and additive type (C); (**c**) JBC content (A) and additive content (D); (**d**) JB particle size (B) and additive type (C); (**e**) JB particle size (B) and additive content (D), (**f**) Additive type (C) and additive content (D).

3.2.3. Pellet Formulation Comparison

Test Results on Pellet Drop Resistance, Pellet Density, and Energy Density

The pellet samples prepared with two optimized formula parameters were compared with their related samples. Table 2 reports the specific sample numbers and formula parameters, while Figure 9 presents the experimental results of the drop resistance, pellet density, and energy density. As revealed in the analysis presented in Section 3.2.1 (Optimization of JBC Contents), the increased JBC content reduced the drop resistance and pellet density of the pellet samples. However, with the use of additives, the drop resistance of the pellet samples containing JBC was restored to more than 98%. Furthermore, for the sample with a JB particle size of 80 mesh, the density was 965 kg/m^3, which slightly differed from the recommended standard of the Ministry of Agriculture of China (NY/T 1878-2010), while that of the 30-mesh samples was exactly 1000 kg/m^3. The pellet density of the optimized formula parameter samples decreased by 16.16% and 10.28%, respectively, compared to the sample prepared from pure JB. However, in terms of fuel performance, the energy density of the optimized formula parameter samples increased by 53.64% and 53.47%, respectively, compared to the pure JB samples. The overall fuel quality of the samples thus improved significantly, indicating the effectiveness of the biomass carbon and additive applications.

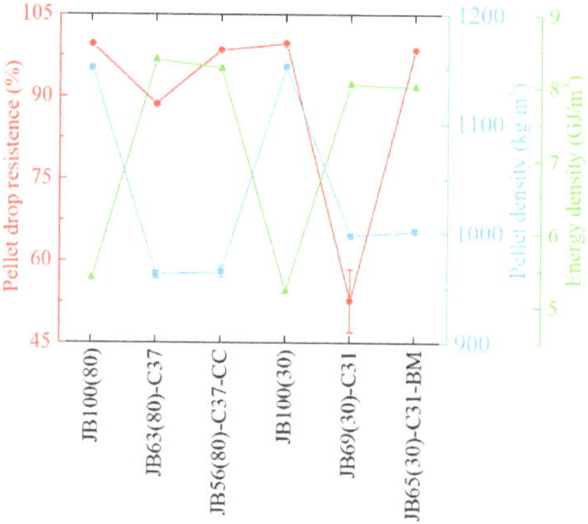

Figure 9. Test results on the drop resistances of the pellets. The samples are named in the form of JB + number 1 + (number 2) + C + number 3 + additive type. For example, JB56(80)-C37-CC means that the sample consists of 80 mesh JB with a mass fraction of 56 wt%, and JBC with a mass fraction of 37 wt%, with CC as the additive.

Hydrophobicity Analysis

Hydrophobicity is an important indicator due to the critical influence of the moisture on the physical stability, energy density, and combustion process of pellet [24]. The hygroscopicity of the prepared pellet should be as low as possible during the storage and transportation processes.

Figure 10 reports the moisture uptake curves of samples with two optimized formula parameters and the corresponding samples. Following approximately 2 h, the moisture uptake of each sample began to change significantly; after about 25 h, the moisture uptake of the samples leveled off and gradually stabilized. The European Granular Council standard [46] indicates that the moisture content of the samples should be less than 10 wt%, which was met by all samples in the test.

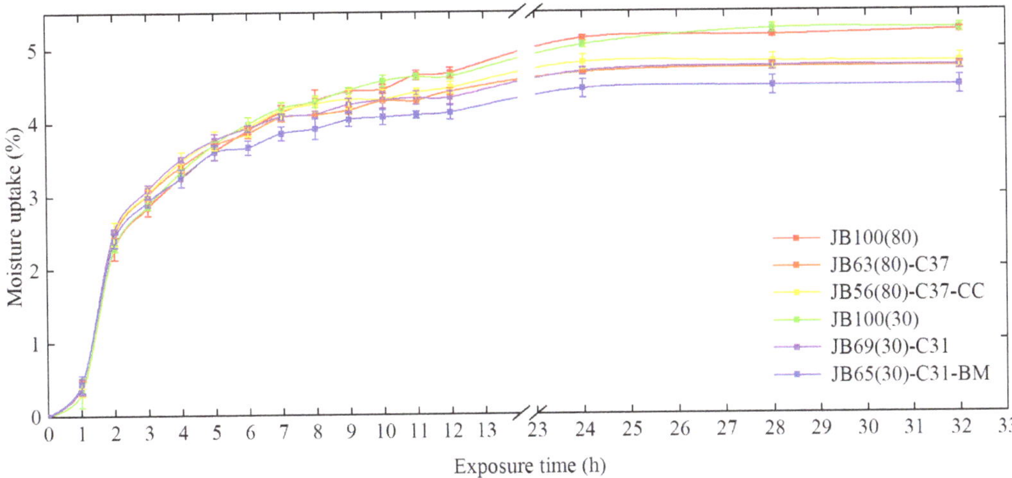

Figure 10. Moisture absorption test results of the different pellet samples. The samples are named in the form of JB + number 1 + (number 2) + C + number 3 + additive type. For example, JB56(80)-C37-CC means that the sample consists of 80 mesh JB with a mass fraction of 56 wt%, and JBC with a mass fraction of 37 wt%, with CC as the additive.

The maximum moisture uptake values of JB100 (80) and JB100 (30), which were densified from pure JB particles of varying meshes, were highly similar, at 5.23 ± 0.03 wt% and 5.25 ± 0.06 wt%, respectively. When 37 wt% and 31 wt% JBC were added, the maximum moisture uptake of JB63(80)-C37 decreased from 5.23 ± 0.03 wt% to 4.37 ± 0.05 wt%, and that of JB69(30)-C31 from 5.25 ± 0.06 wt% to 4.75 ± 0.05 wt%. When CC (7 wt%) and BM (4 wt%) were added, the maximum moisture uptake of JB56(80)-C37-CC increased from 4.75 ± 0.05 wt% to 4.81 ± 0.10 wt%, whereas the maximum moisture uptake of JB65(30)-C37-BM decreased from 4.75 ± 0.05 wt% to 4.48 ± 0.13 wt%.

As described in Section 3.2.2 (Variance Analysis), the added JBC may have wrapped the JB particles, thus hindering the flow of water and the formation of solid bridges. Moreover, the JBC particles are more hydrophobic compared to JB particles; therefore, the added JBC particle may block the water from entering the pellet. This consequently increased the hydrophobicity of the sample. The role of CC was to partially replace the JBC particles on the sample surface, and the CC properties were similar to those of the JB particles, thus slightly increasing the water absorption. The BM also partially replaced the JBC particles on the sample surface. However, its main components were calcium carbonate and calcium phosphate, which are insoluble or slightly soluble in water, further increasing the hydrophobicity of the sample. Although the BM blocked the flow of water, it was difficult to form a solid bridge between the particles [47]. However, due to the filling effects of the BM particles into the void space, it helped to promote the contact of different particles, and further improved the formation of inter-particle bonds in the contact area, thus enhancing the binding [48].

FT-IR Analysis

The composition of functional groups plays an important role in pellet hydrophobicity. Figure 11 presents the FT-IR spectra of two optimized formula parameters with the corresponding samples. The peaks at 2921–2918 cm^{-1} are linked to the stretching vibration of aliphatic C-H; the peaks at 1435–1425 cm^{-1} are characteristic absorption peaks of the C=C in benzene ring; the peaks at 1382–1375 cm^{-1} correspond to the bending vibration of aliphatic C-H, and the peaks at 1055–1034 cm^{-1} correspond to the C-O bond stretching vibrations [49]. These are all hydrophobic functional groups, which could reduce the moisture absorption rate of the pellet during storage and transportation.

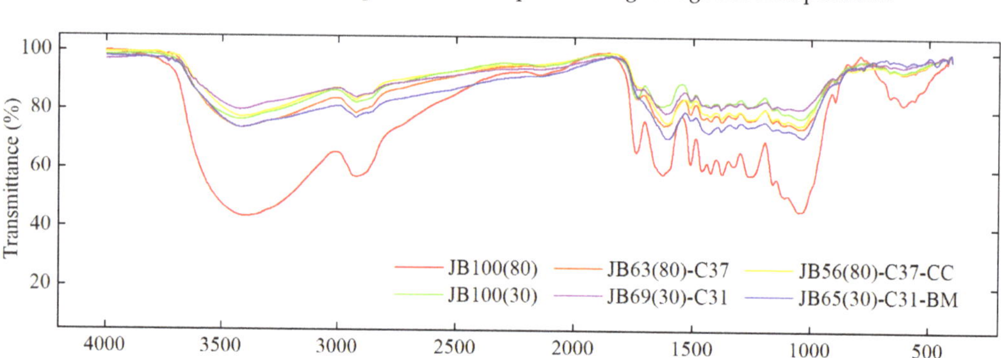

Figure 11. FT-IR spectra of the pellet samples. The samples are named in the form of JB + number 1 + (number 2) + C + number 3 + additive type. For example, JB56(80)-C37-CC means that the sample consists of 80 mesh JB with a mass fraction of 56 wt%, and JBC with a mass fraction of 37 wt%, with CC as the additive.

Thermogravimetric Analysis

In order to study the effects of carbon content and additives on the combustion characteristics of pellets, a thermogravimetric analysis was performed on the pellets before and after optimization. Figure 12 demonstrates the TG and DTG curves of the two samples with optimized formula parameters and their related samples, where JB56(80)-C37-CC corresponds to JB100(80) and JB63(80)-C37, and JB65(30)-C31-BM corresponds to JB100(30) and JB69(30)-C31.

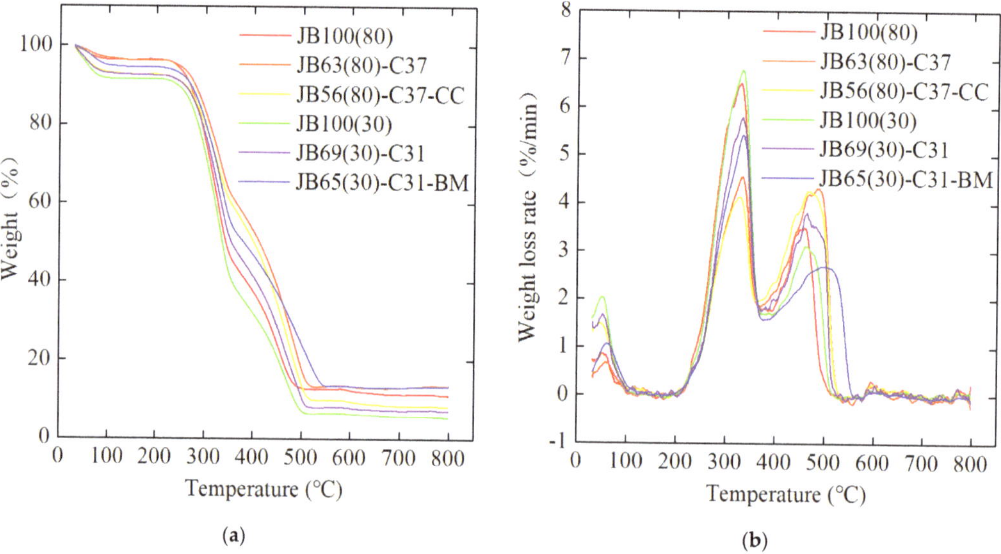

Figure 12. Thermogravimetric analysis of the pellet samples: (**a**) TG curve; (**b**) DTG curve. The samples are named in the form of JB + number 1 + (number 2) + C + number 3 + additive type. For example, JB56(80)-C37-CC means that the sample consists of 80 mesh JB with a mass fraction of 56 wt%, and JBC with a mass fraction of 37 wt%, with CC as the additive.

The curves of six pellets were quite similar, indicating that the addition of the JBC and additives did not significantly affect the overall combustion process of pellets (Figure 12a). Notably, the BM significantly increased the residual content following combustion, and the difference between JB100(30) and JB65(30)-C31-BM reached 8.01 wt%, which was more likely to cause slagging in the furnace.

Figure 12b depicts the DTG curves of the six pellets. Combining the data in both TG and DTG curves, the ignition temperature, burnout temperature, and combustion time were calculated. Although the ignition temperature of the JBC was approximately 100 °C higher than that of the JB, its impact on the ignition temperature after blending and densifying with the JB was limited compared with JB100(80) and JB63(80)-C37, as well as JB100(30) and JB69(30)-C31, which increased by 3.9 °C and 5.9 °C, respectively. Furthermore, the combustion temperature, compared with JB100(80) and JB63(80)-C37, as well as JB100(30) and JB69(30)-C31, increased by 29.4 °C and 8.1 °C, and the combustion time increased by 11.41% and 2.35%, respectively. This indicates the improvement of the pellet combustion performance via the JBC blending, releasing more energy. The adoption of the CC reduced the combustion temperature by 1.1 °C and increased the time by 0.86% compared JB63(80)-C37 between JB56(80)-C37-CC, which had no effect on the combustion characteristics of the pellet. Furthermore, when BM was used as an additive, the combustion temperature and combustion time increased by 32.7 °C and 14.11% compared JB69(30)-C31 between JB65(30)-C31-BP. This may result in combustion residue with the bone meal as an additive for the case of insufficient combustion.

3.3. Cost Evaluation

The two optimized pellet formulations obtained in this study greatly improved the combustion performance of the biomass densified pellet while simultaneously ensuring its storage and transportation requirements. The applicability of the fuel to actual industrial production was dependent on its cost-effectiveness. However, due to the lack of necessary industrial data, it was not possible to obtain the actual production costs of the biomass-densified pellets used in this study. In order to estimate the application value of pellets, the cost-effectiveness of the pellets for laboratory preparation was calculated.

The raw biomass is typically densified into pellets following three processes: raw material collection, raw material pre-treatment, and pellet production by densification. Hence, the costs were divided into raw material treatment costs and pellet densification costs. For the processing cost of raw materials, the purchase data were based on the actual purchase price of the biomass raw materials, namely, the retail market price. The process parameters and production equipment were the same for all raw material pre-treatment processes. However, the yield of the final usable raw material particles differed across biomass types following sufficient grinding. Thus, the relative content of different particle sizes was different (Figure 1). This resulted in different processing costs for different raw materials. The same process parameters and production equipment were used for the preparation of pellets in the pellet densification process. However, the percentage of different raw materials differed among the different formula parameters. Thus, the costs of the raw materials and pre-treatment were distinct. Moreover, the energy density of the densified pellet differed across formula parameters, indicating the variation in the value of different densified pellets.

The cost effectiveness (relative unit cost) between different types of biomass-densified pellets was investigated. To facilitate the analysis, the raw biomass material processing cost was calculated based on the cost of JB particles with a size of 30 mesh (JB(30)) at the base value of 1.00, while the cost of biomass-densified pellet was calculated using JB100(30) as the base (set to 1.00). The calculation formulas are listed as follows:

$$RMCo_{Rmt} = \frac{PP}{PS}PY \; (Yuan/g),\tag{1}$$

$$RUCo_{Rmt} = RMCo_{Rmt}/BCo_{Rmt},\tag{2}$$

$$RUCo_{Pc} = \sum(RUCo_{Rmt} \times RMCt), \tag{3}$$

$$RUED_{Pf} = ED_{Pf}/BED, \tag{4}$$

$$RUECo_{Pf} = RUCo_{Pc} \times RUED_{Pf}. \tag{5}$$

In Formulas (1)–(5), $RMCo_{Rmt}$ represents the raw material treatment cost; PP represents the raw material market purchase; PS represents the raw material purchasing specifications; PY represents the raw material process yield; $RUCo_{Rmt}$ represents the relative unit cost of the raw material treatment; BCo_{Rmt} represents the baseline cost of the raw material treatment; $RUCo_{Pc}$ represents the relative unit cost of the pellet densifying process; $RMCt$ represents the raw material content; $RUED_{Pf}$ represents the relative unit energy density of the pellet; ED_{Pf} represents the energy density of the pellet; BED represents the baseline energy density; and $RUECo_{Pf}$ represents the relative unit energy cost of the pellet.

Table 4 reports the relative unit cost of the raw material processing. Despite the low price when using the JB as the main raw material, its relative unit cost was the highest among raw materials as it was more difficult to crush in the pre-treatment stage and required a small particle size. In contrast, taking the JBC as the secondary raw material and CC and BM as additives resulted in lower relative unit costs due to the ease of crushing and the required high particle size.

Table 4. Relative unit costs of raw material processing.

Raw Materials	Purchase Price (yuan)	Purchase Specifications (g)	Unit Price (Yuan/g)	Relative Content (%)	Raw Materials Relative Unit Cost
JB(30) [a]	3.00	500	0.006	24.59	1.000
JB(80)	3.00	500	0.006	19.82	1.241
JBC	32.04	2500	0.013	98.69	0.532
CC	11.39	2800	0.004	79.69	0.209
BM	10.32	1000	0.010	98.90	0.428

[a] JB(30): 30 mesh pruned jujube tree branches; JB(80): 80 mesh pruned jujube tree branches; JBC: The charcoal of the pruned jujube tree branches; CC: Coco coir; BM: Bone meal.

Table 5 details the relative cost per unit energy of different pellets. The addition of the JBC and additives reduced the cost of the pellet formation, while increasing the energy density of the fuel. The relative cost per unit energy of the pellet, with the 80 mesh JB particles (JB(80)) as the primary raw material, was reduced by 52.50% with the addition of the JBC and CC; the equivalent reduction with 30 mesh JB particles (JB(30)) as the principle raw material and JBC and BM as the additives was 46.00%.

Table 5. Relative cost per unit energy of different pellets.

Sample	Content					Relative Unit Cost	Relative Unit Energy Density	Relative Cost per Unit Energy
	JB(30)	JB(80)	JBC	CC	BP			
JB100(80) [a]	0.00	1.00	0.00	0.00	0.00	1.24	1.03	1.20
JB63(80)-C37	0.00	0.63	0.37	0.00	0.00	0.98	1.61	0.61
JB56(80)-C37-CC	0.00	0.56	0.37	0.07	0.00	0.91	1.58	0.57
JB100(30)	1.00	0.00	0.00	0.00	0.00	1.00	1.00	1.00
JB69(30)-C31	0.69	0.00	0.31	0.00	0.00	0.85	1.54	0.55
JB65(30)-C31-BM	0.65	0.00	0.31	0.00	0.04	0.83	1.53	0.54

[a] The samples are named in the form of JB + number 1 + (number 2) + C + number 3 + additive type. For example, JB56(80)-C37-CC means that the sample consists of 80 mesh JB with a mass fraction of 56 wt%, and JBC with a mass fraction of 37 wt%, with CC as the additive.

The results reveal the improved performance of the formulated pellets compared to the raw biomass densified pellets in terms of the storage and transportation characteristics and combustion heat, as well as the lower production costs.

However, we would like to mention that the cost estimation presented here is based on the data from this bench-scale study. For further industrialization, more rigorous analyses using data from a large quantity of practical cases should be carried out.

4. Conclusions

Biomass-based pellet is an important contributor to the development of alternative fuels, and plays an indispensable role in the promotion of renewable energy. In this study, the formulation of jujube tree (JB)-based pellet using pruned branches was optimized to improve the overall quality. The main conclusions are as follows:

(1) The optimized formulation of JB, JBC, CC or BM in proportion can effectively improve the combustion performance and simultaneously provide the storage and transportation performance.

(2) Cost estimation showed that the two optimized formulations have significant advantages in terms of relative cost per unit energy compared to pellets made from JB alone; therefore, they have potential for commercial applications.

(3) The results revealed that, in the co-densified pellet, the primary raw materials mainly contributed to forming the pellet framework, the secondary raw materials mainly contributed to improving the fuel characteristics, and the additives mainly contributed to enhancing the storage and transportation performance. All these components were found to be critical for complementarily forming the high-quality pellet.

The limitations of this study are: (1) Only the formula parameters were optimized, but the effect of process parameters on pellet quality was not studied; (2) Due to the lack of relevant data, it was impossible to accurately calculate the real cost of pellets. Therefore, it is recommended that future research considers the following two aspects: (1) Considering the influence of process parameters such as pressure, temperature and mold size on pellet quality, in order to further optimize the formula parameters and process parameters; (2) The life cycle assessment method is adopted to analyze the optimization results in order to provide more effective data support for the industrial production of pellet.

Author Contributions: Conceptualization, K.K.; methodology, K.K., Y.L. and W.W.; software, Y.L.; validation, W.W. and Y.L.; formal analysis, K.K. and W.W.; investigation, Y.L.; resources, Y.L.; data curation, Y.L.; writing—original draft preparation, Y.L.; writing—review and editing, K.K. and W.W.; visualization, Y.L.; supervision, K.K. All authors have read and agreed to the published version of the manuscript.

Funding: This research was funded by Natural Science Special Project of Shaanxi Education Department (20JK0499) and Natural Science Basic Research Program of Shaanxi (2020JQ-243).

Data Availability Statement: The is available upon reasonable request from the corresponding author.

Acknowledgments: We thank Longtao Mu and Xiuqing Wang for provide valuable suggestions on this research, and thank Wei Xia, Yan Zhang, Jie Ma and Jingyu Zhang for their help in the tests.

Conflicts of Interest: The authors declare no conflict of interest.

References

1. Martins, F.; Felgueiras, C.; Smitkova, M.; Caetano, N. Analysis of Fossil Fuel Energy Consumption and Environmental Impacts in European Countries. *Energies* **2019**, *12*, 964. [CrossRef]
2. Rehman, A.; Rauf, A.; Ahmad, M.; Chandio, A.A.; Deyuan, Z. The effect of carbon dioxide emission and the consumption of electrical energy, fossil fuel energy, and renewable energy, on economic performance: Evidence from Pakistan. *Environ. Sci. Pollut. Res.* **2019**, *26*, 21760–21773. [CrossRef] [PubMed]
3. D'Adamo, I.; Morone, P.; Huisingh, D. Bioenergy: A Sustainable Shift. *Energies* **2021**, *14*, 5661. [CrossRef]
4. Xia, T.; Ji, Q.; Zhang, D.; Han, J. Asymmetric and extreme influence of energy price changes on renewable energy stock performance. *J. Clean. Prod.* **2019**, *241*, 118338. [CrossRef]

5. Civitarese, V.; Acampora, A.; Sperandio, G.; Assirelli, A.; Picchio, R. Production of Wood Pellets from Poplar Trees Managed as Coppices with Different Harvesting Cycles. *Energies* **2019**, *12*, 2973. [CrossRef]

6. Danish; Wang, Z. Does biomass energy consumption help to control environmental pollution? Evidence from BRICS countries. *Sci. Total Environ.* **2019**, *670*, 1075–1083. [CrossRef] [PubMed]

7. Monarca, D.; Cecchini, M.; Colantoni, A. Plant for the production of chips and pellet: Technical and economic aspects of a case study in central Italy. In Proceedings of the International Conference on Computational Science and Its Applications, Santander, Spain, 20–23 June 2011; pp. 296–306.

8. Tursi, A. A review on biomass: Importance, chemistry, classification, and conversion. *Biofuel Res. J.* **2019**, *6*, 962–979. [CrossRef]

9. Picchi, G.; Lombardini, C.; Pari, L.; Spinelli, R. Physical and chemical characteristics of renewable fuel obtained from pruning residues. *J. Clean. Prod.* **2018**, *171*, 457–463. [CrossRef]

10. Carter, E.; Shan, M.; Zhong, Y.; Ding, W.; Zhang, Y.; Baumgartner, J.; Yang, X. Development of renewable, densified biomass for household energy in China. *Energy Sustain. Dev.* **2018**, *46*, 42–52. [CrossRef]

11. Sun, X.; Wei, X.; Zhang, J.; Ge, Q.; Liang, Y.; Ju, Y.; Zhang, A.; Ma, T.; Fang, Y. Biomass estimation and physicochemical characterization of winter vine prunings in the Chinese and global grape and wine industries. *Waste Manag.* **2020**, *104*, 119–129. [CrossRef] [PubMed]

12. Alves, C.A.; Vicente, E.D.; Evtyugina, M.; Vicente, A.; Pio, C.; Amado, M.F.; Mahía, P.L. Gaseous and speciated particulate emissions from the open burning of wastes from tree pruning. *Atmos. Res.* **2019**, *226*, 110–121. [CrossRef]

13. Nematian, M.; Keske, C.; Ng'Ombe, J.N. A techno-economic analysis of biochar production and the bioeconomy for orchard biomass. *Waste Manag.* **2021**, *135*, 467–477. [CrossRef] [PubMed]

14. Kazimierski, P.; Hercel, P.; Suchocki, T.; Smoliński, J.; Pladzyk, A.; Kardaś, D.; Łuczak, J.; Januszewicz, K. Pyrolysis of Pruning Residues from Various Types of Orchards and Pretreatment for Energetic Use of Biochar. *Materials* **2021**, *14*, 2969. [CrossRef] [PubMed]

15. Pantaleo, A.; Villarini, M.; Colantoni, A.; Carlini, M.; Santoro, F.; Hamedani, S.R. Techno-Economic Modeling of Biomass Pellet Routes: Feasibility in Italy. *Energies* **2020**, *13*, 1636. [CrossRef]

16. Brand, M.A.; Jacinto, R.C. Apple pruning residues: Potential for burning in boiler systems and pellet production. *Renew. Energy* **2020**, *152*, 458–466. [CrossRef]

17. Barbanera, M.; Lascaro, E.; Stanzione, V.; Esposito, A.; Altieri, R.; Bufacchi, M. Characterization of pellets from mixing olive pomace and olive tree pruning. *Renew. Energy* **2016**, *88*, 185–191. [CrossRef]

18. Riva, L.; Wang, L.; Ravenni, G.; Bartocci, P.; Buø, T.V.; Skreiberg, Ø.; Fantozzi, F.; Nielsen, H.K. Considerations on factors affecting biochar densification behavior based on a multiparameter model. *Energy* **2021**, *221*, 119893. [CrossRef]

19. Kang, K.; Qiu, L.; Sun, G.; Zhu, M.; Yang, X.; Yao, Y.; Sun, R. Codensification technology as a critical strategy for energy recovery from biomass and other resources—A review. *Renew. Sustain. Energy Rev.* **2019**, *116*, 109414. [CrossRef]

20. Niu, Y.; Tan, H.; Hui, S. Ash-related issues during biomass combustion: Alkali-induced slagging, silicate melt-induced slagging (ash fusion), agglomeration, corrosion, ash utilization, and related countermeasures. *Prog. Energy Combust. Sci.* **2016**, *52*, 1–61. [CrossRef]

21. Zhou, Y.; Zhang, Z.; Zhang, Y.; Wang, Y.; Yu, Y.; Ji, F.; Ahmad, R.; Dong, R. A comprehensive review on densified solid biofuel industry in China. *Renew. Sustain. Energy Rev.* **2016**, *54*, 1412–1428. [CrossRef]

22. Kang, K.; Qiu, L.; Zhu, M.; Sun, G.; Wang, Y.; Sun, R. Codensification of Agroforestry Residue with Bio-Oil for Improved Fuel Pellets. *Energy Fuels* **2018**, *32*, 598–606. [CrossRef]

23. Kang, K.; Zhu, M.; Sun, G.; Qiu, L.; Guo, X.; Meda, V.; Sun, R. Codensification of Eucommia ulmoides Oliver stem with pyrolysis oil and char for solid biofuel: An optimization and characterization study. *Appl. Energy* **2018**, *223*, 347–357. [CrossRef]

24. Whittaker, C.; Shield, I. Factors affecting wood, energy grass and straw pellet durability—A review. *Renew. Sustain. Energy Rev.* **2017**, *71*, 1–11. [CrossRef]

25. Guan, Y.; Tai, L.; Cheng, Z.; Chen, G.; Yan, B.; Hou, L. Biomass molded fuel in China: Current status, policies and suggestions. *Sci. Total Environ.* **2020**, *724*, 138345. [CrossRef]

26. Zhang, Y.; Zhang, Z.; Zhou, Y.; Dong, R. The Influences of Various Testing Conditions on the Evaluation of Household Biomass Pellet Fuel Combustion. *Energies* **2018**, *11*, 1131. [CrossRef]

27. Oguntunde, P.E.; Adejumo, O.A.; Odetunmibi, O.A.; Okagbue, H.I.; Adejumo, A.O. Data analysis on physical and mechanical properties of cassava pellets. *Data Brief* **2018**, *16*, 286–302. [CrossRef] [PubMed]

28. Gao, W.; Tabil, L.G.; Liu, Q.; Zhao, R.; Zhang, M. Experimental Study on the Effect of Solid-State Fermentation on Pellet Density and Strength of Corn Stover. *J. Biobased Mater. Bioenergy* **2019**, *13*, 840–847. [CrossRef]

29. Aller, D.; Bakshi, S.; Laird, D.A. Modified method for proximate analysis of biochars. *J. Anal. Appl. Pyrolysis* **2017**, *124*, 335–342. [CrossRef]

30. Zaini, I.N.; Novianti, S.; Nurdiawati, A.; Irhamna, A.R.; Aziz, M.; Yoshikawa, K. Investigation of the physical characteristics of washed hydrochar pellets made from empty fruit bunch. *Fuel Process. Technol.* **2017**, *160*, 109–120. [CrossRef]

31. Hu, Q.; Shao, J.; Yang, H.; Yao, D.; Wang, X.; Chen, H. Effects of binders on the properties of bio-char pellets. *Appl. Energy* **2015**, *157*, 508–516. [CrossRef]

32. Pradhan, P.; Arora, A.; Mahajani, S.M. Pilot scale evaluation of fuel pellets production from garden waste biomass. *Energy Sustain. Dev.* **2018**, *43*, 1–14. [CrossRef]

33. Kaliyan, N.; Morey, R.V. Factors affecting strength and durability of densified biomass products. *Biomass Bioenergy* **2009**, *33*, 337–359. [CrossRef]
34. Shah, M.A.; Khan, M.N.S.; Kumar, V. Biomass residue characterization for their potential application as biofuels. *J. Therm. Anal. Calorim.* **2018**, *134*, 2137–2145. [CrossRef]
35. Zheng, Y.; Tao, L.; Yang, X.; Huang, Y.; Liu, C.; Zheng, Z. Comparative study on pyrolysis and catalytic pyrolysis upgrading of biomass model compounds: Thermochemical behaviors, kinetics, and aromatic hydrocarbon formation. *J. Energy Inst.* **2019**, *92*, 1348–1363. [CrossRef]
36. Fan, M.; Li, C.; Sun, Y.; Zhang, L.; Zhang, S.; Hu, X. In Situ characterization of functional groups of biochar in pyrolysis of cellulose. *Sci. Total Environ.* **2021**, *799*, 149354. [CrossRef] [PubMed]
37. Apaydın-Varol, E.; Pütün, A.E. Preparation and characterization of pyrolytic chars from different biomass samples. *J. Anal. Appl. Pyrolysis* **2012**, *98*, 29–36. [CrossRef]
38. El-Hendawy, A.-N.A. Variation in the FTIR spectra of a biomass under impregnation, carbonization and oxidation conditions. *J. Anal. Appl. Pyrolysis* **2006**, *75*, 159–166. [CrossRef]
39. Zçimen, D.; Ersoy-Meriçboyu, A. Characterization of biochar and bio-oil samples obtained from carbonization of various biomass materials. *Renew. Energy* **2010**, *35*, 1319–1324. [CrossRef]
40. Cao, J.; Xiao, G.; Xu, X.; Shen, D.; Jin, B. Study on carbonization of lignin by TG-FTIR and high-temperature carbonization reactor. *Fuel Process. Technol.* **2013**, *106*, 41–47. [CrossRef]
41. El-Sayed, S.A.; Ismail, M.A.; Mostafa, M.E. Thermal decomposition and combustion characteristics of biomass materials using TG/DTG at different high heating rates and sizes in the air. *Environ. Prog. Sustain. Energy* **2019**, *38*, 13124. [CrossRef]
42. Kok, M.V.; Ozgur, E. Characterization of lignocellulose biomass and model compounds by thermogravimetry. *Energy Sources Part A Recover. Util. Environ. Eff.* **2017**, *39*, 134–139. [CrossRef]
43. Kong, L.; Xiong, Y.; Liu, T.; Tu, Y.; Tian, S.; Sun, L.; Chen, T. Effect of fiber natures on the formation of "solid bridge" for preparing wood sawdust derived biomass pellet fuel. *Fuel Process. Technol.* **2016**, *144*, 79–84. [CrossRef]
44. Chen, W.-H.; Peng, J.; Bi, X.T. A state-of-the-art review of biomass torrefaction, densification and applications. *Renew. Sustain. Energy Rev.* **2015**, *44*, 847–866. [CrossRef]
45. Peng, J.H.; Bi, H.T.; Lim, C.J.; Sokhansanj, S. Study on Density, Hardness, and Moisture Uptake of Torrefied Wood Pellets. *Energy Fuels* **2013**, *27*, 967–974. [CrossRef]
46. Duca, D.; Riva, G.; Pedretti, E.F.; Toscano, G. Wood pellet quality with respect to EN 14961-2 standard and certifications. *Fuel* **2014**, *135*, 9–14. [CrossRef]
47. Rumpf, H. The strength of granules and agglomerates. In *Agglomeration, Proceedings of the First International Symposium on Agglomeration, Philadelphia, PA, USA, 12–14 April 1961*; John Wiley & Sons: New York, NY, USA, 1962; pp. 379–418.
48. Grover, P.; Mishra, S. *Biomass Briquetting: Technology and Practices*; Food and Agriculture Organization of the United Nations: Bangkok, Thailand, 1996; Volume 46.
49. Chen, Y.; Liu, B.; Yang, H.; Yang, Q.; Chen, H. Evolution of functional groups and pore structure during cotton and corn stalks torrefaction and its correlation with hydrophobicity. *Fuel* **2014**, *137*, 41–49. [CrossRef]

Article

Biogas Potential Assessment of the Composite Mixture from Duckweed Biomass

Alexander Chusov, Vladimir Maslikov, Vladimir Badenko *, Viacheslav Zhazhkov, Dmitry Molodtsov and Yuliya Pavlushkina

Institute of Civil Engineering, Higher School of Hydraulic and Power Engineering Construction, Peter the Great St. Petersburg Polytechnic University, 195251 St. Petersburg, Russia; chusov_an@spbstu.ru (A.C.); maslikov_vi@spbstu.ru (V.M.); zhazhkov@spbstu.ru (V.Z.); molodtsov_dv@spbstu.ru (D.M.); pavlushkina_yu@spbstu.ru (Y.P.)
* Correspondence: badenko_vl@spbstu.ru

Abstract: The article presents the research results of anaerobic digestion processes in bioreactors of composite mixtures based on initial and residual biomass of Lemna minor duckweed and additives: inoculum (manure), food waste, and spent sorbents to determine biogas potential (biogas volume, methane content). Duckweed Lemna minor, which is widespread in freshwater reservoirs, is one of the promising aquatic vegetation species for energy use. Residual biomass is obtained by chemically extracting valuable components from the primary product. The purpose of the research was to evaluate the possibility of the energy potential of residual biomass of Lemna minor to reduce the consumption of fossil fuels and reduce greenhouse gas emissions. This is in line with the International Energy Agency (IEA) scenarios for the reduction of environmental impact. The obtained results confirm the feasibility of using this type of waste for biogas/biomethane production. The recommendations on the optimal composition of the mixture based on the residual biomass of Lemna minor, which will allow for an increase in biogas production, are given. The obtained data can be used in the design of bioreactors.

Keywords: biogas potential; methane content; composite mixture; bioreactors; common duckweed; residual biomass; organic waste; circular economy

Citation: Chusov, A.; Maslikov, V.; Badenko, V.; Zhazhkov, V.; Molodtsov, D.; Pavlushkina, Y. Biogas Potential Assessment of the Composite Mixture from Duckweed Biomass. *Sustainability* **2022**, *14*, 351. https://doi.org/10.3390/su14010351

Academic Editors: Idiano D'Adamo and Piergiuseppe Morone

Received: 28 November 2021
Accepted: 27 December 2021
Published: 29 December 2021

Publisher's Note: MDPI stays neutral with regard to jurisdictional claims in published maps and institutional affiliations.

1. Introduction

The technology of anaerobic decomposition of wet biomass with the production of biogas, consisting of 55–60% methane, has been widely used to reduce the consumption of fossil organic fuels and greenhouse gas emissions in many countries around the world (including Russia) [1]. Among the potential energy forms to be derived from biowastes, biogas is of great interest [2]. The processing of biogas into biomethane has become a subject of increased technical interest, as biomethane is expected to be introduced into the national gas grid or used as a vehicle fuel. Moreover, the use of biogas derived from organic matter can support the energy transition [3]. The production and usage of biomethane can provide new opportunities for efficient Circular Economy, but the delay in this could cause significant economic losses [2].

Recently, the use of various aquatic plant organisms as an available raw material for biogas/biofuel production has been recognized as promising [4]. The processing of duckweed has been included in the list of promising pathways for biogas/biofuel production. This property is attributed to its simple harvesting method and high protein or starch content, depending on its species and growing environment [5]. One of the promising species for energy use is duckweed (Lemna minor), the most common representative of higher aquatic plants in freshwater bodies [6,7]. This plant has unique properties: short life cycle, high biological productivity and growth rate (biomass doubles within 2–3 days), wide spreading in various climatic zones of the globe, and undemanding to the quality of the water environment (used for wastewater treatment) [8,9]. In natural water bodies, the productivity of the duckweed is 0.7–1.0 kg of biomass from 1 m^2 of surface. Growing duckweed in artificial conditions is not difficult [10,11].

The biochemical composition of duckweed in terms of the amount of nutrients is not inferior to cereals. The biomass of the duckweed contains about 25.8% protein (twice as much as in cereals), 4.7% fat, and 24.6% fiber (11 times more than cereals). The biomass of the duckweed contains starch, organic nitrogen in the form of protein, and free amino acids, which are useful for biogas production [12,13]. Biomass is used to produce food, medicines, fodder, bioenergy, etc. Lemna minor can be considered as a universal aquatic plant [7,10,14–16]. The widespread use of duckweed will contribute to the challenge of meeting the growing biomass requirements of modern society and will protect the environment by removing excess nutrients and heavy metals from surface water and wastewater [4,17]. These unique characteristics of duckweed make it a technical and medicinal raw material for the extraction of valuable components, such as proteins, lipids, pectins, etc. [18,19].

Publications on the energy potential of duckweed biomass mainly focus on bioethanol production [20–23]. Relatively few research papers concern the possibility of using duckweed for biogas production [24–26]. It should be noted that these reports deal with the initial biomass of Lemna minor, which is not always economically efficient.

The aim of this research is a comparative analysis of the biogas potential of the initial and residual biomass of Lemna minor duckweed, as well as composite mixtures based on residual biomass and additives: inoculum, food waste, and waste sorbents.

2. Materials and Methods

2.1. Laboratory Research

The initial biomass of the duckweed was collected from the water body in the Leningrad Region in the north-west of the Russian Federation [27]. Residual duckweed biomass is formed after the extraction of pectin substances by hydrolysis. This process takes place in a citric acid solution at pH 1–2 and at 90 °C for 2 h. Then, the plant material is separated from the solution, and the residual biomass is used as a component of the composite mixture for fermentation.

The sorbents are made from carbonized residual biomass of duckweed, with the addition of a chitosan solution as a binder to produce sorbent granules. Granular sorbents were used for the extraction of cadmium, zinc, and copper ions from model solutions [28]. The use of waste sorbents as fermentation additives can solve the problem of their utilization. Fresh cow manure with a moisture content of 82% and an organic carbon content of 92% was used as an inoculum.

The laboratory experiment to assess the biogas potential of prototypes of organic substrates allows a relatively accessible method, with the required accuracy to simulate the processes of biodegradation, considering the influence of various factors (physical, chemical, biological, etc.) [29–31].

2.2. Conducting the Experiment

The program of laboratory experiments included the preparation of composite mixtures based on the initial and residual biomass of the duckweed for addition to bioreactors:

- Initial duckweed biomass + inoculum
- Residual duckweed biomass + inoculum
- Residual duckweed biomass + inoculum + food waste
- Residual duckweed biomass + inoculum + food waste + waste sorbent
- Inoculum + waste sorbent
- Inoculum (control sample).

Additives from food waste and waste sorbents were used to assess the impact on anaerobic digestion intensity and biogas production.

For each component of the composite mixture, before loading into the bioreactors, the mass fraction of moisture was determined by the drying method, and the content of organic carbon was determined by the calcination method (Table 1). The moisture content of the samples was determined using an Ohaus MB35 moisture analyzer. The temperature of the drying process was maintained at 105 °C. Drying was carried out automatically to a constant value of the sample mass. Then, the content of organic carbon in the samples was determined by weighing before and after calcining in a PT200 muffle oven. Calcination was carried out at 550 °C for 120 min [32]. The data obtained were used to calculate the mass ratios of the components of composite mixtures based on duckweed.

Table 1. Parameters of composite mixtures components.

Component	Organic Carbon, % of Total Carbon	Moisture, %
Initial biomass of Lemna minor duckweed	98.0	16.20
Residual biomass of Lemna minor duckweed	93.0	6.44
Food waste	93.0	70.28
Inoculum (Fresh cow manure)	88.3	82.75
Waste sorbent	87.7	7.26

Eight bioreactors were prepared for the experiment. Data on the loading of bioreactors in terms of organic carbon in grams are presented in Table 2.

Table 2. Contents of composite mixture components (in terms of organic carbon in grams).

Component	Bioreactor No.							
	1	2	3	4	5	6	7	8
Initial biomass of Lemna minor duckweed	4							
Residual biomass of Lemna minor duckweed		4	4	2	4	2		
Food waste			4	2	4	2		
Waste sorbent					2	4		4
Inoculum (Fresh cow manure)	4	4	4	2	4	4	4	4
Total	8	8	12	6	14	12	4	8

2.3. Laboratory Setup

To assesses the biogas potential of composite mixtures based on the initial and residual duckweed biomass, a laboratory setup was created. Bioreactors with a volume of 1 L were placed in a thermobox with a constant temperature. The laboratory setup included a discrete mode of anaerobic digestion, in which the bioreactors were loaded with the composite mixture only at the beginning of the process. The duration of the experiments ranged from 35 to 50 days. The laboratory setup layout for testing laboratory samples of the composite mixtures is shown in Figure 1.

Samples of composite mixtures were loaded into bioreactors (Table 2). Then, 600 mL of filtered water was added to these containers. The bioreactors were blown with inert gas to create an anaerobic mode of organic substance decomposition. The bioreactors were connected to Ritter MilliGascounters using gas lines to determine the volume and intensity of biogas emission. The generated biogas was diverted into airtight gas bags with a 3 L capacity. The bioreactors were placed in the thermostatic box, which automatically maintained the constant temperature of 35 °C required for mesophilic fermentation. The bioreactors were equipped with additional gas lines to enable the connection of a gas analyzer for periodic monitoring of the biogas composition. The contents of the bioreactors were mixed daily to prevent floating crust formation on the liquid surface. For remote control of biodegradation processes and monitoring of the experiment in real time, an information-analytical complex was created. This information-analytical complex consisted of a personal computer connected to the control unit of gas meters by wires.

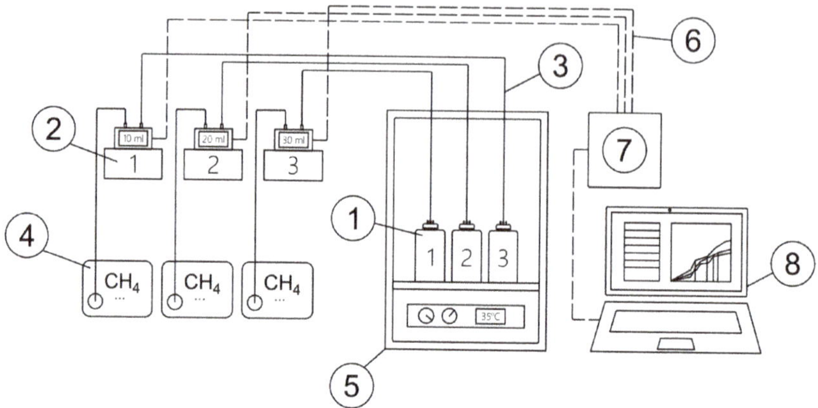

Figure 1. Laboratory setup layout: 1—bioreactors, 2—Ritter MilliGascounters, 3—gas lines, 4—gas bags, 5—thermostatic box, 6—wired connection to a personal computer, 7—control unit of the gas meters, 8—personal computer.

In the bioreactors, a discrete mode of anaerobic digestion was carried out, in which the loading of the bioreactors with the samples under study was carried out only once at the beginning of the experiment. The experiment was carried out until the end of biogas emission from the bioreactors.

2.4. Process Parameters Monitoring

The hydrogen pH value of the liquid phase was periodically controlled. In the case of low pH value and stoppage of gas emission, a buffer solution of baking soda (10%) was added to bioreactors to increase the pH value to 6.5–7.0.

During the experiment, a scheme for monitoring the component composition of the biogas was used directly in the bioreactor (Figure 2) using the portable gas analyzer GA2000 Plus [32]. Measurements were taken at least once a week. This made it possible to carry out the necessary measurements during the experiment without affecting the component composition of the biogas.

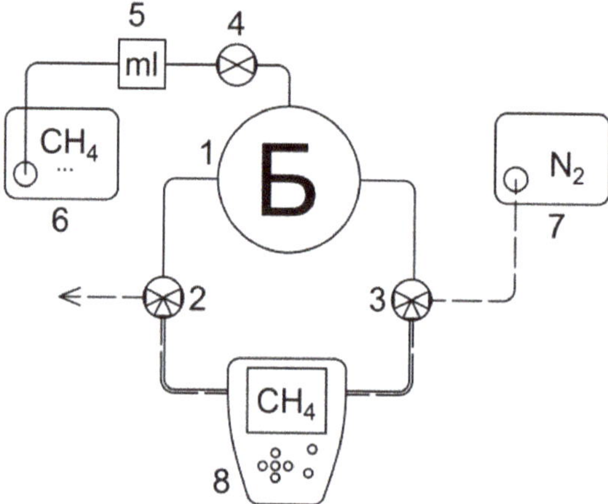

Figure 2. The scheme of connecting and measuring the component composition of biogas with the GA2000 Plus gas analyzer: 1—bioreactor; 2,3—three-way valves; 4—two-way valve; 5—gas meter; 6,7—gas bags; 8—gas analyzer.

3. Results and Discussion

The biogas-specific emission at biodegradation of the initial and residual biomass of *Lemna minor* duckweed is shown in Figure 3.

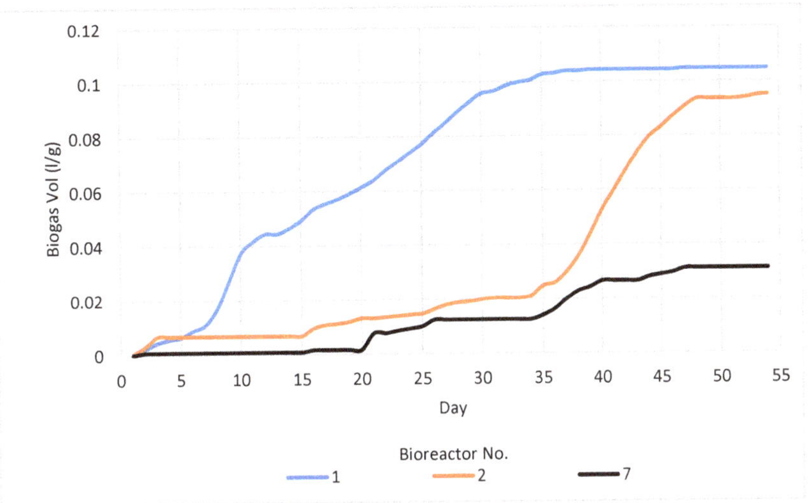

Figure 3. Biogas emissions from bioreactors.

The analysis of the presented graph shows that an intensive biogas yield in bioreactor No. 1 (loaded the initial duckweed biomass 50% + the inoculum 50%) was observed from 7 to 35 days during the 50-day experiment; after that, the gas emission practically stopped. The biogas emission on day 35 was 0.103 L/g of organic carbon.

The biogas emission from bioreactor No. 2 (loaded with the residual duckweed biomass 50% + the inoculum 50%) during the same period was four times less, 0.025 L/g of organic carbon, which could be explained by the presence of inhibitory impurities in the sample. The process of intensive biogas production was observed from day 34 to 48 of the experiment. On the 48th day of the experiment, the biogas emission was 0.093 L/g of organic carbon.

There was almost no biogas emission from bioreactor No. 7, loaded only with the inoculum, during the first 20 days. A slight increase in biogas emission was observed from 20 to 48 days. The biogas emission on the 48th day was 0.031 L/g of organic carbon.

The changes in methane concentration in the biogas during fermentation of the initial and residual duckweed biomass are shown in Figure 4.

The analysis of the graphs shows that the stable stage of methanogenesis was observed in bioreactor No. 1 on day 12, characterized by high methane concentration (35%), and a carbon dioxide content of 20.1%. In the following period, the methane concentration increased and reached its maximum on day 35 (45.2%).

The slow growth of methane concentration (from 0.7% to 12%) was observed in bioreactor No. 2 during the first 35 days, with a carbon dioxide content of 15.5–16.9%. Then, the growth rate increased, and by 48 days, the methane concentration reached 45%.

It should be noted that the biodegradation modes of the initial and residual duckweed biomass are very different. In case of the initial biomass, the experiment time was 30–35 days. In case of the residual biomass, the duration of the experiment was increased (up to 50 days), as it took at least 30 days to start a stable biodegradation process.

The active stage of methanogenesis in fermentation of residual biomass started on day 34, when the intensity of the biogas emission and methane content increased.

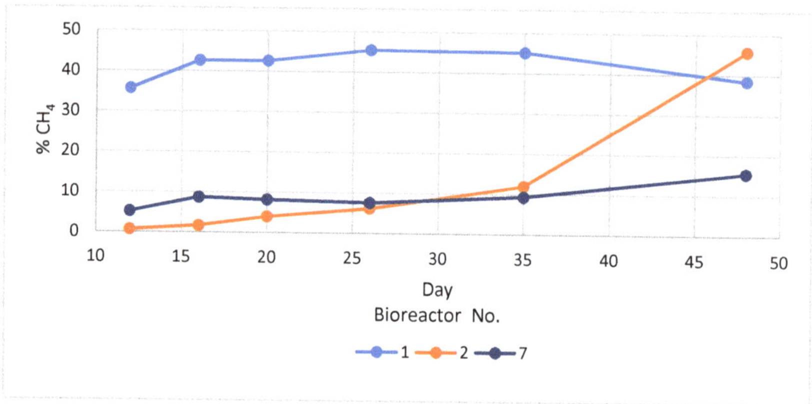

Figure 4. Methane concentration in bioreactors.

Figure 5 shows the specific emissions of biogas from bioreactors Nos. 2–6, loaded with composite mixtures based on the residual biomass of the duckweed Lemna minor, and from the control bioreactors Nos. 7–8 (Table 2).

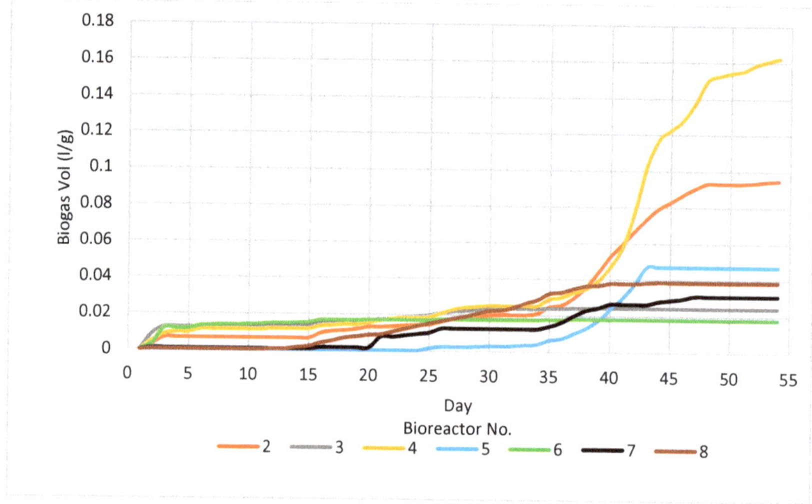

Figure 5. The biogas-specific emissions from bioreactors.

Despite the percentage of residual biomass and additives, the biodegradation processes during the first 35 days were slow. On the 48th day of the experiment, the highest specific biogas yield was observed in bioreactor No. 4 (0.16 L/g of organic carbon), which had the following composition mixture: residual duckweed biomass (33.3%), inoculum (33.3%), food waste (33.3%); the total content of organic carbon was 6 g.

The methane concentration changes in bioreactors Nos. 2–8 are shown in Figure 6.

The highest methane concentration for 48 days was also observed in bioreactor No. 4 (49.8%). Therefore, the optimal composition of the compositional mixture, which provided the greatest biogas potential (0.16 L/g of organic carbon, 49.8% of methane) was achieved by the following ratio: residual duckweed biomass (33.3%), inoculum (33.3%), food waste (33.3%).

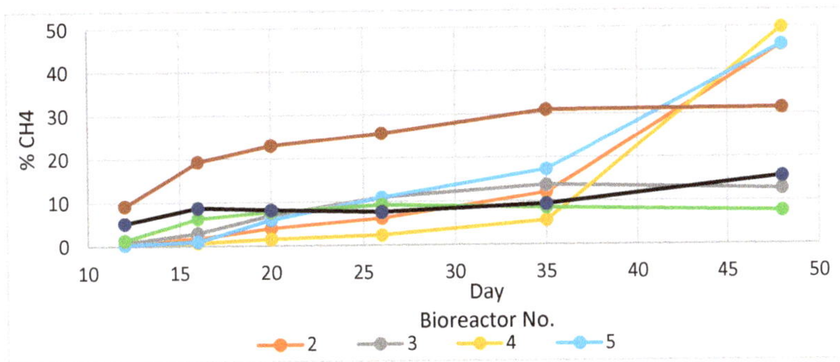

Figure 6. Methane concentration in bioreactors.

Directions for future research include the search for methods to accelerate the beginning of the period of intensive biogas emission during the decomposition of composite mixtures based on the residual biomass of Lemna minor. This direction of research is promising not only for the residual duckweed biomass, which is confirmed by the results of studies, for example, as presented in [33] for corn stover waste biomass.

The most challenging task, of course, is the choice of bio-substrates for biogas production. The use of residual biomass for energy purposes most fully complies with the principles of efficient Circular Economy. In this context, it is essential to broaden and deepen the analyses of bio-energy materials that could be used to help to achieve sustainability objectives [34]. Thus, in the future, this work will be also expanded in terms of research on the energy potential of other aquatic plant residual biomass, formed in large volumes after the extraction of valuable components.

4. Conclusions

The results obtained allowed us to identify the integral effect of the extraction of valuable components from the primary biomass. Prediction of the biogas potential of the residual biomass of Lemna minor duckweed without experimental studies was difficult because, on the one hand, the chemical treatment of the primary biomass causes partial destruction of a number of cellular structures that are difficult to degrade, and on the other hand, causes an inhibitory effect and a decrease in the content of organic matter in the residual biomass.

It was determined that when the bioreactors were loaded with the same amount (in terms of organic carbon) of the primary and residual biomass of Lemna minor duckweed, the highest specific biogas yield from the residual duckweed biomass was slightly lower, by about 9% (does not exceed the error of similar experiments), than from the primary biomass (with a high methane content of about 50%).

The optimal composition of the mixture was determined based on the residual biomass of the duckweed Lemna minor, which provides the highest specific yield of biogas (0.16 L/g carbon).

Considering the possible scale of Lemna minor duckweed processing and a significant amount of waste generation, the use of residual biomass for biogas production will contribute to sustainable development and compliance with the goals of efficient Circular Economy.

The modes of biodegradation of the primary and residual biomass of the duckweed Lemna minor differed sharply. In the case of primary biomass, the experiment took 30–35 days. When using the residual biomass, the duration of the experiment increased (up to 50 days), because it took at least 30 days to start a stable biodegradation process. This must be considered to optimize the specific loading of bioreactors and calculate their parameters.

Author Contributions: Conceptualization, A.C., V.M. and V.B.; methodology, V.M. and D.M.; software, D.M.; validation, V.Z. and Y.P.; formal analysis, A.C., V.M. and V.B.; investigation, V.Z. and Y.P.; writing—original draft preparation, V.M. and D.M; writing—review and editing, A.C. and V.B.; visualization, D.M.; supervision, A.C.; project administration, A.C.; funding acquisition, A.C. and V.B. All authors have read and agreed to the published version of the manuscript.

Funding: This research was partially funded by the Ministry of Science and Higher Education of the Russian Federation under the strategic academic leadership program 'Priority 2030' (Agreement 075–15–2021–1333 dated 30 September 2021).

Institutional Review Board Statement: Not applicable.

Informed Consent Statement: Not applicable.

Data Availability Statement: Not applicable.

Conflicts of Interest: The authors declare no conflict of interest.

References

1. Bioenergy in the Russian Federation: Roadmap 2019–2030. Moscow. 2019. Available online: http://tp-bioenergy.ru/upload/file/dorozhnaya_karta_tp_bioehnergetika.pdf (accessed on 25 November 2021). (In Russian)
2. D'Adamo, I.; Falcone, P.M.; Huisingh, D.; Morone, P. A circular economy model based on biomethane: What are the opportunities for the municipality of Rome and beyond? *Renew. Energy* **2021**, *163*, 1660–1672. [CrossRef]
3. Piechota, G. Multi-step biogas quality improving by adsorptive packed column system as application to biomethane upgrading. *J. Environ. Chem. Eng.* **2021**, *9*, 105944. [CrossRef]
4. Zhang, K.; Zhang, F.; Wu, Y.R. Emerging technologies for conversion of sustainable algal biomass into value-added products: A state-of-the-art review. *Sci. Total Environ.* **2021**, *784*, 147024. [CrossRef] [PubMed]
5. Politaeva, N.; Badenko, V. Magnetic and electric field accelerate Phytoextraction of copper Lemna minor duckweed. *PLoS ONE* **2021**, *16*, e0255512. [CrossRef] [PubMed]
6. Djandja, A.; Prinsen, P.; Vuppaladadiyam, A.K.; Zhao, M.; Luque, R. A review on sustainable microalgae based biofuel and bioenergy production: Recent developments. *J. Clean. Prod.* **2018**, *181*, 42–59. [CrossRef]
7. Djandja, O.S.; Yin, L.; Wang, Z.; Guo, Y.; Zhang, X.; Duan, P. Progress in thermochemical conversion of duckweed and upgrading of the bio-oil: A critical review. *Sci. Total Environ.* **2021**, *769*, 144660. [CrossRef] [PubMed]
8. Tua, C.; Ficara, E.; Mezzanotte, V.; Rigamonti, L. Integration of a side-stream microalgae process into a municipal wastewater treatment plant: A life cycle analysis. *J. Environ. Manag.* **2021**, *279*, 111605. [CrossRef] [PubMed]
9. Ceschin, S.; Crescenzi, M.; Iannelli, M.A. Phytoremediation potential of the duckweeds Lemna minuta and Lemna minor to remove nutrients from treated waters. *Environ. Sci. Pollut. Res.* **2020**, *27*, 15806–15814. [CrossRef]
10. Baek, G.; Saeed, M.; Choi, H.K. Duckweeds: Their utilization, metabolites and cultivation. *Appl. Biol. Chem.* **2021**, *64*, 1–15. [CrossRef]
11. Iqbal, J.; Javed, A.; Baig, M.A. Growth and nutrient removal efficiency of duckweed (*Lemna minor*) from synthetic and dumpsite leachate under artificial and natural conditions. *PLoS ONE* **2019**, *14*, e0221755. [CrossRef]
12. Devlamynck, R.; de Souza, M.F.; Michels, E.; Sigurnjak, I.; Donoso, N.; Coudron, C.; Leenknegt, J.; Vermeir, P.; Eeckhout, M.; Meers, E. Agronomic and Environmental Performance of Lemnaminor Cultivated on Agricultural Wastewater Streams—A Practical Approach. *Sustainability* **2021**, *13*, 1570. [CrossRef]
13. Ekperusi, A.O.; Sikoki, F.D.; Nwachukwu, E.O. Application of common duckweed (*Lemna minor*) in phytoremediation of chemicals in the environment: State and future perspective. *Chemosphere* **2019**, *223*, 285–309. [CrossRef]
14. de Beukelaar, M.F.; Zeinstra, G.G.; Mes, J.J.; Fischer, A.R. Duckweed as human food. The influence of meal context and information on duckweed acceptability of Dutch consumers. *Food Qual. Prefer.* **2019**, *71*, 76–86. [CrossRef]
15. Arefin, M.A.; Rashid, F.; Islam, A. A review of biofuel production from floating aquatic plants: An emerging source of bio-renewable energy. *Biofuels Bioprod. Biorefin.* **2021**, *15*, 574–591. [CrossRef]
16. Appenroth, K.J.; Sree, K.S.; Bog, M.; Ecker, J.; Seeliger, C.; Böhm, V.; Lorkowski, S.; Sommer, K.; Vetter, W.; Tolzin-Banasch, K.; et al. Nutritional value of the duckweed species of the genus *Wolffia* (Lemnaceae) as human food. *Front. Chem.* **2018**, *6*, 483. [CrossRef]
17. Bokhari, S.H.; Mahmood-Ul-Hassan, M.; Ahmad, M. Phytoextraction of Ni, Pb and, Cd by duckweeds. *Int. J. Phytoremediat.* **2019**, *21*, 799–806. [CrossRef] [PubMed]
18. Dvoretsky, D.; Dvoretsky, S.; Peshkova, E.; Temnov, M. Optimization of the process of cultivation of microalgae chlorella vulgaris biomass with high lipid content for biofuel production. *Chem. Eng. Trans.* **2015**, *43*, 361–366. [CrossRef]
19. Zhao, X.; Moates, G.K.; Wellner, N.; Collins, S.R.A.; Coleman, M.J.; Waldron, K.W. Chemical characterisation and analysis of the cell wall polysaccharides of duckweed (*Lemna minor*). *Carbohydr. Polym.* **2014**, *111*, 410–418. [CrossRef]
20. Calicioglu, O.; Brennan, R.A. Sequential ethanol fermentation and anaerobic digestion increases bioenergy yields from duckweed. *Bioresour. Technol.* **2018**, *257*, 344–348. [CrossRef]

21. Nahar, K.; Sunny, S.A. Duckweed-based clean energy production dynamics (ethanol and biogas) and phyto-remediation potential in Bangladesh. *Model. Earth Syst. Environ.* **2020**, *6*, 1–11. [CrossRef]
22. Cui, W.; Cheng, J.J. Growing duckweed for biofuel production: A review. *Plant Biol.* **2015**, *17*, 16–23. [CrossRef] [PubMed]
23. Liu, Y.; Xu, H.; Yu, C.; Zhou, G. Multifaceted roles of duckweed in aquatic phytoremediation and bioproducts synthesis. *GCB Bioenergy* **2021**, *13*, 70–82. [CrossRef]
24. Ren, H.; Jiang, N.; Wang, T.; Mubashar Omar, M.; Ruan, W.; Ghafoor, A. Enhanced biogas production in the duckweed anaerobic digestion process. *J. Energy Resour. ASME* **2018**, *140*, 041805. [CrossRef]
25. Yadav, D.; Barbora, L.; Bora, D.; Mitra, S.; Rangan, L.; Mahanta, P. An assessment of duckweed as a potential lignocellulosic feedstock for biogas production. *Int. Biodeterior. Biodegrad.* **2017**, *119*, 253–259. [CrossRef]
26. Tonon, G.; Magnus, B.S.; Mohedano, R.A.; Leite, W.R.M.; da Costa, R.H.R.; Filho, P.B. Pretreatment of Duckweed Biomass, Obtained from Wastewater Treatment Ponds, for Biogas Production. *Waste Biomass Valorization* **2017**, *8*, 2363–2369. [CrossRef]
27. Smyatskaya, Y.; Politaeva, N.; Toumi, A.; Olshanskaya, L. Influence of extraction conditions on the recovery lipids extracted from the dry biomass of duckweed Lemna minor. In Proceedings of the International Scientific Conference on Energy, Environmental and Construction Engineering (EECE-2018), Saint-Petersburg, Russia, 19–20 November 2018; Volume 245, p. 18004. [CrossRef]
28. Politaeva, N.A.; Smyatskaya, Y.A.; Efremova, S.Y. Production of Sorbents from Residual Biomass of Chlorella Sorokiniana Microalgae and Lemna Minor Duckweed. *Chem. Petrol. Eng.* **2020**, *56*, 543–547. [CrossRef]
29. Wenjing, L.; Chao, P.; Lama, A.; Xindi, F.; Rong, Y.; Dhar, B.R. Effect of pre-treatments on biological methane potential of dewatered sewage sludge under dry anaerobic digestion. *Ultrason. Sonochem.* **2019**, *52*, 224–231. [CrossRef]
30. Ryue, J.; Lin, L.; Liu, Y.; Lu, W.; McCartney, D.; Dhar, B.R. Comparative effects of GAC addition on methane productivity and microbial community in mesophilic and thermophilic anaerobic digestion of food waste. *Biochem. Eng. J.* **2019**, *146*, 79–87. [CrossRef]
31. Maslikov, V.; Korablev, V.; Molodtsov, D.; Chusov, A.; Badenko, V.; Ryzhakova, M. Organization of Organic Waste Samples Tests for Biogas Potential Assessment. *Adv. Intell. Syst.* **2018**, *983*, 440–448. [CrossRef]
32. Chusov, A.; Maslikov, V.; Zhazhkov, V.; Pavlushkina, Y. Determination of biogas potential of residual biomass of microalgae Chlorella Sorokiniana. *IOP Conf. Ser. Earth Environ. Sci.* **2019**, *403*, 012225. [CrossRef]
33. Olugbemide, A.D.; Oberlintner, A.; Novak, U.; Likozar, B. Lignocellulosic Corn Stover Biomass Pre-Treatment by Deep Eutectic Solvents (DES) for Biomethane Production Process by Bioresource Anaerobic Digestion. *Sustainability* **2021**, *13*, 10504. [CrossRef]
34. D'Adamo, I.; Morone, P.; Huisingh, D. Bioenergy: A Sustainable Shift. *Energies* **2021**, *14*, 5661. [CrossRef]

Article

Strategies for the Mobilization and Deployment of Local Low-Value, Heterogeneous Biomass Resources for a Circular Bioeconomy

Fabian Schipfer [1,*], Alexandra Pfeiffer [2] and Ric Hoefnagels [3]

[1] Energy Economics Group, Technische Universität Wien, Gusshausstraße 25-29, 1040 Vienna, Austria
[2] Deutsches Biomasseforschungszentrum, 04347 Leipzig, Germany; alexandra.pfeiffer@dbfz.de
[3] Copernicus Institute of Sustainable Development, Universiteit Utrecht, 3584 CB Utrecht, The Netherlands; r.hoefnagels@uu.nl
* Correspondence: schipfer@eeg.tuwien.ac.at

Abstract: With the Bioeconomy Strategy, Europe aims to strengthen and boost biobased sectors. Therefore, investments in and markets of biobased value chains have to be unlocked and local bioeconomies across Europe have to be deployed. Compliance with environmental and social sustainability goals is on top of the agenda. The current biomass provision structures are unfit to take on the diversity of biomass residues and their respective supply chains and cannot ensure the sustainability of feedstock supply in an ecological, social and economical fashion. Therefore, we have to address the research question on feasible strategies for mobilizing and deploying local, low-value and heterogeneous biomass resources. We are building upon the work of the IEA Bioenergy Task40 scientists and their expertise on international bioenergy trade and the current provision of bioenergy and cluster mobilization measures into three assessment levels; the legislative framework, technological innovation and market creation. The challenges and opportunity of the three assessment levels point towards a common denominator: The quantification of the systemic value of strengthening the potentially last remaining primary economic sectors, forestry, agriculture and aquaculture, is missing. With the eroding importance of other primary economic sectors, including fossil fuel extraction and minerals mining, the time is now to assess and act upon the value of the supply-side of a circular bioeconomy. This value includes the support the Bioeconomy can provide to structurally vulnerable regions by creating meaningful jobs and activities in and strengthening the resource democratic significance of rural areas.

Keywords: bioeconomy strategy; regional development; residues; policy; market; technology; commoditization

Citation: Schipfer, F.; Pfeiffer, A.; Hoefnagels, R. Strategies for the Mobilization and Deployment of Local Low-Value, Heterogeneous Biomass Resources for a Circular Bioeconomy. *Energies* **2022**, *15*, 433. https://doi.org/10.3390/en15020433

Academic Editors: Idiano D'Adamo and Piergiuseppe Morone

Received: 4 December 2021
Accepted: 1 January 2022
Published: 7 January 2022

Publisher's Note: MDPI stays neutral with regard to jurisdictional claims in published maps and institutional affiliations.

1. Introduction

The European Bioeconomy Strategy [1] aims to "strengthen and boost biobased sectors". By definition, the bioeconomy includes "all primary production sectors that use and produce biological resources (agriculture, forestry, fisheries and aquaculture); and all economic and industrial sectors that use biological resources and processes to produce food, feed, bio-based products, energy and services. To be successful, the European bioeconomy needs to have sustainability and circularity at its heart."

Since the 1970s, industrial ecology and industrial metabolism discussions coin the term Circular Economy (C.E.), the C.E. has been used as a guideline in policy-making, especially in China and Europe. Today the C.E. is mainly attributed to electronic waste (see Circular Economy Action plan [2]) and recently also plastics (see Plastic Strategy [3]).

The Annex of the draft proposal for a European Partnership for a Circular Biobased Europe [4] argues why also a bioeconomy is inherently a C.E.; biobased sectors have CO_2 avoidance and retention, reduction, recycling and reuse of wastes and residues as its goals, all traits which have been primarily credited to the circular economy. The success of both the circular bioeconomy and the broader circular economy depend on a sustainable feedstock supply. However, shifting the respective primary economic sectors, i.e., the feedstock supply, to sustainable practices comes with considerable technical, societal and organizational challenges that have to be addressed [5,6].

The current bioenergy provision is mainly based on wood chips, wood pellets or first generation biofuel plantations [7]. The underlying resources are mobilized primarily for material services (e.g., construction wood, pulp and paper) [8], while first generation biofuel resource provision use similar production techniques and supply chains as agricultural production. As a result and in their review on bioenergy supply and demand scenarios and projections, Mandley et al. [9] stress a potential mismatch due to limited modelling and analysis of crucial conversion processes between fresh biomass and end-user services. A plethora of underutilized, non-commodity biomass resources is still not touched upon, which could become the feedstock basis for the circular bioeconomy of tomorrow. These resources can be categorized in energy crops-, forestry and agricultural residues, and biogenic waste [10–13]. However, they are diverse in, e.g., physical properties (energy density, moisture content, ash content but also contamination such as sand/plastic), origin (landscape management, residential garden/kitchen waste) and legal status (waste vs. resources/material). The current biomass provision structures are unfit to take on this diversity and cannot ensure the sustainability of feedstock supply in an ecological, social and economical fashion. Therefore, we have to address the research question on feasible strategies for mobilizing and deploying these local, low-value and heterogeneous biomass resources.

Thus, and for the present paper, we are building upon the work of the IEA Bioenergy Task40 scientists and their expertise on international bioenergy trade and the current provision of bioenergy. To address all sustainability dimensions, mobilization strategies have to respect planetary boundaries [12] and have to be financially viable and contribute to other societal goals. Especially for the provision of local and low-value biomass resources, this means supporting structurally weak and rural regions. The research focuses on the European Union concerning policies but is also inspired by technology- and market developments in the rest of the world.

2. Materials and Methods

This work is based on an extensive discussion on biomass mobilization strategies between International Energy Agency (IEA) Bioenergy Technology Collaboration Program (TCP) Task 40 scientists. The expertise of the authors and discussion participants undoubtedly defines the scope of the presented findings. Task40 initially focused on international bioenergy trade. However, the established supply-chain knowledge proved to be applicable to strategic questions about biomaterials as well (see, e.g., Schipfer et al. [14]). The international consortium specializes by now on the "deployment of biobased value chains" in support of a broader, circular bioeconomy. Systemic assessments, including the utilization of bioenergy, as, e.g., discussed for energy system models in Chang et al. [15], are increasing in spatial, temporal and sectoral resolutions. The IEA Bioenergy Task40 follows this zeitgeist by dedicating a task force to *Regional Transitions* studies. For the sake of this paper, we understand "region" as an area that could have its own characteristics or even administration. We refrain from setting a precise definition, but as a rule of thumb, "regions", "regional" and "local" could span from municipalities, the lowest local administrative unit to groups of districts, or the NUTS 3 level.

For this paper, the IEA Bioenergy Task40 experts focused on transferring and extending their knowledge on current bioenergy carrier provision structures to the local, low-value feedstock base of tomorrows circular bioeconomy. During the discussions within the Task force and based on previous works on mobilization strategies for bioenergy of lower spatial and sectoral resolution (e.g., Junginger et al. [7]), we collect information on respective current developments, barriers and opportunities. The discussion is further complemented by scientific literature on the identified topics and a collection of unpublished research- and development projects. The here presented collection does not claim completeness or indicates any ranking of importance. Instead, it aims at creating a coherent reference work on challenges and opportunities for novel biomass provision structures. It should be used to derive key concepts for follow-up scientific-, market- or patent research. To facilitate the analysis and discussion beyond the project, we cluster the topics into three categories; legislative framework, market structures and technological innovation (see Figure 1).

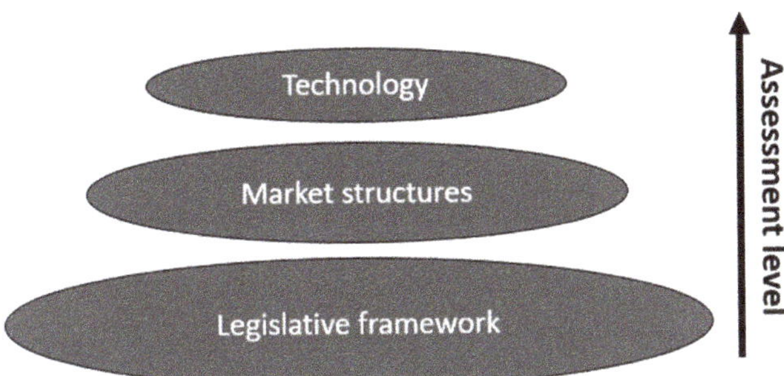

Figure 1. Biomass mobilization strategy categories. The arrow points in the direction of increasing assessment resolution. Source: own illustration.

This paper's Results and Discussion (Section 3) are structured following the outlined categories, starting with the lowest assessment level, highlighting top-down the current developments, opportunities, and barriers in the European legislative framework before zooming into the highest assessment level on bottom-up technological innovation mobilization strategies. The Results and Discussion section is completed with an analysis of biomass markets for energy and material use. The Conclusions section (Section 4) connects the different assessment levels back together and provides recommendations and limitations of the present study.

3. Results and Discussion

3.1. Legislative Framework for Biomass Mobilization

For the basic assessment level of biomass mobilization strategies, we focus on the European Union and its common legislative framework for the 27 Member States (M.S.). We first and foremost are interested in high-level documents labelled "strategy", "blueprint", "roadmap", or "action plan". Even though these terms lack clear definitions and unambiguity, they are often used sequentially (1) with a strategy outlining a general perspective for development, while (2) blueprints and roadmaps are frequently used to illustrate one or several development paths and timelines. (3) Action plans should include concrete steps or action points and preferably quantifiable goals. On an E.U. level, respective documents facilitate the non-juridical discussion of overall trends, even though outlined targets and measures are of non-binding characters eventually to be implemented in regulations and directives. E.U. regulations are binding in their entirety in all Member States/M.S., while directives are to be "transposed" into national laws of the M.S. In contrast, decisions (addressing particular States or organizations) and ordinances on a national, regional or sub-regional level are out of the scope of the present paper.

In the following sub-sections, we provide a top-down mapping of the legislative frameworks of high relevance for mobilizing local, low-value and heterogenous biomass. We explore the E.U. policy landscape (Section 3.1.1), international projects on regional mobilization strategies (Section 3.1.2) and how the novel concept of Multilevel governance tries to bridge local with E.U. governance (Section 3.1.3).

3.1.1. EU Policy Environment Affecting Regional Biomass Mobilisation

The European Green Deal lays down the strategy for a broad set of E.U. policies currently formulated and enacted between 2019 and 2024, building upon the existing policy framework [16]. To achieve climate neutrality by 2050, the Climate Target Plan proposes to cut greenhouse gas (GHG) emissions by 55% in 2030 through a combination of legislation on the Emission Trading System (ETS), Effort Sharing, and Land Use [17]. The current proposal on revising the Renewable Energy Directive includes the amendment of renewable energy targets to 40% by 2030, quantitative sector-specific renewable energy goals for buildings, transport, industry and district heating and the tightening of sustainability criteria for biomass [18]. Biodiversity protection in forests, GHG saving criteria for existing bioenergy installations (as small as 5 MWel), phase-out of electricity-only production from biomass, and enforced cascading principles are proposed by the European Commission.

Coherence with already existing and to-be-revised documents has to be ensured. The primary strategy (or action plan) on the European level for biomass mobilization can be seen in the updated E.U. bioeconomy strategy [1]. This document complements similar objectives to the 2012 bioeconomy strategy with main action areas, including deploying local bioeconomies rapidly across Europe. A Strategic Deployment Agenda (SDA) "for sustainable food and farming systems, forestry and bio-based production in a circular bioeconomy" was envisaged to be finalized by 2021 [1]. This "roadmap" will optimize "synergies between the Common Agricultural Policy (CAP), [maritime and] fisheries [policies], [the] Agricultural Fund for Rural Development (EAFRD), other European Structural and Investment Funds (ESIF)" and mobilize the agricultural European Innovation Partnership (EIP-AGRI). Furthermore, the Covid pandemic brought the relevance of healthy regions off the urban centers to the fore. Considerable increases in grants and loans are proposed in the NextGenerationEU package to be directed to recovery measures but also to rural development [19].

Furthermore, local bioeconomy development is supported for coastal (e.g., Blue Bioeconomy grants), urban (Urban Circular Bieoconomy Strategy funding) and rural areas (in national CAP strategic plans). The strategy also aims at piloting carbon farming initiatives to "make carbon sequestration and emission reduction a profitable farming/forestry activity". Finally, E.U. Bioeconomy policy support facilities (via the BIOEAST initiative) and a European Bioeconomy Forum for M.S. is initiated. Furthermore, and under the Green Deal, the Just Transition Mechanism (JTM), including the Just Transition Fund (JTF), the InvestEU "Just Transition" scheme, and the European Investment Bank (EIB) public sector loan facility will provide support to "reduce regional disparities and to address structural changes in the E.U." related to the transition towards climate neutrality. In addition, the public–private partnership for circular bioeconomy R&D&D (now CBE JU, former bio-based industries joint undertaking—BBI JU) with potential impacts on biomass mobilization has to be mentioned as an essential tool to foster innovation in regions to mobilize biomass for a circular bioeconomy.

Focusing on regional development, the E.U. Long-term Vision for Rural Areas aims at strengthening the provision of "food, homes, jobs, and essential ecosystems services". Proposed measures include a rural revitalization platform, R&I for rural communities, boosting digital connectivity and competencies, establishing carbon-sink focus areas and fostering rural entrepreneurship [20]. The Farm to Fork Strategy addresses sustainability throughout the life-cycle of our nutrient services, including "production, processing/distribution, consumption [and] food loss and waste" prevention [21]. The Forest Strategy for 2030 promotes a sustainable forest bioeconomy, including the use of wood-based resources but also ecotourism while "ensuring forest restoration and reinforced sustainable forest management for climate adaptation and forest resilience" [22]. The Biodiversity Strategy for 2030 aims at the same time to establish and extend an "EU-wide network of protected areas on land and at sea" and announce "binding nature restoration targets" [23].

More specifically, and based on the current CAP (2014–2020), farmers have to set aside a mandatory share of 5% of farmland for Ecological Focus Areas (EFAs), including "grasslands, hedges, buffer strips or nitrogen-fixing crops" [24]. Short rotation plantations (SRP), including short rotation coppice (SRC) and single-stemmed trees (SRF), count towards EFAs. However, implementation of these greening measures is still limited to about 50.000 hectares in Europe, with considerable shares in Sweden due to an unrelated willow plantation trend between 1986–1996 and some measures such as establishment grants in the U.K., Ireland and Germany [25].

This section provides a preliminary list and description of E.U. policies that should be considered when planning for the mobilization of local, low-value and heterogenous biomass feedstock. Strategically deploying these top-down resources is subject to regional, national and international efforts. The next section addresses the evolution of respective projects on regional biomass mobilization.

3.1.2. Regional Strategies Focusing on Regional Biomass Mobilization

Historically, the E.U. Biomass Action Plan from 2005 aimed at setting out "measures to increase the development of biomass energy from wood, wastes and agricultural crops by creating market-based incentives" [26]. To promote regional structures, the E.U. Biomass Action Plan was followed up by several regional biomass action plans, bioenergy action plans and regional development plans. These plans are often developed for specific regions, countries, and sometimes inter-regional partnerships by trans-disciplinary project consortia, including research and regional energy agencies, biomass associations, and regional stakeholders. Renown projects, their funding source and runtimes are listed in Table 1.

Table 1. Projects aiming at creating and implementing regional biomass action plans with multi-regional and international scopes. Source: own elaboration.

Funding Agency	Project Name	Project Years
IEE [1]	REGBIE+	2007–2009
EFRD [2]	4Biomass.eu	2009–2011
IEE [1]	Bioregions.eu	2010–2013
Interreg [3]	Bio-En-Area	2011–2015
IEE [1]	Biomass Policies	2013–2016
FP7 [4]	S2Biom	2013–2016
IEE [1]	Basis	2013–2016
Horizon2020 [5]	BioVill	2016–2019
Interreg [3]	Bio4Eco	2016–2020
Nordic Council of Ministries	Nordic Bioeconomy Prgrm	2018–2022
Horizon2020 [5]	BioEastsUp (initiative)	2019–2022

[1] Intelligent Energy Europe, [2] European Funds for Regional Development, [3] Innovation & Environment Regions of Europe Sharing Solutions, [4] 7th Framework Program for Research of the European Commission, [5] Horizon2020 Funding Program of the European Commission.

The selection of the projects listed in Table 1 is based on their multi-regional and international scope. Project consortia consists of partners from 6–13 Member States, including neighboring countries. Following up on the E.U. Biomass Action plan projects until 2013 mainly focused on creating action plans and action plan templates and improving regional policies for bioenergy uptake and market creation. Older, potentially fitting projects such as "Make-it-BE", "BioMob", "BioCLUS" and "Rok-FOR" are mentioned in the review on regional biomass planning by Kautto and Peck [27]. Still, information on these projects is insufficient for further analysis. Since 2013, sustainable and efficient use of biomass and the interaction between biomaterial, food and feed and bioenergy based on supply chain approaches is in the foreground. For this purpose, especially the "Biomass Policies" and the "S2Biom" projects mapped sustainable supply potentials. They are published as openly accessible and "updated harmonized datasets at a local, regional, national and pan-European level for EU28, Western Balkans, Moldova, Turkey and Ukraine" [28]. More recent projects focus on establishing "knowledge-bio hubs" and "bio villages", exchanging information and know-how between regions. The Nordic Council of Ministries representing Denmark, Finland, Iceland, Norway, Sweden, the Faroe Islands, Greenland and Åland furthermore established the "Nordic Bioeconomy Programme" to collaboratively improve the use of biogenic residues and to remain in a leading position with regard to regional bioeconomy development.

European regions concerned about structural losses in the primary economic sectors through the fossil fuel phase-out also show interest in strengthening local and regional bioeconomies. These regions include mainly coal mining since self-sufficiency rates of the E.U. are at 60% for coal, 16% for natural gas and only 5% for oil and petroleum products (in 2019) [29]. The Horizon 2020 project TRACER supports regions in Bulgaria, Germany, Greece, Czech Republic, Poland, Romania, Serbia, Ukraine and the United Kingdom in designing or re-designing their Research and Innovation strategies. The project "analyses the impact of the energy transition, in terms of social change, communities shrinking, migration, demographic ageing, poverty, high youth unemployment rates and participation to education and training" in intensive coal regions [30]. The mono-industrial character of these regions makes them specifically vulnerable to respective socioeconomic challenges and asks for dedicated measures, especially regarding re-skilling, job creation, a productive re-usage of the industrial landscape, investments in infrastructure and addressing legal issues related to land ownership. Shifting the focus from primary economic sectors to secondary or tertiary sectors might not be feasible for some of these regions; thus, coupling fossil-fuel phase-out with bioeocomy actions are promising strategies. A regional bioeconomy transition example is reported for the chemistry park "Schwarzheide" with leading research parties developing and pioneering in the deployment, together with BASF company and other plastics processing companies, of bioplastics and biodegradable synthetics [31].

The strengthening of primary economic sectors in these regions will have to focus on sustainable management of agriculture and forestry for a circular bioeconomy. The circular bioeconomy sectors will cover multiple services, including electricity, heat, chemicals, bio-based materials, food/feed and services on ecosystems and the carbon budget. Respective bioeconomy concepts stand in stark contrast to a direct substitution of coal with biomass for electricity-only at the same scales of incumbent coal-fired power plants. Subsidies for bioelectricity and electricity-only are already phased out, e.g., in the Netherlands to re-orientate limited biomass resource potentials to economic sectors, which are more challenging to abate [32]. An obligatory phase-out is proposed by the European Commission for all Member States, starting with 2026, except for Bioenergy Carbon Capture and Storage (BECCS) or plants located in a "region identified in a territorial just transition plan" [33].

In summary, we can highlight that numerous regional development and biomass action plans have been developed in dedicated projects over the last decades. The focus and scope of these plans co-evolve with the policy framework. In the last decade, the main driver was the provision of first generation biomass feedstocks for bioenergy and meeting renewable energy targets. Since then, the aim of these plans seems to have shifted to (1) more generic approaches, including quantitative feedstock and market potential assessments (2) with a more holistic view on different bioeconomy sectors. Regions dependent on coal mining now have the opportunity to tap into this evolution of action plans. Creating added value and opportunities for structurally weak and vulnerable regions has been an objective in the projects following up upon the E.U. biomass action plan of 2005. However, and with the enlargement of the E.U., barriers and opportunities multiplied that have to be eventually addressed based on multilevel governance frameworks.

3.1.3. Multilevel Governance for Biomass Mobilization

The E.U. is a valuable case study for biomass mobilization strategies due to its great variety in biomes, economic structures, governmental forms and cultures but with the ambition to agree on the direction forward without inhibiting the diversity in approaches for the actual progress. It is not surprising that the member states themselves have a similar governance philosophy, resulting in a highly dynamic patchwork of legislative structures. The Multilevel Governance (MLG) concept provides a framework for acknowledging the interactions between the different spatial or organizational resolutions while giving room for a more coherent policy mix across these resolutions and across sectors. An MLG view is essential, especially for enabling local and regional energy and climate initiatives. Dobravec et al. [34], for example, "analyses the existing energy planning governance in Austria throughout the MLG-structure by focusing on the alignment between the local energy and climate initiatives and the national and E.U. goals". They find that a "general willingness of Austrian municipalities to take part in local energy actions" as well as "cooperation of different levels of governance from the top-down and bottom-up perspective" via local, regional and inter-regional initiatives such as the e5—Program of energy-efficient municipalities, KEM—Klima- und Energiemodellregionen, CoM—Covenant of Mayors can be observed. Based on the identified shortcomings, especially concerning data availability and spatial energy planning for renewables crossing different jurisdictions and responsibilities, the paper recommends extending the existing governance on multiple levels with a more flexible MLG, including neighborhoods and zones and their interconnection with varying levels up to the European Union (see Figure 2). Exemplary action points in such a framework could include "blueprint[s] for pioneering feasible regional energy initiatives". In contrast, regional sustainable development goals need to be integrated into national energy transition policy [35].

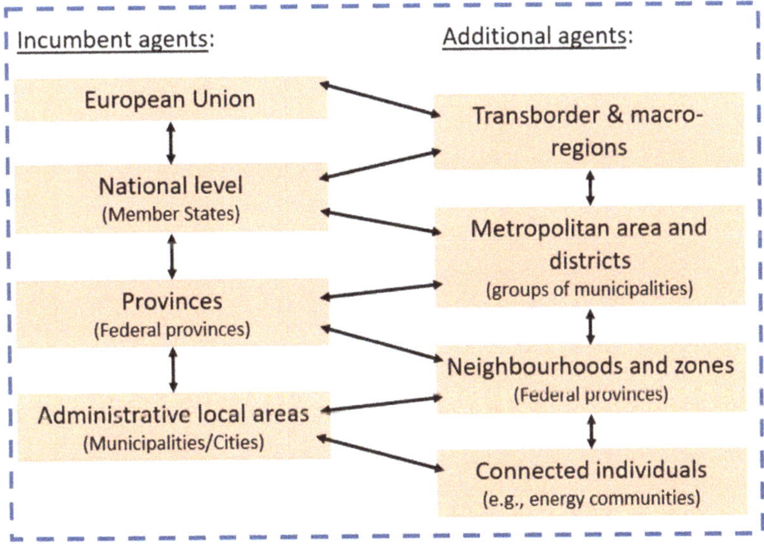

Figure 2. New integrated action space for multilevel governance. Source: own illustration modified from Dobravec et al., 2021 [34].

Renewable local energy initiatives historically focused on tackling problems related to social acceptance, such as the "nimby" (not in my backyard)-phenomenon. Today, especially modularity of renewable electricity generation, prosumer frameworks and demand-side management are rather coined by questions on social participation instead of acceptance. Participatory processes in governance and investments and providing energy production and consumption flexibility or engaging and nudging social networks to enhance energy efficiency or more sustainable consumption hold significant potentials, not yet recognized, e.g., in energy system planning [36–38].

The Renewable Energy Directive acknowledges the manifold "opportunities for growth and employment that investments in the regional and local energy production from renewable sources bring". The regional and local development opportunities include "export prospects, social cohesion and employment opportunities, particularly SMEs and independent energy producers", with decentralization fostering "community development and cohesion by providing income sources and creating jobs locally." The Renewable Energy Directive and the European Economic and Social Committee (EESC) mainly address "the role of civil society in the implementation of [decentralized P.V.- and wind energy]" [39]. Based on the findings of this paper, we think it is time to extend this broader socioeconomic benefits discussion to regional biomass mobilization. For example, a civic power plant ("Bürger* innenkraftwerk") based on P.V. already boosts the possibilities for participation manifold compared to a fossil-based power plant based on energy carrier imports. However, a local CHP-plant connected to a district heating network and supplied by forestry residues from forests primarily cultivated for stem-wood for wood construction and engineered wood products must exhibit an even higher societal participation potential.

Socioeconomic benefits of regional bioeconomies such as income, employment and net-profit for and of the engaged stakeholders are vital parameters to highlight here for policymakers and society (see, e.g., Wang et al. [40]). However,, the "inclusion of unique types of possibilities that each town or location offers" [41] needs to be taken into account, even though this might be more difficult to assess quantitatively. Furthermore, different levels of purchasing power parity (PPP) result in trade between regions [42]. While biomass export can provide economic benefits, the availability of ecologically sustainable options for meeting the own demand of the exporting regions has to be ensured. More indirect societal benefits include "protection against unpredictable energy pricing, improved energy access and security, reduction of transmission and distribution costs, independence from multinational utility interests and strategies and increased feasibility of renewable energy deployment within the framework of decentralized business cases" EESC in McGovern and Klenke [35].

For the present paper, we can outline that the valorization of the outlined socioeconomic benefits is still in its infancy. This observation is based on the fact that the quantification and assessments of the discussed aspects do not even take place in the theoretical energy system- and bioeconomy models today. Following Krumm et al. [37], we urge modelers to take "heterogeneity of actors, public acceptance and opposition, public participation and ownership" into account to at least theoretically explore the social dimension and benefits of regional biomass mobilization quantitatively.

3.2. Mobilization through Technological Innovation

Biomass mobilization will most likely be enhanced through various innovation types; organizational/institutional and social innovations can extend technological innovation. Organizational/institutional innovations address "changes in and among various organizational aspects of functions [of an organization or institution]", e.g., "the idea of networks—involving actors inside and outside [of the organization]" [43]. Social innovation is defined as "preferences of consumers, citizens, and workers for the types of products, services, environmental quality, leisure activities, and work they want" as well as respective changes in their behavior and interactions [43]. It also refers to "new solutions that imply conceptual, process, product, or organizational change, which ultimately aim to improve the welfare and wellbeing of individuals and communities" [44].

Still, and for this assessment level, we first and foremost focus on the market introduction and diffusion of technological innovations. We explore the current frontiers in adopting respective technologies for mobilizing the feedstock base of the circular bioeconomy. Especially biomass pre-treatment technologies (Sections 3.2.1 and 3.2.2), improvements in planning and harvesting (Section 3.2.3) and biomass production (Section 3.2.4) can provide opportunities for the mobilization of local, low-value and heterogenous biomass.

3.2.1. Decentralized Pre-Treatment

Classical, mechanical pre-treatment processes include chipping, pelletization, briquetting and bailing of biomass, which reduces transport and handling costs and better facilitates the storage and trade of densified bioenergy carriers [45]. Torrefaction, a mild form of pyrolysis, can further enhance relevant properties of the bioenergy carriers such as energy density, grindability and hydrophobicity [46]. Pyrolysis to maximize the liquid fraction of the output [47], hydrothermal treatment [48,49], upgrading of biogas from anaerobic digestion [50] or from biomass gasification to biomethane [51] are other strategies to facilitate biomass mobilization. A large number of publications focused on the impact of densification technologies on decreasing supply chain costs [52] to promote biomass commodification and trade [53] and to improve conversion efficiency, for example, through gasification [54].

In contrast to properties related to trade also the value of energy and carbon being reliably stored over a long time becomes particularly relevant in a Circular Bieoconomy [45]. This value is based on improved volumetric energy densities and suitability for storage in existing infrastructures in light of flexibility needs due to increasing shares of intermittent renewable energy production. However, similar to the socioeconomic benefits outlined in Section 3.1, tools to quantify and valorize this flexibility, even in theoretical energy system models, are missing today (see, e.g., Thrän et al. [55]).

3.2.2. Mobile/Portable Pre-Treatment

Investments in conversion- and pre-treatment plants are primarily driven by economies of (unit-) scale. It is recommended to optimize between plant size and the "respective feedstock supply distances for various feedstocks, supply modes and feedstock yield, availability and accessibility combinations" [45]. This obvious connection has far-reaching consequences, including the need for pre-treatment steps based on commodity markets (see Section 3.3), emerging overcapacities with increasing feedstock competition and tendencies to create vertically integrated supply chains.

Still, some niche actors aim to down-scale respective stationary technologies for them to become relocatable, transportable or even mobile. Polagye et al. [56] outline in a detailed cost comparison how the economy of scale results in the issue that "the production of bio-fuels using mobile and transportable facilities is significantly more costly than production at a stationary or relocatable facility." A more recent project, "mobileflip.eu" from VTT and SLU, acknowledges that the added value of the smallest functional unit would be reflected by its flexibility to switch between feedstocks that are scattered spatially but also in time. They discuss mobile pre-treatment facilities of 687 tonnes of forest residues input per year [57]. De la Fuente et al. [58] outline the LCA of mobile pelletization, torrefaction, slow pyrolysis, hydrothermal pre-treatment and carbonization and respective environmental challenges for downscaling pre-treatment. Demonstration projects are furthermore described in Mirkouei et al. [59,60] for mobile bio-oil production (i.e., the Renewable Oil International LCC) based on a relatively old refinery concept from Badger and Fransham, 2005 [61] and another slow pyrolysis for biochar production (Schatz Energy Research Center) as an alternative to slash pile burning [62,63]. Some commercialized concepts exist, such as the relocatable shipping container "PelletBox" by Prodesa and mobile pelletization plants such as the "Krone Premos 5000', the "Schaider Groups Pelletec", the "Gmco mobile pellet plant" and the "Proxipel concept". However, while these concepts could significantly help mobilize local and heterogenous biomass resources, they do not play a considerable role in current provision structures.

3.2.3. GIS Supported Planning and Harvesting

Subramanian et al. [64] classify energy system models regarding their (1) decision-making hierarchy (strategic, tactical, operational) and (2) the level of technology aggregation (unit operation, plant, supply chain, energy sector and whole economy). However, we can observe supply chain innovation and the utilization of Geographic Information

Systems (GIS) on all three hierarchy levels and imagine supply chain management to become the backbone of connecting organizational planning to mobilize biomass for a circular bioeconomy.

However, a notable literature review on biomass supply chain optimization by Ba et al. [65] finds that "GIS [is] mainly used in a strategic context because they lack the short and medium-term temporal dimension that is required for tactical and operational decisions." They are often applied to find economically optimal solutions for mobilizing forestry residues. In contrast to forestry biomass, optimization of agricultural residues and/or based on environmental or social parameters (or a combination thereof) are discussed to be less plentiful [66]. Economic optimization addresses minimizing costs for public endeavors and society, e.g., drafting policy recommendations or maximizing revenues for private projects and investment decisions. The "BeWhere model" (based on Leduc, [67]), for example, "identifies the localization, size and technology of the renewable energy system that should be applied in a specific region". At the same time, Frombo et al. in Ba et al. [65] provide a "GIS-based Environmental Decision Support System (EDSS)" to support investment decisions for location and size of pre-treatment plants for forest residues.

The utilization of GIS data in a strategic context includes decision variables such as the optimal location to construct a biomass densification or conversion facility, its capacity, technological set-up and the biomass supply and distribution between facilities and the end-user. In contrast, tactical models aim at inventory planning and identify optimalharvest quantity, harvest schedules, inventory deployment and optimize transport modes, shipment size and routing. These models are mainly business-oriented, such for example, the "VITO MooV model" (based on De Meyer et al. [68]) optimizing supply chain configurations, including decisions on transport mode planning, storage capacity planning and feedstock- and product variability for medium-term (i.e., next months to years) scheduling. Vopenka et al. [69] furthermore describe a tool for spatial and temporal optimization of forest harvesting in a user-friendly digital map, potentially bridging the gap between tactical and operational planning.

The operational context can be discussed regarding scheduling activities in a temporal granularity below months and weeks. An H [70] aims to optimize the "daily scheduling of [trucks and] mobile [loaders] to transport biomass from satellite storage locations to a bioenergy plant" and present a case study on corn stover. Zamar et al. [71] identify the best daily routing schedule for trucks to collect sawmill residues for energy conversion in the pulp and paper industry. Besides these rather classical travelling salesman problems, the literature on biomass supply chain GIS-modeling for operational decision support is scarce.

The view citations mentioned in the central reviews on biomass supply chain modelling [65,68,72,73] can mostly and arguable be better grouped in the tactical or even strategic context. We acknowledge that thorough market research would be more thankful for this type of model than the scientific literature research performed for the objective of this paper. Innovations in the field of precision agriculture, including optimizing fertilizer- and pesticide application, as well as harvest scheduling, weather forecasts of high temporal and spatial resolutions but also dynamic record-keeping based on data collection from satellites, drones and on the ground (e.g., https://geomarvel.com/, accessed on 31 December 2021), can be used for increasing the mobilization of biogenic residues. Digitally guided forest management, planning for collecting and utilizing damaged wood from extreme weather events or minimizing soil contamination through harvesting after natural washing (rain) and optimized deployment of mobile pre-treatment could be potential applications. These big data strategies can be complemented by further digitalization and mechanization efforts, e.g., in silviculture operations and with soil mechanics fundamentals to assess terrain trafficability as, e.g., currently developed in the "H2020 EFFORTE project".

3.2.4. Next-Generation Primary Sources

Wild cards in the bioenergy and the circular bioeconomy discussions can be seen in novel biogenic carbon sources and -sinks for biogenic carbon. Their particular potentials for regional mobilization strategies supporting a circular bioeconomy are challenging to discuss. Alternatively, we provide a short overview of the different research frontiers that will have to be assessed in follow-up projects in detail.

The production of short rotation coppice (SRC) remains relatively low (see Section 3.1.1). Various projects assess the potential of crops such as Miscanthus and hemp, such as the "BBI GRACE", or agricultural prunings and plantation removal in the "uP_Running project". The EU-Brazil cooperation in the "BECOOL project" addresses different "annual and perennial dedicated lignocellulosic crops, together with crop and process residues such as cereal straw, sugar cane straw, bagasse, and lignin-rich residues." The "MAGIC" and the "ADVANCEFUEL" projects have an even broader scope for abundant oil, lignocellulosic, carbohydrate or specialty crops. Innovation for agricultural management such as optimization of planting density, crop establishment improvements, crop rotation intercropping, multi-purpose cropping, cropping on marginal land and precision farming, and increasing the harvesting frequency, will potentially provide additional biomass. These topics are subject, e.g., to the "LIBBIO" project, the "SeemLA" project and the "FORBIO" project.

In addition, "technological advances in agriculture and forestry can still be expected through improved fertilization, breeding, crop selection, and gene editing, and genetically modified organisms not only for yield improvements but also to provide resilience against temperature tipping points [for biomes] caused by global warming" (see Duffy et al. [74]).

In this line, the production of micro-and macro-algae also have to be mentioned. Algae are "produced in photobioreactors, in open ponds or harvested from the natural environment are also promising primary feedstocks and should be addressed, e.g., in bioeconomy modelling and discussions" [75]. As part of the "blue economy", this is mainly commercially realized for food or specialty food products (e.g., Omega-3 fatty acids). The high water content renders energy or chemicals production, particularly energy and cost-intensive. Furthermore, services such as nutrient recycling and recirculation or urban solutions are still in their infants, often discussed under the umbrella of Nature-Based Solutions [76].

3.2.5. Next-Generation Primary Sinks

The IEA Bioenergy project on the deployment of biocarbon capturing and sequestration published three case studies of large scale BECCS coupled with CHP in Denmark [77] for bioelectricity only in the Drax Power Station in the U.K. [78] and with the waste incineration plant of Fortum Oslo Varme in Norway [79]. In addition to the "centralized" and large-scale BECCS, more decentralized carbon storage solutions, such as halting deforestation and degradation, have the most significant carbon emissions mitigation potential followed by afforestation (non-forest areas to forests) as outlined, e.g., in [80], reforestation (deforested areas to forests) (see Chazdon et al. [81]) and forest restoration (degrading forest to healthy forest) [82]. Furthermore, Fritsche et al. [83] lists biochar addition to soil "improving water holding capacity and nutrient use efficiency" while sequestering carbon. Especially in the light of climate change, we want to stress that these decentralized carbon management strategies will require substantial efforts attracting skilled labor to rural areas and providing mobilization opportunities for circular bioeconomy feedstock.

3.3. Market Creation for Biomass Mobilization

Zooming into the regional context of biomass mobilization, we find that existing legislative frameworks (Section 3.1), readily deployable technologies, and niche innovations (Section 3.2) are often pre-conditions to establish economic activities but do not necessarily result in such. The creation and establishment of dedicated and functioning physical markets, regional-, interregional- and international trade depend on additional factors, such as market competitiveness and -liquidity [83]. In the following pages, we provide and discuss selected strategies on the market creation level for the mobilization of low-value and heterogenous biomass feedstock.

3.3.1. Market Catalysts for Wastes, Residues, Post-Consumer Products and Secondary Raw Materials

Regarding European efforts to transform the economy, the European Bioeconomy Strategy states that "to be successful, the European bioeconomy needs to have sustainability and circularity at its heart" [84]. A comprehensive European policy overview on circularity measures can be found in Milios L. [85] who outlines, that even though the E.U. can be seen as a leader in circularity, its actions mainly focus on the end-of-life phase of consumables so far while avoiding waste through improved quality and repair options are rather novel concepts. Waste collection, processing and treatment gained a high priority in the many E.U. Member States, distinguished by waste fractions, e.g., containing fossil-based plastic, biogenic waste and electronic waste. The E.U. Circular Economy action plan [2] thus goes a step further by aiming to create a secondary feedstock market, primarily focused on fossil-based plastics from waste collection but also including residues from downstream industries. This creates market opportunities for commercial waste and residues plastic collectors and distributors (e.g., see https://polymerstocklist.com/, accessed on 31 December 2021). The collection and mainly energy utilization of waste wood, also coined post-consumer wood, is already better established, resulting in small but quantifiable international trade [7]. Similarly, the collection of used cooking oil and animal fats for international trade and biodiesel production has gained importance in the recent decade [86]. While used cooking oil and animal fats exhibit a limited potential for business creation, post-consumer wood from traditional applications but also from increased utilization, e.g., in a potentially growing wood-based construction sector (see, e.g., Churkina et al. [87]), as well as novel materials such as biopolymers (see, e.g., Schipfer et al. [14]), will demand residues and waste collection, processing, treatment, intermediary storage and distribution to biobased industries.

3.3.2. Physical and Virtual Bio-Hubs

Decentral or regional biomass processing depots, or bio-hubs, are facilities that are discussed to overcome the mismatch between the distributed occurrence of biogenic resources and large-scale centralized conversion plants such as biorefineries [88–90]. Conceptually, by including on-site pre-processing and/or densification technologies, bio-hubs are enabling regional market creation to allow farmers and forest owners to convert their residues into valuable by- or even co-products in the form of bioenergy or biogenic carbon carriers. Residue collection and forwarding can be either done by the farmers themselves or third parties eventually owning mobile pre-treatment equipment (see Section 3.2) or machinery to make the residues accessible. Economically feasible options for residue collection are often highly case sensitive, mainly based on low energy densities and high water content and thus limit economically viable transportation ranges [45]. Different collection options exist and should be compared on a case-by-case basis.

A dedicated project for collecting the best bio-hubs practices was initiated within the IEA Bioenergy TCP. The IEA Bioenergy Biohub project collects and disseminates case studies from different world regions and theoretical considerations regarding agri- and silvicultural residues collection centers (see https://arcg.is/qLqaK, accessed on 31 December 2021). The broad definition of bio-hubs by Nasso et al. [91] valorize opportunities such as "streamlining of processing, storage and transportation, reduction in administrative costs, making a variety of biomass types available at a single location, providing an opportunity for suppliers of biomass products to continue producing in the offseason [...], as well as a place for companies to connect and trade with one another" is used for the projects' assessment. Successful examples include the Tschiggerl Agrar GmbH, a logistics center for processing agricultural residues to feed, animal bedding material and fuel, but also virtual bio-hubs such as the Rosewood Network for knowledge transfer in eastern European countries, discussed in the IEA Bioenergy Biohub workshops [92].

Other virtual bio-hubs have also been initiated several times throughout history. While some aim to facilitate knowledge transfer (see Table 1), others establish online market platforms to trade bioenergy commodities. Projects such as "b2bbioenergy.eu" and "promobio.eu", "pellet-zone.com" and "bioexchange.com" can be named within this realm. While these projects did not succeed and homepages are not maintained anymore, the "Biomass Commodity Exchange (BCEX)" for cellulose biomass trade within the U.S., including energy crops such as willow, poplar and switchgrass, is under construction. For more international trading of biomass and bioenergy commodities, the wood pellets futures from "Euronext" and platforms to prove the origin, especially of liquid and gaseous commodities such as the "German Nabisy" system or national and international renewable gas registries fall in the category of virtual bio-hubs. In the Baltic States, "Baltpool" enables the trade of biomass feedstocks and even heat, with the latter being reported a relatively underdeveloped market in other parts of the E.U. [93,94]. If virtual market platforms can alter regional trade remains to be shown. However, the comparison, implementation, development, and impact of virtual bio-hubs could be an exciting field of energy economic research without considerable publications to date.

Similarly, virtual bio-hubs include the "Electronic Reverse Auction (eRA)" system in Denmark. Via the eRA straw-based thermal power stations source their fuel since 2006. A power station initiates the auctioning by requesting a certain amount and quality of straw. Various potential suppliers can underquote (reverse auctioning) until a determined deadline. Each buyer runs its auctioning model, while the supplier can decide on amounts, contract periods (1–3 years) and prices based on blind auctioning.

The applicability of eRA for the German straw market has been discussed in Pfeiffer et al. [95]. Even though the eRA system is a virtual market platform, it still requires intensive relationship management. This results in additional work for the buyer and significant competition among the sellers. Furthermore, it has to be mentioned that eRA in Denmark mainly was implemented to stabilize a market, not to establish it. For new markets, feasibility studies need to address supply and demand potential, engagement of possible stake- and shareholders and include a thorough risk assessment [95].

3.3.3. Commoditization of Intermediary Products

In the previous years, the IEA Bioenergy TCP Task40 discussed the commodification of bioenergy carriers as a necessary step for market uptake and market creation. For example, in Olsson et al. [83], we present the key features of commodity markets, including trade with standardized and perfectly fungible (interchangeable) units. These product-related properties are complemented with market-related properties, such as the engagement of many buyers and sellers, resulting in high market liquidity. These properties can be expressed by international trade and equilibrating forces on the last remaining product differentiator, the product price. In Schipfer et al. [96], we try to quantify the commoditization of wood pellets based on competitive spatial equilibrium modelling using modern trade theory. The wood pellets market can be seen as a role model for possibly upcoming pre-treated bioenergy carriers such as straw pellets, briquettes, pyrolysis oil and biomethane and biogenic carbon-based (liquid) chemicals, including biodiesel, bioethanol and liquid organic hydrogen carriers (LOHC). Still, we find a relatively inefficient international wood pellets market in central Europe with high margins for traders. The identified low efficiency and low commoditization could result from low market transparency and an "intrinsic valuation of non-quality related properties like pellets color and more regional biomass supply chains". In the article we argue, that internationally harmonized data and sustainability standardization (and awareness thereof) could steer the market towards a joint price benchmark and more stable and on average lower prices. However, we acknowledge, that this development holds the risk to drive smaller and less economic viable producers with regional supply chains out of the market.

A more critical view and direct reaction on this topic are presented in McGovern and Klenke [35]; here, commoditization is deemed "counter to generating viable regional energy projects as it reduces the stakeholder role of local agricultural biomass producers". Commoditization is addressed as a challenge with the potential impact to "seriously decelerate or weaken the movement towards energy democracy and decentralized renewable energy". Provided supporting arguments include that longer and international supply chains give citizens "little-to-no influence over the sustainability of their energy supply" and pose a risk of distorting ecological equilibrium since they undermine efforts for (local) circularity. This critique understands the heterogeneity of regional biomass supply chains as an essential driver for the participation of diverse stakeholders, providing ownership, local revenue and job creation, and independence from multinational utility interests.

Some of these commoditization-critical arguments are not sufficiently substantiated; e.g., sustainability certification provides control over the energy supply chain, even for internationally purchased wood pellets, wood pellet supply chains, especially for residential heating, are by no means centralized. However, the overall controversy and potential trade-off between commoditization and specialization, hence between larger and efficient markets and smaller and less efficient markets, is highly relevant, especially in the light of regional and civic energy initiatives. As discussed under the legislative strategies (see Section 3.1), circular bioeconomic structures have exceptionally high decentralization- and societal participation potentials. We propose that this key performance indicator (KPI) could be better measured by the number of stake- and shareholders part of a specific bioeconomy supply chain rather than by the transport distance of bioenergy or biogenic carbon carriers. Still, the number of stake- and shareholders, similar to the transport distance, contributes negatively (higher costs) to the accumulated costs of the final product/service and for the consumer. Without "giving credits" to this KPI, commoditization and the resulting (international) market efficiency indeed result in (1) less stake-/shareholder diverse supply chains and (2) relocation of supply chains to regions of lower production costs. In addition, and in line with Section 3.1, biomass export from lower PPP-regions does only provide economic and ecological benefits, as long as it does not endanger the availability of sustainable options to meet the demand of the exporting regions.

Sustainability certification can ensure that efficient markets do not race to the lowest environmental standards and even elevate standards and transfer higher standards to some regional markets. Additional socioeconomic certification for smaller producers, as implemented for some products (fair-trade of coffee and tea, chocolate), can be understood as the specialization of the supply chain; however, a critical mass of equitable supply chains could set new standards for commodity markets. Extending sustainability certification of internationally traded biomass commodities with socioeconomic KPIs such as stake-/shareholder variety could be a promising strategy for just and regional biomass mobilization. Research on comprehensive frameworks for sustainable assessments of biobased products is performed, e.g., in the Horizon2020 STAR-ProBio project.

3.3.4. Informative Networks for Knowledge Exchange and Market Creation

Transparency is critical for functioning international markets and is also expected to play an essential role in regional markets to support values (e.g., sustainability, stake-/shareholder diversity) but also stability through the exchange of harmonized price information, spatially and temporally explicit production and consumption data and data on flows and stocks, including storage volumes. Bioenergy storage, and its flexibility service character, is of particular importance when energy systems develop towards higher implementation of intermittent renewable energies such as photo-voltaic and wind power. Market transparency induces fair competition, which can be beneficial not only for the end-users but also for the shareholders of the supply chain if the information on innovation is transparently traded and best practices are shared. Currently, market information networks are collecting, preparing and providing respective knowhow and knowledge, often in cooperation with experts from industry and academia:

The IEA Bioenergy TCP and related TCPs (see Figure 3) is part of the intergovernmental OECD/IEA network. It is committed to providing scientific backed information on the level of markets, developed- and immature technologies and how their status today, opportunities and barriers for market diffusion in the upcoming decades.

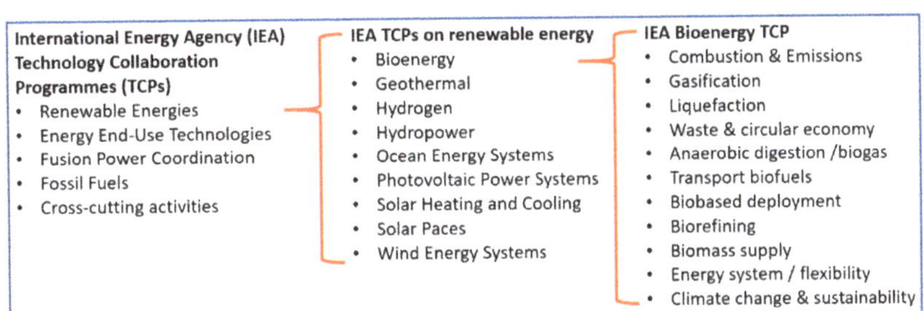

Figure 3. Structure of the IEA network and technology collaboration program. Source: own illustration based on BMK, KG and Eggler, Indinger and Zwieb, 2018 [96].

Other energy-related international and partly intergovernmental information networks include the International Renewable Energy Agency (IRENA), the U.N. initiative Sustainable Energy for All (SE4ALL), the REN21 and many more (e.g., International Energy Forum /IEF, Global Bioenergy Partnership/GBEP, BioFuture Platform, EurObserv'ER, European Renewable Energy Council/EREC, World Energy Council and Food and Agriculture Organisation on bioenergy and food security/FAO BEFS). On a European level, especially the European Energy Research Alliance (EERA) has to be mentioned. While these networks focus on the exchange between countries, also considerable efforts are undertaken for inter-regional exchange of know-how and knowledge. Networks on an inter-regional and regional level often include national and regional governments and regional energy agencies, and interest groups. They are supported by international or national structural and regional development funds (see Section 3.1). Analyzing the unique selling propositions and the effectiveness of the different networks and initiatives is beyond the scope of this project. Still, it could be an exciting research topic for participatory research and research on international consortia, potentially resulting in harmonized performance metrics and ultimately improved impact on regional sustainable energy implementation.

4. Conclusions

We assess mobilization strategies for local, low-value and heterogenous biomass feedstock. Respective feedstocks, including energy crops-, forestry- and agricultural residues and biogenic wastes, are identified as the backbone of the circular bioeconomy of tomorrow. In contrast to currently deployed biomass for energy purposes, novel provision structures face considerable technical-, societal- and organizational challenges.

To explore and discuss these challenges, we are building upon the IEA Bioenergy Task40 expertise on international bioenergy trade and the current provision of bioenergy. For the present paper, we aim at transferring and extending our knowledge on current bioenergy carrier provision structures to the local, low-value feedstock base of the circular bioeconomy.

This approach exhibits limitations on each of the three assessment levels: The (1) legislative framework level limits the scope to top-down frameworks and the E.U. policy landscape. A first attempt to overcome the top-down view is made by bringing together findings from international projects on regional biomass mobilization. The (2) innovation level assessment is limited to technological innovation only. This limitation is mainly due to the available expertise in the consortium. However, social and organizational innovations also visibly coined the findings of all three assessment levels. For the (3) market creation level, the results and discussion section builds on the authors' particularly strong scientific background. The remaining limitations concern the under-researched nature of this area; they include limited scientific, energy-economic publications on electronic bioenergy carrier trading or comparative discussions of different market structures and market instruments.

We find that the E.U. policy landscape, especially under Covid recovery's umbrella, provides significant funds for regional development and biomass mobilization. Most regional action plans shifted to international, quantitative potential assessment approaches to provide biomass to different Bioeconomy sectors. The next frontier can be seen moving towards Multilevel governance, entangling governance levels from neighborhoods to E.U. governance. Respective opportunities for regional biomass mobilization are particularly exciting since the circular bioeconomy exhibits an outstanding decentralization- and thus resource democratization potential.

The niche technological innovations for the mobilization of local, low-value and heterogenous biomass resources already exist. They include mainly pre-treatment technologies, GIS-supported planning and novel primary biomass sources. However, ensuring economic and ecological sustainability while down-scaling pre-treatment technologies to the smallest possible functional unit need to be addressed. In return, mobile or at least portable pre-treatment and densification plants could induce valuable operational flexibility. Mobile pre-treatment, coupled with big-data and GIS support, could overcome the challenges of seasonal fluctuations and in-homogenous geographical feedstock distribution.

However, markets for local, low-value and heterogenous biomass resources are largely underdeveloped. Physical- and virtual bio-hubs and market platforms are required to connect the highly diverse supply side with the demand side for biomaterials and bioenergy. Today, these hubs are rather an exception than the rule; numerous attempts of establishing virtual market platforms have failed over the years. The heterogeneity of market actors and the traded goods can be identified as a major challenge, also for successful platforms such as the eRA-straw market. With this regard, commoditization is addressed as a double-edged sword; without environmental and socioeconomic standards, market creation might be at the expense of biodiversity and stakeholder variety.

The challenges and opportunities of the three assessment levels point towards a common denominator: The quantification of the systemic value of strengthening the potentially last remaining primary economic sectors, forestry, agriculture and aquaculture, is missing. With the eroding importance of other primary economic sectors, including fossil fuel extraction and minerals mining, the time is now to assess and act upon the value of the supply side of a circular bioeconomy. This value includes the support the Bioeconomy can provide to structurally vulnerable regions by creating meaningful jobs and activities in and strengthening the resource democratic significance of rural areas.

Energy system and circular bioeconomy modelling could play an important role, theoretically simulating the systemic value, e.g., of temporal- and spatial flexibility of pre-treatment technologies and of stakeholder diversity in markets and multilevel governance. Therefore, modelling should account for multiple assessment criteria and modelling functions, based on all types of resources, including monetary-, natural-, CO2-budget- but also human resources. Based on the theoretical work, sound recommendations for biomass mobilization action plans, technology investment decisions and market organization should be derived.

Author Contributions: Conceptualization, F.S. and A.P.; methodology, F.S. and A.P.; investigation, F.S. and A.P.; writing—original draft preparation, F.S.; writing—review and editing, F.S. and R.H.; project administration, F.S.; funding acquisition, F.S. and R.H. All authors have read and agreed to the published version of the manuscript.

Funding: This research was funded by the IEA Bioenergy TCP Task40 and the BMK, FFG grant number 876720.

Institutional Review Board Statement: Not applicable.

Informed Consent Statement: Not applicable.

Data Availability Statement: Not applicable.

Acknowledgments: We want to warmly thank the three reviewers and the editor for the valuable comments and the fast but thorough reviewing process. Your input improved this paper considerably. We also want to thank the Open Access Funding by TU Wien library.

Conflicts of Interest: The authors declare no conflict of interest. The funders had no role in the design of the study; in the collection, analyses, or interpretation of data; in the writing of the manuscript, or in the decision to publish the results.

Nomenclature

BBI JU	Biobased Industries Joint Undertaking
BECCS	Bioenergy Carbon Capture and Sequestration
CAP	Common Agriculture Policy
CBE JU	Circular Bioeconomy Joint Undertaking
C.E.	Circular Economy
CHP	Combined heat and power
CO2	Carbon dioxide
EESC	European Economic and Social Committee
EFA	Ecological Focus Areas
eRA	Electronic Reverse Auctioning
E.U.	European Union
GHG	Greenhouse Gas
GIS	Geographic Information System
IEA	International Energy Agency
KPI	Key performance indicator
M.S.	Member States
MLG	Multilevel governance
MWel	Megawatt electric
NUTS	Nomenclature of Territorial Units for Statistics
OECD	Organization for Economic Co-operation and Development
P.V.	Photo voltaic
R&D&D	Research, development and deployment
SME	Small and medium enterprises
TCP	Technology Collaboration Programme
U.K.	United Kingdom
U.S.	United States

References

1. EC. *A Sustainable Bioeconomy for Europe: Strengthening the Connection between Economy, Society and the Environment*; EC: Brussels, Belgium, 2018.
2. EC. *A New Circular Economy Action Plan For a Cleaner and More Competitive Europe*; EC: Brussels, Belgium, 2020.
3. EC. *A European Strategy for Plastics in a Circular Economy*; EC: Brussels, Belgium, 2018.
4. BBI JU. Draft Proposal for aEuropean Partnership under Horizon Europe European Partnership for a Circular Bio-Based Europe: Sustainable Innovation for New Local Value from Bio-Waste and Biomass (CBE). 2020. Available online: https://ec.europa.eu/info/sites/default/files/research_and_innovation/funding/documents/ec_rtd_he-partnerships-circular-biobased-europe.pdf (accessed on 31 December 2021).
5. D'Adamo, I.; Morone, P.; Huisingh, D. Bioenergy: A Sustainable Shift. *Energies* **2021**, *14*, 5661. [CrossRef]
6. Fonseca, L.M.; Domingues, J.P.; Pereira, M.T.; Martins, F.F.; Zimon, D. Assessment of Circular Economy within Portuguese Organizations. *Sustainability* **2018**, *10*, 2521. [CrossRef]
7. Junginger, H.M.; Mai-Moulin, T.; Daioglou, V.; Fritsche, U.; Guisson, R.; Hennig, C.; Thrän, D.; Heinimö, J.; Hess, J.R.; Lamers, P.; et al. The Future of Biomass and Bioenergy Deployment and Trade: A Synthesis of 15 Years IEA Bioenergy Task 40 on Sustainable Bioenergy Trade. *Biofuels Bioprod. Biorefining* **2019**, *13*, 247–266. [CrossRef]
8. Camia, A.; Robert, N.; Jonsson, R.; Pilli, R.; García-Condado, S.; López-Lozano, R.; Van der Velde, M.; Ronzon, T.; Gurría, P.; M'barek, R.; et al. *Biomass Production, Supply, Uses and Flows in the European Union: First Results from an Integrated Assessment*; Publications Office of the European Union: Luxembourg, 2018; 126p.
9. Mandley, S.J.; Daioglou, V.; Junginger, H.M.; van Vuuren, D.P.; Wicke, B. EU Bioenergy Development to 2050. *Renew. Sustain. Energy Rev.* **2020**, *127*, 109858. [CrossRef]
10. Brosowski, A.; Krause, T.; Mantau, U.; Mahro, B.; Noke, A.; Richter, F.; Raussen, T.; Bischof, R.; Hering, T.; Blanke, C.; et al. How to Measure the Impact of Biogenic Residues, Wastes and by-Products: Development of a National Resource Monitoring Based on the Example of Germany. *Biomass Bioenergy* **2019**, *127*, 105275. [CrossRef]
11. Haase, M.; Rösch, C.; Ketzer, D. GIS-based Assessment of Sustainable Crop Residue Potentials in European Regions. *Biomass Bioenergy* **2016**, *86*, 156–171. [CrossRef]

12. Hamelin, L.; Borzęcka, M.; Kozak, M.; Pudełko, R. A Spatial Approach to Bioeconomy: Quantifying the Residual Biomass Potential in the EU-27. *Renew. Sustain. Energy Rev.* **2019**, *100*, 127–142. [CrossRef]
13. Searle, S.; Malins, C. A Reassessment of Global Bioenergy Potential in 2050. *GCB Bioenergy* **2015**, *7*, 328–336. [CrossRef]
14. Schipfer, F.; Kranzl, L.; Leclère, D.; Sylvain, L.; Forsell, N.; Valin, H. Advanced Biomaterials Scenarios for the EU28 up to 2050 and Their Respective Biomass Demand. *Biomass Bioenergy* **2017**, *96*, 19–27. [CrossRef]
15. Chang, M.; Thellufsen, J.Z.; Zakeri, B.; Pickering, B.; Pfenninger, S.; Lund, H.; Østergaard, P.A. Trends in Tools and Approaches for Modelling the Energy Transition. *Appl. Energy* **2021**, *290*, 116731. [CrossRef]
16. EC. *The European Green Deal*; EC: Brussels, Belgium, 2019.
17. EC. *Stepping Up Europe's 2030 Climate Ambition. Investing in a Climate-Neutral Future for the Benefit of Our People*; EC: Brussels, Belgium, 2020.
18. EC. *Directive of the European Parliament and of the Council Amending Directive (EU) 2018/2001 of the European Parliament and of the Council, Regulation (EU) 2018/1999 of the European Parliament and of the Council and Directive 98/70/EC of the European Parliament and of the Council as Regards the Promotion of Energy from Renewable Sources, and Repealing Council Directive (EU) 2015/652*; EC: Brussels, Belgium, 2021.
19. EC. *NextGenerationEU. European Commission-European Commission*; EC: Brussels, Belgium, 2021. Available online: https://ec.europa.eu/info/strategy/recovery-plan-europe_en (accessed on 23 December 2021).
20. EC. *A Long-Term Vision for the EU's Rural Areas-Towards Stronger, Connected, Resilient and Prosperous Rural Areas by 2040*; EC: Brussels, Belgium, 2021.
21. EC. *A Farm to Fork Strategy for a Fair, Healthy and Environmentally-Friendly Food System*; EC: Brussels, Belgium, 2020.
22. EC. *New EU Forest Strategy for 2030*; EC: Brussels, Belgium, 2021.
23. EC. *EU Biodiversity Strategy for 2030*; EC: Brussels, Belgium, 2020.
24. Emmerling, C.; Pude, R. Introducing Miscanthus to the Greening Measures of the EU Common Agricultural Policy. *GCB Bioenergy* **2017**, *9*, 274–279. [CrossRef]
25. Parra-López, C.; Holley, M.; Lindegaard, K.; Sayadi, S.; Esteban-López, G.; Durán-Zuazo, V.H.; Knauer, C.; von Engelbrechten, H.G.; Winterber, R.; Henriksson, A.; et al. Strengthening the Development of the Short-Rotation Plantations Bioenergy Sector: Policy Insights from Six European Countries. *Renew. Energy* **2017**, *114*, 781–793. [CrossRef]
26. EC. *Communication From The Commission. Biomass Action Plan COM(2005) 628*; EC: Brussels, Belgium, 2005.
27. Kautto, N.; Peck, P. Regional Biomass Planning–Helping to Realise National Renewable Energy Goals? *Renew. Energy* **2012**, *46*, 23–30. [CrossRef]
28. Elbersen, B.; de Groot, H.; Staritsky, I.; Dees, M.; Datta, P.; Leduc, S. A Full Technical Description of the Integrated Toolset, Central Database and General User Interface Developed in WP4. S2Biom Project Grant Agreement n 608622. 2017. Available online: https://www.s2biom.eu/images/Publications/D4.10_S2Biom_Technical_description_toolset_Final2.pdf (accessed on 31 December 2021).
29. Eurostat. *Energy from Renewable Sources-Shares*; Eurostat: Luxembourg City, Luxembourg, 2021.
30. Radulov, L.; Nikolaev, A.; Genadieva, V.; Frouz, J. Report on Social Challenges and Re-Skilling Needs of the Workforce Solutions in the TRACER Target Regions. TRACER D3.4. 2020. Available online: https://tracer-h2020.eu/wp-content/uploads/2020/09/TRACER_D3.4_31072020_public-rev.pdf (accessed on 31 December 2021).
31. Knoche, D.; Mergner, R.; Janssen, R.; Doczekal, C. Fact Sheet:Chemistry park "Schwarzheide"Conversion of a Lignite-Basedrefineryinto a Showcase for Industrial Transition. TRACER Fact Sheet. 2019. Available online: https://tracer-h2020.eu/wp-content/uploads/2019/10/TRACER_D2.1-Chemical-Park-Schwarzheide.pdf (accessed on 31 December 2021).
32. Sociaal-Economische Raad. *Biomass in the Balance*; Sociaal-Economische Raad: The Hague, The Netherlands, 2020; p. 30. Available online: https://www.ser.nl/-/media/ser/downloads/engels/2020/biomass-in-the-balance.pdf (accessed on 31 December 2021).
33. EC. *Proposal for a Directive of the European Parliament and of the Council amending Directive (EU) 2018/2001. COM (2021) 557 Final*; EC: Brussels, Belgium, 2021.
34. Dobravec, V.; Matak, N.; Sakulin, C.; Krajačić, G. Multilevel Governance Energy Planning and Policy: A View on Local Energy Initiatives. *Energy Sustain. Soc.* **2021**, *11*, 2. [CrossRef]
35. McGovern, G.; Klenke, T. Towards a Driver Framework for Regional Bioenergy Pathways. *J. Clean. Prod.* **2018**, *185*, 610–618. [CrossRef]
36. Aryanpur, V.; O'Gallachoir, B.; Dai, H.; Chen, W.; Glynn, J. A Review of Spatial Resolution and Regionalisation in National-Scale Energy Systems Optimisation Models. *Energy Strategy Rev.* **2021**, *37*, 100702. [CrossRef]
37. Krumm, A.; Süsser, D.; Blechinger, P. Modelling Social Aspects of the Energy Transition: What Is the Current Representation of Social Factors in Energy Models? *Energy* **2022**, *239*, 121706. [CrossRef]
38. Süsser, D.; Gaschnig, H.; Ceglarz, A.; Stavrakas, V.; Flamos, A.; Lilliestam, J. Better Suited or Just More Complex? On the Fit between User Needs and Modeller-Driven Improvements of Energy System Models. *Energy* **2022**, *239*, 121909. [CrossRef]
39. EESC. *EU Renewable Energy Directive. The Role of Civil Society in the Implementation of the EU Renewable Energy Directive: An Impact Study Across Six Member States*; EESC: Bruxelles, Belgium, 2014.
40. Wang, X.; Li, K.; Song, J.; Duan, H.; Wang, S. Integrated Assessment of Straw Utilization for Energy Production from Views of Regional Energy, Environmental and Socioeconomic Benefits. *J. Clean. Prod.* **2018**, *190*, 787–798. [CrossRef]

41. Lehtonen, O.; Okkonen, L. Socio-Economic Impacts of a Local Bioenergy-Based Development Strategy–The Case of Pielinen Karelia, Finland. *Renew. Energy* **2016**, *85*, 610–619. [CrossRef]

42. Thrän, D.; Schaubach, K.; Peetz, D.; Junginger, M.; Mai-Moulin, T.; Schipfer, F.; Olsson, O.; Lamers, P. The Dynamics of the Global Wood Pellet Markets and Trade–Key Regions, Developments and Impact Factors. *Biofuels Bioprod. Biorefining* **2019**, *13*, 267–280. [CrossRef]

43. Ashford, N.A.; Technological, Organisational, and Social Innovation as Pathways to Sustainability. Innovation the Pathway to Threefold Sustainability 2001. Available online: http://hdl.handle.net/1721.1/38475 (accessed on 31 December 2021).

44. OECD. *LEED Forum on Social Innovations*; OECD: Paris, France, 2021; Available online: https://www.oecd.org/cfe/leed/social-economy/social-innovation.htm (accessed on 27 December 2021).

45. Schipfer, F.; Kranzl, L. Techno-economic evaluation of biomass-to-end-use chains based on densified bioenergy carriers (dBECs). *Appl. Energy* **2019**, *239*, 715–724. [CrossRef]

46. Thrän, D.; Witt, J.; Schaubach, K.; Kiel, J.; Carbo, M.; Maier, J.; Ndibe, C.; Koppejan, J.; Alakangas, E.; Majer, S.; et al. Moving Torrefaction towards Market Introduction–Technical Improvements and Economic-Environmental Assessment along the Overall Torrefaction Supply Chain through the SECTOR Project. *Biomass Bioenergy* **2016**, *89*, 184–200. [CrossRef]

47. Bridgwater, A.V. Review of Fast Pyrolysis of Biomass and Product Upgrading. *Biomass Bioenergy* **2012**, *38*, 68–94. [CrossRef]

48. Gollakota, A.R.K.; Kishore, N.; Gu, S. A Review on Hydrothermal Liquefaction of Biomass. *Renew. Sustain. Energy Rev.* **2018**, *81*, 1378–1392. [CrossRef]

49. Shen, Y. A Review on Hydrothermal Carbonization of Biomass and Plastic Wastes to Energy Products. *Biomass Bioenergy* **2020**, *134*, 105479. [CrossRef]

50. Kapoor, R.; Ghosh, P.; Kumar, M.; Vijay, V.K. Evaluation of Biogas Upgrading Technologies and Future Perspectives: A Review. *Env. Sci. Pollut. Res.* **2019**, *26*, 11631–11661. [CrossRef]

51. Schildhauer, T.J.; Biollaz, S.M.A. *Fluidised Bed Methanation for SNG Production–Process Development at the Paul-Scherrer Institut, in Synthetic Natural Gas from Coal, Dry Biomass, and Power-to-Gas Applications*; Schildhauer, T.J., Biollaz, S.M.A., Eds.; Wiley & Sons: New York, NY, USA, 2016; pp. 221–229.

52. Uslu, A.; Faaij, A.P.C.; Bergman, P.C.A. Pre-Treatment Technologies, and Their Effect on International Bioenergy Supply Chain Logistics. Techno-Economic Evaluation of Torrefaction, Fast Pyrolysis and Pelletisation. *Energy* **2008**, *33*, 1206–1223. [CrossRef]

53. Lamers, P.; Junginger, M.; Hamelinck, C.; Faaij, A. Developments in International Solid Biofuel Trade—An Analysis of Volumes, Policies, and Market Factors. *Renew. Sustain. Energy Rev.* **2012**, *16*, 3176–3199. [CrossRef]

54. Chen, W.-H.; Peng, J.; Bi, X.T. A State-of-the-Art Review of Biomass Torrefaction, Densification and Applications. *Renew. Sustain. Energy Rev.* **2015**, *44*, 847–866. [CrossRef]

55. Thrän, D.; Kjell, A.; Schildhauer, T.; Schipfer, F.; Lange, N. Five Cornerstones to Unlock the Potential of Flexible Bioenergy. 2021. Available online: https://www.ieabioenergy.com/wp-content/uploads/2021/12/IEA-Bioenergy-Task-44-Five-cornerstones-to-unlock-the-potential-of-flexible-bioenergy.pdf (accessed on 31 December 2021).

56. Polagye, B.L.; Hodgson, K.T.; Malte, P.C. An Economic Analysis of Bio-Energy Options Using Thinnings from Overstocked Forests. *Biomass Bioenergy* **2007**, *31*, 105–125. [CrossRef]

57. Tamminen, T. The Horizon2020 Mobile Flip Project. 2018. Available online: http://www.mobileflip.eu/ (accessed on 31 December 2021).

58. de la Fuente, T.; Bergström, D.; González-García, S.; Larsson, S.H. Life Cycle Assessment of Decentralized Mobile Production Systems for Pelletizing Logging Residues under Nordic Conditions. *J. Clean. Prod.* **2018**, *201*, 830–841. [CrossRef]

59. Mirkouei, A.; Haapala, K.R.; Sessions, J.; Murthy, G.S. A Review and Future Directions in Techno-Economic Modeling and Optimization of Upstream Forest Biomass to Bio-Oil Supply Chains. *Renew. Sustain. Energy Rev.* **2017**, *67*, 15–35. [CrossRef]

60. Mirkouei, A.; Mirzaie, P.; Haapala, K.R.; Sessions, J.; Murthy, G.S. Reducing the Cost and Environmental Impact of Integrated Fixed and Mobile Bio-Oil Refinery Supply Chains. *J. Clean. Prod.* **2016**, *113*, 495–507. [CrossRef]

61. Badger, P.C.; Fransham, P. Use of Mobile Fast Pyrolysis Plants to Densify Biomass and Reduce Biomass Handling Costs—A Preliminary Assessment. *Biomass Bioenergy* **2006**, *30*, 321–325. [CrossRef]

62. Puettmann, M.; Sahoo, K.; Wilson, K.; Oneil, E. Life Cycle Assessment of Biochar Produced from Forest Residues Using Portable Systems. *J. Clean. Prod.* **2020**, *250*, 119564. [CrossRef]

63. Sahoo, K.; Bilek, E.; Bergman, R.; Mani, S. Techno-Economic Analysis of Producing Solid Biofuels and Biochar from Forest Residues Using Portable Systems. *Appl. Energy* **2019**, *235*, 578–590. [CrossRef]

64. Subramanian, A.S.R.; Gundersen, T.; Adams, T.A. Modeling and Simulation of Energy Systems: A Review. *Processes* **2018**, *6*, 238. [CrossRef]

65. Ba, B.H.; Prins, C.; Prodhon, C. Models for Optimization and Performance Evaluation of Biomass Supply Chains: An Operations Research Perspective. *Renew. Energy* **2016**, *87*, 977–989. [CrossRef]

66. Cambero, C.; Sowlati, T. Assessment and Optimization of Forest Biomass Supply Chains from Economic, Social and Environmental Perspectives—A Review of Literature. *Renew. Sustain. Energy Rev.* **2014**, *36*, 62–73. [CrossRef]

67. Leduc, S. *Development of an Optimization Model for the Location of Biofuel Production Plants*; Luleå Tekniska Universitet: Luleå, Sweden, 2009.

68. De Meyer, A.; Cattrysse, D.; Van Orshoven, J. Considering Biomass Growth and Regeneration in the Optimisation of Biomass Supply Chains. *Renew. Energy* **2016**, *87*, 990–1002. [CrossRef]

69. Vopěnka, P.; Kašpar, J.; Marušák, R. GIS Tool for Optimization of Forest Harvest-Scheduling. *Comput. Electron. Agric.* **2015**, *113*, 254–259. [CrossRef]

70. An, H. Optimal Daily Scheduling of Mobile Machines to Transport Cellulosic Biomass from Satellite Storage Locations to a Bioenergy Plant. *Appl. Energy* **2019**, *236*, 231–243. [CrossRef]

71. Zamar, D.S.; Gopaluni, B.; Sokhansanj, S. Optimization of Sawmill Residues Collection for Bioenergy Production. *Appl. Energy* **2017**, *202*, 487–495. [CrossRef]

72. Atashbar, N.Z.; Labadie, N.; Prins, C. Modeling and Optimization of Biomass supply Chains: A Review and a Critical Look. *IFAC-PapersOnLine* **2016**, *49*, 604–615. [CrossRef]

73. Mafakheri, F.; Nasiri, F. Modeling of Biomass-to-Energy Supply Chain Operations: Applications, Challenges and Research Directions. *Energy Policy* **2014**, *67*, 116–126. [CrossRef]

74. Duffy, K.A.; Schwalm, C.R.; Arcus, V.L.; Koch, G.W.; Liang, L.L.; Schipper, L.A. How Close Are We to the Temperature Tipping Point of the Terrestrial Biosphere? *Sci. Adv.* **2021**, *7*, eaay1052. [CrossRef]

75. Welfle, A.; Thornley, P.; Röder, M. A Review of the Role of Bioenergy Modelling in Renewable Energy Research & Policy Development. *Biomass Bioenergy* **2020**, *136*, 105542. [CrossRef]

76. Gómez Martín, E.; Giordano, R.; Pagano, A.; van der Keur, P.; Máñez Costa, M. Using a System Thinking Approach to Assess the Contribution of Nature based Solutions to Sustainable Development Goals. *Sci. Total Environ.* **2020**, *738*, 139693. [CrossRef] [PubMed]

77. Bang, C. *Deployment of Bio-CCS: Case Studyon Bio-Combined Heat and Power*; HOFOR Amager CHP: Copenhagen, Denmark, 2021.

78. Marris, Z.M. *Deployment of Bio-CCS: Case Study on Bioelectricity*; DraxPower Station: Selby, United Kingdom, 2021.

79. Becidan, M. *Deployment of Bio-CCS:Case Study onWaste-to-Energy*; Fortum Oslo Varme (FOV): Oslo, Norway, 2021.

80. Houghton, R.A.; Nassikas, A.A. Negative Emissions from Stopping Deforestation and Forest Degradation, Globally. *Glob. Change Biol.* **2018**, *24*, 350–359. [CrossRef] [PubMed]

81. Chazdon, R.L.; Broadbent, E.N.; Rozendaal, D.M.; Bongers, F.; Zambrano, A.M.A.; Aide, T.M.; Balvanera, P.; Becknell, J.M.; Boukili, V.; Brancalion, P.H.; et al. Carbon Sequestration Potential of Second-Growth Forest Regeneration in the Latin American Tropics. *Sci. Adv.* **2016**, *2*, e1501639. [CrossRef]

82. IPCC. *Climate Change and Land: An IPCC Special Report on Climate Change, Desertification, Land Degradation, Sustainable Land Management, Food Security, and Greenhouse Gas Fluxes in Terrestrial Ecosystems*; Shukla, P.R., Skea, J., Calvo Buendia, E., Masson-Delmotte, V., Pörtner, H.-O., Roberts, D.C., Zhai, P., Slade, R., Connors, S., van Diemen, R., et al., Eds.; IPCC: Geneva, Switzerland, 2019.

83. Olsson, O.; Lamers, P.; Schipfer, F.; Wild, M. *Chapter 7-Commoditization of Biomass Markets. Developing the Global Bioeconomy*; Academic Press: New York, NY, USA, 2016; pp. 139–163.

84. Milios, L. Advancing to a Circular Economy: Three Essential Ingredients for a Comprehensive Policy Mix. *Sustain. Sci.* **2018**, *13*, 861–878. [CrossRef]

85. Pelkmans, L.; Goh, C.S.; Junginger, M.; Parhar, R.; Bianco, E.; Pellini, A.; Benedetti, L. *Impact of Promotion Mechanisms for Advanced and Low-iLUC Biofuels on Biomass Markets: Summary Report*; IEA Bioenergy Task 40: Utrecht, The Netherlands, 2014.

86. Churkina, G.; Organschi, A.; Reyer, C.P.; Ruff, A.; Vinke, K.; Liu, Z.; Reck, B.K.; Graedel, T.E.; Schellnhuber, H.J. Buildings as a Global Carbon Sink. *Nat. Sustain.* **2020**, *3*, 269–276. [CrossRef]

87. Lamers, P.; Roni, M.S.; Tumuluru, J.S.; Jacobson, J.J.; Cafferty, K.G.; Hansen, J.K.; Kenney, K.; Teymouri, F.; Bals, B. Techno-Economic Analysis of Decentralized Biomass Processing Depots. *Bioresour. Technol.* **2015**, *194*, 205–213. [CrossRef]

88. Maheshwari, P.; Singla, S.; Shastri, Y. Resiliency Optimization of Biomass to Biofuel Supply Chain Incorporating Regional Biomass Pre-Processing Depots. *Biomass Bioenergy* **2017**, *97*, 116–131. [CrossRef]

89. Soren, A.; Shastri, Y. Optimization Based Design of a Resilient Biomass to Energy System. In *Computer Aided Chemical Engineering*; Friedl, A., Klemeš, J.J., Radl, S., Varbanov, P.S., Wallek, T., Eds.; Elsevier: Amsterdam, The Netherlands, 2018; Volume 43, pp. 797–802. [CrossRef]

90. Nasso, S.; Sweazey, B.; Gagnon, B. Bio-Hubs as Keys to Successful Biomass Supply for the Bioeconomy. In *Report from Joint IEA Bioenergy Task 43 & Natural Resources Canada Workshopheld in Ottawa on 6 March 2020*; IEA Bioenergy Task 43: Ottawa, ON, Canada, 2020.

91. Kulišić, B.; Brown, M.; Dimitriou, I.; Murphy, E.K.; Gagnon, B. Bio-Hubs as Keys to Successful Biomass Supply Integration for Bioenergy within the Bioeconomy. 2020, p. 35. Available online: https://www.ieabioenergy.com/wp-content/uploads/2020/06/Bio-hubs-as-Keys-to-Successful-Biomass-Supply-for-the-Bioeconomy.pdf (accessed on 31 December 2021).

92. Liu, W.; Klip, D.; Zappa, W.; Jelles, S.; Kramer, G.J.; van den Broek, M. The Marginal-Cost Pricing for a Competitive Wholesale District Heating Market: A Case Study in the Netherlands. *Energy* **2019**, *189*, 116367. [CrossRef]

93. Wang, J.; You, S.; Zong, Y.; Cai, H.; Træholt, C.; Dong, Z.Y. Investigation of Real-Time Flexibility of Combined Heat and Power Plants in District Heating Applications. *Appl. Energy* **2019**, *237*, 196–209. [CrossRef]

94. Pfeiffer, A.; Mertens, A.; Brosowski, A.; Thrän, D. *Der Strohmarkt in Deutschland*; Deutsches Biomasseforschungszentrum Gemeinnützige GmbH: Leipzig, Germany, 2019.

95. Schipfer, F.; Kranzl, L.; Olsson, O.; Lamers, P. The European Wood Pellets for Heating Market-Price Developments, Trade and Market Efficiency. *Energy* **2020**, *212*, 118636. [CrossRef]

96. Eggler, L.; Indinger, A.; Zwieb, L.; Mapping of Activities in Technology Collaboraiton Programmes (TCPs) in the Energy Technology Network of the International Energy Agency (IEA). 2018, p. 64. Available online: https://nachhaltigwirtschaften.at/resources/iea_pdf/endbericht_201810_iea-mapping-tcp.pdf (accessed on 31 December 2021).

 *sustainability*

Article

Tree Resin, a Macroergic Source of Energy, a Possible Tool to Lower the Rise in Atmospheric CO_2 Levels

Jaroslav Demko and Ján Machava *

Department of Biology and Ecology, Faculty of Education, Catholic University in Ruzomberok,
Hrabovska Cesta 1, 034 01 Ruzomberok, Slovakia; jaroslav.demko@ku.sk
* Correspondence: jnmchv9@gmail.com; Tel.: +421-905-756824

Abstract: Tree resin is a macroergic component that has not yet been used for energy purposes. The main goal of this work is to determine the energy content of the resin of spruce, pine, and larch and of wood components—pulp and turpentine. The combustion heat of resin from each timber was determined calorimetrically. Approximately 1.0 g of liquid samples was applied in an adiabatic calorimeter. The energy values of the tree resin (>38.0 MJ·kg^{-1}) were 2.2 and 2.4 times higher than that of bleached and unbleached cellulose, and the highest value was recorded for turpentine (>39.0 MJ·kg^{-1}). Due to the high heating values of the resin, it is necessary to develop approaches to the technological processing of the resin for energy use. The best method of resin tapping is the American method, providing 5 kg of resin ha^{-1} yr^{-1}. The tapped resin quantity can be raised by least 3 times by applying a stimulant. Its production cost compared to other feedstocks was the lowest. Tree resin can be applied as a means of mitigating global warming and consequently dampening climate change by reducing the CO_2 content in the atmosphere. One tonne of tree resin burned instead of coal spares the atmosphere 5.0 Mt CO_2.

Keywords: resin; combustion heat; renewable energy source; wood; carbon sequestration; climate change

Citation: Demko, J.; Machava, J. Tree Resin, a Macroergic Source of Energy, a Possible Tool to Lower the Rise in Atmospheric CO_2 Levels. *Sustainability* **2022**, *14*, 3506. https://doi.org/10.3390/su14063506

Academic Editors: Idiano D'Adamo and Marc A. Rorasen

Received: 3 August 2021
Accepted: 25 February 2022
Published: 16 March 2022

Publisher's Note: MDPI stays neutral with regard to jurisdictional claims in published maps and institutional affiliations.

1. Introduction

According to European Union policy (Directive 2009/28/EC), 20% of the overall energy consumption in the EU by 2020 should be provided by renewable sources. In this respect, adequate findings must be obtained for the characterization and identification of specific biomass types [1,2], because forest biomass quality is closely associated with carbon sequestration, lowering its content in the atmosphere [3]. A more detailed study on the energy issue of resin as an important component of tree biomass is therefore necessary.

Resins are related to a family of extractive substances, including waxes, suberin, cutin, glycosides, alkaloids, and others, that protect the plants against unfavorable biotic and abiotic influences [4–9]. Resin can be chemically characterized as a group of aromatic compounds of a terpene nature, formed by two major classes of substances—the first group of compounds is derived from phenylpropane, and the second is from terpene compounds. Resin is a mixture of various substances, of which resin acids, terpene-type hydrocarbon, terpenoid alcohols, waxes, and solutions of relatively light or volatile terpenes are the most important [10–13]. The terpenes are largely monoterpenes that generally comprise about 25% of the total weight of resin. The remainder of the liquid fraction is chiefly sesquiterpenes (0–20%). Thus, the aggregate of all terpenes in the whole resin is in the range 25–45% [14]. It is generally agreed that terpenes are formed by the polymerization of isoprene (C_5H_8) from hemiterpenes, through monoterpenes, and up to polyterpenes (including rubber and gutta) [15,16].

Tree resin is a complex mixture of volatile and non-volatile terpenes. Mono-($C_{10}H_{16}$) and sesquiterpenes ($C_{15}H_{24}$, turpentine; farnesyl diphosphate, FPP) are generally volatile, giving fluidity to the resin, and they usually give resin its characteristic odor. When only volatile mono- and sesquiterpenes occur, they often are called essential oils. This designation, however, is misleading because these terpenoids are neither essential to plant metabolism nor true oils; "essential" refers to their essence or fragrance, and "oil" refers to their feel. Turpentine constitutes the largest group of secondary products and derives from a 5C compound, IPP (isopentenyl pyrophosphate). Non-volatile diterpenic acids ($C_{20}H_{32}$, rosin), the most frequent and abundant diterpenoid resin compounds occurring in rosin, are derivatives of abietane, pimarane, and isopimarane skeletons [17,18]. Diterpene acids are particularly important in resin. The doubling (dimerization) of the $C_{15}H_{24}$ (FPP) leads to $C_{30}H_{52}$ compounds, i.e., triterpenes. Triterpenes include a wide variety of structurally diverse substances. These terpene fractions are a natural source of valuable materials for chemical industries [19].

Plant resin can be defined primarily as a lipid-soluble mixture of volatile and non-volatile terpenoid and/or phenolic secondary compounds that are usually secreted in specialized structures located either internally or on the surface of the plant, and of potential significance in ecological interactions [10]. Resin is sometimes confused with oleoresin. Oleoresins are comparatively fluid terpenoid resins with a higher ratio of volatile to non-volatile terpenes. Hereinafter, the resin will mostly be referred to as oleoresin.

After the collection, crude resin conversion into gum turpentine and gum rosin is carried out by steam distillation [20]. Subsequently, these by-products are processed for use in the fabrication of diverse industrial products such as feedstock, cleaners, pine oil, fragrances and flavoring compounds, pesticides, solvents and thinners for paints, and pharmaceuticals products [21–23]. Most *Pinus* species yield pinenes (α- and/or $\exists$-pinene) as major compounds in their monoterpenic turpentine fraction. Large amounts of pinenes are used in the flavor and fragrance industry [24]. However, due to their strong odor, they cannot be extensively used as additives in flavors or fragrances; pinenes are chemically converted to more valuable products, such as verbenone, a monoterpene that exhibits an ecological role as an anti-aggregating pheromone (tree protection) [25–30]. Besides this, pinenes might also be used in the production of pharmaceuticals, plasticizers, repellents, insecticides, solvents, perfumery, cosmetics, and antiviral and antimicrobial compounds [26–32]. Some investigations have been carried out with the aim of assessing the economic viability of performing resin-tapping operations [21,33–35].

The combustion heat and the calorific value of stump wood, tree crown components, and assimilatory organs are relatively well known [36–38], and fluctuate depending on tree species. For chestnut and pine, the crown fraction generates the highest energetic values, and the wood produces the lowest. The calorific value per tree is lowest for chestnut ($17.419 \ MJ \cdot kg^{-1}$), intermediate for Short-Rotation Coppice crops ($18.185–18.419 \ MJ \cdot kg^{-1}$), and highest for maritime pine ($19.336 \ MJ \cdot kg^{-1}$) [38]. This is caused by the higher lignin and resin contents in coniferous species. These results are consistent with the trends observed by Telmo [39], who reported higher energetic values for different pine species than for native broad-leaved and autochthonous species in northern Portugal. Wood extractives such as resin, waxes, oils, tannins, and other phenolic substances also have much higher heating values than cellulose and hemicelluloses, and they are abundant in the wood of coniferous species [40]. The contents of terpenes and oleoresins significantly affect the energetic values of "lignocellulosic fuels". Furthermore, Howard [41] has calculated the higher heating values of resin as 15,000 to 16,000 btu/lb ($34.890–37.216 \ MJ \cdot kg^{-1}$), but the variation in the results was large, as the resin samples were semiliquid and rather inhomogeneous. In a follow-up on the inquiries stated, this work examines the properties of tree resin and its potential for use as a renewable energy source.

Problems

Resins are macroergic solids (high energy), formed in a manner similar to the essential oils in specific resin canals. Physiological resin is formed by the metabolic activity of trees, and pathological resin is mainly produced as a result of damage to the trees. The amount and composition of the resin in various types of wood depends largely on the biochemical taxon of individual species, environmental effects [42], geographic origin [43], local conditions [44,45], etc.

Significant differences in the composition of resin were found not only between different coniferous trees, but also between trees of the same species [46]. Differences were found between the physiological resin composition coming from a healthy tree and the pathological resin flowing from the bark of the injured tree (similar designations: a constitutive and induced resin). Pathological resin consists substantially of terpenes and resin acids, it is fat-free, it provides excellent resin after distilling turpentine, and it is also referred to as callus resin [47]. The resin obtained by extraction from harvested wood cannot also be considered as physiological in origin, because the solvents used are not indifferent to the oleoresin or to its constituents, and the resin obtained by extraction is chemically modified. From a chemical point of view, the oleoresin obtained from growing trees is, therefore, more acceptable and represents a more appropriate approach.

The importance of an energy analysis of dendromass components, oleoresin in this case, is important for obtaining information on their living nature and their high content in some woods. Due to their high energy value, these findings play an important role, for example, in selecting the biochemical taxon of trees, which ensures the highest possible production of energy biomass at energy forest plantations [48,49].

Generally, four pine resin-tapping techniques are used worldwide. The most effective is the American method (Figure 1, an improved V-shape) [50], often referred to as the bark streak method [51].

Figure 1. American pine resin tapping techniques (V-shaped streaks (2–3 mm wide)). Cut into xylem. Source: [52].

Nowadays, it is well established that tree resin properties depend on key factors, and mainly on climate conditions [53,54], the genetic background, and the environmental effects [42,55]. Additionally, the quality and quantity of resin and its by-products can be influenced by age [56], geographic origin [43], and different stresses, such as low-fertility soils [57], drought [44,58], extreme temperatures [59–61], and the hydrological impact [62]. Further, resin production can be modulated by resin-tapping methods [50], associated with chemical stimulation treatments [63].

Although the history of the tapping and processing of tree resin dates back to the 18th century, energy values, such as the combustion heat and calorific values, are rarely stated for a macroergic substance such as oleoresin. Absent or incomplete findings prompted us to conduct an energy review of the resin of the most economically important conifers in the Central European region: spruce (*Picea abies* (L.) H. Karst), pine (*Pinus sylvestris* L.), and larch (*Larix decidua* Mill.) trees. The main goal of this work is to determine the combustion heats of the oleoresins of these trees, and to compare them with those of further tree wood components. Knowing the energy levels of these substances is important because biomass currently forms the basis of renewable energy sources and forest ecosystem services in Central European conditions.

2. Materials and Methods

2.1. Sampling Material

The resin samples were taken by a modified streaking method from spruce (*Picea abies* (L.) H. Karst), pine (*Pinus sylvestris* L.), and larch (*Larix decidua* Mill.) trees growing on the Cernova research area (3,3 ha), nearby Ruzomberok (A: 49.101501 N, 19.236523 E, B: 49.100178 N, 19.236632 N, C: 49.100344 N, 19.239303 E, D: 49.101986 N, 19.239519 E), at an altitude of approximately 490 ± 90 m asl., and were transported to the laboratory and stored in a freezer prior to energy determination analyses. The tree bark was cut with a groove knife into the sapwood (a modified V-shaped streak), and the resin flowing from there was later collected with attached cups. The resin obtained in this way is not actually physiological resin. However, it is more suitable than that obtained from harvested timber by chemical extraction.

To compare the energy of the resin obtained by the streaking method, a resin component from pulp production, known as turpentine, was used. This is essentially a mixture of resins. The measured energy values of the resin were further compared with the basic constituents of the dendromass, such as bleached and unbleached cellulose and turpentine. These wood constituents were obtained from Mondy SCP, Ltd. (Ružomberok, Slovakia), a pulp and paper complex in Ruzomberok.

2.2. Assessment of Energy—Calorimetry

Approximately 1.0 g of liquid samples was weighted on the Denver Instrument SI-234 scale with a precision of 0.0001 g and placed into a 3 mL metal crucible. After weighing, determination of the combustion heat was accomplished in an adiabatic bomb calorimeter (Model IKA Calorimeter C400). The resin sample was completely incinerated at 3.0 ± 0.3 MPa in a pure oxygen environment. The heat emitted during incineration was transferred to the calorimeter fluid, whereby the heat capacity for each calorimeter was specified. From the difference in temperature (ΔT) between the initial condition and after incineration, the level of energy released from the sample material as well as the combustion heat was calculated by the respective equation [64]. At least one replicate determination was carried out for each resin sample.

2.3. Statistical Evaluation

The data determined were tested for normality using the Kolmogorov-Smirnov test and the SPSS statistical packet. The normal distribution of the combustion heat data of pine and larch was confirmed by the p-values of 0.141 and 0.200. In the spruce data, the first two values were outliers, and the data did not meet the normality presumption ($p = 0.026$). No significant differences between the combustion heat values determined for pine, spruce, and larch were indicated by the non-parametric Kruskal-Wallis test ($p < 0.509$). Statistical parameters of combustion heat values for pine (*Pinus sylvestris* L.), spruce (*Picea abies* (L.) H. Karst), and larch (*Larix decidua* Mill.) are stated in Table 1.

Table 1. Statistical parameter of combustion heat of tree resin (MJ·kg^{-1}).

	Pine	Spruce	Larch
Minimum	36.00	34.60	36.30
Maximum	40.10	40.00	39.70
Range	4.10	5.40	3.40
Mean	38.60	38.37	38.33
Stand. error	0.30	0.35	0.23
Stand. deviation	1.30	1.53	0.99
Variance	1.70	2.33	0.97
Skewness	−0.65	−1.51	−0.57
Kurtosis	0.52	2.05	0.52
Number	19.00	19.00	19.00

3. Results

The renewable biomass resources studied in the past for energy purposes comprise many herbaceous and woody plants. In the composition of the wood, the main components account for 97–98% (saccharide—cellulose 49%; hemicellulose 24%; aromatic—lignin 24%), and the accompanying components account for 2–3% (e.g., resin). This work focuses on the tree resin.

The results of the energy analysis are presented in terms of the combustion heat values of the resins from *Pinus sylvestris* L., *Picea abies* (L.) H. Karst, and *Larix decidua* Mill. (Figure 2), as well as the pulp, an intermediate in wood production, and turpentine, a component of the resin (Table 2). For the other components of tree biomass, the energy parameters are relatively well known [37,65]. The heat values of pine, spruce and larch are quite high, and tree resin is probably the substance with the highest energy content out of all tree components in plants. The mean combustion heat values of pine, spruce, and larch are, respectively, 38.591 ± 1.307 MJ·kg^{-1}, 38.373 ± 1.521 MJ·kg^{-1}, and 38.326 ± 0.975 MJ·kg^{-1}. The differences between the combustion heat values of pine, spruce, and larch are statistically insignificant ($p < 0.509$).

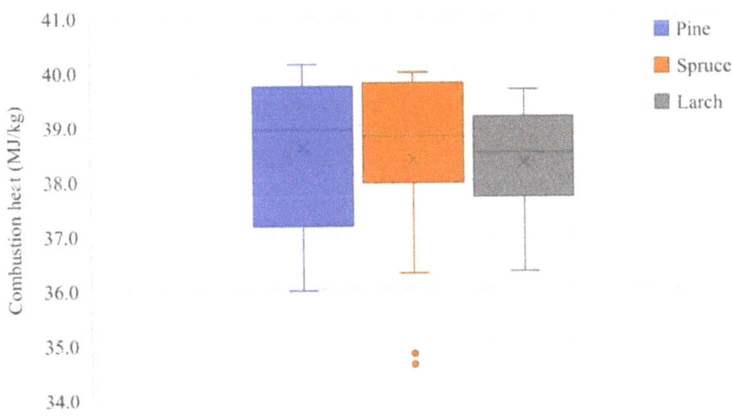

Figure 2. Resin combustion heat of pine, spruce, and larch. $n = 19$.

Table 2. Resin combustion heat of bleached and unbleached pulp and of turpentine.

Component of Wood	Combustion Heat (MJ·kg^{-1})	Ash (%)
Bleached pulp	17.319 ± 0.025	1.29
Unbleached pulp	15.955 ± 0.036	8.21
Turpentine	39.773 ± 0.027	2.83

The resin energy values measured were further compared with cellulose associated with paper production, and can be supposed to be a by-product. The readings of bleached pulp (cellulose) produced by the Kraft sulfate process may serve as the energy reference standard for tree biomass. The mean combustion heat values of bleached (17.319 ± 0.025 MJ·kg^{-1}) and unbleached pulp (15.955 ± 0.036 MJ·kg^{-1}) (Table 2) are 2.2 and 2.4 times lower compared to the lowest value of the investigated tree resin (larch 38.326 ± 0.975 MJ·kg^{-1}).

The combustion heat of turpentine (39.773 ± 0.027 MJ·kg^{-1}), a constituent produced as a by-product in the sulfate technological procedure, is 2.3 and 2.5 times higher than those of white and brown pulp. This figure is even slightly higher than that of tree resin, and this may be due to the different representations of the volatile components of turpentine (Table 3; Figure 3).

Table 3. Composition of turpentine.

Constituent	Content (%)
Dimethylsulfide	25.54
Dimethyldisulfide	0.14
α-pinene	36.28
Camphene	0.72
β-pinene	9.74
Myrcene	0.15
δ-3 carene	9.34
P-cymene	0.29
D-limonene	2.70
α-terpinene	0.93
Γ-terpinene	0.18
Terpinolene	0.94
Unidentifiable	3.95
Unidentifiable (heavier than terpinolene)	9.1

Method: Gas chromatograph—Agilent 7890A Yield and technical characteristics of the tapping operations.

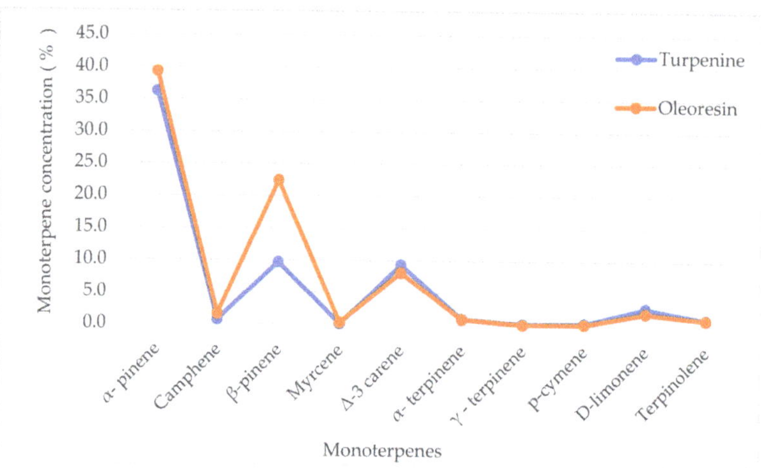

Figure 3. Monoterpene concentrations in turpentine and oleoresin of *Pinus sylvestris* L. (oleoresin values taken from [52]).

The properties of oleoresin are determined mainly by the proportion of volatile monoterpenes. The ratio of volatile to non-volatile components affects the physical and defensive properties of oleoresin. The volatile and more fluid components of resin enable the movement of the more viscous non-volatile components. The non-volatile di- and triterpenoid or phenolic components increase the viscosity and rate of crystallization of the oleoresin. These properties affect the rate of resin flow and the ability of the resin to trap and immobilize enemies or coat wounds in tree trunks, which has been considered a first line of tree defense [10].

In this work, the analysis of the representations of turpentine components was performed, and the concentrations measured were compared to those in the oleoresin of *Pinus sylvestris* L. (Figure 3). The highest levels in turpentine were recorded for α- and β-pinenes (together making up 46% of the total content), but were still lower than that of oleoresin (62%). Δ-3-carene and D-limonene were in the range of 3–10%. Further resin constituents observed were below 1%, namely, terpinolene, camphene, α- and γ-terpinene, p-cymene, and myrcene. Although turpentine is prepared by a distillation process, its composition is rather similar to that of the oleoresin of *Pinus sylvestris* L. [52]. A similar composition of monoterpenes was found in needles and cortical oleoresin [66]. In needles, the major constituent was β-pinene, and it was α-pinene in the latter. In general, the content of monoterpenes in needle oleoresin decreases from winter to summer, while the concentration of sesquiterpenes increases. In cortical oleoresin, the case was the reverse. This descending trend of monoterpene concentrations in needles was most likely caused by their ascending release into the atmosphere, with an increasing temperature up to the summer period. The monoterpene content, which represents the main components of the natural emission load, has its own effect on the course of global warming.

Furthermore, the combustion value of resin was compared to that of the tree wood. The combustion heat of tree wood irrespective of tree species ranges from 18.000 to 20.000 MJ·kg^{-1} [67]. Similar results were stated for softwood [38,68]. The lowest value was found in eastern white cedar branches (18.668 MJ·kg^{-1}) and the highest was found in black spruce treetop (21.562 MJ·kg^{-1}); the mean value of all softwood components is 20.178 MJ·kg^{-1}. For the hardwood, the span is wider—the lowest value is 17.230 MJ·kg^{-1} in Manitoba maple foliage, and the highest one is 21.119 MJ·kg^{-1} in White birch foliage; the mean of all components is 19.146 MJ·kg^{-1}. An even wider span of values was given [69] in a statistical summary of the calorific values of 402 species of wood of 246 genera, and 66 species of bark of 33 genera, based primarily on literature surveys. The calorific values range from 15.584 to 23.723 MJ·kg^{-1} in hardwood and from 18.608 to 28.447 MJ·kg^{-1} in softwood. Excluding the highest values in softwood, the resin combustion heat is almost twice as high as that of wood.

In the case of refined liquid hydrocarbon fuels such as petrol or diesel, the mean energy value of conifer resins is close to the lower end of the combustion heat range of diesel (about 41.900 MJ·kg^{-1}) and petrol (43.500 MJ·kg^{-1}). However, the maximum measured value of pine resin combustion heat (40.109 MJ·kg^{-1}) is only 5% lower than hydrocarbon fuel's combustion heat (1.8 MJ). High-calorific solid fuel such as coke registered only 70% of the heating value measured in the pine resin. Black coal achieved only 65%, and brown coal achieved only 36%.

4. Discussion

The use of renewable energy sources is becoming increasingly important to achieving the changes required to address the impacts of global warming. In the context of current European Union policy, woody biomass is expected to be an important energy resource in the near future. Up to now, tree biomass has been investigated mostly for the energy utilization of tree body components, such as the branches and stumps of broadleaves and conifers [37,68]. However, several studies have specifically focused on the enhancement of the fuel characteristics of woody biomass, such as wood density, volatile matter, the calorific value of wood [70], the fertilization of *Picea abies* stands [71], landscaping, and bioesthetic planning [72]. Further, the individual constituents of tree resin, such as rosin and turpentine, have been studied [73–75]. However, only a limited number of studies have focused on a macroergic material such as tree resin.

This paper reveals that the combustion heat values of pine, spruce, and larch are 38.591 MJ·kg^{-1}, 38.373 MJ·kg^{-1}, and 38.326 MJ·kg^{-1}, respectively, and the difference between figures is statistically insignificant ($p < 0.509$). The analysis of the energy components of the wood biomass shows that the combustion heat of oleoresin (Figure 2) is higher than that of bleached cellulose (2.2-fold) and unbleached cellulose (2.4-fold). Only the calorific value of turpentine is higher, by 4%, but the percentage of turpentine in spruce wood is only 0.1–0.2%. The resin content of the tree's wood is more than 1–2%.

Similar results, albeit showing large variation, were recorded by Howard [41] and Ivask [76]. While Howard [41] reported lower heating values in a range of 34.89–37.22 MJ·kg⁻ in tree pine resin, Ivask [76] recorded higher values in spruce resin, namely, 40.10 ± 0.62 MJ·kg^{-1}. These differences in energy values in tree resin can be explained by differences in the ratio of cellulose and lignin [41], by topographical [77], ecological [78], seasonal [79], and environmental aspects, and naturally by the conditions and methods of determination. The variation in values [41] was large, as the samples were difficult to mix thoroughly, and were therefore not homogeneous. Ivask [76] predominantly investigated the influence of seasonal aspects, as well as the local and climatic conditions, on the energy parameters of tree wood components. Spruce tree resin combustion heat was determined in only six samples, and the author [76] himself admitted that a random estimate of calorific values yields very little information.

Although oleoresin is obtained by tapping in small quantities, it takes place over almost the entire year, and can be obtained from a living tree. The quantity and quality of the oleoresin are also determined by a tapping process. Resin sampling from pine trees has occurred since the mid-19th century [80]. However, it is necessary to mention that, since the 1980s, there has been a decline in oleoresin sampling due to lower purchase prices. This decrease in price was caused by the advent of cheaper competing commodities produced from crude oil. However, nowadays, oleoresins with high calorimetric values might return in economies that utilize green biofuels and bioproducts from non-food feedstocks [81]. There are four methods of oleoresin tapping; the American method is the most effective and provides 2.5 times the yield of the Chinese method (Table 4). The largest quantities of resin have been tapped from pine trees.

Table 4. Yield and technical characteristics of the tapping operations. The values are valid for pine trees.

Location	Brazil	China
Tapping Technique	American	Chinese
Density (trees/ha)	800	700
DBH (cm)	25	15
Season (months)	9	5.6
Years in production	~20	5 to 7
Yield per time (g/day)	19.7	11.2
Trees tapped per worker	7000	1500
Hectares tapped per worker	8.75	2.18
Metric tonnes produced per worker	35	3
Pine resin (kg/year)	5	2

Legend: Modified table [50].

The American sampling method provides 5 kg of resin per year under optimal conditions. This amount fluctuates according to local conditions; for example, in Portugal, this amount was reduced by almost half [82]. Some investigations have been carried out with the aim of assessing the economic viability of performing resin-tapping operations [21,33–35]. However, the highest increase in extractive contents in *Pinus elliottii* biomass was achieved by using a 2% paraquat-cation stimulant [83]. In the low 152 cm bolt, there was an 884% rise in resin acid amounts and a 2360% rise in turpentine values, and these values underwent 273% and 684% increases, respectively, in the whole stem. By applying a 2% paraquat-cation stimulant in the American tapping method, the yield might be as much as 18.65 kg of oleoresin per pine tree. Per hectare, this yields a total amount of

14,920 kg (575.78 GJ). Increasing resin production occurs with damage to the health status of the sample trees.

Tapping, irrespective of the method employed, causes intensive wounds in tree stems, leading to wood deformation. Tapped trees, compared to non-tapped ones, after one tapping, show a decrease in mean tree ring width by 14.1% (an average tree ring width of 2.41 ± 0.85 mm was reduced to 2.07 ± 0.7 mm), but this decrease is only by 6% in late wood [84]. Decreases in the volume of the wood are visible; however, this damage is negligible compared to the energy that is stored in the tapped resin.

Due to its diverse applications, the *Pinus* genus is considered one of the most important commercial timber species [85]. Today, it is well established that resin properties depend mostly on factors such as genetic background and environmental effects [42,55]. Its low technical requirements for planting [19] make *Pinus* one of the most suitable woody species for cultivating and restoring marginal areas, as well as abandoned and degraded agricultural lands [86].

In the 2015–2030 period, incremental abandonment is expected to reach 4.2 Mill. ha net (approximately 280,000 ha per year on average) of agricultural land, bringing the total abandoned land to 5.6 Mill. ha by 2030 (3% of total agricultural land). Arable land is projected to account for the largest share of abandoned land (4.0 Mill. ha; 70%), followed by pasture (1.2 Mill. ha; 20%) and permanent crops (0.4 Mill. ha; 7%). Nearly a quarter ($\approx$1.38 Mill. ha) of all agricultural areas in mountainous areas in the EU will probably be abandoned [87]. This supposes that half of the abandoned agricultural land in mountainous areas (700,000 ha) will be afforested, and pine oleoresin will be tapped from an adult pine stand growing on this land. The pine oleoresin collected by the American method using a 2% paraquat-cation stimulant over one year will yield an energy value of 403.044 PJ, which would provide 0.82% of the fossil coal energy and 0.24% of the total energy (160–180 EJ) required worldwide (as of 2018). This energy is produced by more than 15.5 Mt of coal, with each metric tonne of coal producing 1700–1800 m^3 of CO_2, thus exacerbating the problem of global warming [88]. If this coal is replaced by tapped resin via the combustion process, the atmospheric load will be reduced by 27.1 Gm^3 CO_2, i.e., 53.7 Mt, as its origin is not fossil fuel (Table 5).

Table 5. Resin and carbon dioxide quantities associated with a 700,000-ha pine forest stand.

Area	Resin Tapped Over One Year		C Sequestered by Pine Trees	CO$_2$ Released from Coal (Equivalent of Resin Amount)	
(ha)	(Mt)	Q (PJ)	(Mt)	(Gm^3)	(Mt)
700,000	10.444	403.044	1.179	27.128	53.700

Legend: 4.320 Mt CO_2 is equivalent to 1.179 Mt C; ρ_{CO_2} = 1.98 kg m^{-3}; $m_{coal} = \left(\frac{Q_{resin}}{Q_{coal}} \right) * 10.444 = 15.5$ (Mt).

Further, forests have also been recognized for the other services they offer, such as the ability to store carbon and mitigate the impacts of climate change [89]. Forest plantations are beneficial in mitigating climate change and reducing airborne emissions only through stringent management. A 700,000 ha *Pinus sylvestris* plantation with a net ecosystem exchange (NEE) rate of 1.684 t C ha^{-1} yr^{-1} sequesters 1.179 Mt of carbon per year (4.320 Mt CO_2). A NEE rate of 1.684 t C ha^{-1} yr^{-1} converges with the average annual increase in C in a Scots pine stand equaling 1.234 t yr^{-1} (105.42 t/85.46 years) [90]).

Although the aforementioned example of afforestation is only theoretical, it reveals the possibilities of oleoresin utilization in the environmental field. All constituents of biomass are photosynthesized in plant leaves from CO_2, water, and absorbed solar energy. A specific feature of biomass is that its combustion produces the same amount of the greenhouse gas that was absorbed during photosynthesis. Biomass is neutral in terms of CO_2 emissions. The establishment and management of forests as a source of oleoresin also supports the removal and storage of CO_2 from the atmosphere, and offsets the increase in the anthropogenic emissions of greenhouse gasses (GHGs), consequently reducing the rate of global warming and mitigating the impacts of climate change [52,89]. A plantation system is a good means of climate change mitigation [91,92].

Generally, according to the essence of the production processes, feedstocks are classified as first-, second-, or third-generation. Oleoresin, by its lignocellulosic nature, can be assigned to the second generation of feedstocks. Third-generation feedstocks are produced from algae, sewage sludge, and municipal solid waste (this is omitted) [93]. The prices of these feedstock generations differ. First-generation feedstock supply is supposed to be available at production costs of EUR 5–15 GJ^{-1} compared to EUR 1.5–4.5 GJ^{-1} for second-generation feedstocks [94]. The production cost of first-generation feedstocks is approximately triple that of the second-generation ones. The difference is slightly smaller between the cost of residues in agriculture and forestry (EUR 1–7 GJ^{-1} and EUR 2–4 GJ^{-1}, respectively).

For the 2008–2011 period, the production cost of oleoresin was nearly double that of lignocellulosic crops, i.e., USD 0.2–0.4 kg^{-1} $\equiv$ EUR 4.6–9.2 GJ^{-1} [95]. However, when oleoresin tapping was carried out by the American tapping method with an application of a 2% paraquat-cationic stimulant, the production cost was significantly reduced. If we take into account the ratio between the energy stored in oleoresin tapped from 1 ha of pine stand (575,78 GJ yr^{-1}) and that stored in wood from 1 ha of SRC of willow (172 GJ yr^{-1}), the oleoresin production per unit amount is the cheapest, at EUR 1.3–2.75 GJ^{-1} (Table 6). Further investigations are needed to demonstrate this.

Table 6. Comparison of energy values of renewable energy sources.

Species	Yield	Heating Value	Energy per Hectare		Production Cost
	(t)	(MJ kg^{-1})	(GJ)	(MWh)	(€/GJ)
Feeding [a] sorrel	10.0	16.00	160.00	44.44	
Common [b] reed	12.7	17.45	221.60	61.56	1.5–4.5
Miscanthus [c]	14.0	17.60	246.40	68.45	
Pine resin [d]	14.9	38.59	575.80	159.94	1.3–2.75

Legend: [a] a hybrid of *Rumex patientia* L. (maternal line) and *Rumex tianschanicu* (paternal line) species [96]; [b] *Phragmites australis* (Cav.) Trin. [97]; [c] *Miscanthus gigantheus* [98]; [d] *Pinus sylvestris*.

Biofuels are a potentially low-carbon energy source, but whether biofuels offer carbon savings depends on how they are processed [92]. Tree resin has been proven to be an excellent renewable energy source that has not yet been used for this purpose. Compared to previously investigated energy sources, resin has the highest energy content per hectare (Table 6). The production of oleoresin is not the primary goal, but the volume of the greenhouse gas CO_2 will be reduced with afforestation; consequently, the rate of global warming will be dampened. Afforestation associated with the conversion of marginal agricultural or forestry lands to purpose-grown crops is an important practice used to lower the rise in atmospheric CO_2 concentration, due to a forest's ability to fix carbon in tree biomass and in soil [52,99,100].

While wood reduction in forests requires the removal of woody biomass, utilizing it for power generation reduces overall emissions by 98% in comparison with slash pile burning [101]. Energy production in the forest is more environmentally friendly. For example, regrowth energy in a forest stand generated from wood leads to a production of 0.057 metric tonnes of CO_{2e} per MWh, compared with the average US rate of 0.60 metric tonnes of CO_{2e} per MWh [102]. Striking relations also exist between the annual amounts of carbon sequestered by a pine stand, the resin tapped from pine trees by the American method, and the carbon dioxide released from coal with the same energy content as tapped resin (Figure 4).

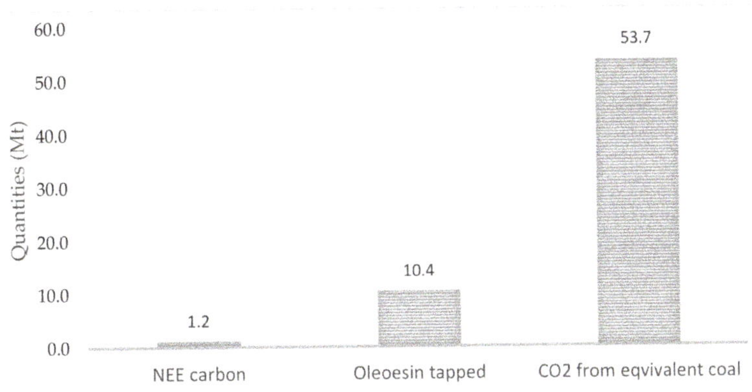

Figure 4. Quantities of net ecosystem exchange (NEE) C accumulated in stand, oleoresin obtained from a pine trees stand and CO_2 from coal (energy equivalent to resin tapped over one year; 700,000 ha).

The lowest amount was recorded for NEE carbon sequestered by a tree stand. The extracted resin and the released CO_2 held, respectively, 9 times and 45 times the quantity of the carbon sequestered. This means that 1 tonne of oleoresin corresponds to approximately 5 tonnes of CO_2. If the oleoresin is burned as a renewable energy source instead of coal, the emission load in the atmosphere would be reduced by roughly 5 tonnes of anthropogenic CO_2. (This is valid provided that CO_2 generated from oleoresin is environmentally neutral)

Tree resin production is directly related to bioeconomic issues. The bioeconomy may be conceived as a prime way of engaging with ecological modernization, i.e., economic and technological modernization that seeks to address perceived environmental issues [103–105]. The bioeconomy may well be an inevitable transition if fossil resources are to be phased away [106]. The bioeconomy is attracting interest as a conceivable win–win solution for green growth. The European Bioeconomy Strategy supports the production of renewable biological resources and their conversion into vital products and bioenergy in order to satisfy the 2030 Agenda and its Sustainable Development Goals [107,108]. It represents a wide range of opportunities for sustainable development in bio-based industries [109,110], which encompass various sectors, including agriculture and forestry.

The bioeconomy aims to substitute fossil resources with bio-based alternatives, such as tree resin applications for energy. This energy utilization can be evaluated by the socio-economic indicator of the bioeconomy (SEIB) [111]. The circular economy (CE) concept aims to reduce resource use and consumption, favoring reuse and recycling activities and aiming to minimize waste and emissions (environmental risks). To make progress toward the CE, it is essential to prepare accurate estimates of the environmental/economic and ethical dimensions of proposals to support this transition. So far, among the potential energy forms to be derived from biomaterials, biogas is of the greatest significance due to its ability to transform organic feedstocks into biomethane (CH_4).

The bio-CH_4 potential of several European cities was estimated, and a large share of this potential can be used as vehicle fuel and can therefore help the European Union (EU) to achieve its Paris Agreement commitments within the transport sector [112,113]. Currently, the transport sector is responsible for a third of the global energy demand, and one-sixth of global GHG emissions. [112,114]. This sector is currently dominated by the usage of fossil fuels in Europe. The experiment with bio-CH_4 yielded the following.

The production of bio-CH_4 in Portugal would only be profitable under potential government incentives in the form of feed-in premia above 15.42 $EUR.GJ^{-1}$ (55.5 $EUR.MWh^{-1}$) [115]. Selling CO_2 for a price of 46 $EUR.t^{-1}$ CO_2 would also be profitable. In 2017, a total of 1.94×10^9 (American billion) cubic meters of bio-CH_4 was produced in Europe. In 2050, the European biomethane potential will be 95×10^9 cubic meters [116,117]. This bio-CH_4 production, showing a trend of 50-fold enlargement from 2017 to 2050, indicates the possible energetic use of tree resin due to its price. The bio-CH_4 production cost is approximately five times the cost of tree resin production (1.3–2.75 EUR.GJ^{-1}). The application of tree resin for energy utilization is associated with environmental issues, and may have social utility in such cases as in southern Brazil [110,118–120].

The objective of the 2015 Paris Climate Agreement is to hold average global warming to well below 2 °C above preindustrial levels. While this objective is formulated at the global level, the success of the agreement depends on the implementation of climate policies at the national level. Each country submits nationally determined contributions (NDCs) [121]. In the EU, in order to achieve conditional NDCs, greater reductions are necessary than those achieved by national policies only. The implementation of conditional NDCs (NDC scenario) is projected to result in 51.9 (50.4–57.4) GtCO$_{2eq}$ GHG emissions by 2030 [121]. When tree resin tapped from 700,000 ha over one year (the previous example) was used for heat purposes, 0.1% of the 51.9 Gt CO_2 would be reduced. For a period of 15 years, it would be 1.55%. The same is valid for the year 2050. Several means to achieve an 80% reduction in GHG emissions, implying an 85% decline in energy-related CO_2 emissions (including those from transport), have been examined (Table 7). This is where the oleoresin application seems appropriate—it relates to reduced GHG emissions and to improved energy efficiency.

Table 7. Plans to reduce greenhouse gas emissions into the atmosphere, to increase the use of renewable energy sources and to improve energy efficiency by 2050.

Parameters Examined	2020 Climate and Energy Package [122,123]	2030 Climate and Energy Framework [124]	Energy Road 2050 [125]
Cut in GHG from 1990 levels	20%	40%	80%
EU energy from renewables	20%	32%	66%
Improvement in energy efficiency	20%	32.5	75%

The use of tree resin for energy is only possible through stringent bioeconomic management. Its benefit is to reduce the additional load of CO_2 into the atmosphere, as aforementioned. On the one hand, oleoresin combustion is associated with the release of neutral CO_2, and on the other hand, the abandoned and degraded land is used for the afforestation of new forest stands that accumulate CO_2 for at least one rotation. However, the technological processing of the resin needs to be developed.

5. Conclusions

This paper focuses on the energy value (calorific or heating value) of certain components of the tree biomass of pine, spruce, and larch. The energy parameters of the trees examined were determined by calorimetry. The components of wood pulp (bleached and unbleached) and turpentine were obtained from the pulp and paper processing plant Mondi SCP (Slovak Cellulose Paper mill), Ltd. in Ruzomberok. Samples of tree resin were taken from V-shaped wounds carved in the bark of trees growing on the Cernova research stand nearby Ruzomberok. The mean combustion heat values of the tree resins (pine: 38.591 ± 1.307; spruce: 38.373 ± 1521; larch: 38.326 ± 0975 MJ·kg^{-1}) are not statistically different from each other. However, these values are double the heating value of forest tree wood and its components.

The combustion heats of bleached pulp (cellulose) produced by the Kraft sulfate process are supposed to be the reference standard for tree biomass. The average combustion heats of bleached (17.319 ± 0.025 MJ·kg^{-1}) and unbleached (15.955 ± 0.036 MJ·kg^{-1}) pulps are 2.2 and 2.4 times lower than those of the investigated tree resin samples. The highest energy value was recorded for turpentine (39.773 ± 0.027 MJ·kg^{-1}), but its technical processing is considerably more complicated than that of the resin.

The quality and quantity of oleoresin was influenced by tapping. The American method was best, providing 5 kg of oleoresin per tree over one year. However, by using the paraquat-cation stimulant, this amount was enhanced to 18.65 kg of oleoresin per tree over a year; thus, a 1 ha pine stand provides 14,920 kg.

Tree resin was shown to be an excellent renewable energy source that has not yet been used for this purpose. Compared to the previously investigated energy sources (feeding sorrel: 160 GJ ha^{-1}; common reed: 221.6 GJ ha^{-1}; Miscanthus *giantheus*: 246.4 GJ ha^{-1}), pine resin, with 575.8 GJ ha^{-1}, showed the highest energy content per hectare.

The production price of oleoresin was compared to that of other feedstocks. The production cost of first-generation feedstocks is triple that of second-generation ones (EUR 5–15 GJ^{-1} compared to EUR 1.5–4.5 GJ^{-1}). When using a 2% paraquat-cation stimulant, the production cost of oleoresin is the lowest (EUR 1.3–2.75).

The potential contribution of forest expansion to the sequestration of carbon and the utilization of renewable forest resources is well known. The tapping of resin and the expansion of its industrial uses can further enhance the carbon sequestration potential of forest resources and expand the forest ecosystem services. An average NEE rate of 1.684 t C ha^{-1} yr^{-1} was calculated at three geographic locations (lat. from 42° N to 62° N). If one tonne of oleoresin was burned instead of the energy equivalent amount of coal, the atmosphere would be spared more than 5 Mt CO_2.

The use of oleoresin for energy is only possible through stringent bioeconomic management. The circular economy aims to reduce resource use and consumption to minimize emissions. So far, among the potential energy forms derived from biomaterials, biogas is of the greatest significance. However, its production is only profitable under government incentives (15.42 ERU.GJ^{-1}), which is about five times the cost of tree resin production. The application of oleoresin for energy use is also associated with environmental security, and has a social role. It can also be helpful to reduce CO_2 GHG emissions and improve energy efficiency.

Tree oleoresin, as a macroergic substance, approaches the heating parameters of liquid hydrocarbon fuels such as oil and petroleum products. Therefore, oleoresin can play an important role in the supersession of fossil fuels. Resin quantity and quality might be controlled by selecting suitable tree taxa of high oleoresin content, or its energy content in a tree can be increased by applying a suitable stimulant. The aim of this work is the cultivation of energy-modified biomass, a revolutionary shift in the production of renewable energy sources.

Author Contributions: The basic idea and the overall scope of this research was under the responsibility of J.D. Chemical processing of biological material, chemical analyses and writing the article managed J.M. All authors have read and agreed to the published version of the manuscript.

Funding: This research was funded by the 2/0013/17 VEGA grant—"Ecosystem services to support landscape protection in conditions of global change" in the period 1.1.2017-31.12.2020.

Data Availability Statement: In this manuscript, all the necessary data were published, except for individual combustion heat values of studied trees, where average values were depicted in a graph.

Acknowledgments: The authors have reviewed and agree with the contents of the manuscript and there is no financial claim against publication. It is confirmed that the paper is an original work and is not subject to review in any other publication.

Conflicts of Interest: The authors have declared no conflict of interest.

References

1. Warburg, C.T.; King'ondu, C.K. Energy characteristics of five indigenous tree species at Kitulangalo Forest Reserve in Morogoro, Tanzania. *Int. J. Renew. Energy Res.* **2014**, *4*, 1078–1084.
2. Constanza, J.K.; Abt, R.C.; McKerrow, A.J.; Collazo, J.A. Bioenergy production and forest landscape change in the south-eastern United States. *GCB Bioenergy* **2017**, *9*, 924–939. [CrossRef]
3. Johnson, J.M.; Coleman, M.D.; Gesch, R.; Jaradat, A.; Mitchell, R.; Reicosky, D.; Wilhelm, W.W. Biomass-bioenergy crops in the United States: A changing paradigm. *Am. J. Plant Sci. Biotechnol.* **2007**, *1*, 1–28.
4. Franceschi, V.R.; Krokene, P.; Christiansen, E.; Krekling, T. Anatomical and chemical defenses of conifer bark against bark beetles and other pests. *New Phytol.* **2005**, *167*, 353–376. [CrossRef]
5. Wallin, K.F.; Raffa, K.F. Effects of folivory on subcortical plant defenses: Can defense theories predict interguild processes? *Ecology* **2001**, *82*, 1387–1400. [CrossRef]

6. Fischer, M.J.; Waring, K.M.; Forstetter, R.W. The resin composition of ponderosa pine (Pinus ponderosa) attacked by the round headed pine beetle (*Dendroctonus adjunctus*) (Coleoptera: Curculionidae, Scolytinae). In Proceedings of the Conference Proceeding (2008), Proceedings RMRS-P-5, USDA FS 2008, Flagstaff, AZ, USA, 7–9 August 2008; pp. 178–182.
7. Raffa, K.F.; Aukema, B.H.; Bentz, B.J.; Carroll, A.L.; Hicke, J.A.; Turner, M.G.; Romme, W.H. Cross-scale drivers of natural disturbances prone to anthropogenic amplification: The dynamics of bark beetle eruptions. *BioScience* **2008**, *58*, 501–517. [CrossRef]
8. Wang, J.; Chen, Y.P.; Yao, K.; Wilbon, P.A.; Zhang, W.; Ren, L.; Zhou, J.; Nagarkatti, M.; Wang, C.; Chu, F.; et al. Robust antimicrobial compounds and polymers derived from natural resin acids. *Chem Commun.* **2012**, *48*, 916–918. [CrossRef]
9. Gottlieb, O.R. Phytochemicals: Differentiation and function. *Phytochemistry* **1990**, *29*, 1715–1724. [CrossRef]
10. Langenheim, J.H. *Chemistry, Evolution, Ecology, Ethnobotany*; Timber Press: Cambridge, UK, 2003; p. 612.
11. Nascimento, M.S.; Santana, A.L.; Maranhao, C.A.; Oliveira, L.S.; Biebe, L. Phenolic extractives and natural resistance of wood. In *Biodegradation—Life of Science*; Chamy, R., Rosenkranz, F., Eds.; BoD—Books on Deman: Nordstedt, Germany, 2013; Chapter 13; pp. 349–370, ISBN 978-953-51-1154-2.
12. Mirov, N. *Composition of Gum Turpentines of Pines*; Tech. Bull. 1239; USDA FS (US Department of Agriculture, Forest Service): Alexandria, VA, USA, 1961; p. 158.
13. Langenheim, J.H. Plant Resins. *Am. Sci.* **1990**, *78*, 16–24.
14. Smith, R.H. *Xylem Monoterpenes of Pines: Distribution, Variation, Genetics, Function*; General Technical Report PSWGTR-177; Pacific Southwest Research Station: Albany, CA, USA; USDA FS (US Department of Agriculture, Forest Service): Alexandria, VA, USA, 2000; p. 454. [CrossRef]
15. Zwenger, S.; Basu, C. Plant terpenoids: Applications and future potentials. *Microbiol. Mol. Biol. Rev.* **2008**, *3*, 1–7.
16. Silvestre, A.J.D.; Gandini, A. Terpenes: Major sources, properties and application. In *Monomers, Polymers and Composite from Renewable Resources*, 1st ed.; Belgacem, M.M., Gandidi, A., Eds.; Elsevier Ltd.: Amsterdam, The Netherlands, 2008; p. 552, ISBN 978-0-08-045316-3.
17. Smith, R.H. *Monoterpenes of Ponderosa Pine Xylem Resin in Western United States*; Tech. Bull. No. 1532; Forest Service, United State Department of Agriculture: Alexandria, VA, USA, 1977; p. 48.
18. Leonhardt, S.D.; Blüthgen, N.; Schmitt, T. Tree resin composition, collection behaviour and selective filters shape chemical profiles of tropical bees (Apidae: Meliponini). *PLoS ONE* **2011**, *6*, e23445. [CrossRef]
19. Bohlmann, J.; Keeling, C.I. Terpenoid biomaterials. *Plant J.* **2008**, *54*, 656–669. [CrossRef]
20. Rezzi, S.; Bighelli, A.; Castola, V.; Casanova, J. Composition and chemical variability of the oleoresin of *Pinus nigra* ssp. Laricio from Corsica. *Ind. Crops Prod.* **2005**, *21*, 71–79. [CrossRef]
21. Jantam, I.; Ahmad, A.S. *Oleoresins of Three Pinus Species from Malaysian Pine Plantations*; ASEAN Review of Biodiversity and Environmental Conservation: Kota Samarahan, Malaysia, 1999; pp. 1–9. Available online: https://www.researchgate.net/publication/238101480_Oleoresins_of_Three_Pinus_Species_from_Malaysian_Pine_Plantations (accessed on 3 November 2020).
22. Lee, H.J.; Ravn, M.M.; Coates, R.M. Synthesis and characterization of abietadiene, levopimaradiene, palustradiene, and neoabietadiene: Hydrocarbon precursors of the abietane diterpene resin acids. *Tetrahedron* **2001**, *57*, 6155–6167. [CrossRef]
23. Mandaogade, P.; Satturwar, P.M.; Fulzele, S.V.; Gogte, B.B.; Dorle, A.K. Rosin derivatives: Novel film forming materials for controlled drug delivery. *React. Funct. Polym.* **2002**, *50*, 233–242. [CrossRef]
24. Swift, K.A.D. Catalytic transformations of the major terpene feedstocks. *Top. Cat.* **2004**, *27*, 143–155. [CrossRef]
25. McMorn, P.; Roberts, G.; Hutchings, G.J. Oxidation of α-pinene to verbenone using silica–titania co-gel catalyst. *Catal. Let.* **2000**, *67*, 203–206. [CrossRef]
26. Gillette, N.E.; Hansen, E.M.; Mehmel, C.J.; Mori, S.R.; Webster, J.N.; Erbilgin, N.; Wood, D.L. Area-wide application of verbenone-releasing flakes reduces mortality of white bark pine Pinus albicaulis caused by the mountain pine beetle *Dendroctonus ponderosae*. *Agric. For. Entomol.* **2012**, *14*, 367–375. [CrossRef]
27. Mercier, B.; Prost, J.; Prost, M. The essential oil of turpentine and its major volatile fraction (α- and ∃-pinenes): A review. *Int. J. Occup. Med. Environ. Health* **2009**, *22*, 331–342. [CrossRef]
28. Adams, T.B.; Gavin, C.L.; McGowen, M.M.; Waddell, W.J.; Cohen, S.M.; Feron, V.J.; Marnett, L.J.; Munro, I.C.; Portoghese, P.S.; Rietjens, I.M.C.M.; et al. The FEMA GRAS assessment of aliphatic and aromatic terpene hydrocarbons used as flavor ingredients. *Food Chem. Toxicol.* **2011**, *49*, 2471–2494. [CrossRef]
29. Limberger, R.P.; Schuh, R.S.; Henriques, A.T. Pine terpene biotransformation. In *Pine Resin: Biology, Chemistry and Applications*; Fett-Neto, A.G., Rodrigues-Corrêa, K.C.S., Eds.; Research Signpost: Kerala, India, 2012; pp. 107–126, ISBN 978-81-308-0493-4.
30. Silva, A.C.R.; Lopes, P.M.; Azevedo, M.M.B.; Costa, D.C.M.; Alviano, C.S.; Alviano, D.S. Biological activities of α-pinene and β-pinene enantiomers. *Molecules* **2012**, *17*, 6305–6316. [CrossRef]
31. Macchioni, F.; Cioni, P.L.; Flamini, G.; Morelli, I.; Maccioni, S.; Ansaldi, M. Chemical composition of essential oils from needles, branches and cones of Pinus pinea, P. halepensis, P. pinaster and P. nigra from central Italy. *Flavour Fragr. J.* **2003**, *18*, 139–143. [CrossRef]
32. Burdock, G.A.; Carabin, I.G. Safety assessment of Ylang-Ylang (*Cananga* spp.) as a food ingredient. *Food Chem. Toxicol.* **2008**, *46*, 433–445. [CrossRef]
33. Tümen, I.; Reunanen, M.A. Comparative study on turpentine oils of oleoresins of *Pinus sylvestris* L. from three districts of Denizli. *Rec. Nat. Prod.* **2010**, *4*, 224–229.
34. Teshome, T. Analysis of resin and turpentine oil constituents of Pinus patula grown in Ethiopia. *Ed-JRIF* **2011**, *3*, 38–48.

35. Bublinec, E.; Kuklová, M.; Oszlányi, J. Fytometria a distribúcia energie v bukových ekosystémoch = Phytometry and energy distribution in beech ecosystems. In *Buk a Bukové Ekosystémy Slovenska*; VEDA, vydavateľstvo SAV: Bratislava, Slovakia, 2011; pp. 289–327, ISBN 978-80-224-1192-9.
36. Liu, M.H.; Yi, L.T. Combustibility of fresh leaves of 26 forest species in China. *J. Trop. For. Sci.* **2013**, *25*, 528–536.
37. Aniszewska, M.; Gendek, A. Comparison of heat of combustion and calorific value of the cones and wood of selected forest tree species. Lesne Prace Badawcze. *For. Res. Pap.* **2014**, *75*, 231–236.
38. Álvarez-Álvarez, P.; Pizarro, C.; Barrio-Anta, M.; Cámara-Obregón, A.; Bueno, J.L.M.; Ávarez, A.; Gutiérrez, I.; Burslem, D.F. Evaluation of tree species for biomass energy production in Northwest Spain. *Forests* **2018**, *9*, 160. [CrossRef]
39. Telmo, C.; Lousada, J.; Moreira, N. Proximate analysis, backwards, stepwise regression between gross calorific value, ultimate and chemical analysis of wood. *Bioresour. Technol.* **2010**, *101*, 3808–3815. [CrossRef]
40. Francescato, V.; Antonini, E.; Bergomi, L.Z. *Wood Fuels Handbook*; AIEL—Italian Agroforestry Energy Association, Agripolis: Rome, Italy; Food and Agriculture Organization: Rome, Italy, 2008; p. 83.
41. Howard, E.T. Heat of combustion of various southern pine materials. *Wood Sci.* **1972**, *5*, 194–197.
42. Aguiar, A.V.; Shimizu, J.Y.; Sousa, V.A.; Resende, M.D.V.; Freitas, M.L.M.; Moraes, M.L.T.; Sebbenn, A.M. Genetics of oleoresin production with focus on Brazilian planted forests. In *Pine Resin: Biology, Chemistry and Applications*; Fett-Neto, A.G., Rodrigues-Corrêa, K.C.S., Eds.; Research Signpost: Kerala, India, 2012; pp. 87–106, ISBN 978-81-308-0493-4.
43. Lachowicz, H.; Sajdak, M.; Paschalis-Jakubowicz, P.; Cichy, W.; Wojtan, R.; Witczak, M. The influence of location, tree age and forest habitat type on basic fuel properties of wood of the Silver birch (*Betula pendula Roth.*) in Poland. *BioEnergy Res.* **2018**, *11*, 638–651. [CrossRef]
44. Turtola, S.; Manninen, A.M.; Rikala, R.; Kainulainen, P. Drought stress alters the concentration of wood terpenoids in Scots pine and Norway spruce seedlings. *J. Chem. Ecol.* **2003**, *29*, 1991–1995. [CrossRef] [PubMed]
45. Ferreira, A.G.; Fior, C.S.; Gualtieri, S.C.J. Oleoresin yield of Pinus elliottii Engelm seedlings. *Braz. J. Plant Physiol.* **2011**, *23*, 313–316. [CrossRef]
46. Blažej, A.; Košík, M. *Fytomasa Ako Chemická Surovina (Phytomass as a Chemical Resource)*; Veda Vydavateľstvo SAV: Bratislava, Slovakia, 1985; p. 404.
47. Holmbom, T.; Reunanen, M.; Fardim, P. Composition of callus resin of Norway spruce, Scots pine, European Larch and Douglas fir. *Holzforschung* **2008**, *62*, 417–422. [CrossRef]
48. Santo, A.M.; Vasconcelos, T.; Mateus, E.; Farrall, M.H.; Gomes da Silva, M.D.R.; Gutiérrez, I.; Burslem, D.F. Characterization of the volatile fraction emitted by phloems of four pinus species by solid-phase microextraction and gas chromatography—Mass spectrometry. *J. Chromathogr. A* **2006**, *1–2*, 191–198. [CrossRef]
49. Marques, F.A.; Frensch, G.; Zaleski, S.R.M.; Nogata, N.; Maia, B.H.L.N.S.; Lazzari, S.M.N.; Lenz, C.A.; Corrêa, G. Differentiation of five pine species cultivated in Brazil based on chemometric analysis of their volatiles identified by gas chromatography-mass spectrometry. *Food Energy Secur.* **2012**, *1*, 81–93. [CrossRef]
50. Cunningham, A. Pine resin tapping techniques used around the world. In *Pine Resin: Biology, Chemistry and Applications*; Fett-Neto, A.G., Rodrigues-Corrêa, C.S., Eds.; Research Signpost: Kerala, India, 2012; pp. 1–8, ISBN 978-81-308-0493-4.
51. Rodrigues-Corrêa, K.L.S.; Lima, J.C.; Fett-Neto, A.G. Oleoresins from Pine: Production and industrial uses. In *Natural Products*; Ramarat, K., Mérillon, J.M., Eds.; Springer: Berlin/Heidelberg, Germany, 2013; pp. 4037–4060.
52. Rodrigues-Corrêa, K.C.S.; Lima, J.C.; Fett-Neto, A.G. Pine oleoresin: Tapping green chemicals, biofuels, food protection, and carbon sequestration from multipurpose trees. *Food Energy Secur.* **2016**, *1*, 81–93. [CrossRef]
53. Juntunen, V.; Neuvonen, S. Natural regeneration of Scots pine and Norway spruce close to the timberline in northern Finland. *Silva Fenn.* **2006**, *40*, 443–458. [CrossRef]
54. McLane, S.C.; Daniels, L.D.; Aitken, S.N. Climate impacts on lodgepole pine (*Pinus contorta*) radial growth in a provenance experiment. *For. Ecol. Manag.* **2011**, *262*, 115–123. [CrossRef]
55. Lombardero, M.J.; Ayres, M.P.; Lorio, P.L.; Ruel, J.J. Environmental effects on constitutive and inducible resin defences of Pinus taeda. *Ecol. Lett.* **2000**, *3*, 329–339. [CrossRef]
56. Wang, Z.; Calderon, M.M.; Carandang, M.G. Effects of resin tapping on optimal rotation age of pine plantation. *J. For. Econ.* **2006**, *11*, 245–260. [CrossRef]
57. Dickens, E.D. Fertilization options for longleaf pine stands on marginal soils with economic implications. In Proceedings of the SOFEW 2001 Proceedings, Atlanta, GA, USA, 27–28 March 2001.
58. Matías, L.; Jump, A.S. Interactions between growth, demography and biotic interactions in determining species range limits in a warming world: The case of *Pinus sylvestris*. *For. Ecol. Manag.* **2012**, *282*, 10–22. [CrossRef]
59. Peñuelas, J.; Llusia, J. Seasonal emission of monoterpenes by the Mediterranean tree Quercus ilex L. in the field conditions. Relations with photosynthetic rates, temperature and volatility. *Physiol. Plant.* **1999**, *105*, 641–647. [CrossRef]
60. Parantainen, A.; Pulkkine, P. Pollen viability of Scots pine (*Pinus sylvestris*) in different temperature conditions: High levels of variation among and within latitudes. *For. Ecol. Manag.* **2002**, *167*, 149–160. [CrossRef]
61. Salmela, M.J.; Cavers, S.; Cottrell, J.E.; Iason, G.R.; Ennos, R.A. Seasonal patterns of photochemical capacity and spring phenology reveal genetic differentiation among native Scots pine (*Pinus sylvestris* L.) populations in Scotland. *For. Ecol. Manag.* **2011**, *262*, 1020–1029. [CrossRef]

62. Maestre, F.T.; Cortina, J. Are Pinus halepensis plantations useful as a restoration tool in semiarid Mediterranean areas? *For. Ecol. Manag.* **2004**, *198*, 303–317. [CrossRef]

63. Rodrigues, K.C.S.; Apel, M.A.; Henriques, A.T.; Fett-Neto, A.G. Efficient oleoresin biomass production in pines using low cost metal containing stimulant paste. *Biomass Bioenergy* **2011**, *35*, 4442–4448. [CrossRef]

64. STN EN ISO 18125. Solid biofuels—Determination of calorific value (ISO 18125:2017). Office for Standardization, Metrology and Testing of the Slovak Republic: Bratislava, Slovakia, 2017.

65. Chakradhar, S.; Pate, K.S. Combustion characteristics of tree woods. *JSBS* **2016**, *6*, 31–43. [CrossRef]

66. Zeneli, G.; Tsitsimpikou, C.; Petrakis, P.V.; Naxakis, G.; Habili, D.; Roussis, V. Foliar and cortex oleoresin variability of Sliver fir (*Abies alba* Mill.) in Albania. *Z. Naturforsch.* **2001**, *56*, 531–539. [CrossRef]

67. Dzurenda, L. *Spaľovanie Dreva a Kôry (Burning of Wood and Bark)*; Technical University in Zvolen: Zvolen, Slovakia, 2005; p. 124, ISBN 80-228-1555-1.

68. Singh, T.; Kostecky, M.M. Calorific value variations in components of 10 Canadian tree species. *Can. J. For. Res.* **1986**, *16*, 1378–1381. [CrossRef]

69. Haarke, A.P.; Sandels, A.; Burley, J. Calorific values for wood and bark and a bibliography for fuel wood. In *Singh and Kostecky*; CAB International: Wallingford, UK, 1986.

70. Jain, R.K.; Singh, B. Fuelwood characteristics of selected indigenous tree species from central India. *Bioresour. Technol.* **1999**, *68*, 305–308. [CrossRef]

71. Rhén, C. Chemical composition and gross calorific value of the above-ground biomass components of young *Picea Abies*. *Scand. J. For. Res.* **2004**, *19*, 72–81. [CrossRef]

72. Seth, M.K. Trees and their economic importance. *Bot. Rev.* **2003**, *69*, 321–376. [CrossRef]

73. Abdel-Raouf, M.E.; Abdul-Raheim, A.R.M. Rosin: Chemistry, Derivatives, and Applications: A review. *BAOJ Chem.* **2018**, *4*, 39.

74. Simpraga, M.; Ghimire, R.P.; Van der Straeten, D.; Blande, J.D.; Kasurinen, A.; Sorvari, J.; Holopainen, T.; Adriaenssens, S.; Holopainen, J.K.; Kivimäenpää, M. Unravelling the functions of biogenic volatiles in boreal and temperate forest ecosystems. *Eur. J. For. Res.* **2019**, *5*, 763–787. [CrossRef]

75. Maaten, E.; Mehl, A.; Wilmking, M.; Maaten-Theunissen, M. Tapping the tree-ring archive for studying effects of resin extraction on the growth and climate sensitivity of Scots pine. *For. Ecosyst.* **2017**, *4*, 7. [CrossRef]

76. Ivask, M. Caloric value of Norway spruce organs and its seasonal dynamics. *Balt. For.* **1999**, *1*, 44–49.

77. Oszlany, J. Wood, bark, needles, leaves and roots energy values of *Pinus sylvestris* L., *Picea excelsa* Link. and *Fagus sylvatica* L. *Ekológia (CSSR)* **1982**, *1*, 289–296.

78. Steubling, I.C.; Ramirez, M.A.; Alberdi, M. Artenzusammensetzung, Lichtgenuss und Energiegehalt der Krautschicht des Valdvianischen Regelwaldes bei St. Martin. *Vegetativo* **1979**, *39*, 25–33. [CrossRef]

79. Hughes, M.K. Seasonal calorific values from a deciduous woodland in England. *Ecology* **1971**, *52*, 923–926. [CrossRef]

80. Genova, M.; Caminero, L.; Dochao, J. Resin tapping in *Pinus pinaster*: Effects on growth and response function to climate. *Eur. J. For. Res.* **2014**, *233*, 323–333. [CrossRef]

81. Vallinayagam, R.; Vedharaj, S.; Yang, W.M.; Lee, P.S.; Chua, K.J.E.; Chou, S.K. Pine oil-biodiesel blends: A double biofuel strategy to completely eliminate the use of diesel in a diesel engine. *Appl. Energy* **2014**, *130*, 466–473. Available online: http://www.fao.org/docrep/005/y4496e/Y4496E27.htm (accessed on 8 October 2020). [CrossRef]

82. Acar, H.H.; Barli, O.; Youshimura, T. The Effect of Harvesting, Transportation and Stockpiling Activities in the Resin Tapping on the Resin Productivity and Quality. Available online: http://www.fao.org/3/y4496e/Y4496E27.htm (accessed on 20 December 2020).

83. Kossuth, S.V.; Roberts, D.; Huffman, J.; Wang, S.C. Resin acid, turpentine, and caloric content of paraquat-treated slash pine. *Can. J. For. Res.* **1982**, *12*, 489–492. [CrossRef]

84. Papadopoulos, A.M. Resin tapping history of an Aleppo Pine forest in Central Greece. *Open For. Sci. J.* **2013**, *6*, 50–53. [CrossRef]

85. Krokene, P.; Nagy, N.E. Anatomical aspects of resin-based defenses in pine. In *Pine Resin: Biology, Chemistry and Applications*; Fett-Neto, A.G., Rodrigues-Corrêa, K.C.S., Eds.; Research Signpost: Kerala, India, 2012; pp. 67–86.

86. Fox, T.R.; Jokela, E.J.; Allen, H.L. The development of pine plantation silviculture in the southern United States. *J. For.* **2007**, *105*, 337–347.

87. Perpiňa, C.C.; Kavalov, B.; Diogo, V.; Jakobs-Crisioni, C.; Batista e Silva, F.; Lavalle, C. *Agricultural Land Abandonment in the EU within 2025–2030*; JRC 113718; European Commission: Brussels, Belgium, 2018; Available online: https://ec.europa.eu/eu/jrc/en/luisa (accessed on 3 March 2021).

88. Ioelowich, M. Thermodynamics of biomass-based solid fuels. *Acad. J. Polym. Sci.* **2018**, *2*, 555577.

89. Creuzburg, M.K.; Scheller, R.M.; Lucash, M.S.; Evers, L.B.; Ludec, S.D.; Johnson, M.G. Bioenergy harvest, climate change, and forest carbon in the Oregon Coast Range. *GCB Bioenergy* **2016**, *8*, 357–370. [CrossRef]

90. Lee, J.; Tolunay, D.; Makineci, E.; Comez, A.; Son, Y.M.; Kim, R.; Son, Y. Estimating the age-dependent changes in carbon stocks of Scots pin (*Pinus sylvestris* L.) stands in Turkey. *Ann. For. Sci.* **2016**, *73*, 523–531. [CrossRef]

91. Pan, Y.; Birdsey, R.A.; Fang, J.; Houghton, R.; Kauppi, P.E.; Kurz, W.A. A large and persistent carbon sink in the world's forests. *Science* **2011**, *333*, 988–993. [CrossRef]

92. Bastin, J.F.; Finegold, Y.; Garcia, C.; Mollicone, D.; Rezende, M.; Routh, D.; Crowther, T.W. The global tree restoration potential. *Science* **2019**, *365*, 76–79. [CrossRef]

93. Nanda, S.; Rana, R.; Sarangi, P.K.; Dalai, A.K.; Kozinski, J.A. A Broad Introduction to First-, Second-, and Third-Generation Biofuels. In *Recent Advancements in Biofuels and Bioenergy Utilization*; Sarangi, P., Nanda, S., Mohanty, P., Eds.; Springer: Singapore, 2018; pp. 1–25. [CrossRef]

94. De Wit, M.; Faaij, A. European biomass resource potential and costs. *Biomass Bioenergy* **2010**, *34*, 188–202. [CrossRef]

95. Palma, A.; Pestana, M.; Azevedo, A. Pine resin sector in Portugal—Weaknesses and challenges. *For. Ideas* **2012**, *1*, 10–18.

96. Maga, J.; Piszczalka, J. *Biomasa Ako Zdroj Obnoviteľnej Energie [Biomass as an Energy Renewable Source]*; Slovenská Poľnohospodárska Univerzita: Nitra, Slovakia, 2006; p. 108.

97. Demko, J.; Machava, J.; Saniga, M. Energy production analysis of Common Reed—*Phragmites australis* (Cav.) Trin. *Folia Oecol.* **2017**, *44*, 107–113. [CrossRef]

98. Weger, J.; Knápek, J.; Bubeník, J.; Vávrová, K.; Strašil, Z. Can Miscanthus fulfill its expectations as an energy biomass source in the current conditions of the Czech republic?—Potentials and barriers. *Agriculture* **2021**, *11*, 40. [CrossRef]

99. Shan, J.; Morris, L.A.; Hendrick, R.L. The effects of management on soil and plant carbon sequestration in slash pine plantations. *J. Appl. Ecol.* **2001**, *38*, 932–941. [CrossRef]

100. Balboa-Murias, M.A.; Rodríguez-Soalleiro, R.; Merino, A.; Àlvarez-González, J.G. Temporal variation and distribution of carbon stocks in aboveground biomass of radiata pine and maritime pine pure stands under different silvicultural alternatives. *For. Ecol. Manag.* **2006**, *237*, 29–38. [CrossRef]

101. Malmsheimer, R.W.; Heffernan, P.; Brink, S.; Crandall, D.; Deneke, F.; Galik, C.S.; Gee, E.; Helms, J.A.; McClure, N.; Mortimer, M.; et al. Preventing GHG emissions through wood substitution. *J. For.* **2008**, *106*, 136–140.

102. Spath, P.L.; Mann, M.K. *Biomass Power and Conventional Fossil Systems with and without CO$_2$ Sequestration—Comparing the Energy Balance, Greenhouse Gas Emissions and Economics*; EERE Publication and Product Library: Washington, DC, USA, 2004; p. 28.

103. Bastos Lima, M.G. *The Politics of Bioeconomy and gb: Lessons from Biofuel Governance, Policies and Production Strategies in the Emerging World*; Springer: Dordrecht, The Netherlands, 2021; p. 227.

104. Sadik-Zada, E.R. Natural resources, technological progress, and economic modernization. *Rev. Dev. Econ.* **2021**, *25*, 381–404. [CrossRef]

105. D'Adamo, I.; Morone, P.; Huisingh, D. Bioenergy: A sustainable shift. *Energies* **2021**, *14*, 5661. [CrossRef]

106. Bastos Lima, M.G. Corporate Power in the Bioeconomy Transition: The Policies and Politics of Conservative Ecological Modernization in Brazil. *Sustainability* **2021**, *13*, 6952. [CrossRef]

107. European Commission. *A New Bioeconomy Strategy for a Sustainable Europe*; European Commission: Brussels, Belgium, 2018.

108. Dietz, T.; Börner, J.; Förster, J.J.; von Braun, J. Governance of the bioeconomy: A global comparative study of national bioeconomy strategies. *Sustainability* **2018**, *10*, 3190. [CrossRef]

109. Morone, P.; D'Amato, D. The role of sustainability standards in the uptake of biobased chemicals. *Curr. Opin. Green Sustain. Chem.* **2019**, *19*, 45–49. [CrossRef]

110. Pehlken, A.; Wulf, K.; Grecksch, K.; Klenke, T.; Tsydenova, N. More Sustainable Bioenergy by Making Use of Regional Alternative Biomass? *Sustainability* **2020**, *12*, 7849. [CrossRef]

111. D'Adamo, I.; Falcone, P.M.; Morone, P. A new socio-economic indicator to measure the performance of bioeconomy sectors in Europe. *Ecol. Econ.* **2020**, *176*, 106724. [CrossRef]

112. D'Adamo, I.; Gastaldi, M.; Rosa, P. Assessing Environmental and Energetic Indexes in 27 European Countries. *Int. J. Energy Econ. Policy* **2021**, *11*, 417–423. [CrossRef]

113. Prussi, M.; Padella, M.; Conton, M.; Postma, E.D.; Lonza, L. Review of technologies for biomethane production and assessment of Eu transport share in 2030. *J. Clean. Prod.* **2019**, *222*, 565–572. [CrossRef]

114. Lorenzi, G.; Baptista, P. Promotion of renewable energy sources in the Portuguese transport sector: A scenario analysis. *J. Clean. Prod.* **2018**, *186*, 918–932. [CrossRef]

115. Baeno-Moreno, F.M.; Malico, I.; Marques, I.P. Promoting sustainability: Wastewater treatment plants as a source of biomethane in regions far from a high-pressure grid. A real Portuguese case study. *Sustainability* **2021**, *13*, 8933. [CrossRef]

116. Terlouw, W.; Peters, D.; van Tilburg, J.; Schimmel, M.; Berg, T.; Cilhar, J.; Mir, G.U.R.; Spöttle, M.; Staats, M.; Lejaretta, A.V.; et al. *Gas for Climate. The Optimal Role for Gas in a Net-Zero Emissions Energy System. Gas for Climate, a Path to 2050*; Navigant Consulting Inc.: Utrecht, The Netherlands, 2019; p. 219.

117. D'Adamo, I.; Falcone, P.M.; Huisingh, D.; Morone, P. A circular economy model based on biomethane: What are the opportunities for the municipality of Rome and beyond? *Renew. Energy* **2021**, *163*, 1660–1672. [CrossRef]

118. Galende-Sánchez, E.; Sorman, A.H. From consultation toward co-production in science and policy: A critical systematic review of participatory climate and energy initiatives. *Energy Res. Soc. Sci.* **2021**, *73*, 101907. [CrossRef]

119. Rodrigues, K.C.S.; Azevedo, P.C.N.; Sobreiro, L.E.; Pelissari, P.; Fett-Neto, A.G. Oleoresin yield of Pinus elliottii plantations in a subtropical climate: Effect of tree diameter, wound shape and concentration of active adjuvants in resin stimulating paste. *Ind. Crops. Prod.* **2008**, *27*, 322–327. [CrossRef]

120. Roelfsema, M.; van Soest, H.L.; Harmsen, M.; van Vuuren, D.P.; Bertram, C.; den Elzen, M.; Höhne, N.; Iacobuta, G.; Krey, V.; Kriegler, E.; et al. Taking stock of national climate policies to evaluate implementation of the Paris Agreement. *Nat. Commun.* **2020**, *11*, 2096. [CrossRef]

121. Yoon, S.H.; van Heiningen, A. Green liquor extraction of hemicelluloses from southern pine in an Integrated Forest Biorefinery. *J. Ind. Eng. Chem.* **2010**, *16*, 74–80. [CrossRef]

122. ECA (European Commission). Climate Action, 2020 Climate & Energy Package. Available online: https://ec.europa.eu/clima/policies/strategies/2020_en (accessed on 1 August 2021).
123. Alsaleh, M.; Abdulwakil, M.M.; Abdul-Rahim, A.S. Does Social Businesses Development Affect Bioenergy Industry Growth under the Pathway of Sustainable Development? *Sustainability* **2021**, *13*, 1989. [CrossRef]
124. ECB (European Commission). Climate Action, 2030 Climate & Energy Framework, Greenhouse Gas Emissions—Raising the Ambition. Available online: https://ec.europa.eu/clima/policies/strategies/2030_en (accessed on 1 August 2021).
125. ECC (European Commission). Climate Action, 2050 Long-Term Strategy. Available online: https://ec.europa.eu/clima/eu-action/climate-strategies-targets/2050-long-term-strategy_en (accessed on 1 August 2021).

MDPI

St. Alban-Anlage 66

4052 Basel

Switzerland

Tel. +41 61 683 77 34

Fax +41 61 302 89 18

www.mdpi.com

Sustainability Editorial Office

E-mail: sustainability@mdpi.com

www.mdpi.com/journal/sustainability